Neutron Physics

NUCLEAR ENGINEERING

Neutron Physics

Paul Reuss

Institut national des sciences et techniques nucléaires

17, avenue du Hoggar
Parc d'activités de Courtabœuf, BP 112
91944 Les Ulis Cedex A, France

The author would like to thank Nova Traduction (K. Foster) and Chris Latham for the translation of his book.

Cover illustrations: Jules Horowitz (1921-1995), a highly talented physicist, founded the French school of neutron physics. In 2014, the Jules Horowitz reactor being built at Cadarache will become the main irradiation reactor in the world (100 MWth) for research on materials and nuclear fuels.
In the background, the meshing for a neutron physics core calculation and in the foreground the power distribution, result of this calculation. (Documents courtesy of CEA.)

Cover conception: Thierry Gourdin

Printed in France

ISBN: 978-2-7598-0041-4

Introduction to the *Nuclear Engineering* Collection

Within the French Atomic Energy Commission (CEA), the National Institute of Nuclear Science and Technology (INSTN) is a higher education institution operating under the joint supervision of the Ministries of Education and Industry. The purpose of the INSTN is to contribute to disseminating the CEA's expertise through specialised courses and continuing education, not only on a national scale, but across Europe and worldwide.

This mission is focused on nuclear science and technology, and one of its main features is a Nuclear Engineering diploma. Bolstered by the CEA's efforts to build partnerships with universities and engineering schools, the INSTN has developed links with other higher education institutions, leading to the organisation of more than twenty five jointly-sponsored Masters graduate diplomas. There are also courses covering disciplines in the health sector: nuclear medicine, radiopharmacy, and training for hospital physicists.

Continuous education is another important part of the INSTN's activities that relies on the expertise developed within the CEA and by its partners in industry.

The Nuclear Engineering course (known as 'GA', an abbreviation of its French name) was first taught in 1954 at the CEA Saclay site, where the first experimental piles were built. It has also been taught since 1976 at Cadarache, where fast neutron reactors were developed. GA has been taught since 1958 at the School for the Military Applications of Atomic Energy (EAMEA), under the responsibility of the INSTN. Since its creation, the INSTN has awarded diplomas to over 4400 engineers who now work in major companies or public-sector bodies in the French nuclear industry: CEA, EDF (the French electricity board), AREVA, Cogema, Marine Nationale (the French navy), IRSN (French TSO)... Many foreign students from a variety of countries have also studied for this diploma.

There are two categories of student: civilian and military. Civilian students will obtain jobs in the design or operation of nuclear reactors for power plants or research establishments, or in fuel processing facilities. They can aim to become expert consultants, analysing nuclear risks or assessing environmental impact. The EAMEA provides education for certain officers assigned to French nuclear submarines or the aircraft carrier.

The teaching faculty comprises CEA research scientists, experts from the Nuclear Safety and Radiation Protection Institute (IRSN), and engineers working in industry (EDF, AREVA, etc.). The main subjects are: nuclear physics and neutron physics, thermal hydraulics, nuclear materials, mechanics, radiological protection, nuclear instrumentation, operation and safety of Pressurised Water Reactors (PWR), nuclear reactor systems, and the nuclear fuel cycle. These courses are taught over a six-month period, followed by a final project that rounds out the student's training by applying it to an actual industrial situation.

These projects take place in the CEA's research centres, companies in the nuclear industry (EDF, AREVA, etc.), and even abroad (USA, Canada, United Kingdom, etc.). A key feature of this programme is the emphasis on practical work carried out using the INSTN facilities (ISIS training reactor, PWR simulators, radiochemistry laboratories, etc.).

Even now that the nuclear industry has reached full maturity, the Nuclear Engineering diploma is still unique in the French educational system, and affirms its mission: to train engineers who will have an in-depth, global vision of the science and the techniques applied in each phase of the life of nuclear installations from their design and construction to their operation and, finally, their dismantling.

The INSTN has committed itself to publishing all the course materials in a collection of books that will become valuable tools for students, and to publicise the contents of its courses in French and other European higher education institutions. These books are published by EDP Sciences, an expert in the promotion of scientific knowledge, and are also intended to be useful beyond the academic context as essential references for engineers and technicians in the industrial sector.

The European Nuclear Education Network (ENEN) fully supported INSTN, one of it founder members, in publishing this book. For ENEN this book constitutes the first of a series of textbooks intended for students and young professionals in Europe and worldwide, contributing to the creation of the European Educational Area.

Joseph Safieh
Nuclear Engineering Course Director
ENEN President

Contents

Chapter 6: One-group/diffusion theory

Exercises

Chapter 7: Neutron slowing down

Exercises

Chapter 8: Resonant absorption of neutrons (physical aspects)

Exercises

Chapter 9: Thermalisation of neutrons

Exercises

Chapter 16: Perturbation theory

Exercises

Chapter 17: Overview of the "Calculation Scheme"

Exercises

Chapter 18: Overview of core design problems

Exercises

Appendice A: Annotated Bibliography

Appendice B: Physical tables and constants

Appendice C: Mathematical supplement

Appendice D: Handbook

Foreword

This guide to neutron physics is intended to provide a basic knowledge of this area of science to Nuclear Engineering students.

This book is laid out according to the syllabus of the Nuclear Engineering diploma as taught at Saclay, Cadarache and Cherbourg.

Its contents are partly culled from my previous publications in the field of neutron physics and partly inspired by the documents and photocopied teaching aids used by my fellow instructors. I would therefore like to thank all of them for these very helpful materials.

I strongly advise anyone who has had no instruction in nuclear engineering to start by reading my little book *L'Énergie nucléaire* (*Que sais-je ?* No. 317, PUF, 2006), which introduces the basic concepts of energy and the structure of matter, and gives a description of nuclear power plants and an analysis of the issues involved with this type of energy.

I can also recommend another book in the same series: *La Neutronique* (PUF, *Que sais-je ?* No. 3307, 1998), which introduces the same concepts as the present *Guide*, but in a more accessible way.

Finally, I wish to mention the book I co-wrote with Jean Bussac: *Traité de neutronique* (Hermann, 1978 and 1985). Although it is not very recent, it can still be a useful reference for anyone seeking further information or additional details about the physical aspects. In the current volume, I have not reviewed certain analytical theories that were used at the time: the fast fission factor theory, the ABH (Amouyal-Benoist-Horowitz) theory, the Cadilhac secondary thermalisation model, etc. — but which have now fallen out of use. On the other hand, I have given more detail about the techniques for solving the Boltzmann equation, although I confine myself to a discussion of the main principles because this book is intended more for those who use calculation software than for the specialists who develop the software.

This guide to neutron physics is the English translation of my books published by EDP Sciences in the *Génie atomique* series, *Précis de neutronique* (2003) and (partly) *Exercices de neutronique* (2004).

For the French version of this guide and the associated exercises, we used the CGS system of units as almost all the neutron physicists did since the very beginnings of neutronics and are still doing in their books, articles, reports or communications. Nevertheless, when an English translation was decided by INSTN, it seemed to us important to follow the recommandations of the Academies and Universities, and the usage in the other branches of physics, for the SI (international system of units). So we converted all the numerical examples and applications into this last system. The main parameters which are concerned are the lengths (m or mm instead of cm), the macroscopic cross-sections (m^{-1} instead of cm^{-1}) and the fluxes ($m^{-2}\,s^{-1}$ instead of $cm^{-2}\,s^{-1}$). The mass burn-ups must be expressed in $J\,kg^{-1}$ (or more conveniently in $TJ\,kg^{-1}$ [i.e. $10^{12}\,J\,kg^{-1}$]) if this system is used. However we kept also the values with the usual units (MWd/t or GWd/t) in order to avoid the reader mentally converting.

Acknowledgements: I would like to offer my most sincere thanks to Nova traduction who translated the *Précis* and to Christopher Latham who greatly improved my own translation of the exercices, as well as ENEN and INSTN who participated to the financial backing. Many thanks also to EDP Sciences –particularly France Citrini – for agreeing to publish this book and for making improvements to its presentation. Last but not least, I express all my gratitude to Laurent Turpin and Joseph Safieh (INSTN) who let me have the possibility to finalize this project.

P. R.
Gif-sur-Yvette, February 2008.

About the Author

Paul Reuss is an alumnus of the *École Polytechnique* and has a doctorate in physical sciences.

He has spent his entire career at the French Atomic Energy Commission (CEA) at Saclay and Fontenay-aux-Roses, dividing his time between research and development, teaching, and training.

His research focuses on improving, validating, and qualifying computer programs used by engineers to design and monitor nuclear reactor cores, most notably CORÉGRAF (natural uranium and graphite reactors) and APOLLO (all reactor types, particularly water reactors). He has participated in some important developments, such as the physical study of plutonium recycling in water reactors (the topic of his doctoral thesis), the generalisation of the theory of resonant neutron absorption, and "trend research", i.e. the use of neutron physics measurements performed on critical experiments and power reactors to gain increased knowledge about nuclear data. He has over a hundred technical publications to his name.

After taking the DEA graduate degree in nuclear reactor physics, Paul Reuss soon became a lecturer, and finally the professor in charge of this DEA. He has also taught many other courses. He is currently the coordinator in charge of neutron physics teaching for the *Nuclear Engineering* diploma. In addition to the present work *(Neutron Physics: A Guide)*, he is the author of several texts on neutron physics and nuclear physics; most notably the co-author with Jean Bussac of *Traité de neutronique*, which is considered to be the key reference text for neutron physics students and specialists.

Paul Reuss has supervised the thesis work of about twenty doctoral candidates, and has been a member of over a hundred thesis committees. His other education-related activities have included two years in charge of training at the Nuclear Safety and Protection Institute (formerly IPSN, now the IRSN), as well as many seminars given at the CEA and at *Électricite de France* on neutron physics, neutron transport theory, and neutron absorption; he also wrote very complete course notes for all of the above.

Books by Paul Reuss:

- *Traité de neutronique*, Hermann, 1978 and 1985, 670 pages (with Jean Bussac).
- *Éléments de physique nucléaire à l'usage du neutronicien*, in the "Enseignement" series, CEA/INSTN, 1981, 1987 and 1995, 91 pages.
- *Éléments de neutronique*, in the "Enseignement" series, CEA/INSTN, 1986 and 1995, 175 pages.
- *Clefs pour la neutronique des réacteurs à eau*, in the "Enseignement" series, CEA/INSTN, 1990, 348 pages.
- *L'Énergie nucléaire*, in the "Que sais-je ?" series, No. 317, PUF, 1994, 1999 and 2006, 128 pages.
- *La Neutronique*, in the "Que sais-je ?" series, No. 3307, PUF, 1998, 128 pages.
- *Précis de neutronique*, Génie atomique series, EDP Sciences, 2003, 533 pages.
- *Exercices de neutronique*, Génie atomique series, EDP Sciences, 2004, 334 pages.
- *L'épopée de l'énergie nucléaire, une histoire scientifique et industrielle*, Génie atomique series, EDP Sciences, 2007, 167 pages.

Part I

FUNDAMENTALS OF NEUTRON PHYSICS

1 Introduction: general facts about nuclear energy

1.1. A brief history

1.1.1. Fermi's pile

The date was December 2nd, 1942, and for the very first time, Man created a fission chain reaction. The credit for this achievement goes to a Chicago team led by Enrico Fermi (1901–1954). On that day, the neutron population scattering in the pile amplified very gradually, even after the source was withdrawn. When the nuclear power level reached about half a watt, the cautious Fermi ordered the insertion of the cadmium control rod to stop the divergence.

What a shame that nobody had thought to invite a photographer for the occasion. The event was immortalised by a table and a drawing, reproduced below (see Figure 1.1). They show that the *critical condition* (the configuration allowing the chain reaction to be self-sustaining) was reached when 400 tonnes of graphite, 6 tonnes of uranium metal and 37 tonnes of uranium oxide were piled up[1] in a carefully planned arrangement.

Some of the main principles later to be applied in all reactors, both research reactors and power plants, were already used in Fermi's pile.

1/ Monitoring and control, symbolised by the two operators at the bottom: on the left, the operator monitoring the detector display represents the monitoring function. On the right, the operator in charge of the cadmium control rod represents the control function. Cadmium is an efficient neutron-capturing material. When the rod is pushed in, the number of neutrons captured by the cadmium increases. This reduces the number of neutrons causing fission in the uranium. The chain reaction is then stifled. Conversely, if the rod is pulled out slightly, more neutrons become available to cause fission reactions. The chain reaction is then amplified. To control the system according to requirements, the monitoring and control functions must talk to each other (in this case, simply a verbal dialogue between the two operators).

2/ Safety depends first and foremost on good monitoring and control. It also requires an emergency stop mechanism in the event of an incident. In this experiment, the emergency stop function is provided by an unseen operator located above the pile.

[1] This explains the origin of the term *atomic pile*, which we often use to refer to a nuclear reactor. It is now a slightly archaic term.

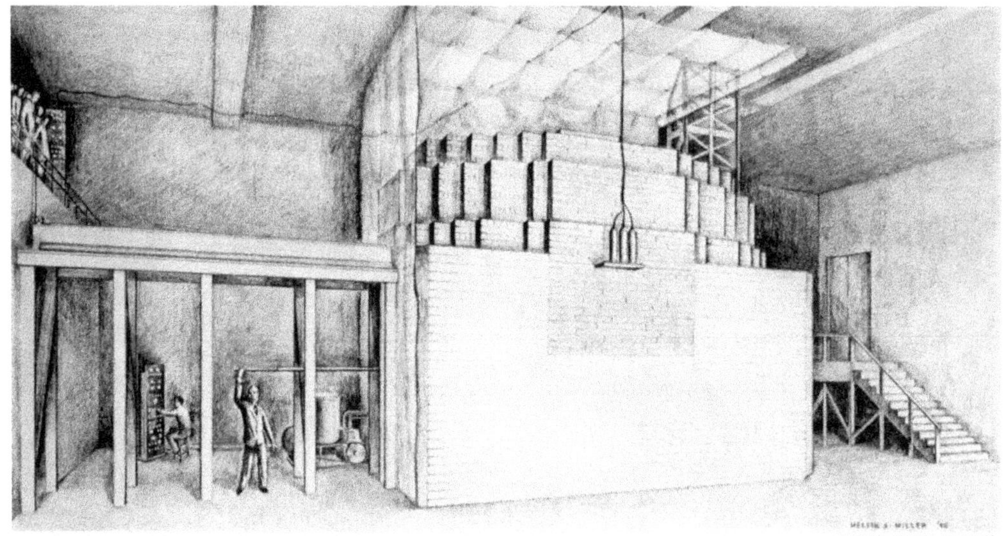

Figure 1.1. Fermi's pile (courtesy of Argonne National Laboratory).

This person is armed with an axe, and on Fermi's signal can cut the rope holding an emergency cadmium control rod. The last line of defense consisted in a tank of cadmium salt solution to release the solution into the pile.

3/ **Radiation shielding** is provided in this case by a detector hanging in front of the pile to measure the ambient radiation level. The signal passes through the cable running along the ceiling to a display placed in view of Fermi himself, on the balcony. Fermi can thereby ensure that he and his colleagues do not run the risk of excessive irradiation and can trigger the emergency stop if necessary.

1.1.2. The end of a long search...

The divergence of Fermi's pile concluded half a century of very active research in nuclear physics.

Nuclear physics is generally considered to have begun in 1896 with the discovery, almost by chance, of radioactivity by Henri Becquerel (1852–1908). Becquerel was intrigued by some photographic plates that were clouded even though they had been kept in a drawer away from sunlight. (It turned out that they had been placed in the vicinity of some uranium samples.)

A brief chronology with a few milestones:

1898: Discovery of polonium and radium by Marie Sklodowska (1867–1934) and her husband Pierre Curie (1859–1906).

1913: First atomic models developed by Ernest Rutherford (1871–1937) and Niels Bohr (1885–1962).

The very concept of the atom had already pervaded physics and chemistry for many years. It was thought of by Democritus (ca. 460-370 B.C.), and appears very clearly in the work of such chemists as John Dalton (1766–1844) and Louis Joseph Gay-Lussac (1778–1850). It is the only model that makes the periodic table of the elements suggested in 1869 by Dmitri Ivanovich Mendeleev (1834–1907) comprehensible.

1932: Discovery of the neutron by James Chadwick (1891–1974).

1934: Artificial radioactivity discovered by Frederic Joliot (1900–1958) and his wife Irene Curie (1897–1956): by the action of alpha particles on aluminium 27 (common aluminium), a neutron and a phosphorus 30 are produced. The phosphorus takes two and a half minutes to disintegrate by beta radioactivity, as this experiment revealed.

1934–1938: Study of neutron-induced reactions.

As soon as it was known how to create neutron radiation, particularly following the work of Chadwick, nuclear physicists became interested in the reactions between these particles and the various elements in Mendeleev's table. Because the neutron has no electrical charge, it can easily approach the nuclei of atoms, and experimental evidence shows that neutrons are quite often captured. This process creates an isotope of the initial nucleus, which is sometimes radioactive and is transformed into another element by beta decay (these radioactive processes will be described in greater detail in the next chapter). Fermi in particular was interested in these reactions. He thought that by bombarding uranium — element number 92, the last one in Mendeleev's table — he should be able to create new, artificial elements and extend the list of known elements.

Experiments did indeed show that the reaction gave off radioactive products, but there were clearly more of them than expected.

1938: Discovery of fission.

Fermi's experiment was repeated in other laboratories. It took four years for a correct explanation of the phenomenon to be found. When Otto Hahn (1879–1968) and Fritz Strassmann discovered that barium was present among the reaction products, and because barium was an element with an intermediate mass, they concluded that the uranium nucleus had split in two after absorbing the neutron. The discovery of fission was announced by Lise Meitner (1878–1968), who calculated that a considerable quantity of energy must be released when this splitting occurs, which lends credence to the theory.

1939: Patents for an *energy production device*.

As soon as the discovery of fission was announced, the experiment was repeated in other laboratories, and obviously there was a great deal of activity in the field of nuclear physics in the 1930s. In particular, Joliot and his colleagues Hans von Halban and Lew Kowarski (1907–1979) discovered the emission of secondary neutrons during fission, and measured the average number of secondary neutrons to be approximately three (which turned out to be slightly optimistic, as subsequent measurements would show). They quickly realised that this should make a self-sustaining chain reaction possible, because each fission would release neutrons that would in turn induce new fission reactions.

Along with their colleague Francis Perrin (1901–1992), who introduced the concept of critical mass, they designed and submitted patent applications for a device that would later be known as a nuclear reactor. The research team was disbanded within a few weeks of the German invasion, and these patents remained secret throughout the War.

Research continued on a very active basis in Great Britain, Canada, and the USA. Many feared that Nazi Germany might gain a decisive advantage by developing an atomic weapon[2]. This fear was most notably expressed in the letter sent by Albert Einstein

[2] This fear turned out to be unfounded.

(1879–1955) on the initiative of Leo Szilard (1898–1964) and Eugene Wigner (1902–1995) to President Roosevelt on 2 August 1939, and led the United States to undertake the gigantic Manhattan Project.

1945: Hiroshima and Nagasaki.

This project, led by Robert Oppenheimer (1904–1967), explored the two possible routes (which will be discussed later): uranium 235 and plutonium 239. Plutonium is element 94, which Fermi was seeking but which was finally discovered by Glenn Seaborg (1912–1999) in 1940. The plutonium route led to the Trinity test at Alamogordo (New Mexico) on 16 July 1945, and then the bomb that was dropped on Nagasaki (9 August 1945). The uranium 235 route led to the bomb that was dropped on Hiroshima (6 August 1945).

1.1.3. ... and the beginning of a great adventure

There is no doubt that this "original sin" left an impression that affected public opinion about the post-war developments in the field of nuclear energy. This was all the more true because the surrender of Japan did not spell the end of the arms race; on the contrary. The participants in this race were the United States and some of the other developed countries, particularly the USSR. It only took a few years (1952 and 1953, respectively) for these two superpowers to develop an even more terrifying weapon: the fusion H-bomb.

Let us concentrate, however, on peaceful developments in the field of nuclear energy. As soon as the war ended, most of the large industrialised nations took an interest in this new energy and, in some cases, undertook very ambitious programmes.

Nuclear electricity was first produced in 1951 in Chicago in a small fast neutron reactor called EBR-1.

In France, the CEA (Atomic Energy Commission) was created by a decree signed by Charles de Gaulle on 18 October 1945. Just three years later (15 December 1948), French atomic scientists made the Zoe pile diverge (natural uranium and heavy water) at Fort Chatillon at Fontenay-aux-Roses (Hauts-de-Seine). After Fontenay, three other major nuclear research centres were created: Saclay (Essonne), Grenoble (Isère), and Cadarache (Bouches-du-Rhône), as well as several centres for military applications. After Zoe, France developed the UNGG procedure (natural uranium, graphite, gas) with prototypes G1, G2, and G3, followed by six high-power reactors which have now been shut down. France was also interested in the heavy water procedure (Brennilis, which has now been dismantled) and fast neutron reactors (Phenix and Superphenix; Superphenix was shut down in 1997). At the end of the 1960s, there was a decision to redirect all efforts towards pressurised water reactors with slightly enriched uranium, and now five reactors rated at 900, 1300 or 1450 MWe account for about three quarters of French electricity production.

Table 1.1, which gives a breakdown of nuclear electric power plant worldwide (more than 400 reactors producing approximately 16% of the electricity), shows that France is remarkably well equipped for the production of nuclear power. France adopted this strategy because it has practically no other sources of energy (except hydro, which is already saturated). Japan is in a similar situation because it also lacks energy sources. Very different situations are observed in other countries, particularly within the European Union.

Table 1.1. Nuclear Electric Power Plant Worldwide: Installed Power in GWe (Number of Units) at the end of 2006 and Proportion of Nuclear in Electricity Production in 2006. (Source: ELECNUC, 2007 Edition).

COUNTRY	Installed	Under construction	Shutdown	Nuclear electricity
USA	99.3 (103)	–	9.8 (28)	19%
France	63.3 (59)	–	4.0 (11)	78%
Japan	47.6 (55)	0.9 (1)	0.3 (3)	30%
Russia	21.7 (31)	4.5 (5)	0.8 (5)	16%
Germany	20.3 (17)	–	5.9 (19)	31%
South Corea	17.5 (20)	1.0 (1)	–	39%
Ukraine	13.1 (15)	1.9 (2)	3.5 (4)	48%
Canada	12.6 (18)	–	0.5 (3)	16%
United Kingdom	10.2 (19)	–	3.3 (26)	18%
Sweden	9.0 (10)	–	1.2 (3)	48%
China	7.6 (10)	3.6 (4)	–	2%
Spain	7.5 (8)	–	0.6 (2)	20%
Belgium	5.8 (7)	–	0.0 (1)	54%
Taiwan	4.9 (6)	2.6 (2)	–	20%
India	3.6 (16)	3.1 (8)	–	3%
Czech Republik	3.5 (6)	–	–	32%
Switzerland	3.2 (5)	–	–	37%
Finland	2.7 (4)	1.6 (1)	–	28%
Slovakia	2.0 (5)	–	0.5 (2)	57%
Bulgaria	1.9 (2)	1.9 (2)	1.6 (4)	44%
Brazil	1.9 (2)	–	–	3%
South Africa	1.8 (2)	–	–	4%
Hungary	1.8 (4)	–	–	38%
Mexico	1.4 (2)	–	–	5%
Lithuania	1.2 (1)	–	1.2 (1)	72%
Argentina	0.9 (2)	0.7 (1)	–	7%
Slovenia	0.7 (1)	–	–	40%
Romania	0.7 (1)	0.7 (1)	–	9%
Netherlands	0.5 (1)	–	0.1 (1)	4%
Pakistan	0.4 (2)	0.3 (1)	–	3%
Armenia	0.4 (1)	–	0.4 (1)	42%
Iran	–	0.9 (1)	–	–
Italy	–	–	1.4 (4)	–
Kazakhstan	–	–	0.1 (1)	–
TOTAL	368.9 (435)	23.6 (30)	35.2 (119)	19%

In addition to these reactors there are hundreds of other, smaller reactors with a wide variety of characteristics. These include reactors used to power various systems (naval propulsion, desalination, space applications, etc.), research reactors (technological irradiation, neutron imaging, etc.), industrial reactors (production of radioactive elements), and reactors used in teaching.

1.2. Principle of a nuclear power plant

The operation of a nuclear electric power plant is very similar in principle to that of a classic coal-, oil-, or gas-fired power plant. A heat source heats a gas to a high temperature. The gas then expands in a turbine and converts its energy to mechanical energy, which is then converted to electricity via an alternator connected to the turbine. When cooled, this gas is usually cycled back to the heat source, thus completing a thermodynamic cycle. The efficiency of the conversion of heat to mechanical energy (not to mention the conversion to electricity) cannot exceed a maximum value known as the *Carnot efficiency*. This is given by the following formula:

$$r_{\max} = 1 - \frac{T_{cold}}{T_{hot}}, \tag{1.1}$$

where the two (absolute) temperatures shown are those of the heat source and the cold source. In a nuclear power plant, the heat source is not the product of the combustion of a fossil fuel with oxygen, but of nuclear "combustion"[3]: heavy nuclei undergoing fission by neutrons and releasing large amounts of energy in the form of heat.

There are several different materials and layouts that can be used to create a chain reaction, and therefore there are several different types of reactor. There are also different versions of the thermodynamic cycles to convert the heat to electricity. Figure 1.2, for example, illustrates pressurised water reactors (PWR)[4]. In France, all but one of the reactors producing nuclear electric power are of this type.

The part of the reactor where nuclear reactions take place and where heat is released is called the core. In a PWR, the core is enclosed in a vessel that can withstand the 150-bar pressure of the primary circuit. This is a closed circuit of water which the high pressure, regulated by the pressuriser, maintains in the liquid state in spite of its high temperature (approximately 280 °C at the core inlet, and 320 °C at the outlet).

Water leaving the core is divided between three loops in 900 MWe reactors and four loops in 1300 or 1450 MWe reactors. Each loop has a steam generator and a primary coolant pump. Steam generators are heat exchangers in which the water in the primary circuit transfers its heat to the water in the secondary circuit. The water enters in liquid form and is converted to vapour; the vapour produced by each loop is collected to be sent to a series of turbines, high-pressure and then low-pressure, directly coupled to the same shaft as the alternator. Finally, the steam is condensed and then taken up (in liquid form) by the secondary pumps and sent back to the steam generators.

The condenser is also a heat exchanger. It is cooled by a third water circuit that is open to the environment. In some cases, water is tapped from a river or the sea, or this third circuit might itself be cooled by a cooling tower. Because of the temperatures involved, particularly the temperature of the pressurised water in the primary circuit, which affects all the others, the overall efficiency of a nuclear reactor is not very high: about 33%. This means that, for every 3 joules released by fission reactions, 1 joule is converted to electricity, and 2 joules are dissipated in the environment as heat. Compare this to a 50% efficiency rating for classic thermal power plants, where steam is produced at a higher temperature. This drawback is not redhibitory; it is possible to design more efficient nuclear reactors, such as fast neutron reactors and graphite high-temperature reactors (HTR).

[3] By analogy with normal combustion, technicians in the nuclear industry talk about nuclear "combustion" and nuclear "fuels", but these are actually misnomers.

[4] For details, refer to the book in this series about pressurised water reactors.

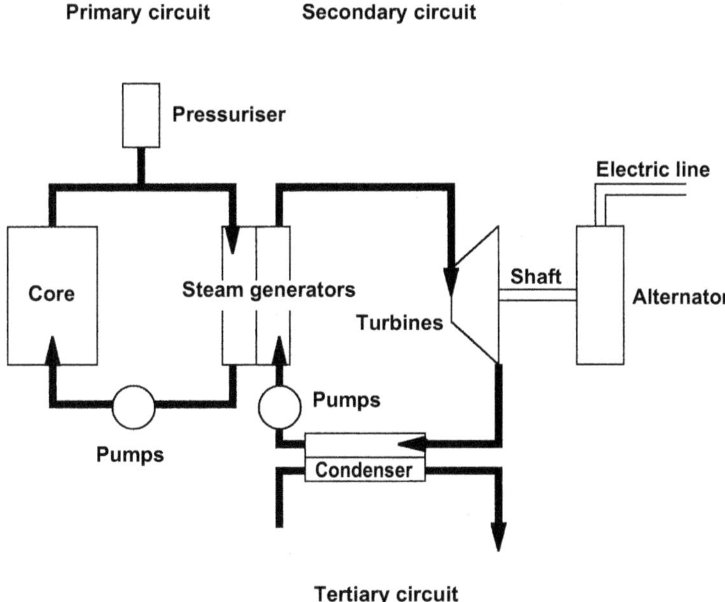

Figure 1.2. Block diagram of a PWR (Pressurised Water Reactor).

1.3. Fission

There are two types of force acting between the protons and neutrons (known collectively as "nucleons") that constitute the nucleus of an atom. The first of these is well known on a macroscopic scale: the electrical or Coulomb force. In this case, the force acts between protons only, and is repulsive because only positive charges are involved (neutrons do not carry a charge). It is a long-range force obeying an inverse square law ($1/r^2$). The second force is the only thing that can explain the stability of nuclear structures. It is called the nuclear force, and acts between all nucleons of any type. It is an attractive, short-range force. In very approximate terms, it can be compared to a type of strong glue that makes nucleons stick together very tightly whenever they are placed in contact. (In fact, the nuclear force is the result of more fundamental interactions — strong and weak interactions — between the constituents of nucleons, which are called quarks. For the purposes of studying neutron physics, however, a classic and rather approximate model like this is generally adequate.)

The proportion of protons and neutrons observed in nuclei is a result of the best possible balance between these two forces. If the proportion is not optimum, the nucleus is radioactive. The neutron/proton ratio is close to 1 for light nuclei, and gradually increases to about 1.5 for heavy nuclei. Despite the fact that this provides more dilution of the electrical charges that repel each other, the average bond energy per nucleon is lower for heavy nuclei because, all other things being equal, the negative energy due to Coulomb repulsion of the protons is proportional to the square of the number of charges. Two consequences immediately arise from these two observations:

1/ Heavy nuclei are slightly less strongly bonded than intermediate nuclei; in other words, the fission of a heavy nucleus into two fragments increases the bond energy and therefore releases energy to the outside. (Note that the bond energy is the energy released when the bond is formed or, looking at it from the opposite perspective, the energy that would need to be supplied to the system to break the bond.) Experiments have shown that a fission reaction releases approximately 200 MeV (200 million electron volts) — a huge amount of energy. Compare this to the energy released by a chemical reaction: on the order of a few electron volts per atom (e.g., 4.08 eV for the combustion of a carbon atom).

2/ Because of the ratio of neutrons to protons, less balanced for medium nuclei than heavy nuclei, it is logical that some neutrons would "evaporate" during fission, i.e. would be emitted in a free state. Like the distribution of nucleons between fragments, the number of neutrons emitted varies between zero and seven. It is the average value, v, that is important. For example, for the fission of uranium 235 (neutron-induced), the measured value of v is in the region of 2.4.

1.4. Principle of chain reactions

Chain reactions are an everyday concept. For example, fire is a chain reaction in which heat causes a chemical reaction (combustion) that produces heat, which causes combustion to continue, producing more heat, and so forth. As mentioned above, when physicists discovered that neutron-induced fission also emitted a few neutrons, they realised that chain reactions were a possibility:

$$\text{Neutrons} \Rightarrow \text{Fissions} \Rightarrow \text{Neutrons} \Rightarrow \text{Fissions} \Rightarrow \text{Neutrons} \Rightarrow \text{etc.}$$

Such a reaction would release a phenomenal amount of energy, which could be used either for peaceful purposes or to create a formidable weapon.

To produce energy for peaceful purposes, the rate of reaction must be controlled, as it would be in a classic boiler. For a weapon, as in a bomb using chemical explosives, fast amplification of the reaction is required. The behaviour of the reaction will depend on the factor k, defined below.

Let ω be the probability of a neutron placed in the system causing a fission reaction (the complementary probability $1 - \omega$ is the probability that the neutron is either captured in the system without causing a fission reaction or escapes, i.e. is captured on the outside). If a fission reaction occurs, it emits ν new neutrons on average.

The product $k = \omega\nu$ is the average number of neutrons that are direct descendants of a neutron placed in the system. By applying this reasoning in reverse on these two factors (a fission reaction releases ν neutrons, each with a probability ω of causing a fission reaction), we see that k is also the average number of fission reactions that result from an initial fission reaction. In other words, applying this argument to a large number N of fission reactions would give the following scheme:

This shows that the behaviour of the chain reaction will depend on the value of this factor k with respect to 1:

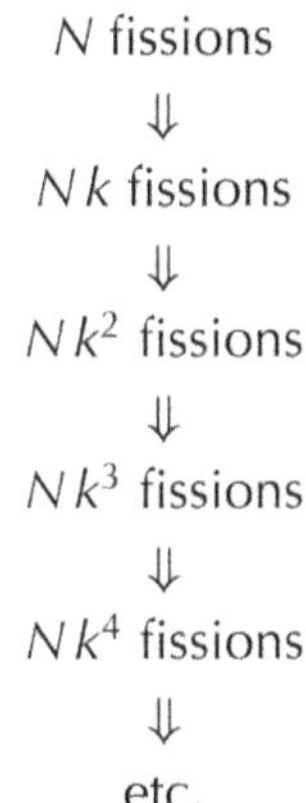

Figure 1.3. Progression of a chain reaction.

- if $k > 1$, the reaction accelerates;
- if $k < 1$, the reaction is stifled;
- if $k = 1$, the reaction is self-sustaining at a constant rate.

The first of these configurations is the suitable for a weapon. The third configuration, known as the *critical configuration*, where the factor k is equal to 1, is the configuration that exists in a reactor in stable operation. To start up a reactor or to increase its power level, it is temporarily placed in a slightly supercritical state ($k > 1$), and to decrease the power level or shut down the reactor, it is placed in a subcritical configuration ($k < 1$).

Fermi adjusted the position of a control rod to perform fine adjustments to the value of this factor. This is still the most common method in use today.

1.5. Main moderators and coolants; types of reactor

Because the factor ν is in the region of 2.4 for uranium 235, a probability ω of approximately $1/2.4 \cong 42\%$ must be achieved to attain criticality (or a little higher to create a weapon). Is this actually possible? This is what physicists were wondering in the late 1930s.

The information required to find the answer to this question was, on the whole, already known at the time, i.e.:

1/ Uranium is the only element found in nature liable to undergo fission on impact by neutrons.

2/ Uranium has two isotopes: uranium 238 and uranium 235. Uranium 238 cannot undergo fission (except in very rare cases with very energetic neutrons). Uranium 235, on the other hand, undergoes fission very readily no matter what the kinetic energy of the incident neutron.

3/ Unfortunately, the fissile 235 isotope constitutes only about 0.72% of all naturally-occurring uranium (1/139 in terms of number of nuclei), and the other isotope constitutes 99.28% of the total[5].

4/ Neutrons emitted by fission are emitted at an energy of approximately 2 MeV, i.e. around 20 000 km/s.

5/ At this energy, the cross-sections[6] of both uranium isotopes are of the same order of magnitude.

6/ By successive scatterings[7] in materials with low capture, neutrons can be decelerated until they reach approximate thermal equilibrium with matter; thermal neutrons have energy in the region of 1/40 eV, i.e. 2 or 3 km/s, if the matter that thermalised them is at room temperature.

7/ For neutrons in this speed range, the cross-section of uranium 235 is much larger than that of uranium 238 (by a factor of approximately 250).

Thus, there are two possible ways of creating chain reactions:

1/ Enriched uranium and fast neutrons: This uses neutrons at the energy at which they are produced by fission reactions, without decelerating them, and using a fuel that is highly enriched in fissile isotopes (uranium 235 or an artificial substitute such as plutonium 239).

2/ Unenriched uranium and thermal neutrons: This uses neutrons that have been slowed and thermalised by a suitable material called a *moderator*. A fuel poor in fissile isotopes can then be used; even natural uranium might suffice. Even in this case, a slow neutron passing through uranium has a higher probability of being absorbed by the 235 isotope and causing its fission than of being captured (without fission) by the 238 isotope, because the factor of 250 between the cross-sections more than compensates for the handicap of the factor 139 on the concentrations.

These methods were both introduced during the War; the second was used by Fermi's pile, with natural uranium, as already described. This method can be used for a nuclear reactor but not directly for a bomb. It takes too long to decelerate and thermalise neutrons for an efficient explosion to take place; moreover, the need to include a moderator would make the device too large and cumbersome.

This method does however produce plutonium from the fissionless capture of neutrons by uranium 238. This by-product of the chain reaction can be recovered by chemical reprocessing of the fuel, and then used instead of uranium 235 to make a weapon. (Note that the explosions at Alamogordo and Nagasaki were caused by plutonium bombs.)

The first method was also initiated by the Manhattan project, in which several procedures for the separation of uranium isotopes were explored; in particular, an electromagnetic separation, and gradual enrichment by gaseous diffusion. (The weapon that

[5] Traces of the 234 isotope are also found in natural uranium, but they are negligible.

[6] The concept of *cross-section* will be defined in the next chapter. For the present it suffices to say that the cross-section represents the *probability* (to within a factor) that an incident neutron will interact with a nearby nucleus.

[7] The scattering of a neutron by a nucleus is a collision analogous to the collision between two billiard balls.

destroyed Hiroshima was a uranium bomb that was very highly enriched in the 235 isotope.)

A moderator must consist of a material with *low neutron capture* to avoid wasting the neutrons supplied by fission. It must also contain *light nuclei,* which are most effective at slowing neutrons[8]. The moderator material must also be sufficiently dense, i.e. it must contain enough decelerating nuclei. Because of this consideration, *a liquid or a solid* is chosen in preference to a gas. In practice, these criteria lead to quite a restricted choice of moderators:

- Liquid or solid *hydrogenated materials,* such as *water* in particular.
- *Heavy water* (water in which all the hydrogen is deuterium).
- *Beryllium* or beryllium oxide **BeO**, known as *beryllia.*
- Carbon in the form of *graphite.*

Because hydrogen has a slight capturing ability, natural uranium cannot be used with hydrogenated materials (a content of about 2% of 235 isotope is the minimum requirement). In spite of the need for enrichment, therefore, these low-cost materials are often chosen because hydrogen is very effective at slowing neutrons.

Natural uranium can however be used with the three other moderators.

Graphite is the least effective choice from the point of view of neutron deceleration, but it is relatively easy to obtain[9] and not too expensive, so this is what Fermi chose. Beryllium and beryllia are rarely used because of their poor metallurgical properties. Heavy water is the best neutron-slowing moderator, but it is expensive because its production requires the separation of hydrogen isotopes (natural hydrogen contains approximately one heavy hydrogen (*deuterium*)[10] atom in 6500).

In a power reactor, a fluid must be made to circulate to extract the heat produced by the fission reactions. This fluid is called the *coolant,* and it can be a gas (carbon dioxide, helium, etc.) or a liquid (water[11], heavy water, liquid metal, etc.).

The choice of moderator (or the absence of a moderator, for a fast neutron reactor), coolant, and fuel (fissile material, physical-chemical form, geometry) as well as its cladding, define the concept of a nuclear reactor. This concept can then lead to a technology: a *type* of reactor.

1.6. Monitoring and control of reactors

Fermi understood the need for monitoring and control of reactors. For a chain reaction to be stable, the multiplication factor k must be set to precisely 1. This *control* of the chain reaction is the primary function the control equipment must perform. This often involves one or more bars containing a material that captures neutrons (such as boron or cadmium), as used by Fermi.

[8] For the same reason that the slowing of a billiard ball colliding with another ball placed on the table is more efficient if the balls have the same (or similar) mass.

[9] High chemical purity must be obtained, particularly for boron, which captures a lot of neutrons.

[10] Deuterium nuclei comprise a proton and a neutron. A normal hydrogen nucleus has one proton only.

[11] In pressurised water reactors or boiling water reactors, the water serves as both a moderator and a coolant.

Note that, in power reactors, there are *counter-reactions* related to temperature variations.

These variations can modify the intensity of neutron reactions and therefore the multiplication factor k. In practice, counter-reactions lower this factor if the power increases, providing *self-regulation of the system*. Under these conditions, control is required only to modify the equilibrium power level or to start or stop the chain reaction.

In fact, reactor controls have several functions:

- *Control.*
- *Compensating for long-term changes* in the multiplication factor due to variations in the concentrations (fission products and heavy nuclei).
- *Flattening the power distribution* (where necessary).
- *Safety*: In the event of an incident, the operator must be able to stop the chain reaction very quickly by inserting a material with a strong neutron capture capability. In practice, however, safety does not depend solely on the alertness of the operator. Every reactor also has fast *automatic shutdown systems* to be used as soon as the monitoring systems detect a malfunction.

These functions can all be performed by a single system, or each function can be performed by a dedicated system. The emergency shutdown system, however, is always run by a dedicated system.

1.7. Nuclear fuel cycle

Irradiation in a reactor is a small but essential part of the fuel's life story. Upstream from this, there are several steps leading up to the manufacture of the fuel element to be loaded into the core of a reactor. Uranium is extracted from a mine, concentrated and purified, chemical formulae are changed and, where necessary, enrichment is performed.

The downstream process can involve interim storage of the irradiated assemblies for a period pending a permanent storage decision. This is the policy currently applied in the USA, for example. (In this case, the fuel is not actually going through a "cycle".) In other countries, such as France in particular, the fuel used in nuclear reactors is usually reprocessed after a few years of interim storage that allows the radioactivity to attenuate. During reprocessing, any energetic matter can be separated and recovered for recycling, and radioactive waste can be separated according to type. In particular, fission products consisting of intermediate-mass nuclei with beta radioactivity[12] and "minor" (non-recyclable) actinides[13] which mainly have alpha radioactivity are separated for special packaging for appropriate interim storage and then final storage. Figure 1.4 is a simplified illustration of the fuel cycle for French pressurised water reactors. It shows that uranium containing a significant quantity (about 1%) of the 235 isotope can be recycled and re-enriched, although this is not currently done on a large scale for *Électricité de France*. Even more significant is the recycling of plutonium created by neutron capture in uranium 238, which is also not

[12] The various radioactivity modes will be described in Chapter 2.

[13] Actinide: element in the series of actinium, i.e. atomic number 89 and above.

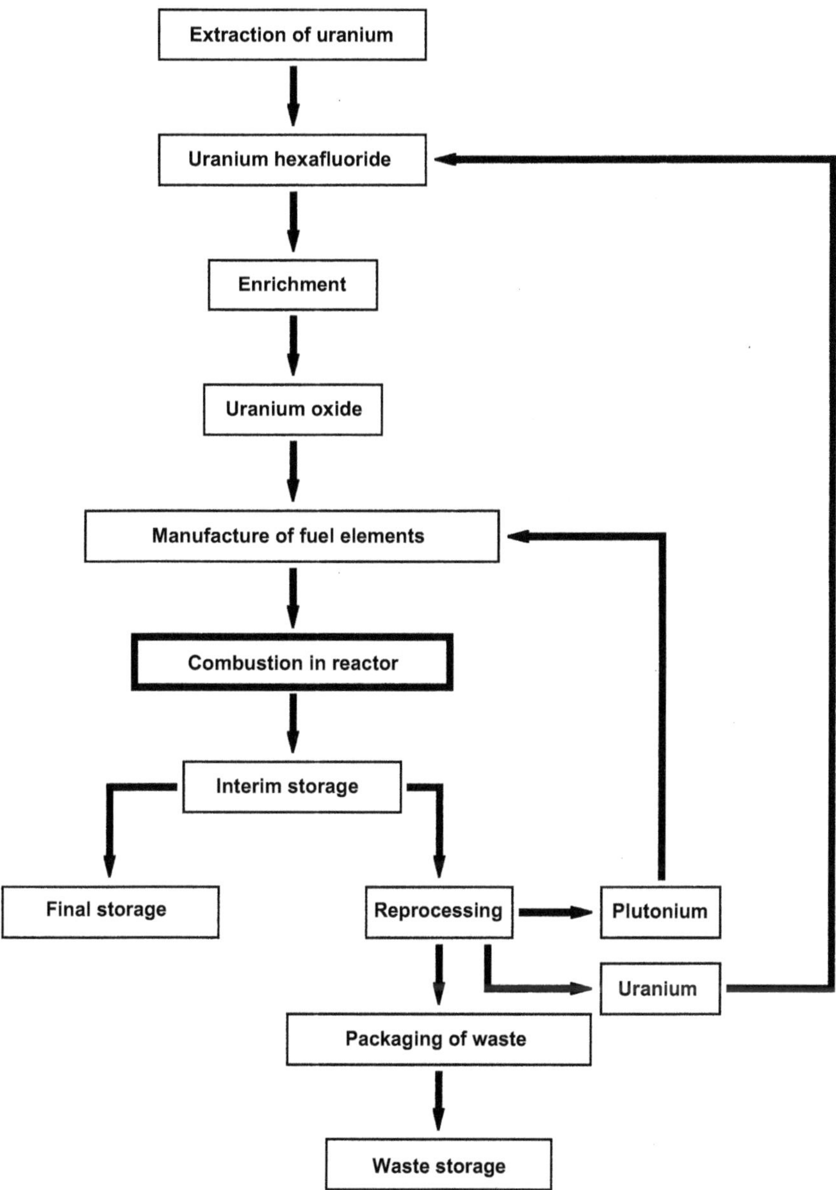

Figure 1.4. Nuclear fuel cycle of pressurised water reactors.

fully consumed during irradiation. In France, most of the plutonium obtained via uranium fuel reprocessing is recycled as a fuel called MOX, a mixed oxide of plutonium and uranium with a low uranium 235 content[14]. Currently, however, irradiated MOX assemblies are not reprocessed.

[14] The plutonium problem will be discussed in detail in Chapters 12 and 18.

1.8. Nuclear safety and radiation protection

Like all other industries, the nuclear industry has risks that need to be analysed and controlled. As we have seen, Fermi was already perfectly aware of the problems in this area.

The unique aspect of nuclear risk is obviously due to the radioactive nature of the materials used, particularly the radioactivity of the "ashes" of nuclear reactions, fission products and actinides.

The fact that radioactivity has harmful effects is now well known; the only remaining uncertainty is about the effect of low doses, because there is not enough statistical evidence to determine whether there is no risk or whether the probability of radiation-induced cancer is proportional to the dose. In practice, caution dictates applying the latter assumption. The coefficient is determined by the observation of cohorts that have been subjected to low, but not very low doses, such as the survivors of Hiroshima and Nagasaki. For the general population and all risks combined (fatal cancer, non-fatal cancer, and hereditary effects), the coefficient has been estimated at 7.3×10^{-2} per sievert. As an example, applying this assumption, a person subjected to the average natural background radiation in France (2.4 mSv/year) for fifty years, i.e. $50 \times 0.0024 = 0.12$ Sv, has a probability of $7.3 \times 10^{-2} \times 0.12 = 0.00876$, or less than 1%, of contracting a radiation-induced cancer. (By way of comparison, 25% of deaths in France are due to cancer. Because the type of cancer is independent of the cause that triggered it, it is virtually impossible to determine which cancers are due to radioactivity and which are due to other causes.)

Maximum allowable doses can be defined using this type of coefficient and by setting an *acceptable risk* threshold. From this, *annual limits on intake (ALI)* for the various radionuclides can be set, and strict radiation protection regulations can be applied to the nuclear industry in particular.

The matter of safety is approached in different terms because it concerns accidental situations rather than normal ones. Risk exists in reactors and other installations involving the fuel cycle and material transport. Safety is not among the main topics of this book, but, very briefly, the philosophy of safety has two aspects:

- *Prevention:* limiting the *occurrence* of accidents as much as possible, which has implications for the design of installations, as well as their construction, operation, maintenance, and therefore the training of personnel.

- *Mitigation:* limiting the *consequences* of an accident as much as possible whenever an accident occurs despite all the precautions taken. This leads firstly to the concept of *defence in depth* such as placing a series of barriers (at least three) between the radioactivity and the environment[15], followed by emergency plans that are tested by staging drills.

[15] For example, in water reactors, the fuel cladding forms the first barrier. It is designed to confine almost all the radioactive products of nuclear reactions. The primary circuit, especially the vessel, is the second barrier; in the event that the cladding should burst, the vessel should confine the active products because this circuit is fully isolated from the others. In the event of a rupture in this second barrier (breach in a primary pipe), the third barrier comes into play. This third barrier is the reactor containment. The accident at Three Mile Island demonstrated the effectiveness of the third barrier after the rupture of the first two.

1.9. Nuclear programmes: prospects

Table 1.1 summarises the situation with regard to nuclear equipment in various countries. The current involvement of different countries in nuclear energy varies widely. France is a leader in terms of the proportion of nuclear energy used in the production of electricity. This is without a doubt because all things "atomic" or "nuclear" generated tremendous public enthusiasm in the pre-war and post-war decades, and this enthusiasm was then very effectively reflected in government policy.

This situation is now observed in certain countries in the Far East, specially Japan, South Corea, China and India which are currently the only ones that seem to be moving forwards in this area. Public opinion in these countries is now, however, beginning to have some reservations about nuclear power, in the same way that this happened in Western countries a few years earlier. These doubts are aggravated by lobbying groups that often base their arguments on irrational ideas, but their views are by no means universal. Opinion polls have shown that many members of the general public and their elected representatives understand that fission energy is not evil, as some people might suggest, and that it even offers some tremendous advantages for the preservation of the environment; in many countries it has become practically indispensable.

It therefore seems possible that there will be renewed enthusiasm for nuclear energy in the medium term. There are early signs of this in the USA, where they made an ambitious start but then did not build any new reactors for nearly thirty years. In France, public opinion will have to be prepared for the time when the reactors currently in operation need to be replaced, whether by nuclear or non-nuclear power plants. It is obviously difficult to predict what will happen in the long term. In one or two centuries' time, the supplies of fossil fuels (except coal) will be almost exhausted. It is likely that "new" sources of energy (which have in fact been used since the dawn of time), such as solar energy (thermal or photovoltaic), wind energy, biomass, geothermics, etc., will be used increasingly but will continue to represent a relatively small proportion of the total energy used, for both technical and environmental reasons. Perhaps thermonuclear fusion will finally be brought into use, although it will probably remain at a modest level because of its tremendous technological complexity.

Perhaps new forms of energy will be discovered or invented.

Whatever happens in this uncertain future, nuclear fission energy is a resource that will remain available almost indefinitely. If the known supplies of uranium are divided by the annual consumption, it would appear that a uranium shortage will arise within a century, but in fact this is not true, for two reasons. Firstly, there is the possibility of breeding. In the types of reactor usually operated at present, particularly water reactors, uranium 235 is used almost exclusively. We say "almost" exclusively because, on the one hand, not all of the uranium 235 is consumed (because the reactor must contain a certain mass of fissile material at all times to remain critical) and, on the other hand, a little bit of uranium 238 is converted to plutonium and then fissioned. In fast breeder reactors such as *Superphenix*, for example, the rate of conversion of uranium 238 to plutonium can be increased with respect to the consumption of fissile matter until the breeding threshold, where the mass of fissile matter produced exceeds consumption, is reached or exceeded. Under these conditions, all uranium, not only uranium 235, could be used, disregarding any reprocessing losses. This means that our uranium resources could produce fifty to sixty times more energy, thereby postponing any shortage until the very distant future.

Secondly, the estimate of available resources (approximately four million tonnes) does not include all existing uranium, but only sources that can demonstrably or probably be exploited under current economic conditions (at a cost of $130 per kg). In fact, however, uranium is a relatively abundant element, and a great deal more of it could be made available if we were willing to pay a higher price for it. For example, there is an enormous amount of uranium dissolved in sea-water, although in a very low concentration. Under current conditions, this source of uranium would be far too costly to exploit, but it could be used in the event of a crisis. It would even be cost-effective to use with the breeding method, where energy production is 50 times greater per unit mass.

* * *

This introductory chapter began with Fermi, but in conclusion we should point out that Fermi did not invent the fission reactor as such.

In 1972 when the uranium deposit at Oklo (Gabon) was being mined, isotopic abundance anomalies were found. Initial investigations quickly concluded that a spontaneous chain reaction was the only possible explanation. In-depth analyses were then performed, and showed that about fifteen nuclear reaction cores had "ignited" and had been gently "simmering", probably for hundreds of thousands of years. This must have occurred almost two thousand million years ago, just after the creation of the deposit.

This phenomenon is the result of an exceptional combination of circumstances; above all the extreme age of the deposit (the further back in time you go, the higher the content of 235 isotope in uranium) and the high ore content, as well as the remarkable geological stability that allowed these remains of "fossil reactors" to survive until now.

Exercises

Exercise 1.1: the mass-energy relationship

Let us consider a nuclear reactor producing 1000 MWe by the fisson of uranium-235 atoms, with an average load factor of 70%, and with an efficiency for the conversion of heat into electricity equal to 33%. Using the Einstein formula, calculate the mass of matter converted into energy during one year, and the mass of uranium-235 that is fissioned. (*Recall:* one fission yields 200 MeV, and 1 eV equals 1.602×10^{-19} J.)

Compare with the mass of coal necessary to provide the same amount of electricity with an efficiency equal to 50%. (*Recall:* one ton equivalent of coal is 29.3 GJ.)

Exercise 1.2: fuel efficiency in a PWR

A typical pressurized water reactor (PWR) extracts about 40 000 MWd/t from its fuel (1 MWd = $86\,400 \times 10^6$ joules of heat, and where the relevant mass is the initial mass of the heavy nuclides in the reactor core). What is the fuel efficiency of the reactor, i.e. the proportion of heavy nuclei that are consumed?

N.B: for the purposes of this exercise it is sufficient to assume that all actinide species possess the same atomic mass, i.e. 235 *atomic mass units per atom.*

Exercise 1.3: which nuclei have been fissioned?

The uranium fuel described in the previous exercise initially contains 3.7% of the fissile isotope 235. Why is the number on heavy nuclei consumed greater than the number of uranium-235 nuclei? Note here that the spent fuel still contains about 1% uranium-235.

Exercise 1.4: consumption of natural uranium

How much natural uranium (0.71% ^{235}U) is needed to produce 1 kg of enriched uranium (3.7% ^{235}U) if the remaining depleted uranium contains 0.25% uranium-235? (*N.B: concentrations here are expressed in terms of mass, not number.*)

Exercise 1.5: fuel efficiency with regard to natural uranium

Use the results of exercises **1.2** and **1.4** to calculate the proportion of the atoms of the natural uranium which have been consumed in the present example.

Exercise 1.6: uranium reserves

The identified economically viable reserves of uranium (valued at less than 130 USD per kg) are estimated to be around 4 million tons. Using the results of the exercises **1.1** and **1.5**, evaluate the number of reactor × years which are available if the reactors are of the type considered in exercise **1.1**.

What result is obtained if breeder reactors are employed instead which have a fuel efficiency about equal to 50%?

Exercise 1.7: control of a chain reaction

a) If the multiplication factor of a reactor is 1.001 (positive reactivity equal to 100 pcm)[16], how many neutron generations are necessary to double the power output? How much time is necessary to achieve this increase if the generation time is:

- 0.1 μs (fast neutron reactor without taking into account the delayed neutrons);
- 20 μs (water reactor without taking into account the delayed neutrons);
- 0.1 s (any reactor taking into account the delayed neutrons)?

b) If the multiplication factor of a reactor is decreased to 0.99 (negative reactivity equal to −1000 pcm), how many neutron generations are necessary to decrease the power by a factor of 2? How many are needed for a factor of 1000?

Exercise 1.8: probability to initiate a chain reaction

Let ω be the probability that a neutron emitted by fission induces a new fission, and let p_n be the probability that this fission yields n neutrons (this number n of secondary neutrons can be between 0 and 7).

What is the mean number ν of neutrons emitted per fission? What is the multiplication factor k?

Let α be the probability that a neutron entering the reactor initiates an unlimited fission chain reaction (probability that the chain reaction starts), and let $\varepsilon = 1-\alpha$ be the probability that the fission chain reaction stops after $0, 1, 2...$ or any finite number of generations (probability that the chain reaction ceases).

Expressing this last probability according to the various possible events, write the so-called Hansen equation,

$$\varepsilon = F(\varepsilon),$$

giving ε.

What can you say about $F'(\varepsilon)$, $F(0)$, $F(1)$, $F'(1)$ and $F(\infty)$; consequently, what form does the solution of the Hansen equation take?

Evaluate by a first order calculation the value of α for a slightly overcritical system characterized by its reactivity ρ.

Determine the minimum number N of neutrons needed for a chain reaction to start with a probability greater than a given value P.

Data for the example problems (fictitious values giving a reasonable order of magnitude for the uranium-235 ν-factor)

- $p_0 = 0.09$
- $p_1 = 0.15$
- $p_2 = 0.30$
- $p_3 = 0.26$
- $p_4 = 0.10$
- $p_5 = 0.05$
- $p_6 = 0.03$
- $p_7 = 0.02$

Hint: use $\rho = 100$ pcm and $P = 0.999$.

[16] pcm means *'per cent mille'*, i.e. 10^{-5}

Solutions

Exercise 1.1: the mass-energy relationship

During one year the reactor produces

$$1000 \times 10^6 \times 0.7 \times 3 \times 365 \times 86\,400 = 6.62 \times 10^{16}\ \text{J}.$$

Dividing by $c^2 = (3 \times 10^8\ \text{m s}^{-1})^2$, we get a mass converted into energy equal to 0.736 kg.

One fission yields $200 \times 10^6 \times 1.602 \times 10^{-19} = 3.20 \times 10^{-11}$ J. In this reactor 2.07×10^{27} fissions occur each year. As the mass of one uranium-235 atom is $235 \times 10^{-3}/6.02 \times 10^{23} = 3.90 \times 10^{-25}$ kg, this number of fissions corresponds to $3.90 \times 10^{-25} \times 2.07 \times 10^{27} = 806$ kg of uranium consumed. Notice that about 0.1% of the mass is converted into energy (i.e. about 10% of the binding energy of the nucleus).

For coal, we get $6.62 \times 10^{16}/29.3 \times 10^9 = 2.26 \times 10^6$ tons per year. Thus, in terms of mass, there is a factor of order a few million between nuclear and chemical energy.

Exercise 1.2: fuel efficiency in a PWR

40 000 MWd corresponds to $4 \times 10^4 \times 10^6 \times 86\,400/3.20 \times 10^{-11} = 1.08 \times 10^{26}$ fissions, which represents $1.08 \times 10^{26} \times 3.90 \times 10^{-25} = 42.1$ kg of uranium consumed per ton of heavy nuclei in the reactor. Therefore, the fuel efficiency is 4.21%.

Exercise 1.3: which nuclei have been fissioned?

Part of the energy comes from the fission of plutonium atoms, following the transmutation of uranium-238 into plutonium-239. The contribution from plutonium to the fuel efficiency is 4.2 – (3.7 – 1) = 1.5%, i.e. more than one third.

Exercise 1.4: consumption of natural uranium

Let m be the masses and e the uranium-235 concentrations, and let 0, 1, and 2 be the subscripts for natural uranium, enriched uranium, and depleted uranium, respectively. The mass balances for all the uranium and for uranium-235 are

$$m_0 = m_1 + m_2,$$
$$m_0 e_0 = m_1 e_1 + m_2 e_2.$$

Therefore,

$$\frac{m_0}{m_1} = \frac{e_1 - e_2}{e_0 - e_2} = 7.5.$$

Exercise 1.5: fuel efficiency with regard to natural uranium

Using the previous results, we get: 4.21%/7.5 = 0.56%.

Exercise 1.6: uranium reserves

With these reserves and this fuel efficiency, there is $4 \times 10^9 \times 0.56\% = 2.24 \times 10^7$ kg of fuel available. As 806 kg of uranium are consumed per reactor, per year, this corresponds to 27 800 reactor × years. At present there are about 400 nuclear reactors in the world; hence, the reserves will be exhausted in 70 years.

Breeder reactors provide about 100 times improvement in fuel efficiency; thus, the known reserves of uranium would in this case provide sufficient fuel for 70 centuries instead of 70 years at the present rate of consumption.

Exercise 1.7: control of a chain reaction

a) The number n of generations is given by the equation $1.001^n = 2$, therefore $n = 693$. For each of the three generation times the power doubles in 69 µs, 14 ms, and 69 s, respectively.

b) The equation giving n is now $0.99^n = 1/2$, hence $n = 69$. To get the factor 1000, about 10 'half lives' are necessary, or about 700 generations.

Exercise 1.8: probability to initiate a chain reaction

The number ν is the average value of n with the weights p_n, i.e. 2.5. The multiplication factor is $k = \nu\omega$.

The reaction ceases either when the neutron is captured, or escapes from the core, or when it induces a fission with the emission of n neutrons where none of them subsequently initiates another fission:

$$\varepsilon = 1 - \omega + \omega\left(p_0 + p_1\varepsilon + p_2\varepsilon^2 + \cdots + p_7\varepsilon^7\right).$$

The function F is increasing for all ε; for $\varepsilon = 0$, its value is

$$1 - \omega(1 - p_0) < 1;$$

for $\varepsilon = 1$, its value is 1; if $\varepsilon \to \infty$—which is clearly not physical—then $F \to \infty$. The derivative of F equals k for $\varepsilon = 1$. Consequently, the function $F(\varepsilon)$ possesses the following properties.

- For $k < 1$, $\varepsilon = F(\varepsilon)$ when $\varepsilon = 1$, and at a second point where $\varepsilon > 1$ (which has no physical meaning). The probability α equals zero.
- For $k > 1$, $\varepsilon = F(\varepsilon)$ at a point where $\varepsilon < 1$, and again when $\varepsilon = 1$ (which has no physical meaning). The probability α is finite.
- For $k = 1$, a double root of the Hansen equation occurs: $\varepsilon = 1$; $\alpha = 0$.

If the reactivity is positive, then a Taylor expansion gives

$$\alpha \simeq \frac{2\nu}{\xi}\rho,$$

where the number $\xi = 6.1$ is the average value of $n(n-1)$ with the weights p_n.

The number N is obtained by the equation

$$\varepsilon^N = (1 - \alpha)^N < 1 - P.$$

With $\rho = 100$ pcm and $P = 0.999$, we get $\alpha = 0.00082$ and $N > 8400$.

Nuclear physics for neutron physicists

Introduction

Neutron physics is the study of the travel of neutrons through matter and the resulting reactions, most notably power generation via the fission of nuclei of heavy atoms[1].

Neutron physics is unusual in that it lies between macroscopic and microscopic physics. In general terms, it describes the interaction of elementary particles — neutrons — with the nuclei of atoms that form matter. Neutron physics is thus derived from nuclear physics. Because the neutron population is very large, however, it can be treated globally by likening it to a continuous fluid and applying the techniques of fluid mechanics. To do this, we use an equation formulated by Ludwig Boltzmann (1844–1906) in the context of his work on statistical mechanics in 1879 — more than half a century before the discovery of the neutron. The study and numerical processing of the Boltzmann equation for neutrons is the main challenge faced by neutron physicists.

The equation has *two* mathematical operators to represent both aspects of neutron migration: a *transport operator* and a *collision operator*.

The path of a neutron from emission to absorption consists of a combination of straight lines, without interaction with matter (in technical terms, this is known as neutron *transport*), and interactions with an atomic nucleus (these are "collisions", analogous to a projectile hitting a target).

The microscopic aspect of the problem becomes important where collisions are concerned. To write the Boltzmann equation and to try to solve it, we use only a phenomenological model, essentially in terms of cross-section. Neutron physics thus goes beyond nuclear physics, since it deals with the transport of particles, but on the other hand it falls far short of covering all of nuclear physics, since it considers only the input and output of reactions, without examining the details of what happens inside the "black box".

This is the approach taken in the present chapter. The phrase "for Neutron Physicists" in the title is intended as a reminder that this chapter does not deal with all aspects of nuclear physics. We will take a quick peek into the black box for a better understanding of the inputs and outputs of nuclear reactions that govern neutron physics, but only the most basic and essential points will be discussed.

Note: This chapter contains some numerical values and orders of magnitude; additional and more accurate values are given in the appendix entitled Physical Constants and Tables.

---

[1] See P. REUSS, *La Neutronique, Que sais-je?*, n° 3307, PUF, 1998.

A. STRUCTURE OF MATTER AND NUCLEAR BINDING ENERGY

2.1. Structure of matter

2.1.1. The classical atomic model

The structure of atoms gradually came to be understood through the discovery of radioactivity and the many nuclear physics experiments conducted over the first half of the twentieth century. The first model of interest is the Bohr–Rutherford atomic model (1913), where the atom looks like a miniature solar system, with a nucleus that has electrons gravitating around it like planets orbiting the sun. This model introduces two fundamental concepts of neutron physics:

1/ The nucleus is very small compared to the overall size of the atom (on the order of 10^{-15} to 10^{-14} m compared to 10^{-10} m; the equivalent of a small marble in the middle of a football pitch).

2/ Although small, the nucleus contains almost the entire mass of the atom. This means that it is extremely dense: on the order of 10^{17} kg/m^3.

Electrons orbit the nucleus. Electrons are lightweight particles, each carrying a negative unit charge. The nucleus carries one positive unit charge for each electron in the atom so that, in its normal state, the atom is neutral overall.

2.1.2. Elements and isotopes

The number of electrons, Z, and therefore the number of electric charges in the nucleus, defines a chemical element. Chemical bonds involve the electronic structure of atoms only. The electrons arrange themselves in a series of layers, which explain the patterns observed by Mendeleev as he created his Periodic Table of the Elements.

The structure of the nucleus was unknown when Bohr proposed his atomic model. It soon became apparent, however, with experimental proof arriving thanks to Chadwick in 1932, that nuclei comprised two types of particle: protons and neutrons, collectively known as *nucleons*.

- The *proton* is 1836 times heavier than the electron, and has a positive electric charge.
- The *neutron* has almost the same mass (1839 times heavier than the electron), but carries no electric charge.

Each element is characterised by the number Z of protons (which is also the number of electrons), and we often find that different atoms of the same element have a different number N of neutrons accompanying the protons in the nucleus. These are *isotopes*[2]. A *nuclide* is a nuclear species characterised by Z and N, or, according to preference,

[2] This word means "same place", and indicates that these different atoms occupy the same position in the Periodic Table.

by Z (*atomic number*) and $A = Z + N$ (*mass number*). The mass number is in fact approximately the mass of an atom. The presence of isotopes explains why certain elements have a non-integer (average) mass.

Although isotopes appear to be the same from a chemical point of view, their nuclear properties can be completely different. Two examples were seen in the previous chapter: hydrogen ($Z = 1$) comprises two isotopes: ordinary hydrogen ($N = 0$, $A = 1$) and heavy hydrogen (*deuterium*) ($N = 1$, $A = 2$); there is also superheavy hydrogen (*tritium*) ($N = 2$, $A = 3$), which is radioactive. At the other end of the Periodic Table, there is uranium ($Z = 92$), which essentially has two natural isotopes: uranium 235 ($N = 143$, $A = 235$) and uranium 238 ($N = 146$, $A = 238$).

2.1.3. Nuclide notation

Nuclides are designated in full by the name of the element (without capitals) followed by the mass number (without a hyphen), for example: **uranium 235**.

The abbreviated notation uses the symbol of the element with the mass number A placed as a superscript to the left, for example: $^{235}\mathbf{U}$. Sometimes it is also useful to write the atomic number Z as a reminder, even though this information is already known for any given element. In this case, Z is placed as a subscript to the left, e.g.: $^{235}_{92}\mathbf{U}$.

Exceptions: For the heavy isotopes of hydrogen, the symbols **D** (deuterium) and **T** (tritium) are normally used instead of $^{2}\mathbf{H}$ and $^{3}\mathbf{H}$.

2.1.4. Stable and unstable nuclei

The nuclides found in nature show that not all N-Z combinations are possible. Most of them are stable nuclei that will endure forever unless destroyed by a nuclear reaction. Some, such as potassium 40, are radioactive: after a certain time, characterised by the radioactive half-life, which is defined below, they are spontaneously transformed into a different nuclide.

All elements beyond bismuth ($Z = 83$) are radioactive.

A helpful way to visualise all the nuclides is to represent each one by a point on a Cartesian diagram (Z-N) as in Figure 2.1, which shows the 267 stable nuclides found on Earth and 19 natural nuclides that are almost stable (i.e. have a very long half-life), such as the two main natural isotopes of uranium.

Note that there is no point at $Z = 43$ (technetium) or $Z = 61$ (promethium). These two elements have no stable isotope and are not found in nature.

Hundreds of new, man-made nuclides have been created using nuclear reactions, and all are radioactive. (In other words: all the possible stable nuclei are found in nature). All are located in the immediate neighbourhood of the cloud of points in the figure.

This cloud follows an approximate line that seems to correspond to an optimum neutron/proton ratio for each value of Z. This line is called the *valley of stability*. The liquid drop model, described below, attempts to account for this.

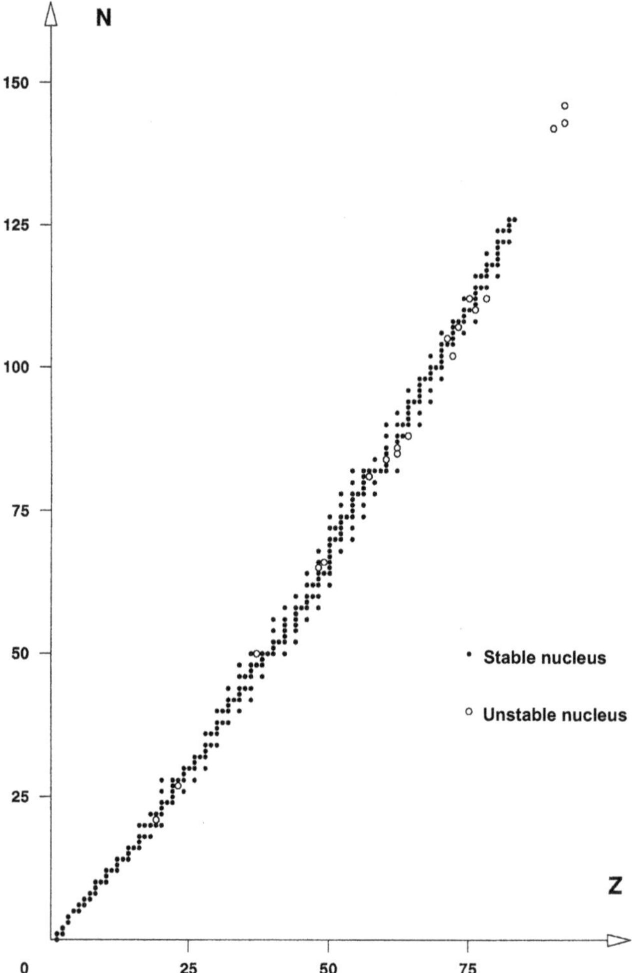

Figure 2.1. Diagram of *Z-N* combinations giving stable nuclei (includes natural unstable nuclei with half-life greater than 100 million years).

2.1.5. Pattern of stable nuclei

A careful study of Figure 2.2 shows that there are more points at the even abscissas and ordinates than at the odd ones. Out of 286 stable or nearly-stable nuclides, there are:

- 167 (58%) with an even number of protons and of neutrons,
- 57 (20%) with an even number of protons but an odd number of neutrons,
- 53 (19%) with an odd number of protons but an even number of neutrons,
- only 9 (3%) with an odd number of protons and of neutrons.

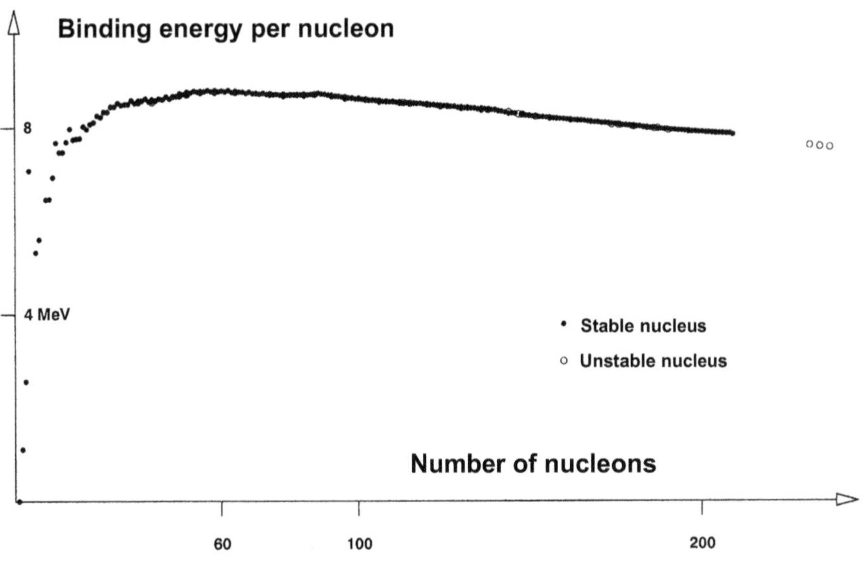

Figure 2.2. Average binding energy per nucleon of stable nuclei as a function of mass number (including natural unstable nuclei with a very long half-life).

If we exclude four light nuclei, ${}^{2}_{1}\mathbf{H}$, ${}^{6}_{3}\mathbf{Li}$, ${}^{10}_{5}\mathbf{B}$, and ${}^{14}_{7}\mathbf{N}$, then only ${}^{40}_{19}\mathbf{K}$, ${}^{50}_{23}\mathbf{V}$, ${}^{138}_{57}\mathbf{La}$, ${}^{176}_{71}\mathbf{Lu}$, and ${}^{180}_{73}\mathbf{Ta}$ are odd–odd; and these five are all radioactive.

These differences will be explained by the binding energy and the beta radioactivity.

2.2. Nuclear binding energy

2.2.1. Mass defect and nuclear binding energy

One would expect the mass of a nucleus A containing Z protons and N neutrons to be the sum of the masses of its constituents, but measurements show that this is not so. There is in fact a mass defect

$$\Delta m = Zm_P + Nm_N - m_A, \tag{2.1}$$

which is of the order of a percent, and therefore accurately measurable.

According to Einstein's principle of the equivalence of mass and energy, this mass defect corresponds to the nuclear binding energy:

$$W = \Delta mc^2. \tag{2.2}$$

This energy, W, was taken from the mass of the constituents and released when the nucleus was formed.

Conversely, this is the amount of energy that would have to be provided to separate the nucleons that are together in the nucleus. (Imagine this as the work that must be supplied to "pull apart" nucleons that were stuck to each other.)

There is a similar mass defect for any bond; for example, that of electrons about a nucleus, or atoms in a molecule. In these cases, however, the mass defect is negligible (i.e. the binding energy is very small with respect to the mass of the constituents).

By contrast, the binding energy of nuclei is enormous. As an order of magnitude, nuclear binding energies are typically a million times greater than chemical bonds.

The mass defect measurements, and therefore the binding energies W of the various nuclides can therefore be shown on a diagram. For practical reasons, W/A (*average binding energy per nucleon*) instead of W is usually plotted against A (mass number). The resulting diagram is shown in Figure 2.2.

2.2.2. Nuclear units

SI units are not very suitable for the orders of magnitude encountered in atomic and nuclear physics, so two new units have been introduced for use in this field.

- The *atomic mass unit* (u) is defined as one twelfth the mass of a carbon 12 atom, i.e. approximately the mass of one nucleon: $1\ u = 1.66054 \times 10^{-27}$ kg.
- The *electron volt* (eV) is defined as the energy acquired by a unit electric charge passing through a potential difference of 1 volt: $1\ \text{eV} = 1.60218 \times 10^{-19}$ J.

  Like other units, the electron volt has multiples and sub-multiples, such as *meV*, *keV*, *MeV*, *GeV*, and *TeV*.

The Einstein equivalence between these two units is: $1\ \text{u} \sim 931.49$ MeV.

2.2.3. Nuclear forces

Classically, the energy of a bond is the result of the work performed by the binding forces. The forces operating in nuclei are as follows:

1/ *Electrical* (Coulomb) forces are well known on a macroscopic scale. These are long-range forces that diminish as a function of $1/r^2$. In the nucleus, these forces operate between protons only, since neutrons have no charge, and are *repulsive* since they are operating between charges of the same sign. These forces can certainly not explain the stability of nuclei.

2/ The cohesion of nuclei is the result of a second type of force: the *nuclear* force. This is an attractive force between all nucleons of any type. It is a very strong, short-range force that can be compared to a type of strong glue that holds nuclear particles together as soon as they come into contact.

This (macroscopic) analogy is obviously very approximate. A better model, although still a phenomenological one, was introduced in the 1930s: the resonant exchange of mesons[3] between nucleons. Nowadays, in the "standard model", it is considered that each nucleon is composed of three quarks (*uud* for a proton and *udd* for a neutron), and that the nuclear forces are the result of interactions between quarks, governed by quantum chromodynamics and thanks to gluons, which are the mediators of the strong interaction.

[3] Particles with an intermediate mass between that of electrons and that of nucleons.

These concepts are not essential for physicists who study reactors, which involve low-energy nuclear physics.

2.2.4. Liquid drop model

The Liquid Drop Model is a completely phenomenological nuclear model that helps to grasp the basics of nuclear physics required for our purposes. It was proposed in 1935 by Hans Albrecht Bethe (1906-2005) and Carl von Weizsäcker (1912-2007).

Bethe and von Weizsäcker began by assuming that the nucleus, as a relatively large number of "marbles" joined by contact forces, is similar to a drop of liquid, which is a collection of molecules joined by short-range forces (Van der Waals forces). They then used the formulation of binding energy for a drop and added some special terms. This led them to the following expression for binding energy W as a function of mass number A and atomic number Z:

$$W = a_v A - a_s A^{2/3} - a_a \frac{(A/2 - Z)^2}{A} a_c \frac{Z^2}{A^{1/3}} + \delta a_p A^{-1/2}. \qquad (2.3)$$

The first term is the *volume term*: for each nucleon, a certain quantity of energy corresponds to the bonds with nearest neighbours. With an appropriate coefficient, this gives a total proportional to the number of nucleons.

The second term is the *surface term*: nucleons located at the surface of the nucleus have no outside neighbours and are therefore less bonded, giving a negative correction. Assuming that the "marbles" are incompressible, as in a drop of water, (their radius is on the order of 1.2×10^{-15} m), and assuming the drop to be spherical, the volume is proportional to A, the radius is proportional to $A^{1/3}$, and the surface — and therefore the number of nucleons concerned — is proportional to $A^{2/3}$.

For the case of a normal liquid drop, these are the only two terms required. For a "drop" of nuclear matter, however, three other corrective terms must be introduced. The *asymmetry term* states that, all other things being equal, maximum stability (and therefore maximum binding energy) is attained when the number of nucleons is equally divided between protons and neutrons. This term goes to zero when $Z = N = A/2$, and gives a lower binding energy when the proton/neutron distribution is not equal.

The *Coulomb term* expresses the Coulomb repulsion between the positive electric charges. Its form comes from a simple electrostatics calculation assuming that the positive charges are uniformly distributed in the spherical drop.

The final term is the *parity term*, which reflects the fact that protons on the one hand and neutrons on the other hand tend to pair up, independently of whether the two types of nucleon are present in equal proportions. The creation of each pair corresponds to a binding energy. By convention, the case Z even and N odd, or vice-versa, is used as the reference ($\delta = 0$). For the even–even case, a term for the bond of the additional pair must be added by taking $\delta = +1$. For the odd–odd case, the term for the bond of the missing pair is removed by taking $\delta = -1$. The coefficient a_p is chosen so that the expression will correspond to plus or minus this bond. (Some authors suggest a slightly different relationship to the mass number, i.e. a function of $A^{-1/2}$.) The presence of this parity term explains why more nuclides are characterised by even numbers than odd numbers.

Certain coefficients a can be calculated theoretically. In practice, to compensate for the approximations of this model and to obtain the most accurate formula possible, the coefficients are adjusted by a least squares method. This produces the closest possible values to actual mass measurements. For example, the coefficients proposed by Luc Valentin are as follows (expressed in MeV):

$$a_v = 15.56, \quad a_s = 17.23, \quad a_a = 94.4, \quad a_c = 0.70, \quad a_p = 12.$$

The formula obviously does not take minor irregularities into account, particularly for light nuclei, which are not really comparable to a drop of liquid. It does, however, fit quite well with the curves approximated by the sets of points in Figure 2.1 and Figure 2.2.

The *valley of stability* can be seen quite clearly by identifying the value of Z that gives the highest binding energy for a fixed value of A. It is worth noting that this valley lies approximately along the line of neutron–proton symmetry for light nuclei. For heavier nuclei, the relative proportion of neutrons must increase to approximately 50% more than protons. This reduces the asymmetry term, but the electric charges are "diluted", and so the Coulomb term decreases. The valley curves to achieve the best possible trade-off between these two effects. The curve of *binding energy per nucleon W/A* is then found by writing the equation for the valley of stability into the Bethe–Weizsäcker formula.

Figure 2.3 shows the contribution of the various terms (except the parity term) to W/A as a function of A. In particular, note how the optimum trade-off between the Coulomb term and the asymmetry term changes as a function of A.

2.2.5. Magic numbers and the layer model

The liquid drop model, then, does not account for small irregularities. In particular, slightly higher binding energies are observed in nuclei possessing a *magic number* of protons and/or neutrons: 8, 20, 28, 50, 82, or 126.

Most notably, helium 4 ($^4_2\mathbf{He}$ also known as an *alpha particle*) and oxygen 16 ($^{16}_{8}\mathbf{O}$), which are doubly magical, are much more strongly bonded than their nearest neighbours in Figure 2.3.

This is the result of "layers" related to the quantum aspect of nuclear physics. These layers are comparable to the electron layers of atoms.

The layer model complements the liquid drop model and attempts to take this aspect into account (with a potential that goes as r^2, it explains the first three magic numbers.) This model will not be explained in detail here, but it is similar to the atomic model with layers of electrons: we take a simple, empirical form of the nuclear potential $V(r)$ in which the nucleons are swimming, and find the stationary solutions (eigenfunctions) of the Schrödinger equation:

$$\left[-\frac{\hbar^2}{2m}\Delta + V(r)\right]\Psi(\vec{r}) = E\Psi(\vec{r}). \qquad (2.4)$$

2.2.6. Spin and parity

Spin and parity are also quantum concepts. The spin characterises the intrinsic angular momentum in terms of $\hbar$ (reduced Planck's constant: $\hbar = h/2\pi$). A proton and a neutron both have spin 1/2. The two nucleons in a pair have opposite spin, so that the overall contribution to the spin of the nucleus is zero.

Even–even nuclei also have zero spin and can be considered as approximately spherical.

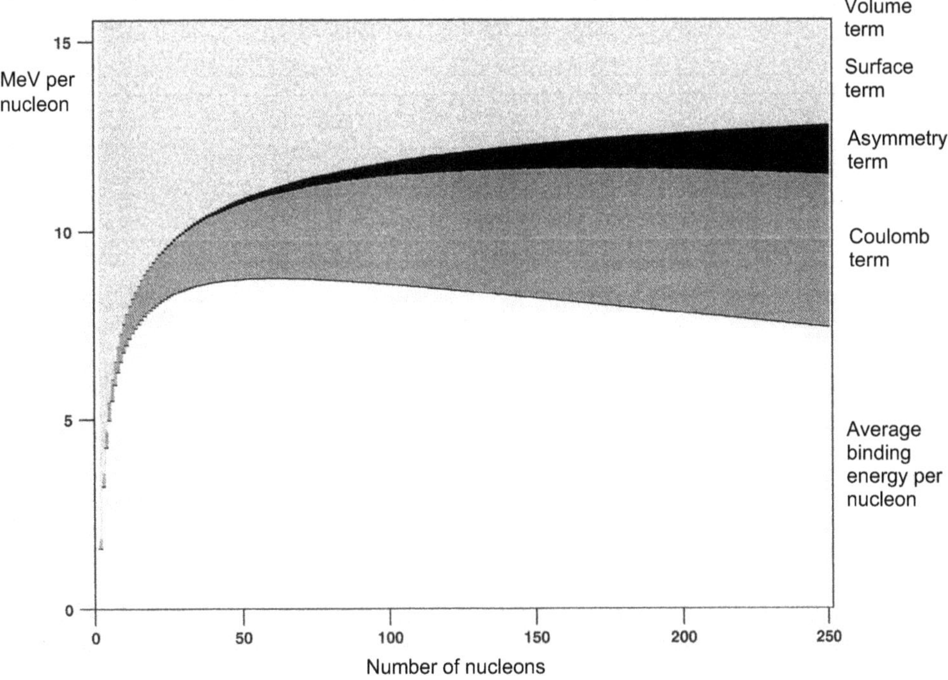

Figure 2.3. Average binding energy per nucleon of stable nuclei as a function of mass number according to the liquid drop model. The volume term is the constant value forming the top boundary of the box, the three negative corrections are represented by the shaded areas, and the binding energy is the curve delineating the white area.

For even–odd nuclei, the spin is of the form $n + 1/2$ (with n not necessarily zero) and for odd–odd nuclei, the spin is a whole number. Deviation from the spherical shape can be demonstrated by measuring a quadripole moment.

Parity (+ or –) tells whether the wave function associated with the nucleus is symmetric or antisymmetric with respect to the origin of the coordinate system.

2.2.7. Excited levels of nuclei (isomeric states)

Another consequence of quantum nuclear physics is the existence of excited levels. In its normal state, a nucleus is at the lowest energy level, or *ground state*, but after a nuclear reaction, for example, a nucleus may acquire an excitation energy. This means that the nucleus goes to a higher energy level. An excited nucleus will release this additional energy, usually quickly, in one or more stages. This most often occurs in the form of gamma radiation (high-energy photons).

As in the case of atoms, the excited states of nuclei take the form of a structure with discrete levels, but the distance between the levels is six orders of magnitude greater.

The width Γ of these levels is related to their lifetime τ by the Heisenberg uncertainty relation: $\Gamma\tau \cong \hbar$.

As a general rule, the space between levels decreases as you climb the ladder of excitation energies, until a continuum zone where the levels overlap. Note also that the overall structure is weaker for light nuclei, where the energy separating the ground state from the first level is on the order of a few MeV, than for heavy nuclei, where the separation is on the order of only a few tens of keV.

2.2.8. Other nuclear models

There are two difficulties involved in modelling the atomic nucleus. The first arises from having only approximate knowledge of the nuclear interaction, and the second arises from the practical impossibility of solving the N-body problem, particularly for large N. This is why all the models proposed have limitations. The two models mentioned above are not the only ones; they merely illustrate two possible approaches.

The first approach is purely phenomenological, and is based on a classical analogy. It is illustrated by the liquid drop model, which will be the most useful one for our purposes. Other examples include the Fermi model, where the nucleus is likened to a gas of fermions (protons and neutrons) in a nuclear potential well, and the optical model, which is useful for handling very high-energy reactions.

The second approach attempts to treat each nucleon individually, but by handling all the interactions with others as an average. This is how the layer model operates, for example. The mathematical representation of the potential field can be improved, as in the best-known example: the Woods-Saxon potential, which finds all the magic numbers:

$$V(r) = \frac{V_0}{1 + \exp\left(\dfrac{r - R}{0.228\,a}\right)}, \tag{2.5}$$

where R is the radius of the nucleus, r is the distance to the centre, and V_0 (on the order of 50 MeV) and a are adjustable parameters.

Let us also mention the compound nucleus model, which will be introduced with nuclear reactions.

2.3. Principle of release of nuclear energy

2.3.1. Nuclear recombination

It is well known that chemistry is the art of recombining atoms into new molecules. The nature and number of the atoms themselves do not change; only the bonds and the associated energies change.

If the binding energy *increases*, the supplement is evacuated to the outside (usually in the form of heat). This is called an *exothermic* (or *exoenergetic*) reaction. For example, the combining of two $\mathbf{H_2}$ hydrogen molecules with an $\mathbf{O_2}$ oxygen molecule to form two water molecules ($\mathbf{H_2O}$) is an exothermic reaction (used in blowlamps). If the binding energy *decreases*, the difference must come from outside (for example, taken up in the form of heat). This is an *endothermic* reaction. The dissociation of water into hydrogen and oxygen is one example.

Nuclear reactions obey the same general principle, except that they involve *nucleons* rather than atoms[4]. Like chemical reactions, they are either exoenergetic or endoenergetic according to the sign of the change in binding energy.

Because nuclear binding energies are typically a million times greater than chemical bond energies, the reaction energies also differ by this order of magnitude.

2.3.2. Reaction energy

A reaction energy can easily be calculated using the accurate tables that are available giving the masses of all nuclides (or the associated atoms). The mass *difference* between the initial reactants and the final products is calculated, and this difference is converted into energy. The reaction is exoenergetic if the mass decreases and endoenergetic if it increases.

For example, consider the following reaction[5]:

$$\text{neutron} + \text{nitrogen 14} \Rightarrow \text{proton} + \text{carbon 14}.$$

The following values are found in a mass table[6]:

- neutron: 1.008665 u,
- nitrogen 14: 14.003074 u,
- light hydrogen: 1.007825 u,
- carbon 14: 14.003242 u.

The mass totals are therefore:

- initial reactants: 15.011739 u,
- final products: 15.011067 u,

this gives a mass defect of 0.000672 u, equivalent to 0.626 MeV. This is therefore an exoenergetic reaction.

2.3.3. Principle of fusion and fission

For the purpose of producing nuclear energy, exoenergetic reactions, i.e. those giving an increase in binding energy, are obviously required. The overall shape of the curve showing binding energy per nucleon (Figure 2.4), with a maximum near mass 60 (iron and neighbouring elements) hints at two possible approaches:

---

4 Like the atoms in a chemical reaction, the number of nucleons does not change in a nuclear reaction. Protons can, however, be transformed into neutrons, and vice-versa, by beta decay.

5 This reaction occurs in the upper atmosphere, and explains the presence of small quantities of radioactive carbon 14 in the biosphere. The neutrons are produced by reactions induced by the protons from cosmic radiation.

6 Except for the neutron, these are atomic masses; this is equivalent to using the masses of nuclei while neglecting the electronic bond energies, since there is the same number of electrons on both sides of the reaction.

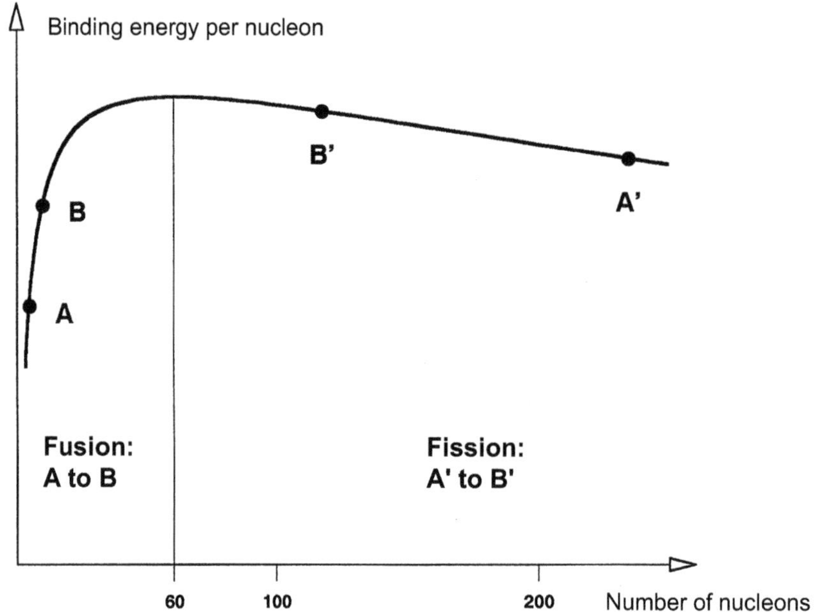

Figure 2.4. Average binding energy per nucleon of stable nuclei and principle of fusion and fission.

- *Fusion* is the joining of small nuclei to form larger ones, which means going from A to B, for example, with an increase of W/A.
- *Fission* involves splitting a large nucleus in two (for example), which means going from A′ to B′, also with an increase of W/A.

Stars produce their energy by fusion. The main mechanism (but not the only one) is a complicated series of reactions that can be summarised as follows:

$$4 \text{ hydrogen } 1 \Rightarrow \text{helium } 4,$$

generating approximately 28 MeV, i.e. the binding energy of helium 4.

Engineers find that a fusion reaction is difficult to produce because it requires placing two nuclei in contact with each other to bring the nuclear binding forces into play, which means overcoming the Coulomb repulsion between the (positively charged) nuclei.

The fusion of two heavy hydrogen isotopes is the reaction usually considered because it is the least problematic:

$$\text{deuterium (hydrogen 2)} + \text{tritium (hydrogen 3)} \Rightarrow \text{helium 4} + \text{neutron}.$$

Note that deuterium is one of the natural isotopes of hydrogen. Tritium (radioactive nuclide with a half-life of 12 years), however, must be manufactured. The reaction used to do this is neutron capture by lithium. The main reaction is:

$$\text{neutron} + \text{lithium 6} \Rightarrow \text{tritium} + \text{helium 4}.$$

(It is advantageous to combine these reactions by recovering the neutron from the fusion reaction to regenerate the tritium that was consumed.)

To perform a fusion reaction in practice, the reactants must be heated to a very high temperature (typically, 100 million degrees Celsius) to give them sufficient kinetic energy to overcome the Coulomb repulsion barrier when a collision occurs.

There are three possible ways of imparting the required temperature. The first is used in hydrogen bombs (known as *H-bombs*), and the other two are being studied for the purposes of peaceful production of thermonuclear energy (another name for fusion energy):

- the explosion of an atomic fission bomb (uranium 235 or plutonium 239),

- magnetic fusion: magnetic fields are used to confine a plasma of reactants far from the wall of the combustion chamber (usually in the shape of a torus and known as a *Tokamak*). Various heating methods can be used and are often combined: resistive heating (the Joule effect), injection of neutral energetic particles, and cyclotron heating using high-frequency waves,

- inertial fusion: the reactants are enclosed in a little ball whose diameter is of the order of a millimetre and are rapidly heated to a very high temperature using a set of high-power laser beams (or beams of other particles). This causes high compression and extreme heating that can trigger the fusion reaction.

In the last two cases, once the fusion reaction had begun, it would generate enough heat to be self-sustaining. More development work has been devoted to the magnetic concept, but the inertial concept also has advantages. In the short term, neither of these methods seems likely to allow fusion to make a significant contribution to the world's energy requirements. The *Iter* project would make it possible to assess the feasibility of a magnetic fusion reactor.

Fission seems far easier, and has been in use for over half a century. One of its essential features is that it was possible to start with machines that were not very powerful (note that the Fermi reactor generated only half a watt) and subsequently to work on perfecting the technology. By contrast, it is impossible to perform small-scale exoenergetic fusion by any means whatsoever.

Moreover, the neutrons released at the same time can induce new fissions, thereby maintaining the chain reaction.

The binding energy curve (Figure 2.4) shows that fission, assumed to be symmetrical as a simplification, increases the binding energy by about 0.85 MeV per nucleon. Assuming 236 nucleons (uranium 235 plus the incident neutron), this gives approximately $0.85 \times 236 \cong 200$ MeV of energy. This figure has been confirmed experimentally, and is huge compared to "classic" forms of energy.

This brings us to the subject of neutrons and the reactions they can induce. This calls firstly for some general remarks about nuclear reactions and, in particular, about the spontaneous reactions known as radioactive decay.

B. RADIOACTIVITY

2.4.1. Regions of instability

The *Z-N* diagram of natural nuclides (Figure 2.1) shows that not all combinations lead to stable nuclei (there are no stable nuclei other than those found on Earth). This does not mean that no other combinations exist; merely that they are *unstable;* after a certain time, whose duration depends on the specific example, an unstable nucleus will spontaneously change to produce a stable nucleus, or even another unstable nucleus that will, in turn, undergo a similar transformation. This spontaneous mechanism is *radioactivity*, and this type of transformation is *radioactive decay*. There are a few tens of natural radioactive nuclides, and there are a few thousand other, artificial, radioactive nuclides whose characteristics are known. All of these nuclei are represented by points close to the valley of stability; if this were not the case, their existence would be too ephemeral to be discovered.

The word "valley" implies a surface located above the *Z-N* plane representing the binding energy of the nucleus (or, more accurately, its mass) with a depression along the curve concerned. Thus, any point not located on the curve or in its immediate neighbourhood would tend to "slide down the slope" and fall to the bottom of the valley.

With this image in mind, three regions of instability can be identified (see Figure 2.5):

- Region **A** contains nuclei located approximately in the axis of the valley, but too high up (imagine the valley sloping gradually upwards in the area corresponding to medium-sized nuclei and then rising along an increasingly steep slope): this is the region of *oversized nuclei*.

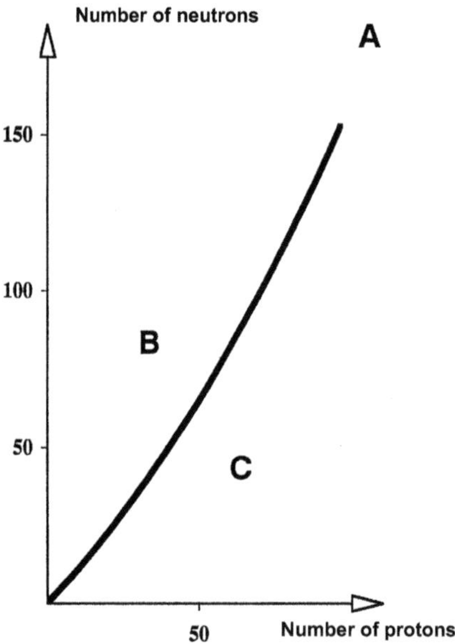

Figure 2.5. The three regions of instability.

- Region **B** contains nuclei located on the "right bank"; this is the region of nuclei having *excess neutrons* with respect to protons.
- Region **C** contains nuclei located on the "left bank"; this is the region of nuclei having *excess protons* with respect to neutrons.

2.4.2. Main types of radioactivity

Each of the main modes of radioactive decay is associated with one of these three regions of instability.

For nuclei in Region **A**, decay most often occurs[7] by *alpha radioactivity* (α): within the (large) nucleus, an alpha particle, which is a helium 4 nucleus consisting of two protons and two neutrons, isolates itself and is then ejected. Using **X** and **Y** to represent the chemical symbols of the initial and final nuclei, the reaction can be written as:

$$ {}^{A}_{Z}\mathbf{X} \Rightarrow {}^{A-4}_{Z-2}\mathbf{Y} + {}^{4}_{2}\,\mathbf{He}. $$

For Region **B** nuclei, which have excess neutrons, one of the neutrons is transformed into a proton by a weak interaction process:

$$ {}^{1}_{0}\mathbf{n} \Rightarrow {}^{1}_{1}\mathbf{p} + {}^{0}_{-1}\mathbf{e} + {}^{0}_{0}\,\bar{\nu}, $$

where the electron and the antineutrino are ejected. For an observer, this reaction, known as *beta-minus* (β^-) decay, is written as:

$$ {}^{A}_{Z}\mathbf{X} \Rightarrow {}^{A}_{Z+1}\mathbf{Y} + {}^{0}_{-1}\mathbf{e} + {}^{0}_{0}\,\bar{\nu}. $$

For Region **C** nuclei, which have excess protons, the symmetric process can occur:

$$ {}^{1}_{1}\mathbf{p} \Rightarrow {}^{1}_{0}\mathbf{n} + {}^{0}_{1}\bar{\mathbf{e}} + {}^{0}_{0}\,\nu. $$

where the positron (antielectron) and neutrino are ejected. For an observer, this reaction, known as *beta-plus* (β^+) decay, is written as:

$$ {}^{A}_{Z}\mathbf{X} \Rightarrow {}^{A}_{Z-1}\mathbf{Y} + {}^{0}_{1}\bar{\mathbf{e}} + {}^{0}_{0}\,\nu. $$

For these Region **C** nuclei with excess protons, there is another process, without a symmetric equivalent, that can also take place: *orbital electron capture* (EC). This is the capture of a nearby electron by a proton in the nucleus:

$$ {}^{1}_{1}\mathbf{p} + {}^{0}_{-1}\mathbf{e} \Rightarrow {}^{1}_{0}\mathbf{n} + {}^{0}_{0}\nu. $$

For an observer, the reaction is written as:

$$ {}^{A}_{Z}\mathbf{X} + {}^{0}_{-1}\mathbf{e} \Rightarrow {}^{A}_{Z-1}\mathbf{Y} + {}^{0}_{0}\nu, $$

and is different from the previous reaction in that only a neutrino (which would not be detected in practice) is ejected. Some X-rays, caused by the rearrangement of the electron cloud, may also be observed.

7 Other rare processes include spontaneous fission, which will be discussed later.

For these various decay modes, the nucleus **Y** is often obtained not in its ground state, but in an excited state **Y***, which is an *isomer* of the ground state.

With few exceptions, isomers have a very short lifetime, because decay occurs by the emission of a gamma photon (if the nucleus returns directly to ground state), or several gamma photons (if the nucleus passes through one or more intermediate levels). This decay, known as *gamma radioactivity* (γ), accompanies (or, more accurately, follows) most other types of radioactive decay. The emitted particles constitute types of *radiation* named according to the decay that produced them:

- Alpha radiation is stopped in normal matter within a distance on the order of a tenth of a millimetre, and deposits its energy (approximately 5 MeV or more) in the matter. Beta radiation is stopped in normal matter within a distance on the order of a millimetre. The energy carried by the particle and deposited in the matter can be between a few keV and a few MeV. In the case of beta-plus radioactivity, the positron is annihilated with an electron by emitting two 511-keV photons at 180 degrees. This energy corresponds to the mass of each of these particles.

- Antineutrinos or neutrinos share the energy of radioactive beta decay involving electrons or positrons[8]. These particles have practically no interaction with matter and they escape undetected.

- Gamma photons from radioactivity can have energies ranging from a few keV to several MeV. The higher their energy, the longer their average trajectory (typically on the order of a decimetre). Certain photons can travel a very long distance because the trajectory length is random according to an exponential distribution, unlike charged particles, whose trajectory is more or less fixed once the energy of the particle and the type of material through which it travels have been specified. Gamma radiation is therefore the most difficult type of radiation to provide protection against.

2.4.3. Law of radioactive decay

Although there are several decay processes, radioactivity is governed by a universal exponential law. This law arises from the fact that the transformation occurs at a random instant, and it expresses the fact that a radioactive nucleus does not "age", but rather "dies" at an unpredictable moment. The key idea, then, is the concept of *probability of radioactive decay* during the upcoming unit time interval dt, i.e. $\lambda\, dt$. This infinitesimal probability is obviously proportional to the infinitesimal interval dt. The proportionality constant λ is known as the *radioactive decay constant*. The fact that an unstable nucleus does not "age", but is only liable to "die", is reflected in the fact that, as long as we know that the nucleus is still "alive", λ is independent of the age of the nucleus, i.e. of when it was produced. Moreover, experiments have shown that this constant does not depend on the physical-chemical conditions of the nucleus concerned; it is truly a nuclear characteristic. This constant differs, however, according to the process and the unstable nucleus.

[8] Long before these particles were found experimentally (1953), their existence was postulated by Pauli, and then by Fermi in his theory of beta radioactivity, in accordance with the principle of conservation of energy. They are also required in reaction equations to conserve angular momentum; their spin is 1/2. Their mass is very small, if not zero.

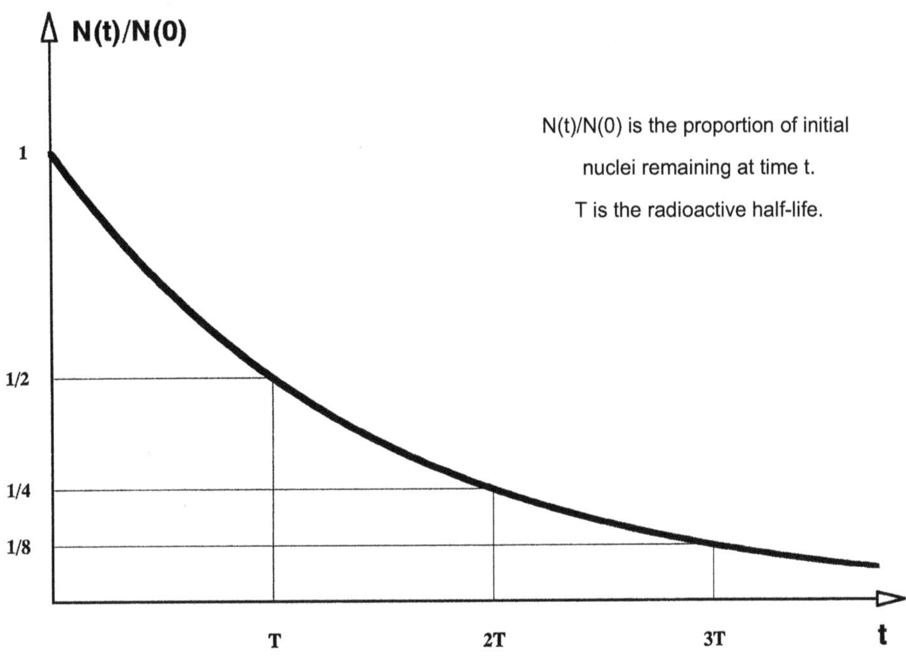

Figure 2.6. Law of radioactive decay.

Consider a large population of radioactive nuclei of a given type. Let $N(t)$ be the number of nuclei at instant t. Between t and $t + dt$, each nucleus has probability $\lambda\, dt$ of disappearing; therefore:

$$dN = -N(t)\lambda dt.$$

The exponential law is then obtained by integration:

$$N(t) = N(0)e^{-\lambda t}. \tag{2.6}$$

Figure 2.6 shows the curve representing this exponential function, and introduces the concept of *radioactive half-life*. The half-life is the time after which half of the population has disappeared (thus, after two half-lives the population is down to one quarter, after three half-lives it is reduced to one eighth, after ten half-lives it is approximately one thousandth, and so on).

The half-life is given by the following formula:

$$T = \frac{\ln(2)}{\lambda} \simeq \frac{0.693}{\lambda}. \tag{2.7}$$

This should be distinguished from the *mean life* $1/\lambda$, which is the average amount of time after which an unstable nucleus observed at a given instant will disintegrate.

The *activity* of a radioactive material is the number of decays per unit time, i.e.:

$$A = \lambda N. \tag{2.8}$$

This is expressed in *becquerels* (Bq), where 1 Bq = 1 decay per second. This replaces an old unit, the curie (Ci), where 1 Ci = 3.7×10^{10} Bq. Note that activity decreases over time according to the same law as the number N of unstable nuclei.

2.4.4. Examples of radioactive decay

We will give just a few examples to illustrate the wide variety of processes, the range of half-lives, and the diverse areas of scientific and medical endeavour that involve radioactivity:

- ^{226}**Ra**: Radium 226, an indirect descendant of uranium 238, is found in small amounts in uranium ores. This *alpha* radionuclide with a half-life of *1620 years* is famous for having been discovered by Marie and Pierre Curie in 1898 and then isolated by Marie Curie and André Debierne in 1910. Note that the activity of 1 g of radium is equal to 1 curie; this was the initial definition of this unit.

- ^{235}**U**: Uranium 235 is the only fissile nucleus of all the natural nuclides. It is an *alpha* emitter with a long half-life: *710 million years.*

- ^{238}**U**: The other isotope, uranium 238, is an *alpha* emitter with an even longer half-life: *4.5 thousand million years*. These very long half-lives mean that these nuclides are still found in significant quantities on Earth since their creation by the explosion of a supernova five thousand million years ago. Uranium 235 is less abundant because of its shorter half-life. Note also that these very heavy nuclei can undergo spontaneous fission, but at a very slow rate. For example, 26 spontaneous fissions occur per hour per gramme of uranium 238.

- ^{232}**Th**: Natural thorium comprises the thorium 232 isotope only. This is an *alpha* emitter with a very long half-life: *14 thousand million years.* This and the two previous nuclides (the uranium isotopes) are the only nearly-stable nuclei, i.e. those with a very long half-life, of the elements above bismuth (Z = 83). Intermediate elements, with shorter half-lives, however, are found in nature because they are continually regenerated from one of these three nuclides.

- ^{1}**n**: Paradoxically, the neutron is unstable in its free state, but stable when inserted into a stable nucleus. It has a half-life of *12 minutes*, and disintegrates by beta-minus decay. In reactors, the lifetime of a free neutron is about a millisecond or less. The probability of decay during this period is infinitesimal, and this radioactivity is negligible in reactor physics.

- ^{3}**H**: *Tritium* is one of the reactive substances in the fusion reaction D + T (the other one, deuterium, is stable). Its half-life is *12 years.* It is converted to helium 3 by *beta-minus* decay. Once it is produced, it must be used promptly (Tokamak) or renewed regularly (weapons).

- ^{11}**C**: There are two stable isotopes of carbon: carbon 12 (99%) and carbon 13 (1%). With its neutron deficit, the carbon 11 isotope is a *beta-plus* emitter, as one would expect. It has a half-life of *20 minutes*. Carbon 11 is one of the radioisotopes used in medical imaging (positron emission tomography, or PET), which is used in particular for studying the metabolism of the brain. The positron is annihilated with an electron, almost at the very spot where the decay takes place, and emits two 511 keV gamma particles that are detected at 180 degrees and in coincidence. This gives the axis on which the decay occurred. After a large number of such events, mathematical processing provides a map of the carbon 11 concentration. The PET process is carried out by manufacturing the carbon 11 using an accelerator on site, marking the molecules of interest (e.g., glucose), and then injecting them into the patient before performing the tomography. This entire procedure must be done in the few tens of minutes that are available before the radionuclide disappears.

- ^{14}**C**: This isotope has excess neutrons and is therefore a *beta-minus* emitter. Its half-life is *5700 years*. As we have already seen § 2.3.2, carbon 14 is found in small but fairly constant quantities in the natural carbon in the biosphere because it is continually generated by nuclear reactions induced by solar radiation in the upper atmosphere. Because of this, every living thing, whether plant or animal, contains a certain amount of carbon 14. After death, however, the carbon 14 is not renewed, and starts to disappear by radioactive decay. The date of death can be estimated by measuring the remaining concentration. Given the half-life of carbon 14, this method is suitable for determining ages from a few hundred to a few tens of thousands of years, which makes it useful for archaeological research.

- ^{97}**Tc**: Technetium 97, with a half-life of *2.7 million years,* is an example of a radionuclide that decays mostly by *electron capture*. It has the longest half-life of all the isotopes of this element. Because this half-life is short compared to the age of the earth (4.5 thousand million years), there is no natural technetium.

- ^{16}**N**: This is an emitter particularly known for its *gamma* emissions following *beta-minus* decay. It has a short half-life: *7 seconds*. This emitter is produced in water reactors by an (n,p) reaction on common oxygen (oxygen 16). Unfortunately, because of the half-life, most of the gamma emission occurs outside the core, particularly in the steam generators of pressurised water reactors, but this also makes it possible to measure the primary water flow rate.

- ^{60}**Co**: Another example of a *beta-minus/gamma* emitter. Its half-life is *5.7 years*. In the past, this radionuclide was used a great deal in *cobalt bombs* for cancer therapy. When cancer is irradiated, more malignant cells than healthy cells are destroyed, because the malignant ones are more sensitive to radiation. (Today, other radionuclides are used, and are selected according to the type of cancer.) This radionuclide is also the main contributor to the activation of steel structures subjected to a neutron flux.

2.4.5. Alpha instability

The above examples lead to a few general remarks about the alpha and beta processes.

Concerning the alpha process, it is interesting to note that the half-life is almost always long, except for elements 84 to 89 and the heaviest (artificial) nuclei, and that the decay energies are of the order of 5 MeV or a little more. Neglecting the recoil of the nucleus, this energy is transmitted to the alpha particle if it is alone, or shared between the alpha particle and the gamma radiation. Finally, note that only heavy nuclei undergo alpha decay.

This is a consequence of the reaction energy:

$$Q = W(A-4, Z-2) + W(4,2) - W(A,Z),$$
$$Q = [M(A,Z) - M(A-4, Z-2) - M(4,2)]c^2, \qquad (2.9)$$

where the binding energy of the alpha particle, $W(4, 2)$, is equal to 28.3 MeV. Using the Bethe–Weizsäcker formula to perform a calculation for various points on the axis of the valley of stability, we find that Q is positive only when A is greater than about 150. As one would intuitively guess, quantum physics calculations show that the half-life decreases as Q increases. In practice, alpha decay only becomes significant when Q exceeds 4 or 5 MeV, corresponding to a mass number of about 220.

2.4.6. Beta instability

For beta instability, the most important consideration is not the mass number, but the ratio of neutrons to protons with respect to the optimum ratio. Once again, the Bethe–Weizsäcker formula is used to perform the analysis. We set A[9] — not modified in a beta decay or an electron capture — and look at the changes in the binding energy (or the mass M) as a function of Z.

If A is odd, the parity term δ is zero for all proton-neutron distributions; all the points representing M as a function of Z are located on a *parabola* whose equation is given by the liquid drop model. In principle, only the nucleus corresponding to the point nearest the bottom of the parabola is stable. The nuclei represented by points further to the left are unstable by beta-minus radioactivity, and the points further to the right are unstable by beta-plus radioactivity and/or electron capture. Given the small difference in mass between the neutron and the proton that must be taken into account to calculate the reaction exactly, there are sometimes two stable nuclei.

If A is even, the parity term δ can have the value +1 or −1, depending on the combination. In this case, the points are located alternately on *two parabolas* set apart (in the ordinate direction) by twice the energy associated with the parity term. There can be up to three stable isobars.

As a general rule, beta half-lives are shorter than alpha half-lives (with some exceptions), and are even shorter with increasing decay energy. (Parity also has an effect, however: transitions without a parity change occur more easily than transitions with a parity change.)

In terms of number of isotopes (i.e. for fixed Z), these considerations show that even elements have more isotopes than odd elements. A systematic analysis shows that:

- If Z is *even*, there are *at least* two stable isotopes, with the sole exception of beryllium ($Z = 4$), whose only stable isotope is beryllium 9.

- If Z is *odd*, there are *at most* two stable isotopes.

[9] Nuclei with the same mass number are called "isobars".

2.4.7. Gamma instability

Note (*see.* § 2.2.7) that nuclei have *excited states* or *isomeric states* (isomer: same A and same Z). Alpha and beta decays can produce different isomers of the same nuclide. Isomers generally disintegrate quickly by emission of gamma photons until they reach the ground state.

The spectra of these gamma emissions are made additionally complicated by the fact that the alpha or beta decay might have placed the daughter nucleus[10] on a higher level. This means that the structure of gamma emissions is often more complicated for heavy nuclei than for light nuclei.

2.4.8. Radioactive series

We often find that a nucleus **Y**, obtained by the decay of a radioactive nucleus **X**, is itself radioactive and disintegrates to produce a nucleus **Z**. In particular, under each of the three almost-stable heavy nuclei found in nature, there is a long chain of descendants, sometimes with branches in the chain. For example, Figure 2.7 shows the series starting from uranium 238, with fifteen unstable nuclides preceding the final, stable product: $^{206}\mathbf{Pb}$.

These series involve alpha decay, which reduces the number A by four, and beta decay, which does not alter A. All the values of A in a series are therefore of one of the forms $4n$ (thorium 232 series), or $4n+1$ (a series that does not exist in nature), or $4n+2$ (uranium 238 series), or $4n+3$ (uranium 235 series).

2.4.9. Radioactive series equations

The equations governing the number of atoms X, Y, Z, etc. related by a radioactive decay chain:

$$\mathbf{X} \Rightarrow \mathbf{Y} \Rightarrow \mathbf{Z} \Rightarrow \ldots,$$

are a generalisation of the equation for decay of a nuclide: except for the first link, a production term must be added to the disappearance term[11]:

$$\begin{aligned} \frac{dX}{dt} &= -\lambda_X X, \\ \frac{dY}{dt} &= +\lambda_X X - \lambda_Y Y, \\ \frac{dZ}{dt} &= +\lambda_Y Y - \lambda_Z Z. \end{aligned} \tag{2.10}$$

The general solution is a combination of exponentials of the form $e^{-\lambda t}$. The coefficients are obtained by substituting back into the system of equations and by taking the initial conditions into account (if the series is linear, the equations can be solved from one coefficient to the next).

10 The French say *"fils"* (*son*).

11 The chain can be imagined as a series of tanks where each is pouring its contents out into the next. The equations are obtained by listing all the incoming flows (+ sign) and outgoing flows (– sign).

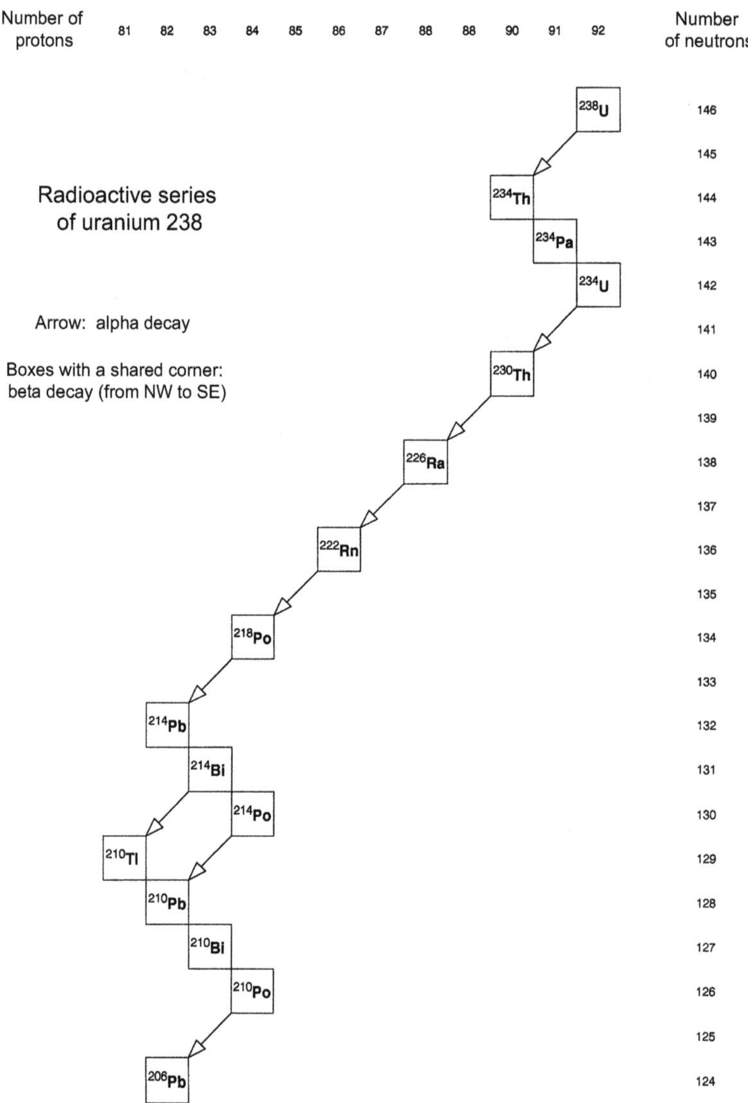

Figure 2.7. Radioactive series of uranium 238.

If the half-life of the leading nuclide is much longer than all the others, which is the case for the thorium 232, uranium 235, and uranium 238 series, then the concentrations tend towards *secular equilibrium*. This equilibrium is reached when all nuclides have remained unchanged for a time that is long compared to all half-lives except the first, and every type of activity affects it equally:

$$\lambda_X X = \lambda_Y Y = \lambda_Z Z = \ldots, \tag{2.11}$$

The concentrations are therefore inversely proportional to the half-lives, and are practically time-independent over periods that are short compared to the first half-life.

C. NUCLEAR REACTIONS

2.5. General information about nuclear reactions

2.5.1. Spontaneous reactions and induced reactions

Radioactivity could be described as a spontaneous nuclear reaction, but the term "reaction" is normally used when the reaction is *induced*. In practice, this is done by a *projectile* hitting a *target*. (In fact, the projectile and the target play completely symmetrical roles, as can be seen if we place ourselves in the centre of mass system.) Let the projectile be **a** and let the target be **A**. Assume, for example, that the reaction has two products: a large **B** and a small **b**. We write:

$$\mathbf{a} + \mathbf{A} \Rightarrow \mathbf{B} + \mathbf{b},$$

or, more concisely:

$$\mathbf{A}(\mathbf{a}, \mathbf{b})\mathbf{B}.$$

2.5.2. Nuclear reaction examples

The following examples are useful for anyone interested in nuclear energy (most have already been mentioned):

- The reaction most often used in *neutron sources* (obtained by mixing any alpha emitter with beryllium):

$$^{4}\mathbf{He} + {}^{9}\mathbf{Be} \Rightarrow {}^{12}\mathbf{C} + {}^{1}\mathbf{n}.$$

- The reaction that led to the *discovery of artificial radioactivity* by Joliot:

$$^{4}\mathbf{He} + {}^{27}\mathbf{Al} \Rightarrow {}^{30}\mathbf{P} + {}^{1}\mathbf{n}.$$

- The reaction producing *carbon 14*:

$$^{1}\mathbf{n} + {}^{14}\mathbf{N} \Rightarrow {}^{14}\mathbf{C} + {}^{1}\mathbf{p}.$$

- The *thermonuclear fusion* reaction:

$$^{2}\mathbf{H} + {}^{3}\mathbf{H} \Rightarrow {}^{4}\mathbf{He} + {}^{1}\mathbf{n}.$$

- Production of the *tritium* required for this reaction:

$$^{1}\mathbf{n} + {}^{6}\mathbf{Li} \Rightarrow {}^{4}\mathbf{He} + {}^{3}\mathbf{H}.$$

- *Fission* reaction:

$$^{1}\mathbf{n} + {}^{235}\mathbf{U} \Rightarrow \textbf{two fission fragments} + \textbf{a few neutrons.}$$

– *Radiative capture of a neutron,* which can occur on all nuclei:

$$^{1}\mathbf{n} + {}^{238}\mathbf{U} \Rightarrow {}^{239}\mathbf{U} + \textbf{gamma photon(s)}.$$

(In this example, the nucleus obtained is uranium 239, and after two short half-life beta-minus decays it becomes fissile plutonium 239.)

– Another *neutron capture* reaction:

$$^{1}\mathbf{n} + {}^{10}\mathbf{B} \Rightarrow {}^{7}\mathbf{Li} + {}^{4}\,\mathbf{He}.$$

(This reaction is used in reactors, particularly pressurised water reactors, to regulate the reactivity.)

2.5.3. Laws of conservation

As in all physical processes, there are certain parameters that are conserved in these reactions. The main ones are as follows:

– *Number of nucleons,*

– (Algebraic) *number of electric charges,*

– *Energy,*

– *Momentum,*

– *Angular momentum.*

By applying the first two of these laws we can, for example, find **B** if **a**, **A**, and **b** are known. In this way we can discover (in case it had been forgotten) that the (n,α) neutron capture reaction by boron 10 gives lithium 7.

In the third law, the mass must be counted along with the other forms of energy (kinetic and excitation). In practice, it allows us to calculate the balance of energy exchanges in the reaction using nuclide mass tables.

The fourth law is used to find the distribution of the energy output: for example, for the fusion reaction D+T, the 17.6 MeV energy from the reaction (third law) is distributed as follows: 3.5 MeV for the α particle, and 14.1 for the neutron. This can be seen by assuming that the momentum is negligible at the input and therefore also at the output of the reaction. The third and fourth laws suggested the presence of a phantom particle other than the electron in the beta decay process: the *neutrino*. The conservation of angular momentum also imposes the condition that the neutrino must have spin 1/2.

2.5.4. Cross-section

The concept of cross-section is necessary if we wish to quantify the *number of reactions* between a flux of particles **a** and targets **A**, or the *probability of an interaction*.

A cross-section can be microscopic, meaning that it is characteristic of an individual target, or macroscopic, meaning that it is characteristic of a material containing a large number of targets. There are various possible ways of presenting these concepts.

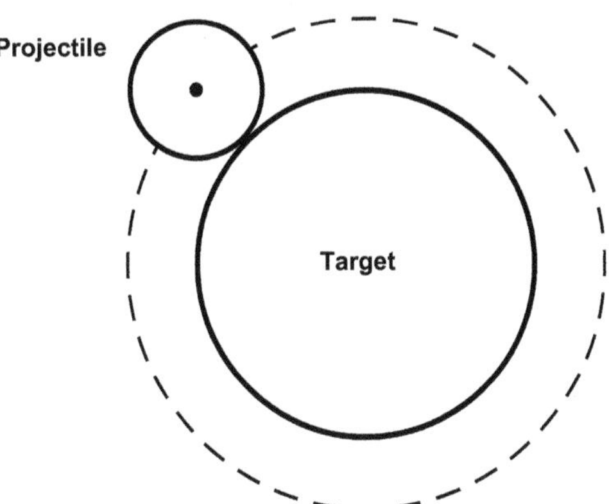

Figure 2.8. Intuitive concept of cross-section.

We will start by introducing both ideas and then looking at how they are related. To do this, we need to use some models from classical mechanics that require the use of intuition. The resulting model will obviously be an oversimplification (although not wrong), and so we must also qualify it with the necessary details.

Figure 2.8 represents the definition of cross-section in an intuitive manner (in this case with a target and projectile that are assumed to be spherical): a collision will occur if and only if the trajectory of the projectile takes its centre near the centre of the target to within a distance less than the sum of the radii, i.e. it must cross the circle in the plane of the figure whose radius is the sum of the radii of the two objects. The cross-section is the surface area σ of this circle.

This surface area has a probability attached to it; if we place in the plane of the figure not one target, but a number n of targets distributed in a contour of surface area S and then send the projectile at random through S, the probability of its hitting one of the targets is the surface area of the entire set of dashed circles divided by the total surface area, i.e. $n\sigma/S$.

In nuclear physics and neutron physics, the intuitive definition is too simplistic. This becomes obvious if we notice that the neutron cross-section σ can differ considerably between nuclei that, at first glance, appear not to be very different (e.g. the uranium 235 and 238 isotopes), and if we also notice that it can vary greatly for a given nucleus according to the speed of the neutron. This image does show, however, that the order of magnitude of the cross-sections should be somewhere in the region of 10^{-28} m^2, since the radius of a nucleus is on the order of 10^{-14} m. Measurements have confirmed this estimate. This is why, in nuclear physics, cross-sections are expressed in barns:

$$1 \text{ barn } (b) = 10^{-24} \text{ cm}^2 = 10^{-28} \text{ m}^2.$$

The probabilistic definition of σ does however hold true in quantum physics.

2.5.5. Macroscopic cross-section

Imagine a neutron travelling in matter, which the neutron "sees" as if it were a vacuum, since the neutron is unaware of electrons. It therefore travels in a straight line at a constant speed until it hits a nucleus in its path. For a short path dx, this collision probability is infinitesimal and proportional to dx (for example, the neutron is twice as likely to hit a nucleus on a 2 µm path as on a 1 µm path): this can be written as Σdx where Σ is the appropriate proportionality coefficient, or *macroscopic cross-section*. Neutron physicists generally use the centimetre as the unit of length for their calculations, and so cross-sections are expressed in cm^{-1}. Here we shall use the SI units, m and m^{-1}.

Using this definition, we can calculate the probability relationship governing the distance x between the starting point of the neutron and the point where it will have its *first* collision. This first collision takes place at distance x to within dx, i.e. between x and $x + dx$:

1/ *if* the neutron has had no collision between 0 and x; this probability is denoted $Q(x)$,

2/ ***and*** *if* the neutron has a collision between x and $x+dx$; by definition, this probability is Σdx.

The probability of this event is therefore: $p(x)\,dx = Q(x) \times \Sigma dx$.

To calculate $Q(x)$, note that $Q(x + dx)$, the probability of *no collision* over the distance $x + dx$, is the product of:

1/ the probability of *no collision* between 0 and x, i.e. $Q(x)$;

2/ the probability of *no collision* between x and $x + dx$, i.e., by definition, $1 - \Sigma dx$.

This gives: $Q(x + dx) = Q(x) \times (1 - \Sigma dx)$.

Simplify, integrate, and recall that $Q(0)$ is equal to 1 by definition, to obtain:

$$Q(x) = \exp(-\Sigma x),$$

which gives:

$$p(x)dx = Q(x) \times \Sigma dx = exp(-\Sigma x)\Sigma dx. \tag{2.12}$$

The inverse of the macroscopic cross-section, $\lambda = 1/\Sigma$, is the mean free path of the neutrons, i.e. the average value of the distance x at which the first collision occurs. We have:

$$\lambda = \langle x \rangle = \int_0^\infty x p(x) dx = \int_0^\infty x \exp(-\Sigma x)\Sigma dx = \frac{1}{\Sigma}\cdot \tag{2.13}$$

In typical materials and for neutrons, measurements have shown that macroscopic cross-sections are often on the order of cm^{-1} (100 m^{-1}) and mean free paths are on the order of a cm.

The probability element Σdx of interaction with matter for a path dx is obviously proportional to the number of obstacles the neutron is likely to encounter, and therefore to the number N of nuclei per unit volume. With σ denoting the proportionality coefficient, we can write:

$$\Sigma = N\sigma. \tag{2.14}$$

This coefficient σ will be known as the *microscopic cross-section* (as opposed to the *macroscopic* one). It is a very small number if expressed in the usual units, since N is of the order of magnitude of Avogadro's number. The barn would therefore be the appropriate unit. It is also important to note that this second definition of microscopic cross-section is consistent with the intuitive definition given above.

To illustrate this, we can use the image of the projectile and target to evaluate the probability of interaction with matter of a particle travelling along a path dx. We can associate with this path the small cylinder whose height is dx and whose base is the dashed circle in Figure 2.8, i.e. surface area σ. The volume of this cylinder is $\sigma\, dx$. The particle has a collision on path dx if and only if the centre of an atomic nucleus is located in this volume; if there are N nuclei per unit volume, the probability of this being true is $N \times \sigma dx$. By introducing this probability into the expression Σdx, this is exactly what we find: $\Sigma = N\sigma$.

2.6. Neutron reactions

2.6.1. General remarks

Of the nuclear reaction examples given earlier (*see* § 2.5.2), those involving neutrons show that several processes exist. Before listing those of interest for reactor physics, i.e. those where the neutron's energy is between zero and about 10 MeV, we should mention two points that are important in neutron physics:

1/ Neutron–electron interactions are negligible; firstly, because the cross-section is infinitesimal (only a weak interaction is possible between these two particles) and secondly, because the ratio of their masses is such that the collision of a neutron with an electron would hardly alter the neutron's trajectory.

Consequently, neutrons have a very long trajectory (on their scale); typically on the order of a centimetre. This means that a neutron passes approximately a hundred million atoms before accidentally meeting a nucleus with which it has an interaction.

(To return to the image evoked in § 2.1.1, imagine that the neutron is crossing a football pitch, but can only "see" the little marble at the centre. It must therefore go across a large number of these football pitches before having any significant chance of colliding with a marble.)

Note: A neutron physicist does not need to introduce a precise model of the electron clouds of atoms, because neutron–electron interactions are not important in this case. The "solar system" model, for example, can suffice.

2/ Neutron–neutron interactions are also negligible, not for reasons related to the cross-section[12], but because of the density. Expressed in orders of magnitude, there are 10^{28} atomic nuclei per m^3 in normal matter, and only 10^{14} neutrons per m^3 in a high-power reactor, so a neutron travelling in the system has roughly a 10^{14} times greater probability of meeting a nucleus than of meeting another neutron. The likelihood of meeting a neutron can therefore be neglected.

12 The scattering cross-sections for p-p, n-p and n-n interactions have the same order of magnitude.

The consequence of this is that the Boltzmann equation governing neutron density is *linear*. This simplifies the mathematical analysis and numerical processing. If neutron–neutron interactions had to be considered, they would be represented by a term proportional to the *square* of the density, i.e. a nonlinear term.

2.6.2. Scattering and "real" reactions

When considering the interactions of neutrons with matter, it is important to distinguish real reactions from potential scattering. In *potential scattering* there is no actual contact between the neutron and the target nucleus (i.e. nuclear forces are not brought into action); the wave associated with the neutron is scattered by the nuclear field. This process is phenomenologically similar to the elastic collision between two billiard balls, with conservation of momentum and conservation of kinetic energy[13]. In a "real" reaction, on the other hand, the neutron penetrates the nucleus. In the energy range of interest here, the reaction can be described by the *compound-nucleus model*. This model has three stages:

1/ The *entrance channel*: The target nucleus incorporates the incident neutron, producing the isotope of the next rank up, and giving this isotope an excitation energy equal to the sum of the *binding energy* of the extra neutron (work of the nuclear binding forces) and the *kinetic energy* contributed by the neutron,

2/ The actual life of the *compound nucleus*, an isotope of the target nucleus: its lifetime is brief on a human scale — on the order of 10^{-14} s — but long on a nuclear scale, i.e. compared to the time required for the new nuclear bond to be created, which is on the order of 10^{-22} s. This means that the excitation energy can become "uniform" within the compound nucleus. In other words, it lives long enough to "forget" that it was created via the entrance channel. What happens next will be independent of the process that created the compound nucleus (absorption of a neutron, a proton, a gamma particle, etc.),

3/ The *exit channel*: the excited compound nucleus will very quickly (on a human scale) disintegrate by a radioactive-type process. With the energy acquired when the neutron was absorbed, there are several possible mechanisms which compete with each other (these are described below).

The excitation energy of the compound nucleus is significant; the binding energy of the extra neutron is of the order of the average binding energy per nucleon, which is approximately 5 to 10 MeV[14], and the kinetic energy supplied by the neutron can be from zero to a few MeV.

2.6.3. Main reactions induced by neutrons in reactors

Table 2.1 summarises the main reactions seen in reactors.

Other than potential scattering, two reactions are always possible for any target nucleus and for any neutron energy: 1/ *Resonant elastic scattering*, which involves the formation

[13] This image will be used in Chapter 7 to establish the laws of impact.

[14] The binding energy of the last nucleon is slightly greater than average in the region where the *W/A* vs. *A* curve is increasing, and slightly less than average in the region where the curve is decreasing, particularly in the region of heavy nuclei such as uranium.

Table 2.1. Main reactions undergone by neutrons in reactors.

Interaction without formation of a compound nucleus		
Potential scattering (elastic)	$n + A \Rightarrow n + A$	Always possible
Interactions occurring via formation of a compound nucleus		
Elastic resonant scattering	$n + A \Rightarrow n + A$	Always possible
Inelastic resonant scattering	$n + A \Rightarrow n + A^*$ $A^* \Rightarrow A + \gamma$	Threshold: first level of A
Reaction (n,2n)	$n + A \Rightarrow n + n + (A - 1)$	Threshold: separation energy of a neutron from A
Radiative capture	$n + A \Rightarrow \gamma + (A + 1)$	Always possible
Reaction (neutron, charged particle)	$n + A \Rightarrow p + B$ $n + A \Rightarrow \alpha + C$ etc.	Usually with threshold; sometimes without threshold
Fission	$n + A \Rightarrow$ $PF_1 + PF_2$ + a few neutrons (On average, ν, from 2 to 3)	Heavy nuclei: without threshold for odd N, with threshold for even N; tunnel effect.

of a compound nucleus (the energy contributed by the incident neutron is restored in full —part of it to break the bond, and the rest in the form of kinetic energy— to an ejected neutron), 2/ *Radiative capture* (the excitation energy of the compound nucleus is ejected in full in the form of a photon or photons).

Certain reactions usually have a threshold, but not always: 1/ *Fission* (see § 2.9), 2/ Reactions in which a *charged particle* is ejected.

Other reactions always have a threshold: 1/ *Inelastic (resonant) scattering* leaving the residual nucleus in an excited state after ejection of the neutron (the threshold is the energy of the excited state concerned, and therefore at least the energy of the first level), 2/ *Reactions of the type* $(n, 2n)$ (ejection of two neutrons: the threshold is the binding energy of the second neutron, which needs to be picked up to separate it), *(n,np)*, $(n, 3n)$, etc.

2.6.4. Partial cross-sections and additivity of cross-sections

The microscopic cross-section σ and macroscopic cross-section Σ we have just defined characterise the whole interaction of a neutron in matter; that is why they are called "total" (the index t can be added to indicate this unambiguously).

As we have seen, there are always several types of possible interaction for neutrons. Each of these processes is therefore characterised by a partial cross-section σ_r or Σ_r, where r denotes the reaction. By definition, the ratio σ_r/σ or Σ_r/Σ is the *probability* of the neutron-matter interaction occurring by the process r if it occurs at all. Consequently, *the sum of the partial cross-sections is the total cross-section.*

In neutron physics, indices are used to distinguish *scattering* (s) from *absorption* (a), depending on whether the neutron is re-emitted after the reaction ((n,2n) processes are considered as scattering):

$$\sigma_s + \sigma_a = \sigma, \quad \Sigma_s + \Sigma_a = \Sigma, \tag{2.15}$$

and, for absorption processes, *fission* and any other type of absorption will be known as *captures*:

$$\sigma_f + \sigma_c = \sigma_a, \quad \Sigma_f + \Sigma_c = \Sigma_a. \tag{2.16}$$

In compound materials, it is possible to distinguish the type of nucleus with which the neutron has interacted, for example uranium 235, uranium 238 and oxygen in the case of uranium oxide. The macroscopic cross-section of the *mixture* (total or partial) will be the sum of the macroscopic cross-sections $\Sigma_k = N_k\sigma_k$ of each of the components:

$$\Sigma = \Sigma_1 + \Sigma_2 + \cdots \tag{2.17}$$

In other words, Σ_k/Σ is the *probability* that the reaction, if it occurs, concerns component k of the mixture.

2.6.5. Neutron cross-section curves

The general shape of the cross-section curves depends on whether scattering or absorption has occurred and on the nuclides concerned. Scattering cross-sections are often more or less constant and on the order of a few barns, but absorption cross-sections are generally larger for slower neutrons and their order of magnitude varies greatly according to nuclide. Many irregularities are also observed.

A typical neutron absorption cross-section curve is given as an example in Figure 2.9: the (n,γ) reaction of gold 197 (a unique natural isotope), plotted between 0 and 20 MeV. (This cross-section has been measured very carefully because it is often used as a standard for relative capture measurements. It is easier to measure the ratio of two cross-sections than to perform an absolute measurement of one cross-section.) Note the general behaviour that is almost always found in absorption cross-section curves:

1/ General "1/v" behaviour, i.e. which is inversely proportional to the speed of the neutrons or to the square root of their kinetic energy. *Absorption* cross-sections (radiative capture, fission, (n,p) and (n,α) reactions) often follow this rule in the domain of "thermal" neutrons, i.e. below one electron volt.

2/ Complex behaviour with a curve showing several fairly irregular peaks in the epithermal domain, located between the fast domain and the thermal domain, typically between a few eV and a few keV. These peaks are called *resonances* of the cross-section. For gold, a spectacular resonance is observed around 5 eV.

The next figures show three other cross-section curves:

- the (n,α) reaction on boron 10 (one of the rare neutron–charged particle reactions without a threshold) is an example of a cross-section that follows the 1/v rule almost perfectly over the entire domain of the energies of interest (Figure 2.10);
- the fission reaction on uranium 235 obviously plays an essential role in reactors (Figure 2.11);

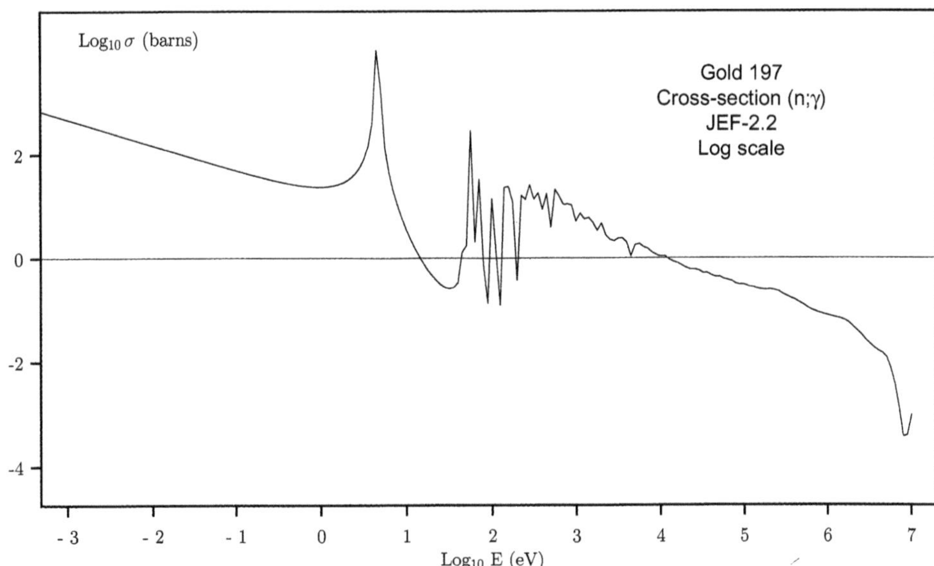

Figure 2.9. Cross-section of the (n,γ) reaction on gold 197.

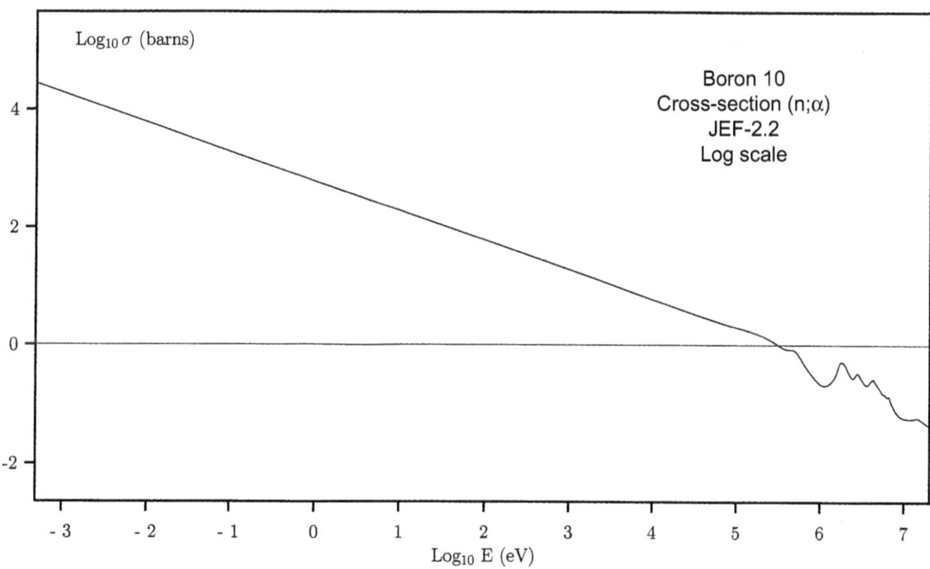

Figure 2.10. Cross-section of the (n, α) reaction on boron 10.

- the radiative capture reaction on uranium 238 (which, after two beta-minus decays, will give plutonium 239) is characterised by many large resonances that cause a great deal of trouble for neutron physicists, as we shall see (Figure 2.12).

Thus, even though absorption cross-sections keep a similar shape, particularly in the thermal domain, the order of magnitude can change considerably from one example to

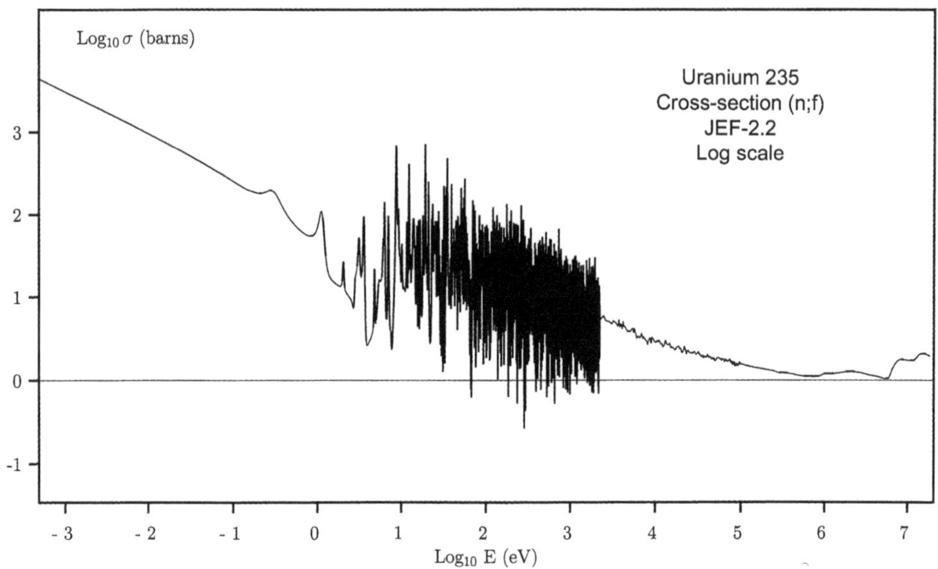

Figure 2.11. Cross-section of the (n,f) reaction on uranium 235.

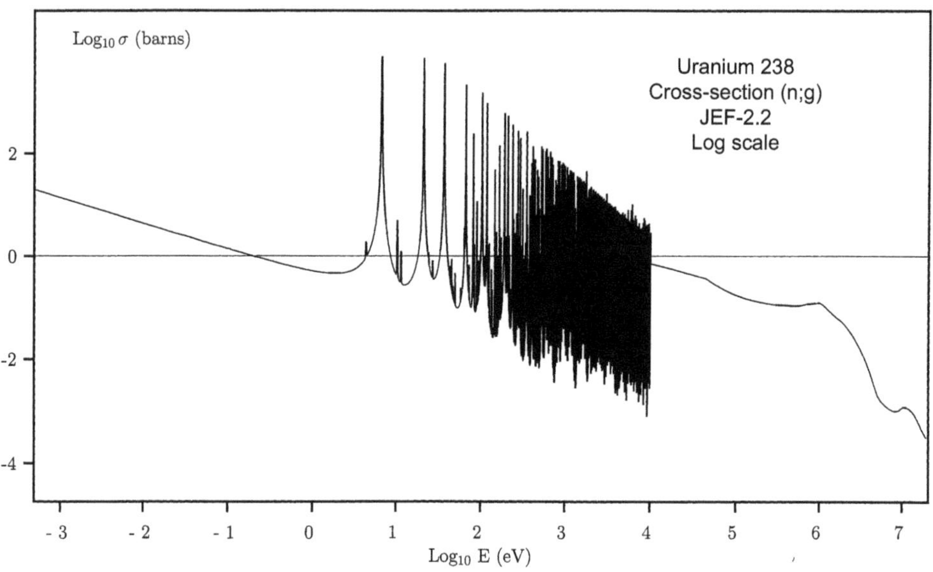

Figure 2.12. Cross-section of the (n, γ) reaction on uranium 238.

the next. The following tables (Tables 2.2 and 2.3) illustrate this for a few values of cross-section for 2200 m/s neutrons (this is often used as a reference speed). A nuclide has a large cross-section for slow neutrons if it happens to have a resonance centred on an energy close to zero (possibly negative); this is what happens in the examples shown in the next two figures (Figures 2.13 and 2.14), in particular the four main isotopes of plutonium.

Table 2.2. A few absorption cross-sections (in barns) for 2200 m/s neutrons.

Nucleus or element	Cross-section
Natural hydrogen	0.332
Deuterium	0.000506
Natural beryllium	0.0076
Boron 10	3840
Natural carbon	0.00337
Natural oxygen	0.000191
Natural zirconium	0.184
Xenon 135	2 650 000
Natural gadolinium	48 600
Samarium 149	40 500
Natural lead	0.178

Table 2.3. Characteristics of the main heavy nuclei for 2200 m/s neutrons (fission, capture: cross-sections in barns; ν: average number of neutrons emitted per fission).

Nucleus	Fission	ν	Capture
Thorium 232	–	–	7.40
Uranium 233	525.2	2.498	45.9
Uranium 235	582.6	2.439	98.9
Uranium 238	–	–	2.719
Plutonium 239	747.3	2.874	270.4
Plutonium 240	–	–	288.8
Plutonium 241	1 012.3	2.939	363.0

Scattering cross-sections can also have resonances, especially for intermediate and heavy nuclides.

Contrary to the 1/v rule for absorption cross-sections, however, their overall behaviour is a *constant*. This constant simply corresponds to the potential scattering. This potential scattering cross-section is of the order of a few barns, i.e. the geometric value of the image of the target and the projectile.

The only exception is light hydrogen which, despite being the smallest atomic nucleus, has the largest scattering cross-section: 20 barns. For very low-energy neutrons, quantum effects lead to higher scattering cross-sections than the plateau values seen in most of the domain of interest. This is illustrated in Figure 2.15 for the three main moderator nuclei.

2.7. Why resonances?

The presence of resonances comes from the structure in levels of excited states of the *compound nucleus* obtained by absorption of the incident neutron (see Figure 2.16).

Note that the excitation energy acquired by the compound nucleus is the sum of the binding energy of the incident neutron (the work of the nuclear forces) and the kinetic energy provided by this neutron. If this excitation energy is located exactly on one of the levels of the compound nucleus, as in Figure 2.16, or in its immediate neighbourhood,

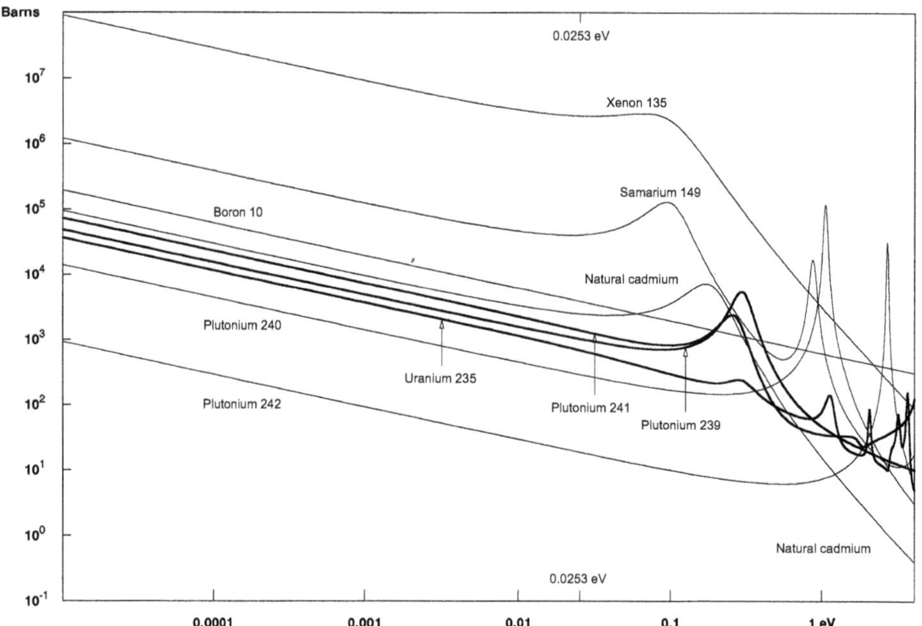

Figure 2.13. A few absorption cross-sections in the thermal neutron domain.

the reaction will occur easily and a large cross-section will be observed. If the excitation energy is not located on one of the levels of the compound nucleus, however, then the reaction will occur with greater difficulty. This will be reflected in a small cross-section. Thus the neutron cross-section can change by several decades for a very small change in the energy of the neutron, as the preceding figures show.

The general structure becomes increasingly crowded for heavier nuclei; that is why few or no resonances are seen on the cross-section curves for light nuclei, in contrast to a dense forest of peaks on the curves for heavy nuclei, particularly the actinides.

The binding energy of the incident neutron is higher if the target nucleus has an odd number of neutrons, because a pair is formed, than when it has an even number of neutrons, because the newcomer remains "single". All other things being equal, in the first case, the excitation energy, which is higher, is in a domain where the levels are denser, because this density increases as the excitation energy becomes higher.

This is why more resonances are observed for nuclides with an odd number of neutrons, such as uranium 235, plutonium 239, etc., (approximately one per electron volt, when they can be separated) than for nuclides with an even number of neutrons, such as uranium 238, thorium 232, etc., (approximately one resonance every 20 eV on average).

For the low kinetic energies of neutrons, and therefore low excitation energies, the levels are clearly separated, and it is easy to identify all the resonances when performing measurements. This is known as the *resolved* domain. For higher energies, the resonances remain, but they can no longer be distinguished by measurement; this is the *statistical* domain. At even higher energies, the resonances end up overlapping because of their width. This is the *continuum* domain.

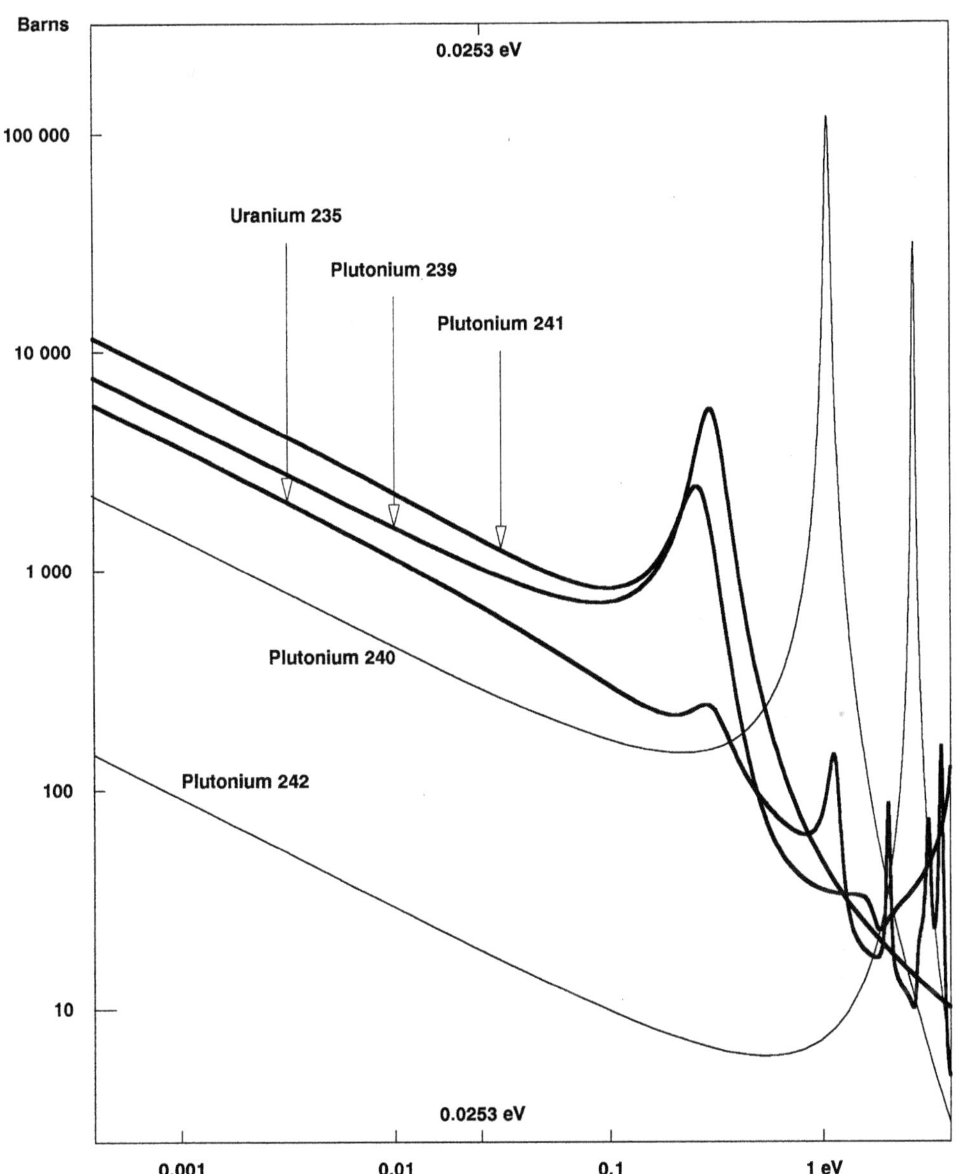

Figure 2.14. Comparison of absorption cross-sections of plutonium isotopes to uranium 235 in the thermal neutron domain.

In view of the above comment, the (approximate) limits between these domains are lower for nuclei with an odd number of neutrons than for the nuclei of neighbouring weights with an even number of neutrons. This is why the statistical domain is located (very approximately) around one keV in the first case, and around ten keV in the second case, for actinides.

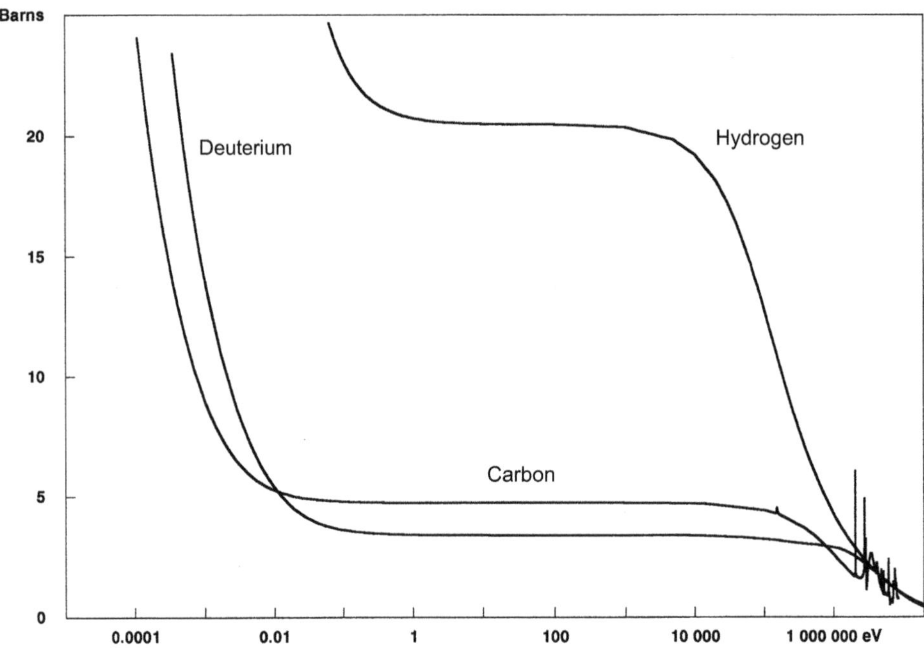

Figure 2.15. Elastic scattering cross-sections for the main nuclei used as moderators: ordinary hydrogen, deuterium, and carbon.

Note also that resonances tend to be more oriented towards either absorption or (resonant) scattering, according to the case. Examples that tend to concern absorption include the three big resonances of uranium 238 at 6.7, 20.9, and 36.7 eV. Examples concerning resonant scattering include the resonance of sodium at 5 keV and resonances of oxygen around one MeV. In either case, however, both components always exist. The same observation holds true for absorption processes if there are several of them, e.g. fission and radiative capture.

Finally, note that there are *negative resonances* or, more accurately, resonances at negative energies. These correspond to the levels located *below* the binding energy in Figure 2.16. The peaks of these resonances can obviously not be detected by measurement. It is possible, however, to detect the "wings" (in the positive energy region) if the level is close to the origin.

2.7.1. Resonant cross-sections: Breit–Wigner law

Using the quantum mechanical formalism and the compound nucleus model, i.e. assuming there is no correlation between the entrance and exit channels, it is possible to model a nuclear reaction and to formulate the cross-section relationships. In practice, the general, or **R**-matrix, formalism must be simplified. There are different degrees of approximation and therefore different models. Of these, the one most often used in nuclear physics calculation codes for neutron physics is the Reich–Moore formalism. The simplest model,

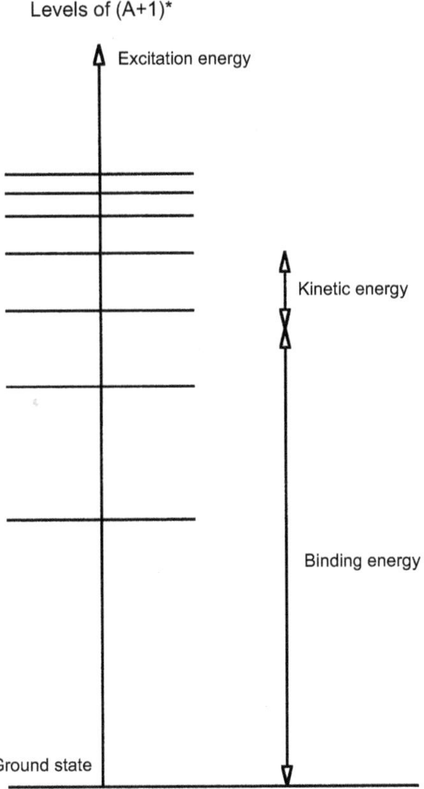

Figure 2.16. A resonance is observed for each kinetic energy value of the neutron that brings the excitation energy of the compound nucleus to one of the levels.

which is amply sufficient for most applications and, in any case, for the main qualitative aspects, is the Breit–Wigner one-level model[15].

Before giving details of the formulae, there are three considerations to note:

1/ Like any microscopic object, the neutron not only has a bodily aspect, but also a wave aspect. The wavelength associated with a neutron is calculated as follows:

$$\lambda = \frac{h}{mv} = \frac{2.86 \times 10^{-11}}{\sqrt{E}}, \tag{2.18}$$

where λ is in m and E is in eV (non-relativistic De Broglie equation). For fast neutrons, this wavelength is of the order of the size of a nucleus (e.g., 2.9×10^{-14} m for 1 MeV neutrons). For thermal neutrons, however, this wavelength is of the order of the size of an atom (e.g., 1.8×10^{-10} m for 1/40 eV neutrons). This has two important consequences:

15 There is an extension: the Breit–Wigner multi-level model. This allows for interference effects, not only between potential scattering and resonant scattering (see below), but also between the excited levels of the compound nucleus.

- in general, slow neutrons see a larger portion of space than fast neutrons, which means that slow neutrons often have larger cross-sections. More specifically, this also leads to the 1/v rule for absorption.
- Very low-energy neutrons in crystals and in certain molecules can undergo interference effects when scattered. This allows them to be used as a tool to study the structure of matter, providing a complement to the X-ray and electron methods.

2/ For a billiard ball hitting another ball more or less tangentially, an orbital angular momentum must be taken into account. This moment, $L = mvd$, is the product of the momentum mv of the projectile and the impact parameter d, the minimum distance between the centre of the projectile and the centre of the target. In quantum physics, this moment is quantified as $L = \ell\hbar$. By substituting one formula into the other, we see that the whole number ℓ must be on the order of $mvd/\hbar$, with d less than or equal to the sum of the radii of the neutron ($r_0 \cong 1.2 \times 10^{-15}$ m) and target ($r_0 A^{1/3}$). This evaluation (although obviously an oversimplification) shows that, for thermal and epithermal neutrons, the wave $s\,(\ell = 0)$ dominates, but at high energy[16], the next waves, p, d, f, g, h, etc. ($\ell = 1, 2, 3, 4, 5$, etc.) play an increasingly important part.

As we have seen, resonances are most prominent in the lower part of the epithermal domain; this means that s waves (which have no angular moment and are therefore spherically symmetric) play an essential role, although in practice the other waves must sometimes be considered. This is why we write the Breit–Wigner formulae for s waves only[17].

3/ Neutrons and certain nuclei have an intrinsic angular momentum (spin). A neutron has spin $s = 1/2$ (the unit being $\hbar$). The spin of even–even nuclei is $I = 0$, and the spin of other nuclei is nonzero. For s waves, the spin J of the compound nucleus is $1/2$ if $I = 0$ and $I - 1/2$ or $I + 1/2$ if I is nonzero. A statistical factor must be introduced to allow for the different angular momentum orientations:

$$g = \frac{2J + 1}{(2s + 1)(2I + 1)}, \tag{2.19}$$

where $g = 1$ if I is zero, and $g = (I + 1)/(2I + 1)$ or $g = I/(2I + 1)$ if I is nonzero.

The other parameters involved in the formulae are as follows:

- *normalised mass* μ defined by $1/\mu = 1/m_{\text{target-nucleus}} + 1/m_{\text{neutron}}$;
- *kinetic energy of the neutron in the centre of mass system*: E;
- *momentum*: $i = \sqrt{2\mu E}$;
- *normalised wavelength*: $\lambda\!\!\!^{-} = \hbar/i$;
- *energy of resonance peak*: E_0 (same definition as for E);

[16] For uranium, the formula gives $\ell = 1$ for 300 keV neutrons.

[17] In reality, the general formulae are not much more complicated; coefficients need to be added to the resonant and interference terms.

- *width of resonance*: $\Gamma = \hbar/\tau$ where τ is the average lifetime of the compound nucleus (the inverse of its decay constant), the width Γ has the dimensions of energy and is expressed in electron volts or, more often, meV,
- *partial widths*: Γ_i. As we have seen, the compound nucleus can disintegrate according to several modes (exit channels): neutron emission ($i = n$), photon emission ($i = \gamma$), fission ($i = f$), etc. Associated with each mode is a partial width whose sum is Γ; in other words, the ratio Γ_i/Γ is the probability that the exit will occur along channel i;
- *potential scattering cross-section*: p (over the energy interval covered by a resonance, this cross-section can be considered as a constant).

The Breit–Wigner equations describe the partial cross-sections for *one* resonance assumed to be isolated and characterised by its *resonance parameters* E_0 and Γ_i ($i = n, \gamma$ ...). In practice, there are several resonances, and *the expressions must be summed* for all resonances. The equations are as follows:

- **for absorption cross-sections:**

$$\sigma_i = \pi \lambda\!\!\!^{-2} g \frac{\Gamma_n \Gamma_i}{(E - E_0)^2 + \Gamma^2/4}, \tag{2.20}$$

with $i = \gamma$ for radiative capture, $i = f$ for fission, etc.

- **for the scattering cross-section:** to the resonant process whose cross-section is given by the above equation with $i = n$, we must add the potential scattering whose cross-section is p and an *interference term* for the interference between the two processes[18]:

$$\sigma_s = \pi \lambda\!\!\!^{-2} g \frac{\Gamma_n^2}{(E - E_0)^2 + \Gamma^2/4} + 2\sqrt{\pi \lambda\!\!\!^{-2} g p} \frac{\Gamma_n (E - E_0)}{(E - E_0)^2 + \Gamma^2/4} + p; \tag{2.21}$$

- **for the total cross-section:**

$$\sigma = \pi \lambda\!\!\!^{-2} g \frac{\Gamma_n \Gamma}{(E - E_0)^2 + \Gamma^2/4} + 2\sqrt{\pi \lambda\!\!\!^{-2} g p} \frac{\Gamma_n (E - E_0)}{(E - E_0)^2 + \Gamma^2/4} + p. \tag{2.22}$$

Let:

$$r = g \frac{h^2 (A + 1)}{2\pi E_0 m A} = g \frac{2\,603\,911}{E_0} \cdot \frac{A + 1}{A}, \quad q = \sqrt{pr}, \tag{2.23}$$

(where A is the ratio of the target nucleus mass to the mass of the neutron, i.e., very close to the mass number) and:

$$x = \frac{E - E_0}{\Gamma/2}, \quad \Psi = \frac{1}{1 + x^2}, \quad \chi = \frac{2x}{1 + x^2}, \tag{2.24}$$

we can write these equations more simply:

18 In quantum mechanics, complex wave functions are added, and since the square of the norm is taken to calculate the cross-section, this generates some "crossed" terms between those that are added.

- **for absorption cross-sections:**

$$\sigma_i = r\psi \cdot \frac{\Gamma_n \Gamma_i}{\Gamma^2}; \tag{2.25}$$

The expression for the function ψ shows that a resonant cross-section is represented by a *symmetrical* bell curve centred on the resonance energy and vanishing at infinity. The width Γ is the energy interval delineated by taking a value at mid-height with respect to the peak (x between -1 and $+1$). The curves for the various absorption reactions differ from each other by a simple factor[19],

- **for the scattering cross-section:**

$$\sigma_s = (r\psi + q\chi) \cdot \frac{\Gamma_n^2}{\Gamma^2} + p; \tag{2.26}$$

- **for the total cross-section:**

$$\sigma = (r\psi + q\chi) \cdot \frac{\Gamma_n}{\Gamma} + p. \tag{2.27}$$

In addition to the bell curve representing absorption, the scattering and total cross-sections contain a constant (potential) term and an interference term that is asymmetric with respect to the resonance peak: it is positive at energies above E_0, and negative below.

These equations give the cross-sections in the *centre of mass* system, i.e. the system in which nuclear physicists such as Breit and Wigner perform their calculations. When neutron physicists write the Boltzmann equation and try to solve it, however, they are working under laboratory conditions (i.e. using a reactor). Assuming the target nucleus to be initially at rest, the coordinate system change formulae give:

$$E_{\mathrm{lab}} = \frac{A+1}{A} E_{cm}, \tag{2.28}$$

When introduced into the Breit–Wigner equations, this gives similar expressions. To take into account the thermal agitation of the target nuclei (which can be significant if there are resonances), a mathematical convolution between the Breit–Wigner relationship and the thermal agitation relationship must be performed. This "Doppler effect" problem will be examined in Chapter 8.

2.7.2. Resonant cross-sections: statistical aspects

No nuclear model can give a detailed prediction of a resonant cross-section structure. Such information can only be supplied by measurements. A statistical analysis of measurement results shows the following:

- the Γ_n widths fluctuate greatly from one resonance to the other,
- the Γ_γ widths, however, are roughly the same for all resonances,

[19] Neglecting the variation of widths with energy, which is not strictly correct.

Table 2.4. First resonances of uranium 238 (*s* waves only; JEF-2 evaluation).

Peak energy	Neutr. width	Gamma width	Total width	Maximum section
6.674	1.493	23	24.493	23 992
20.871	10.26	22.91	33.17	38 923
36.682	34.13	22.89	57.02	42 849
66.032	24.6	23.36	47.96	20 399
80.749	1.865	23	24.865	2448
102.56	71.7	23.42	95.12	19 295
116.902	25.49	22.99	48.48	11 811
145.66	0.847	23	23.847	651
165.29	3.367	23	26.367	2036
189.67	173.2	22.38	195.58	12 253
208.51	51.11	23.94	75.05	8573
237.38	27.16	24.54	51.70	5812
273.66	25.78	22.1	47.88	5166
291	16.87	22.12	38.99	3907

- the average distance $\langle D \rangle$ between two resonances varies slightly with changes of energy domain, according to the following approximate rule:

$$\langle D \rangle = \alpha \exp\left(\beta \sqrt{E_{ex}}\right), \tag{2.29}$$

where α and β are constants, and E_{ex} is the excitation energy. This relationship explains the series of domains (starting from the bottom): resolved, statistical, continuum,

- The spaces between consecutive resonances D approximately follow the *Wigner probability distribution*:

$$p(w)dw = \frac{\pi}{2} w \exp\left(-\frac{\pi}{4} w^2\right) dw, \tag{2.30}$$

where $w = D/\langle D \rangle$.

Some of these statistical aspects are shown in Table 2.4, which gives the parameters of the first fourteen resonances of uranium 238.

The energy of the peak is given in eV; widths are in meV; total resonant cross-section at peak is in barns; the potential cross-section for this nuclide is $p = 8.9$ barns.

2.7.3. Cross-sections in the thermal domain

At low energy, the absorption cross-sections (fission and capture) approximately obey a $1/v$ relationship. This can be demonstrated using the Breit–Wigner formula for absorption:

- Γ_f, Γ_γ, Γ_α, etc. are independent of the energy E;
- Γ_n is proportional to the square root of this energy (for *s* waves);

- $\lambda\!\!\!^{-2}$ is inversely proportional to this energy;

- The denominator is approximately equal to the constant E_0^2 assuming that E and Γ are small compared to E_0.

Thus σ_f and σ_c are inversely proportional to the square root of E, or proportional to $1/v$.

Even if several resonances make a contribution, the reasoning remains valid.

By the same reasoning, the diffusion cross-sections are more or less constant at low energy.

These approximations do not remain valid if the fourth assumption fails, i.e. if the peak energy E_0 is close to zero. Unless if there is an exception, such as boron 10, this is what happens with neutrons that have a large cross-section for thermal neutrons because, by sheer coincidence, they have a resonance peak near the origin. Examples include xenon 135, samarium 149, uranium 235 (negative resonance close to zero), plutonium 239 (resonance at 0.3 eV)[20], etc.

2.8. Neutron sources

Fission is obviously the main source of neutrons in a reactor.

Other neutron sources can be used in other applications of neutron physics; however, even in a reactor, the chain reaction must be initialised by a source other than neutron-induced fission. That is why this subsection devoted to nuclear reactions ends with a brief introduction to neutron sources.

2.8.1. Spontaneous sources

In our environment there are some neutrons produced by nuclear reactions due to cosmic rays. In reactors there are also some neutrons emitted by spontaneous fissions. This effect is mainly seen, although at a very slow rate, among even–even heavy nuclei, particularly uranium 238 (*see* § 2.4.4) and plutonium 240. In a reactor that has operated and that contains irradiated fuel —and therefore alpha emitters— there are also some neutrons emitted by (α,n) and (γ,n) reactions on oxygen.

In a subcritical system (reactor or other installation of the cycle) characterised by a multiplication factor k, these sources are amplified because of the induced fissions by a factor G that increases as the system approaches criticality: $G = 1+k+k^2+k^3+\cdots = 1/(1-k)$.

---

[20] As we have seen (*see* § 2.7), heavy nuclei with an odd number of neutrons have a high resonance density (spacing on the order of 1 eV); this means that there is a high probability of finding a resonance close to zero, and therefore a large cross-section and a deviation from the 1/v rule. Conversely, heavy nuclei with an even number of neutrons have fewer resonances (spacing on the order of 20 eV); there is only a small chance of finding a resonance close to zero, and therefore the cross-section is usually not very large and the 1/v rule is respected; examples: uranium 238, thorium 232; counter-example: plutonium 240 (resonance at 1 eV).

2.8.2. Reactions induced by radioactivity

These sources, which are very weak[21], are theoretically sufficient to start up a chain reaction in a reactor[22]. In practice, a much more intense source is introduced so that the divergence can be monitored by measuring the neutron flux until a significant power level is reached. The source can then either be removed or left in place, because the induced fission chain reaction becomes dominant. The most commonly used start-up sources use the (α,n) reaction or the (γ,n) reaction on beryllium. They are manufactured from a mixture of beryllium and an α emitter (such as radium) or a γ emitter (such as antimony 124; if the source is left in the reactor, this isotope with a relatively short half-life of 60 days can be regenerated by neutron irradiation).

2.8.3. Fusion reactions

For applications requiring more intense sources, such as neutron physics measurements outside a reactor or in subcritical systems, the D+T fusion reaction is most often used. The usual method is to accelerate deuterons (obtained via the ionisation of deuterium) to hit a target containing tritium.

2.8.4. Spallation reactions

The techniques mentioned above cannot be extrapolated for far more intense sources, for example with a view to generating energy from subcritical reactors, or the large-scale incineration of nuclear waste[23].

The most promising method for this type of application seems to be the method involving spallation of heavy nuclei by high-energy protons.

Although spallation reactions are now mainly of interest to nuclear physicists, they were in fact initially discovered and studied by astrophysicists.

These reactions can occur at the surface of stars during stellar eruptions and, in particular, between galactic cosmic radiation and the nuclei of the (rare) interstellar atoms. This interaction modifies the composition of this cosmic radiation and thereby provides information about its origins; cosmic radiation also provokes spallation reactions in the matter in meteorites: observing the products of these reactions gives an indication of how long the meteorites have been in space.

Spallation reactions are induced by nucleons or small nuclei (such as alpha particles) at high speed, if their kinetic energy exceeds approximately 10 MeV and, preferably, if it is on the order of a GeV. (The protons of galactic cosmic radiation have an average energy of 4 GeV, and that is why they cause spallation reactions if they interact with matter.)

Spallation reactions take place on intermediate or heavy nuclei. By pulling fragments off the nuclei, these reactions eject nucleons or light nuclei: isotopes of hydrogen, helium, lithium, or even beryllium or boron. Sometimes the fission of the nucleus that has been struck occurs.

---

[21] Despite being very weak, these sources must be taken into account when planning radiation protection for the handling of irradiated nuclear fuels.

[22] In plutonium weapons it is important not to exceed a few percent of plutonium 240, an emitter of neutrons by spontaneous fission, because too much could trigger a premature and less efficient start-up of the chain reaction.

[23] See Chapter 18.

There are two stages to these spallation reactions:

- the first stage lasts only as long as the transit time of the incident particle through the nucleus it encounters, i.e. between 10^{-22} and 10^{-21} s; a few nucleons are ejected by a series of "billiard ball"-type collisions: this is *intranuclear cascade*;

- the second stage is longer (approximately 10^{-16} s): the residual nucleus, which is very energetic because it has kept some of the energy from the incident particle, releases this excess energy by emitting a few more nucleons: this is *evaporation*.

In dense matter, the nucleons or small nuclei thus emitted have a certain probability of colliding with other nuclei and of repeating these two phases if they still have enough energy: this is *extranuclear cascade* (see Figure 2.17).

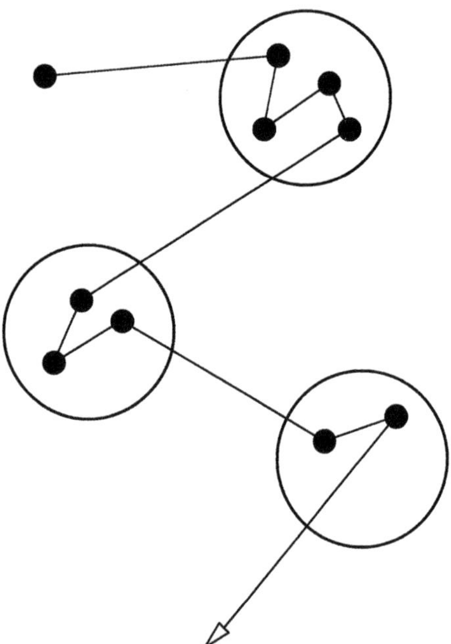

Figure 2.17. Spallation: intranuclear and extranuclear cascades (diagram).

These reactions became of interest to reactor physicists when they discovered that firing a proton with energy on the order of a GeV at a target of heavy nuclei (tungsten, lead, bismuth, thorium, uranium, etc.), could produce about *thirty secondary neutrons*. (In this range, the number of neutrons obtained by spallation is approximately proportional to the kinetic energy of the incident proton.) These neutrons can induce fissions (which themselves supply more energy than was required to produce the incident proton) or transmute nuclei: to convert fertile matter into fissile matter, destroy radioactive waste, etc.

D. NUCLEAR FISSION

2.9. Spontaneous fission and induced fission

2.9.1. The fission barrier

When a nuclear process is exoenergetic, it can be expected to occur spontaneously; this is what happens in radioactivity, for example. The fission of a heavy nucleus is very exoenergetic, since it releases approximately 200 MeV, as we have seen. Spontaneous fission is sometimes seen in certain actinides, but only at a ridiculously slow rate that is negligible in practice.

The reason that fission does not occur spontaneously (and that heavy nuclei exist in nature) is that a certain amount of energy needs to be added, as Figure 2.18 suggests to the intuitive observer.

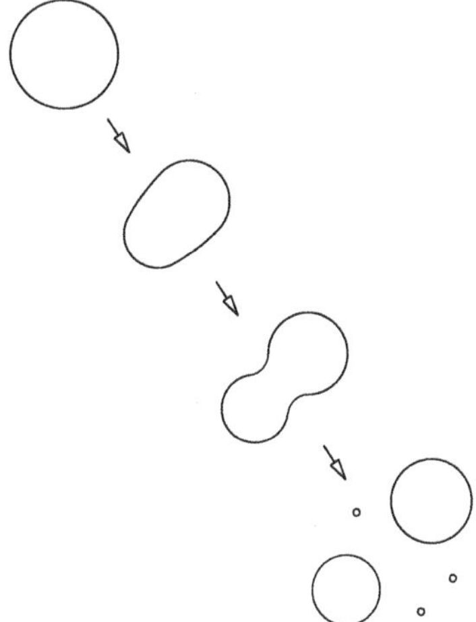

Figure 2.18. Deformation before fission (diagram).

Like a drop of liquid, the nucleus, which is initially almost spherical, can only be split in two if it is first sufficiently deformed for constriction to occur. This deformation, however, requires an energy input; without this input, there can be no fission.

More precisely, and with reference to the liquid drop model (see § 2.2.4), it can be said firstly that the necessary deformation corresponds essentially to an increase in surface area, and therefore to a decrease in binding energy (or a mass increase or an energy input). Next, the work of the Coulomb forces will do the opposite, i.e. release energy by accelerating the constriction effect until splitting occurs, and then make the two residual nuclei repel each other violently. Finally, most of the energy released by fission will end up as kinetic energy in these two fragments. These two stages are illustrated in Figure 2.19 (not to scale): this plot is shown as a function of a parameter called *deformation*, which is defined such that it increases over the successive stages and quantifies them. It shows how the surface term varies: it increases to the splitting point (two tangential spheres), and afterwards does not

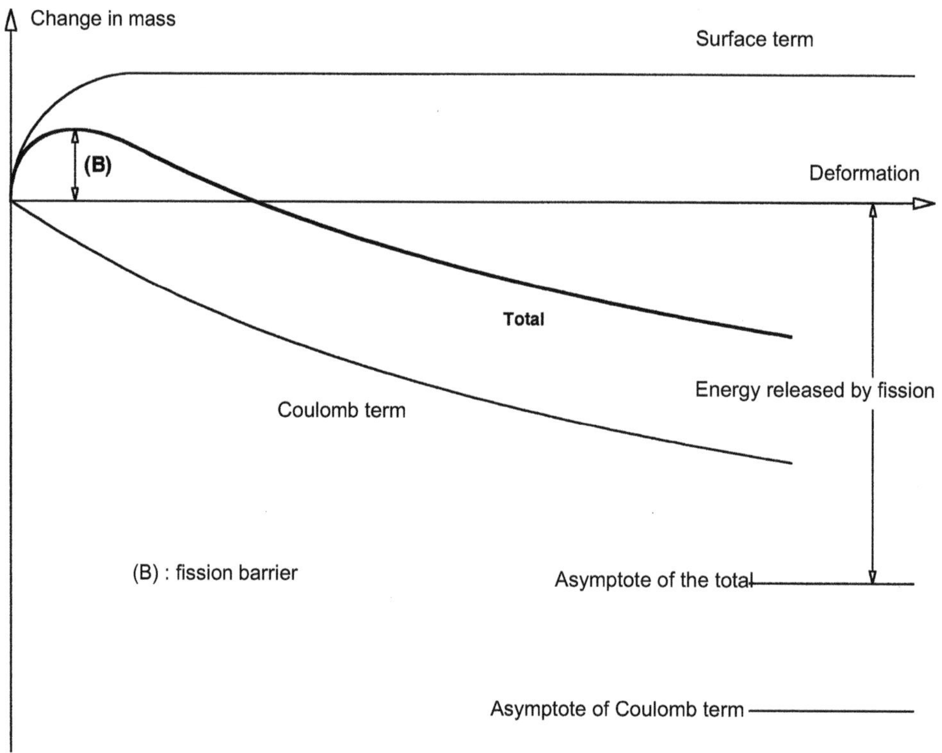

Figure 2.19. Concept of fission barrier (diagram).

change. It also shows how the Coulomb term continuously decreases. The sum of these two terms is initially positive, passes through a maximum, and then decreases towards an asymptote at a large negative value. The height of the "hill" (barrier) to be overcome is the energy contribution required to cause fission; the ordinate of the asymptote is the energy finally released by the fission. Note that this diagram is only approximate; not only because the variable along the abscissa axis has not been defined, but also because the curve might be more complicated. It could start out at a small positive value and go through zero before approaching the barrier (meaning that the nucleus in ground state is not spherical, but ellipsoidal), or it could have two or three humps, etc.

2.9.2. Fission-related thresholds

Following these very general considerations, we should define two thresholds. The first is the threshold from which fission becomes exoenergetic, and the other is the threshold where the barrier disappears, i.e. from where nuclei are unstable with respect to spontaneous fission.

Assuming that fission takes place symmetrically, the first threshold is defined as follows:

$$2W\left(\frac{A}{2}, \frac{Z}{2}\right) - W(A, Z) > 0. \tag{2.31}$$

Using the Bethe–Weizsäcker equation and neglecting its parity term, we obtain:

$$\frac{Z^2}{A} > \frac{(2^{1/3} - 1)a_s}{(1 - 2^{-2/3})a_c} = 17.3, \tag{2.32}$$

On the stability line, this corresponds approximately to $A = 90$.

An approximate evaluation of the fission barrier disappearance threshold can be found by starting at the splitting point (two tangential spheres) and by setting the condition that the changes in the surface and Coulomb terms exactly compensate each other. The calculation performed under the same conditions gives:

$$\frac{Z^2}{A} > \frac{(2^{1/3} - 1)a_s}{(1 - 2^{-2/3} - 5.2^{1/3}/24)a_c} = 59.5, \tag{2.33}$$

The stability line does not reach this value[24].

Between these two limits, the height of the fission barrier (in MeV) can be estimated using an empirical formula proposed by G. Seaborg:

$$B = 19.0 - 0.36\frac{Z^2}{A} + \varepsilon, \tag{2.34}$$

with $\varepsilon = 0$ for even–even nuclei, $\varepsilon = 0.4$ for even–odd or odd–even nuclei, and $\varepsilon = 0.7$ for odd–odd nuclei. (This formula gives a slightly lower limit value of Z^2/A, around 54, which the stability line still does not reach.)

2.9.3. Parity effect

Seaborg's formula shows that the barrier is a little lower for even–even nuclei, even though these are not the nuclei that fission most easily. The other consideration involved is the energy input that can overcome the barrier.

In the present case, this input comes from the absorption of an incident neutron.

Note that the excitation energy of the compound nucleus resulting from this absorption is the sum of the kinetic energy supplied by the neutron and the binding energy of the additional neutron. This binding energy is much greater when the initial target has an odd number of neutrons than an even number, because a pair is formed.

Figure 2.20 and Table 2.5 illustrate these differences for the two main isotopes of uranium (values are expressed in MeV)[25].

There is a large positive energy balance for isotope 235, which suggests that this nucleus will undergo fission even if the neutron it absorbs has very little kinetic energy. The energy balance for the other isotope, however, is a negative value of 1.8 MeV, meaning that at least this amount of energy must be provided to provoke fission. This turns out to be more or less true, apart from a few additional complications due to quantum effects.

---

24 If the stability line is extrapolated beyond the real nuclei, this parameter passes through a maximum of 49 at approximately mass 600.

25 The values for the barrier height are measurements that differ slightly from the results of the Seaborg formula. (Note that, for fission induced by absorption of a neutron, the formula must be applied to the *compound* nucleus.) The diagrams show the mass (or energy) curve limited to low deformations, i.e. around the barrier.

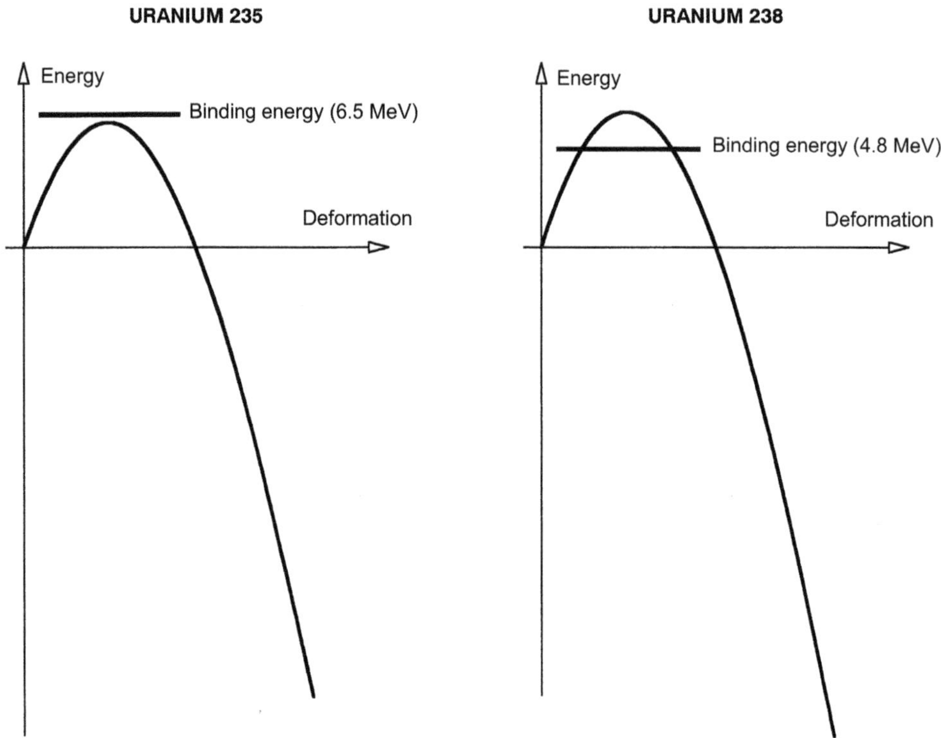

Figure 2.20. Comparison of two uranium isotopes with regard to fission.

Table 2.5.

Isotope	235	238
Binding energy	6.5	4.8
Fission barrier	6.1	6.6
Energy balance	+0.4	−1.8

2.9.4. *Quantum effects: tunnel effect and anti-tunnel effect*

In classical physics, a projectile will reach the other side of a hill if it was launched over the top of the hill, but will fail to reach the other side if it was not. In quantum physics, this all-or-nothing situation must be replaced by a probability that varies continuously from 1 to 0.

This means that, even if the excitation energy of the compound nucleus is insufficient to take it over the barrier, fission can still occur (the more negative the energy balance, the smaller the probability of this event). This is called the *tunnel effect,* to convey the image that the system is going through the metaphorical hill as if through a tunnel. Thus, the fission cross-section of uranium 238 is not strictly zero for slow neutrons, but only negligible in practice (on the order of a microbarn); and although the (classic) threshold is located near 1.8 MeV, this cross-section becomes significant around 0.8 MeV. Note also that spontaneous fission occurs thanks to the tunnel effect.

Conversely, if the excitation energy exceeds the threshold, fission does not necessarily occur — there is merely a strong probability that it will. This explains why uranium 235 fission occurs only about six times out of seven after absorption of a slow neutron, even though there is a positive energy balance of 0.4 MeV.

2.10. Fission products

2.10.1. Neutrons

There are three types of fission product: energy (the desired product), fragments (by-products for which no use has yet been found), and neutrons, which allow the chain reaction to take place.

Most neutrons are emitted by "evaporation" by the two fragments before they have been stopped by the matter, approximately 10^{-17} s after the split, which seems instantaneous on a human scale. These *prompt neutrons* vary in number, from zero to seven. They are emitted according to a spectrum extending roughly between 0 and 10 MeV, called the *fission spectrum*.

There are analytical approximations of the fission spectrum. The most commonly-used ones are:

- *The Maxwell spectrum*:

$$\chi(E)dE = \frac{2\pi}{(\pi T)^{3/2}}\sqrt{E}\exp\left(-\frac{E}{T}\right)dE, \tag{2.35}$$

  (number of neutrons emitted between E and $E + dE$) with T in the region of 1.33 MeV for uranium (a little more for plutonium), i.e. an average energy 3/2 T on the order of 2 MeV (Figure 2.21);

- *The Cranberg spectrum:*

$$\chi(E)dE = \frac{2\exp(-AB/4)}{\sqrt{\pi A^3 B}}\exp\left(-\frac{E}{A}\right)sh\sqrt{BE}dE, \tag{2.36}$$

  with $A = 0.965$ MeV and $B = 2.29$ MeV^{-1} for uranium.

In addition to prompt neutrons, there are other neutrons that are emitted with a delay after fission[26]. Chapter 4 explains why delayed neutrons play an essential role in reactor kinetics even though they make up only a small proportion of the neutrons emitted (0.7% for induced fission in uranium 235 and 0.2% in plutonium 239). The delay between fission and the emission of these neutrons can be between a second and a minute, i.e. a very long time compared to the life of a neutron (measured in microseconds).

There are about a hundred different processes for delayed neutron emission, and all are similar: a beta-minus decay of a fission fragment, followed by a neutron emission

[26] The values of ν given on § 2.6.5 (Table 2.3) are total values for prompt neutrons + delayed neutrons.

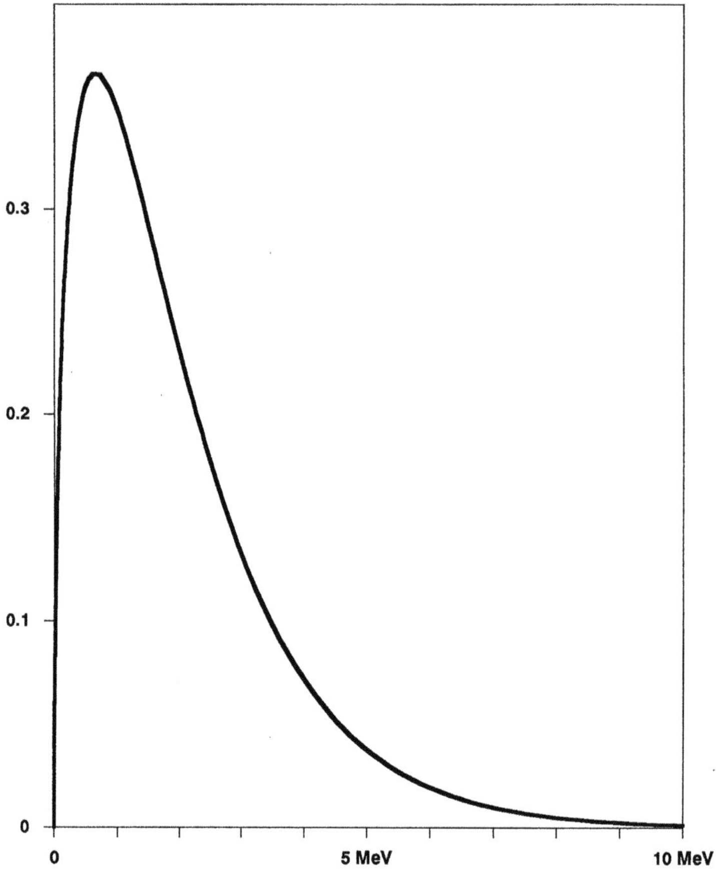

Figure 2.21. Fission Spectrum (uranium, prompt neutrons).

(which is possible if the nucleus obtained following beta decay has an excitation energy greater than the separation energy of a neutron). The delay between fission and neutron emission is caused by the beta decay; the neutron emission that follows, if any[27], is almost instantaneous. Examples:

$$^{87}\mathbf{Br} \Rightarrow {}^{87}\mathbf{Kr}^{*} \Rightarrow {}^{86}\mathbf{Kr} + {}^{1}\mathbf{n} \text{ (beta half – life : 54.5 s),}$$

$$^{137}\mathbf{I} \Rightarrow {}^{137}\mathbf{Xe} \Rightarrow {}^{136}\mathbf{Xe} + {}^{1}\mathbf{n} \text{ (beta half – life : 21.8 s).}$$

Delayed neutrons are emitted at a slightly lower energy (0.2 to 0.6 MeV) than prompt neutrons (2 MeV on average).

[27] De-excitation can also occur via gamma emission.

2.10.2. Fission fragments

Fission is always binary, except when neutrons and, very rarely, light nuclei (tritium, alpha particle) are emitted during *ternary* fission. The two fragments are nuclei with a mass that is approximately half the mass of the fissioning nucleus. There is a wide variety of possible fragments, because there are many different ways in which the available nucleons can be arranged: several hundred nuclides, which are isotopes of over thirty elements, can be obtained.

Figure 2.22 summarises the statistical distribution of these fragments, each of which is characterised by a *fission yield (y)* (the probability of its production when fission occurs). Note that the sum of all fission yields is 200%, because every fission is binary.

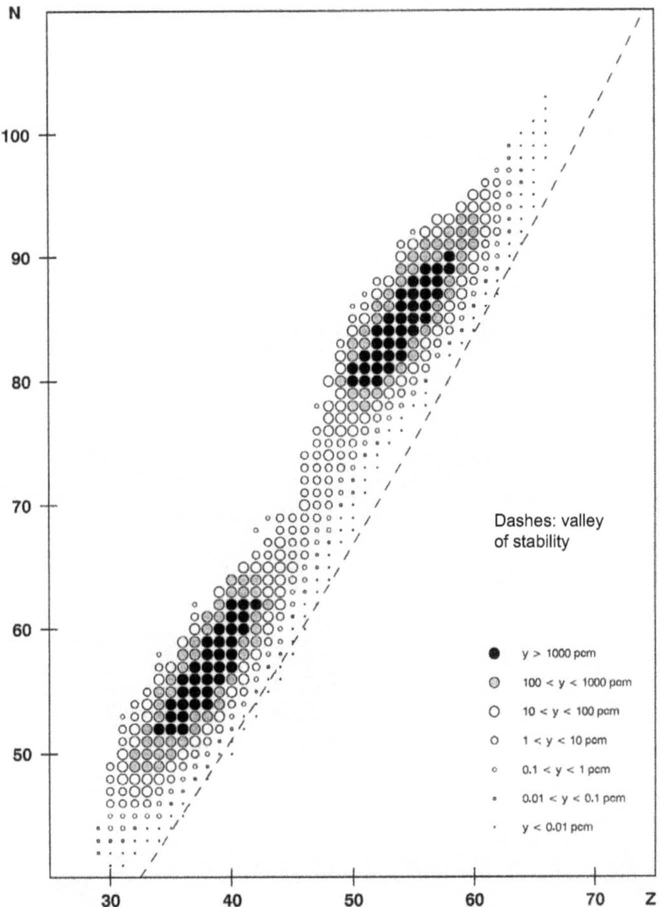

Figure 2.22. Fragments of the thermal neutron-induced fission of uranium 235.

Note the wide spread of the numerical values of yield and the fact that fission is usually *asymmetrical*, since one of the fragments (the heavier one) is located in the higher cloud, and the other fragment is in the lower cloud.

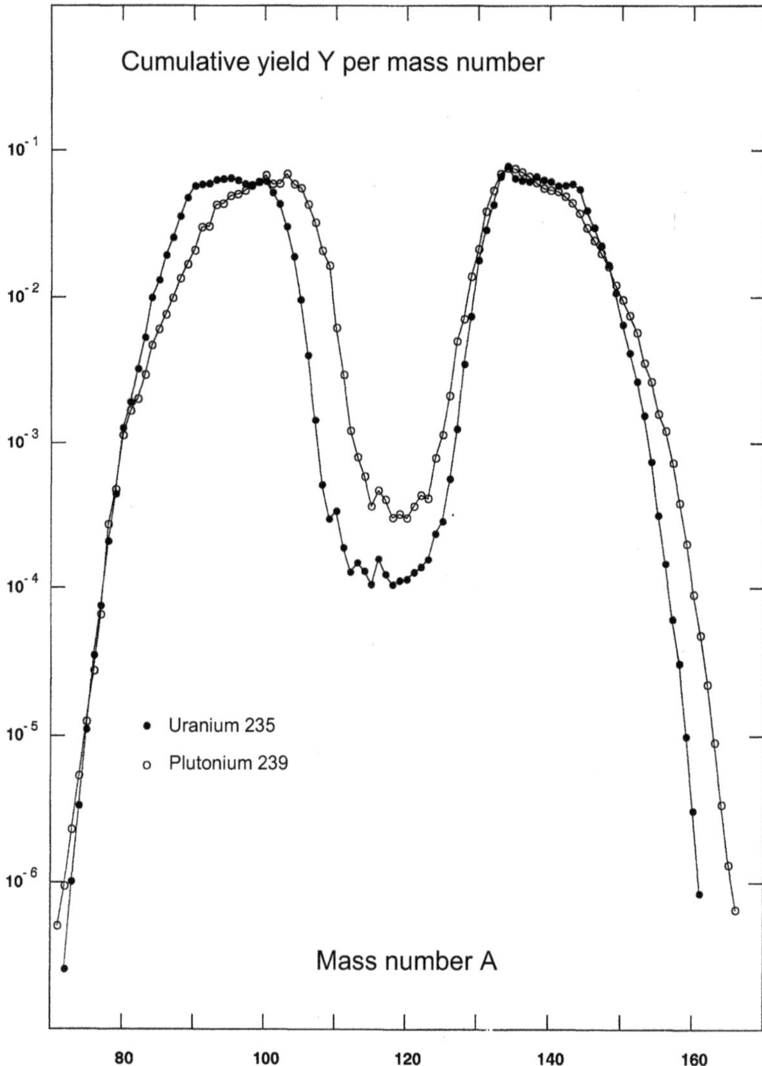

Figure 2.23. Cumulative yields for thermal neutron-induced fission: comparison of uranium 235 and plutonium 239.

This asymmetry is even more obvious on a plot of the *cumulative* yields *Y* for each mass number *A* (Figure 2.23). This diagram also illustrates how yields differ according to the fissioned nucleus, i.e. the number of nucleons to be distributed.

Finally, note that almost all of these fragments are beta-minus radioactive because they are located *above* the stability line (which is explained by the curvature of this line: see Figure 2.1). Moreover, some of them have a medium or high neutron capture rate. *Fission products* is the name given to all the nuclides obtained in reactors following fission reactions, i.e. fragments plus all nuclides (which might or might not be included in the list of fragments) obtained by radioactive decay and/or neutron capture.

2.10.3. Energy

The energy released during fission and the relative proportions of its contributing components tend to be the same to within a few percent for all fissile nuclei. The sample breakdown given below refers to the energy released on thermal neutron-induced fission of uranium 235 (in MeV):

• Fission fragments:	166.2
• Instantaneous gamma photons:	8.0
• Neutrons:	4.8
• Beta radiation (electrons):	7.0
• Antineutrinos accompanying beta emission:	9.6
• Gamma photons after beta emission:	7.2
Total:	202.8

Over eighty percent of this energy is taken away by the two fragments and dispersed within the fuel where the fission takes place (mean free path of fragments in metallic uranium: 7 μm).

The energy of the electrons is also deposited over a short distance in the fuel.

The mean free path of gamma photons is on the order of several centimetres. The corresponding energy is mainly deposited in heavy materials: here again, in the fuel, but over a wider area.

The energy of the neutrons (approximately 2.4 times 2 MeV), mostly prompt neutrons, is mostly deposited in the moderator, which decelerates them.

The energy of the antineutrinos, which do not interact with matter, is not recovered.

Thus, when calculating the total amount of energy actually recovered, the "antineutrinos" line should be left out. Another energy should be added, however: the energy produced by the capture (usually radiative) of the $\nu - 1$ neutrons which, on average, in a chain reaction at equilibrium, do not cause fission. The gamma energy of the capture reactions depends on the materials, and therefore the reactor, concerned. An average, order-of-magnitude figure, usually used for water reactors, would be 6 MeV per capture, i.e. $6 \times 1.4 = 8.4$ MeV total. The final total (for uranium 235 would be:

• Energy released:	202.8
• Antineutrinos accompanying beta decay:	−9.6
• Gamma photons after neutron capture:	+8.4
Energy recovered by fission:	201.7

Three other examples are given for comparison:

• Uranium 235 (thermal neutron):	201.7
• Uranium 238 (fast neutron):	205.0
• Plutonium 239 (thermal neutron):	210.0
• Plutonium 241 (thermal neutron):	212.4

Overall, 3.1×10^{10} fissions are required to produce one joule.

Finally, note that 14.2 MeV, i.e. 7% of the energy labelled as "recovered", is released into the system a certain time after the fission; this time can be anywhere from a few seconds to several years[28]. This is all the energy from delayed beta and gamma (post-beta) radiation.

28 In this case, the energy will probably not be recovered.

E. EVALUATION AND PROCESSING OF BASIC NUCLEAR DATA

Strictly speaking, the problems of nuclear data used in reactor calculations concern specialist nuclear physicists rather than neutron physicists. That is why we do not intend to expand on this topic here. It is important, however, for a neutron physicist to have a basic knowledge of this subject, to be able to discuss it and develop synergies with nuclear specialists.

To determine cross-sections and other nuclear data is not simply a measurement problem; the measurement results need to be *evaluated* (selected and/or weighted), any missing information must be *filled in* using nuclear models, the data must be placed in a standard format and processed for the purpose of use in neutron physics calculations and, finally, the calculations must be qualified by neutron physics experiments. These experiments are called *integral* experiments to distinguish them from *differential* experiments, and they can be a valuable source of additional information beyond that provided by direct nuclear measurements.

2.11. Measuring basic neutron physics data

2.11.1. Neutron sources

When measuring the nuclear parameters of interest in neutron physics, the neutron source is the obvious place to start. Of the various possible techniques, the following two are most often used:

For *differential measurements*, an accelerator in continuous or pulsed operation supplies neutrons via a carefully-chosen nuclear reaction caused by accelerated particles hitting a target. For example: Linear electron accelerator: electrons are brought to an energy of about a hundred MeV and, by bremsstrahlung in a thick target, produce gamma photons which in turn will supply neutrons via a (γ,n) or (γ,f) reaction; Van de Graaff electrostatic proton or deuteron accelerator giving neutrons by reaction on a target. For example:

$$\mathbf{d}\,(^{2}\mathbf{He},{}^{3}\mathbf{He})\mathbf{n} + 3.27\ \mathrm{MeV},$$

$$\mathbf{d}\,(^{3}\mathbf{He},{}^{4}\mathbf{He})\mathbf{n} + 17.60\ \mathrm{MeV},$$

$$\mathbf{d}\,(^{9}\mathbf{Be},{}^{10}\mathbf{B})\mathbf{n} + 4.36\ \mathrm{MeV},$$

$$\mathbf{p}\,(^{7}\mathbf{Li},{}^{7}\mathbf{Be})\mathbf{n} - 1.65\ \mathrm{MeV}.$$

If the source is pulsed, it is possible to distinguish between the neutron energies using the *time-of-flight method* (Figure 2.24).

The particles (neutrons) are emitted at an initial instant according to an energy spectrum; the beam is collimated and sent down a pipe that can be up to a hundred metres long. The target containing the material to be measured is located at the other end. The time between emission and detection of the event in the target is measured, and then the speed (and hence the energy) of the neutron that provoked the reaction can be calculated.

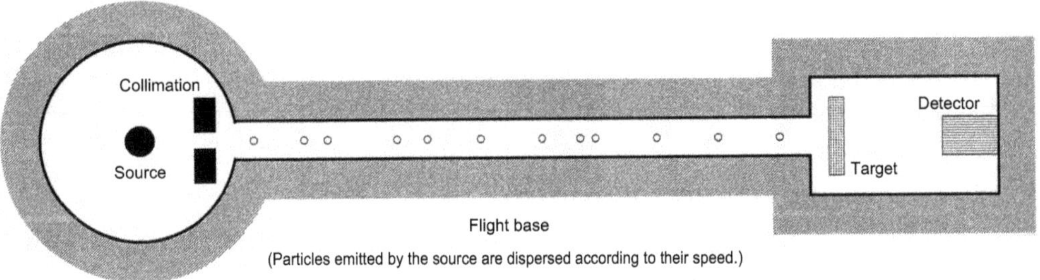

Figure 2.24. Principle of the time-of-flight method.

This entire process takes only a few milliseconds, so the measurements can be repeated many times[29];

For *integral measurements*, neutrons from an experimental reactor are usually used.

2.11.2. Detection of neutrons

The most commonly used neutron detector is the NaI scintillator associated with a photo-multiplier that detects the 470 keV photon resulting from a capture by boron. The following reactions are involved:

$$\mathbf{n} + {}^{10}\mathbf{B} \Longrightarrow {}^{7}\mathbf{Li}^* + \alpha,$$
$${}^{7}\mathbf{Li}^* \Longrightarrow {}^{7}\mathbf{Li} + \gamma\ (470\ \text{keV}).$$

2.11.3. Measurement of total cross-section

The total cross-section σ is the easiest to measure. An *absolute* measurement is performed. This is performed by *transmission* through a sample of the material to be studied, which has a carefully-chosen thickness and contains N atoms per unit volume. The transmission coefficient, i.e. the ratio of the experimental signals (neutron absorption in the detector shown in the diagram) with and without a target in the beam is $\exp(-eN\sigma)$: with known e and N, the measurement of this ratio gives σ.

2.11.4. Measurement of partial cross-sections and number of neutrons emitted per fission

It is more difficult to measure partial cross-sections (diffusion, capture, fission, etc.) because to do this requires detecting the reaction products (neutrons, gamma photons, etc.). To catch as many as possible, the technique is to place a set of detectors around the target in the maximum solid angle.

[29] Note that the choice of flight base requires a trade-off; a longer one provides better energy discrimination, but a lower count rate (a consequence of the solid angle through which the target is seen from the source).

The problem lies in knowing both the number of incident neutrons and the number of events. This normalisation problem can often be solved by taking *relative* measurements, i.e. by comparing the results obtained with a sample of the material being studied on the one hand, and a sample of a known material (standard) on the other hand.

2.11.5. Integral measurements

Integral measurements are performed either in power reactors or in purpose-built experimental reactors. They are called integral measurements because they concern parameters (rate of reaction, multiplication factor, etc.) that are expressed mathematically by integrals containing cross-sections. These measurements are of interest for two reasons: 1/ Firstly, they concern parameters that are of direct interest to the engineer, who will then be able to assess the quality of the calculations, 2/ Secondly, they are often very accurate, and thus provide relevant (although indirect) information about the nuclear data.

The principles of these measurements and how they are used will be discussed in a little more detail in Chapter 17 (calculation scheme).

2.12. Evaluation and libraries of nuclear data

There is now a huge body of data collected from nuclear measurements performed by specialists over the past half-century on various nuclides, for different reactions, and according to the energy of the incident neutrons. Because of the large amount of data and the necessity to choose between redundant measurements (or to average them using appropriate weighting) and to fill in any gaps, it was necessary to organise this information and to standardise the way it is presented and the procedures for its use. Evaluation bodies are responsible for this. The main evaluations currently used by reactor physicists are listed below (covering the entire range: fission, fusion and protection):

- **ENDF/B-7** (Evaluated Nuclear Data File), USA (Brookhaven National Nuclear Data Center) ;
- **JEF-2** (Joint European File), Europe (Nuclear Energy Agency of the OECD);
- **EFF-3** (European Fusion File), Europe (EU);
- **EAF-2003** (European Activation File), Europe (EU) ;

  These three last librairies are now gathered together into **JEFF-3**;
- **JENDL-3** (Japanese Evaluated Nuclear Data Library), Japan;
- **CENDL-2** (Chinese Evaluated Nuclear Data Library), China;
- **FENDL** (Fusion Evaluated Nuclear Data Library), International (IAEA);
- **BROND**, Russia (Obninsk Data Center);
- **EPDL** (Evaluated Photon Data Library), USA (Lawrence Livermore National Laboratory);

- **ENDL** (Evaluated Nuclear and Atomic Reaction Data Library), USA (Lawrence Livermore National Laboratory);
- **IAEA PDL** (IAEA Photonuclear Data Library), International (IAEA);
- **ESTAR, PSTAR, ASTAR** (Stopping-Power and Range Tables for Electrons, Protons and Helium Ions), USA (National Institute of Standards and Technology, Physics Laboratory, Ionizing Radiation Division).

2.13. Processing of nuclear data for neutron physics codes

In these evaluations, nuclear data are presented so that all the details, if known, can be included. This presentation is not necessarily the most appropriate, however, for reactor physics codes. For example, these codes often do not operate using point data (continuous energy curves), but multiple group data (curves that are approximated by steps; see Chapter 10).

Likewise, the rather complicated processing of the resonances of heavy nuclei generally involves pre-tabulation (see Chapter 15); Doppler broadening must also be processed (see Chapter 8), as well as thermalisation (see Chapter 9), etc. Special utility software must be developed to handle all of these aspects, and provides an interface between the files supplied by evaluators and the actual reactor physics code.

Exercises

A. Structure of Matter and Nuclear Binding Energy

Exercise 2.1: the sizes of atoms and of nuclei

In this exercise we assume that an atom is a sphere with a diameter D contained in a cube of volume $V = D^3$.

a) Size of a hydrogen atom: in liquid water (density 1000 kg m^{-3}), assume that the hydrogen and oxygen atoms have the same sizes; and, therefore, that two thirds of the total volume is occupied by hydrogen atoms. Evaluate V and D for a hydrogen atom.

b) Size of a uranium atom: evaluate V and D for a uranium atom in metallic uranium which has a density 18 950 kg m^{-3}; compare the sizes of hydrogen and uranium atoms.

c) Size of nuclei: given that a nucleus containing A nucleons is a sphere of radius $R = r_0 A^{1/3}$ with $r_0 = 1.2\times10^{-15}$ m, estimate the diameters of H- and U-nuclei, and compare with the sizes of the corresponding atoms.

Exercise 2.2: equation for the stability line of nuclei

Neglecting the parity term in the Bethe-Weizsäcker formula, find the value of Z corresponding to the minimum mass of the nuclei with a given mass number A, using the assumption that Z is a continuous variable, and express the result in the following form:

$$N - Z = f(A).$$

In practice, this equation can be approximated by the simpler expression:

$$N - Z = kA^{5/3}.$$

Evaluate the constant k.

Deduce from this equation analytical expressions for the volume, surface, Coulomb, and asymmetry terms as functions of A.

Exercise 2.3: fusion reaction

In this exercise, and those that follow it, the masses are given in atomic mass units for the neutral ***atoms*** *(except the free neutron); the binding energies of the electrons can be neglected.*

Consider the fusion reaction

$$^{2}_{1}\mathbf{H} + ^{3}_{1}\mathbf{H} \Longrightarrow ^{4}_{2}\mathbf{He} + ^{1}_{0}\mathbf{n}.$$

Calculate the amount of energy yielded by this reaction. Neglecting the initial momentum of deuterium and tritium, calculate the kinetic energies of the alpha particle and the neutron.

Masses:

- Deuterium: 2.014102
- Tritium: 3.016049
- Helium: 4.002603
- Neutron: 1.008665

Exercise 2.4: fission reaction

Next, consider the symmetrical fission reaction of uranium-235 induced by a neutron, and the following beta decays:

- ${}^{235}_{92}\mathbf{U} + {}^{1}_{0}\mathbf{n} \Longrightarrow {}^{236}_{92}\mathbf{U} \Longrightarrow 2\ {}^{117}_{46}\mathbf{Pd} + 2\ {}^{1}_{0}\mathbf{n}$;
- ${}^{117}_{46}\mathbf{Pd} \Longrightarrow {}^{117}_{47}\mathbf{Ag} + {}^{0}_{-1}\mathbf{e} + {}^{0}_{0}\bar{\nu}$;
- ${}^{117}_{47}\mathbf{Ag} \Longrightarrow {}^{117}_{48}\mathbf{Cd} + {}^{0}_{-1}\mathbf{e} + {}^{0}_{0}\bar{\nu}$;
- ${}^{117}_{48}\mathbf{Cd} \Longrightarrow {}^{117}_{49}\mathbf{In} + {}^{0}_{-1}\mathbf{e} + {}^{0}_{0}\bar{\nu}$;
- ${}^{117}_{49}\mathbf{In} \Longrightarrow {}^{117}_{50}\mathbf{Sn} + {}^{0}_{-1}\mathbf{e} + {}^{0}_{0}\bar{\nu}$.

Calculate the total energy released by all these reactions.

Masses:

- Uranium-235: 235.043924
- Tin-117: 116.902956
- Neutron: 1.008665

Compare with the result obtained when the analytical expressions of exercise **2.2** are used instead.

Exercise 2.5: evaluation of the Coulomb term

a) Calculate the electrostatic energy of a charge Q uniformly distributed in a sphere of radius R.

b) Deduce from this calculation the expression of the Coulomb coefficient a_C in the Bethe-Weizsäcker formula. *Hint*: use $r_0 = 1.2 \times 10^{-15}$ m.

c) Evaluate a_C for the 'mirror' nuclei nitrogen-15 and oxygen-15 by comparing their binding energy. In what way does it differ from the previous result?

Masses:

- Nitrogen-15: 15.000109
- Oxygen-15: 15.003065
- Hydrogen: 1.007934
- Neutron: 1.008665

B. Radioactivity

Exercise 2.6: radioactive activity

Calculate in becquerels and curies the activity of one gram of radium-226 (half life: 1599 years) and one gram of tritium (half life: 12.32 years).

Exercise 2.7: secular equilibrium

Estimate the mass of radium-226 per kilogram of uranium-238 for natural uranium ore from a mine.

Radioactive half lives:

- Uranium-238: 4.47×10^9 years
- Radium-226: 1599 years

Exercise 2.8: decay of uranium isotopes in the geological Oklo reactors

The present average concentration of uranium-235 in natural uranium by number of atoms is 0.72%; what was the concentration when the Oklo reactors formed about two billion years ago?

Radioactive half lives:

- Uranium-235: 7.04×10^8 years
- Uranium-238: 4.47×10^9 years

Exercise 2.9: beta decay of tritium and alpha decay of plutonium

Calculate and compare the energies released by the decay of tritium,

$$^3_1\mathbf{H} \Longrightarrow {}^3_2\mathbf{He} + \mathbf{e} + \bar{\nu},$$

and the decay of plutonium:

$$^{239}_{94}\mathbf{Pu} \Longrightarrow {}^{235}_{92}\mathbf{U} + \alpha + (\gamma).$$

Masses:

- Tritium: 3.01604927
- Helium-3: 3.01602931
- Plutonium-239: 239.052158
- Uranium-235: 235.043924
- Helium-4: 4.002603

Exercise 2.10: limit of the alpha instability

Using the Bethe-Weizsäcker formula (and, possibly, the simplified analytical formulae obtained in exercise **2.2**), estimate approximately from which mass alpha decay becomes exoenergetic.

Exercise 2.11: three body relationship

Study the evolution of the numbers, expressed with respect to their initial values, for the nuclei X, Y, and Z, with decay constants λ and μ for X and Y, respectively, and where Z is assumed to be stable.

Apply the previous result to the xenon-135 chain:

$$^{135}\mathbf{I} \Longrightarrow {}^{135}\mathbf{Xe} \Longrightarrow {}^{135}\mathbf{Cs}$$

Radioactive half lives:

- Iodine-135: 6.53 hours
- Xenon-135: 9.17 hours
- Caesium-135: 2.6×10^6 years (effectively infinite)

C. Nuclear reactions by neutrons

Exercise 2.12: production of neutrons by a Van de Graff machine

One of the reactions used to obtain neutrons from a Van de Graff accelerator for performing measurements of neutron cross-sections is

$${}^1_1\mathbf{p} + {}^7_3\mathbf{Li} \Longrightarrow {}^7_4\mathbf{Be} + {}^1_0\,\mathbf{n}.$$

What is the threshold for this reaction?

Masses:

- Hydrogen: 1.007825
- Lithium-7: 7.016003
- Beryllium-7: 7.016929
- Neutron: 1.008665

Exercise 2.13: reaction thresholds

Among the following reactions, which ones present a threshold, and if so, how much is the threshold?

Tritium production:

$${}^1_0\mathbf{n} + {}^6_3\mathbf{Li} \Longrightarrow {}^4_2\mathbf{He} + {}^3_1\mathbf{H}.$$

Atmospheric carbon-14 production:

$${}^1_0\mathbf{n} + {}^{14}_7\mathbf{N} \Longrightarrow {}^{14}_6\mathbf{C} + {}^1_1\mathbf{p}.$$

Reactivity control of the PWRs:

$${}^1_0\mathbf{n} + {}^{10}_5\mathbf{B} \Longrightarrow {}^7_3\mathbf{Li} + {}^4_2\mathbf{He}.$$

Deuterium dissociation:

$${}^1_0\mathbf{n} + {}^2_1\,\mathbf{H} \Longrightarrow {}^1_1\mathbf{H} + 2\,{}^1_0\mathbf{n}.$$

Masses:

- Hydrogen: 1.007825
- Deuterium: 2.014102
- Tritium: 3.016049
- Helium-4: 4.002603
- Lithium-6: 6.015121
- Lithium-7: 7.016003
- Boron-10: 10.012937
- Carbon-14: 14.003242
- Nitrogen-14: 14.003074
- Neutron: 1.008665

Exercise 2.14: neutron capture by Xenon-135

Compare and comment on the energies yielded by the radiative capture of neutrons by xenon-135 and xenon-136 isotopes.

Masses:

- Xenon-135: 134.907130
- Xenon-136: 135.907214
- Xenon-137: 136.911557
- Neutron: 1.008665

Exercise 2.15: neutron scattering by hydrogen

Show that after the scattering of a neutron by a proton, which is initially at rest, the two particles move on perpendicular paths in the laboratory frame, based on the assumption that the proton and neutron have identical mass.

Exercise 2.16: extrema in cross-sections

For a nucleus presenting a unique resonance described by the Breit-Wigner law (with $g = 1$), calculate the maximum and minimum values of the total cross-section. *Hints*: the variations of the resonance parameters and of λ with the energy of the incident neutron should be neglected. One barn (b) is 10^{-28} m^2.

Apply the previous result to the strong resonance of iron-56 using the following parameters:

$$E_0 = 27\,600 \text{ eV}$$

$$\Gamma_n = 1409 \text{ eV}$$

$$\Gamma_\gamma = 1 \text{ eV}$$

$$\sigma_p = 11 \text{ b}$$

Exercise 2.17: limit of the absorption cross-section

For a nucleus presenting a unique resonance described by the Breit-Wigner law (with $g = 1$), and for a neutron of a given energy E, how large is the highest value of the absorption cross-section which can be observed? *Apply this using the following values of E in eV:* 0.0253; 1; 10; 100; 1000.

Exercise 2.18: pratical width of a resonance

The practical width Γ_p can be defined as the energy interval where the absorption cross-section exceeds the scattering potential cross-section. For a nucleus characterized by a unique resonance described by the Breit-Wigner formula (with $g = 1$), calculate Γ_p, and compare it with the nuclear width Γ.

Apply the previous result to the first resonance of uranium-238:

$$E_0 = 6.674 \text{ eV}$$
$$\Gamma_n = 1493 \text{ meV}$$
$$\Gamma_\gamma = 23 \text{ eV}$$
$$\sigma_p = 8.90 \text{ b}$$

D. Nuclear fission

Exercise 2.19: energy released by a symmetrical fission without neutron emission

Let us consider symmetrical fission without neutron emission:

$$^{236}_{92}\mathbf{U} \Longrightarrow 2\ ^{118}_{46}\mathbf{Pd}.$$

Using the Bethe-Weizsäcker formula, and neglecting the parity term, calculate the amount of energy released by this reaction, and the velocity of each fission fragment.

Exercise 2.20: radioactive decays of the fission products

Using the equation of the stability line obtained in exercise **2.2**, calculate the number of beta decays that each fission fragment of the previous symmetrical fission undergoes until a stable nucleus is obtained. What will this number of decays be when 2, 4, or 6 neutrons are emitted, and the fission remains symmetrical?

Exercise 2.21: precursor of delayed neutrons

Fission of uranium-235 nuclei induced by thermal neutrons yields a significant quantity of bromine-87 (1.3%). The latter undergoes β^- decay with a half life of 56 seconds into krypton-87, which usually goes into an excited state with an energy of 5.4 MeV. Why is this excitation energy sufficient for the nucleus to emit a neutron?

A similar process occurs in the chain reaction,

$$^{137}_{53}\mathbf{I} \Longrightarrow {}^{137}_{54}\mathbf{Xe} \Longrightarrow {}^{136}_{54}\mathbf{Xe}.$$

What is it?

Solutions

A. Structure of matter and nuclear binding energy

Exercise 2.1: the sizes of atoms and of nuclei

The following diameters are obtained:

Hydrogen atom:	2.15×10^{-10} m
Uranium atom:	2.75×10^{-10} m
Hydrogen nucleus:	2.40×10^{-15} m
Uranium nucleus:	1.49×10^{-14} m

Paradoxically, the sizes of the smallest and largest atoms are almost the same. The ratio atom/nucleus is 90 000 for hydrogen and 18 000 for uranium.

Exercise 2.2: equation for the stability line of nuclei

If the parity term is neglected, then the Bethe-Weizsäcker formula for the mass of a nucleus X can expressed as

$$m_X = Z\,m_P + N\,m_N - a_v\,A + a_s\,A^{2/3} + a_a\,\frac{(A/2 - Z)^2}{A} + a_c\,\frac{Z^2}{A^{1/3}}.$$

Assuming A is constant, replace N with $A - Z$, and set the derivative with respect to Z equal to zero:

$$m_P - m_N - 2\,a_a\,\frac{A/2 - Z}{A} + 2\,a_c\,\frac{Z}{A^{1/3}} = 0.$$

Thus,

$$Z = \frac{1}{2}\,\frac{1 + (m_N - m_P)/a_a}{1 + a_c\,A^{2/3}/a_a},$$

and,

$$N - Z = \frac{a_c}{a_a}\,\frac{1 - (m_N - m_P)\,A^{-2/3}/a_c}{1 + a_c\,A^{2/3}/a_a}\,A^{5/3}.$$

It can be seen that the value of the fraction varies very little with A; hence, it is reasonable to make the following approximation.

$$N - Z \simeq k\,A^{5/3} \simeq 0.006\,A^{5/3}.$$

Substituting into the Bethe-Weizsäcker formula, without the parity term, the average binding energy per nucleon can be expressed as

$$\frac{W}{A} \simeq a_v - \frac{a_s}{A^{1/3}} - \frac{a_a\,k^2}{4}\,A^{4/3} - \frac{a_c}{4}\,A^{2/3}\,(1 - k\,A^{2/3})^2.$$

Exercise 2.3: fusion reaction

When the masses of the atoms are used (except, of course, for the free neutron), the same numbers of electrons appear on both sides of the reaction; hence, the electron mass need not be taken into account in the mass defect.

The fusion reaction yields 17.6 MeV. If we assume that the momentum is zero, then it can be shown that this energy is split in proportion to the inverse of the masses, i.e. 3.5 MeV for the alpha particle and 14.1 MeV for the neutron.

Exercise 2.4: fission reaction

The complete set of the reactions can be summarized by

$$^{235}_{92}\mathbf{U} + ^{1}_{0}\mathbf{n} \Longrightarrow 2\,^{117}_{50}\mathbf{Sn} + 8\,^{0}_{-1}\mathbf{e} + 8\,^{0}_{0}\bar{\nu} + 2\,^{1}_{0}\mathbf{n}.$$

If the masses of the atoms are used, then there are 92 electrons on each side; hence,

$$\text{Atom}\,^{235}_{92}\mathbf{U} + ^{1}_{0}\mathbf{n} \Longrightarrow 2\,\text{Atom}\,^{117}_{50}\mathbf{Sn} + 8\,^{0}_{0}\bar{\nu} + 2\,^{1}_{0}\mathbf{n}.$$

The energy released by the complete set of the reactions (mass defect) is 213.6 MeV.

Using the formula obtained in exercise **2.2** with:

$$a_v = 15.56\text{ MeV},$$
$$a_s = 17.23\text{ MeV},$$
$$a_a = 94.4\text{ MeV},$$
$$a_c = 0.70\text{ MeV},$$

we obtain:

$$A_{235} = 235 \times 7.5691 = 1778.73\text{ MeV},$$
$$A_{117} = 117 \times 8.4802 = 992.18\text{ MeV}.$$

Therefore, 205.6 MeV total energy is released. This value is near the previous exact one.

Exercise 2.5: evaluation of the Coulomb term

a) If Q is the total charge, then the charges dq and dq' in volume elements dV and dV' are $Q\,dV/V$ and $Q\,dV'/V$, respectively. Integrating $dq\,dq'/(4\pi\varepsilon_0 d)$, where d is the distance between dV and dV', and dividing by two to eliminate double counting of each elementary interaction, gives

$$|E| = \frac{3}{5}\frac{Q^2}{4\pi\varepsilon_0 R}.$$

b) Taking $Q = Ze$ and $R = r_0 A^{1/3}$, the expression $a_c Z^2/A^{1/3}$ for the Coulomb term is obtained using

$$a_c = \frac{3}{5}\frac{e^2}{4\pi\varepsilon_0 r_0}.$$

With $e = 1.602 \times 10^{-19}$ C, $r_0 = 1.2 \times 10^{-15}$ m, and $4\pi\varepsilon_0 = 1.113 \times 10^{-10}$ F m^{-1}, the result is $a_c = 0.72$ MeV. (The empirical value obtained by a mean square adjustment is 0.70 MeV.)

c) The difference in mass between $^{15}\mathbf{O} + \mathbf{n}$ and $^{15}\mathbf{N} + {}^{1}\mathbf{H}$ (0.003796 u) is, according to the Bethe-Weizsäcker formula, $(64 - 49)\,a_C/15^{1/3}$; therefore, $a_C = 0.58$ MeV. This result is not very precise because the oxygen and nitrogen atoms are rather small 'drops'.

B. Radioactivity

Exercise 2.6: radioactive activity

- One gram of radium-226 $\Longrightarrow 3.7 \times 10^{10}$ Bq = 1 Ci (from the definition of the curie).
- One gram of tritium $\Longrightarrow 3.6 \times 10^{14}$ Bq $\simeq$ 10 000 Ci.

Exercise 2.7: secular equilibrium

In natural uranium ore, the activities of all the elements in the radioactive chain are equal in the so-called secular equilibrium. In particular,

$$\frac{N_{226}}{N_{238}} = \frac{T_{226}}{T_{238}} = \frac{1599}{4.47 \times 10^{9}} = 3.577 \times 10^{-7}.$$

Therefore,

$$\frac{M_{226}}{M_{238}} = \frac{226 \times N_{226}}{238 \times N_{238}} = 3.397 \times 10^{-7},$$

or 0.34 milligrams of radium per kilogram of uranium.

Exercise 2.8: decay of uranium isotopes in the geological Oklo reactors

For every 100 nuclei of uranium now, two billion years ago there were:

$$0.72 \times 2^{(2\times10^{9}/7.04\times10^{8})} = 5.159 \text{ nuclei of uranium-235, and}$$

$$99.28 \times 2^{(2\times10^{9}/4.47\times10^{9})} = 135.4 \text{ nuclei of uranium-238.}$$

This means that when the reactor formed, the concentration of $^{235}_{92}\mathbf{U}$ was 3.67% by number of atoms.

Exercise 2.9: beta decay of tritium and alpha decay of plutonium

The energy equivalents of the mass defects for these two reactions are 18.6 keV and 5.25 MeV, respectively; thus, there is almost 300 times less energy in the tritium decay than in the plutonium decay.

Exercise 2.10: limit of the alpha instability

Assuming that, as was done previously in exercise **2.2**, all the nuclei are at the optimum value of Z, and if we use the simplified Bethe-Weizsäcker formula (without the parity term), then the binding energy is

$$W \simeq a_v A - a_s A^{2/3} - a_a \frac{k^2}{4} A^{7/3} - \frac{a_C}{4} A^{5/3} (1 - kA^{2/3})^2.$$

The energy balance of an alpha decay is

$$Q = W(A-4) + W_\alpha - W(A),$$

where $W_\alpha = 28.3$ MeV is the binding energy of an alpha particle. The following table gives some numerical examples which show that the limit of the alpha instability (Q positive) is in the region of $A = 150$.

A	50	100	150	200	250
Q	−7.75	−3.72	−0.09	+3.07	+5.89

Exercise 2.11: three body relationship

The equations governing the numbers of nuclei X, Y and Z are

$$\frac{dX}{dt} = -\lambda X, \qquad \frac{dY}{dt} = \lambda X - \mu Y, \qquad \frac{dZ}{dt} = \mu Y.$$

Solving the two first equations gives X and Y; the third number Z can be deduced by knowing that $X + Y + Z$ is constant:

$$X(t) = X(0)\, e^{-\lambda t},$$

$$Y(t) = X(0) \frac{\lambda}{\mu - \lambda} e^{-\lambda t} + \left[Y(0) - X(0) \frac{\lambda}{\mu - \lambda} \right] e^{-\mu t},$$

$$Z(t) = X(0) \left[1 - \frac{\lambda}{\mu - \lambda} e^{-\lambda t} + \frac{\lambda}{\mu - \lambda} e^{-\mu t} \right] + Y(0)\,(1 - e^{-\mu t}) + Z(0).$$

The function $X(t)$ decreases monotonically to zero while the function $Z(t)$ increases monotonically to the total number of the initial nuclei (*cf.* the signs of the derivatives in the differential equations). The function $Y(t)$ is extremal when $\lambda X(t)$ is equal to $\mu Y(t)$. This condition can be verified or not according to the values of λ and μ, and to the initial conditions. For instance, a peak in the amount of xenon-135 always occurs when a reactor is shut down after operating at equilibrium. When an extremum occurs, it is a maximum, as the second derivative of Y is then equal to $-\lambda^2 X(t)$. The value of Y at this time is

$$Y_m = \frac{\left[\frac{\lambda}{\mu} \right]^{\mu/(\mu-\lambda)}}{\left[1 - \frac{\mu-\lambda}{\lambda} \frac{Y(0)}{X(0)} \right]^{\lambda/(\mu-\lambda)}}.$$

C. Nuclear reactions by neutrons

Exercise 2.12: production of neutrons by a Van de Graff machine

The reaction is endothermal:

$${}^{1}_{1}\mathbf{p} + {}^{7}_{3}\mathbf{Li} \Longrightarrow {}^{7}_{4}\mathbf{Be} + {}^{1}_{0}\mathbf{n} - 1.64 \text{ MeV}.$$

The threshold energy, in the centre of mass system, is 1.64 MeV.

Exercise 2.13: reaction thresholds

The net energies for each reaction are as follows.

$$ {}^{1}_{0}\mathbf{n} + {}^{6}_{3}\mathbf{Li} \Longrightarrow {}^{4}_{2}\mathbf{He} + {}^{3}_{1}\mathbf{H} + 4.78 \text{ MeV}. $$
$$ {}^{1}_{0}\mathbf{n} + {}^{14}_{7}\mathbf{N} \Longrightarrow {}^{14}_{6}\mathbf{C} + {}^{1}_{1}\mathbf{p} + 0.63 \text{ MeV}. $$
$$ {}^{1}_{0}\mathbf{n} + {}^{10}_{5}\mathbf{B} \Longrightarrow {}^{7}_{3}\mathbf{Li} + {}^{4}_{2}\mathbf{He} + 2.79 \text{ MeV}. $$
$$ {}^{1}_{0}\mathbf{n} + {}^{2}_{1}\mathbf{H} \Longrightarrow {}^{1}_{1}\mathbf{H} + 2\,{}^{1}_{0}\mathbf{n} - 2.22 \text{ MeV}. $$

Exercise 2.14: neutron capture by xenon-135

The energies released by the radiative capture (n,γ) reactions, i.e. the binding energies of the added neutron, are:

- for xenon 135: 7.99 MeV,
- for xenon 136: 4.03 MeV.

There is almost a factor of two between these binding energies!

The nucleus ${}^{136}_{54}\mathbf{Xe}$ is magic for the number of neutrons (82), which explains why it forms readily from ${}^{135}_{54}\mathbf{Xe}$, implying that there is a large cross-section for the process. However, neutron capture by ${}^{136}_{54}\mathbf{Xe}$ destroys this magic state which hinders the reaction; hence, it has a small cross-section.

Exercise 2.15: neutron scattering by hydrogen

Let $\vec{v}$ and $\vec{0}$ be the neutron and proton velocities in the laboratory frame before the collision, and $\vec{v'}$ and $\vec{v''}$ be their velocities after the collision. Since the particles possess equal mass, conservation of momentum and kinetic energy give

$$ \vec{v} = \vec{v'} + \vec{v''}, $$
$$ v^2 = v'^2 + v''^2. $$

Substracting the second equation from the first equation squared, gives $\vec{v'}.\vec{v''} = 0$; hence, the vectors are orthogonal.

Exercise 2.16: extrema in cross-sections

Using the Breit-Wigner formalism, the total cross-section can be expressed as

$$ \sigma = \frac{A + Bx}{1 + x^2} + \sigma_p, $$

where

$$ x = \frac{2(E - E_0)}{\Gamma}, $$

$$ A = 4\pi\lambda\!\!\!^{-2}\frac{\Gamma_n}{\Gamma}, \qquad B = 4\sqrt{\pi\lambda\!\!\!^{-2}\sigma_p}\frac{\Gamma_n}{\Gamma}. $$

If the derivative of σ with respect to x is zero, then the equation giving the values of x for the extrema of the cross-section is

$$Bx^2 + 2Ax - B = 0.$$

Therefore,

$$x = \frac{-A \pm \sqrt{A^2 + B^2}}{B},$$

$$x = \frac{\sqrt{\pi}\lambda\!\!\!^{-}}{\sigma_p}\left[-1 \pm \sqrt{1 + \sigma_p/\pi\lambda\!\!\!^{-2}}\right],$$

and

$$\sigma = 2\pi\lambda\!\!\!^{-2}\frac{\Gamma_n}{\Gamma}\left[1 \pm \sqrt{1 + \sigma_p/\pi\lambda\!\!\!^{-2}}\right] + \sigma_p.$$

Applying this result to iron-56:

$x_- = -3.26$; $E_- - E_0 = -2299$ eV; $\sigma_- = 1.04$ b.

$x_+ = +0.31$; $E_+ - E_0 = +216$ eV; $\sigma_+ = 116.9$ b.

Exercise 2.17: limit of the absorption cross-section

The absorption cross-section is maximum for $E = E_0$, and then is equal to

$$\sigma_{a,max} = 4\pi\lambda\!\!\!^{-2}\frac{\Gamma_n\Gamma_\gamma}{\Gamma}.$$

This expression is maximum when $\Gamma_n = \Gamma_\gamma = \Gamma/2$, and then is equal to $\pi\lambda\!\!\!^{-2}$. Consequently,

$$\sigma_a \leq \pi\lambda\!\!\!^{-2} = \frac{\lambda^2}{4\pi}.$$

Example application:

When E_0 is expressed in electronvolts (1.602×10^{-19} J) and σ_a in barns (10^{-28} m^2),

$$\sigma_a \leq \frac{A+1}{A}\,\frac{650\,978}{E_0}.$$

For a heavy nucleus, $A \gg 1$, which gives the following result.

Energy (eV)	Limit (barns)
1000	651
100	6510
10	65 100
1	651 000
0.0253	25 700 000

Exercise 2.18: practical width of a resonance

If the interference term is neglected, then the total cross-section is

$$\sigma = \frac{\sigma_{max}}{1 + x^2} + \sigma_p,$$

where

$$\sigma_{max} = 4\pi\lambda\!\!\!^{-2}\frac{\Gamma_n}{\Gamma}, \qquad x = \frac{E - E_0}{\Gamma/2}.$$

The resonant cross-section exceeds the potential cross-section when $x^2 < \sigma_{max}/\sigma_p - 1$, i.e. approximately, since the peak resonant cross-section is much greater than the potential cross-section:

$$x < \sqrt{\sigma_{max}/\sigma_p}.$$

Therefore,

$$\Gamma_p = \Gamma\sqrt{\sigma_{max}/\sigma_p}.$$

Applying this result to the first resonance of uranium-238 gives

$$\sigma_{max} = 23\,783 \text{ b};$$
$$\Gamma_p = 51.7 \times \Gamma = 1.266 \text{ eV}.$$

Remark: if uranium is mixed with a diffusing material, then it would be better to take the latter value for the potential cross-section when comparing it with the resonant cross-section, and, of course, using the macroscopic rather than the microscopic cross-sections.

D. Nuclear fission

Exercise 2.19: energy released by a symmetrical fission without neutron emission

The Bethe and Weizsäcker formula gives the result,

$$W(236, 92) = 1785.01 \text{ MeV},$$
$$W(118, 46) = 984.38 \text{ MeV}.$$

Therefore an energy equal to 183.75 MeV is released by the reaction (i.e. 91.87 MeV for each of the fragments). The associated velocity given by the kinetic energy formula $v = \sqrt{2E/m}$ is equal to 12 300 km s^{-1}.

Exercise 2.20: radioactive decays of the fission products

The stability line is given by the equation (*cf.* exercise **2.2**)

$$Z = \frac{A}{2} - 0.003\,A^{1/3}.$$

The following table gives, as a function of A, the numerical values for the fragments constituted with 46 protons and $A - 46$ neutrons:

- value of Z given by this equation;
- rounded up values of Z;
- value of Z of the stable nucleus actually obtained;
- number of β^- decays predicted;
- number of β^- decays really observed (to be doubled in order to take into account *two* fission fragments).

It can be seen that, due to the parity effect, Z is always even.

Mass of the fragments	Values of Z			Number of β	
	Formula	Rounded up	Exact	Predicted	Exact
118	50.48	50	50	4	4
116	49.72	50	48	4	2
114	48.96	49	48	3	2
112	48.19	48	48	2	2

Exercise 2.21: precursor of delayed neutron

The nuclei krypton-87 and xenon-137 have 51 and 83 neutrons, respectively, which are magic numbers (50 and 82) plus one. Therefore, it is expected that they may readily eject the 'extra' neutron. It turns out, in fact, that the binding energies of these extra neutrons are about only 5 MeV, compared with 8 MeV for the average binding energy per nucleon.

3 Introduction to neutron physics

3.1. Neutron–matter interactions

3.1.1. Cross-sections (review)

The concept of cross-section, introduced in the previous chapter (§ 2.5.4 and § 2.5.5), is not exclusive to the field of neutron physics. Here is a quick review of some of the jargon involving neutron-induced reactions:

- *Scattering* indicates any reaction that re-emits at least one neutron,
- *Absorption* indicates any reaction that terminates the neutron's travel in a free state,
- *Fission* (induced) for an absorption leading to the fission of the compound nucleus formed,
- *Capture* for any other absorption.

We therefore have:

- total cross-section: $\sigma_t = \sigma_s + \sigma_a$,
- absorption cross-section: $\sigma_a = \sigma_f + \sigma_c$.

The interaction probability element for a path dx is Σdx, with $\Sigma = N\sigma$ (the index t is understood), where N is the number of atoms per unit volume. (For a homogeneous mixture of several types of atom, the $(N\sigma)_k$ terms must be added);

The probability that the first collision occurs at a distance x from the starting point, to within dx, is $e^{-\Sigma x}\Sigma dx$. One of the results is that the mean free path of the neutrons (average distance at which the first impact occurs) is $\lambda = 1/\Sigma$.

The mean free path of neutrons in most materials is of the order of a centimetre. That is why neutron physicists measure distances in centimetres (instead of the SI unit, the metre). Macroscopic cross-sections are therefore measured in cm^{-1}, and concentrations are measured in atoms per cm^3 (often expressed in 10^{24} per cm^3, i.e. atoms per barn.centimetre, to simplify the multiplication of N by microscopic cross-sections expressed in barns, where $1\ b = 10^{-24}\ cm^2$). However we shall use SI units in this book.

3.1.2. Neutron density, neutron flux, reaction rate

Even though the neutron population is very dilute compared to the population of atoms, it is still very large; on the order of $10^{14}/m^3$ in a power reactor. We therefore handle it by statistical methods using the concept of *density*. The neutron density n is the number of neutrons observed per unit volume. More specifically, imagine a small volume element d^3r placed at a point $\vec{r}$ in the system. The *average* number of neutrons found in this volume[1] would be $n(\vec{r})\,d^3r$. This density could obviously vary from one point to another and over time. It could also be possible to distinguish the speeds of the neutrons in terms of their modulus and direction.

As a simplification, we assume n to be constant with respect to these variables (which does not affect the following reasoning). Let v be the speed of the neutrons. During a time interval dt, each neutron travels a path $dx = v\,dt$ and therefore has probability $\Sigma dx = \Sigma v\,dt$ of interacting with matter. Multiply this by $n(\vec{r})\,d^3r$, the number of neutrons present in the volume element d^3r, to obtain the number $n(\vec{r})\Sigma v d^3r\,dt$ of neutron-matter interactions in d^3r during dt. The parameter

$$R = n\Sigma v,$$

representing the number of interactions per unit volume and per unit time, is the *reaction rate*.

In practice it is useful to distinguish between different *types of reaction* by breaking down Σ according to partial cross-sections Σ_r (see § 2.6.4) and R according to partial reaction rates R_r:

$$R_r = n\Sigma_r v.$$

The product nv often appears in formulae, and so we set:

$$\Phi = nv, \qquad (3.1)$$

a legitimate substitution. This value is known as *flux*. This is now the established term, but it is an unfortunate choice of word, because this is not a flux in the usual sense of a quantity passing through a surface[2], since Φ is defined on the basis of the *volumetric* concept of density.

Finally, reaction rates are represented by the following equation:

$$R_r = \Sigma_r\Phi, \qquad (3.2)$$

where Σ represents matter, and Φ is the population of neutrons travelling through the matter.

---

[1] Imagine taking a photograph of this and examining it later.

[2] This flux does nonetheless have the dimensions of a number of neutrons passing through a unit surface per unit time, i.e. the same units as a "real" flux (what a neutron physicist would call "current").

3.1.3. Concept of phase flux

Like density, the flux can depend on $\vec{r}$ and t; and, as in the case of density, it can also be useful to distinguish the neutron speeds $\vec{v}$.

In practice, the materials placed in reactors are *isotropic*, which means that they have the same properties no matter which angle they are seen from[3]. Consequently, *cross-sections are not dependent on the direction of the incident neutron,* but only on its speed. That is why it is important in neutron physics to distinguish the scalar variable v (speed) — or any variable related to it, such as kinetic energy E — and $\vec{\Omega} = \vec{v}/v$ (unit velocity vector, which in practice is identified by two angular coordinates: usually the colatitude θ and longitude φ) (see Figure 3.1).

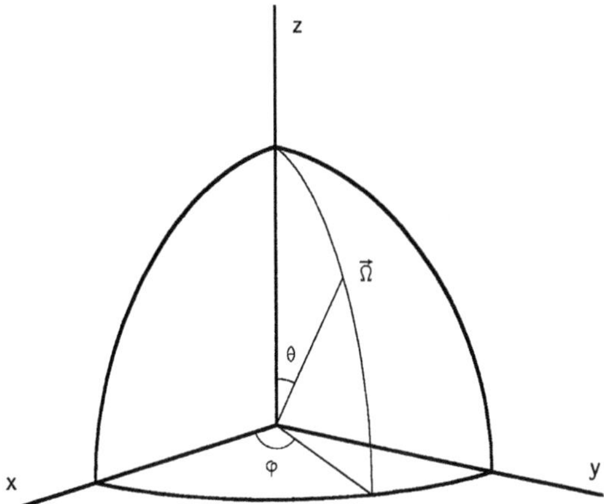

Figure 3.1. Identification of a direction in space by two angles.

When directions are distinguished in density or flux space, this is known as phase *density* (or *flux*). For example, $n(\vec{\Omega}, ...)\, d^2\Omega$ is the number of neutrons that can be counted in the solid angle element $d^2\Omega$ obtained by changing the angle θ by $d\theta$ and the angle φ by $d\varphi$ (a small "rectangle" on the unit sphere).

Note the following useful formulae:

1/ Cartesian components of the vector $\vec{\Omega}$:

$$\Omega_x = \sin\theta\cos\varphi, \quad \Omega_y = \sin\theta\sin\varphi, \quad \Omega_z = \cos\theta; \tag{3.3}$$

2/ Solid angle element:

$$d^2\Omega = \sin\theta\, d\theta\, d\varphi. \tag{3.4}$$

[3] In practice, this applies even to crystalline materials such as metals, because the size of their crystals is on the order of a micrometre, which is tiny compared to the mean free path of neutrons, and their orientation is variable and completely random.

3.1.4. Concept of current

Neutron physicists use the term "current" to denote what is known as "flux" in other branches of physics: the number of neutrons passing through a surface element, normalised per unit surface and unit time.

Consider a surface element $d\vec{S} = dS\vec{N}$, with surface area dS and located perpendicular to the unit vector $\vec{N}$ (normal). Let us examine the neutrons with a given direction $\vec{\Omega}$ to within $d^2\Omega$. Those that go through dS during the time interval from t to $t+dt$ are those that are, at the instant t, in the unit cylinder adjacent to the contour of the surface element and with parallel generators $\vec{\Omega}$ and length $v\,dt$ (see Figure 3.2).

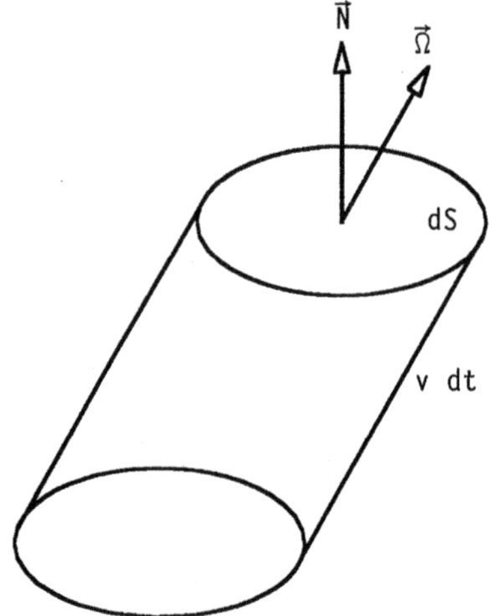

Figure 3.2. Calculation of neutron current.

By the definition of density, the number of neutrons concerned is the product of $nd^2\Omega$ by the volume of this cylinder, i.e. $dSv\,dt \cos\theta$, where θ is the angle formed by the vectors $\vec{\Omega}$ and $\vec{N}$. This number can also be written as:

$$\vec{J} \cdot d\vec{S} d^2\Omega dt,$$

by setting:

$$\vec{J}(\vec{\Omega}) = \vec{v}n(\vec{\Omega}) = v\vec{\Omega}n(\vec{\Omega}) = \vec{\Omega}\Phi(\vec{\Omega}), \tag{3.5}$$

(the point denotes a scalar product of the vectors; the variables $\vec{r}$, v and t are understood).

This parameter is the *current vector*; the scalar product of this vector $\vec{J}$ by the unit vector $\vec{N}$ is the number of neutrons by unit solid angle around the direction $\vec{\Omega}$, passing through the unit surface area element per unit time.

Note that the number of neutrons is *positive* if θ is smaller than $\pi/2$, i.e. *if the neutrons pass through the surface element in the normal direction,* and *negative* if θ is greater than $\pi/2$, i.e. *if the neutrons pass through the surface element in the opposite direction from the normal.*

Currents *integrated* on the phase are also defined, which pass through in the normal direction:

$$J_+ = \int_{0<\theta<\pi/2} \vec{J}(\vec{\Omega}) \cdot \vec{N} d^2\Omega, \tag{3.6}$$

and in the other direction (absolute value):

$$J_- = -\int_{\pi/2<\theta<\pi} \vec{J}(\vec{\Omega}) \cdot \vec{N} d^2\Omega, \tag{3.7}$$

as well as the net algebraic current:

$$J_{net} = J_+ - J_-. \tag{3.8}$$

For an *isotropic phase flux*, i.e. one whose value is independent of $\vec{\Omega}$, and therefore a phase current in cos θ, we obtain:

$$J_+ = J_- = \frac{\Phi}{4}, \quad J_{net} = 0. \tag{3.9}$$

3.1.5. Concept of opacity

In neutron physics, it is always important to evaluate whether an area of space, for example a fuel element, is "large" or "small". If it is large, heterogeneities will affect the flux distribution, but if it is small, the heterogeneities will be "erased".

The size of a zone must be evaluated using the neutron mean free path $\lambda = 1/\Sigma$ as the unit of measurement. It is useful to adopt the *mean chord*, i.e. the average distance separating the point of exit from the point of entry of a neutron crossing the area under consideration, as the "size" of a zone[4]. According to a theorem proposed by Augustin Cauchy (1789–1857), this chord is given by the following very simple formula:

$$\langle X \rangle = \frac{4V}{S}, \tag{3.10}$$

where V denotes the volume of the area and S denotes its surface. By normalising this mean chord to λ (or multiplying it by Σ), we define the *opacity of the area*:

$$\omega = \frac{4V\Sigma}{S}. \tag{3.11}$$

By comparison to the unit, ω is used to specify the adjectives "large" and "small".

To provide examples, Table 3.1 shows the opacity of a cylinder of diameter (and mean chord) 1 cm for a few materials (moderators, coolants, fuels, absorbents) and for thermal neutrons (the [total] cross-sections are in barns and the concentrations in number per b.cm, i.e. 10^{-30} m^{-3}):

[4] If convex.

Table 3.1.

Material	Main cross-section and value		Number of atoms or molecules per unit volume	Opacity
Water	s	107	0.03337	3.59
Graphite	s	4.95	0.0802	0.40
Sodium	s	3.63	0.0254	0.09
Lead	s	11.4	0.0328	0.37
U (natural)	a	17.03	0.0483	0.82
U 235	a	696.6	0.0483	33.6
Gadolinium	a	48 900	0.0303	1 482

3.1.6. The Boltzmann equation: a first approximation

When a reactor has been described in terms of its geometry, composition, and cross-sections, the purpose of a neutron physics calculation is to determine the reaction rates and therefore the neutron density or flux. The flux is the product of sources that are given in certain problems; they are usually sources of neutron-induced fission, and are therefore proportional to the flux and are also unknown.

We intend to introduce this problem using a simple example before presenting the more general case. Our example applies three simplifying assumptions:

1/ The neutrons are monokinetic, with speed v,

2/ The sources and the neutron population are stationary in time,

3/ The sources are isotropic.

a) We begin with a simple case: a point source placed alone in a vacuum emits S neutrons per unit time (see Figure 3.3).

To estimate the density, let us imagine the unit volume delineated by two spheres of radius R and $R + dR$. Because neutrons take an amount of time $dt = dR/v$ to

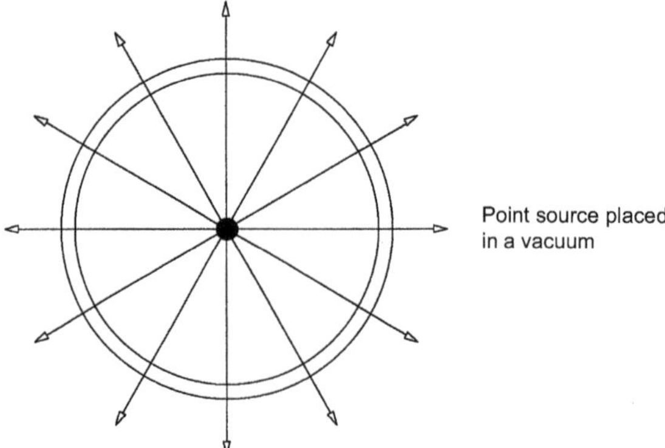

Figure 3.3. Basic problem of neutron "transport".

pass through this volume, we continuously observe what has been emitted during this time, i.e. $Sdt = SdR/v$. The density is obtained by dividing this number by the volume $4\pi R^2\, dR$, and the flux is obtained by multiplying by v:

$$n = \frac{S}{4\pi R^2 v}, \quad \varphi = \frac{S}{4\pi R^2}. \tag{3.12}$$

b) We now place this source in an absorbent material. The only neutrons present will be those that have passed through this material without interaction through a distance R separating the source from the volume element. The probability of making such a crossing is $\exp(-\Sigma R)$, which gives:

$$\varphi = \frac{S\exp(-\Sigma R)}{4\pi R^2}. \tag{3.13}$$

c) Let us now suppose that there is not one source, but a set of sources distributed at $S(\vec{r}')\,d^3r'$ in the volume element d^3r'. The calculation is *linear*, which means that if there are several sources, the flux Φ is obtained by adding the fluxes φ from each of the sources. In this case, we can represent this calculation using an integral:

$$\Phi(\vec{r}) = \int \frac{\exp(-\Sigma R)}{4\pi R^2} S(\vec{r}')d^3r'. \tag{3.14}$$

In this equation, we must integrate over the entire system and remember that R is the distance from the source to the point where the flux is measured, i.e., $R = |\vec{r} - \vec{r}'|$.

d) In Paragraph **b**, we considered that all neutrons interacting with matter are "lost", but in fact, they may be scattered and therefore re-emitted. A re-emission is strictly equivalent to an emission[5], and so, to account for scattering, we add the scattering sources to the "real" sources. Scattering sources are represented by $\Sigma_s\Phi$ according to the general formula for reaction rates. This gives:

$$\Phi(\vec{r}) = \int \frac{\exp(-\Sigma R)}{4\pi R^2} \left[S(\vec{r}') + \Sigma_s\Phi(\vec{r}')\right] d^3r'. \tag{3.15}$$

e) The system might consist of a heterogeneous material. In this case, as can immediately be seen by combining the probabilities of non-impact, we can simply replace ΣR in the exponential by the straight line integral from $\vec{r}'$ to $\vec{r}$ of the total cross-section. If we denote this integral by τ, we obtain:

$$\Phi(\vec{r}) = \int \frac{e^{-\tau}}{4\pi R^2} \left[S(\vec{r}') + \Sigma_s(\vec{r}')\Phi(\vec{r}')\right] d^3r'. \tag{3.16}$$

f) Finally, if the sources originate entirely from fission, they are expressed as follows in terms of flux: $S = \nu\Sigma_f\Phi$, which gives:

$$\Phi(\vec{r}) = \int \frac{e^{-\tau}}{4\pi R^2} \left[\nu\Sigma_f(\vec{r}')\Phi(\vec{r}') + \Sigma_s(\vec{r}')\Phi(\vec{r}')\right] d^3r'. \tag{3.17}$$

Note that this equation, unlike the preceding ones, becomes homogeneous; if it has a solution, this solution is only defined to within a factor.

[5] If we assume, as we do here, that scattering does not affect the speed of the neutron and re-emits it isotropically.

3.2. General representation of a neutron population

3.2.1. Variables to introduce

To represent fully (but statistically) a population of neutrons in a system, *seven variables* must be used:

- *three spatial variables*, such as x, y and z, to identify the *position* of particles;
- *three speed variables*, such as v, θ and φ, to identify the *state* of the particles[6];
- *the time variable* t, to specify the instant at which the observation is made.

3.2.2. General concept of neutron flux

Density, n, is the number of neutrons ***per*** unit volume, ***per*** unit speed, ***per*** unit solid angle and ***at*** the instant considered. Note the importance of the choice of preposition:

- If the argument u of a function f calls for "per", it means that it is a *density function*; a change of variable (or unit) $u \Longrightarrow v$ must be done according to the following formula:

$$|f(u)du| = |f(v)dv|,$$

i.e.:

$$f(v) = f(u)|du/dv|,$$

by expressing the second term with the variable v;

- If the argument u of a function f calls for "at", it means that it is a *"real" function*; a change of variable $u \Rightarrow v$ must be done according to the following formula:

$$f(v) = f(u).$$

The neutron flux $\Phi = nv$ is a density function with respect to all variables, including time[7], as can be seen by examining the reaction rate $R = \Sigma\Phi$, where R is a number of reaction per unit volume and per unit time, and Σ is a "real" function[8].

3.2.3. Boltzmann equation

Note: To talk about neutron physics, it is not essential to introduce the Boltzmann equation immediately. The formalism is rather daunting, and the details of it are required only to explain the solving methods and the principles of the calculation codes (to be introduced in Chapter 14). The purpose of briefly introducing this equation here is simply to make

[6] Note that spherical coordinates are used for the velocity vector because in practice cross-sections are independent of neutron direction.

[7] Speed is actually a density with respect to time, since it is a distance travelled ***per*** unit time.

[8] The flux itself can be considered as a reaction rate: it is the sum of the paths travelled by all neutrons, per unit volume and per unit time.

the reader aware of its complexity and to highlight the difficulty, or even the practical impossibility, of finding a detailed numerical solution, even with a very fast computer. The reader may skip this paragraph if desired.

In the simple case above, we can identify on the one hand the *transport operator* linking the flux Φ to the emission density Q (neutrons leaving a point after emission or scattering) in the following integral:

$$\Phi(\vec{r}) = \int \frac{e^{-\tau}}{4\pi R^2} Q(\vec{r}\,')d^3r',$$

and, on the other hand, the *collision operator* linking $Q = D + S$ to Φ:

- for scattering, $D = \Sigma_s\Phi$,
- for sources (in the case of fissions), $S = \nu\Sigma_f\Phi$.

Here is what happens to these three equations in the general case (assuming that the reactor is fixed).

a) Emission density

(Neutrons leaving an element of volume × speed × solid angle)

$$Q = D + S. \tag{3.18}$$

b) Scattering operator

(Neutrons changing speed and direction)

$$D(\vec{r}, v, \vec{\Omega}, t) = \int_0^\infty dv' \int_{(4\pi)} d^2\Omega' \Sigma_s \left[\vec{r}, (v', \vec{\Omega}') \to (v, \vec{\Omega})\right] \Phi(\vec{r}, v', \vec{\Omega}', t). \tag{3.19}$$

The double integral expresses both the change in speed and direction of the neutron when scattering occurs; the double differential scattering cross-section quantifies the probability of the events.

c) Production operator

(Neutrons emitted by fission)

$$S(\vec{r}, v, \vec{\Omega}, t) = \frac{1}{4\pi}\chi(v) \int_0^\infty dv' \int_{(4\pi)} d^2\Omega' \nu\Sigma_f(\vec{r}, v')\Phi(\vec{r}, v', \vec{\Omega}', t). \tag{3.20}$$

The double integral indicates that fission can be induced by neutrons of all speeds and all directions. In this case we assume that the neutrons emitted by fission are emitted isotropically on the 4π steradians (which is always valid), and that the fission spectrum χ is the same irrespective of the fissioned nucleus or the energy of the neutron that induced the fission. (It is of course possible not to apply these approximations).

d) Transport operator

(Neutrons travelling in a straight line without interaction)

Integral form

$$\Phi(\vec{r}, v, \vec{\Omega}, t) = \int_0^{\infty} ds \exp(-\tau) Q(\vec{r} - s\vec{\Omega}, v, \vec{\Omega}, t - s/v), \tag{3.21}$$

with:

$$\tau = \int_0^{s} \Sigma(\vec{r} - s'\vec{\Omega}, v) ds' \quad \text{(optical path)}. \tag{3.22}$$

The integral expresses the fact that any neutron leaving a point of the half-line parallel to $\vec{\Omega}$, and ending up at a point where the flux is evaluated contributes to the evaluation *if and only if* it does not undergo a collision on its path (an absorption would make it disappear, and scattering would send it in a different direction with a different speed).

Differential form

The transport operator can be written in differential form, either by applying physical arguments or by deriving the above equation with respect to point of observation $\vec{r}$ and in direction $\vec{\Omega}$ (to be described in detail in Chapter 14). This means that it would be written using space and time derivatives instead of an integral over s:

$$-\text{div}\left[\vec{\Omega}\Phi(\vec{r}, v, \vec{\Omega}, t)\right] - \Sigma(\vec{r}, v)\Phi(\vec{r}, v, \vec{\Omega}, t) + Q(\vec{r}, v, \vec{\Omega}, t) = \frac{1}{v}\frac{\partial}{\partial t}\Phi(\vec{r}, v, \vec{\Omega}, t). \tag{3.23}$$

e) Boltzmann equation

The Boltzmann equation governing the flux Φ is obtained by combining these formulae, i.e. by replacing the emission density expressed in integral form as a function of flux in one of the transport operator expressions. The reader is welcome to try it!

3.2.4. Probabilistic and deterministic solutions of the Boltzmann equation

The two forms of the transport operator are strictly equivalent from a mathematical point of view, which means that the theoretical solutions are the same.

In practice, however, there are differences: firstly, because we often make approximations, and not necessarily the same ones, in the fully integral approach and in the integral/differential approach (integral for v and $\vec{\Omega}$, and differential for $\vec{r}$ and t), and secondly because the solutions are rarely analytical, and so we must settle for numerical processing, which obviously takes different forms according to the operator to be processed. These numerical processes are called "deterministic" as opposed to "probabilistic" processes, which are increasingly used by engineers because they are more accessible thanks to the greater computing power that is available.

Engineers apply the "Monte Carlo" method to neutron physics (an allusion to gambling). The principle of this technique is to simulate neutron paths as accurately as possible by randomly selecting events as real neutrons do, i.e. according to the laws of nuclear physics: emission point, collision points, type of atom impacted, type of reaction, exit channel, etc. This approach remains statistical, like the deterministic approach, since the greatest possible number of neutron histories will be simulated in order to obtain the rates in which we are interested with the greatest possible statistical accuracy.

This method not only avoids the need to write the Boltzmann equation explicitly; its other advantage is that it does not necessarily require simplifications to be applied to the geometry, composition, and distribution of nuclei. In this sense, it can be described as "exact". In practice, however, it is only a reference method at best, because statistical uncertainties can never be reduced to zero. They vary as an inverse square root with the number of events; for example, to reduce the uncertainly by a factor of 10, 100 times more histories must be processed, and therefore the computer must be 100 times more powerful for the calculation time to remain the same.

3.3. Neutron spectra and energy balances

3.3.1. Fast neutron reactors and thermal neutron reactors

Note (see § 1.5) that, schematically, two main channels are possible to create a chain reaction:

- *Fast neutron reactors* which avoid slowing down the neutrons to optimise their use *above* the capture resonance region.

- *Thermal neutron reactors* which do the opposite by adding a moderator to slow down the neutrons and optimise their use *below* the capture resonance region, in practice in the thermal domain where the neutrons are more or less at the temperature of the system.

The spectra of the neutrons, emitted by fission in both cases at an energy in the region of 2 MeV, are therefore completely different below this range.

- In *fast neutron reactors*, the spectrum is always more or less degraded with respect to the fission spectrum, because a certain amount of slowing down by inelastic scattering is inevitable (particularly for heavy nuclei), as well as some elastic scattering (for example, on sodium if this coolant is used). This spectrum is relatively different according to the reactor design. Typically, for a high-power sodium-cooled reactor, the flux[9] has a maximum around a hundred keV, and becomes negligible below about a hundred electron volts.

- In *thermal neutron reactors*, the spectrum always has a somewhat similar appearance: a "hump" at high energy reflecting the fission spectrum (see Figure 2.21) but slightly degraded because of scattering.

- A slight decrease in the epithermal region, reflecting resonant capture losses, particularly by uranium 238 (see Figure 2.12).

[9] This is the lethargy flux, which will be defined in Chapter 7.

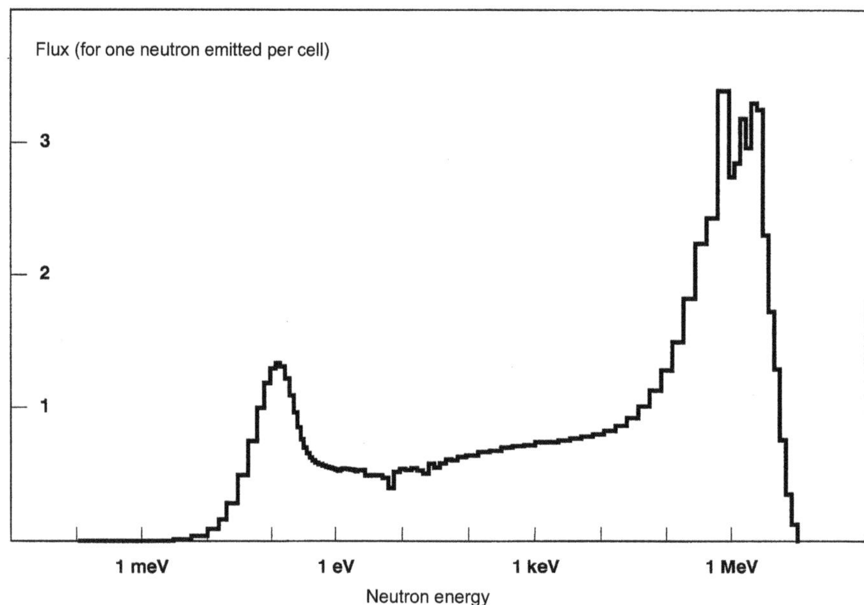

Figure 3.4. Flux for a pressurised water reactor (fresh fuel).

- A new "hump" at low energy reflecting the Maxwell distribution of the thermal agitation but a little "harder" (shifted towards the higher energies) because temperature equilibrium has not been perfectly achieved. As an illustration, Figure 3.4 shows a typical neutron spectrum in a PWR. (The "staircase" representation reflects the "multiple group" calculation that was performed; in this case by the APOLLO code and 99-group library.)

3.3.2. Neutron balances: the four-factor formula and variants

Enrico Fermi, who chose the thermal neutron reactor route for the Chicago pile, proposed a breakdown of the neutron balance into four factors: the formulae expressing each of them allowed him to optimise the lattice and evaluate the critical mass. With the modern-day use of computers in neutron physics, these formulae are no longer used for reactor calculations, but they still help us to understand reactor physics and evaluate various effects. Details of these will be given in Chapters 8 and 9, but for the moment it is of interest to give definitions and orders of magnitude for an example (we chose the case of PWRs) to give some idea of what happens to neutrons emitted by fission[10].

For the purpose of simplification, Fermi does not take neutron leaks outside the reactor core into account, i.e. he reasons according to the "infinite" multiplication factor k_∞ that

[10] Here we shall restrict the discussion to the classical definitions from Fermi's work. Other breakdowns of the neutron balance, which are more detailed or can be generalised to all reactor types, have been suggested, most notably by Roger Naudet, Alain Santamarina, and the present author.

the system would have if it extended to infinity. (The real factor k is obviously lower than k_∞, since some of the neutrons escape. An infinite multiplication factor greater than 1 should therefore be sought.)

The neutron balance is established by starting from one fission and analysing events until the next fission, and therefore by following the "history" of the neutron. The "four-factor" formula is written as follows:

$$k_\infty = \varepsilon p f \eta. \tag{3.24}$$

Figure 3.5 summerizes the analysis.

Conventionally, the analysis is performed for *one emitted neutron;* the following values are therefore *averages*. The fuel of the system under consideration is assumed to consist of a mixture of a fissile material (such as uranium 235) and a non-fissile but capturing material (such as uranium 238). For the first, and for other capturing materials (moderator, cladding), it is possible to accept that all absorptions occur in the thermal region[11]. For the second, on the other hand, fission induced by fast neutrons must be taken into account, as well as (and especially) captures in the many large resonances in the epithermal region.

Also conventionally, our reasoning applies to one neutron emitted by the main fission flux, i.e. *a neutron emitted by induced fission by a thermal neutron* (for example, uranium 235)[12].

That is why a correction factor ε must immediately be introduced to allow for the *additional* neutrons (a few percent) due to fast fission (induced by a fast neutron very quickly after its birth). This factor ε, known as the *fast fission factor,* is defined as the total number of neutrons that will enter the next stage — slowing down — for one neutron from a "thermal" fission (induced by a thermal neutron).

The next factor, p, accounts for the "survivors" at the end of slowing down, which is a very dangerous region for neutrons because they can be caught in the many (fissionless) capture resonance "traps". Despite the spectacular nature of these resonances, approximately three quarters of neutrons escape from them. The factor p, called the *resonance escape probability,* is the probability that a neutron reaching the slowing down domain will cross it and reach the entrance to the thermal domain (roughly 1 eV).

The next two factors concern *thermal* neutrons. The factor f, known as the *thermal utilisation factor,* is the probability that the neutron, when thermal, will be absorbed usefully, i.e. in the fuel rather than in some sterile material such as the moderator or cladding. In general, over 90% of neutrons are "well utilised". The last factor, η, is the *reproduction factor,* defined as the number of neutrons emitted by fission for one thermal neutron absorption in the fuel. It is therefore the product of the probability that the absorption is a fission (rather than a sterile capture by one or other of the materials constituting the fuel) and the mean number ν of neutrons emitted per fission. The value of this factor is directly dependent on the fissile content of the fuel. For the chosen example, where the fuel is uranium enriched to slightly over 3%, the values of η, and therefore finally of k_∞, are relatively high.

For this reactor type, a value of approximately 1.3 should be obtained for k_∞; on the one hand, to compensate for leaks that will reduce the multiplication factor by a few percent, and on the other hand (and in particular) to anticipate the changes that can lower

[11] This assumption is flawed in the case of pressurised water reactors...

[12] In the calculation codes, however, all fissions are put on the same plane.

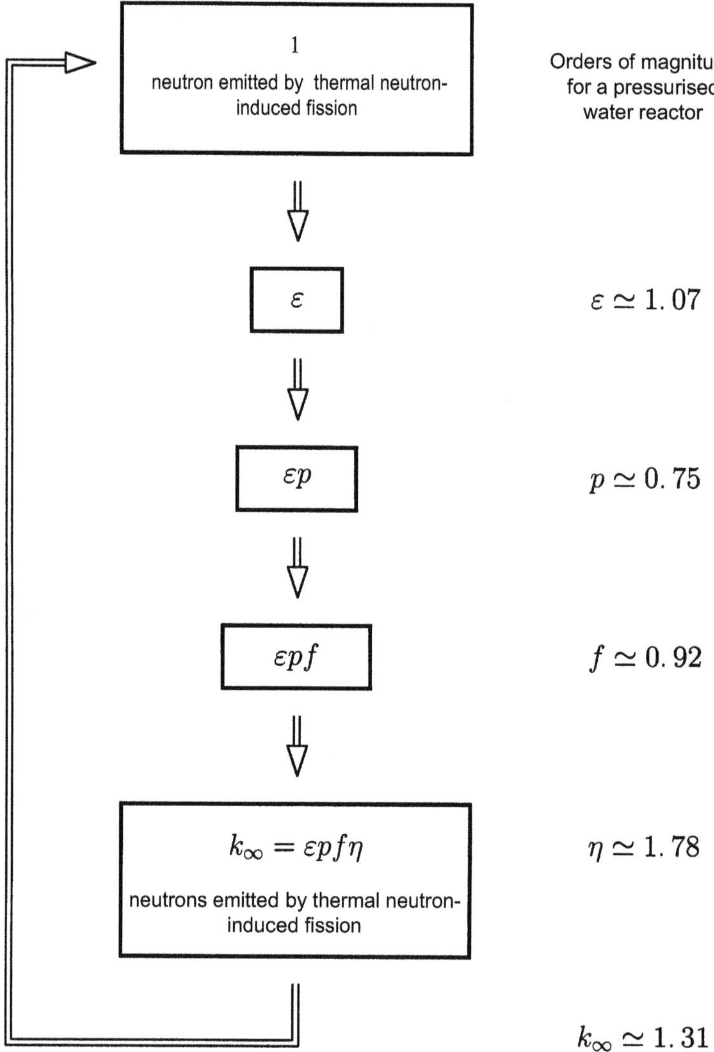

Figure 3.5. Neutron cycle in a thermal neutron reactor and classic four-factor formula (without leaks). Orders of magnitude for a pressurised water reactor.

the multiplication factor by several tens of percentage points through the degradation of the fissile material and the accumulation of fission products. Very irradiated fuel assemblies can thus be characterised by k_∞ less than 1, which is compensated by the neighbouring assemblies which are less irradiated and have k_∞ greater than 1. We shall also see that a factor k greater than 1 should be aimed for at the beginning of the reactor operating cycles, so that criticality will be achieved at the end of the cycle. The excess at the beginning of the cycle is compensated by absorbent "poisons".

Exercises

Exercise 3.1: absorption mean free path

For monokinetic neutrons emitted in an infinite, homogeneous medium characterized by the macroscopic cross-sections Σ_t, Σ_s, and Σ_a. Derive expressions for,

a) the mean number $\bar{n}$ of elementary paths between emission and absorption;

b) the absorption mean free path (i.e. expanded mean path between emission and absorption).

Exercise 3.2: diffusion length

Following on from the previous exercise, next let R be the distance in a straight line between the point of emission and the point of absorption. Derive expressions for the average value of R^2, based upon the definition of the diffusion area L^2, which has an average value of $6L^2$, where L is the diffusion length. Assume that the neutron scattering is isotropic; therefore, the mean values of the cosines of the angles between two elementary paths are zero.

Exercise 3.3: comparison of moderators

For the main moderators — light water, heavy water, beryllium (metal), beryllia (beryllium oxide) and graphite — and for monokinetic 2200 m s^{-1} neutrons, calculate the macroscopic cross-sections, the mean expanded absorption path (*cf.* exercise **3.1**), the mean duration of the migration, and the diffusion length (*cf.* exercise **3.2**), using the data provided in the following tables[13]. The atomic masses are given in atomic mass units, the cross-sections in barns (10^{-28} m^2) and the densities in kg m^{-3}.

Nuclide	Atomic mass (u)	Absorption (barns)	Diffusion (barns)
Hydrogen	1.00794	0.322	30.3
Deuterium	2.01410	0.00051	4.25
Beryllium	9.01218	0.0076	6.34
Carbon	12.0107	0.00337	4.94
Oxygen	15.9994	0.000191	3.76

Material	Density (kg m^{-3})
Light water	998
Heavy water	1105
Beryllium	1850
Beryllia	3010
Graphite	1600

Exercise 3.4: an example nuclear power plant

Consider a 900 MWe PWR.

a) The efficiency for the conversion of heat into electricity is equal to 33%; how much is the nominal thermal power? How many fissions per second are necessary to produce this thermal power?

[13] The density of ideal graphite is about 2200 kg m^{-3}; however, in practice the material is porous. The figure given here is that for the early 1960s experimental research reactor *Marius* at Cadarache in southern France. Since then, the technology for making reactor-grade graphite has improved so that modern material is denser.

b) The core contains 157 fuel assemblies (0.215 m × 0.215 m square cross-section, and 3.658 m height); what size is its volume and its equivalent radius if it is transformed into a cylinder? It contains 82 t of uranium oxide (density: 10 300 $\mathrm{kg\,m^{-3}}$); how large is the volume of fuel, and what fraction of the total is occupied by the fuel?

c) How large is the mean specific power per unit of volume of core, and per unit of volume of fuel?

d) The uranium of the fuel is enriched up to 3%. Calculate the number of uranium-235 atoms per unit of volume. Using 582 barns for the uranium-235 fission microscopic cross-section, and neglecting fission of the uranium-238 atoms, calculate the macroscopic fission cross-section of the fuel. Consequently, how large is the mean neutron flux in the fuel?

e) Assuming that the neutrons are monokinetic, with a velocity equal to 3100 $\mathrm{m\,s^{-1}}$, calculate the mean neutron density in the fuel. Compare this with the number of atoms per unit of volume.

Exercise 3.5: 'peak' and 'hole' of iron main resonance

Natural iron (density: 7860 $\mathrm{kg\,m^{-3}}$) consists of four stable isotopes, of which about 92% are $^{56}_{26}\mathbf{Fe}$ atoms. Based on the results of exercise **2.16**, calculate the neutron mean free paths for isotopically pure iron-56 at the energies of the maximum and minimum cross-sections.

Exercise 3.6: how should the mean free path be defined?

Paradox of the mean free path: in a homogeneous material, consider a neutron passing by a point *A*. This neutron then continues to a point *N* where the next collision is observed. It seems reasonable to assert that the path *AN* will, on average, be equal to $\lambda = 1/\Sigma$, i.e. the mean free path. However, this neutron started from a point *M* located *before* point *A*. Since the path *MA* is not nil, the total path $MN = MA + AN$ is greater than λ. What is the explanation for this apparent paradox?

Can you gives other examples of similar paradoxes?

Exercise 3.7: mean chord

Cauchy's formula: the mean chord $\langle X\rangle$ of any convex body with surface area S enclosing a volume V is given by the formula

$$\langle X\rangle = \frac{4V}{S}.$$

In order to demonstrate this formula, consider a uniform, isotropic flux Φ of monokinetic particles (for instance neutrons) in otherwise empty space.

a) How many particles are there inside the volume *V* at any given time?

b) How many particles pass through the surface *S* both into and out of the volume *V* per unit of time?

c) Consequently, how long is the mean duration $\langle t\rangle$ that a particle takes to cross the volume *V* upon entering it?

d) What is the expression for the time $\langle t \rangle$ as a function of the mean chord $\langle X \rangle$, and the velocity of the particles v.

e) Derive the Cauchy formula from **(c)** and **(d)**.

Which statistical law is used to define the mean value of the chord X, in order to apply this theorem?

Exercise 3.8: neutron current density and flux per unit solid angle

a) Find expressions for the neutron current density and flux per unit solid angle passing through an infinitesimal surface element dS, located in a vacuum at a distance x from a plane source isotropically emitting E particles per unit of surface and time, where dS is parallel to the source.

b) Repeat part **(a)** for an isotropic flux and compare the result.

c) Repeat part **(a)** for a point source isotropically emitting E particles per unit of time, where dS is located at a distance r from the source and its normal lies on the line to the source.

Exercise 3.9: fission spectrum

The fission spectrum can be described approximately by a Maxwell distribution:

$$\chi(E)\,dE = C\sqrt{E}\,e^{-E/T}\,dE,$$

where C is a constant and T a characteristic parameter equal to two thirds of the mean energy $\bar{E}$.

Using $\bar{E} = 2$ MeV (approximately the energy of the neutrons released by the fission of uranium-235 atoms induced by thermal neutrons), calculate the following.

a) The proportion of neutrons emitted by fission above 0.8 MeV, which are able possibly to induce fission of uranium-238 atoms;

b) the proportion of neutrons emitted above 10 MeV, which are often neglected in calculations employing, for example, the APOLLO program package with the standard 99 group library.

Exercise 3.10: activation by neutron irradiation

By neutron irradiation, sulphur-32 is transmuted into phosphorus-32 via the reaction

$${}^{32}_{16}\mathbf{S} + {}^{1}_{0}\mathbf{n} \Longrightarrow {}^{32}_{15}\mathbf{P} + {}^{1}_{1}\mathbf{p}.$$

The neutron cross-section for this process is $\sigma = 0.3$ b.

Phosphorus-32 subsequently decays back into sulphur-32 by β^- radioactivity:

$${}^{32}_{15}\mathbf{P} \Longrightarrow {}^{32}_{16}\mathbf{S} + {}^{0}_{-1}\mathbf{e} + {}^{0}_{0}\bar{\nu},$$

with a half life $T_{1/2} = 14.3$ d.

a) Assuming that there is initially no phosphorus at the start of the irradiation, and that the neutron flux Φ is constant, derive and solve the equations that describe the system.

b) How long does it take for the activity of the phosphorus to reach 90% of its asymptotic value when $\Phi = 4 \times 10^{13}\ \mathrm{m^{-2}\,s^{-1}}$?

Solutions

Exercise 3.1: absorption mean free path

Let $c = \Sigma_s/\Sigma_t$ be the number of 'secondaries' (re-emitted particles) per collision, and p_n be the probability that a neutron undergoes exactly n collisions, i.e. performs exactly n elementary paths; then p_n is the probability to observe $n-1$ scatterings, before absorption:

$$p_1 = 1 - c, \qquad p_2 = c \times (1 - c), \qquad p_3 = c \times c \times (1 - c),$$

$$p_n = c^{n-1} \times (1 - c).$$

Therefore,

$$\bar{n} = \sum_{n=1}^{\infty} n\, p_n = \frac{1}{1 - c} = \frac{\Sigma_t}{\Sigma_a},$$

and

$$d = \bar{n}\lambda = \frac{1}{\Sigma_a}.$$

Exercise 3.2: diffusion length

First, consider the neutrons which have performed exactly n elementary paths. If $\vec{\rho}_i$ ($i = 1, 2, \cdots n$) are these paths, then

$$\vec{R}_n = \vec{\rho}_1 + \vec{\rho}_2 + \cdots + \vec{\rho}_n.$$

Hence, the average of their squares is

$$\langle \vec{R}_n^{\,2} \rangle = \langle \vec{\rho}_1^{\,2} \rangle + \langle \vec{\rho}_2^{\,2} \rangle + \cdots + \langle \vec{\rho}_n^{\,2} \rangle \; + \; 2\langle \rho_1 . \rho_2 . \cos\theta_{12} \rangle + \cdots ,$$

where θ_{ij} is the angle between the vectors $\vec{\rho}_i$ and $\vec{\rho}_j$.

The probability law of the length of each elementary path is $P(\rho)\, d\rho = \exp(-\Sigma_t \rho)\, \Sigma_t\, d\rho$. Thus,

$$\langle \vec{\rho}_i^{\,2} \rangle = \frac{2}{\Sigma_t^2} \qquad \text{(for any } i\text{)}.$$

The averaged values of all the terms containing a cosine are nil when the scattering is isotropic. Therefore,

$$\langle \vec{R}_n^{\,2} \rangle = \frac{2n}{\Sigma_t^2}.$$

Taking the average of the $\langle \vec{R}_n^2 \rangle$ terms weighted by their probabilities p_n gives

$$\langle \vec{R}^{\,2} \rangle = \frac{2}{\Sigma_t \Sigma_a},$$

and

$$L^2 = \frac{1}{3\Sigma_t\Sigma_a}.$$

Exercise 3.3: comparison of moderators

Macroscopic cross-sections in m^{-1}, distances in m, and durations in milliseconds for the five materials used in this exercise are given in the following table.

Material	Σ_a	Σ_s	$\bar{n}$	Abs. path	Duration	Diff. length
Light water	2.15	215	101	0.465	0.212	0.0267
Heavy water	0.00402	40.7	10100	249.0	113	1.43
Beryllium	0.0940	78.4	835	10.6	4.84	0.213
Beryllia	0.0565	73.2	1300	17.7	8.05	0.284
Graphite	0.0270	39.6	1470	37.0	16.8	0.558

Exercise 3.4: an example nuclear power plant

a) With 200 MeV per fission, the thermal power (2700 MWth) is generated by 8.4×10^{19} fissions per second.

b) The volume of the core is 27 m^3 which represents an equivalent radius of 1.52 m when in the form of a cylinder; the volume of the fuel is 8 m^3, i.e. 30% of the core volume.

c) The specific power is 100 MW m^{-3} of core, or 340 MW m^{-3} of fuel.

d) $N_U = 230\times10^{26}$ atoms per m^3; $N_{U\text{-}235} = 7\times10^{26}$ atoms per m^3. Therefore, $\Sigma_f = 40$ m^{-1}. Using $P = \kappa\Sigma_f\Phi$, means that the neutron flux is $\Phi = 2.6 \times 10^{17}$ m^{-2} s^{-1}).

e) Using $\Phi = nv$, the neutron density in the fuel is $n = 0.8 \times 10^{14}$ m^{-3}; this represents one neutron for 2.7×10^{14} atoms of uranium, which is equivalent to one neutron, on average, in a cube that has 65 000 $\mathbf{UO_2}$ molecules along each edge!

Exercise 3.5: 'peak' and 'hole' of iron main resonance

The concentration of atoms is $N = 8.48 \times 10^{28}$ m^{-3}.

- Peak: $\sigma = 117$ b $\lambda = 1.01$ mm.
- Hole: $\sigma = 1.04$ b $\lambda = 113$ mm.

This means that neutrons with an energy of about 23.5 keV are practically not stopped by an iron shield!

Exercise 3.6: how should the mean free path be defined?

The populations 'neutrons leaving the point M, towards A', and 'neutrons passing by the point A' are *not* equivalent from a statistical point of view. The second population is the first one *unless* the neutrons suffered a collision between M and A, *i.e. a population where the short-path neutrons have been elimitated*. As λ is the mean path of the first population neutrons, the mean path of the second population neutrons is necessarily greater.

An equivalent paradox is obtained with radioactivity: the radioactive nuclei observed at a given moment will, on average, survive longer than the mean life time τ, since they have already not decayed during a given time, and will remain intact during the mean duration τ.

Note that these processes are known as *Markovian*, meaning that the future will depend only on the present and not on the past: a neutron carries only the information about its *present* state (position, momentum, spin), and ignores what happened to it in the past. The same it true for a radioactive nucleus.

However, the paradox still applies for non-Markovian processes: for instance, a 70 year old man will certainly have a longer mean life than a baby, because we know that he did not die before 70!

Exercise 3.7: mean chord

a) If n is the density, then the number of particles in the volume V at any instant is simply $N = nV$.

b) The flux is $\Phi = nv$; the partial currents everywhere are $J_{\pm} = \Phi/4$ per unit area, and in particular at each point of the surface S. This result is rigourous because the flux is uniform and isotropic. Therefore, the number of particles entering per unit time is $E = SJ_{-} = \Phi S/4 = nvS/4$.

c) The mean time that it takes for a particle to pass through the volume V is $\langle t \rangle = N/E = 4V/(vS)$.

d) Expressed in terms of the mean chord $\langle X \rangle$, and the velocity of the particles v, this duration is $\langle t \rangle = \langle X \rangle / v$.

e) Simply equate the two formulae obtained previously to derive the Cauchy formula.

For this theorem, it is assumed that the particles enter uniformly and isotropically, i.e. the mean chord X is defined by:

- choosing the entry point uniformly on the surface S;
- choosing the entry direction isotropically over 2π steradians.

Exercise 3.8: neutron current density and flux per unit solid angle

a) In the following figure, where the surface element dS is represented by the segment AB (transverse section), it can be seen that the current through dS is independent of x, due to the absence of any intervening matter. The neutrons emitted in the direction $\vec{\Omega}$ by the source element dS' (shifted from dS in the direction $-\vec{\Omega}$, and represented by $A'B'$) will cross dS. Hence, per unit of time and solid angle, there are $(E/4\pi)\,dS$ such neutrons. Identifying with $\left[\vec{\Omega}\,\Phi(\vec{r},\vec{\Omega})\right].\vec{N}\,dS$, it can be seen that

$$\Phi(\vec{r},\vec{\Omega}) = \frac{E}{4\pi\vec{\Omega}.\vec{N}} \qquad \text{(for } \vec{\Omega}.\vec{N} \text{ positive only when } x \text{ is positive).}$$

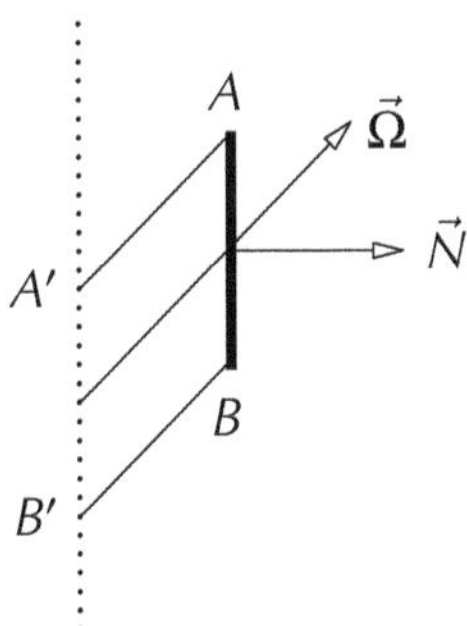

Figure 1

b) Note that this phase flux proportional to $1/\mu$ (with $\mu = \cos\theta = \vec{\Omega}.\vec{N}$) is infinite for the directions parallel to the source plane. The scalar flux is also infinite everywhere! Of course, if we add matter, then the flux becomes finite — unless inside the source plane — as the previous expression of the phase flux must be reduced by a factor $\exp(-\Sigma x/\mu)$. For an isotropic flux, Φ would be independent of $\vec{\Omega}$.

c) For a point source (fig. 2), the neutrons cross dS only in the direction of the vector $\vec{u} = \vec{r}/r$. Therefore,

$$\Phi(\vec{r}, \vec{\Omega}) = \frac{E\,\delta(\vec{\Omega} - \vec{u})}{4\pi r^2}.$$

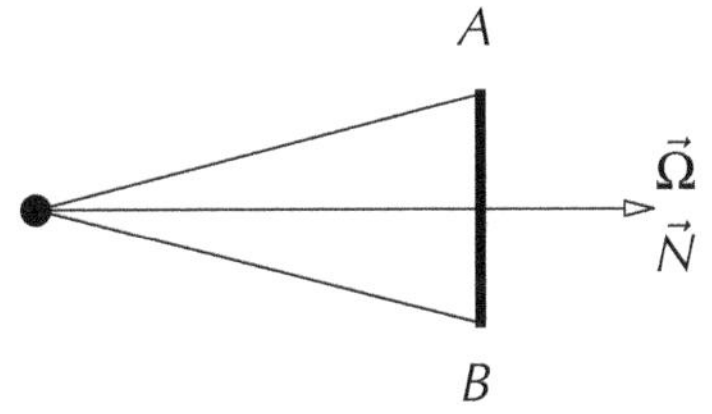

Figure 2

Both phase distributions are very anisotropic (compare with an isotropic phase flux, by definition independent of $\vec{\Omega}$). Note that in the first example the *current* — not the flux — is independent of $\vec{\Omega}$.

Exercise 3.9: fission spectrum

The function describing the spectrum can be normalized by calculating the constant as follows.

$$I = \int_0^E \chi(E)\,dE = \Theta(x) - \frac{2}{\sqrt{\pi}}\,x\,e^{-x^2}, \qquad x = \sqrt{E/T},$$

$$I = 1 - \frac{2}{\sqrt{\pi}}\,x\,e^{-x^2}\left[1 + \frac{1}{2x^2} - \frac{1}{4x^4} + \cdots\right] \qquad (x \longrightarrow \infty),$$

where Θ is the error function (also written 'erf'). Numerical values of this integral I are as follows.

a) $E = 0.8$ MeV $x^2 = 0.6$ $I = 0.24669$ $1 - I = 0.75331$;
b) $E = 10$ MeV $x^2 = 7.5$ $I = 0.99818$ $1 - I = 0.00182$.

Exercise 3.10: activation by neutron irradiation

a) Let S and P be the concentrations, then the dynamical equations that describe the system are as follows.

$$\frac{dS}{dt} = -\sigma\Phi S + \lambda P,$$

$$\frac{dP}{dt} = -\lambda P + \sigma\Phi S.$$

They can be solved by taking the expression

$$S + P = C^{te} = S_0,$$

and substituting it into either the first or second equation:

$$P = \frac{\sigma\Phi}{\lambda + \sigma\Phi} S_0 \left[1 - e^{-(\lambda+\sigma\Phi)t}\right].$$

Since $\sigma\Phi$ is very small relative to λ, the following approximation may be made.

$$P \simeq \frac{\sigma\Phi}{\lambda} S_0 \left[1 - e^{-\lambda t}\right].$$

b) 90% of the asymptotic value is reached when $e^{-\lambda t} = 0.1$, i.e. when $t = 47.5$ days.

4 Point kinetics

Introduction

As we have seen (Figure 1.3), in a chain reaction, the number of fissions or the number of neutrons is multiplied by a factor k from one generation to the next, and therefore the kinetic behaviour of the reactor is exponential. We have not yet evaluated the duration of a generation, however: this is what we aim to do in the present chapter, to determine the speed of the exponential growth according to the value of the multiplicative factor k.

Moreover, even though our reasoning had the advantage of being intuitive, it was too simple, because the increase in the rate of fissions and the number of neutrons does not occur in a discrete way (with a whole number n), but continuously over a time t.

Above all, this reasoning neglects the fact that there are two neutron populations: prompt neutrons and delayed neutrons (see § 2.10.1), and that delayed neutrons completely change the kinetic behaviour of the system, as we shall see.

When studying problems in kinetics, it can generally be accepted that the neutron flux varies in amplitude without changing its spatial distribution, i.e. that it is factorised:

$$\Phi(\vec{r}, v, \vec{\Omega}, t) = \varphi(\vec{r}, v, \vec{\Omega}) \times n(t). \tag{4.1}$$

Using this assumption, the kinetic analysis can be limited to the time parameter only, i.e. the analysis can be performed as if the reactor were reduced to a point. In this chapter, we handle the kinetics problem in this "point" context.

4.1. Kinetics without delayed neutrons

4.1.1. First approach

Referring back to the chain reaction diagram (Figure 1.3), we see that the neutron population is multiplied by k^n after n generations. The evaluations we are able to perform show that the duration of a generation, i.e. the average time between one fission and the next, is extremely short: of the order of 10^{-7} s in a fast neutron reactor, and between 10^{-5} s and 10^{-3} s in thermal neutron reactors according to the choice of moderator, e.g. approximately 2.5×10^{-5} s in pressurised water reactors. (It is a short time because neutrons travel quickly. Even a thermal neutron covers 2 to 3 km per second.)

For the present example, 40 000 generations pass by in one second. To grasp the significance of this, let us take a value of k that is 0.01% different from criticality. We can see that the power varies in one second by the following factors:

- $1.0001^{40\,000} = 55$ if the difference is positive,
- $0.9999^{40\,000} = 0.018$ if the difference is negative.

Under these conditions, it is difficult to imagine how the chain reaction could be monitored and regulated[1].

4.1.2. Chain reaction equations

As has been remarked, this reasoning is too simple, because the progress of the reaction is continuous. Let $n(t)$ be the number of neutrons as a function of time t.

Let θ be the average life of a neutron from emission to absorption. Assuming that the neutron has the same probability of being absorbed irrespective of its "age"[2], the probability of each neutron disappearing during a unit time interval dt is dt/θ; thus the average number of neutrons disappearing during dt is $n\,dt/\theta$.

Each neutron disappearance has probability ω of being a fission (notation used in § 1.4) and, if fission actually occurs, it gives rise to ν new neutrons on average. Each disappearance therefore contributes an average of $\omega\nu = k$ new neutrons; thus, the $n\,dt/\theta$ disappearances observed during dt contribute $kn\,dt/\theta$ new neutrons.

Finally, the change in number of neutrons during dt is:

$$dn = -n\,dt/\theta + kn\,dt/\theta,$$

i.e.:

$$\frac{dn}{dt} = \frac{k-1}{\theta}n;$$

and, by integration:

$$n(t) = n(0)\exp\left(\frac{(k-1)t}{\theta}\right). \qquad (4.2)$$

By trying this out with the same numbers as in the previous numerical example, the reader can confirm that the results are almost identical; this improved model does not change the conclusions!

[1] Joliot *et al.* did not know about the existence of delayed neutrons. After this type of order of magnitude calculation, they wrote patents in May 1939 describing a very complicated pulsed mechanism...

[2] This is strictly correct in monokinetic theory only. The assumption is allowable in practice, particularly in thermal neutron reactors, because the slowing down duration is short compared to the duration of the thermal phase, which is more or less monokinetic. Also note that, in spectral theory, it would be strictly necessary to distinguish the lifetime (from neutron emission to absorption) from the generation time (from one fission to the next), because there is no reason why the energy spectrum of fissions and the energy spectrum of captures (and therefore of absorptions) should be identical.

4.1.3. Reactivity

Before addressing this question again whilst taking delayed neutrons into account, we introduce the concept of reactivity, which is often used in kinetics. The reactivity ρ is defined by:

$$\rho = \frac{k-1}{k}, \tag{4.3}$$

and is expressed in pcm[3], or 10^{-5}.

Reactivity becomes zero at criticality. Its sign defines the direction of the trend:

- increasing trend if reactivity is positive (supercritical state);
- decreasing trend if reactivity is negative (subcritical state).

Note that the reactivity is very close to $k-1$ in the neighbourhood of criticality.

The reactivity of a reactor often varies because of a movement of the absorbent, because an increase in absorption reduces the multiplication factor in inverse proportion.

If absorption increases, the neutrons disappear more quickly, and the lifetime also decreases in inverse proportion to absorption. We can therefore see that θ varies with k, giving:

$$\theta = \ell k, \tag{4.4}$$

where ℓ is the neutron lifetime in a *critical* reactor.

By introducing reactivity, with this assumption, we can simplify the kinetics equation:

$$\frac{dn}{dt} = \frac{\rho}{\ell} n,$$

and, by integration:

$$n(t) = n(0) \exp\left(\frac{\rho t}{\ell}\right). \tag{4.5}$$

4.2. Kinetics with delayed neutrons

4.2.1. Parameters of delayed neutrons

As we have seen (§ 2.10.1), some neutrons are emitted with a delay with respect to fission because of a beta decay that precedes neutron emission. The two channels leading to the longest delays have been mentioned. The others, about a hundred similar processes, are grouped in practice into four other "pseudo-chains" (i.e. a total of *six groups of delayed neutrons*) whose characteristics are adjusted according to the measurement results.

For a given fissile nucleus, each group i of delayed neutrons is characterised by two nuclear parameters:

- the proportion β_i of delayed neutrons in this group as compared to all neutrons emitted by fission (prompt and delayed), expressed in pcm;

[3] The unit pcm is an abbreviation for the French term *pour cent mille*, meaning "per one hundred thousand". In English, the alternative term "millinile" is sometimes used.

Table 4.1. Characteristics of the six groups of delayed neutrons (thermal neutron-induced fission of uranium 235). The unique "average" group is characterised by the sum of all β_i and the average of all T_i *weighted by* β_i.

Group	β (pcm)	*T* (s)	τ (s)
1	24	54.5	78.6
2	123	21.8	31.5
3	117	5.98	8.62
4	262	2.23	3.22
5	108	0.495	0.714
6	45	0.179	0.258
Average	679	7.84	11.31

Table 4.2. Total proportion of delayed neutrons for a few fissile nuclei (cases of thermal neutron-induced fission or fast neutron-induced fission).

Nucleus	Fission	β (pcm)
Thorium 232	Fast	2 433
Uranium 233	Thermal	296
Uranium 235	Thermal	679
Uranium 238	Fast	1 828
Plutonium 239	Thermal	224
Plutonium 240	Fast	292
Plutonium 241	Thermal	535

- the radioactive decay constant λ_i of the *precursor,* i.e. the beta decay that will lead to the nucleus emitting the neutron (almost instantaneously) or, equivalently, the radioactive period $T_i = \ln 2/\lambda_i$.

These periods are not exactly the same for all fissile nuclei because they are obtained by adjustment but, in practice, the values are near neighbours. The same applies to the ratios β_i/β where β is the *total* proportion of delayed neutrons, i.e. the sum of all β_i.

That is why the tables 4.1 and 4.2 give β_i, T_i and mean lifetime $\tau_i = 1/\lambda_i$ for one example only (uranium 235), and only the value of β for the others.

4.2.2. Qualitative aspects

As a first approximation, the above reasoning can continue to be applied if we simply take delayed neutrons into account in the *mean* generation time calculation, i.e. by making the following substitution for the delayed neutrons: ℓ replaced by $\tau_i + \ell$:

$$\bar{\ell} = (1 - \beta) \times \ell + \sum_i \beta_i \times (\tau_i + \ell) = \ell + \beta\bar{\tau}. \tag{4.6}$$

Taking the example of the uranium 235 pressurised water reactor, ℓ goes from 2.5×10^{-5} s to $\bar{\ell} = 2.5\times10^{-5} + 679\times10^{-5}\times11.31 = 0.077$ s (the term $\beta\bar{\tau}$ added by the delayed neutrons overwhelmingly dominates ℓ). There are no longer 40 000, but 13 "effective generations" per second to be taken into account. If the multiplication factor differs from criticality by 0.01%, the power does not evolve by a factor of 55, but only 1/10 of a percent. It is therefore plain to see that the presence of delayed neutrons completely changes the kinetic behaviour of the reactor, making it easy to control, notwithstanding any fears raised by the initial calculation.

4.2.3. Chain reaction equations

This evaluation is correct in qualitative terms, but not in quantitative terms if the reactivity is too high. That is why a more accurate model is required. Remaining in the point model, let n be the number of neutrons, and c_i the number of precursor nuclei[4] in group i:

- For **neutrons**, the rate of disappearance is the same as before, but a distinction must be made between the direct channel (prompt neutrons) for the proportion $1 - \beta$ of neutrons produced and the delayed channel resulting from radioactive decay and expressed by a λc term;
- For **precursors**, the rate of disappearance is the radioactivity rate (λc term), and the rate of production is equal to the number of neutrons to be emitted with a delay by the chain concerned:

$$\frac{dn}{dt} = \frac{k(1-\beta)-1}{\theta} n + \sum_i \lambda_i c_i,$$
$$\frac{dc_i}{dt} = \frac{k\beta_i}{\theta} n - \lambda_i c_i. \tag{4.7}$$

By setting $\theta = k\ell$ as before and by introducing the reactivity, we can simplify these equations as follows:

$$\frac{dn}{dt} = \frac{\rho-\beta}{\ell} n + \sum_i \lambda_i c_i,$$
$$\frac{dc_i}{dt} = \frac{\beta_i}{\ell} n - \lambda_i c_i. \tag{4.8}$$

4.2.4. Inhour equation

All of the coefficients in this system are physical constants, except the reactivity, which can change if the system is acted upon or if spontaneous changes occur (*see* Chapters 11 to 13). In the case where reactivity does not vary, the system is a "constant coefficient" system, and its solution consists of linear combinations of $g+1$ exponential functions, if there are $g+1$ equations, i.e. g groups of precursors of delayed neutrons plus the neutron equation. The time constants ω of the exponentials are obtained by substituting this type of expression into the system, i.e.:

$$n(t) = a \exp(\omega t),$$
$$c_i(t) = b_i \exp(\omega t), \tag{4.9}$$

[4] In the decay chain **A** $\Rightarrow$ **B*** $\Rightarrow$ **C** + **n**, intermediate nucleus **B** can also decay by gamma emission. By convention, "delayed neutron precursors" will refer only to nuclei **A** that will give a neutron; in other words, the number c of precursor nuclei is the real number multiplied by the probability that the de-energising of **B** occurs by neutron emission.

where a and b_i are constants. After simplification by $\exp(\omega t)$, the equations are reduced to the following algebraic system:

$$\omega a = \frac{\rho - \beta}{\ell} a + \sum_i \lambda_i \beta_i,$$

$$\omega b_i = \frac{\beta_i}{\ell} a - \lambda_i \beta_i. \tag{4.10}$$

The equations for each of the groups give b_i as a function of a. By substituting into the first equation, we obtain a homogeneous equation. After simplification using a, this gives the condition that ω must satisfy:

$$\rho = \omega \left[\ell + \sum_i \frac{\beta_i}{\lambda_i + \omega} \right]. \tag{4.11}$$

In France, this condition is known as the Nordheim Equation. In English-speaking countries, however, it is called the "inhour equation" because it gives a quantity that can be expressed in hour-1 (inverse hour).

The plot as a function of ω of the expression in the second term (see Figure 4.1) shows (if the graph is cut by a horizontal line at the ordinate ρ) that there are always $g + 1$ real solutions, no matter what the reactivity.

Each of the $g + 1$ roots ω_k must be associated with an integration constant a_k (the other constants $b_{i,k}$ are expressed in terms of the a_k, once ω_k has been determined, as we have just seen). These $g + 1$ integration constants must be determined from $g + 1$ conditions: in general, these will be the initial values of n and c_i.

The figure shows that there are g transient exponentials (ω always negative) corresponding to the g (six in this case) left-hand branches and one asymptotically dominant exponential characterised by the largest value of ω, corresponding to the right-hand branch. *The dominant value of* ω *is of the same sign as* ρ: if the reactivity is positive, the functions grow asymptotically (subcritical system). If the reactivity is negative, the functions decrease asymptotically (supercritical system). If the reactivity is zero, the functions tend toward an asymptotic value (critical system).

4.2.5. Low reactivities

If the reactivity is low (in terms of absolute value), the dominant value of ω is small. In the denominator of the inhour equation, ω can be ignored in comparison to λ_i, which makes the curve the same as its tangent at the origin. The result is:

$$\omega \simeq \rho / \bar{\ell}. \tag{4.12}$$

We thus find the simple behaviour described at the beginning: the elementary formula with an average generation time allowing for emission delays. This reasoning was therefore acceptable, provided that the system remains close to criticality.

The following is a better approximation:

$$\omega \simeq \frac{\rho}{(\beta - \rho)\bar{\tau} + \ell}. \tag{4.13}$$

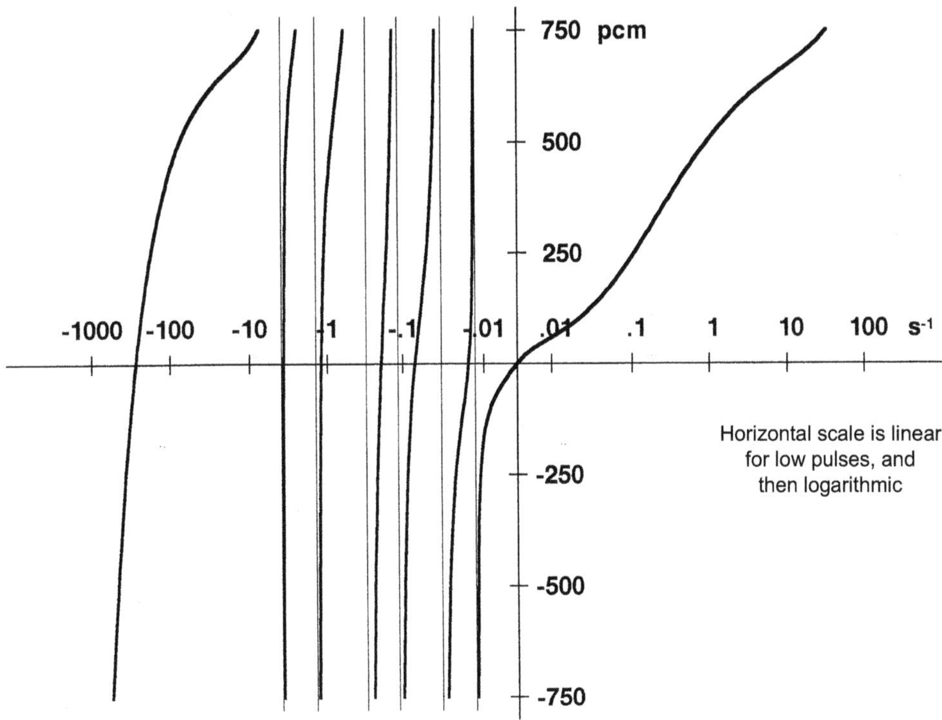

Figure 4.1. Graphical representation of the second term of the inhour equation (Fission of uranium 235 induced by thermal neutron, processed to six groups of delayed neutrons).

4.2.6. High reactivities

If the reactivity is positive and high, ω becomes large compared to each λ_i, as the preceding figure shows. In this case, we can neglect λ_i compared to ω in the denominator of the inhour equation. It then becomes:

$$\omega \simeq \frac{\rho - \beta}{\ell}. \tag{4.14}$$

This brings us back to the elementary formula *without delayed neutrons* but with reactivity $\rho - \beta$ instead of ρ. *This regime is supercritical with the prompt neutrons alone*[5], and therefore has the same reactivity as if delayed neutrons did not exist, as well as the "real" lifetime.

Obviously, this situation can only be accidental. In normal operation, operators are instructed to remain far below the prompt neutron criticality threshold to allow a power "doubling time", $T = \ln 2/\omega$, long enough to control the installation. This means at least fifteen seconds for an experimental installation (therefore a reactivity threshold of approximately 250 pcm for uranium) and a far lower reactivity threshold for a power reactor.

[5] This is known as the prompt critical regime; the term "prompt criticality" is also used.

4.2.7. The "natural" unit of reactivity: the "dollar"

These considerations show that pcm is a convenient but conventional unit; for physics, and in particular to identify such a threshold not to be approached, the natural unit of reactivity is the proportion β of neutrons emitted with a delay. The Americans suggested calling this unit a "dollar", and one-hundredth of it a "cent"[6].

The value of a dollar in terms of pcm (real reactivity) depends on the fuel. For example, it is three times lower for a plutonium-based fuel than a uranium-based fuel[7].

4.2.8. Effective proportion of delayed neutrons

To summarise: we can see that this proportion β of delayed neutrons appears to be the essential parameter to characterise a system in kinetic terms and, more precisely, in terms of the risk of a criticality accident (uncontrolled insertion of a high positive reactivity[8]). This goes to show that this parameter must be evaluated with the greatest of care. Note two points in particular:

a/ The parameters of the point model must be defined on the basis of the spatio-energetic part of the flux, $\varphi(\vec{r}, v, \vec{\Omega})$, as well as the neutron importance function (which will be mentioned in Chapter 16).

b/ This problem applies particularly to β, the parameter that is most sensitive to these aspects. On the one hand, careful weighting must be applied according to the fission rate of each nuclide (because the individual β values are very different from each other). On the other hand, it is important to allow for the fact that delayed neutrons are emitted at a lower energy than prompt neutrons (400 keV on average instead of 2 MeV). In the case of a thermal neutron reactor, the delayed neutrons are at less risk of escaping in the fast region before being slowed down and likely to provoke a fission. They are therefore more "efficient" than prompt neutrons. They will not, however, cause rapid fissions. This leads to the use of an "effective beta" in calculations; this β_{eff} is slightly different from the "mean beta" $\bar{\beta}$ obtained by weighting the "nuclear betas" β_k of the various fissile nuclides by the fission rates.

4.2.9. Fast kinetics model

As we have just seen, if ρ is greater than β, the kinetics can be handled by the model without delayed neutrons, provided that ρ is replaced by $\rho - \beta$. This simple model is obviously valid only when delayed neutron emission is negligible or at least does not vary significantly. For example, it makes it possible to study a criticality accident (caused by prompt neutrons) qualitatively if not quantitatively, when the essential phase of this event lasts at least one tenth of a second. (The main difficulty of this problem is to evaluate the change in ρ, which is also fast because of the counter-reactions related to temperature effects; see Chapter 13.)

---

[6] In a similar spirit, they often use the "inhour" (inverse hour) to express ω.

[7] Except for the branches the furthest to the left and right on Figure 4.1, which depend mainly on the lifetime of prompt neutrons and therefore the reactor type, the use of the dollar to graduate the vertical axis makes it possible to have a graphical representation of the inhour equation that is not very dependent on the type of fuel.

[8] In pressurised water reactors, for example, this could come from the fast ejection of a control bundle following the rupture of its mechanism, the introduction of clear water instead of boron water, etc.

4.2.10. Slow kinetics model

The slow kinetics model is far more useful for all reactor control problems. In this model, it is supposed that the real lifetime of neutrons ℓ tends towards zero. This is valid because, in this type of problem, delayed neutrons play the essential part in kinetics (see, for example, the calculation of $\bar{\ell}$ in § 4.2.2).

As the lifetime tends to zero, so does the number of neutrons[9]. The rate of absorption (number of absorptions per unit time) $\alpha = n/\ell$, however, remains finite. Re-writing the equations with this new function gives:

$$\ell\frac{d\alpha}{dt} = (\rho - \beta)\alpha + \sum_i \lambda_i c_i,$$

$$\frac{dc_i}{dt} = \beta_i\alpha - \lambda_i c_i, \tag{4.15}$$

and by setting $\ell = 0$, we obtain:

$$\alpha = \frac{\sum_i \lambda_i c_i}{\beta - \rho},$$

$$\frac{dc_i}{dt} = \beta_i\alpha - \lambda_i c_i. \tag{4.16}$$

The first equation becomes explicit; it says that if growth is not too fast, the neutron population is constantly in equilibrium with what is imposed, the "source" level which the precursors represent. This population comprises all (infinite) prompt generations of neutrons from the source. Such equilibrium is obviously possible only if the series is convergent, i.e. if the situation is *prompt neutron subcritical*; in other words, this model applies only if $\beta - \rho$ is positive.

Because of this equilibrium, the order of the differential system is lowered by one unit. If the reactivity is constant, exponential solutions can be sought as before, and the inhour equation can be written (the preceding equation is obviously also found without the ℓ term).

In particular, if we adopt a theory with only one group of delayed neutrons, the problem is reduced to solving just one differential equation: the equation governing the (unique) concentration of precursors:

$$\frac{dc}{dt} = \frac{\rho\lambda c}{\beta - \rho},$$

$$\alpha = \frac{\lambda c}{\beta - \rho}. \tag{4.17}$$

[9] Note that, in a reactor, the number of precursors is much greater than the number of neutrons. A calculation could be performed for a critical equilibrium situation with the numerical data from a pressurised water reactor with uranium.

4.3. A few specific problems

4.3.1. Kinetics with source term

Until now we have assumed that fissions were the only source of neutrons (prompt source and delayed sources. If there is also an "external" source (i.e. outside of the fissions) S, for example a start-up source, this must be added to the second term of the neutron equations:

$$\frac{dn}{dt} = \frac{\rho - \beta}{\ell} n + \sum_i \lambda_i c_i + S,$$
$$\frac{dc_i}{dt} = \frac{\beta_i}{\ell} n - \lambda_i c_i. \tag{4.18}$$

Here we assume that this source and the reactivity are constant in time.

- If the system is *subcritical*, a time-independent solution is added to the transient solution given by the inhour equation (refer to the usual technique for solving this type of problem: general solution = particular solution + general solution for the equation without second term). This time-independent particular solution is:

$$n_{as} = \frac{\ell S}{-\rho}, \quad c_{i,as} = \frac{\beta_i S}{-\rho \lambda_i}. \tag{4.19}$$

- If the system is *just critical*, the asymptotic solution grows linearly with time:

$$\begin{aligned} n_{as} &= \frac{\ell}{\bar{\ell}} St + \mathrm{C^t}, & c_{i,as} &= \frac{\beta_i \tau_i}{\bar{\ell}} St + \mathrm{C^t}, \\ \textstyle\sum_i c_{i,as} &= \frac{\bar{\ell} - \ell}{\bar{\ell}} St + \mathrm{C^t}, & n_{as} + \textstyle\sum_i c_{i,as} &= St + \mathrm{C^t}. \end{aligned} \tag{4.20}$$

- If the system is *supercritical*, the asymptotic solution is qualitatively similar (exponential growth) whether or not there is a source.

4.3.2. Emergency shutdown

Every reactor has a safety system, usually comprising a set of highly absorbent rods used to insert a strong antireactivity very quickly and to halt the chain reaction. For simplification, we assume that this negative reactivity ρ is inserted instantaneously.

If we refer to the slow kinetic model and, in fact, with a time constant ω of the order of $(\rho - \beta)/\ell$ if we refer to the exact inhour equation, the power falls by a factor of $\beta/(\beta - \rho)$. It then continues to fall according to the g exponential modes, and finally according to the slowest one, with a time constant of the order of $-\lambda_1$ (corresponding to the vertical asymptote the farthest to the right on the inhour graph), i.e. approximately a factor of 2 in 55 seconds.

4.3.3. Reactivity window

When we wish to change the power level of the reactor, we briefly insert a (modest) positive reactivity to increase the power, or a negative (and also modest) reactivity to reduce it. In practice, reactivity insertions are never brutal but, to simplify the calculations, it is possible to assume that this takes place via a reactivity "window", i.e. the instantaneous insertion of a reactivity ρ, kept constant during a certain time T, and followed by a return, which is also instantaneous, to zero reactivity.

At the instant of reactivity insertion, the power varies almost instantaneously by a factor of $\beta/(\beta - \rho)$(greater than 1 if the reactivity introduced is positive, and less than 1 if it is negative). It then changes according to the g main exponential modes. When the reactivity vanishes, a sharp change by the inverse factor occurs, followed by a more gradual convergence to the asymptote. Figure 4.2 gives two examples of windows lasting 11.3 seconds (the value chosen here is the average lifetime of the precursors), obtained via exact calculations. The reader is invited to examine the problem using the slow kinetic model.

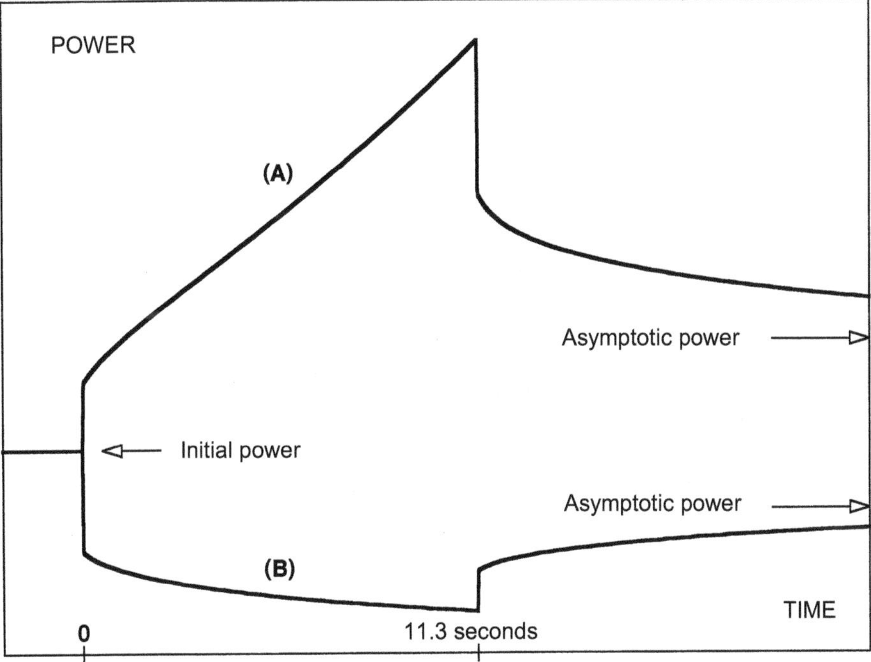

Figure 4.2. Two examples of power variation during reactivity "windows" (fission of uranium 235; six-group delayed neutron theory; A: $\rho = \beta/4$; B: $\rho = -\beta$).

4.3.4. Reactivity ramp

As has been mentioned, reactivity insertions are gradual in practice. For example, they are applied through the raising or lowering of absorbent rods. The reactivity ramp is therefore a better model than the window. The ramp is a linear change over time from a critical state:

$$\rho(t) = \pi t,$$

where the coefficient π (slope) is chosen to be positive if an increase in power is desired, or negative if a decrease is desired.

In this case it is more difficult to solve the problem analytically. To simplify it, we present it here in the framework of the "slow kinetic model with one group of delayed neutrons", whose equations are as follows:

$$\frac{dc}{dt} = \frac{\pi t \lambda c}{\beta - \pi t}, \qquad \alpha = \frac{\lambda c}{\beta - \pi t}. \tag{4.21}$$

The differential equation has separate variables and can therefore be integrated as follows:

$$\ln \frac{c(t)}{c(0)} = -\lambda t + \frac{\lambda \beta}{\pi} \ln \frac{\beta}{\beta - \pi t}. \tag{4.22}$$

We deduce:

$$\frac{c(t)}{c(0)} = e^{-\lambda t} \left(\frac{\beta}{\beta - \pi t} \right)^{\gamma}, \quad \frac{p(t)}{p(0)} = e^{-\lambda t} \left(\frac{\beta}{\beta - \pi t} \right)^{\gamma + 1}, \tag{4.23}$$

with:

$$\gamma = \frac{\lambda \beta}{\pi}. \tag{4.24}$$

(The power p is proportional to the absorption rate α.)

Figure 4.3 shows a few examples of power curves obtained with this formula.

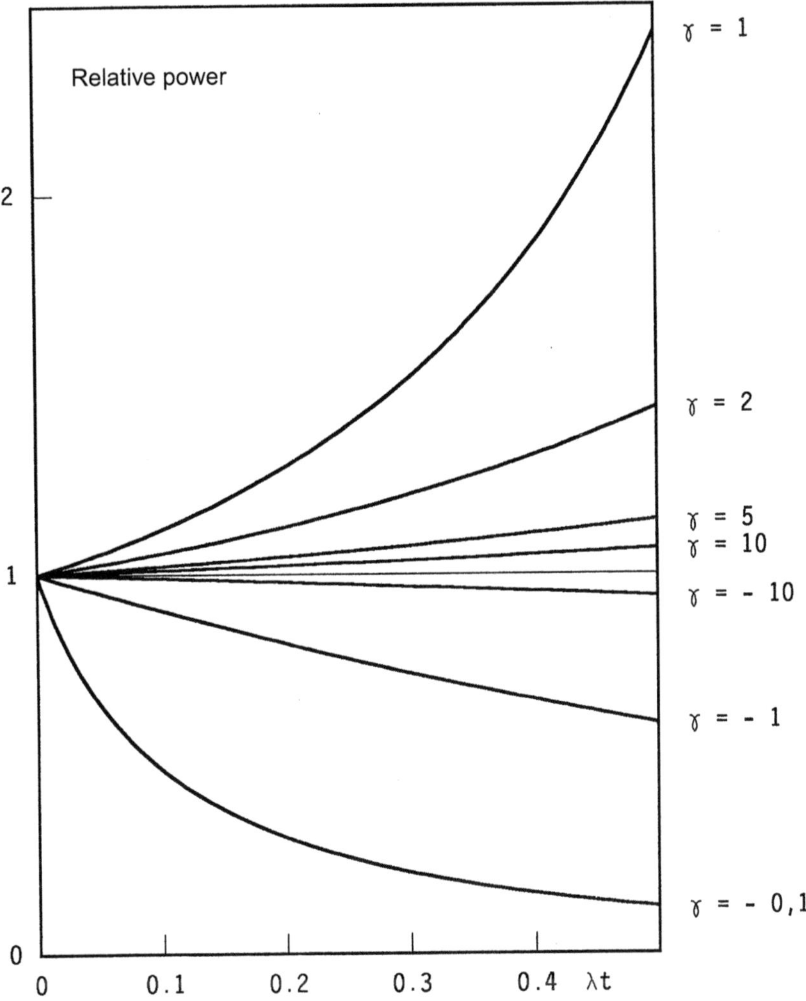

Figure 4.3. A few examples of power variation during reactivity ramps (slow kinetics with one group of delayed neutrons).

Exercises

Exercise 4.1: first order calculation of the power doubling time

Considering only one group of delayed neutrons, compare the exact expression for the time to double or half the power with the approximate one given by a first order expansion around the critical case. Show how the results vary for typical values of the reactivity ρ in the cases of uranium-235 and plutonium-239 reactors where the neutron lifetime is negligible.

Exercise 4.2: cancelling out the precursor concentrations

Write down the second order differential equation that governs the neutron number for a single group of delayed neutrons.

Derive the Nordheim equation for a single group of delayed neutrons from the previous expression, using the assumption that the reactivity is constant.

What form does the Nordheim equation take in the general case of g groups?

Exercise 4.3: reactivity window

The so called 'reactivity window' occurs during a critical equilibrium situation, and consists in the following steps.

- inject instantaneously a positive or negative reactivity ρ at a given time (taken as time zero);
- maintain this reactivity until time T;
- at time T, instantaneously reduce the reactivity to zero and maintain it.

Examine this scenario with the point kinetics model, assuming there is a single group of delayed neutrons, and a negligible neutron lifetime.

Is this theory qualitativly correct?

Exercise 4.4: control rod insertion experiment

Starting from a critical equilibrium situation, this experiment consists of inserting very rapidly a control rod or bundle into the reactor core, then measuring how the neutron density subsequently varies with time. In this exercise, use the point kinetics model with a single group of delayed neutrons that have a negligible neutron lifetime, and assume that the control rod insertion is instantaneous.

a) Determine how the number of neutrons varies with time, $n(t)$.

b) Calculate the integral of $n(t)$ from the instant that the control rod is inserted at time zero to infinity.

c) Show how the measurements of the impulse and of this integral give both the inserted negative reactivity and the mean lifetime of the precursors.

Exercise 4.5: source ejection experiment

In a kind of inverse to the previous situation, the source ejection experiment involves the rapid removal of a source of neutrons from a subcritical reactor instead of using control rods to quench it. How does this affect the answers to questions **a**, **b**, and **c** of the previous exercise?

Exercise 4.6: pulsed neutron experiment

In a subcritical or even non-multiplying system, this experiment consists in injecting a pulse of neutrons, and subsequently measuring the decay constant of the exponential decrease in the neutron number. This decay is rapid; hence, the experiment can be repeated a great number of times in order to improve its statistical precision. To simplify matters, a non-multiplying system is considered here.

a) Assume that the neutron pulse is sufficiently short to be represented by a Dirac distribution $S\,\delta(t)$. Show how the number of neutrons varies with time $n(t)$ following neutron injection. What can be deduced by making pairs of measurements of $n(t)$ over two short time intervals near times t_1 and t_2 after the burst?

b) In order to improve the precision, it is better to repeat the pulses regularly and recurrently with a time interval T and to perform a measurement at a time t after each burst ($t < T$). Compare this procedure with the previous one and describe the consequenses of varying t and T.

Exercise 4.7: oscillation experiments

The previous exercises show that the measurements of dynamic situations can be rich sources of information. In the control rod insertion or source ejection experiments, we only go from one state to a different one. In the pulsed neutron experiments, it is useful to repeat the pulses periodically in order to improve the statistical precision. In oscillation experiments, the idea is similar: periodic behaviour is generated in a part of the system, and subsequently is propagated via the medium throughout the whole system, with changes of phase and level. Measurements of the phase and level correlations can give useful, precise information about the neutron characteristics of the reactor medium. Furthermore, since the period of the oscillation can be varied, better discrimination and a greater precision can be obtained for the measured parameters.

Several variants can be imagined; e.g. various modes of the periodical excitation; under-critical or critical systems; compensation or not of the perturbation; distance between the excitation mechanism and the detector; etc.

For example, the French Atomic Commission (CEA) performed numerous so-called 'oscillation experiments'. In these experiments a sample of fissile or absorbing material is periodically inserted into and extracted from the central channel of the core of a critical reactor. The perturbation is compensated by an automatic mechanism (absorbing control rod at the periphery of the core), whose movement is the *global signal*: its level is approximately proportional to the reactivity effect of the sample insertion. A flux measurement near the sample is a *local signal* which provides further information that is related to the absorption perturbation.

The following exercice represents a highly simplified model for such an experiment. The reactor is treated using the point kinetics model with a single group of delayed neutrons. It is also assumed that the system is in a subcritical state with an external source of neutrons. This source has two components: one is constant, s_0, while the other varies in a sinusoidal manner, $s_1 \exp(i\omega t)$. Notice here that, similar to other problems with oscillatory behaviour, it is convenient to employ complex exponential functions.

a) Write down the equations of this model, assuming that the reactivity does not vary.

b) The source term is

$$s(t) = s_0 + s_1 \exp(i\omega t),$$

where s_0 and s_1 are constants. In order to obtain a positive real result, make s_0 positive and require that $| s_1 | < s_0$.

The solution of the equations then take the form

$$n(t) = n_0 + n_1 \exp(i\omega t),$$
$$c(t) = c_0 + c_1 \exp(i\omega t).$$

Show how the four constants n_0, n_1, c_0, and c_1 are related to one another.

c) The measured oscillating term (flux) as a function of the imposed oscillation term (source) can be characterized by the complex parameter, $z = n_1/(\ell s_1)$, which can be written as $z = x/y$ with $x = \lambda + i\omega$. Find y.

d) What are the moduli and the arguments of the complex numbers x, y, and z?

e) Find approximate expressions for the variations of the modulus and argument of z as a functions of the period $T = 2\pi/\omega$, based on a typical uranium-fuelled PWR that is subcritical by one dollar.

Solutions

Exercise 4.1: first order calculation of the doubling time

If the lifetime is assumed to be zero, then the first order $\omega_{(1)}$ and exact ω_{ex} values of the rate, respectively, are

$$\omega_{(1)} = \frac{\rho}{\beta\tau} \qquad \text{and,} \qquad \omega_{ex} = \frac{\rho}{(\beta - \rho)\tau}.$$

The following table gives some numerical values of $T = \ln 2/ | \omega |$ (in seconds).

ρ (pcm)	Uranium reactor		Plutonium reactor	
	Order 1	Exact	Order 1	Exact
−200	26.6	34.5	8.78	16.6
−150	35.4	43.3	11.7	19.5
−100	53.2	61.1	17.6	25.4
−50	106	114	35.1	43.0
−10	532	540	176	183
+10	532	524	176	168
+50	106	98.6	35.1	27.3
+100	53.2	45.4	17.6	9.72
+150	35.4	27.6	11.7	3.87
+200	26.6	18.8	8.78	0.0940

Exercise 4.2: cancelling out the precursor concentrations

The initial equations are

$$\frac{dn}{dt} = \frac{\rho - \beta}{\ell} n + \lambda c,$$

$$\frac{dc}{dt} = \frac{\beta}{\ell} n - \lambda c.$$

Take the derivative of the first equation, and the sum of the equations:

$$\frac{d^2 n}{dt^2} = \frac{\rho - \beta}{\ell} \frac{dn}{dt} + \lambda \frac{dc}{dt} + \frac{d\rho}{dt} \frac{n}{\ell},$$

$$\frac{dn}{dt} + \frac{dc}{dt} = \frac{\rho}{\ell} n.$$

Cancelling dc/dt by a linear combination, gives

$$\frac{d^2 n}{dt^2} = \frac{\rho - \beta}{\ell} \frac{dn}{dt} - \lambda \frac{dn}{dt} + \frac{d\rho}{dt} \frac{n}{\ell} + \frac{\lambda \rho}{\ell} n.$$

If ρ is independent of the time, then the solution is an exponential of the form $a \exp(\omega t)$. The Nordheim equation is thus obtained for this case where there is a single group of delayed neutrons.

When there are g groups of delayed neutrons, provided that the reactivity is constant, the Nordheim equation takes the form of a polynomial of degree $g + 1$, equal to zero. This polynomial is obtained from the usual form of the Nordheim equation by placing all the terms in the same denominator. Now the equation for n is expected take the form of a rather complicated differential equation of order $g + 1$.

Exercise 4.3: reactivity window

The evolution of the number of precursors is described by a continuous function, described as follows.

- $t < 0$: $c = c_0$;
- $0 < t < T$: $c = c_0\, e^{\omega t}$, $\omega = \beta\lambda/(\beta - \rho)$;
- $t > T$: $c = c_1 = c_0\, e^{\omega T}$.

However, the evolution of the number of neutrons is described by a function discontinuous for $t = 0$ and for $t = T$, because ℓ is assumed to be zero:

- $t < 0$: $n = n_0$;
- $t = 0$: $n = n_0 \Longrightarrow n'_0 = n_0\, \beta/(\beta - \rho)$;
- $0 < t < T$: $n = n'_1 = n'_0\, e^{\omega t}$;
- $t = T$: $n = n'_0\, e^{\omega T} \Longrightarrow n_1 = n_0\, e^{\omega T}$;
- $t > T$: $n = n_1 = n_0\, e^{\omega T}$.

If the neutron lifetime is assumed to be zero, then this means that finite jumps are replaced by instantaneous ones. The consequence of assuming that there is only a single group of precursors, is it eliminates a transient lasting a few tens of second after these jumps, due to the changes in the *relative* proportions of the precursors. The transient following the second jump can be seen easily in figure 4.2, paragraph 4.3.3.

Exercise 4.4: control rod insertion experiment

Upon insertion of the control rod into the reactor's core, the number of neutrons jumps from its initial value n_0 to $n_1 = n_0\beta/(\beta - \rho)$. Following this jump, exponential decay occurs:

$$n = n_1\, e^{\omega t}, \qquad \omega = \frac{\beta\lambda}{\beta - \rho}.$$

Notice here that ρ is *negative*; therefore, the integrated intensity is

$$I = \int_0^\infty n(t)\, dt = \frac{\beta}{-\rho\lambda} n_0.$$

Thus, the reactivity in dollars $\$ = \rho/\beta$, and the decay constant of the precursors λ is given by

$$\frac{n_1}{n_0} = \frac{1}{1 - \$}; \qquad \frac{I}{n_0} = \frac{1}{-\$\lambda}.$$

Exercise 4.5: source ejection experiment

The final formulae, and the conclusions, are identical to those in the previous exercise.

Exercise 4.6: pulsed neutron experiment

a) Now $n(t) = a\exp(-t/\ell)$ where a is a constant proportional to the source. The ratio between two measurements, $n_2/n_1 = \exp\left[-(t_2 - t_1)/\ell\right]$, gives the lifetime ℓ.

b) Once the equilibrium is reached, the number of neutrons observed is

$$N = n(t) + n(t+T) + n(t+2T) + n(t+3T) + n(t+4T) + \cdots;$$

$$N = a\,e^{-t/\ell}\left[1 + e^{-T/\ell} + e^{-2T/\ell} + e^{-3T/\ell} + e^{-4T/\ell} + \cdots\right];$$

$$N = a\,\frac{e^{-t/\ell}}{1 - e^{-T/\ell}}.$$

By varying the periods t and T, the statistical precision can be improved beyond that which can be achieved by simply making repeated pulses, and ℓ can be estimated.

Exercise 4.7: oscillation experiments

a) The equations for the chosen model are

$$\frac{dn}{dt} = \frac{\rho - \beta}{\ell}n + \lambda c + s,$$

and

$$\frac{dc}{dt} = \frac{\beta}{\ell}n - \lambda c.$$

b) For the constant terms, the usual relationships are obtained:

$$n_0 = \frac{\ell s_0}{-\rho}, \qquad c_0 = \frac{\lambda\beta s_0}{-\rho}.$$

The oscillating terms are

$$i\omega n_1 = \frac{\rho - \beta}{\ell}n_1 + \lambda c_1 + s_1,$$

and

$$i\omega c_1 = \frac{\beta}{\ell}n_1 - \lambda c_1.$$

c) Therefore,

$$z = \frac{n_1}{\ell s_1} = \frac{1}{-\rho + \frac{i\omega\beta}{\lambda + i\omega} + i\ell\omega} = \frac{x}{y},$$

with

$$x = \lambda + i\omega, \qquad \text{and} \qquad y = -\rho\lambda - \ell\omega^2 + (\beta - \rho + \ell\lambda)\,i\omega.$$

In practice, the term $\ell\lambda$ can be neglected; however, the term $\ell\omega^2$ may be not negligible if the frequency of the oscillation is rapid.

d) Using,

$$x = \xi e^{ia}, \qquad y = \eta e^{ib}, \qquad z = \zeta e^{ic},$$

with,

$$\zeta = \frac{\xi}{\eta}, \qquad c = a - b,$$

gives (neglecting $\ell\lambda$)

$$\xi = \sqrt{\lambda^2 + \omega^2}; \qquad \zeta = \sqrt{(-\rho\lambda - \ell\omega^2)^2 + (\beta - \rho)^2\,\omega^2};$$

$$a = \arctan\left(\frac{\omega}{\lambda}\right); \qquad b = \arctan\left[\frac{(\beta - \rho)\,\omega}{-\rho\lambda - \ell\omega^2}\right].$$

e) For a typical uranium-fuelled PWR that is subcritical by one dollar, the following approximations are applicable.

- If $\omega \ll \lambda$, i.e. $T \gg 80$ s, then

$$\zeta \simeq \frac{1}{-\rho}; \qquad c \simeq 0.$$

  - If $\lambda \ll \omega \ll \beta/\ell$, i.e. $0.02 \text{ s} \ll T \ll 80$ s, then

$$\zeta \simeq \frac{1}{\beta - \rho}; \qquad c \simeq 0.$$

  - If $\omega \ll \beta/\ell$, i.e. $T \ll 0.02$ s, then

$$\zeta \simeq \frac{1}{\ell\omega}; \qquad c \simeq \frac{\pi}{2}.$$

By varying the frequency and measuring ζ (and possibly c), it is possible to estimate the reactivity in dollars (from the two first 'plateaus'), the decay constant of the precursors (from the transition between them), and the prompt neutron lifetime (from the behaviour at high frequencies).

5 Diffusion equation

Introduction

Because the Boltzmann equation is so complex, there is not much hope of using it to find analytical solutions. The only way to make any progress in that direction would be to simplify the operators in the equation. Such simplifications are often necessary even if the calculations are to be performed numerically using a computer.

The diffusion approximation is a simplification very often used for the transport operator. It is useful because it eliminates the need to take the "phase" variable $\vec{\Omega}$ (neutron direction) into account and allows us to express particle transport using a simple, well-known mathematical operator: the Laplace operator. This approximation obviously has its limitations, however.

In this chapter, our primary concern is to study *migration* in space, and so we shall simplify matters by restricting ourselves to a monokinetic theory. This simplification is not essential, however, as we shall see in Chapter 10.

In the present chapter, we present this diffusion approximation and examine, within this context, the calculation of flux when the sources are assumed to be known. In practice, sources are often fission sources which are therefore expressed in terms of the unknown flux. The problem arising in that case will be discussed in the next chapter.

This chapter has two parts: 1/ Establishing the diffusion approximation and a discussion about the assumptions, 2/ A few problems are studied as examples.

5.1. Establishing the diffusion equation

5.1.1. Neutron balance

Let us consider any domain in space, D, in an effort to determine the number of neutrons it contains. By definition of the density n, at an instant t this number is:

$$X(t) = \int_D n(\vec{r}, t)d^3r. \tag{5.1}$$

As usual, we would use the flux $\Phi = nv$ rather than density, giving:

$$X(t) = \int_D \frac{1}{v}\Phi(\vec{r}, t)d^3r. \tag{5.2}$$

(Here, the neutrons are assumed to be monokinetic, and so v is not an argument; because we do not distinguish the directions $\vec{\Omega}$ of the neutrons, we use a density and a flux that are integrated in this variable.)

We wish to express the variation of this number between t and $t + dt$

$$dX = X(t + dt) - X(t) = \int_D \frac{1}{v} \frac{\partial \Phi(\vec{r}, t)}{\partial t} d^3r\, dt, \tag{5.3}$$

by analysing the various phenomena likely to affect it.

Three phenomena must be identified:

1/ ***Sources:*** during dt, they contribute to making X *increase* by:

$$d_{(1)}X = \int_D S(\vec{r}, t) d^3r\, dt; \tag{5.4}$$

2/ ***Absorptions:*** during dt, they contribute to making X *decrease* by:

$$d_{(2)}X = \int_D \Sigma_a(\vec{r})\Phi(\vec{r}, t) d^3r\, dt; \tag{5.5}$$

(*Note:* Scattering is not considered here because it does not change the speed [monokinetic theory] and, even if it changes the direction, it does not affect the number X.)

3/ ***Entrances and exits:*** Entrances contribute to *increasing* X and exits contribute to *decreasing* it. As we have seen (§ 3.1.4), the vector $\vec{J}(\vec{\Omega})$ (variables $\vec{r}$ and t being understood) expresses the *net* balance of crossings of a surface element (a positive value indicates the number of crossings in the normal direction, and a negative value indicates the number of crossings in the opposite direction) for neutrons whose direction is $\vec{\Omega}$. If we consider the vector $\vec{J}$ (variables $\vec{r}$ and t understood) — i.e. the current vector

$$\vec{J} = \int \vec{J}(\vec{\Omega}) d^2\Omega \tag{5.6}$$

integrated over the phases — we can likewise express the net number of crossings in all directions. By orienting the normal towards the outside of the domain D and integrating over its entire surface S, we can express the third variation of X (decreasing if positive, increasing if negative):

$$d_{(3)}X = \int_D \vec{J}(\vec{r}, t) \cdot \vec{N} dS\, dt. \tag{5.7}$$

To write this integral in an analogous form to the others, i.e. as a volume integral, we apply the *divergence theorem:*

$$d_{(3)}X = \int_D \operatorname{div} \left[\vec{J}(\vec{r}, t)\right] d^3r\, dt. \tag{5.8}$$

Finally:

$$dX = d_{(1)}X - d_{(2)}X - d_{(3)}X, \tag{5.9}$$

expresses the balance in D during dt. Because D is any ordinary domain, it can be reduced to the volume element d^3r (i.e. the "summation" signs can be eliminated) and then simplified by $d^3r\,dt$:

$$\frac{1}{v}\frac{\partial\Phi(\vec{r},t)}{\partial t} = S(\vec{r},t) - \Sigma_a(\vec{r})\Phi(\vec{r},t) - \mathrm{div}\left[\vec{J}(\vec{r},t)\right]. \tag{5.10}$$

Note that, although the variable $\vec{\Omega}$ does not appear in it, this equation is absolutely rigorous (in monokinetic theory). Unfortunately, it alone is not sufficient because it contains not one, but two unknown functions: flux Φ and current $\vec{J}$. That is why another relationship is required, to link $\vec{J}$ with Φ, which will only be approximate because we do not wish to calculate the phase flux.

5.1.2. Evaluating the current: Fick's law

The second relationship to be adopted is known as "Fick's Law"[1].

Fick's Law is used in chemistry, and expresses the fact that a flux (in the usual sense of the word) of matter in a solution is established in the direction opposite to the concentration gradient and proportionally to the modulus of the gradient[2]. This is transposed into neutron physics by replacing the term "flux" with "current" and by replacing the "concentration" n of neutrons by the parameter that is proportional to it, Φ; as in chemistry, the coefficient of proportionality is called "diffusion coefficient" and written D, i.e.:

$$\vec{J}(\vec{r},t) \simeq -D(\vec{r})\,\overrightarrow{\mathrm{grad}}\,\Phi(\vec{r},t). \tag{5.11}$$

So far we have applied intuitive arguments to this law, but it can also be justified by more specific means. Firstly, we present a physical justification (some possible mathematical justifications will be presented subsequently).

Let us assume that the situation changes little — or not at all — over time, which allows us to apply steady-state reasoning.

We place a surface element dS, oriented by its normal vector $\vec{N}$ (see Figure 5.1).

The neutrons passing through it will be all those that have left a volume element dV (after being emitted or scattered), leaving in the solid angle element under which dS is seen from dV, that have travelled the path without any collisions. Reasoning per unit time:

- neutrons leaving dV: $Q(\vec{r})dV$ (where Q is the emission density);
- solid angle under which dS is seen, normalised to the total of 4π steradians: $dS\cos\theta/(4\pi R^2)$;
- probability of travelling the path without collision: $e^{-\Sigma R}$.

1 The term "law" should be understood here in its physical sense, as in "Fourier's Law" for heat, for example. A physical law is always more or less approximate.

2 In other words, these fluxes tend to make their concentrations more uniform; thus the sugar initially accumulated at the bottom of a cup of coffee ends up being evenly distributed even if you do not stir your coffee.

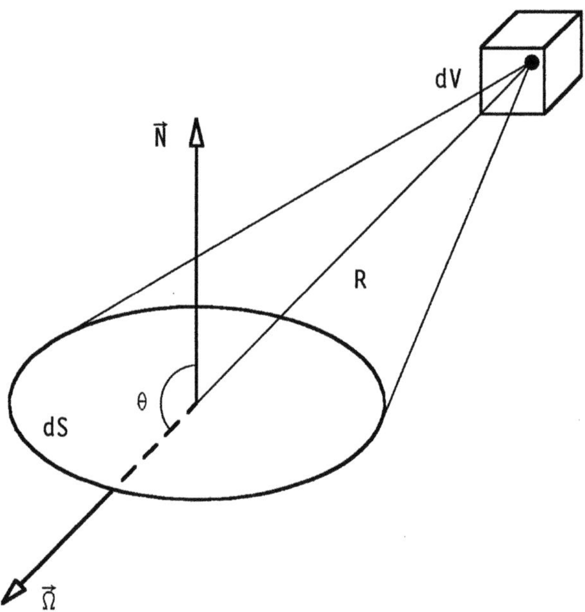

Figure 5.1. Current evaluation.

By summing for all volume elements *above dS*, we evaluate $J_- dS$ (§ 3.1.4) and by summing for all volume elements *below dS*, we evaluate $J_+ dS$:

$$J_+ dS = \int_{(0<\theta<\pi/2)} Q(\vec{r}\,')dV \frac{dS\cos\theta}{4\pi R^2} e^{-\Sigma R}, \tag{5.12}$$

$$J_- dS = \int_{(\pi/2<\theta<\pi)} Q(\vec{r}\,')dV \frac{dS\cos\theta}{4\pi R^2} e^{-\Sigma R}. \tag{5.13}$$

To calculate the integrals we make three approximations:

1/ The medium is homogeneous in the neighbourhood of dS; in other words, Σ is a constant. Note that "in the neighbourhood" means "within a few mean free paths", because the exponential becomes negligible beyond that distance;

2/ In steady state ($\partial\Phi/\partial t = 0$), the balance equation is reduced to $S = \Sigma_a\Phi + \mathrm{div}\vec{J}$, i.e. $S \simeq \Sigma_a\Phi$ if we neglect the second term, i.e. if the variations in flux are not too fast[3]. We deduce that:

$$Q = S + \Sigma_s\Phi \simeq \Sigma_a\Phi + \Sigma_s\Phi \simeq \Sigma\Phi;$$

3/ Maintaining this assumption of a small variation of flux and therefore of Q, a first-order Taylor expansion should suffice:

$$Q(\vec{r}\,') \simeq Q(\vec{r}) + x\frac{\partial Q(\vec{r})}{\partial x} + y\frac{\partial Q(\vec{r})}{\partial y} + z\frac{\partial Q(\vec{r})}{\partial z}, \tag{5.14}$$

[3] Fick's Law, which we may temporarily adopt, indicates that the current is weak if the flux gradient is low.

where $\vec{r}$ is the point where dS is placed, $\vec{r}'$ is the point where dV is placed, and $\vec{R}$, with components x, y and z, is the vector connecting $\vec{r}$ to $\vec{r}'$. In the second term, the function Q and its derivatives are in terms of $\vec{r}$.

By performing the calculation in spherical coordinates, i.e. by setting:

$$x = R\cos\varphi\sin\theta,\quad y = R\sin\varphi\sin\theta,\quad z = R\cos\theta,\quad dV = R^2 dR\sin\theta\, d\theta\, d\varphi,$$

and replacing Q by $\Sigma\Phi$, we obtain:

$$J_+ = \frac{\Phi}{4} - \frac{1}{6\Sigma}\frac{\partial\Phi}{\partial z},\quad J_- = \frac{\Phi}{4} + \frac{1}{6\Sigma}\frac{\partial\Phi}{\partial z},\tag{5.15}$$

from which we obtain:

$$J_{\text{net}} = J_+ - J_- = -\frac{1}{3\Sigma}\frac{\partial\Phi}{\partial z}.\tag{5.16}$$

These equations can be re-written by setting:

$$D = \frac{1}{3\Sigma}.\tag{5.17}$$

Note that the currents we have calculated are relative to the z component of the current vector; for any component N:

$$J_{+,N} = \frac{\Phi}{4} - \frac{D}{2}\frac{\partial\Phi}{\partial N},\quad J_{-,N} = \frac{\Phi}{4} + \frac{D}{2}\frac{\partial\Phi}{\partial N},\tag{5.18}$$

and for the current vector itself:

$$\vec{J} = -D\,\overrightarrow{\text{grad}}\,\Phi,\tag{5.19}$$

i.e. the law introduced above.

5.1.3. Diffusion equation

By substituting Fick's Law into the balance equation, we obtain the so-called diffusion equation:

$$\frac{1}{v}\frac{\partial\Phi(\vec{r},t)}{\partial t} = S(\vec{r},t) - \Sigma_a(\vec{r})\Phi(\vec{r},t) + \text{div}\left[D(\vec{r})\,\overrightarrow{\text{grad}}\,\Phi(\vec{r},t)\right].\tag{5.20}$$

In practice, the medium in which this equation operates is homogeneous: Σ_a and D are therefore independent of $\vec{r}$, and this is simplified by the observation that $\text{div}(\overrightarrow{\text{grad}})$ is the Laplace operator Δ, i.e.:

- in Cartesian coordinates:

$$\Delta = \frac{\partial^2}{\partial x^2} + \frac{\partial^2}{\partial y^2} + \frac{\partial^2}{\partial z^2};\tag{5.21}$$

- in cylindrical coordinates:

$$\Delta = \frac{\partial^2}{\partial\rho^2} + \frac{1}{\rho}\frac{\partial}{\partial\rho} + \frac{1}{\rho^2}\frac{\partial^2}{\partial\varphi^2} + \frac{\partial^2}{\partial z^2};\tag{5.22}$$

– in spherical coordinates[4]:

$$\Delta = \frac{\partial^2}{\partial r^2} + \frac{2}{r}\frac{\partial}{\partial r} + \frac{1}{r^2}E, \qquad (5.23)$$

with:

$$E = \frac{1}{\sin\theta}\frac{\partial}{\partial\theta}\left(\sin\theta\frac{\partial}{\partial\theta}\right) + \frac{1}{\sin^2\theta}\frac{\partial^2}{\partial\varphi^2}. \qquad (5.24)$$

5.1.4. Initial condition, boundary conditions, interface conditions

In addition to the equation itself and the geometric and physical description of the media, we require the following information to demonstrate that this diffusion equation has one and only one solution:

1/ for a time-dependent situation: the *initial condition*, i.e. the spatial distribution of the neutrons at the instant of origin, $\Phi(\vec{r}, 0)$;

2/ in every case: *boundary conditions*.

In practice, the boundary conditions consist of a relationship to be satisfied involving the flux and/or its derivative with respect to the normal vector.

In general, a reactor consists of one or more homogeneous regions. We must then consider the diffusion equation in each of these regions, with constant values of D and Σ_a, and express the continuity of flux and current at the interface between two regions, 1 and 2:

$$\Phi_1 = \Phi_2, \quad D_1\frac{\partial\Phi_1}{\partial N} = D_2\frac{\partial\Phi_2}{\partial N}. \qquad (5.25)$$

Because the neutrons do not "know" that they are crossing an interface, the phase density must be continuous. The same applies to the flux (integrated) and the current (integrated), which are integrals over $\vec{\Omega}$ of this phase density[5].

5.1.5. External boundary: black body extrapolation distance; extrapolated surface

In general, it can be considered that neutrons leaving the reactor are lost because they are located beyond the absorbent material where they disappear (steel, concrete, etc.). The following boundary condition expresses this:

$$J_- = 0,$$

(neutrons can escape, but none can return). Such materials that capture all neutrons are known as "black bodies". Note that a vacuum is also a black body, because in this case also, neutrons that go out will not come back!

---

[4] Here, the angles θ and φ denote the angular components of the space variable, not the velocity of the neutron.

[5] If the diffusion coefficients are different, the flux curve is continuous but has an "angle" at the interface. This is an artefact caused by the diffusion approximation.

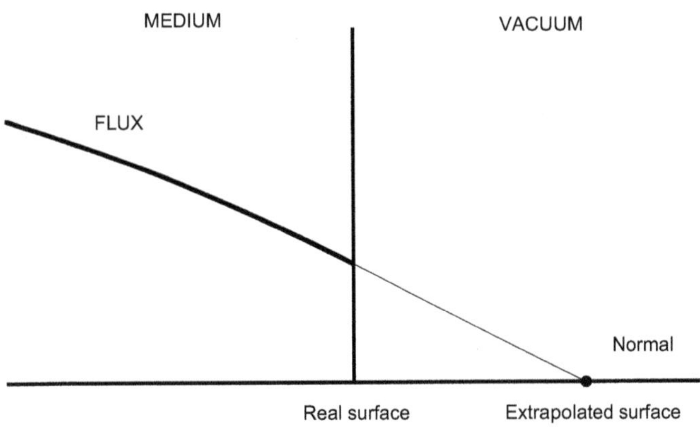

Figure 5.2. Extrapolation distance.

In the context of the diffusion approximation, this boundary condition is expressed as:

$$J_{-} = \frac{\Phi}{4} + \frac{D}{2}\frac{\partial \Phi}{\partial N} = 0, \tag{5.26}$$

i.e. by a value imposed on the *logarithmic derivative* of the flux:

$$\frac{\partial \Phi / \partial N}{\Phi} = -\frac{1}{2D} = -\frac{3\Sigma}{2} = -\frac{3}{2\lambda} = -\frac{1}{d}. \tag{5.27}$$

A basic geometric calculation (Figure 5.2) shows that this is equivalent to considering that the tangent to the flux curve at the surface cuts the axis beyond the surface at a distance d equal to two-thirds of the mean free path of the neutrons.

This *extrapolation distance* is of the order of a centimetre, i.e. small compared to the size of a reactor. Replacing the flux by its tangent over this distance, which means replacing the actual boundary condition with the zero-flux condition at this distance, i.e. on the *surface extrapolated* towards the outside of d with respect to the actual surface, only creates a very small error.

This is what is generally done, because it is simpler to explain the condition $\Phi = 0$ at the extrapolated boundary than the logarithmic derivative condition at the real surface. Because it does not cause additional complications, we can also replace the coefficient 2/3 in the expression for d with a more accurate coefficient obtained by performing an exact calculation of the black body problem[6], without using the diffusion approximation:

$$d = 0.7104\lambda. \tag{5.28}$$

[6] This problem is called the "*Milne problem*".

5.1.6. Approach based on the integral equation

As an exercise, here is another way of obtaining the diffusion equation (in steady state, for the sake of simplicity). The reader is invited to perform the detailed calculations:

- starting with the isotropic diffusion integral Boltzmann equation (see § 3.2.3):

$$\Phi(\vec{r}) = \int \frac{e^{-\Sigma R}}{4\pi R^2} Q(\vec{r}\,')d^3r'; \tag{5.29}$$

- assume that the material is homogeneous around $\vec{r}$ in the region contributing to the integral (constant Σ);
- to calculate the integral, replace $Q(\vec{r}\,')$ by its second-order Taylor expansion around $\vec{r}$;
- to calculate the Laplace operator that then appears, we apply the approximation for $Q \simeq \Sigma\Phi$, i.e. $\Delta Q \simeq \Sigma\Delta\Phi$.

5.1.7. Conditions for validity of the diffusion approximation

The "demonstration" of Fick's law presented above shows that the diffusion approximation is valid provided that *the variations (in space and time) are slow*. In concrete terms, this holds true when:

- there is little geometric heterogeneity,
- the absorption cross-section is small compared to the scattering cross-section,
- the position is not too close to the interfaces (at least a few mean free paths away),
- the position is not too close to the "concentrated" sources.

5.1.8. Transport correction

Much of the error due to the diffusion approximation comes from the assumption of isotropic scattering. (This assumption is necessary to reach the simple form of the integral transport equation used above; it was also made to establish Fick's Law when we assumed that a neutron leaving dV had probability $d^2\Omega/4\pi$ of hitting dS, where $d^2\Omega$ was the solid angle under which dS is seen from the point of emission.) The transport correction, established from the differential form of the transport operator, can compensate for most of this error. Because it is simple to introduce, it is always used in practice. It consists of replacing the total cross-section Σ with the so-called "transport" cross-section in the expression for the diffusion coefficient.

$$\Sigma_{tr} = \Sigma - \bar{\mu}\Sigma_s, \tag{5.30}$$

where $\bar{\mu}$ is the average cosine of the neutron deviation angle in a collision: this is a nuclear parameter that is simply added to the others. (In Chapter 7 we shall see that a good approximation for a nucleus A times heavier than the neutron is $\bar{\mu} = 2/(3A)$.) The following formulae are therefore to be used:

$$D = 1/3\Sigma_{tr}, \quad d = 0.7104/\Sigma_{tr}. \tag{5.31}$$

5.2. Example problems

5.2.1. Kernels of the diffusion equation in a homogeneous, infinite medium

The "kernels" of an equation (the diffusion equation in this case) are the elementary solutions obtained for the simplest media and the simplest "second terms" (the sources in this case).

Here we shall examine the simplest medium seen in neutron physics problems: a boundaryless (infinite), homogeneous material. We shall examine three elementary sources that schematically represent real sources: a point source emitting S neutrons per unit time; the filiform source (along a line) emitting S neutrons per unit length and time, and the planar source emitting S neutrons per unit surface and time.

1/ **"Point" source:** We place this source at the origin of the coordinate system and use spherical coordinates; for symmetry, the angular variables have no effect and the equation to be solved is reduced to:

$$D\left[\frac{d^2\Phi}{dr^2}+\frac{2}{r}\frac{d\Phi}{dr}\right]-\Sigma_a\Phi+S=0. \tag{5.32}$$

Outside of the origin, the source is nil. By seeking a solution of the form $\Phi = f/r$, we can see that the general solution is expressed with two constants A and B:

$$\Phi(r)=A\frac{e^{-\kappa r}}{r}+B\frac{e^{+\kappa r}}{r}, \tag{5.33}$$

with:

$$\kappa^2=\frac{\Sigma_a}{D}. \tag{5.34}$$

The constant B must be zero because the flux cannot increase indefinitely with distance from the source. The constant A remains to be determined.

- *First method:* The net number $4\pi\varepsilon^2 J(\varepsilon)$ of neutrons leaving a sphere of radius ε must tend towards the intensity S of the source if ε tends to zero. By performing the calculation with:

$$\Phi(r)=A\frac{e^{-\kappa r}}{r},\quad J(r)=-D\frac{d\Phi(r)}{dr}, \tag{5.35}$$

we find $A = S/4\pi D$, i.e.:

$$\Phi(r)=S\frac{e^{-\kappa r}}{4\pi D r}. \tag{5.36}$$

- *Second method:* The source can be represented mathematically by $S\delta(\vec{r})$ where δ is the three-dimensional Dirac distribution; near the origin, the flux is equivalent to A/r, because the exponential is approximately equal to 1. We can use the following formula:

$$\delta(\vec{r})=-\frac{1}{4\pi}\Delta\frac{1}{r}. \tag{5.37}$$

By balancing the coefficients of the two Dirac distributions that thus appear in the equation, we see that A must be equal to $S/4\pi D$.

2/ **"Line" source:** we place this source along the z axis and use cylindrical coordinates. Only the variable ρ is involved, and the equation to be solved is reduced to:

$$D\left[\frac{d^2\Phi}{d\rho^2}+\frac{1}{\rho}\frac{d\Phi}{d\rho}\right]-\Sigma_a\Phi+S=0. \tag{5.38}$$

Outside of the *z* axis, the source is zero. The general solution is expressed with two constants *A* and *B*:

$$\Phi(\rho)=AK_0(\kappa\rho)+BI_0(\kappa\rho), \tag{5.39}$$

where K_0 and I_0 are two Bessel functions (see appendices).

The constant *B* must be zero, because the function I_0 increases exponentially, and the flux cannot increase indefinitely with increasing distance from the source. The constant *A* can be determined by a similar method to that used for the "point" source.

- *First method:* Analogous to the previous one: considering a cylinder of radius ε and unit height, we find $A = S/2\pi D$, i.e.:

$$\Phi(\rho)=S\frac{K_0(\kappa\rho)}{2\pi D}. \tag{5.40}$$

- *Second method:* The source can be represented mathematically by $S\delta(\vec{\rho})$ where δ is the two-dimensional Dirac distribution bearing in mind that, near the origin, the function $K_0(u)$ is equivalent to $-\ln u$ and that we have the formula:

$$\delta(\vec{\rho})=\frac{1}{2\pi}\Delta\ln\rho, \tag{5.41}$$

the value of *A* is found by balancing the coefficients of the two Dirac distributions appearing in the equation.

3/ **"Plane" source:** We place this source in the plane $x = 0$ and use Cartesian coordinates. The variable *x* is involved, and the equation to be solved is reduced to:

$$D\frac{d^2\Phi}{dx^2}-\Sigma_a\Phi+S=0. \tag{5.42}$$

Outside of the plane of origin, the source is zero. Let *x* be strictly positive. The general solution is expressed with two constants *A* and *B*:

$$\Phi(x)=Ae^{-\kappa x}+Be^{+\kappa x}. \tag{5.43}$$

The constant *B* must be zero because the flux cannot increase indefinitely with increasing distance from the source. For strictly negative *x*, the flux is the symmetric function, which can be dealt with by introducing the absolute value of *x*:

$$\Phi(r)=Ae^{-\kappa|x|}. \tag{5.44}$$

The constant *A* can be determined by a method similar to the one already applied.

- *First method:* Consider two planes that are infinite neighbours of the source plane at $x = -0$ and $x = +0$; because half the neutrons are emitted to the right and the other half to the left, we have:

$$J_+(+0) = J_+(-0) + \frac{S}{2}, \quad J_-(-0) = J_-(+0) + \frac{S}{2},$$

which gives:

$$J(+0) = J(-0) + S. \tag{5.45}$$

Using Fick's Law, we obtain $A = S/2\kappa D$ and:

$$\Phi(r) = S\frac{e^{-\kappa|x|}}{2\kappa D}. \tag{5.46}$$

- *Second method:* the source can be represented mathematically by $S\delta(x)$ where δ is the one-dimensional Dirac distribution. Using the following formula:

$$\delta(x) = \frac{1}{2}\Delta|x|, \tag{5.47}$$

we obtain the value of A by balancing the coefficients of the two Dirac distributions that appear in the equation.

Mean squares of crow-fly distances

After multiplication by Σ_a, these kernels represent the absorption density, for example per unit volume for the first; it would be possible to verify that the integral of $\Sigma_a\Phi$ respectively over r (from zero to infinity), over ρ (from zero to infinity) and over x (from $-\infty$ to $+\infty$) is equal to S, and that there are therefore as many neutrons absorbed as emitted per unit time. By weighting r^2 — respectively ρ^2 and x^2 — with this absorption density, we obtain the mean of the square of the crow-fly distance travelled by a neutron from its emission to its absorption — respectively from the projection of the path on the x-y plane and on the $\vec{x}$ axis — we thus obtain (by setting $S = 1$ to normalise the absorption density):

$$\begin{aligned}
\langle r^2\rangle &= \int_0^\infty r^2\Sigma_a\frac{e^{-\kappa r}}{4\pi Dr}4\pi r^2\,dr = 6L^2,\\
\langle \rho^2\rangle &= \int_0^\infty \rho^2\Sigma_a\frac{K_0(\kappa\rho)}{2\pi D}2\pi\rho\,d\rho = 4L^2,\\
\langle x^2\rangle &= \int_0^\infty x^2\Sigma_a\frac{e^{-\kappa x}}{2\kappa D}2dx = 2L^2,
\end{aligned} \tag{5.48}$$

with $L^2 = 1/\kappa^2 = D/\Sigma_a$; this parameter has the dimensions of a surface, and is called a "diffusion area". Note that the second formula, followed by the first, can be deduced from the third by applying the Pythagoras theorem.

5.2.2. Generalisation: the Green function

A filiform source can be considered as a set of point sources located on an axis, and a planar source can be considered in a similar way. Because neutron physics is linear, the flux from a set of sources can be calculated by adding the elementary fluxes from each of the sources. This will give the "line" and "plane" kernels if the appropriate integrals of the "point" kernel are calculated.

This can be generalised to the case of any heterogeneous and/or finite reactor. Any source $S(\vec{r})$ can be considered as an infinite set of point sources: in the volume element d^3r' there is an elementary source $S(\vec{r}\,')\,d^3r'$ that can be likened to a point source. If the flux $g(\vec{r},\vec{r}\,')$ at point $\vec{r}$ due to a unit point source placed at point $\vec{r}\,'$ could be calculated, the flux due to the set $S(\vec{r}\,')$ of sources can be deduced from this by the following integral:

$$\Phi(\vec{r}) = \int g(\vec{r},\vec{r}\,')S(\vec{r}\,')d^3r'. \tag{5.49}$$

The elementary flux $g(\vec{r},\vec{r}\,')$ is the *Green function* of the diffusion equation[7].

In monokinetic theory, neutron physics is not changed by inverting the direction of time flow. The Green function is therefore *symmetric*:

$$g(\vec{r},\vec{r}\,') = g(\vec{r}\,',\vec{r}), \tag{5.50}$$

This means that the same result can be achieved by placing a source emitting one neutron per unit time at a point $\vec{r}\,'$ and measuring the flux at a point $\vec{r}$, and placing a source emitting one neutron per unit time at a point $\vec{r}$ and measuring the flux at a point $\vec{r}\,'$.

This result is obviously incorrect in spectrum theory: in practice, neutrons slow down in reactors. If the direction of time were reversed, they would accelerate instead!

5.2.3. The "albedo" concept

A layer of reflective material is generally placed around the core of a reactor. The purpose of this material is to send back to the core any neutrons that are leaving it and that would otherwise be lost. The multiplication factor is thus enhanced. The materials used as reflectors are often the same as those chosen as moderators, because the property of a good reflector is to scatter neutrons until they eventually return to the core, without capturing too many of them.

A reflector is characterised by its coefficient of reflection, or albedo, defined as the proportion of neutrons leaving the core that are sent back towards the core, i.e.:

$$\beta = \frac{J_-}{J_+}, \tag{5.51}$$

a formula in which the currents are taken at the core/reflector interface with the normal vector oriented from the core towards the reflector. (In problems where the currents depend upon the point on the interface, these currents are taken either locally or as an average.)

[7] The concept of the Green function is general, and applies to any *linear* equation connecting an unknown function Φ to a given function S.

5.2.4. Calculating the albedo of a plate

Note that the albedo can be expressed with the flux of the core region (at its limit) or with the flux of the reflector zone (also at its limit) because of the continuity between the two currents. In practice, the albedo is *calculated* by taking position in the reflector, and is *used* to calculate the flux of the core.

As an example, let us calculate the albedo of an infinite plate along y and z and contained between the planes $x = 0$ and $x = e$. We assume that the neutrons enter uniformly via the face $x = 0$, and therefore that the flux is dependent on x only. There are no sources within the reflector. The equation to be solved is:

$$D\frac{d^2\Phi}{dx^2} - \Sigma_a\Phi = 0. \tag{5.52}$$

The general solution is expressed with two constants A and B:

$$\Phi(x) = Ae^{-\kappa x} + Be^{+\kappa x}.$$

If we assume that the thickness e includes the extrapolation distance beyond the reflector, the condition on this limit is the cancellation of the flux. To take this into account in this example, it is best to write the general solution in an equivalent form using the hyperbolic functions instead of the exponential ones:

$$\Phi(x) = A\sinh[\kappa(e - x)] + B\cosh[\kappa(e - x)],$$

because the condition at $x = e$ immediately leads to $B = 0$.

The constant A can be calculated given the incoming current at $x = 0$. Here, this calculation is not necessary because we wish to obtain a ratio. Using this flux expression and the formulae giving the partial currents (§ 5.1.2), we find:

$$\beta = \frac{1-\gamma}{1+\gamma}, \quad \text{with:} \quad \gamma = 2\kappa D \coth \kappa e. \tag{5.53}$$

(The parameter γ is minus twice the current/flux ratio at the interface.)

The albedo improves as the thickness of the reflector increases, which is natural because there will be less leakage to the outside. An asymptote is reached, however, from a thickness of 2 or 3 diffusion lengths $L = 1/\kappa$. Table 5.1 gives some orders of magnitude for this asymptotic value of the albedo:

$$\beta_{as} = \frac{1 - 2\kappa D}{1 + 2\kappa D}. \tag{5.54}$$

5.2.5. Use of the albedo as boundary condition

We now place ourselves on the core side and assume that the reflector albedo is known. We can use this information to express the boundary condition:

$$\frac{J_-}{J_+} = \beta, \tag{5.55}$$

where the currents are to be expressed with the flux in the core. In practice, it is easier to explain this condition in a form involving the logarithmic derivative of the flux:

$$2D\frac{d\Phi/dN}{\Phi} = -\gamma, \quad \text{with:} \quad \gamma = \frac{1-\beta}{1+\beta}. \tag{5.56}$$

We shall see an example of the application of this technique in the next chapter (reflector economy).

Table 5.1. Albedos for the thermal neutrons of a few common materials (large thickness).

Material	D (cm)	L (cm)	β
Graphite	0.8	55	0.94
Beryllium	0.5	21	0.91
Beryllia	0.6	30	0.92
Heavy water	0.8	130	0.97
Ordinary water	0.2	2.8	0.80

5.2.6. Calculation of configurations described by a single space variable

The calculation examples seen so far could all be handled analytically because the problems involved *only one space variable*: r, ρ, or x.

In general, all problems of this type, even heterogeneous ones, can be solved provided that the general solution of the diffusion equation can be found in each of the regions: i.e. if the function S is not too complicated (or is zero) and allows a particular solution to be found. When there is only one space variable, the diffusion equation is reduced to a second-order differential equation instead of a partial differential equation. The general solution in each region of the system is then expressed as the sum of a particular solution with the source and the linear combination of two solutions to the "equation without a second term" (without source). The conditions at the interfaces and the boundaries make it possible to calculate, from one to the next, the two integration constants that appear for each region (coefficients of the linear combination).

5.2.7. Example of configuration where flux is factorised

This type of analytical calculation can be extended to two- or three-dimensional configurations *if* the system geometry and the distribution of sources allow the *flux to be factorised*. This is a very rare situation.

As an example of a problem where the flux is factorised, we take the case of a vertical cylindrical reactor located between the planes $z = -H/2$ and $z = +H/2$ (including extrapolation distances), comprising various materials arranged in concentric rings and fed by a source proportional to $\cos \pi z/H$ with a coefficient of proportionality that can vary from one region to another or can be zero.

5.2.8. Homogeneous bare reactor: eigenfunctions of the Laplace operator

For a homogeneous bare reactor, the problem can be handled for any source, theoretically at least, using the method of decomposition on the eigenfunctions of the Laplace operator. Before using these functions we shall define them and indicate (without demonstrating) their main properties[8].

• **Definition:** The reactor under consideration, assumed to be homogeneous and bare (or surrounded by a black body), occupies a region D of space limited by its surface S. On S, the function of interest (flux) must be cancelled (we assume that S is the *extrapolated* surface).

This leads to the following purely geometric (and therefore mathematical) problem: to find functions $f(\vec{r})$ of the point in the space defined in D such that:

a) They are *eigenfunctions* of the Laplace operator Δ, i.e. solutions of the following equation:

$$\Delta f + \mu f = 0,$$

where μ is a number known as the *eigenvalue*[9] associated with f;

b) They satisfy the *boundary conditions,* i.e. vanishing on S.

• **Existence of solutions:** there is an infinite series of solutions. This series is enumerable, which means that the terms can be counted using an index n: f_n; μ_n. (Because the equation defining f is homogeneous, two functions that differ from each other only by a multiplication factor can be treated as identical).

• **Properties of eigenvalues:** Eigenvalues are all real and positive. They can be degenerate or not (the order of degeneracy is the number of linearly independent eigenfunctions associated with one eigenvalue).

• **Fundamental mode:** As the smallest eigenvalue, μ_0 is *non degenerate* and the associated eigenfunction f_0 keeps the same sign, for example positive, throughout the region D. The other eigenfunctions f_n are called "harmonics".

• **Orthogonality and normalisation of eigenfunctions:** Two eigenfunctions f_j and f_k associated with two *different eigenvalues are* orthogonal (their *scalar product* is zero):

$$\langle f_j, f_k \rangle = \int_D f_j^* f_k d^3r = 0 \quad (j \neq k). \tag{5.57}$$

[8] Most of these properties are general: different linear operator, different boundary conditions. Here we choose the operator that appears in the equation that interests us (Laplace operator) and we adopt the conditions at the boundaries of the physical problem we wish to analyse (equal to zero at the surface).

[9] In mathematics, the number $\lambda = -\mu$ is generally called an eigenvalue, but here we refer to μ as the eigenvalue for practical reasons. This number is always positive.

(If the functions have complex values, the asterisk denotes the complex conjugate.)

For an m-fold degenerate eigenvalue μ_i, it is always possible to choose a set of two-by-two orthogonal functions $f_{i,j}$:

$$\langle f_{i,j}, f_{i,k}\rangle = 0, \quad (j \neq k). \tag{5.58}$$

Under these conditions, the eigenfunctions are all two-by-two orthogonal, whatever their associated eigenvalues.

Finally, because eigenfunctions are defined to within a factor, this factor can always be chosen so that each eigenfunction is normed (its scalar product with itself is equal to 1); under these conditions, the orthonormalisation of eigenfunctions is summarised by:

$$\langle f_j, f_k\rangle = \delta_{jk}, \tag{5.59}$$

where δ_{jk} is the Kronecker symbol.

• **Completeness of the basis:** The series of functions f_n is complete, i.e. "rich" enough to perform a series expansion of any function or distribution used in physics. This means that for any function[10] $\varphi(\vec{r})$ defined in D, we can have an associated set of coefficients φ_n such that:

$$\varphi(\vec{r}) = \sum_{n=0}^{\infty} \varphi_n f_n(\vec{r}). \tag{5.60}$$

If we multiply this formula by the complex conjugate of one of the eigenfunctions and integrate on D (i.e. taking a scalar product with one of the eigenfunctions), and assuming that the basis has been orthonormed, we find the expression for the coefficients of the expansion:

$$\varphi_n = \int_D f_n^*(\vec{r})\varphi(\vec{r})d^3r. \tag{5.61}$$

• **Closure relation:** Note that the Dirac distribution $\delta(\vec{r}-\vec{r}\,')$ (considered as a function of $\vec{r}$ and therefore $\vec{r}\,'$ is considered to be a parameter) is defined by the following formula:

$$\int_D \delta(\vec{r}-\vec{r}\,')\varphi(\vec{r})d^3r = \varphi(\vec{r}\,'), \tag{5.62}$$

where φ is any function that is continuous in $\vec{r}\,'$; by applying these formulae to $\delta(\vec{r}-\vec{r}\,')$, we obtain:

$$\sum_{n=0}^{\infty} f_n^*(\vec{r})f_n(\vec{r}\,') = \delta(\vec{r}-\vec{r}\,'). \tag{5.63}$$

This is called the "closure relation"[11]; the existence of the closure relation is equivalent to completeness of the basis.

The appendix gives formulae for the eigen elements of the Laplace operator for common geometries. The general properties described above can be checked on these examples.

[10] Even one not satisfying the boundary condition. In this case the expansion has a discontinuity at the crossing of S.

[11] Note the formal similarity between the closure relation and the orthonormalisation relation.

5.2.9. Steady-state problem: flux calculation by decomposition on the eigenfunctions of the Laplace operator

Let us consider the diffusion equation for a bare, homogeneous reactor in steady state.

If the sources $S(\vec{r})$ are distributed randomly, it might be difficult to find the analytical solution, but the eigenfunction decomposition method can be used. (In concrete terms, it might be necessary to calculate the integrals numerically instead of analytically, and it will almost certainly be necessary to truncate the expansions to the finite order N, which must be high enough to obtain the required accuracy.)

The equation to be solved in the domain D of the reactor is:

$$D\Delta\Phi - \Sigma_a\Phi + S = 0,$$

with the condition that the flux is 0 on its surface S.

For a known source, this can be expanded into a series of eigenfunctions of the Laplace operator:

$$S(\vec{r}) = \sum_{n=0}^{\infty} S_n f_n(\vec{r}), \tag{5.64}$$

where the coefficients are calculated by:

$$S_n = \int_D f_n^*(\vec{r})S(\vec{r})d^3r. \tag{5.65}$$

Let us find a similar expansion for the flux:

$$\Phi(\vec{r}) = \sum_{n=0}^{\infty} \Phi_n f_n(\vec{r}). \tag{5.66}$$

By substituting into the diffusion equation, replacing Δf_n by $-\mu_n f_n$ and identifying with zero term by term[12], we obtain:

$$-D\mu_n\Phi_n - \Sigma_a\Phi_n + S_n = 0,$$

which gives the solution in series form:

$$\Phi(\vec{r}) = \sum_{n=0}^{\infty} \frac{S_n}{D\mu_n + \Sigma_a} f_n(\vec{r}). \tag{5.67}$$

We can show that this series converges to a continuous function that vanishes on the surface and therefore satisfies the boundary condition.

---

12 To show that this is valid, simply calculate the scalar product by one of the eigenfunctions, taking orthonormalisation into account.

5.2.10. Study of kinetics after injecting a burst of neutrons

Here is another example of the use of eigenfunctions of the Laplace operator[13]: so-called "pulsed neutron" experiments. The idea is to send a burst of neutrons into a block consisting of the material to be tested, and then measuring the decrease in flux at a point. Because this decay occurs very quickly, as we shall see, the experiment can be repeated many times to improve the statistics. The problem to be solved is the diffusion equation problem in kinetics; here, without a source:

$$D\Delta\Phi - \Sigma_a\Phi = \frac{1}{v}\frac{\partial\Phi}{\partial t}, \tag{5.68}$$

in a domain D assumed to be homogeneous with values vanishing at its surface S. We have the initial flux $\Phi(\vec{r}, 0)$ (satisfying the boundary condition).

1/ Let us find the flux in the form of an eigenfunction expansion of the Laplace operator:

$$\Phi(\vec{r}, t) = \sum_{n=0}^{\infty} \Phi_n(t) f_n(\vec{r}). \tag{5.69}$$

2/ Knowing $\Phi(\vec{r}, 0)$, we can deduce the values $\Phi_n(0)$ of the coefficients at the instant of origin by setting $t = 0$ in this formula:

$$\Phi_n(0) = \int_D f_n^*(\vec{r})\Phi(\vec{r}, 0)d^3r. \tag{5.70}$$

3/ Moreover, by substituting the flux expansion into the diffusion equation and then identifying term by term, we obtain the following differential equations:

$$-D\mu_n\Phi_n(t) - \Sigma_a\Phi_n(t) = \frac{1}{v}\frac{d\Phi_n(t)}{dt}. \tag{5.71}$$

4/ These are easily integrable:

$$\Phi_n(t) = \Phi_n(0)\exp[-v(D\mu_n + \Sigma_a)t]. \tag{5.72}$$

We can set $L^2 = D/\Sigma_a$ and $\ell = 1/(v\Sigma_a)$: these are, respectively, the diffusion area and the neutron lifetime in the medium in the absence of leaks.

5/ Finally:

$$\Phi(\vec{r}, t) = \sum_{n=0}^{\infty} \Phi_n(0) f_n(\vec{r}) \exp\left[-\frac{(1 + L^2\mu_n)t}{\ell}\right]. \tag{5.73}$$

By measuring the decrease in flux at different points to separate the modes (the fundamental eigenfunction and the harmonics), we can obtain the two parameters that characterise the material, ℓ and L^2.

[13] We note in passing that a multi-dimensional problem can sometimes be handled by a mixed technique: analytical solution by one of the variables, and expansion on the eigenfunctions by the other variables. For example, the problem of the "exponential pile" in cylindrical or parallelepiped form, into which a current of neutrons is introduced via one of the bases according to a distribution that is known on that surface. The problem presented here is similar: analytical with respect to time, and eigenfunction expansion with respect to space.

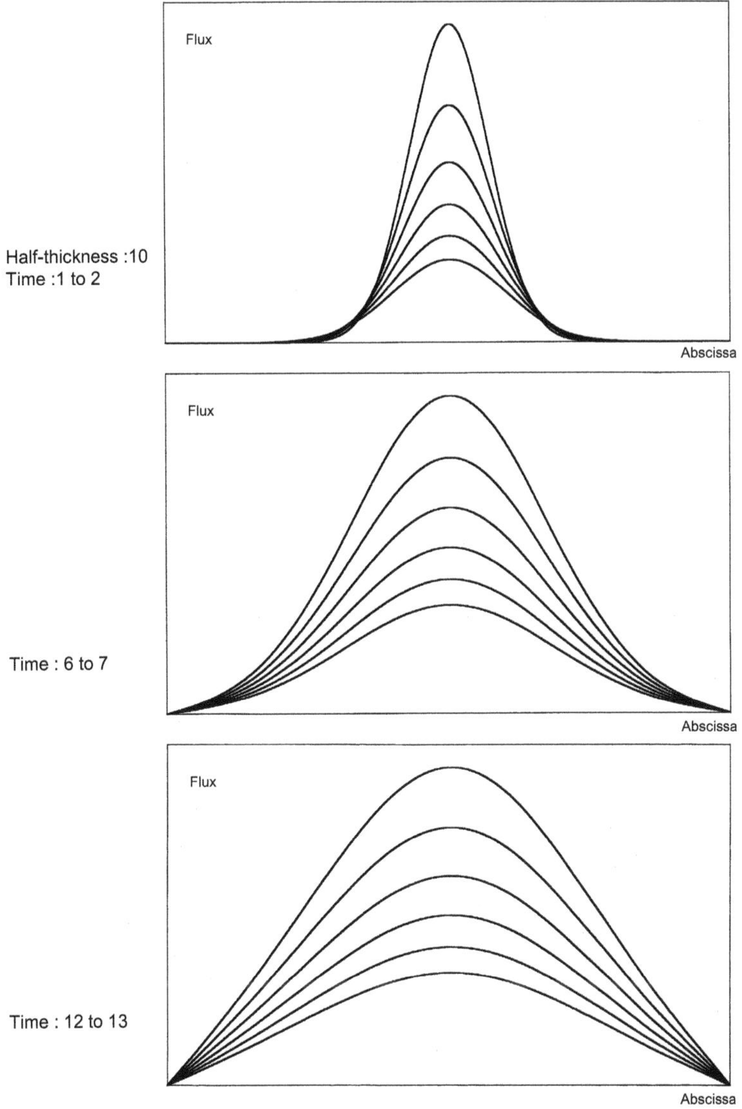

Figure 5.3. Flux curve after a burst of neutrons (unit of length: *L*; unit of time: ℓ).

As an example, we give the flux curves in space and for successive instants for a block in the form of an infinite plate with an initial burst of neutrons on the median plane, represented by $\delta(x)$. Figures 5.3 show the successive curves for three time intervals. For clarity, the fluxes have been renormalised to a value given at the centre of the plate and at the beginning of the time interval. Note that the curves gradually approach the fundamental mode (a cosine in this case) which decreases more slowly than the harmonics.

Exercises

Exercise 5.1: condition at the origin

Derive expressions for the normalisation constants of the plane, line, and point kernels in an infinite, homogeneous medium according to the condition,

$$\int_{(\infty)} \Sigma_a \Phi \, dV = S = 1.$$

Exercise 5.2: 'shell' sources

In an infinite, homogeneous medium, find expressions for the flux distribution resulting from a source emitting S neutrons per unit of area and time that is the surface of,

a) a cylinder of radius a, and

b) a sphere of radius a.

Exercise 5.3: point source in a finite medium

N.B. in this exercise and the ones that follow it, which concern a finite medium, assume that the flux level at the surface falls to zero, i.e. the black body condition with negligible extrapolation distance applies.

Find the expression for the flux distribution in a sphere of radius R, resulting from a point source lying at its centre, and emitting S neutrons per unit of time.

Exercise 5.4: line source in a finite medium

Find the expression for the flux distribution in a cylinder of radius R, and infinite height, resulting from a line source lying along its axis, and emitting S neutrons per unit of length and time.

Exercise 5.5: plane source in a finite medium

Find the expression for the flux distribution in an infinite slab of thickness $2a$, resulting from a source lying on its median plane, and emitting S neutrons per unit of surface and time.

Exercise 5.6: uniform sources in a finite medium

Find expressions for the flux distribution resulting from a uniform source emitting S neutrons per unit of volume and time, and the 'form factor' $F = \Phi_{maximum}/\Phi_{average}$ in,

a) an infinite slab of thickness $2a$;

b) a cylinder of radius R and infinite height; and

c) a sphere of radius R.

Exercise 5.7: expression using an expansion in eigenfunctions

Derive expressions for the flux distributions, in terms of eigenfunction expansions of the Laplace operator, in each of the situations described in exercises **5.3–5.6**.

Exercise 5.8: line source in an infinite slab

Using an expansion in eigenfunctions, find an expression for the flux distribution in an infinite slab of thickness $2a$, resulting from a line source lying on the median plane, and emitting S neutrons per unit of length and time.

Exercise 5.9: point source in an infinite cylinder

Using an expansion in eigenfunctions, find an expression for the flux distribution in a cylinder of radius R and infinite height, resulting from a point source on its axis, and emitting S neutrons per unit of time.

Exercise 5.10: exponential pile

Using an expansion in eigenfunctions, find an expression the flux distribution in a cylinder of radius R and height H, resulting from a beam of S neutrons per unit of surface and time, entering uniformly through its base.

Exercise 5.11: reflected pile

In an infinite, homogeneous medium, a uniform source emitting S neutrons per unit of volume and time lies in a region V. Find expressions for the flux distribution and 'form factor' $F = \Phi_{maximum}/\Phi_{average}$ in V, when it is,

a) an infinite slab of thickness $2a$;

b) a cylinder of radius R with infinite height; and

c) a sphere of radius R.

Compare these results with those obtained in exercise **5.6**.

Exercise 5.12: albedo concept

Consider a planar interface between two homogeneous media possessing albedos α and β, respectively. On average, how many times will a neutron that originates in the first medium, and which enters the second, cross the interface before being absorbed?

Exercise 5.13: expressions for albedos

Derive and compare expressions for the albedos of a homogeneous medium occupying a semi-infinite region that lies,

a) beyond a plane;

b) outside an infinitely long cylinder of radius R; and

c) outside a sphere of radius R.

Recalculate the results of the exercise **5.11** using these albedos.
Examine the limits of these albedos when the absorption cross-section goes to zero.

Exercise 5.14: probability of absorption

Consider a homogeneous medium occupying a sphere of radius R in a vacuum.

a) Find an expression for the probability distribution $p(r)$ for a neutron emitted within the sphere at a distance r from its centre, being absorbed by the medium.
b) Therefore, what fraction of neutrons emitted randomly and uniformly within the sphere fail to escape from it?
c) Reformulate the previous result using the flux obtained in exercise **5.6** part **c**.

Exercise 5.15: kinetic kernel in plane geometry

In an infinite, homogeneous medium, let $\varphi(x, t)$ be the flux resulting from the emission at the time $t = 0$ of one neutron per unit of surface of the plane $x = 0$, i.e. the source is $S = \delta(x)\,\delta(t)$.

a) Solve the integral

$$\int_{-\infty}^{\infty} \varphi\, dx,$$

and check that

$$\int_{0}^{\infty} dt \int_{-\infty}^{\infty} \Sigma_a\, \varphi\, dx = 1.$$

b) Find an expression for the mean square $\langle x^2(t)\rangle$ of the distance from the source plane of the neutrons absorbed between t and $t + dt$. Comment the result.

Exercise 5.16: concept of migration area

If scattering is assumed to be isotropic, then which diffusion coefficient must be chosen in order to respect the migration area for diffusion theory (see exercise **3.2**)?

Exercise 5.17: problem of the 'black hole'

Consider a uniform source of neutrons in a diffusive, non-absorbing medium located between two concentric spheres of radii a and $b > a$. The neutron mean free path λ can be used as the unit of length. The inner part of the sphere of radius a contains a perfectly absorbing material ('black hole'); the outer surface of the sphere of radius b is a perfect mirror (albedo equal to one). Derive an expression for the flux between a and b, and deduce the mean number of collisions suffered by a neutron before it is captured by the 'black hole'. How does this number vary with respect to a and b?

Exercise 5.18: relaxation length

In a homogeneous region where no sources of neutrons exist, if there is a solution of the neutron equation of the form $C^t \exp(-x/L^*)$, then the parameter L^* is called 'relaxation length'. Show using diffusion theory that the relaxation length is equal to the diffusion length L.

Solutions

Exercise 5.1: condition at the origin

The result can be obtained using the following integrals.
Plane:

$$\int_0^\infty e^{-u}\, du = 1.$$

Line:

$$\int_0^\infty u\, K_0(u)\, du = 1.$$

Point:

$$\int_0^\infty u\, e^{-u}\, du = 1.$$

Exercise 5.2: 'shell' sources

a) *Cylindrical shell:*

– *Inner part:*

$$\Phi = \frac{Sa}{D}\, K_0(\kappa a)\, I_0(\kappa \rho).$$

– *Outer part:*

$$\Phi = \frac{Sa}{D}\, I_0(\kappa a)\, K_0(\kappa \rho).$$

b) *Spherical shell:*

– *Inner part:*

$$\Phi = \frac{Sa}{D}\, e^{-\kappa a}\, \frac{\sinh \kappa r}{r}.$$

– *Outer part:*

$$\Phi = \frac{Sa}{D}\, \sinh \kappa a\, \frac{e^{-\kappa r}}{r}.$$

Exercise 5.3: point source in a finite medium

$$\Phi = \frac{S}{4\pi D r}\, \frac{\sinh \kappa (R - r)}{\sinh \kappa R}.$$

Exercise 5.4: line source in a finite medium

$$\Phi = \frac{S}{2\pi D}\left[\mathrm{K}_0(\kappa\rho) - \frac{\mathrm{K}_0(\kappa R)}{\mathrm{I}_0(\kappa R)}\mathrm{I}_0(\kappa\rho)\right].$$

Exercise 5.5: plane source in a finite medium

$$\Phi = \frac{S}{2\kappa D}\,\frac{\sinh\kappa\,|\,a-x\,|}{\cosh\kappa a}.$$

Exercise 5.6: uniform sources in a finite medium

a) *Slab:*

$$\Phi = \frac{S}{\Sigma_a}\left[1 - \frac{\cosh\kappa x}{\cosh\kappa a}\right]; \qquad F = \frac{1 - \frac{1}{\cosh\kappa a}}{1 - \frac{\sinh\kappa a}{\kappa a\cosh\kappa a}}.$$

b) *Cylinder:*

$$\Phi = \frac{S}{\Sigma_a}\left[1 - \frac{\mathrm{I}_0(\kappa\rho)}{\mathrm{I}_0(\kappa R)}\right]; \qquad F = \frac{1 - \frac{1}{\mathrm{I}_0(\kappa R)}}{1 - \frac{2}{\kappa R}\frac{\mathrm{I}_1(\kappa R)}{\mathrm{I}_0(\kappa R)}}.$$

c) *Sphere:*

$$\Phi = \frac{S}{\Sigma_a}\left[1 - \frac{R\sinh\kappa r}{r\sinh\kappa R}\right]; \qquad F = \frac{1 - \frac{\kappa R}{\sinh\kappa R}}{1 - \frac{3(\kappa R\coth\kappa R - 1)}{\kappa^2R^2}}.$$

Exercise 5.7: expression using an expansion in eigenfunctions

- *Normalised eigenfunctions:*
  - *Slab:*

$$f_n(x) = \frac{1}{\sqrt{a}}\cos n\frac{\pi x}{2a}; \qquad \mu_n = \frac{n^2\pi^2}{4a^2};$$

$$(-a < x < +a;\ n\ \text{odd}).$$

  - *Cylinder:*

$$f_n(\rho) = \frac{1}{\sqrt{\pi}\,R\,\mathrm{J}_0(j_{0n})}\,\mathrm{J}_0\left(j_{0n}\frac{\rho}{R}\right); \qquad \mu_n = \frac{j_{0n}^2}{R^2}.$$

  - *Sphere:*

$$f_n(r) = \frac{1}{\sqrt{2\pi R}}\,\frac{\sin n\frac{\pi r}{R}}{r}; \qquad \mu_n = \frac{n^2\pi^2}{R^2}.$$

- *Flux expansion:*

$$\Phi = \sum_{n=1}^{\infty}\varphi_n f_n; \qquad \varphi_n = \frac{s_n}{D\mu_n + \Sigma_a}.$$

- *Source expansion:*
  - *Point source at the centre of a sphere:*

$$s_n = n\,S\,\sqrt{\pi/(2R^3)}.$$

  - *Line source along the axis of a cylinder:*

$$s_n = \frac{S}{\sqrt{\pi}R\,J_1(j_{0n})}.$$

  - *Plane source in a slab:*

$$s_n = \frac{S}{\sqrt{a}} \qquad (n \text{ odd}).$$

  - *Uniform source in a slab:*

$$s_n = \frac{4S\sqrt{a}}{n\pi}.$$

  - *Uniform source in a cylinder:*

$$s_n = 2\sqrt{\pi}\,S\,\frac{R}{j_{0n}}.$$

  - *Uniform source in a sphere:*

$$s_n = -(-)^n\,2n\pi\,S\,\sqrt{2\pi/R}.$$

Exercise 5.8: line source in an infinite slab

When the source is put along the $\vec{z}$ axis, the flux does not depend on z:

$$\Phi(x,y) = \frac{S}{2\pi a}\sum_{n\,(\text{odd})=1}^{\infty}\frac{e^{-\kappa_n|y|}}{\kappa_n}\cos n\frac{\pi x}{2a},$$

where

$$\kappa_n^2 = \kappa^2 + \mu_n; \qquad \mu_n = \frac{n^2\pi^2}{4a^2}.$$

Exercise 5.9: point source in an infinite cylinder

If the origin of z is chosen to be at the source, then

$$\Phi(\rho,z) = \frac{S}{2\pi DR^2}\sum_{n=1}^{\infty}\frac{J_0\left(j_{0n}\frac{\rho}{R}\right)}{\kappa_n\left[J_1(j_{0n})\right]^2}\,e^{-\kappa_n|z|};$$

$$\kappa_n^2 = \kappa^2 + \mu_n; \qquad \mu_n = \frac{j_{0n}^2}{R^2}.$$

Exercise 5.10: exponential pile

Expansion of the radial eigenfunctions yields the following result.

$$\phi(\rho, z) = \sum_{n=1}^{\infty} a_n \sinh \kappa_n (H - z)\, f_n(\rho);$$

$$a_n = \frac{8\sqrt{\pi}\, S\, R/j_{0n}}{\sinh \kappa_n H + 2D\kappa_n \cosh \kappa_n H}; \qquad \kappa_n^2 = \kappa^2 + \mu_n;$$

f_n, μ_n : see exercise **5.7** – Cylinder.

Exercise 5.11: reflected pile

Let 1 be the index of the zone where is the source and 2 the index of the external reflector.

a) *Slab (abscissa origin in the medium plane):*

$$\Phi_1 = \frac{S}{\Sigma_a}\left(1 - e^{-\kappa a} \cosh \kappa x\right); \qquad \Phi_2 = \frac{S}{\Sigma_a} \sinh \kappa a\, e^{-\kappa x};$$

$$F = \frac{\kappa a\,(1 - e^{-\kappa a})}{\kappa a - e^{-\kappa a} \sinh \kappa a}.$$

b) *Cylinder:*

$$\Phi_1 = \frac{S}{\Sigma_a}\left[1 - \kappa R\, K_1(\kappa R)\, I_0(\kappa\rho)\right]; \qquad \Phi_2 = \frac{S}{\Sigma_a} \kappa R\, I_1(\kappa R)\, K_0(\kappa\rho);$$

$$F = \frac{1 - \kappa R\, I_0(\kappa R)\, K_1(\kappa R)}{1 - 2\, I_1(\kappa R)\, K_1(\kappa R)}.$$

N.B: the results are simplified by virtue of the Wronskian formula: $I_0(x)\,K_1(x) + K_0(x)\,I_1(x) = 1/x$.

c) *Sphere:*

$$\Phi_1 = \frac{S}{\Sigma_a}\left[1 - (1 + \kappa R)\, e^{-\kappa R} \frac{\sinh \kappa r}{\kappa r}\right];$$

$$\Phi_2 = \frac{S}{\Sigma_a}\left(\kappa R \cosh \kappa R - \sinh \kappa R\right) \frac{e^{-\kappa r}}{\kappa r};$$

$$F = \frac{1 - (1 + \kappa R)e^{-\kappa R}}{1 - \dfrac{3\,(1 + \kappa R)\left(\kappa R \cosh \kappa R - \sinh \kappa R\right) e^{-\kappa R}}{\kappa^3 R^3}}.$$

Exercise 5.12: albedo concept

Assuming that the neutron makes its first transit of the boundary starting from the medium possessing albedo α, then

$$\bar{n} = \frac{1+\beta}{1-\alpha\beta}.$$

Exercise 5.13: albedo calculations

All the albedo formulae are in the form

$$\beta = \frac{1-u}{1+u},$$

with,

a) *Slab:*

$$u = 2\kappa D.$$

b) *Cylinder:*

$$u = 2\kappa D \frac{K_1(\kappa R)}{K_0(\kappa R)}.$$

c) *Sphere:*

$$u = 2D(\kappa + \frac{1}{R}).$$

When the absorption cross-section goes to zero, the slab and cylinder albedos increase to one, while the sphere albedo becomes

$$\beta_0 = \frac{1-2D/R}{1+2D/R}.$$

Exercise 5.14: probability of absorption

First, notice that for the calculation of the absorption probability in the sphere, a point source located at distance r from the centre can be replaced by a 'shell' source on the surface of the sphere of radius r. The flux Φ resulting from shell can then be found by the same method as in exercise **5.2.b**, except that the flux is constrained to be zero at R. From this, the probability $p(r)$ can be deduced either by integration of $\Sigma_a\Phi$ in the sphere, or by calculation of the total outgoing current from the sphere at the radius R (complement to one):

$$p(r) = 1 - \frac{R \sinh \kappa a}{a \sinh \kappa R}.$$

Either by averaging this probability, or using the result from **5.6.c** (giving the same integral), gives

$$P = 1 - \frac{3(\kappa R \coth \kappa R - 1)}{\kappa^2 R^2}.$$

Exercise 5.15: kinetic kernel in plane geometry

The time dependent kernel of the diffusion equation can be obtained either by a Laplace transform on t, or a Fourier transform on x:

$$\varphi(x,t) = \sqrt{\frac{v}{4\pi Dt}}\,\Upsilon(t)\,\exp(-v\Sigma_a t - \frac{x^2}{4vDt}).$$

a) The solution to the integral is

$$\int_{-\infty}^{+\infty} \varphi(x,t)\,dx = v\,\Upsilon(t)\,e^{-v\Sigma_a t};$$

therefore,

$$\int_0^{\infty} dt \int_{-\infty}^{+\infty} \Sigma_a\,\varphi(x,t)\,dx = 1 = \text{Source}.$$

b) The mean square of the distance from the source plane of the neutrons absorbed at time t is

$$\langle x^2(t)\rangle = 2vDt.$$

Notice how the normal diffusion law applies, where the direct distance increases as the *square root* of the time.

Exercise 5.16: concept of migration area

The exact value is (see exercise **3.2**)

$$\langle R^2\rangle = \frac{2}{\Sigma_a\Sigma_t}.$$

Using the point kernel of the diffusion equation gives

$$\langle R^2\rangle = \int_0^{\infty} r^2\,\Sigma_a\,\frac{e^{-r/L}}{4\pi Dr}\,4\pi r^2\,dr = 6L^2 = 6\frac{D}{\Sigma_a}.$$

Therefore, the expression for the migration area is fulfiled provided we use

$$D = \frac{1}{3\Sigma_t}.$$

Remark: it can similarly be shown that $D = 1/3\Sigma_{tr}$ (with $\Sigma_{tr} = \Sigma_t - \bar{\mu}\Sigma_s$, where $\bar{\mu}$ is mean cosine of the scattering angle) satisfies the migration area expression for *any* scattering law.

Exercise 5.17: problem of the 'black hole'

The equation to solve is

$$D\left(\Phi'' + 2\frac{\Phi'}{r}\right) + S = 0.$$

The general solution is

$$\Phi(r) = A + \frac{B}{r} - \frac{S}{6D}r^2.$$

The constants A and B are obtained from the two boundary conditions

$$J_+(a) = \frac{\Phi(a)}{4} - \frac{D}{2}\Phi'(a) = 0, \qquad J(b) = -D\,\Phi'(b) = 0.$$

It can eventually be shown that

$$\Phi(r) = \frac{S}{6D}\left[\frac{2b^3}{a} + a^2 + 4D\left(\frac{b^3}{a^2} - a\right) - \frac{2b^3}{r} - r^2\right].$$

The mean number of collisions suffered by a neutron is the ratio of the number of collisions per unit of time to the number of emissions per unit of time:

$$n = \frac{\int_a^b \Sigma\,\Phi(r)\,4\pi r^2\,dr}{\int_a^b S\,4\pi r^2\,dr}.$$

Taking into account the relationship $D = 1/3\Sigma$, provided that the collisions are assumed to be isotropic, then

$$n = \frac{\left(b^3 - a^3\right)\left[2b^3/a + a^2 + 4D\left(b^3/a^2 - a\right)\right] - 3b^3\left(b^2 - a^2\right) - 3\left(b^5 - a^5\right)/5}{18D^2\left(b^3 - a^3\right)}.$$

Some numerical examples expressed in terms of the neutron mean free path are given as follows. Notice how n increses when the ratio increases.

b/a	**1.5**	**2**	**3**	**4**	**5**
a = 1	1.96	6.70	29.6	78.8	164
a = 2	4.68	17.4	83.7	231	492
a = 4	12.4	51.1	265	736	1637

Exercise 5.18: relaxation length

Clearly, a flux in the form of $C^t \exp(-x/L)$ satisfies the diffusion equation. Therefore, in the diffusion approximation, $L^* = L$, where *the diffusion length L* is defined as the migration length, square root of one sixth of the mean squared direct distance between the neutron emission and its absorption — see exercise **5.16** — *is equal to the relaxation length.* This result is not general. For instance, in monokinetic transport theory, the relaxation length differs from the diffusion length when the absorption cross-section is finite.

One-group/diffusion theory

Introduction: what is one-group theory?

In the previous chapter, it was assumed that the neutron sources were known *a priori*, and we wished to calculate the resulting flux. We have seen, however, that the sources in reactors are almost all sources of neutrons emitted by flux-induced fission. Like the flux, therefore, they are initially unknown. This is the problem we wish to address here[1].

In the context of the preceding chapter, we could assume that it was natural to compare the thermal neutron population of interest to a population of monokinetic neutrons, because the thermal neutron spectrum is relatively narrow[2]. This assumption now seems far more risky because the neutrons are emitted at an energy of the order of 2 MeV and, in the case of a thermal neutron reactor, mostly disappear at an energy of the order of 0.025 eV, i.e. 10^8 times lower.

Test calculations show, however, that reasoning according to *one group of neutrons*, i.e. by grouping them without distinguishing their energies and therefore handling them *as if* they were monokinetic, will produce results that are qualitatively satisfactory and even relatively accurate in quantitative terms.

The reason for this apparently paradoxical success (both in thermal neutron reactors and fast neutron reactors) is that, in a given area of the reactor, such as the core, the neutron *spectrum* (i.e. the energy distribution of the neutrons) is broadly the same over the entire area. If nuclear data (cross-sections, diffusion coefficients, etc.) averaged over this spectrum[3] are introduced into the one group, we can adequately handle the "space" factor of a flux that is more or less factorised energy x space.

Let $\Phi(\vec{r})$ denote this "space" factor of the flux and $\Sigma_r(\vec{r})$ the associated cross-sections used to express the rates of reaction[4]. In particular, the fission rate can be written as $\Sigma_f\Phi$ and therefore the source is:

$$S = \nu\Sigma_f\Phi. \tag{6.1}$$

[1] Even though it is not much more difficult to handle the general case, with both types of source (spontaneous and induced fission), in this chapter we restrict the discussion to cases with induced fission sources only.

[2] We have also implicitly accepted that the sources were emitting neutrons at this same speed, which is probably not very realistic.

[3] Note that a spectral flux average does not involve the *level* of the flux, but only its *energy distribution*.

[4] The reaction rate is integrated over the spectrum. According to the usual conventions, the flux is also an integral over the spectrum. Under these conditions, the cross-section is a flux-weighted *average* over the spectrum.

Here a second approximation will be made (one that is often, but not necessarily, used for theories with a small number of groups or one-group theory): *the diffusion approximation.*

Because only steady-state situations will be studied here, time will be left out of the equation. The one-group/diffusion equation we shall study therefore takes the form:

$$D\Delta\Phi + \nu\Sigma_f\Phi - \Sigma_a\Phi = 0. \tag{6.2}$$

Instead of the three parameters that appear in this equation and define matter, we shall also use:
– the *infinite multiplication factor*: $k_\infty = \nu\Sigma_f/\Sigma_a$, representing the production/disappearance ratio of neutrons in the absence of leaks (and therefore in a medium extending *ad infinitum*),
– the *migration area*: $M^2 = D/\Sigma_a$, representing a sixth of the mean square of the distance travelled by neutrons from emission to absorption in the supposedly infinite medium. (The term "diffusion area" [or "length"] is generally used in the monokinetic context, and "migration area" [or "length"] is used in one-group theory. That is why this parameter was written L^2 in the previous chapter, but is now written M^2.)
– the *material buckling*: $\chi^2 = (\nu\Sigma_f - \Sigma_a)/D = (k_\infty - 1)/M^2$, so called because it characterises matter and appears as an *eigenvalue of the Laplace operator* if the one-group/diffusion equation is rewritten in canonical form:

$$\Delta\Phi + \chi^2\Phi = 0. \tag{6.3}$$

The first part of this chapter gives a few examples of analytical solutions of this equation. The second part briefly describes the main numerical methods used in calculation codes for this problem.

6.1. A few problems in one-group/diffusion theory

6.1.1. Shape of solutions

If the medium is multiplying, i.e. if its infinite multiplication factor k_∞ is greater than 1, the material buckling χ^2 is positive, and so the solutions are of the type $\cos\chi x$ and $\sin\chi x$ or equivalent for the other geometries. The flux curve has its *concave side facing downwards.*

If the medium is characterised by an infinite multiplication factor k_∞ less than 1 and, in particular, if no fission occurs in it ($k_\infty = 0$), the material buckling is negative and is written $-\kappa^2$. The solutions are of the type $\cosh \kappa x$ and $\sinh \kappa x$ or equivalent for the other geometries. The flux curve has its *concave side facing upwards.*

6.1.2. Bare homogeneous spherical pile

Let us take a spherical pile of radius R, including the extrapolation distance, as an example of a bare homogeneous pile. We assume that χ^2 is positive, which means that k_∞ is greater than 1. (Note that if k_∞ is less than 1, a steady-state solution without an independent source is impossible. This is obvious from a physical standpoint, because the system must be subcritical.)

Taking spherical symmetry into account, the equation to be solved is:

$$\frac{d^2\Phi(r)}{dr^2} + \frac{2}{r}\frac{d\Phi(r)}{dr} + \chi^2\Phi(r) = 0, \tag{6.4}$$

with the following boundary condition:

$$\Phi(R) = 0. \tag{6.5}$$

Seeking a solution of the form $\Phi = f/r$, the equation is simplified as follows:

$$\frac{d^2f(r)}{dr^2} + \chi^2 f(r) = 0,$$

and the solutions are the trigonometric functions with argument χr. This gives:

$$\Phi(r) = A\frac{\sin\chi r}{r} + B\frac{\cos\chi r}{r},$$

where A and B are constants.

Applying a condition of regularity at the origin (the flux remains finite) imposes $B = 0$:

$$\Phi(r) = A\frac{\sin\chi r}{r}.$$

The boundary condition imposes:

$$A\sin\chi R = 0.$$

Ruling out $A = 0$, which would give zero flux, this condition requires χR to be a multiple of π:

$$\chi R = k\pi,$$

with k a whole number. We rule out $k = 0$, which would also give zero flux, and the negative values that give the same functions (with a different sign) as the positive values.

The flux Φ must be positive or zero (note that, except for its speed, flux is a neutron density). Of the whole, positive values of k, only $k = 1$ is physically acceptable. In any other case, the sine would change sign when r goes from 0 to R.

Finally, we reach two conclusions:

1/ The problem has a solution only if χR is equal to π;

2/ This solution takes the form (Figure 6.1):

$$\Phi(r) = A\frac{\sin\pi\frac{r}{R}}{r}. \tag{6.6}$$

The condition will be written as follows:

$$\chi^2 = \frac{\pi^2}{R^2}, \tag{6.7}$$

and specifies that a steady-state flux can exist *if and only if the reactor is critical.* This condition involves the various system characteristics; in the suggested form, the left-hand

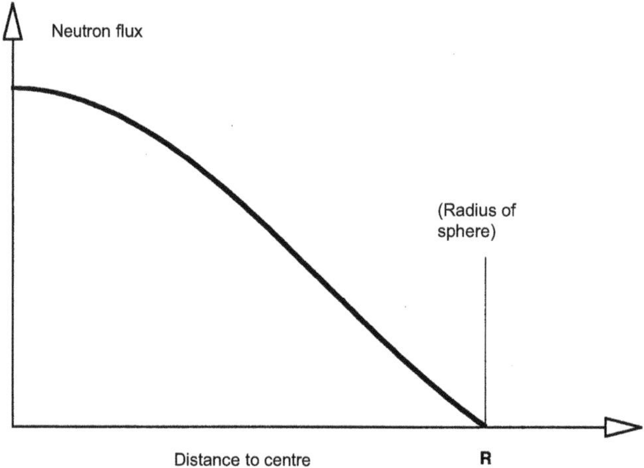

Figure 6.1. Flux in a spherical reactor.

side is a property of the *material* used, and the right-hand side is a *geometric* characteristic (shape and size). This is often summarised as follows:

material buckling = geometric buckling.

If the system is critical, both of these parameters are equal to the *fundamental eigenvalue of the Laplace operator* (the smallest one, denoted μ_0 in the previous chapter, § 5.2.8).

If the reactor is critical, the flux is the *associated fundamental eigenfunction.* This function is defined only to within an arbitrary factor A. Mathematically, this comes from the homogeneous nature (no "second term") of one-group theory. Physically, it arises from the concept of criticality, which means that neutron production and neutron elimination rates are equal; but this equality can occur at any level (a thousand neutrons per second, a million, a thousand million, etc.).

6.1.3. Other homogeneous bare pile examples

These considerations apply to all bare homogeneous piles. Only the formulae of the geometric buckling and flux change. The formulae for two simple types of pile where the flux is factorised according to the variables are given here as examples:

– *cylinder* of radius R and height H:

$$\mu_0 = \frac{j^2}{R^2} + \frac{\pi^2}{H^2}, \quad \Phi(\rho, z) = A J_0\left(j\frac{\rho}{R}\right)\cos\left(\frac{\pi}{2}\frac{z}{H}\right); \tag{6.8}$$

(origin of the coordinates at the centre of the cylinder; J_0: Bessel function; j = 2.40483: the first zero of this function);

– *rectangular parallelepiped* with sides a, b and c:

$$\mu_0 = \frac{\pi^2}{a^2} + \frac{\pi^2}{b^2} + \frac{\pi^2}{c^2}, \quad \Phi(x, y, z) = A\sin\left(\pi\frac{x}{a}\right)\sin\left(\pi\frac{y}{b}\right)\sin\left(\pi\frac{z}{c}\right); \tag{6.9}$$

(origin of the coordinates at a vertex).

6.1.4. Interpretation of critical condition

The geometric buckling μ_0 is often written as B^2 by symmetry with the material buckling χ^2; let us now use this notation[5].

Returning to the expression for χ^2, the critical condition for the bare homogeneous pile can be written as follows:

$$k = \frac{k_\infty}{1 + M^2 B^2} = 1. \tag{6.10}$$

This ratio is the real multiplication factor of the finite-size reactor, often called *"effective"* and written k_{eff} to distinguish it from the "infinite" multiplication factor.

(***Summary of the proof:*** by definition, we have:

$$k = \frac{P}{A + \Lambda}, \tag{6.11}$$

where P, A and Λ are, respectively, the number of neutrons produced in the pile, absorbed in the pile, and the leakage via its surface per unit time. The first two rates are expressed as a function of the integral of the flux in the pile, and the third is expressed as a function of the integral on the surface of the normal derivative of the flux. Using Green's theorem and the flux equation $\Delta\Phi + B^2\Phi = 0$, this third integral becomes the integral of flux in the pile. The formula shown is then found by simplifying with this integral and using k_∞ and M^2.)

This factor k is the product of k_∞ by the *probability of non-leakage* of a neutron emitted in the pile:

$$P_{\text{non-leakage}} = \frac{1}{1 + M^2 B^2}. \tag{6.12}$$

This value is obviously zero if the reactor is reduced to a point, and tends to 1 if the reactor is infinite. Between these limits, and if k_∞ is greater than 1, the reactor is critical for a certain size when its shape is fixed. The concept of critical mass can be associated with this concept of critical size, which is illustrated below (see Figure 6.2) for the case of a sphere.

6.1.5. Reflector saving

The concept of a reflector was introduced in the previous chapter (§ 5.2.3). A reflector sends some of the escaped neutrons back into the core. These neutrons, which would otherwise be lost, return to the multiplying material and might cause fission there. The improved multiplication factor resulting from the better probability $P_{\text{non-leakage}}$ leads to an increase in the effective multiplication factor and, finally, a reduction of the critical size.

The albedo, which has already been defined, and the various parameters just mentioned, can characterise the "efficiency" of a reflector. The parameter that is often preferred is the economy of the multiplying material resulting from the reflector, or "reflector saving".

This concept is defined by Figure 6.3[6].

The flux curve in the core (when normalised) does not depend on the presence of a reflector (the equation has not changed). Without a reflector, this curve must be traced

[5] The notation B_g^2 and B_m^2 is also often used.

[6] The curves traced in Figure 6.3 were calculated for an "infinite plate" geometry.

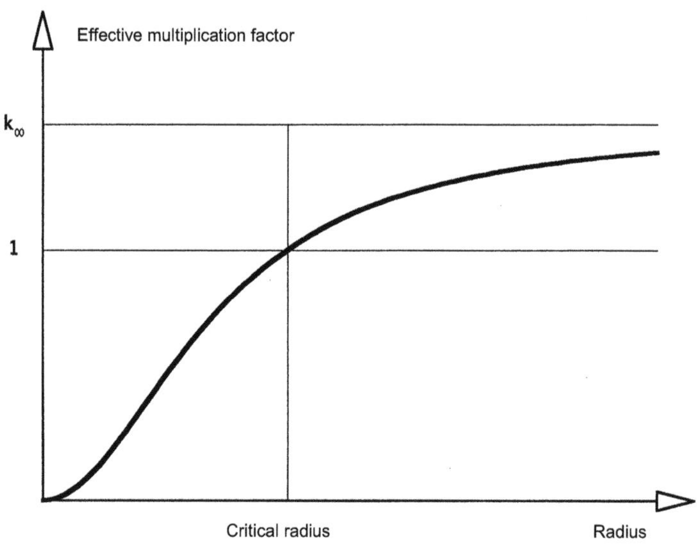

Figure 6.2. Concept of critical size for a bare homogeneous pile (spherical in this case).

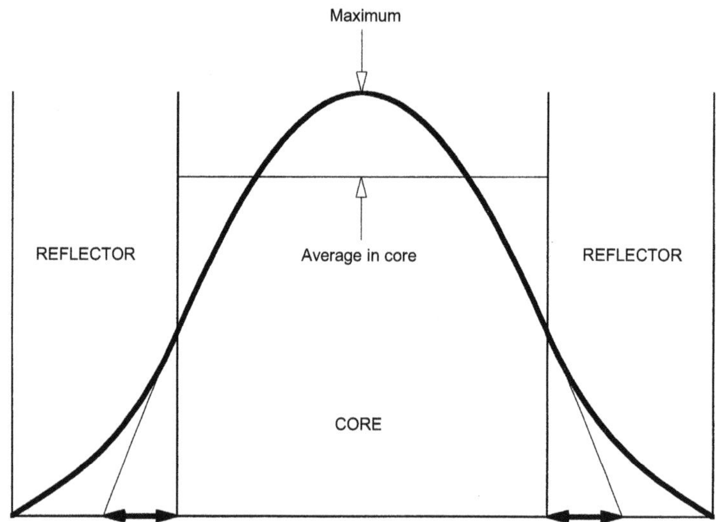

Figure 6.3. Concept of reflector saving. The curve drawn with a thick line represents flux. In the core, power is proportional to flux; in the reflector, it is zero.

to the point where it cuts the axis, which defines the boundary of the pile if it is bare (neglecting the extrapolation distance). If a reflector is then added, the curve must be stopped at the point where the logarithmic derivative of the flux has the required value $-\gamma/2D$ (§ 5.2.5). This point must be closer to the centre of the core than the surface of the bare pile. The difference (arrows) is, by definition, the reflector saving.

Note that inserting a reflector has an additional advantage that can be even more important in practice than the saving of fissile material: the improvement in the shape factor of the power, $F = P_{max}/P_{av}$, maximum power normalised to the average power which, in this case, is equal to the equivalent ratio calculated with the flux, $F = \Phi_{max}/\Phi_{av}$. The maximum temperature threshold imposed for safety reasons sets the value of P_{max}; under these conditions, the power output of the power plant, the product of P_{av} by the volume of the core, is inversely proportional to this shape factor. This is an incentive to seek the lowest possible value of F. The figure shows that the presence of a reflector lowers this factor. (Normalising to P_{max} in the calculation of the average eliminates the part of the curve with the lowest values.) Note also that "flattening" the power distribution allows greater uniformity of fluid temperatures leaving the core, as well as a higher average value. In addition to increasing the thermal power, this also improves the thermodynamic efficiency of the conversion to electricity.

6.1.6. Reflector saving calculation for a "plate" reactor

Consider the geometry described above: a reactor consisting of a core in the shape of a plate of thickness $2a$ enclosed on both sides by a reflector plate of thickness e.

The equation for diffusion in the reflector was solved in the previous chapter (§ 5.2.4). We obtained:

$$\gamma = 2\kappa_r D_r \coth \kappa_r e,$$

(the index "r" is added to denote the reflector).

In the core, allowing for symmetry and taking the origin of the abscissas at the median plane, the flux is:

$$\Phi(x) = A \cos \chi_c x,$$

where χ_c is the square root of its material buckling.

The condition at the right-hand interface[7] at $x = a$:

$$2D_c \frac{d\Phi/dN}{\Phi} = -\gamma, \tag{6.13}$$

gives the critical condition:

$$2\chi_c D_c \tan \chi_c a = 2\kappa_r D_r \coth \kappa_r e.$$

(*Calculation variant:* write that the flux in the (right) reflector, given the boundary condition, is: $\Phi[x] = B \sinh \kappa_r [a + e - x]$ and express the continuity of the flux and current at the interface.)

Write the condition that flux at the core (extrapolated beyond the interface) is zero at $x = a+\delta$ (the point where the argument of the cosine is $\pi/2$) to find the following expression for reflector saving[8] δ:

$$\delta = \frac{1}{\chi_c} \mathrm{Arctg} \left[\frac{D_c \chi_c}{D_r \kappa_r} \mathrm{th} \kappa_r e \right]. \tag{6.14}$$

[7] The condition at the left-hand interface gives the same equation with opposite signs for each of the terms.

[8] In planar geometry, this is not explicitly dependent on a, but this is not general.

– If *the reflector is thin*, the reflector saving is roughly proportional to the thickness:

$$\delta \simeq \frac{D_c}{D_r} e. \tag{6.15}$$

– If *the reflector is thick* (at least two or three migration lengths):

$$\delta \simeq \delta_{as} = \frac{1}{\chi_c} \text{Arctg} \left[\frac{D_c \chi_c}{D_r \kappa_r} \right]. \tag{6.16}$$

– If the *core* size is also *large*, i.e. if $\chi_c \delta$ is small compared to 1:

$$\delta_{as} \simeq \frac{D_c}{D_r} L_r. \tag{6.17}$$

(See Table 5.1 for the orders of magnitude of $L_r = 1/\kappa_r$.)

6.1.7. Geometries described by a single space variable

The method used in this reflected reactor calculation can be generalised to any geometry described by a single space variable x, ρ or r (stacking of plates or concentric spherical or cylindrical shells).

In one of the regions, the one-group/diffusion equation is solved analytically; the general solution is the linear combination of two particular solutions. Both constants of the combination, A and B, can be expressed as a function of the flux and current at one of the interfaces, and vice-versa. This means that, for example, by expressing these constants as a function of the flux and current on the left, and then calculating the flux and current on the right as a function of the constants, we obtain linear expressions that can be written as a 2×2 matrix.

According to the situation, for the left-hand boundary of the system or the central region, where either a boundary condition or a symmetry condition (zero derivative) must be fulfilled, by choosing both solutions carefully we can observe that one of the constants must be zero. The other one, for example A, can be used as a normalisation constant.

We then take successive layers from left to right, using the matrices to calculate the flux and current from one to the next as a function of A.

At the right-hand boundary, the boundary condition must apply. This will be an equation of the form $E(...)A = 0$, where $E(...)$ is an expression that might be complicated but can be calculated based on the product of all the matrices, and involves all the information about the geometry and materials. We do not want A to be zero because we are seeking a non-trivial steady-state solution, and so the other factor must be zero. The *critical condition* of the system is therefore expressed:

$$E(...) = 0.$$

Note: If it is more convenient, the recurrence calculation can be performed from right to left instead of from left to right.

6.1.8. Example of a problem where flux is factorised

In a multi-dimensional case, the flux can be factorised according to variables if the geometry allows. In practice, this is possible only if the system is multi-layered according to one variable, and homogeneous according to the others. The solution is then the fundamental mode of the Laplace operator according to the homogeneous ones, and is calculated by the matrix method according to the multi-layered one[9].

For example, the flux of a cylindrical reactor reflected radially but not axially (or vice-versa) is factorised and can be calculated analytically, but this does not hold true for a reactor reflected *both* radially and axially.

This means that only purely numerical methods can be considered. Because this problem almost always arises for reactors studied by engineers, reactor calculations require the use of computers[10]. Before such calculations can be programmed, the numerical techniques to be used must be chosen. This opens the door to a vast area of applied mathematics where neutron physicists have made significant contributions. An entire book could be written on the subject of numerical analysis (some excellent ones already exist: refer to the bibliography), and so it will not be discussed in detail here. To give the reader an idea of what goes into calculation codes, we provide a very brief review of the most common numerical methods used to solve the diffusion equation and to handle one-group theory[11].

6.2. Main numerical methods used for the diffusion equation

6.2.1. Introduction to numerical processing

The basic difficulty involved in the numerical processing of an equation whose unknown is a function, is that a parameter representing an infinite number of numerical values — a function — must be replaced by a *finite* number of numerical values with the least possible loss of information.

There are two possible approaches:

- *Discretisation*: This consists of creating a table of values of the function concerned for a finite number of values of the argument(s): for example, a table of logarithms or other transcendental functions;

- *Series representation*: This consists of representing the function concerned by a finite sum of carefully-chosen basic functions, known as "test functions": for example, representation in the form of a truncated Fourier series, to a certain rank, by a polynomial of finite degree, etc.

---

[9] In the example given in § 5.2.7, the flux could be factorised exactly because we had chosen an axial source arranged according to this fundamental mode in z. Note that factorising also assumes suitable boundary conditions.

[10] This is all the more true because in practice we *also* wish to deal with the neutron spectrum.

[11] These methods are all generalised to "diffusion/multigroup" theory. In Chapter 14 we introduce the numerical methods used to process the transport equation.

First, the form in which the unknown function will be represented must be chosen, and then the appropriate mathematical process must be defined in order to go from the equation concerned to an algebraic system giving either the discretised values or the coefficients placed before the test functions. Because the equation for neutron flux is always linear, this algebraic system will also be linear. In practice, the problem is therefore reduced to inverting a matrix, often a very large one[12], which will generally be done by iteration[13].

Of the methods mentioned below, the finite difference method is a "discretisation" method, and the others are series representations by test functions.

6.2.2. Source problem and critical problems

The problems examined in the previous chapter, for which the source is known, are called "source problems". After discretisation or representation by a truncated series, these problems are presented in the following form:

$$MV = B,$$

where M is a matrix (replacing the equation) and B is a vector (deduced from the source), which are known when the numerical processing is performed. V is the vector of unknowns.

This system is solved by inverting the matrix M, giving V:

$$V = M^{-1}B.$$

The problems considered in the present chapter are called "critical problems". They are different from the previous ones partly because the source is related to the unknown flux and because the solution exists only if the critical condition is ensured (we could also say that the operator represented by the equation to be solved must have a zero eigenvalue).

In practice, the first difficulty is solved by an "external" iteration applied to invert M. Starting from an initial approximation $S^{(0)}$ of the fission source, a flux approximation $\Phi^{(0)}$ is deduced by treating the problem as a source problem, i.e. by inverting M; from this flux, a better approximation of the source $S^{(1)}$ can be calculated by applying the fission production operator (in this case, multiplication by $\nu\Sigma_f$); $\Phi^{(1)}$ can be deduced from $S^{(1)}$, and so on, until an iteration convergence criterion is satisfied (several tens of "external" iterations are often necessary).

For the second aspect — the critical condition — we introduce a "critical parameter" which is also determined during the iterations for the problem to have a solution: this is any parameter that applies a fictitious modification to the reactor described in the calculation data to make it critical: outside boundary, concentration or position of a poison, fissile material content of a fuel, etc. The parameter most often used is the effective multiplication factor k of the reactor. Returning to its definition, we can see that this is the factor such that *if neutron production were divided by this factor k*, i.e. if ν were replaced by ν/k each time this number appears in the equations, *the reactor would become critical.* [In the "external" iteration the factor k is evaluated, with increasing accuracy as n increases, by comparing the intensity of sources $S^{(n)}$ and $S^{(n+1)}$.]

---

12 Neutron physics problems can often involve hundreds of thousands of unknowns.

13 The choice of the most efficient iterative method is also an interesting and difficult mathematical problem.

Finally note that, in a critical problem, the flux is obtained only to within a factor, A. Before the results are obtained, the designer or user of the code must specify how this factor is chosen. For example, it is possible to normalise to a neutron emitted by the source or to a total power of a unit, etc.

6.2.3. Finite differences

The finite difference method consists of choosing a discretisation according to each of the variables of the problem, and then approximating the derivatives, which are quotients of infinitesimal differences, with quotients of differences that are finite but small. For example, if the variable x has been discretised, i.e. if only the discrete abscissas x_0, x_1, x_2, x_3, etc. are considered, and, for a function $\Phi(x)$, only the associated values $\Phi_0 = \Phi(x_0)$, $\Phi_1 = \Phi(x_1)$, $\Phi_2 = \Phi(x_2)$, $\Phi_3 = \Phi(x_3)$, etc., the derivative at the abscissa x_i is approximated by the following quotient:

$$\frac{d\Phi}{dx}(x = x_i) \simeq \frac{\Phi_{i+1} - \Phi_i}{x_{i+1} - x_i} = \frac{\Phi_{i+1} - \Phi_i}{h}, \tag{6.18}$$

(derivative on the right), or by the following quotient:

$$\frac{d\Phi}{dx}(x = x_i) \simeq \frac{\Phi_i - \Phi_{i-1}}{x_i - x_{i-1}} = \frac{\Phi_i - \Phi_{i-1}}{h}, \tag{6.19}$$

(derivative on the left). To simplify the presentation, we assume that discretisation is regular (performed according to a constant step h), but this restriction is not essential.

By applying this approximation twice (once with the first formula and once with the second to arrive at a symmetric formula), we express a second-order derivative:

$$\frac{d^2\Phi}{dx^2}(x = x_i) \simeq \frac{\Phi_{i+1} - 2\Phi_i + \Phi_{i-1}}{h^2}. \tag{6.20}$$

For a two-dimensional problem processed in Cartesian coordinates x and y, which we simplify by assuming it is regularly discretised with the same step h, a similar approach leads to an approximation of the Laplace operator:

$$(\Delta\Phi)_{i,j} \simeq \frac{\Phi_{i+1,j} + \Phi_{i-1,j} + \Phi_{i,j+1} + \Phi_{i,j-1} - 4\Phi_{i,j}}{h^2}. \tag{6.21}$$

We also sometimes use a nine-point formula:

$$\begin{aligned}(\Delta\Phi)_{i,j} \simeq \Big[&4(\Phi_{i+1,j} + \Phi_{i-1,j} + \Phi_{i,j+1} + \Phi_{i,j-1}) \\ &+ (\Phi_{i+1,j+1} + \Phi_{i-1,j-1} + \Phi_{i+1,j-1} + \Phi_{i-1,j+1}) \\ &-20\Phi_{i,j}\Big] /(6h^2).\end{aligned} \tag{6.22}$$

Similar formulae have been established for three-dimensional problems and/or other coordinate systems.

When the formula has been chosen, it must simply be substituted into the equation to be solved, written at one of the discrete points. For example, for the diffusion equation:

$$D\Delta\Phi - \Sigma_a\Phi + S = 0,$$

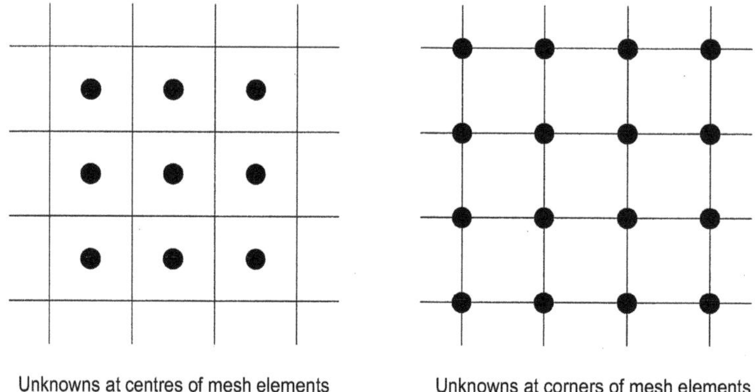

Figure 6.4. ***x-y*** **discretisations for a square mesh (the nodes are those where the functions are considered).**

in a two-dimensional case, we write:

$$D_{i,j}(\Delta\Phi)_{i,j} - \Sigma_{a,i,j}\Phi_{i,j} + S_{i,j} = 0.$$

Varying the indices i and j over the entire reactor domain produces a linear system with as many equations as unknowns. Note that this system is extremely "hollow" (i.e. its matrix M contains many zeros), since each equation involves only five or nine unknowns, according to the formula used to discretise the Laplace operator[14]. The equations cannot however be solved from one to the next: an iterative inversion of the matrix M is necessary.

Note: In practice, the discretisation step is of the order of the mean free path of the neutrons, i.e. a centimetre. For calculations on water reactors, this means that in the (x, y), plane we use one mesh element or $2 * 2$ mesh elements[15] per unit cell (after homogenisation of the fuel rod, its cladding and the associated portion of water); this discretisation can be used if necessary to assign different neutron physical properties to each of the cells[16]. Under these conditions, one might wonder where the points of the discretisation should be placed. There are two variants, which are also symmetrical, each with its proponents and detractors. These are shown in a diagram below (Figure 6.4).

6.2.4. Finite elements

At first glance, the finite element method looks similar to the finite difference method because the first step is to cut the system up into mesh elements of finite volume. In fact, however, the approach is different, because this is a series representation by test functions.

[14] As an example for comparison, a domain discretised along 300 points in x and 300 points in y involves $300^2 = 90\,000$ unknowns.

[15] The formulae just seen are of order h^2, meaning that the accuracy is improved by a factor of 4 if the step is reduced by a factor of 2.

[16] A distinction can also be drawn between fuel cells (possibly of different compositions), absorbent cells, "water holes", etc.

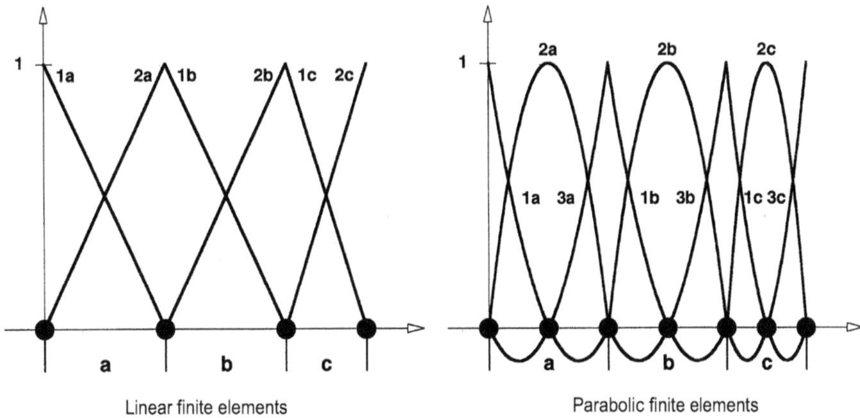

Figure 6.5. Test-function examples for a one-dimensional case (a, b, c denote the elements, and 1, 2, 3 denote the test functions).

Each test function is specific to *one* element and zero for the others. In general, linear, quadratic or cubic functions are used, and defined such that they:

a/ Ensure continuity of the functions at the interfaces between elements; the functions will be represented by a sum of these test functions;

b/ Approximate the functions in each element by a polynomial of degree 1, 2 or 3.

Figure 6.5 gives examples of such test functions for elements with one dimension x and linear and quadratic approximations.

Note that these functions are the zero lines or parabolas, respectively, at all the nodes of the element (bold points in Figure 6.5) except for one, which makes it possible to fulfil both criteria[17].

If these test functions are called φ_k, the unknowns of the problem are the coefficients A_k of the representation of the unknown function:

$$\Phi(\vec{r}) = \sum_k A_k \varphi_k(\vec{r}). \tag{6.23}$$

(In practice, with the choice of test functions made, it is strictly equivalent to say that these unknowns are the values of Φ at the "nodes", which shows how similar this approach is to the finite element method.)

The technique most often used to obtain the linear system giving these unknowns is based on a variational formulation of the problem, and then a Galerkin criterion. Let us show this on an example with one dimension, x, i.e. a plane and possibly heterogeneous geometry:

$$\frac{d}{dx}\left[D(x)\frac{d\Phi(x)}{dx}\right] - \Sigma_a(x)\Phi(x) + S(x) = 0, \quad a \leqslant x \leqslant b, \quad \Phi(a) = \Phi(b) = 0. \tag{6.24}$$

[17] For the purpose of continuity, the same coefficient must be taken for both functions which have the value 1 at the interface, which can be done by grouping the functions concerned; for example, for the linear elements, $2a$ with $1b$, $2b$ with $1c$ and so on.

This problem of solving a differential equation is subject to a "variational" (or "weak") formulation. Let us associate this equation with the following functional[18]:

$$J(v) = \int_a^b \left\{ -D(x) \left[\frac{dv(x)}{dx} \right]^2 + \Sigma_a(x)[v(x)]^2 - 2S(x)v(x) \right\} dx, \qquad (6.25)$$

where $v(x)$ is any function satisfying the boundary conditions.

We have the following *variational theorem*: the functional $J(v)$ is an extremum (in this case, a minimum) if and only if the function v is a solution of the diffusion equation.

[*Overview of the proof*: calculate $J(v + \varepsilon w)$ where v and w are functions fulfilling the limit conditions, and ε is a parameter (which is not necessarily assumed to be small in the present case); note that J is an extremum if the ε term is zero for any w; note that this requires the w factor term to be identically zero, and note that this term is exactly the first term of the diffusion equation.]

Using this theorem, the diffusion equation problem can be reformulated in the following terms: *determine the function v creating the extremum of J*.

If applied in the context of finite elements, this formulation can be considered as follows: the *best approximation* of the solution of the diffusion equation in the form:

$$\Phi = \sum_k A_k \varphi_k, \qquad (6.26)$$

is obtained with the set of coefficients A_k giving the extremum of:

$$F(A_1, A_2, A_3, \ldots) = J\left(\sum_k A_k \varphi_k \right). \qquad (6.27)$$

In practice, the coefficients will be determined by taking the function F of coefficients A_k to be an extremum, i.e. all its partial derivatives are zero:

$$\frac{\partial F}{\partial A_1} = 0, \quad \frac{\partial F}{\partial A_2} = 0, \quad \frac{\partial F}{\partial A_3} = 0, \quad \text{etc.} \qquad (6.28)$$

Because F is a quadratic function, by construction, these equations take the form of a linear system. In this case again, a matrix inversion will need to be performed.

In view of the similarity between the finite element method and the finite difference method, it would be natural to imagine that these two numerical approaches would perform in a comparable way, but in fact they do not. The finite element method, particularly in a parabolic or cubic approximation, is far more efficient that the finite difference method. It is more accurate with the same meshing (or a coarser meshing can be used to achieve the same accuracy as the other method).

As an illustration, let us take the example of a parabolic approximation, where the mesh elements can be twenty times wider for finite elements than for finite differences. Because the parabolic elements use an internal node, the factor is only ten. Moreover, because the system of equations to be inverted is slightly more complex, a further reduction by a factor of two can be applied, i.e. *the finite element method has an overall advantage over the finite difference method by a factor of five*. There are two other important considerations:

[18] In mathematics, a "functional" is an operation that associates a *number* with a *function*. There are other possible choices of functional to reach a variational formulation of the diffusion problem.

- this gain is observed for a one-dimensional problem. For a two-dimensional problem, the gain becomes $5 \times 5 = 25$ and, for a three-dimensional problem, it becomes $5 \times 5 \times 5 = 125$.
- This very desirable gain can only be achieved if it is actually possible to describe the geometry with large, homogeneous mesh elements. For water reactors, this is not possible if a "cell by cell" description is used. In this case, the implementation of the finite element method assumes complete homogenisation of the assemblies.

6.2.5. Nodal methods

Nodal methods also allow a reactor core to be handled with large mesh elements. This is particularly useful for three-dimensional calculations. In this case, the mesh elements are called "nodes"[19].

The approach is similar to finite elements in that the flux is approached by polynomial approximations. The main differences are as follows:

- Not only the flux *in* the mesh elements, but also *on* their faces is considered.
- The degree of the polynomial approximations is not necessarily the same for the internal flux and the flux on the faces.
- Internal flux is calculated from the neutron balance:

$$\text{Absorptions} + \text{Net leakage} = \text{Sources}$$

- The outgoing current is deducted from the incoming current by analytical integration of the diffusion equation (this can be performed using the polynomial approximations of flux values).
- Currents leaving a node are used as the incoming currents in the adjacent node.

Nodal methods perform comparably to finite elements. It is difficult to determine whether one approach is better than the other because there are many possible variants in each case, and because different laboratories might use different programming methods.

6.2.6. Synthesis methods

Unlike the previous methods, the synthesis method is based more on physical intuition than on mathematical rigour. It can provide excellent performance in terms of cost/accuracy ratio, but it can also lead to disappointing results.

The principle is outlined below, using the example of a three-dimensional core calculation:

1/ Note the characteristic geometries of the various axial sections of the core (lower reflector, area without control bundles, area with control bundles, upper reflector, etc.) and perform a two-dimensional calculation for each section, giving values for flux $\varphi_k(x, y)$.

[19] Not to be confused with the "nodes" of finite elements.

2/ Look for the three-dimensional flux in the form of a linear combination of these flux expressions with coefficients that depend on the third variable z:

$$\Phi(x, y, z) \simeq \sum_k \psi_k(z)\varphi_k(x, y), \tag{6.29}$$

(the synthesis operation).

3/ Determine the coefficients using a given criterion, such as a Galerkin technique (error orthogonal with each φ_k).

This method has certain advantages because it is more economical to calculate a few flux values in two dimensions than one flux in three dimensions. (The third step does not use much processing time.)

The synthesis method can obviously be applied to other multi-dimensional problems: for example, a space × time synthesis can be performed for a kinetics calculation[20].

[20] This was done in the example in the previous chapter (§ 5.2.10) concerning pulsed neutron experiments.

Exercises

N.B: for the following exercises, the extrapolation distance beyond a pile will be neglected.

Exercise 6.1: calculation of geometrical buckling factors

Consider a homogeneous, critical, bare pile. This pile is divided into two half-piles, each isolated from the other. Calculate the effective multiplication factor of each half-pile for each of the following four geometrical bodies. Use $k_\infty = 2$.

a) A cube cut along a plane containing its centre and parallel with a pair of opposing faces.

b) A sphere cut into hemispheres.

c) An orthocylinder (diameter = height) cut along a plane containing its axis.

d) An orthocylinder cut along a plane parallel to and equidistant from its ends.

Exercise 6.2: the optimal form of a pile

For a given multiplying, homogeneous material, characterized by its buckling factor B^2,

a) determine the dimensions a, b and c of the cuboid for a critical, bare pile having the minimum volume;

b) determine the radius R and height H of a cylindrical critical, bare pile having the minimum volume; and

c) compare these with the volume of a spherical, critical, bare pile.

Exercise 6.3: a hollow sphere

A critical, spherical, homogeneous, bare pile is replaced by a hollow sphere containing the same amount of material with the same density, and with a spherical void at its centre. What is the boundary condition for the flux at the surface of the inner sphere? Calculate the effective multiplication factor for hollow spheres with several values of inner and outer radii, using $k_\infty = 2$.

Exercise 6.4: the effect of compression and an expansion

The material of a critical, spherical, homogeneous, bare pile is compressed uniformly into a smaller sphere. Calculate the effective multiplication factor as a function of the ratio of the density d to the initial density d_0, using $k_\infty = 2$. Similarly, show the effect of uniform expansion.

Exercise 6.5: sandwich pile

Consider a critical, planar pile consisting of a slab of a homogeneous material labelled 1, extending in thickness from $-a$ to $+a$, sandwiched between two slabs of another homogeneous material labelled 2 of thickness b. Assume that the characteristics of the materials are otherwise identical except for k_∞ which differs.

a) Determine the critical condition for each of the following cases.

0) $k_{\infty 1} = 1$;

1) $k_{\infty 1} > 1$ and $k_{\infty 2} > 1$;

2) $k_{\infty 1} < 1$ and $k_{\infty 2} > 1$;

3) $k_{\infty 1} > 1$ and $k_{\infty 2} < 1$.

b) For each of these cases, where is the maximum value of the flux?

c) Show, by qualitative reasoning, that the form factor defined as $F = \Phi_{\text{maximum}}/\Phi_{\text{average}}$, is greater in case **3** than in case **1**.

d) Calculate the form factor for cases **0**, **1**, and **2**.

e) Assume that the multiplication factor $k_{\infty 1}$ is very nearly one, and let u be $\chi_1 a$ (small) with $\chi_1^2 = \mid k_{\infty 1} - 1 \mid / M^2$. Employ a limited expansion to compare F for cases **1** and **2** with the value F_0 for case **0**. Under which conditions is the latter one smallest?

Exercise 6.6: reflector saving

Derive the relationships giving the reflector savings for a cylindrical pile of infinite height and a spherical pile, when the thickness of the reflector is infinite.

Exercise 6.7: reflected piles

Show how the critical condition and form factor $F = \Phi_{\text{maximum}}/\Phi_{\text{average}}$ (where the maximum and average terms are considered in the core part only) of a homogeneous reflected pile, depend on the reflector albedo for

a) an infinite homogeneous slab of thickness $2a$;

b) a cylinder of radius R and infinite height; and

c) a sphere of radius R.

Exercise 6.8: a pile possessing uniform power-density distribution

A fissile material characterized by σ_a and $\nu\sigma_f = \eta\sigma_a$, with a radial concentration $N(\rho)$ out to a maximum distance R, is added to an infinite, homogeneous moderator material. Assume that the concentration of the moderator and the diffusion coefficient of the mixture do not depend on the concentration of the fissile material. Determine the concentration $N(\rho)$ required to make a critical system characterized by a uniform power-density distribution in the cylindrical region $\rho < R$.

Exercise 6.9: multilayer piles

Consider piles in planar, cylindrical, and spherical geometries that are described in terms of a single spatial coordinate, x, ρ, and r, respectively. Establish what the 2×2 matrices are that describe a homogeneous layer situated between a and b along the relevant coordinate, which enable the flux and current at b to be determined from their values at a.

Exercise 6.10: effect of a small, absorbing body

A small, spherical, absorbing body with a radius ε is located at the centre of a spherical, bare, homogeneous reactor with a larger radius R, that would be critical without the presence of this perturbation. Criticality is artificially restored by replacing $\nu\Sigma_f$ with $\nu\Sigma_f/k = \nu\Sigma_f(1-\rho)$, where $k < 1$ is the multiplication factor of the perturbed reactor, and $\rho < 0$ its reactivity.

a) Show that outside the absorbing body the flux is

$$\Phi(r) = C\left[f(r) - \lambda g(r)\right], \quad \text{with} \quad \Phi(R) = 0,$$

where C is a normalization constant, and λ is a parameter which characterizes the absorption rate of the body. Find expressions for the two functions f and g.

b) Write down the general expression for the critical condition with abritrary λ, using the following notations.

$$\chi^2 = \frac{k_\infty(1-\rho)-1}{M^2}, \quad \text{and} \quad \chi R = \pi + u.$$

Assuming that λ and u are small, simplify the previous result to show that $u \approx \lambda$. Deduce the expression for ρ as a function of λ.

c) Give expressions for the number of neutrons

- produced in the multiplying zone;
- absorbed in the multiplying zone;
- absorbed in the absorbing body;
- escaping the system through its external surface;

and establish the neutron balance.
Show using first order expansions with respect to λ and $\chi\varepsilon$

- the relationship giving ρ as a function of λ; and
- that the antireactivity arising from the presence of the absorbing body is due to two effects of equal magnitude; namely, the neutron absorption by the sample itself, and the increase of the neutron leakage through the external surface.

d) Let γ be the probability for a neutron entering the body to be absorbed; thus, $1-\gamma$ is the probability that the neutron subsequently escapes. Again using first order expansions, express λ as a function of γ; however, note now that ε and D may be of the same order of magnitude.

e) Repeat this exercise for an infinite cylinder containing a small, coaxial, absorbing region, and for an infinite slab with a thin, absorbing layer in the middle.

Exercise 6.11: gradient of the power-density distribution

Consider a reactor made from three infinite slabs of material stacked together. (This could apply to a cylindrical reactor if it is assumed that the values of k_∞ include the radial leakages.) The middle slab, where $k_\infty = 1$, is width $2a$ thick, centred at the origin. The outer pair of slabs both have equal thickness $b-a$, and are characterized by the parameters $k_{\infty 1}$ and $k_{\infty 2}$ for the upper and lower slabs, respectively. All three slabs possess the same values for D and M^2.

a) Describe how the flux behaves in the slabs, and state the critical condition when $k_{\infty 1} = k_{\infty 2} = k_{\infty s}$.

b) Describe how the flux behaves in the slabs, and state the critical condition when $k_{\infty 1} \neq k_{\infty 2}$. *Hint:* use $\lambda = \Phi(+a)$ and $\mu = \Phi(-a)$ to represent the unknown fluxes.

c) Derive an expression for the ratio of the average fluxes in the outer slabs, $\bar{\Phi}_1/\bar{\Phi}_2$.

d) The values of $k_{\infty 1} - k_{\infty s}$ and $k_{\infty 2} - k_{\infty s}$ are assumed to be small. Use a first order expansion to find an expression that describes the relationship between these two parameters, and show that $\bar{\Phi}_1/\bar{\Phi}_2 \approx \lambda/\mu$. *Hint:* use $u = \pi/2 - \chi_1 b$ and $v = \pi/2 - \chi_2 b$ to represent auxiliary unknows.

e) Taking $a = 1$ m, $b = 0.5$ m and $M^2 = 5.6 \times 10^{-3}$ m^2, evaluate $\bar{\Phi}_1/\bar{\Phi}_2$ when $k_{\infty 1}$ exceeds $k_{\infty s}$ by the amounts 100, 250, 500, and 750 pcm.

Exercise 6.12: propagation of a wave

This exercise describes a very simplified oscillation experiment.

a) A point source, whose intensity varies sinusoidally is located at the origin in an infinite, homogeneous moderator medium; thus, $S(\vec{r}, t) = \delta(\vec{r})\, e^{i\omega t}$. Find an expression which describes the flux resulting from this source. The distance over which the magnitude of the flux is attenuated by a factor e is known as the relaxation distance; derive the formula for it. Show what happens when ω is very large or very small. *Hint:* use complex functions to treat the time-varying contribution in a manner analagous to that done for an alternating current in electromagnetism theory, taking the real part to be the solution with physical meaning, and add a constant contibution to obtain a result which is always positive.

b) Repeat the problem for an exactly critical, multiplying medium ($k_\infty = 1$). Include only a single group of delayed neutrons, and assume that the lifetime ℓ of the free neutrons is zero.

Solutions

Exercise 6.1: calculation of geometrical buckling factors

For the initial pile

$$k = \frac{k_\infty}{1 + M^2 B^2} = 1; \qquad (k_\infty = 2;\ M^2 B^2 = 1).$$

For the half-piles, the geometrical buckling factor becomes B'^2, and

$$k = \frac{k_\infty}{1 + M^2 B'^2} = \frac{k_\infty}{1 + (B'^2/B^2) M^2 B^2} = \frac{2}{1 + B'^2/B^2}.$$

The buckling factors, together with their corresponding k-values, for each of the geometrical bodies in this exercise are as follows.
Cube with edge a:

$$B^2 = \frac{3\pi^2}{a^2}; \qquad B'^2 = \frac{6\pi^2}{a^2}; \qquad k = 0.66667.$$

Sphere with radius R:

$$B^2 = \frac{\pi^2}{R^2}; \qquad B'^2 = \frac{\alpha^2}{R^2}; \qquad k = 0.65665.$$

Orthocylinder with radius R cut vertically:

$$B^2 = \frac{4j_{01}^2 + \pi^2}{4R^2}; \qquad B'^2 = \frac{4j_{11}^2 + \pi^2}{4R^2}; \qquad k = 0.64965.$$

Orthocylinder with radius R cut horizontally:

$$B^2 = \frac{4j_{01}^2 + \pi^2}{4R^2}; \qquad B'^2 = \frac{j_{01}^2 + \pi^2}{R^2}; \qquad k = 0.69033,$$

where $\alpha = 4.49340946$ is the first nonzero root of the equation $\tan x = x$, and where j_{n1} is the first nonzero root of the Bessel function J_n.

Exercise 6.2: study of the optimal form of a pile

For the cuboid

$$B^2 = \frac{\pi^2}{a^2} + \frac{\pi^2}{b^2} + \frac{\pi^2}{c^2}; \qquad V = abc.$$

The derivatives must be zero, because in the first instance B^2 is constant, and the in second instance the extremum of V is wanted.

$$\frac{da}{a^3} + \frac{db}{b^3} + \frac{dc}{c^3} = 0; \qquad bc\,da + ac\,db + ab\,dc = 0.$$

The term dc cancels by taking a linear combination; then, by making the expression equal to zero for any da and db gives two equations:

$$a = c; \qquad b = c.$$

Therefore, the best cuboid is a cube where its volume is

$$V = \frac{3\sqrt{3}\pi^3}{B^3} = \frac{161.11}{B^3}.$$

Similarly, it can be shown that for the best cylinder

$$H = \frac{\sqrt{2}\pi}{j_{01}} R = 1.8475\,R,$$

where its volume is

$$V = \frac{3\sqrt{3} j_{01}^2 \pi^2}{2\,B^3} = \frac{148.29}{B^3}.$$

The volume of a sphere possessing the same buckling factor is

$$V = \frac{4\pi^4}{3\,B^3} = \frac{129.88}{B^3}.$$

Exercise 6.3: a hollow sphere

Let a and b be the internal and external radii. The boundary conditions are then $\Phi'(a) = 0$ and $\Phi(b) = 0$. The flux is in the form

$$\Phi = A \frac{\sin B(b - r)}{r},$$

where the geometrical buckling factor is given by

$$\tan B(b - a) + aB = 0.$$

Let R be the radius of a solid sphere, then the radii a and b are linked by

$$b^3 - a^3 = R^3.$$

The multiplication factor is given by

$$k = \frac{k_\infty}{1 + \pi^2 C^2 M^2/R^2},$$

with

$$\alpha = \frac{a}{R}; \qquad \beta = \frac{b}{R}; \qquad \beta = \sqrt[3]{1 + \alpha^3}; \qquad \varepsilon = \frac{\alpha}{\beta - \alpha};$$

where x is the solution to the equation $\tan x + \varepsilon x = 0$; and

$$C = \frac{x}{\pi(\beta - \alpha)}.$$

Some numerical examples obtained using $k_\infty = 2$ are given below.

α	β	ε	x	C	k
0	1	0	π	1	1
0.10	1.00033	0.11107	2.83640	1.00280	0.99721
0.25	1.00518	0.33105	2.45846	1.03624	0.96441
0.50	1.04004	0.92585	2.05475	1.21111	0.81077
0.75	1.12449	2.00275	1.83629	1.56083	0.58204
1.00	1.25992	3.84732	1.72072	2.10726	0.36761

Exercise 6.4: study of a compression and an expansion

Let $r = d/d_0$, then

$$k = \frac{k_\infty}{1 + M_0^2 B_0^2 / r^{4/3}}; \qquad \text{with} \qquad M_0^2 B_0^2 = k_\infty - 1.$$

Some numerical values obtained using $k_\infty = 2$ are given below.

r	k	r	k
1	1	1	1
1.1	1.06345	0.9	0.92987
1.2	1.12095	0.75	0.81053
1.5	1.26391	0.5	0.56821
2	1.43179	0.25	0.27212
5	1.79057	0	0
10	1.91129		
∞	2		

Exercise 6.5: sandwich pile

Define

$$\alpha = \frac{a}{a+b}; \qquad \chi_1^2 = \frac{|k_{\infty 1} - 1|}{M^2}; \qquad u = \chi_1 a;$$

$$\beta = \frac{b}{a+b}; \qquad \chi_2^2 = \frac{|k_{\infty 2} - 1|}{M^2}; \qquad v = \chi_2 b.$$

In case **0**, the flux is constant in the inner region, then decreases toward the surfaces of the outer regions.

In cases **1** and **3**, the flux is maximum at the centre and decreases toward the surfaces. This effect is more important for case **3** which, therefore, cannot be the optimum form factor.

In case **2**, the flux increases in the inner region from its centre, reaching maxima in the outer regions, before decreasing to zero at the surface.

In case **4**, with two negative buckling factors, the pile cannot be critical.

The critical condition and the inverse of the form factor are given by the following relationships for the cases likely to lead to an optimum form factor.

Case **0**

$$u = 0; \; v = \frac{\pi}{2}; \qquad \frac{1}{F} = \alpha + \frac{\pi}{2}\beta.$$

Case **1**

$$\frac{1}{\alpha} u \tan u = \frac{1}{\beta} v \cot v; \qquad \frac{1}{F} = \alpha \frac{\sin u}{u} + \beta \frac{1}{\cos u} \frac{1 - \cos v}{v \sin v}.$$

Case **2**

$$-\frac{1}{\alpha} u \sinh u = \frac{1}{\beta} v \cot v; \qquad \frac{1}{F} = \alpha \frac{\tanh u}{u} \sin v + \beta \frac{1 - \cos v}{v}.$$

If u is small, then a limited expansion, with $\lambda = \beta/\alpha$, gives

Case **1**

$$\frac{1}{F} = \alpha + \frac{2}{\pi}\beta - \left[\frac{1}{6} - \frac{1}{\pi}\lambda + \frac{4}{\pi^2}\left(1 - \frac{2}{\pi}\right)\lambda^2\right] \alpha u^2.$$

Case **2**

$$\frac{1}{F} = \alpha + \frac{2}{\pi}\beta - \left[\frac{1}{3} - \frac{4}{\pi^2}\left(1 - \frac{2}{\pi}\right)\lambda^2\right] + \alpha u^2.$$

The two first terms correspond to the value for case **0**. This case is optimum ($F > F_0$) when the expressions between brackets are positive. For the first one, this is the true when λ

lies outside the interval [0.890; 1.271] between the zeros, and for the second case, when λ is inside the interval [−1.504; +1.504] between the zeros. Since λ is positive, it can be concluded that

- if $\lambda < 0.890$, then case **0** is optimum;
- if $0.890 < \lambda < 1.271$, then the optimum exists among the possibilities for case **1**;
- if $1.271 < \lambda < 1.504$, then case **0** is optimum;
- if $\lambda > 1.504$, then the optimum exists among the possibilities for case **2**.

Example results are presented below, where exact calculations are employed for these four cases, all with $u = 0.25$.

λ	1/3	1/2	3/4	1	4/3	2
α	3/4	2/3	4/7	1/2	3/7	1/3
β	1/4	1/3	3/7	1/2	4/7	2/3
F_0	**1.0999**	**1.1378**	**1.1845**	1.2220	**1.2621**	1.3197
F_1	1.1042	1.1401	1.1849	**1.2217**	1.2623	1.3244
F_2	1.1178	1.1539	1.1972	1.2312	1.2663	**1.3141**

N.B: in practice, it is the power form factor instead of the flux form factor that is examined. The problem is then more difficult because the ratio power/flux is not the same in the two materials. Moreover, it depends on the values of k_∞.

Exercise 6.6: reflector saving

Define

$$\chi^2 = \left(\frac{\nu\Sigma_f - \Sigma_a}{D}\right)_{\text{core}}; \qquad \kappa^2 = \left(\frac{\Sigma_a}{D}\right)_{\text{reflector}}.$$

The formulae giving the reflector saving δ are

Slab:

$$D_{\text{core}}\,\chi\,\cot\chi\delta = D_{\text{reflector}}\,\kappa.$$

Cylinder:

$$D_{\text{core}}\,\chi\,\frac{J_1(j-\chi\delta)}{J_0(j-\chi\delta)} = D_{\text{reflector}}\,\kappa\,\frac{K_1(\kappa r)}{K_0(\kappa r)}.$$

Sphere:

$$D_{\text{core}}\,\chi\left(\frac{1}{\pi-\chi\delta} + \cot\chi\delta\right) = D_{\text{reflector}}\,\kappa\left(1 + \frac{1}{\kappa R}\right).$$

Exercise 6.7: reflected piles

As in exercise **5.13**, when $D = D_{\text{reflector}}$, albedos take the form

$$\beta = \frac{1-u}{1+u}.$$

When $D = D_{\text{core}}$

a) *Slab:*

$$2D\chi \tan \chi a = u; \qquad F = \frac{\chi a}{\sin \chi a};$$

b) *Cylinder:*

$$2D\chi \frac{J_1(\chi R)}{J_0(\chi R)} = u; \qquad F = \frac{\chi R\, J_0(\chi R)}{2\, J_1(\chi R)};$$

c) *Sphere:*

$$2D\chi \left(\frac{1}{\chi R} - \cot \chi R \right) = u; \qquad F = \frac{\chi^3 R^3}{3(\sin \chi R - \chi R \cos \chi R)}.$$

It can be shown for the three cases that $\chi a, \chi R \to 0$ and $F \to 1$ when $\beta \to 1$.

Exercise 6.8: a pile possessing uniform power-density distribution

For a given radius R, the concentration is governed by the relationship

$$N(\rho) = \frac{\Sigma_{am}}{(\eta - 1)\sigma_a} \frac{1}{1 - \kappa R\, K_1(\kappa R)\, I_0(\kappa \rho)},$$

where Σ_{am} and $\kappa^2 = \Sigma_{am}/D$ describe the moderator.

Exercise 6.9: multilayer piles

Define

$$\chi^2 = \frac{|\, k_\infty - 1\,|}{M^2},$$

and $t = \chi x$, $\chi\rho$, or χr according to the geometry; then $u = \chi a$, $v = \chi b$, and $f(t)$ and $g(t)$ are particular solutions of the equation giving the flux.

Provided the function f has the property that its derivative at the origin is zero, then the matrices giving the flux and the current at b from their values at a can be expressed as

$$M = \frac{1}{W} \begin{vmatrix} \alpha & \frac{-\beta}{D\chi} \\ D\chi\gamma & \delta \end{vmatrix},$$

with

$$W = g(u)\dot{f}(u) - f(u)\dot{g}(u),$$
$$\alpha = g(v)\dot{f}(u) - f(v)\dot{g}(u),$$
$$\beta = f(v)\, g(u) - g(v)\, f(u),$$
$$\gamma = \dot{f}(v)\,\dot{g}(u) - \dot{g}(v)\,\dot{f}(u),$$
$$\delta = \dot{f}(v)\, g(u) - \dot{g}(v)\, f(u).$$

For the first layer, where, owing to the symmetry condition, the current is zero:

$$\Phi(a) = \varphi\,\Phi(0), \qquad J(a) = j\,\Phi(0),$$

with

$$\varphi = f(u), \qquad j = -D\chi\,\dot{f}(u).$$

According to the geometry and sign of $k_\infty - 1$, the function f and g, their derivatives, and the Wronskian W are as follows.

Slab, $k_\infty > 1$:

$$f = \sin, \qquad \dot{f} = \cos,$$

$$g = \cos, \qquad \dot{g} = -\sin,$$

$$W = 1.$$

Slab, $k_\infty < 1$:

$$f = \sinh, \qquad \dot{f} = \cosh,$$

$$g = \cosh, \qquad \dot{g} = \sinh,$$

$$W = 1.$$

Cylinder, $k_\infty > 1$:

$$f = J_0, \qquad \dot{f} = -J_1,$$

$$g = Y_0, \qquad \dot{g} = -Y_1,$$

$$W = -\frac{2}{\pi u}.$$

Cylinder, $k_\infty < 1$:

$$f = I_0, \qquad \dot{f} = I_1,$$

$$g = K_0, \qquad \dot{g} = -K_1,$$

$$W = \frac{1}{u}.$$

Sphere, $k_\infty > 1$:

$$f = \frac{\sin t}{t}, \qquad \dot{f} = -\frac{-t\cos t + \sin t}{t^2},$$

$$g = \frac{\cos t}{t}, \qquad \dot{g} = -\frac{t\sin t + \cos t}{t^2},$$

$$W = \frac{1}{u^2}.$$

Sphere, $k_\infty < 1$:

$$f = \frac{\sinh t}{t}, \qquad \dot{f} = \frac{t\cosh t - \sinh t}{t^2},$$

$$g = \frac{\cosh t}{t}, \qquad \dot{g} = \frac{t\sinh t - \cosh t}{t^2},$$

$$W = \frac{1}{u^2}.$$

Exercise 6.10: effect of a small, absorbing body

a) In spherical geometry,

$$f(r) = \frac{\sin \chi r}{r}, \qquad g(r) = \frac{\cos \chi r}{r}.$$

b) The critical condition is

$$\tan u = \lambda, \qquad u \simeq \lambda;$$

therefore,

$$\rho \simeq -\frac{2}{\pi}\frac{k_\infty - 1}{k_\infty}\lambda.$$

c) The integrated flux is

$$I = \frac{4\pi C R^2}{(\pi + u)^2}\left[F(\pi + u) - F(\chi\varepsilon)\right],$$

where

$$F(z) = -z\cos z + \sin z - \lambda\,(z \sin z + \cos z).$$

First order expansion:

$$I \simeq 4CR^2.$$

The total leakage through the external surface is

$$J = 4\pi DC\, F(\pi + u) \simeq 4\pi DC\,(\pi + \lambda).$$

The net leakage through the surface of the absorber is

$$a = -4\pi DC\, F(\chi\varepsilon) \simeq 4\pi DC\lambda.$$

Therefore, the balance is
Production:

$$P = \nu\Sigma_f I(1 - \rho) \simeq 4CR^2\nu\Sigma_f(1 - \rho).$$

Absorption:

$$A = \Sigma_a I + a \simeq 4CR^2\Sigma_a + 4\pi DC\lambda.$$

Leakage:

$$J \simeq 4\pi DC(\pi + \lambda).$$

This means that for the main terms

$$k_\infty = 1 + M^2\frac{\pi^2}{R^2},$$

and for the first order terms

$$-\rho = \frac{k_\infty - 1}{\pi k_\infty}\lambda + \frac{k_\infty - 1}{\pi k_\infty}\lambda,$$

which confirms the result obtained in **b**. The first contribution comes from the absorbing body (internal leakage) and the second contribution comes from the external leakage. Both of these effects make equal contributions to the reactivity effect.

d) Define

$$\frac{J_+(\varepsilon)}{J_-(\varepsilon)} = 1 - \gamma,$$

then it follows that

$$\lambda \simeq \frac{\gamma\chi\varepsilon}{2D(2-\gamma)/\varepsilon + \gamma}.$$

e) In cylindrical and planar geometries, the relationships are not so simple. In particularly, there is no longer an equality between the effects of the internal and external leakages.

Exercise 6.11: gradient of the power-density distribution

a) The flux is constant in the central region, and varies in a sinusoidal manner across the outer regions. The critical condition is $\chi_s b = \pi/2$.

b) The flux is linear in the central region,

$$\Phi = \frac{\lambda+\mu}{2} + \frac{\lambda-\mu}{2a}x,$$

and matched to a sinusoidal variation in the outer regions,

$$\Phi = A_1 \sin\chi_1(a+b-x) \qquad \text{and} \qquad \Phi = A_2 \sin\chi_2(a+b+x).$$

The continuity of the functions and their derivatives determines the critical condition:

$$\text{when } x = +a: \qquad \frac{\mu}{\lambda} = 1 + 2\chi_1 a \cot\chi_1 b;$$

$$\text{when } x = -a: \qquad \frac{\lambda}{\mu} = 1 + 2\chi_2 a \cot\chi_2 b;$$

provided that for both outer regions $k_\infty > 1$. Therefore,

$$(1 + 2\chi_1 a \cot\chi_1 b)(1 + 2\chi_2 a \cot\chi_2 b) = 1.$$

This can also be expressed as

$$\frac{\tan\chi_1 b}{2\chi_1 a} + \frac{\tan\chi_2 b}{2\chi_2 a} + 1 = 0.$$

c) The gradient is

$$\frac{\bar{\Phi}_1}{\bar{\Phi}_2} = \frac{\lambda \sin\chi_2 b\,(1-\cos\chi_1 b)/(\chi_1 b)}{\mu \sin\chi_1 b\,(1-\cos\chi_2 b)/(\chi_2 b)}.$$

d) Employing a limited expansion gives

$$v = -u + \frac{4u^2}{\pi} + \lambda\pi u^2 + \cdots,$$

where

$$u \simeq -\frac{\pi}{4}\frac{k_{\infty s}}{k_{\infty s}-1}\rho_1, \qquad v \simeq -\frac{\pi}{4}\frac{k_{\infty s}}{k_{\infty s}-1}\rho_2,$$

and

$$\frac{\bar{\Phi}_1}{\bar{\Phi}_2} \simeq \frac{\lambda}{\mu}.$$

e) Due to u's relatively large coefficient, is better to use the exact formula for the latter ratio. This is

$$\frac{\lambda}{\mu} = \frac{1}{1 + (2a/b)(\pi/2 - u)\tan u}$$

The ratio of the average fluxes is as follows.

ρ (pcm)	u	λ/μ
100	−0.0150	1.105
250	−0.0375	1.318
500	−0.0750	1.979
750	−0.1125	4.179

Exercise 6.12: propagation of a wave

a) The equation to be solved is

$$D\,\Delta\Phi - \Sigma_a\,\Phi + \delta(\vec{r})\,e^{i\omega t} = \frac{1}{v}\frac{\partial\Phi}{\partial t}.$$

Due to the spherical symmetry and time dependence of the system, the flux is

$$\Phi(\vec{r}, t) = \varphi(r)\,e^{i\omega t},$$

where φ can have complex values. Substituting this into the flux equation, and simplifying, gives

$$D\,\Delta\varphi - \left(\Sigma_a + i\frac{\omega}{v}\right)\,\varphi + \delta\left(\vec{r}\right) = 0.$$

This equation is the point kernel in diffusion theory with a complex 'absorption cross-section' $\Sigma_a + i\omega/v$. Its solution is

$$\varphi(r) = \frac{e^{-\kappa r}}{4\pi D r}.$$

Here we must use a complex κ coefficient,

$$\kappa^2 = \frac{1 + i\omega\ell}{M^2},$$

where $\ell = 1/(v\Sigma_a)$ is the neutron lifetime.

It can be shown that

$$\Phi(r, t) = \frac{e^{-r/L + i(\omega t - \theta)}}{4\pi D r},$$

where the relaxation length is

$$L = \frac{M}{(1 + \omega^2\ell^2)^{1/4}\cos(\alpha/2)}, \qquad \text{with} \qquad \alpha = \arctan\omega\ell,$$

and where the phase displacement is

$$\theta = \frac{r}{M}\left(1 + \omega^2 \ell^2\right)^{1/4} \sin\,(\alpha/2)\,.$$

b) In this case, the formulae are

$$L = M \frac{\left(\lambda^2 + \omega^2\right)^{1/4}}{\sqrt{\beta\omega}\cos\,(\alpha/2)}, \quad \text{and} \quad \theta = \frac{r}{M}\frac{\sqrt{\beta\omega}\sin\,(\alpha/2)}{\left(\lambda^2 + \omega^2\right)^{1/4}}, \quad \text{with} \quad \alpha = \arctan\frac{\lambda}{\omega}.$$

7 Neutron slowing down

Introduction

After examining neutron physics with respect to time and then with respect to space, we shall now approach the subject from the point of view of neutron energy.

This aspect of neutron physics is important for the study of any type of reactor.

- In *thermal neutron reactors*, a difficult problem arises: the crossing of capture resonance traps, particularly those of uranium 238 between a few electron volts and a few thousand electron volts. The resonance escape probability p (see § 3.3.2) in particular must be correctly evaluated, because it has a major effect on the feasibility of a reactor concept.

- In *fast neutron reactors*, there are significant variations in cross-section in the region covered by the neutron spectrum. The neutron balance is therefore very sensitive to the shape of this spectrum, which means that it too must be calculated carefully by taking into account the details of neutron slowing down and the competition between slowing down and absorption.

Neutron slowing down occurs by a series of scattering events whose mechanism we must examine in detail. We shall see that, unlike the laws of absorption, which are simple at high energy and complicated at lower energies (because of resonance), the scattering laws are relatively complicated at high energy (the importance of inelastic and anisotropic aspects), but simpler otherwise. Fortunately, this means that some of the difficulties can be decoupled.

This chapter mainly discusses scattering. The problem of resonant absorption will be presented in the next chapter. We shall then discuss the thermal domain, and then Chapter 10 will present the multi-group processing that is used for almost all neutron physics calculations involving the spectrum.

The present chapter is divided into three parts: the laws of neutron scattering, the slowing down equation, and an examination of a few "academic" problems.

7.1. Scattering collision laws

7.1.1. Elastic and inelastic scattering

Note that a collision is *"elastic"* when kinetic energy is conserved, and *"inelastic"* otherwise, i.e. if some of the energy has gone towards modifying the internal state of the "target"[1]. In the present case, scattering is inelastic if the target nucleus, initially at its fundamental energy level, reaches an excited state after interaction with the neutron. This nucleus will later decay by gamma emission.

Chapter 2 (Table 2.1) explained the difference between several scattering mechanisms:

- *Potential scattering* (always *elastic*) corresponds to a single diffusion of the wave associated with the neutron by the potential field of the nucleus. This reaction can be seen on all nuclei with neutrons of any energy, characterised by a cross-section of the order of a few barns.

- *Resonant scattering* corresponds to the absorption of the incident neutron, the formation of a compound nucleus, and then the re-emission of a neutron[2]; this reaction is characterised by a resonant structure, and therefore a cross-section that can change quickly according to the energy between values that can be very high or very low. There are several possible exit channels. Concerning scattering:

  - If, after ejection of the neutron, the target nucleus is at the fundamental level (same as the initial state), the scattering is *elastic*.

  - If, after ejection of the neutron, the target nucleus is excited, the scattering is *inelastic*.

Elastic scattering has no threshold, which means that it can occur with neutrons of any energy. Inelastic scattering, however, has a reaction threshold: the incident neutron must contribute at least the energy required to take the target nucleus from the fundamental level to the first excited level. This threshold is a few MeV for light nuclei, and a few tens of keV for heavy nuclei. This means that, in reactors, inelastic scattering will mainly be observed in the fuel materials, particularly uranium 238. If necessary, reactions and the associated cross-sections can be distinguished according to the excited level (discrete or continuous) of the impacted nucleus.

Note also that $(n, 2n)$ reactions, essentially on uranium 238, slightly improve the neutron balance (in practice, between 100 and 200 pcm): this is allowed for by the calculation codes, but can be neglected here.

Elastic scattering will play the most important role in neutron slowing down, particularly in thermal neutron reactors containing a moderator. This is why we mention inelastic scattering for information only, and we shall concentrate on elastic slowing down.

For this, note that in neutron physics it does not matter whether the mechanism is potential or resonant, as long as the cross-section for the sum of the two processes is correct. We shall therefore no longer make this distinction.

---

[1] The collision between two ivory billiard balls is nearly elastic, but if the balls were made of modelling clay, the collision would be highly inelastic.

[2] Using a classic image, we could say that, in potential scattering, the incident neutron leaves, while in resonant scattering, any neutron of the compound nucleus is ejected.

7.1.2. Laws of elastic collision

The laws of elastic collision can be established using the assumptions of a purely classical mechanics problem.

In the laboratory system, i.e. a reactor (upper diagram, Figure 7.1), we have a neutron (little black ball) that is initially travelling towards a nucleus (large black ball), which we can consider to be at rest. After the collision, the two objects move away, each with a certain velocity, in different directions. The neutron goes from an initial speed V_{ni} to a final speed V_{nf} with a deflection angle ψ note that this process occurs in three-dimensional space: the figure represents the plane of vectors $\vec{V}_{ni}$ and $\vec{V}_{nf}$; the problem has a rotational symmetry with respect to the axis containing $\vec{V}_{ni}$, which means that all the events characterised by a rotation by an angle φ about this axis are equiprobable. The relationships between these velocities and this angle ψ result from the laws of conservation of momentum and of kinetic energy. The calculation can be performed in the laboratory system, but it is far simpler to use the centre of mass system.

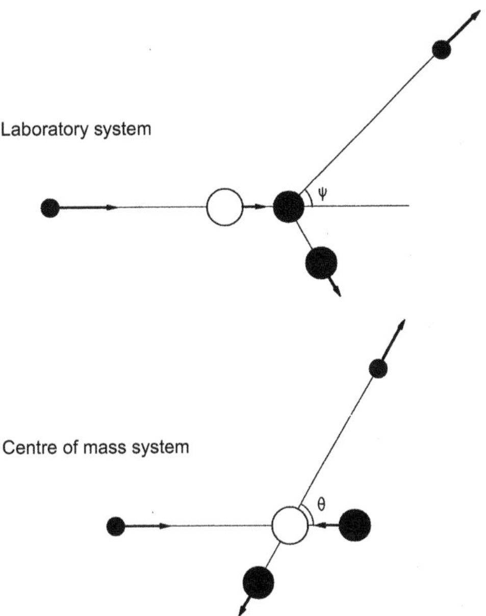

Figure 7.1. Elastic scattering of a neutron by a nucleus initially at rest.

The centre of mass (or centre of gravity, or barycentre) is the point located on the segment joining the two particles at distances inversely proportional to their mass. This point is represented by a fictitious white ball, which has all the mass allocated to it.

In the centre of mass system (whose origin is fixed at this point), scattering occurs according to the diagram at the bottom of Figure 7.1: before the reaction, the two balls are heading towards each other with speeds that are inversely proportional to their masses. After the reaction, they move away from each other with speeds that are also inversely proportional to their masses. This simply means that the momentum of this system is initially zero and remains zero. Because the kinetic energy has not changed, the moduli of the velocity vectors have not changed.

In this system, a simple rotation through an angle θ has occurred (ignoring the angle φ, which does not enter into this discussion).

Let A be the ratio of the mass of the target nucleus to the mass of the neutron (because the mass of the neutron is close to one atomic mass unit, A is close to the mass number of the nucleus). The elastic collision formulae are obtained by applying the following three statements:

1/ In the laboratory system, the velocity of the centre of mass is given by:

$$\vec{V}_{mi} = \vec{V}_{mf} = \vec{V}_m = \frac{1}{A+1}\vec{V}_{ni} + \frac{A}{A+1}\vec{V}_{Ni} = \frac{1}{A+1}\vec{V}_{ni}, \tag{7.1}$$

(i: initial; f: final; n: neutron; N: target nucleus; m: centre of mass; $\vec{V}_{Ni}$ is zero).

2/ The coordinate system change formulae (before and after the collision, and for each ball) are:

$$\vec{v} = \vec{V} - \vec{V}_m, \tag{7.2}$$

(V: laboratory system; v: centre of mass system).

3/ In the centre of mass system, the velocity components of the neutron along the axis of the initial speed and the perpendicular axis in the plane of the figure are $v_{ni}\cos\theta$ and $v_{ni}\sin\theta$.

Note that, if the initial velocity of the neutron V_{ni} and its kinetic energy E_{ni} are fixed, the post-collision variables are all random variables that depend on the impact parameter. In classical mechanics, the value of θ depends on the distance d from the centre of the target to the line containing $\vec{V}_{ni}$; in quantum mechanics, θ and φ can be kept as the main random variables. The other random variables are dependent on θ only, because φ has no effect for reasons of symmetry.

Without giving details of the calculations, here are the two main formulae thus obtained[3]:

1/ *Final kinetic energy of the neutron in the laboratory system:*

$$\frac{E_{nf}}{E_{ni}} = \frac{A^2 + 1 + 2A\cos\theta}{(A+1)^2} = \frac{1}{2}[1 + \alpha + (1-\alpha)\cos\theta]; \tag{7.3}$$

2/ *Neutron deflection angle in the laboratory system:*

$$\cos\psi = \frac{1 + A\cos\theta}{\sqrt{A^2 + 1 + 2A\cos\theta}}. \tag{7.4}$$

In the first formula, we set:

$$\alpha = \frac{(A-1)^2}{(A+1)^2}; \tag{7.5}$$

[3] This discussion is restricted to formulae concerning the neutron, but obviously the formulae for the nucleus could be determined in a similar way if required.

This parameter is the minimum ratio between the final energy and the initial energy of the neutron, obtained when θ is equal to π (180°). Note that this value decreases as the mass of the target nucleus decreases, which shows that these nuclei are better at slowing down neutrons. In particular, for hydrogen[4] α is zero, which means that it would be possible for a neutron to be completely halted by a single collision.

7.1.3. Laws of elastic and isotropic collision

It might be obvious that all values of φ are equiprobable, but it is more difficult to be sure about the probability distribution governing θ. Experience and certain theoretical considerations (refer to the overview in § 2.7.1), indicate that scattering can be considered as *isotropic in the centre of mass system*[5], at least in the epithermal domain where most of the higher resonances of heavy nuclei are found, but possibly not at very high energy. This assumption is valid in particular for the moderator materials that are most often used in thermal neutron reactors.

This isotropy means that, in this system, the post-collision neutron speed vector $\vec{V}_{nf}$[6] has the same probability of pointing in any direction of space; the probability of having φ to within $d\varphi$ and θ to within $d\theta$ is the solid angle element bounded by $d\varphi$ and $d\theta$, normalised to 4π steradians, i.e. $\sin\theta\, d\theta\, d\varphi/(4\pi)$. By integrating over the 2π radians through which φ can vary, we see that, with this assumption of isotropy, the random variable θ is governed by the following probability distribution:

$$P(\theta)\, d\theta = \frac{1}{2} \sin\theta\, d\theta = \frac{1}{2} d|\cos\theta|. \tag{7.6}$$

Because the post-collision neutron energy E_{nf} varies linearly with $\cos\theta$, and because $\cos\theta$ is uniform[7], the same applies to E_{nf} in the interval that can be reached, i.e. between αE_{ni} and E_{ni}:

$$P(E_{nf})dE_{nf} = \frac{dE_{nf}}{(1-\alpha)E_{ni}}. \tag{7.7}$$

The probability distribution for the deflection angle ψ in the laboratory system is not quite so simple: because ψ is always less than θ (except for the limit values 0 and π), isotropy in the centre of mass means an *anisotropy in the forward direction* in the reactor system (scattering towards the front is favoured). In particular, when the mean λ of $\cos\theta$ is zero, the mean of $\cos\psi$ is:

$$\mu = \langle \cos\psi \rangle = \frac{2}{3A}. \tag{7.8}$$

This forward anisotropy increases with decreasing nucleus mass; for example, μ is 2/3 for hydrogen, but close to zero for uranium.

4 Taking the same mass for the proton and the neutron gives A equal to 1.

5 In the formalism introduced in § 2.7.1, scattering is said to occur according to the "*s wave*".

6 As well as the nucleus speed vector.

7 Probability density equal to 1/2 in the interval [−1, +1].

7.1.4. Lethargy

For both practical reasons (drawing cross-section or neutron spectrum curves, etc.) and theoretical reasons (the law of elastic collision directly governs an energy *ratio*), neutron physicists use the *logarithm* of neutron kinetic energy. This is why a variable known as "lethargy" has been introduced. It is defined as follows:

$$u = \ln \frac{E_{ref}}{E}. \tag{7.9}$$

- Because a logarithm refers to a dimensionless number, the energy E must be normalised to a reference energy E_{ref}.
- Neutrons slow down, at least during the first part of their "life", and so it is convenient to put E in the denominator so that lethargy increases with time; in a way, u then becomes a measure of the "age" of the neutron as it slows down.
- The choice of reference energy is completely arbitrary. In calculation codes such as APOLLO, 10 MeV is often used. If neutrons with energy above 10 MeV are neglected, all the neutrons in the calculation then have positive lethargy. Different choices can be made for the sake of convenience.

After a change of variable (refer to the equation in § 3.2.2), the collision law for neutron energy becomes a *lethargy gain* relationship:

$$u_f - u_i = -\ln \frac{A^2 + 1 + 2A\cos\theta}{(A+1)^2} = -\ln \left\{ \frac{1}{2}[1 + \alpha + (1-\alpha)\cos\theta] \right\}. \tag{7.10}$$

This lethargy gain is written as Δu or w. We can see that it varies between 0 (for $\theta = 0$) and a maximum value (for $\theta = \pi$):

$$w_{max} = \varepsilon = -\ln \alpha,$$

which is finite, except for hydrogen.

If scattering is assumed to be isotropic in the centre of mass system, then the uniform distribution for energy becomes a decreasing exponential distribution for lethargy gain, to be taken between 0 and ε and normalised:

$$P(w)dw = \frac{e^{-w}}{1-\alpha} dw. \tag{7.11}$$

The energy and lethargy gain distributions are compared in Figure 7.2 (curves calculated for $A = 4$).

Average values can be deduced from these probability distributions as follows:

$$\langle E_{nf} \rangle = \frac{1+\alpha}{2} E_{ni}, \quad \langle w \rangle = \xi = 1 - \frac{\alpha\varepsilon}{1-\alpha}. \tag{7.12}$$

Note: Because energy and lethargy are not related in a linear fashion, the average post-collision energy and the average post-collision lethargy do not correspond to each other.

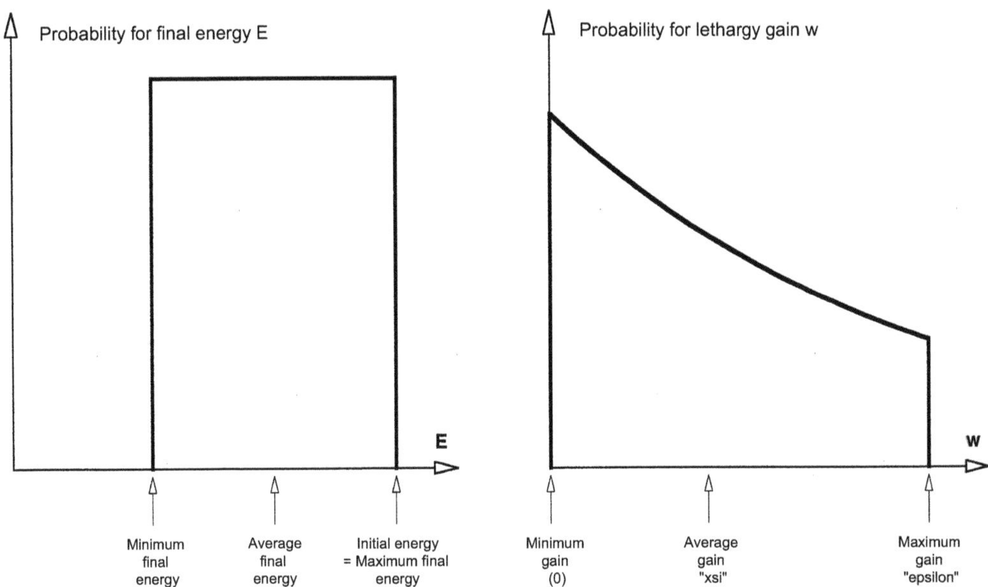

Figure 7.2. Probability distributions for energy and lethargy for an isotropic elastic collision in the centre of mass system.

7.1.5. Evaluating the number of collisions required to slow down a neutron

The parameter ξ turns out to be the most useful one for characterising the "efficiency" of slowing down by a nucleus; it is the average "progress" of the neutrons, in terms of lethargy, on the "path" to slowing down. In particular, it allows us to evaluate the average number of scattering events[8]) required to slow down a neutron from energy E_0 to energy E_1.

Because neutrons advance by ξ lethargy units on average at each collision, in order to overcome the total lethargy interval $U = \ln E_0/E_1$, they need an average number n of collisions such that $n\xi = U$. Table 7.1 gives a few examples for a few values of A, and $E_0 = 2$ MeV and $E_1 = 1$ eV, i.e. 14.51 lethargy units.

As we have mentioned, in hydrogen a single collision can stop a neutron completely. This can only occur in very rare circumstances, however, because about fifteen collisions are required on average to slow down the neutron. It seems that deuterium, despite being twice as heavy as hydrogen, is almost as efficient in terms of the number of collisions required to slow down a neutron. Even carbon 12 is reasonably efficient: in the absence of other materials, 92 scattering events in graphite would be a realistic figure in view of its low capturing ability. On the other hand, it is obvious that a neutron emitted in a block of uranium would be absorbed long before slowing down could occur.

[8] Assuming all collisions are elastic and isotropic in the centre of mass system.

Table 7.1. Slowing down parameters for a few nuclei.

Noyau	A	α	ε	ξ	n
H	1	0	∞	1	15
D	2	0.111	2.197	0.725	20
C	12	0.716	0.334	0.158	92
U	236	0.9832	0.0170	0.0084	1717

7.1.6. Comparison of the main moderators

These considerations show that light nuclei should be used to moderate neutrons. These nuclei should also have low capture, because otherwise many neutrons would disappear rather than being slowed down.

The light nuclei with low capture are as follows[9]:

^{2}**H**, ^{4}**He**, **Be**, **C**, and **O**.

Helium should be removed from this list because it exists only as a gas (the first of the rare gases), and therefore its density is too low for it to be used as a moderator.

Ordinary hydrogen, however, should be considered despite it relatively high capture, because it is the best at slowing down neutrons.

The main liquid or solid materials that contain these nuclei and that are therefore candidates for use as moderators are listed in Table 7.2 [10]. The neutron physics properties of these materials in terms of their ability to slow neutrons and their capture rate are summarised in columns 4 and 5. If both criteria are favourable, a natural uranium reactor could be possible. This holds true for the last three examples, but not water (and other hydrogenated materials) because of its mediocre capture properties.

Table 7.2. Comparison of the main characteristics of the most common moderator materials.

Moderator	Formula	State	Slowing	Capture	Cost	Natural U
Water	H_2O	Liquid	Excellent	Mediocre	Null	Impossible
Heavy water	D_2O	Liquid	Excellent	Excellent	High	Possible
Glucine	BeO	Solid	Average	Good	Average	Possible
Graphite	C	Solid	Average	Good	Average	Possible

The reason that most present-day electronuclear reactors are moderated by water, however, is that besides this relatively high capture cross-section for thermal neutrons, hydrogen, particularly in the form of water, provides three key advantages:

1/ very low cost (only a purity check is required);

2/ useful thermodynamic properties that are well controlled under industrial conditions; in particular, the possibility of using water not only as a moderator, but also as a coolant, and possibly, in the form of vapour, as the fluid sent to the turbine;

[9] The natural helium found on Earth contains practically no helium 3, which has very high capture. Helium 3 is obtained via radioactive decay of tritium. Natural beryllium has only one isotope. Both natural carbon isotopes and the three natural oxygen isotopes have low absorption.

[10] We could also have included hydrogenated carbon compounds such as organic liquids. Their neutron physics properties are similar to those of water (comparing equal numbers of hydrogen atoms per unit volume) because hydrogen plays the main role in all of these compounds, in terms of both slowing and capture.

Table 7.3. Comparison of the main moderator nuclei (the cross-section is taken for epithermal neutrons [a few eV] and the absorption cross-section for thermal neutrons [0.0253 eV]).

Nucleus	Mass	Average lethargy gain	Scattering cross-section	Absorption cross-section
Hydrogen	1.00	1.0000	20.4	0.332
Deuterium	2.00	0.7261	3.40	0.00051
Beryllium	8.93	0.2080	6.00	0.0076
Carbon	12.01	0.1589	4.74	0.00337
Oxygen	15.86	0.1209	3.89	0.000191

Table 7.4. Comparison of the main moderator materials

Material	Density	Concentration	Moderating power	Relative value (material/water)
Ordinary water	998	0.03337	137.72	1
Heavy water	1105	0.0332	17.95	0.130
Beryllium	1850	0.124	15.48	0.112
Beryllia	3010	0.0725	12.46	0.090
Graphite	1600	0.0802	6.04	0.044

3/ exceptional *moderating power*: this parameter turns out to provide the best measure of a material's ability to slow down neutrons. It is the product $\xi \times \sigma_s$ if calculated on a per-atom basis, or $\xi \times \Sigma_s$, if applying the reasoning per unit volume. For a material to have good slowing down properties, it is not sufficient for the average lethargy gain at each scattering to be as large as possible; there must also be the greatest possible number of scattering events (with respect to absorptions, particularly in the resonances); therefore the scattering cross-section must be as large as possible. Tables 7.3 and 7.4 show that hydrogen and hydrogenated materials are clearly distinct from other materials. The reason for this is the exceptional value of the scattering cross-section of the smallest nucleus: hydrogen[11]. Finally, thanks to this tremendous moderating power, water reactors can also be very compact[12].

The use of hydrogen entails one very serious disadvantage, however: the unavoidable requirement to use an enriched fuel[13].

7.1.7. Laws of inelastic collision

The properties of moderators are related to elastic scattering. On heavy materials, in particular the uranium in the fuel, inelastic scattering is responsible for most of the decay in the neutron spectrum, because the lethargy gain by elastic collision is practically zero.

[11] Masses are normalised to the mass of a neutron. Cross-sections are expressed in barns, and the moderating power $\xi\Sigma_s = \xi\sigma_s N$ in m^{-1}; (cold) densities are stated in kg/m^3 and concentrations N in 10^{30} molecules per m^3.

[12] For example, the ratio of the moderator volume to the fuel volume is in the region of 50 in graphite and natural uranium reactors, and in the region of 2 in pressurised water reactors.

[13] Approximate values for the minimum content of isotope 235 required to reach the critical condition in an infinite, water-moderated uranium reactor: 1.04% in homogeneous configuration; 0.80% in heterogeneous configuration.

The relative role of inelastic scattering is low in thermal neutron reactors, but non-negligible in fast neutron reactors.

The proof is similar to that used to establish the laws of elastic collision, so here we simply give the two main formulae for inelastic collision:

$$\frac{E_{nf}}{E_{ni}} = \frac{\gamma^2 + 1 + 2\gamma\cos\theta}{(A+1)^2}, \tag{7.13}$$

and:

$$\cos\psi = \frac{1 + \gamma\cos\theta}{\sqrt{\gamma^2 + 1 + 2\gamma\cos\theta}}, \tag{7.14}$$

with:

$$\gamma = A\sqrt{1 - [(A+1)/A]Q/E_{ni}}, \tag{7.15}$$

where Q is the excitation energy of the target nucleus. It is possible to verify that this parameter γ is reduced to A if Q is zero.

Note that the minimum energy of the neutron for an inelastic collision is:

$$E_{threshold} = \frac{A+1}{A}Q. \tag{7.16}$$

Inelastic scattering can be isotropic in the centre of mass system, and therefore also in the laboratory system, if it occurs on a heavy nucleus.

7.1.8. Slowing down equation

The slowing down equation is simply the reduction of the Boltzmann equation (see § 3.2.3) to the case involving only one variable: the neutron velocity v — or a variable related to it, kinetic energy E or lethargy u — taken in the slowing down domain. To obtain a better grasp of the physical aspects of this slowing down equation, let us examine the problem from its starting point. To obtain a pure slowing down problem, we place ourselves in the situation of an *infinite, homogeneous medium fed by a source that is uniform in space and constant over time* [14]. We shall use the lethargy variable, which turns out to be the most relevant one from a physical standpoint.

The Boltzmann equation in general and the slowing down equation in particular are obtained by performing a *neutron count*. With regard to slowing down, we can imagine two ways of counting the neutrons: this leads to two forms of the equation that are equivalent from a mathematical point of view but not from a physical or problem-handling point of view. The first form is the one most commonly encountered; most notably, it is the equation that is processed numerically in most calculation codes, such as APOLLO. The second form is of interest for certain problems and for a better understanding of the slowing down mechanism, and so it will also be presented, although the reader can omit this section if desired.

[14] A similar, pure slowing-down problem would be obtained by taking the integral over all space of the neutron functions of any system fed by a source that is constant over time.

7.1.9. First form of the slowing down equation

The first form of the slowing down equation involves the *"arrival density"*, written as $\rho(u)$. This density[15] is defined as follows: $\rho(u)du$ is the number of neutrons arriving per unit time and per unit volume in the lethargy interval du situated between u and $u + du$, following a scattering event to another lethargy u' (prior to u, since slowing down is occurring).

The scattering rate[16] to lethargy u' is written as $\Sigma_s(u')\Phi(u')$ using the reaction rate equation; we also write $P(u' \to u)du$ to denote the probability that a neutron scattered at lethargy u' will be transferred in the lethargy interval du between u and $u + du$. These parameters can be used to express the arrival density based on its definition:

$$\rho(u) = \int_{-\infty}^{u} \Sigma_s(u')\Phi(u')du'P(u' \to u), \tag{7.17}$$

(simplifying by du on both sides, and expressing the fact that u' is prior to u).

Note: We often set:

$$\Sigma_s(u' \to u) = \Sigma_s(u')P(u' \to u). \tag{7.18}$$

This parameter, which retains the dimensions of a cross-section, is known as the "(lethargy[17]) *differential scattering cross-section*". The arrival density can now be written more simply:

$$\rho(u) = \int_{-\infty}^{u} \Sigma_s(u' \to u)\Phi(u')du'. \tag{7.19}$$

To the neutrons *arriving* at lethargy u (to within du) after a scattering event, must be added the neutrons created directly at this lethargy u (to within du) by emission from the source. This second density[18] is written as S.

In the present problem, there can be only one outcome for neutrons placed by scattering or by emission at lethargy u (to within du): a collision (scattering or absorption). The sum $\rho du + S du$ is therefore equal to the total collision rate $\Sigma\Phi du$ (the subscript "t" for total cross-section is understood). The slowing-down equation giving the flux $\Phi(u)$ resulting from the sources $S(u)$ is therefore:

$$S(u) + \rho(u) = S(u) + \int_{-\infty}^{u} \Sigma_s(u' \to u)\Phi(u')du' = \Sigma(u)\Phi(u). \tag{7.20}$$

Unless the transfer equation has special properties, this type of *integral equation* can only be solved numerically.

In the specific case of isotropic (in the centre of mass system), monatomic (by nuclei all having the same mass A), elastic slowing down, the transfer probability is:

$$P(u' \to u) = P(u - u') = \frac{e^{-(u-u')}}{1 - \alpha}, \tag{7.21}$$

[15] In the mathematical sense of the term (see § 3.2.2).

[16] This is also a density: $\Sigma_s(u')\Phi(u')du'$ is the number of neutrons travelling in the lethargy interval du' and scattered per unit time and per unit volume.

[17] To distinguish it from the solid angle differential cross-section.

[18] Same dimensions as ρ.

if the difference $u - u'$ is between 0 and ε and zero otherwise. The equation is then written:

$$S(u) + \int_{u-\varepsilon}^{u} \Sigma_s(u')\Phi(u')P(u-u')du' = \Sigma(u)\Phi(u). \tag{7.22}$$

Applying the following function change:

$$f(u) = \Sigma_s(u)\Phi(u), \tag{7.23}$$

the equation takes the following form:

$$S(u) + \int_{u-\varepsilon}^{u} f(u')P(u-u')du' = \frac{\Sigma(u)}{\Sigma_s(u)} f(u). \tag{7.24}$$

The integral is the arrival density ρ. This is clearly a *convolution product*, written $\rho = f * P$, which might point to an analysis using the Laplace transform[19].

7.1.10. Second form of the slowing-down equation

To characterise transfers, a second count can be performed by counting not the neutrons that *"fall to"* a lethargy u (to within du), but the neutrons that *"jump over"* a lethargy u. This number is known as the *"slowing down current"* and is written as $q(u)$. It is analogous to the number of cubic metres of water per second that could be estimated by standing on a bridge over a river. Unlike ρ, which is a density and therefore a differential concept [if the interval du tends to zero, the number of neutrons arriving, $\rho(u)du$, is vanishing], the current $q(u)$ is a true function[20].

By writing out this definition in the form of an equation, we can see that q is given by a double integral that counts all the transfers of a lethargy u' prior to u towards a lethargy u'' subsequent to u:

$$q(u) = \int_{-\infty}^{u} \int_{u}^{\infty} \Sigma_s(u')\Phi(u')du'P(u' \to u'')du'', \tag{7.25}$$

or:

$$q(u) = \int_{-\infty}^{u} \int_{u}^{\infty} \Sigma_s(u' \to u'')\Phi(u')du'du''. \tag{7.26}$$

Let us now compare the currents seen by two observers placed at u and $u + du$. Only the second observer will count the neutrons emitted by the source in this interval, and only the first observer will count the neutrons absorbed in this interval:

$$q(u+du) - q(u) = \frac{dq(u)}{du}du = S(u)du - \Sigma_a(u)\Phi(u)du,$$

or, after simplification by du:

$$\frac{dq(u)}{du} = S(u) - \Sigma_a(u)\Phi(u). \tag{7.27}$$

[19] This *translation invariance* along the axis of the lethargies appears with the choice of this variable instead of energy.

[20] This current is a true function with respect to lethargy, but remains a density with respect to time and space. It is a number of neutrons passing u per unit time and per unit volume. That is why this parameter used to be called the "slowing down density".

Combined with the definition of q, this equation constitutes the second form of the slowing down equation.

By deriving the equation defining q with respect to u, (it must be derived with respect to two of the integral boundaries), we obtain the following *identity*:

$$\frac{dq(u)}{du} = \Sigma_s(u)\Phi(u) - \rho(u). \tag{7.28}$$

By substituting this identity into the second form of the slowing down equation, we obtain the first form. This demonstrates that the two forms are mathematically equivalent.

By setting:

$$R(u' \to u) = \int_u^{+\infty} P(u' \to u'')du'', \tag{7.29}$$

the current can be represented by a simple integral:

$$q(u) = \int_{-\infty}^{u} \Sigma_s(u')\Phi(u')du' R(u' \to u). \tag{7.30}$$

For monatomic, isotropic, elastic slowing down, R, like P, is a function only of the difference $u - u'$:

$$R(u' \to u) = R(u - u') = \frac{e^{-(u-u')} - \alpha}{1 - \alpha}. \tag{7.31}$$

(Since the lethargy gain is between 0 and $\varepsilon = -\ln\alpha$, R is between 0 and 1).

Note that, like ρ, q is expressed as a function of the scattering density f by a *convolution product*:

$$q(u) = \int_{u-\varepsilon}^{u} f(u')R(u - u')du' = (f * R)(u). \tag{7.32}$$

7.2. Analysis of a few specific problems

7.2.1. General remarks

It has already been stated that there is no analytical solution for the general case of the slowing down equation. Even for the simplest case of monatomic, isotropic, elastic slowing, rigorous solutions are rare. Those that have been obtained required mathematical manipulations to go from the integral equation to a differential equation. What follows is an almost complete list of the known analytical solutions.

Unless otherwise indicated, the equation studied below refers to the monatomic, isotropic, elastic case.

7.2.2. Decay of the neutron spectrum by successive scattering events

We have seen that neutrons emitted at an energy E_0, which therefore undergo a collision at this energy $E_{ni} = E_0$, under the assumption of scattering have an energy $E_1 = E_{nf}$ that is uniformly distributed in the interval between αE_0 and E_0.

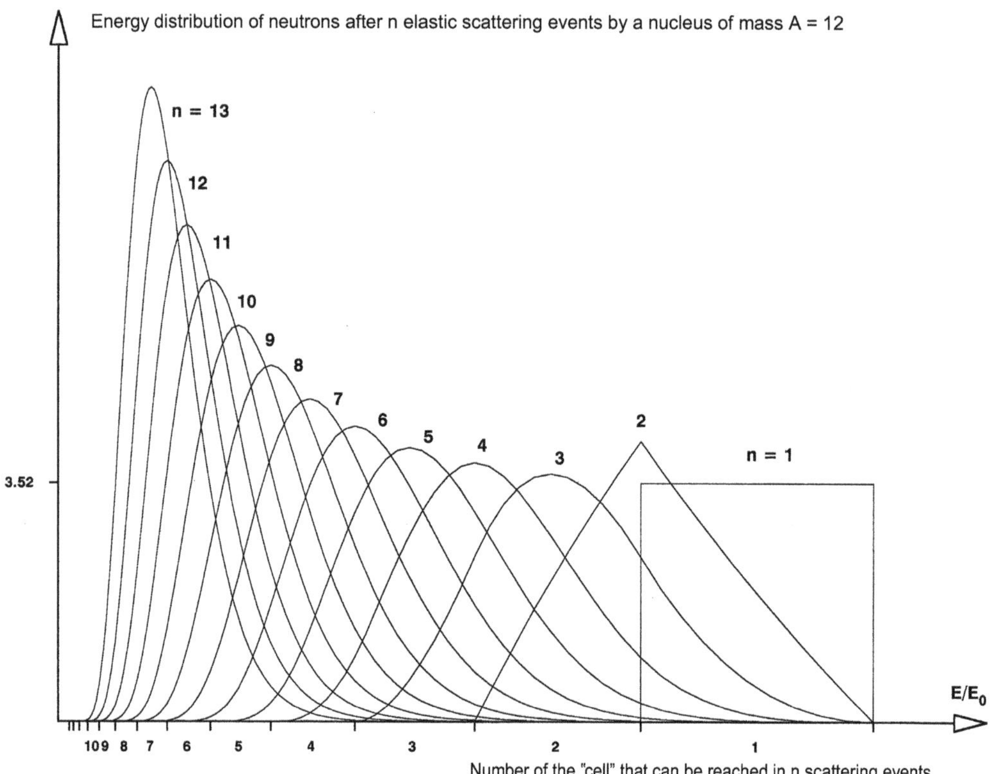

Figure 7.3. Energy distribution, according to *n*, of neutrons emitted at an energy E_0 and scattered *n* times by graphite (the "cell" *n* is the interval that can be reached only after at least *n* collisions).

Still under the assumption of scattering, after two collisions the energy E_2 is divided between α times the minimum of E_1, i.e. $\alpha^2 E_0$ and the maximum of E_1, i.e. E_0. The distribution relationship is calculated by taking the lethargy variable and by convoluting the distribution of E_1 with the transfer equation.

Similarly, and maintaining the assumption of scattering, the energy E_3, after three collisions, is divided between α times the minimum of E_2, i.e. $\alpha^3 E_0$ and the maximum of E_2, i.e. E_0. The distribution relationship is calculated by convoluting the distribution of E_2 with the transfer equation. And so on.

These successive convolutions are somewhat long and tedious to calculate[21]. A relatively simple, easily programmable recurrence formula can then be found. Figure 7.3 shows the successive curves thus obtained for the case of carbon (similar networks of curves can be drawn with the variables v and u). Two interesting characteristics of the slowing down mechanism can be observed: gradual degradation of the spectrum and, at the same time, its increasingly accentuated lethargy dispersion as n, the number of collisions, increases.

[21] These calculations were performed by Philippe Jourdan (private communication).

7.2.3. Slowing down without absorption

Still in the absence of absorption, what is the flux resulting from a source of neutrons placed at a given energy E_0, i.e. the sum over n of all the preceding curves? The answer has been known ever since George Placzek performed the calculation in 1946 and obtained the now-famous "Placzek transient".

To calculate this flux, Placzek did not list the distributions according to n, but solved the slowing down equation directly.

Firstly, note that the second slowing down equation gives the following relationship if we take the origin of the lethargies at E_0, normalise the source to S and represent it by $S\delta(u)$, and cancel the absorption:

$$q(u) = S\Upsilon(u), \tag{7.33}$$

where Υ is the Heaviside step function. This simply means that, at a higher energy than that of the source, there are obviously no neutrons, and at any lower energy, there is a current of S neutrons slowing down, since they do not disappear by absorption. This observation might allow us to simplify the calculation in the following ways (without entering into details):

- by calculating the flux analytically from one term to the next using the solution of a differential equation in the interval $[0, \varepsilon]$, followed by the interval $[\varepsilon, 2\varepsilon]$, and then the interval $[2\varepsilon, 3\varepsilon]$, and so on;

- by determining the asymptotic value of the flux:

$$\Phi_{as}(u) = \frac{S}{\xi\Sigma_s(u)}. \tag{7.34}$$

Figure 7.4 shows the flux curve calculated in this way for a heavy nucleus (large A). (The figure is similar for all values of A, except $A = 1$, where the asymptotic behaviour is observed immediately.)

A discontinuity in the flux is observed at $u = \varepsilon$, a discontinuity in its derivative at $u = 2\varepsilon$, a discontinuity in its second derivative at $u = 3\varepsilon$, etc. The physical explanation for this is as follows: all neutrons have their first collision at the lethargy of origin; at $\varepsilon - 0$, there are neutrons that have undergone 1, 2, 3, or more collisions (refer to the distribution curves as a function of the number of collisions); at $\varepsilon + 0$, it is possible to find neutrons that have undergone 2, 3, or more collisions, but no first-collision neutrons that were not able to exceed ε at the first collision, which explains the discontinuity of the flux. The subsequent discontinuities, which are not as strong, can be explained in a similar way.

The asymptotic behaviour of the scattering density $\Sigma_s(u)\Phi(u)$ has practically been reached by the fourth interval. It is easy to understand why the value is S/ξ: because each neutron advances by an average of ξ at each scattering, and if the scattering density $\Sigma_s\Phi$ is constant in terms of lethargy, the current is the product of these two terms; as we have seen, this current is equal to S in this case: giving this result[22].

[22] This image will be used again in the next chapter: imagine a troop of kangaroos passing in front of you at a rate of q kangaroos per second. The kangaroos jump an average of ξ metres at each bound, and therefore, in each ξ-metre interval, you can count q kangaroos per second touching the ground. This gives $f = q/\xi$ per second per metre.

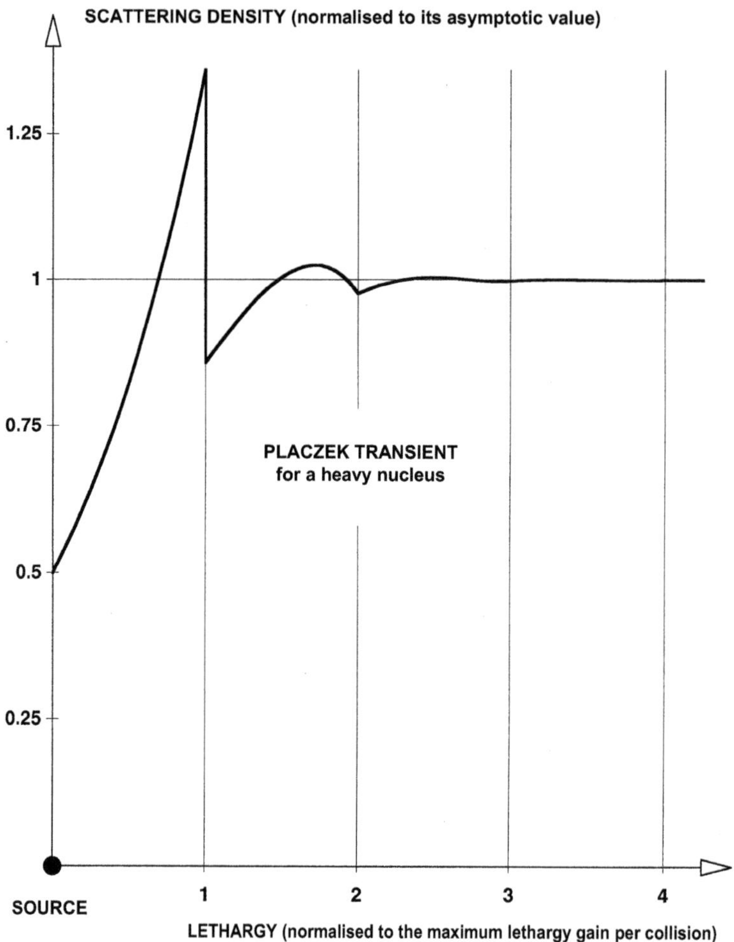

Figure 7.4. Placzek transient for a heavy nucleus.

A general form of the equation can be obtained by neglecting the Placzek transient, giving the following approximate relationship between flux and current:

$$\Phi(u) \simeq \frac{q(u)}{\xi\Sigma_s(u)}. \tag{7.35}$$

Because the flux is a density and the current is a true function, this relationship is written using E and v:

$$\Phi(E) \simeq \frac{q(E)}{\xi\Sigma_s(E)E}, \qquad \Phi(v) \simeq \frac{2q(v)}{\xi\Sigma_s(v)v}. \tag{7.36}$$

7.2.4. Slowing down in hydrogen

Slowing down is simpler to study in hydrogen than other materials, because the lethargy gain exponential transfer distribution is not truncated at a maximum value of ε, but goes up to infinity. The slowing down equation, in particular, can be solved analytically in this case no matter which equations are used for the scattering and absorption cross-sections.

In practice, in the epithermal domain, the scattering cross-section of hydrogen is approximately constant, and it capture cross-section is negligible. This problem is, however, a good representation of a *homogeneous mixture of hydrogen and a heavy material* (such as uranium), because the scattering of the heavy material has no effect (it does not change the energy of the neutrons); only its absorption comes into play.

The first form of the slowing down equation is then written:

$$S(u) + \int_{-\infty}^{u} f(u')e^{-(u-u')}du' = \frac{\Sigma_t(u)}{\Sigma_s(u)} f(u), \tag{7.37}$$

with $f(u) = \Sigma_s(u)\Phi(u)$.

This is multiplied by e^u to produce a first-order differential equation governing:

$$F(u) = \int_{-\infty}^{u} f(u')e^{u'}du'. \tag{7.38}$$

This differential equation can be integrated; f is deduced from this, and then Φ, by differentiating F. After performing all the calculations, we find:

$$\Phi(u) = \int S(u_0)\varphi(u; u_0)du_0, \tag{7.39}$$

where $\varphi(u; u_0)$ (Green's function) is the solution for the source $\delta(u - u_0)$ (source of one neutron per unit time and volume, emitting at lethargy u_0):

$$\varphi(u; u_0) = \frac{1}{\Sigma_t(u)} \left\{ \delta(u - u_0) + \frac{\Sigma_s(u_0)}{\Sigma_t(u_0)} \Upsilon(u - u_0)\, p(u; u_0) \right\}, \tag{7.40}$$

with:

$$p(u; u_0) = \exp\left[-\int_{u_0}^{u} \frac{\Sigma_a(u)}{\Sigma_t(u)} du' \right]. \tag{7.41}$$

Comments about this Green's function:

- the Dirac distribution compensates for the source in the equation. Physically, all neutrons are assumed to be emitted at the same lethargy u_0, and therefore the rate of (first) collision is infinitely "concentrated" at this lethargy, and the same applies to the flux (flux of neutrons before their first collision);
- neutrons that have undergone at least one collision are scattered beyond, not below, the original lethargy u_0, giving the Heaviside step function;
- these neutrons must have been scattered, not absorbed, at their first collision, at u_0, which gives the probability $\Sigma_s(u_0)/\Sigma_t(u_0)$ that this collision was a scattering event;

– similarly, at the lethargies u beyond there are only the neutrons that have not been absorbed at the intermediate lethargies u' between u_0 and u; that is the reason for this quotient Σ_s/Σ_t that is present implicitly in the integral in the argument of the exponential, since $\Sigma_a/\Sigma_t = 1 - \Sigma_s/\Sigma_t$. This integral represents the *probability that a neutron will be slowed down from u_0 to u without having been absorbed*: this can be regarded as a resonance escape probability. The decreasing expression reflects the gradual decrease in the number of "surviving" neutrons when they are made to travel through an increasingly long dangerous passage. The exponential form reflects the fact that these "survivors" had to go through a series of dangerous passages and avoid the danger each time. To multiply the elementary probabilities, we add the arguments of the exponential terms expressing them.

7.2.5. Slowing down in the presence of resonance traps

The equation for the probability of escaping from absorption during slowing down is interesting, because it applies to any distribution of cross sections, but unfortunately it assumes a "hydrogen" distribution for the slowing down. Because it is not possible to calculate a more general slowing down relationship, approximate slowing down models have been constructed. We shall now present the formulae obtained from these models, but without giving the full details of their derivation. One of the models, called the Greuling–Goertzel approximation, is very well suited to the problem of "gentle" (slow and gradually varying) absorption. The second model, called the Wigner approximation, is well suited to the problem of resonance traps (resonances leading to a cross-section that is intense, but only in certain narrow energy domains). We shall begin with this one.

a) The black resonance trap problem

This problem is based on a rather "academic" representation of resonance, but is simple to solve: assuming that the absorption cross-section is zero everywhere except in an interval of width γ, where it is infinite, we wish to calculate the probability p that a slowing down neutron escapes from the resonance trap.

Let us place the lethargy origin at the entrance to the trap, and take a source normalised to one neutron placed far upstream. At the lethargies located just before the resonance trap, the scattering density $\Sigma_s(u)\Phi(u)$ has its asymptotic value $1/\xi$. The number $1 - p$ of neutrons *falling* into the resonance trap can be found by summing over all possible starting lethargies (before the resonance trap) and arrival lethargies (in the trap), separated by ε at the most. The double integral can be calculated analytically:

$$1 - p = \int_0^\gamma du \int_{u-\varepsilon}^0 du' \frac{1}{\xi} \frac{e^{-(u-u')}}{1-\alpha} = \frac{1 - e^{-\gamma} - \alpha\gamma}{\xi(1-\alpha)}. \tag{7.42}$$

Figure 7.5 gives two examples of curves showing how the resonance escape probability varies with γ/ε (when γ is greater than ε, it is obviously zero because no neutron is able to "jump over" the trap.)

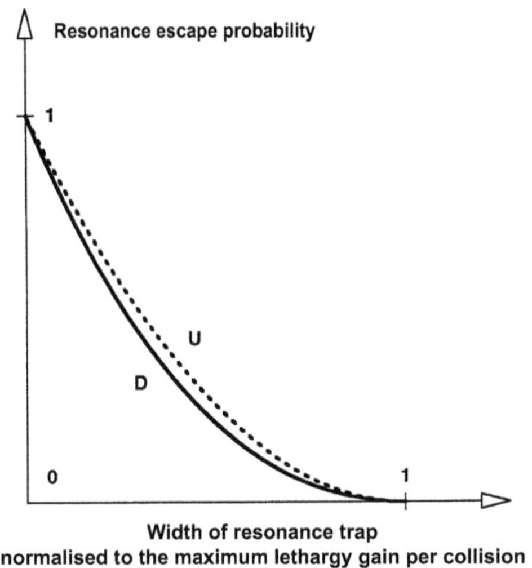

Figure 7.5. Probability of escaping from a "black resonance trap" for a neutron slowed down by deuterium or uranium.

b) Narrow black resonance trap

If the trap is narrow, a limited expansion for small γ gives:

$$1 - p \simeq \frac{\gamma}{\xi}. \tag{7.43}$$

It is important to understand the physical significance of this equation: if the resonance trap is narrow, the number of neutrons falling into it (arrival density) is practically equal to the number of neutrons that would reach the interval concerned if the trap were removed, i.e. $1/\xi$ per lethargy unit. In other words, it makes little difference whether the resonance trap is present or absent. In fact, when the trap is present, the only *missing* neutrons are the very few (for a narrow trap) that would have fallen *twice or more* in this interval.

Note that the unit used to measure the "narrowness" of a resonance trap is ε (or ξ which is approximately equal to $\varepsilon/2$ for intermediate and heavy nuclei).

c) Narrow grey resonance trap

We now consider a narrow grey resonance trap, where "grey" means that it has a *finite* absorption cross-section. As before, the number of neutrons arriving in an interval du located in the trap is approximately equal to the number there would be without the trap, i.e. du/ξ, because the trap is assumed to be narrow. Because the trap is grey, the proportion absorbed is only Σ_a/Σ_t: the ratio of the absorption cross-section to the total cross-section at this lethargy. The other neutrons are scattered and therefore "saved", since most are transferred beyond the trap. We therefore have the following summation:

$$1 - p \simeq \int_{\text{Trap}} \frac{\Sigma_a(u)}{\Sigma_t(u)} \frac{du}{\xi}. \tag{7.44}$$

d) A set of narrow grey traps

Finally, let us consider a series of narrow grey traps, each separated from the next by a sufficient distance that the Placzek transient induced by capture ("negative" source) at one trap is negligible at the subsequent traps. The overall resonance escape probability is then the product of the resonance escape probabilities of each one:

$$p \simeq \prod p_i. \tag{7.45}$$

Each probability can also be expressed approximately in exponential form:

$$p_i \simeq 1 - \int_{(i)} \frac{\Sigma_a(u)}{\xi\Sigma_t(u)} du \simeq \exp\left[-\int_{(i)} \frac{\Sigma_a(u)}{\xi\Sigma_t(u)} du\right]. \tag{7.46}$$

This simplifies the expression for the product:

$$p \simeq \exp\left[-\sum_i \int_{(i)} \frac{\Sigma_a(u)}{\xi\Sigma_t(u)} du\right]. \tag{7.47}$$

e) Resonance escape probability formula

Because the integrated function is zero outside the resonance traps, we can simply write:

$$p \simeq \exp\left[-\int \frac{\Sigma_a(u)}{\xi\Sigma_t(u)} du\right]. \tag{7.48}$$

This approximate formula for the resonance escape probability is a general form of the exact formula we found in the hydrogen case. For hydrogen, ξ is equal to 1.

7.2.6. Slowing down in the presence of low, slowly varying absorption

The resonance trap problem is representative of the low-energy part of the slowing down domain, but this problem of low, slowly varying absorption concerns the high-energy part. We shall present it concisely, without providing details of the mathematical derivation.

a) Slowing down with constant absorption

To begin with, we assume that the ratio $c = \Sigma_a/\Sigma_s$ is constant in lethargy (it is not essential to assume that the cross-sections themselves are constant, although in practice this is the only way that c could be constant). The second form of the slowing down equation for lethargies located beyond the source is written:

$$\frac{dq(u)}{du} = -c f(u), \tag{7.49}$$

where the current q is related to the scattering density f by the convolution product:

$$q = R * f. \tag{7.50}$$

The solution is the following exponential function[23]:

$$q(u) = Ae^{-mu}, \tag{7.51}$$

where the constant A expresses the normalisation of the source: the exponential e^{-mu} is the probability of escaping absorption for a neutron slowing down in a lethargy interval of length u. The constant m is the root of an algebraic equation obtained by substituting into the equations (note that this algebraic equation involves an integral of R that is expressed analytically).

b) Slowing down with constant, low absorption

For small c, the constant m can be calculated by limited expansion. This gives:

$$m \simeq \frac{\Sigma_a}{\xi\Sigma_s + \gamma\Sigma_a}. \tag{7.52}$$

Note that ξ is the average of w, the lethargy gain per collision:

$$\langle w \rangle = \xi. \tag{7.53}$$

The parameter γ (usual notation; not to be confused with the width of a resonance trap, used above) is defined by:

$$\langle w^2 \rangle = 2\gamma\xi. \tag{7.54}$$

c) Slowing down with low, slowly varying absorption

If we now assume the cross-sections to be slowly varying (rather than constant), an approximate expression for the factor p can be obtained by replacing the constant m by its average value over the lethargy interval under consideration:

$$p \simeq \exp\left[-\int \frac{\Sigma_a(u)}{\xi\Sigma_s(u) + \gamma\Sigma_a(u)} du\right]. \tag{7.55}$$

Note how similar this expression is to the previous one; the only difference is that ξ has been replaced by γ in the denominator, before the absorption cross-section. We could show numerically that γ is not very different from ξ. In particular, for hydrogen, γ is equal to 1, and once again we have the exact formula for p.

d) Practical formulae for the resonance escape probability

Unless a numerical calculation for the slowing down is to be performed, one of these formulae will be used to evaluate the resonance escape probability for a homogeneous configuration.

The first will be used for a resonance-type problem, and the second will be used in the case of low absorption. Because in practice the media concerned contain a mixture of different types of atom, the macroscopic cross-sections calculated by the usual additive rule will have to be used and, to respect their definition, the parameters ξ and $2\gamma\xi$ of the mixture must be calculated by weighting the individual value of each nuclide by the various macroscopic scattering cross-sections.

[23] This can be checked by substituting into the equations and demonstrated using the Laplace transform.

7.2.7. Space-energy coupling: Age Theory

In this chapter, we have examined only the "energy" aspect of the neutron problem. Clearly, however, in reality there is a coupling with the "space" aspect. This is a difficult problem which is usually handled using the multigroup theory to be presented in Chapter 10.

There is, however, a theory attributed to Fermi himself, which is called "Age Theory". This is a very physically intuitive theory that enables certain phenomena to be visualised very clearly. Unfortunately, in addition to the fact that this theory is only approximate, it can only be applied if the reactor is homogeneous. This is a very serious limitation.

To give an overview[24], let us say that this theory draws a strict parallel between a variable called "age", which measures the progress of slowing down (broadly speaking, age increases in proportion to lethargy) and time in a kinetic diffusion problem. For example, if a fast neutron source is placed on the median plane of a homogeneous plate reactor, as age (lethargy) increases, we observe a dispersion in space and an attenuation very similar to what was seen in the context of pulsed neutron experiments (§ 5.2.10). The following curves (Figure 7.6) are for an infinite reactor and a non-absorbing material.

The following equation is obtained using Age Theory:

$$q(x, \tau) = S\frac{e^{-x^2/4\tau}}{\sqrt{4\pi\tau}}, \tag{7.56}$$

where τ is the age (this variable has the dimension of a surface) and x is the distance to the source plane.

For Figure 7.6, the convention $S = (0.32\pi)^{1/2}$ was applied.

[24] A few details will be seen in the "exercises" section.

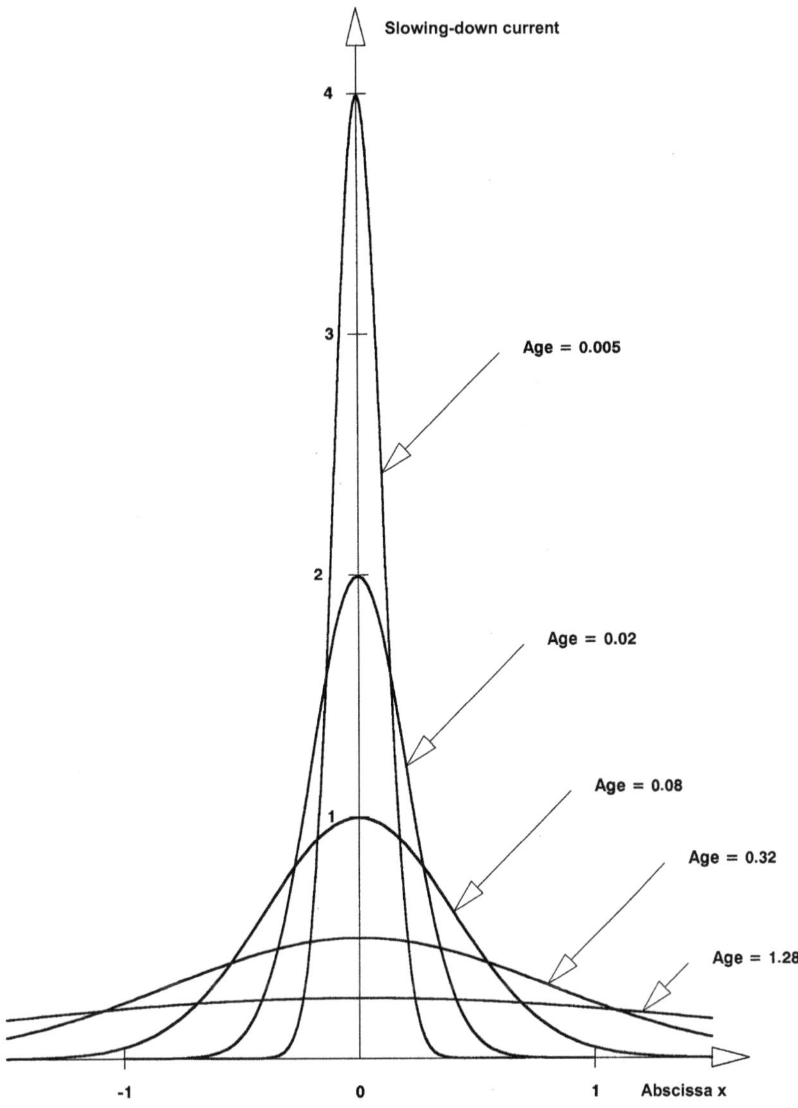

Figure 7.6. Slowing down current at various values of age for a neutron emitted by a planar source placed in an infinite, homogeneous medium.

Exercises

Exercise 7.1: slowing-down by hydrogen

Assume that the mass of the neutron is exactly equal to the mass of the proton.

a) Show that, after the scattering of a non-relativistic neutron by an initially motionless proton, the directions in which the particles move are mutually perpendicular.

b) Simplify the various formulae that describe elastic scattering for this particular case.

Exercise 7.2: slowing-down by a heavy nucleus

For an elastic, isotropic collision in the centre of mass frame occurring with a heavy nucleus

a) perform the limited expansion of ε and ξ in powers of $1/A$; and

b) evaluate the error in ξ when the exponential law for the gain in lethargy between 0 and ε is replaced by a uniform law between the same limits. This approximation is sometimes made, for instance, when calculating the strength of resonant absorption by heavy nuclides.

Exercise 7.3: deflection angle in the laboratory system

a) Write down the formula giving $\mu = \cos\psi$ as a function of $\lambda = \cos\theta$; then give it in terms of the ratio $r = E_f/E_i$ between the neutron energy after collision in the laboratory system, and the initial energy; and finally, express it as a function of the lethargy gain Δu.

b) Write down the probability law for μ as a function of λ when the scattering is isotropic in the centre of mass system.

c) Simplify these formulae for the case of hydrogen, assuming the neutron and the proton possess equal mass.

Exercise 7.4: inelastic collisions

A collision is called *inelastic* when kinetic energy is not conserved. During neutron slowing-down, inelastic collisions occur when a target nucleus—initially in the ground state—temporarily remains excited following neutron emission, before finally undergoing gamma emission. In this exercise, let Q be the difference between the sum of the kinetic energies in the centre of mass system, before and after the collision. All other notations follow the usual conventions.

a) Write down the formulae giving E_f and $\cos\psi$ as a function of $\cos\theta$, using the definition

$$\gamma = A\sqrt{1 - [(A+1)/A]\,Q/E_i}.$$

b) What size is the minimum value $E_{\text{threshold}}$ of the initial neutron energy E_i in the laboratory system, when inelastic scattering occurs?

c) Determine the energy E_{90} giving $\psi = \pi/2$ and show that the deflection is always forward when $E_{threshold} < E_i < E_{90}$.

d) Show that, in this interval for E_i, there are two possible values for θ and E_f for any value of ψ.

Exercise 7.5: number of collisions required to slow down a neutron

There appear to be two plausible ways to estimate the mean number n of collisions required to slow down a neutron. The problem might be formulated in terms of either *energy* using

$$\frac{E_1}{E_0} \simeq \rho^n,$$

or *lethargy* using

$$-\ln\frac{E_1}{E_0} \simeq n\xi,$$

where ρ is the average ratio of the energies E_f/E_i before and after scattering, and $-\xi$ is the average of the logarithm of E_f/E_i. Also, provided that the scattering is elastic and isotropic in the centre of mass system, $\rho = (1-\alpha)/2$, and $\xi = 1 - \alpha\varepsilon/(1-\alpha)$.

a) Compare both expressions for the case of hydrogen.

b) Which one is correct and why?

Exercise 7.6: slowing-down time

The slowing-down time for neutrons can be calculated by describing the process in terms of a deterministic sequence instead of a stochastic one. After being scattered n times, the energy of a neutron that started with energy E_0 is $E_n = E_0\,\mathrm{e}^{-n\xi}$. It is assumed that the path length λ between one collision and the next is always the same. On the basis of this model, find an expression giving the average time taken for the energy of a neutron to become E_N. Using $E_0 = 2$ MeV and $E_N = 1$ eV, calculate this time for the following two materials with the data provided: **a)** water, $\xi \simeq 1$, $\lambda = 2.8$ mm; **b)** graphite, $\xi = 0.158$, $\lambda = 26$ mm.

Exercise 7.7: Placzek's transitory

In a material that does not absorb neutrons, described by the parameter A, assume that slowing-down occurs via collisions that are both elastic and isotropic in the centre of mass frame. The initial energy for the lethargy of the neutrons is a constant E_0, and they are emitted at a normalized rate of one per unit of time.

a) Show that, according to this model, the scattering density f, the arrival density ρ, and the slowing-down current are related by the equation

$$q(u) = \rho(u) - \frac{\alpha}{1-\alpha}\int_{u-\varepsilon}^{u} f(u')\,du'.$$

b) Show that, the current $q(u)$ is equal the Heaviside step function $\Upsilon(u)$.

c) Using **a** and **b** in the slowing-down equation, derive the equation giving f.

d) Give the asymptotic value of f when u is large.

e) Taking the derivative of the equation obtained in **c**, explain how to generate a sequence of elementary differential equations that, when integrated in successive lethargy intervals starting from zero, each of equal width ε, give f for each interval. Show the result for the first few intervals.

Exercise 7.8: slowing-down in the case of a constant absorption

Consider slowing-down in a material where the probability of absorption during a collision $a(u) = \Sigma_a(u)/\Sigma_t(u)$ is independent of the lethargy u. Assume also that the law for the transfer of lethargy by scattering $P(\Delta u)$ depends only on the lethargy gain Δu.

a) Write down the integral equation giving the scattering density $f(u) = \Sigma_s(u)\,\Phi(u)$.

b) Show that, outside the domain of the source, an asymptotic solution of the form $f(u) = A\,e^{-mu}$ satisfies the equation (A and m are constants). Find the expression giving m.

c) Show that, when a is small, the solution of the previous equation is approximately given by

$$m \simeq \frac{a}{P_1 + mP_2/2} \simeq \frac{a}{P_1 + aP_2/(2P_1)} = \frac{\Sigma_a(u)}{\xi\Sigma_s(u) + \gamma\Sigma_a(u)},$$

where P_1 and P_2 are the moments of orders 1 and 2 of the lethargy transfer law (these moments are generally written ξ and $2\gamma\xi$).

d) Derive the equation giving m, and the expressions for ξ and γ, for elastic, isotropic slowing-down in the centre of mass frame by a monoatomic material.

Exercise 7.9: age theory

The age theory, devised by Enrico Fermi, concerns the space-energy coupling during the slowing-down of neutrons. It is based on two approximations:

1) concerning space, the diffusion approximation, and

2) concerning energy, the so-called Fermi model, for which the Placzek transitory is neglected.

In other words, it assumes that neutrons loose energy continuously rather than in finite, discrete amounts. The functions which describe the neutron population—the flux Φ and the slowing-down current Q—depend on space and lethargy.

- In the relationship giving the derivative of the current with respect to lethargy, the leakage rate, by virtue of Fick's law, is added giving

$$\frac{\partial Q}{\partial u} = S - \Sigma_a\,\Phi + D\Delta\Phi. \tag{7.1}$$

- The asymptotic relationship between the flux and the current is assumed to be

$$Q = \xi\Sigma_s\,\Phi. \tag{7.2}$$

a) By assuming that the medium is homogeneous, and eliminating the flux Φ between (7.1) and (7.2), derive the equation for the current Q.

b) Find the solution $p(u)$ (probability for a neutron with lethargy between zero and u to escape absorption) for the case where the functions are uniform in space, and where the source is $\delta(u)$ (emission of one neutron per unit of volume and time at zero lethargy).

c) Define $Q = pq$ where q is a function of $\vec{r}$ and u, and the source is then assumed to be $S(\vec{r})\,\delta(u)$. Find the equation giving q for $u \neq 0$ and the initial condition at $u = 0$.

d) Simplify the previous equation for q by using, instead of u, the variable

$$\tau = \int_0^u \frac{D(u')}{\xi\Sigma_s(u')}\,du'.$$

This parameter is called 'Fermi's age'. What dimensions does it have?

e) Establish the kernel of this 'Fermi age equation' which has solutions in an infinite, homogeneous medium for point, line, and plane sources.

f) Neglecting the extrapolation length, determine the critical condition of a bare, homogeneous pile for 'age + diffusion' theory. 'Age' applies to fast and epithermal neutrons, and monokinetic diffusion is applicable to thermal neutrons. Fissions induced by fast neutrons will be taken into account owing to a factor ε, while fissions induced by epithermal neutrons will be neglected.

g) A linear monokinetic source, emitting one neutron per unit of time and length, is located along the central axis of a square-section column, of infinite height, where each of its four faces is a symmetry mirror-plane, and thereby forms an infinite lattice of identical cells. In addition to the source, each cell contains a homogeneous, non-absorbing material. The energy of the emitted neutrons is determined from the initial value of the slowing-down area at the source. Determine the spatial and energetic distribution $q(x, y, \tau)$ of the neutrons expressed as an expansion of eigenfunctions of the Laplace operator satisfying the boundary conditions. For a 0.15×0.15 m column and a slowing-down area equal to 3×10^{-2} m^2 (typical values for a gas-cooled, graphite-moderated, natural uranium reactor where the neutrons are at the end of their slowing-down), show that this distribution is nearly uniform.

Solutions

Exercise 7.1: slowing-down by hydrogen

a) Conservation of energy and momentum imply that in the laboratory system

$$\vec{V}_{ni} = \vec{V}_{nf} + \vec{V}_{pf},$$

$$V_{ni}^2 = V_{nf}^2 + V_{pf}^2.$$

Subtracting the square of the first equation from the second gives $\vec{V}_{nf} \cdot \vec{V}_{pf} = 0$; hence, the directions are mutually perpendicular.

b) The relationships for the energy and the angle in the laboratory frame in this case are

$$r = \frac{E_f}{E_i} = \cos^2\frac{\theta}{2}; \qquad \cos\psi = \cos\frac{\theta}{2} = \sqrt{r}.$$

Exercise 7.2: slowing-down by a heavy nucleus

a) The expansions to second order are

$$\varepsilon = \frac{4}{A} + \frac{8}{A^2} + \cdots; \qquad \xi = \frac{2}{A} + \frac{8}{3A^2} + \cdots.$$

b) Let ξ' be the approximate value of ξ, then the error is

$$\frac{\xi'}{\xi} = 1 + \frac{2}{3A} + \cdots.$$

Exercise 7.3: deflection angle in the laboratory system

a) The cosine of the deflection angle in the laboratory system is

$$\mu = \frac{A\lambda + 1}{\sqrt{A^2 + 2A\lambda + 1}} = \frac{\left[(A+1)\sqrt{r} - (A-1)\sqrt{1/r}\right]}{2} = \frac{\sinh(\varepsilon/2 - \Delta u)/2}{\sinh(\varepsilon/4)}.$$

b) Provided that the collision is assumed to be isotropic in the centre of mass system, the probability distribution for the scattering angle is obtained by making a change of variables $P(\mu)\,d\mu = P(\lambda)\,d\lambda = d\lambda/2$, which gives

$$P(\mu)\,d\mu = \frac{(A^2 + 2A\lambda + 1)^{3/2}}{2A^2(A+\lambda)}.$$

The expression in terms of μ is complicated because the relationship between λ and μ is itself complicated.

c) However, for the case of hydrogen, the formulae can be simplified:

$$\mu = \sqrt{(\lambda+1)/2} = \sqrt{r} = e^{-\Delta u/2}; \qquad P(\mu) = \sqrt{2(1+\lambda)} = 2\mu.$$

Exercise 7.4: inelastic collisions

a) The formulae are the same as the elastic collision ones except that A is replaced by γ:

$$\frac{E_f}{E_i} = \frac{\gamma^2 + 2\gamma\lambda + 1}{(A+1)^2}; \qquad \mu = \frac{\gamma\lambda + 1}{\sqrt{\gamma^2 + 2\gamma\lambda + 1}}.$$

Notice that $\gamma = A$ in the limit of an elastic collision occuring.

b, **c** and **d)** The relationships are as follows.

$$E_{\text{threshold}} = \frac{A+1}{A}Q; \qquad E_{90} = \frac{A}{A-1}Q.$$

There are two solutions when λ is expressed as a function of μ, and $\gamma < 1$ (i.e. an inelastic collision).

Exercise 7.5: number of collisions required to slow down a neutron

a) For hydrogen, $\rho = 1/2$ and $\xi = 1$. When slowing-down from 2 MeV to 1 eV, $n_{\text{energy}} = 20.9$ and $n_{\text{lethargy}} = 14.5$.

b) The average value of a product is *not* equal to the product of the average values; hence, it is incorrect to formulate this problem in terms of the energy ratio. However, the average of a sum *is* equal to the sum of the averages; therefore, this properly accounts for each random gain in lethargy, and thus the second method is correct. This argument represents the motivation for using *lethargy* as a parameter.

Exercise 7.6: slowing-down time

The velocity during each path n is $v_0 \exp(-n\xi/2)$ and lasts for a duration that is $\lambda/\left[v_0 \exp(-n\xi/2)\right]$. Taking the sum from 0 to N, and using $\lambda/\left[v_0 \exp(-N\xi/2)\right] = \lambda/v_N$ for the duration of the final path gives

$$T = \frac{1 - e^{-(N+1)\xi/2}}{1 - e^{-\xi/2}} \frac{\lambda}{v_N} \simeq \frac{1}{1 - e^{-\xi/2}} \frac{\lambda}{v_N}.$$

The physical interpretation of the last approximation is that the time taken for the slowing-down steps near the begining is negligible in comparison with the ones near the end.

Numerical example: $v_N = 13\,000$ m s^{-1}; $T_{\text{water}} = 0.51$ µs; $T_{\text{graphite}} = 25$ µs.

Exercise 7.7: Placzek's transitory

a) This relation is obtained when combining the expressions for ρ and q with the integral operators expressing the convolutions by the law of the collision in the first and the second forms, respectively.

b) This result is general provided that there is no absorption and the source is monokinetic and normalized.

c) The result is

$$\delta(u) + \Upsilon(u) + \frac{\alpha}{1-\alpha} \int_{u-\varepsilon}^{u} f(u')\, du' = f(u).$$

d) When u is large it can be seen that f can be replaced by a constant which is $1/\xi$.

e) The derivative of the preceding equation gives

$$\delta'(u) + \delta(u) + \frac{\alpha}{1-\alpha} \left[f(u) - f(u-\varepsilon)\right] = f'(u).$$

Over each interval between $(n-1)\varepsilon$ and $n\varepsilon$, provided that $f(u-\varepsilon)$ for the previous interval is known—it is zero for the first interval—an elementary differential equation can be constructed. This can be solved by the normal 'variation of the constant method'. Hence,

- *for the first interval*

$$f(u) = \delta(u) + \exp\left(\frac{\alpha u}{1-\alpha}\right) \Upsilon(u);$$

- *and for the second interval*

$$f(u) = \frac{1-\alpha-\alpha(1-\alpha+u-\varepsilon)\exp\left[-\alpha\varepsilon/(1-\alpha)\right]}{(1-\alpha)^2}\exp\left(\frac{\alpha u}{1-\alpha}\right).$$

For the following intervals, the calculations are progressively more tedious. A graphical representation shows that an asymptotic constant value ($1/\xi$) is essentially obtained beyond 3ε.

Exercise 7.8: slowing-down in the case of a constant absorption

a) With the chosen notations, the slowing-down integral equation to solve is

$$\int_{-\infty}^{u} f(u')\, P(u-u')\, du' + S(u) = (1+a)\, f(u).$$

b) The equation giving m is obtained by replacing $f(u)$ with $A\,e^{-mu}$, and the normalization constant A cancels:

$$\int_0^{\infty} P(w)\, e^{mw}\, dw = 1 + a.$$

c) By an expansion of the exponential, we get

$$m\, P_1 + \frac{m^2}{2!} P_2 + \cdots = a.$$

To first order $m \simeq a/P_1$, and to second order

$$m = \frac{a}{P_1 + \frac{m}{2!}P_2 + \cdots} \simeq \frac{a}{P_1 + aP_2/(2P_1)}.$$

d) For elastic, isotropic slowing-down in the centre of mass frame by a monoatomic material, $P(w) = e^{-w}/(1-\alpha)$ between zero and ε, and the equation for m becomes

$$\frac{1-\alpha^{1-m}}{(1-\alpha)(1-m)} = 1 + a.$$

From the moments of P we obtain

$$\xi = \frac{1-\alpha(1+\varepsilon)}{1-\alpha}; \qquad \gamma = \frac{1-\alpha(1+\varepsilon+\varepsilon^2/2)}{1-\alpha(1+\varepsilon)}.$$

Exercise 7.9: age theory

a) The complete equation is

$$\frac{\partial Q}{\partial u} = S - \frac{\Sigma_a}{\xi\Sigma_s}Q + \frac{D}{\xi\Sigma_s}\Delta Q.$$

b) The solution when the functions in space are uniform is

$$p(u) = \Upsilon(u)\exp\left(-\int_0^u \frac{\Sigma_a(u')}{\xi\Sigma_s(u')}\, du'\right).$$

c) The distribution in space and energy when absorption is neglected is

$$\frac{\partial q}{\partial u} = \frac{D}{\xi \Sigma_s} \Delta q, \qquad \text{with} \qquad q(\vec{r}, 0) = S(\vec{r}).$$

d) The canonical form of the equation is

$$\frac{\partial q}{\partial \tau} = \Delta q.$$

Fermi's 'age' τ has dimensions L^2, i.e. is an area. For this reason, use of the name is deprecated, and is of historic interest only. It was called 'age' by Fermi because it is proportional to the duration between the instant of emission of a neutron and its subsequent observation at a given time, i.e. its age.

e) The kernels of the Fermi age equation are

$$q(d, \tau) = \frac{e^{-d^2/(4\tau)}}{(4\pi\tau)^{n/2}}.$$

where n is equal to 1, 2, or 3, and where d means x, ρ, or r for the plane, line, and point kernels, respectively.

f) The equations of the 'age + diffusion' theory are

$$\frac{\partial q(\vec{r}, \tau)}{\partial \tau} = \Delta q(\vec{r}, \tau), \qquad \text{with} \qquad q(\vec{r}, 0) = k_\infty \Sigma_a \Phi(\vec{r}),$$

$$D \Delta \Phi(\vec{r}) - \Sigma_a \Phi(\vec{r}) + q(\vec{r}, \tau_{th}) = 0,$$

where τ_{th} is the slowing-down area of the neutrons entering the thermal range.

For a bare, homogeneous pile, the functions Q and Φ are proportional to the fundamental eigenfunction of the Laplace operator $f(\vec{r})$ characterized by B^2. By inspection of the equations, it can be seen that the critical condition is

$$\frac{k_\infty \exp(-\tau_{th} B^2)}{1 + L^2 B^2} = 0.$$

g) Let p be the edge of the column, then an expansion in eigenfunctions gives

$$q(\tau, x, y) = \frac{1}{p^2} \sum_{j=0}^{\infty} \sum_{k=0}^{\infty} \varepsilon \exp\left[-4\pi^2 \tau (j^2 + k^2)/p^2\right] f_{jk}(x, y),$$

where $\varepsilon = 1$ for the term $(0, 0)$, $\varepsilon = 2$ for the terms with one null index, and $\varepsilon = 4$ for the other terms. The eigenfunctions are

$$f_{jk}(x, y) = \cos \frac{2\pi j x}{p} \cos \frac{2\pi k y}{p}.$$

The ratio between the coefficients of the first harmonics $(1, 0)$ and $(0, 1)$, and the coefficient of the fundamental mode $(0, 0)$, $2 \exp(-4\pi^2 \tau/p^2)$, is very small: $2\,e^{-52.6}$. The other harmonics are even smaller; therefore, the uniformity of the neutron distribution with this slowing-down area is nearly perfect.

Resonant absorption of neutrons (physical aspects)

Introduction

The tangled "forest" of resonances of uranium 238, with cross-sections of up to tens of thousands of barns (Figure 8.1) gives the impression that trying to slow down and thermalise neutrons in the presence of uranium would be an impossible mission.

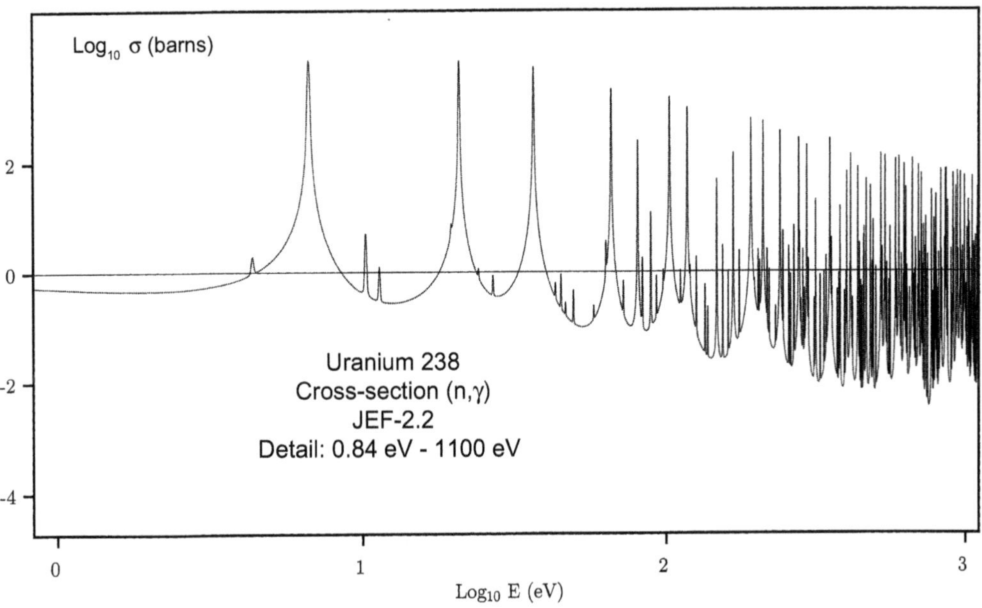

Figure 8.1. Neutron capture cross-section of uranium 238 in the domain of resolved resonances.

In fact, it turns out to be possible after all, thanks to a physical phenomenon known as *self-shielding*. In this chapter we shall attempt to explain the meaning of this term.

The primary reason for resonant capture of neutrons remaining limited despite the very large cross-sections was seen in the resonance trap problem we solved in the previous chapter: even with an infinite cross-section, the probability of falling in the trap is limited, or even small, if the trap is narrow. In fact, this probability does not exceed the ratio γ/ξ

where γ is the width in lethargy of the trap, and ξ is the average lethargy gain acquired by a neutron when a scattering event occurs. Irrespective of the moderator used to scatter and slow down neutrons, the value of ξ is always far greater than the width of the resonances. In other words, *compared to slowing by the moderator, the resonances of capture by the fuel are always narrow.*

This can easily be visualised using the image of the kangaroos (Figure 8.2): if the kangaroos travel along a path that has a trap, even a very deep one, many of them —especially the one whose path is represented— will avoid the trap if it is narrower than the kangaroos' hops.

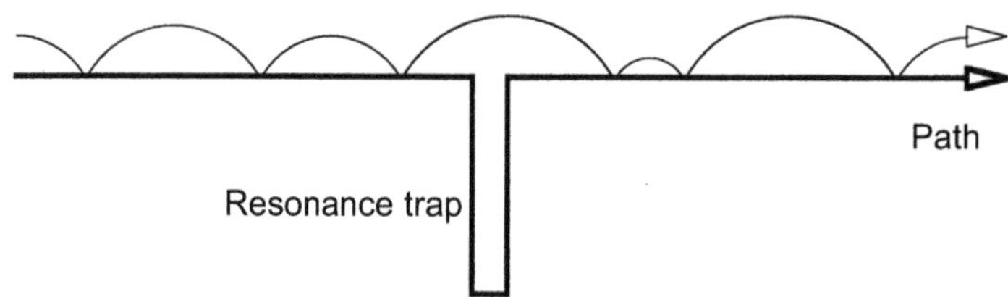

Figure 8.2. Like kangaroos travelling on a path with a trap, many neutrons slow down without disappearing in the resonances.

Note also that the heterogeneous structure generally used in reactors for technological reasons is advantageous in neutron physics terms because it reduces resonant capture (Figure 8.3). After emission by fission, the neutrons usually leave the fuel elements and travel around in the more spacious moderator. As long as they remain there, they can slow down without any risk; when they have become thermal, their paths can take them back into the fuel, where they can cause new fission events.

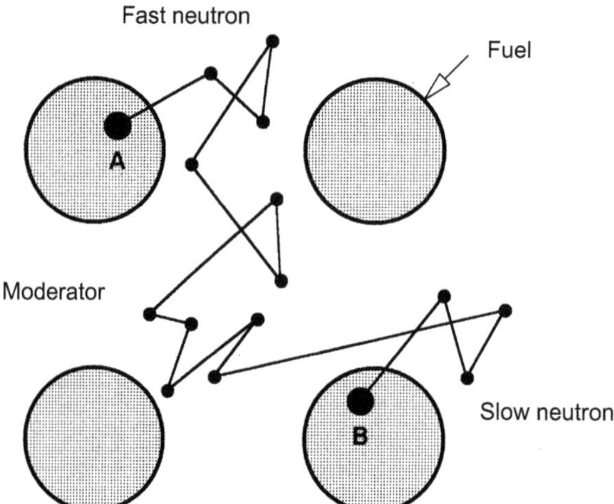

Figure 8.3. A heterogeneous structure reduces absorption in the resonances of the fuel at intermediate energies.

In more technical terms, the absorption rate of neutrons, which is the product $\Sigma_a \times \Phi$ of a cross-section and a flux, is limited, because if the cross-section Σ_a becomes large, the flux Φ falls, approximately in inverse proportion. This is the self-shielding effect.

It occurs at energies (or lethargies) where the cross-section becomes large and, if the structure is heterogeneous, in the region containing resonant material, i.e. the fuel. Figures 8.4a and 8.4b represent both the energetic and spatial aspects of self-shielding[1].

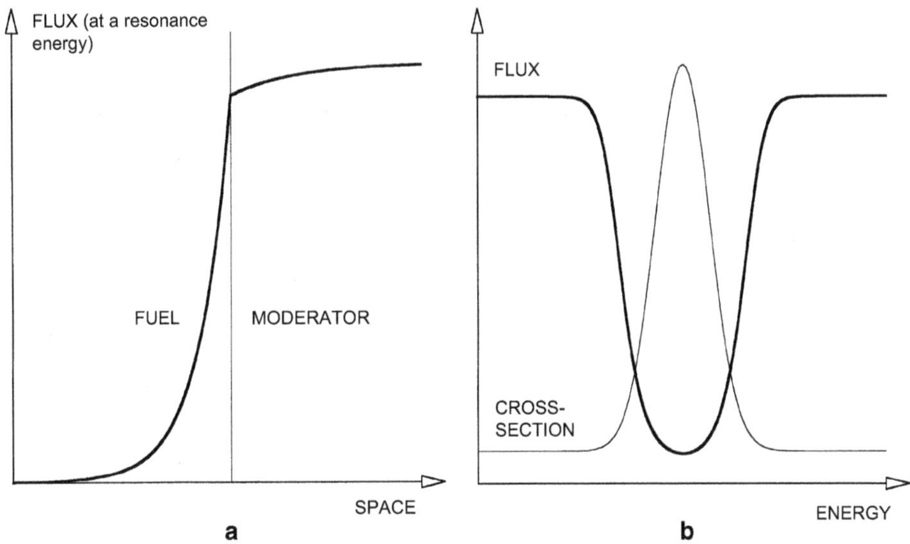

Figure 8.4. The self-shielding phenomenon with respect to energy (left) and space (right).

8.1. Resonant absorption in a homogeneous case by a purely absorbent material

8.1.1. Self-shielding in a homogeneous case

Take the formula for p, the probability of escaping from absorption during slowing through a resonant structure[2]:

$$p \simeq \exp\left[-\int \frac{\Sigma_a(u)}{\xi\Sigma_t(u)}du\right]. \tag{8.1}$$

In practice, we can observe that the scattering cross-section and average lethargy gain concern the *moderator,* which slows down the neutrons but does not capture them, and that the absorption cross-section concerns the *fuel,* which has absorption resonances for neutrons but does not scatter them. More precisely, we can at least provisionally apply the simplifying assumption of neglecting any scattering by this material which, as a first

[1] With regard to space, it can be said that the external layers of the fuel protect the internal layers, giving rise to the term "self-shielding".

[2] An expansion similar to the one we shall present can also be performed on the "low absorption" formula.

approximation, does not change the energy of the neutrons[3]. These two materials are also totally antinomic: the moderator is purely scattering and has a more or less constant cross-section, and the fuel is purely absorbent, with a cross-section that is negligible everywhere except in certain narrow lethargy intervals where it can be very high.

To analyse the structure of the flux and the absorption rate in this problem, let us take the first form of the slowing down equation (§ 7.1.9):

$$\rho(u) + S(u) = \Sigma_t(u)\Phi(u). \tag{8.2}$$

Here, we can neglect the source S (the neutrons are emitted by fission at energies well above the resonance domain). The arrival density:

$$\rho(u) = \int_{-\infty}^{u} \Sigma_s(u')\Phi(u')du'P(u' \to u), \tag{8.3}$$

is an integral that applies in practice to an interval ε that is very large compared to the width of the resonances. Whether u is taken in a resonance, or just before the resonance, practically does not change the integral, which essentially involves values of u' located before the resonance. On other words, this arrival density is roughly constant in the resonance and its neighbourhood. Its value can even be specified, because the situation ***before*** the resonance is asymptotic without absorption:

$$\Phi(u) \simeq \frac{q(u)}{\xi\Sigma_s(u)} \simeq \frac{q}{\xi\Sigma_s} \simeq \mathrm{C^t}, \quad \rho(u) \simeq \Sigma_s(u)\Phi(u) \simeq \frac{q}{\xi} \simeq \mathrm{C^t}. \tag{8.4}$$

Using the slowing-down equation, we deduce the following ***in*** the resonance:

$$\Phi(u) \simeq \frac{q}{\xi\Sigma_t(u)} = \frac{\mathrm{C^t}}{\Sigma_t(u)}. \tag{8.5}$$

This is the essential feature of self-shielding: *in a resonance, the flux decreases approximately in inverse proportion to the total cross-section.*

This result will also make it possible to interpret the formula for p, and then to introduce the concept of an effective resonance integral.

8.1.2. Interpretation of the resonance escape probability formula[4]

If we normalise to a neutron slowing down at a lethargy u ($q = 1$), we can see that the absorption rate in the lethargy interval du is:

$$\Sigma_a(u)\,\Phi(u)\,du \simeq \frac{\Sigma_a(u)du}{\xi\Sigma_t(u)}. \tag{8.6}$$

With this normalisation, this rate is a probability. Its complement can also be written as an exponential:

$$1 - \frac{\Sigma_a(u)du}{\xi\Sigma_t(u)} \simeq \exp\left(-\frac{\Sigma_a(u)du}{\xi\Sigma_t(u)}\right), \tag{8.7}$$

[3] In the homogeneous medium considered here, simply changing the direction of the neutron path does not alter the subsequent events, but in a heterogeneous situation this might no longer be true.

[4] Here we again use the argument introduced in the previous chapter in the context of resonance traps.

and is the probability of non-absorption in the interval du. For a finite lethargy interval, the probability p of non-absorption is calculated by multiplying all of the elementary probabilities, i.e. by adding the arguments of the exponentials, which amounts to taking the integral.

8.1.3. Flux factorisation and the concept of effective cross-section

Outside a resonance, the total cross-section is reduced to the scattering cross-section of the moderator (m), and the flux is:

$$\Phi(u) = \Psi(u) \simeq \frac{q(u)}{\xi\Sigma_{s,m}}. \tag{8.8}$$

This is a slowly decreasing function, in the same way as the current q, which remains constant between resonances, and decreases slightly on passing through each resonance. It is called the *macroscopic flux*, because it has the dimensions of flux, is equal to the flux outside resonances, and has the general form of the actual flux if the perturbations at each resonance are ignored.

At a resonance, the flux undergoes a depression that can be characterised by a second factor $\varphi(u)$:

$$\Phi(u) = \Psi(u)\,\varphi(u) \simeq \frac{q(u)}{\xi\Sigma_t(u)} = \frac{q(u)}{\xi\{\Sigma_{s,m} + \Sigma_{a,f}(u)\}}. \tag{8.9}$$

This abrupt depression is due to the absorption cross-section of the fuel (f), which is added to the scattering cross-section of the moderator (m). The following dimensionless function:

$$\varphi(u) = \frac{\Sigma_{s,m}}{\Sigma_{s,m} + \Sigma_{a,f}(u)}, \tag{8.10}$$

describes this depression (see Figure 8.4a). This function varies very rapidly at a resonance: starting from 1 before the resonance, it "collapses" in the resonance, increasingly for higher resonances, and returns to the value 1 after the resonance. This function is called the *fine structure* or *self-shielding factor*.

It is convenient to normalise the absorption rate not to the true flux Φ, which is complicated and therefore difficult to calculate, but to the macroscopic flux Ψ, which is far smoother. This leads to the introduction of the so-called *"effective" cross-section*, defined by:

$$\Sigma_{a,f}(u)\,\Phi(u) = \Sigma_{a,\text{eff}}(u)\,\Psi(u), \tag{8.11}$$

i.e.:

$$\Sigma_{a,\text{eff}}(u) = \Sigma_{a,f}(u)\,\varphi(u), \tag{8.12}$$

(the subscript f being understood for the effective cross-section). Because this effective cross-section is relative to the fuel, at a concentration N_0, it can also be defined microscopically:

$$\sigma_{a,\text{eff}}(u) = \Sigma_{a,\text{eff}}(u)/N_0 = \sigma_{a,f}(u)\,\varphi(u). \tag{8.13}$$

8.1.4. Practical formula for the resonance escape probability in a homogeneous situation

The integral over the entire resonance domain of this effective cross-section is called the *effective resonance interval*:

$$I_{\text{eff}} = \int \sigma_{a,\text{eff}}(u)\, du. \tag{8.14}$$

Neglecting variations in $\Sigma_{s,m}$, this integral makes it possible to express the resonance escape probability:

$$p \simeq \exp\left[-\frac{N_0 I_{\text{eff}}}{(\xi\Sigma_s)_m}\right]. \tag{8.15}$$

This formula, whilst very simple, is also very physically intuitive. It expresses the fact that the resonance escape probability is the result of the competition between absorption in the resonance traps of the fuel (an unfavourable element, found in the numerator) and slowing by the moderator (a favourable element, found in the denominator).

If we define the *dilution cross-section* as "the number of barns of moderator" associated with each atom of the fuel:

$$\sigma_d = \frac{\Sigma_{s,m}}{N_0}, \tag{8.16}$$

the formula for the effective resonance integral can be written in the following form:

$$I_{\text{eff}} = \int \sigma_{a,f}(u)\, \varphi(u)\, du = \int \sigma_{a,f}(u) \frac{\sigma_d}{\sigma_d + \sigma_{a,f}(u)} du. \tag{8.17}$$

This relationship shows that, for a given resonant nucleus, the effective resonance integral is a function of a single parameter: σ_d. This function can therefore be *tabulated once and for all, and then re-used for each resonance escape probability calculation.* As we shall see, this interesting observation can be applied in a wider context, by allowing for slowing down by the resonant material and/or for a heterogeneous situation.

In this integral, the fraction φ represents the self-shielding factor. This factor tends to 1 if the dilution cross-section tends to infinity, i.e. if the quantity of resonant material likely to depress the flux tends to zero. The limit of the effective resonance integral is the integral of the absorption cross-section, known as the *resonance integral.* As an example, we give approximate values for uranium 238 in the entire resonance domain and for the (equivalent) dilution in a pressurised water reactor:

- dilution cross-section: 50 barns;
- resonance integral: 280 barns;
- effective resonance integral: 20 barns.

On average, self-shielding reduces the integral characterising resonant capture by a factor of 14.

8.2. Slowing down by the absorbing material

The idea of neglecting any slowing down by the resonant material itself was justified by saying that it is a heavy material that leaves the neutron energy practically unchanged after scattering. The slowing down interval certainly is small, but the resonances are narrow too, and it is the relative order of magnitude that matters.

To measure the slowing down interval, ε or ξ can be used. To measure the width of a resonance in the same unit, lethargy, we may consider $\gamma = \Gamma/E_0$, where E_0 is the energy of the resonance peak, and Γ is its width at mid-height (Breit Wigner distribution, § 2.7.1). Even at half the height of the peak, however, the cross-section is still very large. That is why it is more convenient to consider the *practical width* Γ_p defined as the interval where the resonance cross-section exceeds the potential cross-section, and the associated value γ_p. The Breit–Wigner formula gives:

$$\Gamma_p \simeq \sqrt{\sigma_{max}/\sigma_p}, \qquad \sigma_{max} = r\frac{\Gamma_n}{\Gamma}, \tag{8.18}$$

(σ_{max}: resonant cross-section at the peak; σ_p: potential cross-section).

With this criterion, six resonances of uranium 238 (situated below 200 eV, in particular the first four from the bottom) are wide or fairly wide (γ_p greater than ε), and all the others are narrow or fairly narrow.

This means that, even though the resonances of uranium are always narrow when compared to slowing down by the moderator, the situation is more complicated when they are compared to slowing down by this heavy material itself. This is why the simplifying assumption we made must be abandoned.

8.2.1. Equation for the fine structure of the flux in a homogeneous situation

The fine structure φ allowed us to characterise self-shielding and introduce the concept of the effective resonance integral that can be tabulated once and for all. We shall try to keep this general formalism without making any prior assumption about slowing down by resonant nuclei. In so doing, we pursue the developments begun in the late 1960s by Michel Livolant and his doctoral student, Françoise Jeanpierre. (This resonant absorption theory, known in France as the Livolant–Jeanpierre formalism, is used in the APOLLO code; the basic principles are presented in this chapter, and additional details are given in Chapter 15.)

Consider an infinite, homogeneous medium consisting of a mixture of a heavy resonant material (index 0) and one or more light, non-capturing materials (index 1). Fission sources are negligible in the resonance domain; with u understood, the slowing down equation is reduced to:

$$R_0\Phi + R_1\Phi = (\Sigma_0 + \Sigma_1)\Phi. \tag{8.19}$$

We have distinguished the two materials and used R (not to be confused with the kernel introduced in § 7.1.10) to designate the slowing down operators giving the arrival density:

$$(R_i\Phi)(u) = \int_{u-\varepsilon_i}^{u} \Sigma_{s,i}(u')\,\Phi(u')\frac{e^{-(u-u')}}{1-\alpha_i}du'. \tag{8.20}$$

As before, we attempt to write the flux in a factorised form:

$$\Phi = \Psi\varphi,$$

where Ψ is the "flux outside of resonances" (macroscopic flux) and φ is the factor taking the resonance self-shielding (fine structure) into account. The appropriate choice of first factor is:

$$\Psi = \frac{R_1\Phi}{\Sigma_1}. \tag{8.21}$$

Because this function varies little (the denominator is approximately constant, and the long-range integral operator smoothes all the irregularities of the flux) and because it is effectively equal to the flux outside of resonances (if there is no resonance, the total cross-section Σ is equal to the potential cross-section Σ_p, the flux is constant in lethargy, and $R\Phi$ is equal to $\Sigma_p\Phi$ for both materials).

In the neighbourhood of a resonance peak, the flux Φ varies quickly, but Ψ does not. Over the interval of width ε_0, which is more or less on the same scale, the variation of Ψ can also be neglected. An approximation can therefore be applied to the expression for the heavy material slowing down operator: replacing $\Psi(u')$ by $\Psi(u)$ and taking this term out of the integral, as follows:

$$R_0(\varphi\Psi) \cong \Psi R_0\varphi. \tag{8.22}$$

By substituting these last three relationships into the flux equation, we can simplify by Ψ; the remaining equation concerns the fine structure φ only:

$$R_0\varphi + \Sigma_1 = (\Sigma_0 + \Sigma_1)\varphi. \tag{8.23}$$

We usually divide it by the concentration N_0 of the resonant material, introduce the dilution cross-section:

$$\sigma_d = \frac{\Sigma_1}{N_0}, \tag{8.24}$$

and set $r_0 = R_0/N_0$ (operator written with $\sigma_{s,0}$ instead of $\Sigma_{s,0}$). It then has the canonical form:

$$r_0\varphi + \sigma_d = (\sigma_0 + \sigma_d)\varphi. \tag{8.25}$$

A calculation code written to solve this equation numerically, taking all the resonances into account, can be used to calculate the effective resonance integral associated with φ and to *tabulate once and for all* the result for a few values of σ_d. We can then interpolate in the table as a function of σ_d to determine the effective resonance integral I_{eff} and the resonance escape probability p for any (homogeneous) situation that might be of interest.

8.2.2. Slowing-down models for resonant nuclei

Slowing-down models were created to avoid having to solve this fine structure integral equation (similar to the slowing down equation in material "0"). In the present context, they would seem to be obsolete, because the computers of today should be able to solve the fine structure equation exactly and tabulate the effective resonance integral. We shall show, however, that models of this type are still useful for transforming a heterogeneous situation to a homogeneous situation. We limit our discussion to the two classic models[5], which are well suited to the two extreme cases: a very narrow or very wide resonance compared to the scattering lethargy gain[6].

1/ *The narrow resonance* (NR) model is applied preferentially for resonances located at high energy. If u is placed in the resonance, almost all values of u' contributing to the integral $r_0\varphi$ are located *before* the resonance, where $\sigma_{s,0}$ is reduced to the potential cross-section $\sigma_{p,0}$ and φ is equal to 1; therefore:

$$r_0\varphi \simeq \sigma_{p,0}, \quad \varphi(u) \simeq \varphi_{NR}(u) = \frac{\sigma_{p,0} + \sigma_d}{\sigma_0(u) + \sigma_d}, \tag{8.26}$$

which gives the associated effective resonance integral by a simple numerical quadrature. Using the Breit–Wigner formalism, we can express the result analytically:

$$I_{NR} = \frac{I_{\text{res}}}{\sqrt{1 + (1 - i)\sigma_{\max}/(\sigma_{p,0} + \sigma_d)}}, \qquad i = \frac{\Gamma_n}{\Gamma}\frac{\sigma_{p,0}}{\sigma_{p,0} + \sigma_d}, \tag{8.27}$$

with:

$$I_{\text{res}} = \frac{\pi}{2}\frac{\sigma_{\max}\Gamma_\gamma}{E_0}. \tag{8.28}$$

2/ *The wide resonance* (WR) model is applied preferentially for resonances located at low energy. If u is placed in the resonance, all the values of u' contributing to the integral $r_0\varphi$ are located very close to u; by equating $\sigma_{s,0}(u')$ with $\sigma_{s,0}(u)$, we obtain:

$$r_0 \simeq \sigma_{s,0}\varphi, \quad \varphi(u) \simeq \varphi_{WR}(u) = \frac{\sigma_d}{\sigma_{a,0}(u) + \sigma_d}, \tag{8.29}$$

which gives the associated effective resonance integral by a simple numerical quadrature. Using the Breit–Wigner formalism, we can also express this result analytically:

$$I_{WR} = \frac{I_{\text{res}}}{\sqrt{1 + (\Gamma_\gamma/\Gamma)(\sigma_{\max}/\sigma_d)}}. \tag{8.30}$$

8.3. Resonant absorption in a heterogeneous situation

8.3.1. Flux and fine structure equations in the heterogeneous case

It has already been remarked that the structure of a nuclear reactor is rarely homogeneous. In most cases, the fuel and the moderator are physically separated and arranged in a fairly regular pattern of "*cells*", called a "*lattice*". The calculation of the resonance escape probability must therefore be extended to this heterogeneous situation. Here, to simplify, we consider the case of the infinite, regular lattice consisting of cells with two regions

[5] Extensions are presented in Chapter 15.

[6] Unfortunately, most major resonances of uranium 238 tend to be intermediate in this respect.

— fuel and moderator[7] — and we shall see in Chapter 15 that it is possible to generalise this model to more complicated geometries. The two regions are identified by the subscripts f and m; to begin with, we assume that the fuel medium contains "0" resonant material only.

Because of the heterogeneity, the fluxes now depend not only on the lethargy u, but also on the point in space. To express the reaction rates in each of the regions, we need only consider the *average values* at each lethargy, $\Phi_f(u)$ and $\Phi_m(u)$, in the fuel zone and moderator zone respectively. To write the equations governing this flux, we must express not only the lethargy transfers, but also the exchanges between the two zones. These exchanges will be taken into account via the *"first collision probabilities"*. These parameters will be defined and studied in detail in Chapter 14 (they are used to solve the Boltzmann equation); here we simply mention them to point out the essential aspects of the resonant absorption problem.

In any heterogeneous situation, divided into regions, P_{ji} is the *probability that a neutron created or placed in region j will undergo its first collision in region i.*

In the approach to the Boltzmann equation (§ 3.1.6), we saw that the "flux without collision[8]" at $\vec{r}$ for a unit point source placed at $\vec{r}'$ is:

$$\frac{e^{-\tau}}{4\pi R^2}, \tag{8.31}$$

where R is the distance from $\vec{r}'$ to $\vec{r}$ and τ is the integral of the total cross-section on the segment of the line connecting these two points. Multiplying by the total macroscopic cross-section at the point $\vec{r}$ gives the first collision density around the point $\vec{r}$ (number of collisions per unit volume) for a neutron emitted at the point $\vec{r}'$. By integrating over $\vec{r}$ in the entire region i (where the cross-section is equal to Σ_i), we can count all the collisions in this region. Finally, by averaging the result for all the points $\vec{r}'$ of the region j, i.e. by integrating over the entire region and then dividing by its volume V_j, we obtain:

$$P_{ji} = \frac{\Sigma_i}{V_j} \int_{V_j} d^3r' \int_{V_i} d^3r \frac{e^{-\tau}}{4\pi R^2}. \tag{8.32}$$

Note that the double integral is symmetrical, and therefore:

$$P_{ji} \frac{V_j}{\Sigma_i} = P_{ij} \frac{V_i}{\Sigma_j}, \tag{8.33}$$

(reciprocity) and note also that, because the lattice is infinite, the first collision must take place in one of the media:

$$\sum_i P_{ji} = 1, \tag{8.34}$$

(conservation of neutrons).

In this case, there are only two regions. The four probabilities coupling them allow us to write the two equations governing the two fluxes. With u understood, by distinguishing the possibilities concerning the next collision that a neutron placed at this lethargy by

[7] In particular, we neglect the cladding, or assume it to be homogeneous with the moderator.
[8] This refers to neutrons emitted directly, excluding any neutrons re-emitted after scattering.

scattering in one of the regions will undergo, and by expressing the collision rates in the fuel and the moderator, we obtain:

$$V_f R_f \Phi_f \times P_{ff} + V_m R_m \Phi_m \times P_{mf} = V_f \Sigma_f \Phi_f,$$

$$V_f R_f \Phi_f \times P_{fm} + V_m R_m \Phi_m \times P_{mm} = V_m \Sigma_m \Phi_m. \tag{8.35}$$

Resonant absorption occurs in the fuel. We therefore attempt to factorise Φ_c:

$$\Phi_c = \varphi \Psi. \tag{8.36}$$

The definition used earlier for Ψ is still suitable, for the same reasons (we can use either the subscript 1 or m):

$$\Psi = \frac{R_m \Phi_m}{\Sigma_m}. \tag{8.37}$$

As before, we can make the following approximation:

$$R_c(\varphi\Psi) \cong \Psi R_c \varphi. \tag{8.38}$$

By substituting these three relationships into the equation expressing the collision rate in the fuel, we see once again that we can simplify by Ψ and obtain the fine structure equation:

$$V_f R_f \varphi \times P_{ff} + V_m \Sigma_m \times P_{mf} = V_f \Sigma_f \varphi. \tag{8.39}$$

By observing that $V_m \Sigma_m \times P_{mf} = V_f \Sigma_f \times P_{fm}$ (reciprocity) and that $P_{fm} = 1 - P_{ff}$ (conservation), this can be written more simply:

$$R_f \varphi \times P_{ff} + \Sigma_f \times (1 - P_{ff}) = \Sigma_f \varphi. \tag{8.40}$$

Here, we can replace the notation P_{ff} by P_{00}, because the emission volume j (fuel) can be called either f or 0, and the volume of the first collision (fuel also) is assumed to contain only type 0 nuclei:

$$R_0 \varphi \times P_{00} + \Sigma_0 \times (1 - P_{00}) = \Sigma_0 \varphi. \tag{8.41}$$

By dividing by P_{00} and the volume concentration N_0 of the resonant material, we find *a fine structure equation that is formally identical to the one found for the homogeneous case*:

$$r_0 \varphi + \sigma_e = (\sigma_0 + \sigma_e)\varphi, \tag{8.42}$$

and by simply replacing the dilution cross-section by an *equivalent dilution cross-section:*

$$\sigma_e = \frac{\sigma_{00}(1 - P_{00})}{P_{00}}. \tag{8.43}$$

This formal observation hardly seems practical, however: we have seen that the interest of this approach lay in the fact that σ_d was a constant, and therefore a pre-tabulation of the effective resonance integral performed once and for all could be envisaged.

Now, the parameter σ_e that replaces σ_d is no longer a constant, but no doubt a complicated function of lethargy, since its expression contains the resonant cross-section σ_0 (note that P_{00} also depends on this cross-section amongst other things).

If we examine it carefully, however, we notice that σ_e is *almost a constant*. If we therefore make the approximation of replacing it by a constant $\bar{\sigma}_e$, we can calculate the *homogeneous* effective resonance integral for this value using the table, and thereby obtain an approximation of the heterogeneous effective resonance integral. We shall now discuss this further.

8.3.2. Wigner and Bell-Wigner approximations: the concept of homogeneous-heterogeneous equivalence

Wigner noticed that the curves giving P_{ff} as a function of Σ_f for a few common geometries could be approximated fairly well by the following expression:

$$P_{ff} \simeq \frac{\ell\Sigma_f}{1+\ell\Sigma_f}, \tag{8.44}$$

where $\ell = 4V_f/S_f$, four times the volume of the fuel divided by its surface, is the average chord of this region, i.e. the average distance between two points on the surface chosen at random (this result is Cauchy's theorem). By introducing this approximation into the formula for σ_e, we find:

$$\sigma_e \simeq \frac{1}{\ell N_0}, \tag{8.45}$$

i.e., a value that is constant to all intents and purposes.

We might wonder how accurate this approximation is. To assess this, we can plot σ_e as a function of $\ell\Sigma_f$ for a few examples, as in Figure 8.5. It can be shown that the asymptotic value $\sigma_{e,\infty}$ is the Wigner expression; that is why $b = \sigma_e/\sigma_{e,\infty}$ was plotted along the ordinate axis (the Bell factor or, more accurately, the Bell function).

Replacing the function b by a constant b should not lead to a very large error, especially if this constant is well chosen (this problem will be examined in Chapter 15). This so-called Bell-Wigner approximation, giving:

$$\sigma_e \simeq \frac{b}{\ell N_0}, \tag{8.46}$$

or equivalently:

$$P_{ff} \simeq \frac{\ell\Sigma_f}{b+\ell\Sigma_f}, \tag{8.47}$$

makes it possible to establish a *heterogeneous-homogeneous equivalence*: the actual geometry is normalised to an equivalent (in terms of self-shielding) homogeneous geometry whose effective resonance integral can simply be looked up in a table.

8.3.3. Fuel containing a mixture

In addition to the resonant nucleus, the fuel often contains another material inseparably mixed in with it, such as oxygen with uranium in an oxide. To allow for this in the equations, $R_f\Phi_f$ must be replaced by $R_0\Phi_f + R'_f\Phi_f$ and Σ_f must be replaced by $\Sigma_0 + \Sigma'_f$; in these sums the first term represents the resonant material concerned, and the second term represents all the other materials (assumed to be non-resonant) that are mixed with the resonant material in the fuel.

If we note that $R'_f\Phi_f/\Sigma'_f$ is approximately equal to Ψ (the macroscopic flux is more or less "flat" both in space and in lethargy) and that collisions in the fuel are distributed according to proportions of the total cross-sections, i.e. that:

$$P_{00} = P_{ff}\frac{\Sigma_0}{\Sigma_0+\Sigma'_f}, \tag{8.48}$$

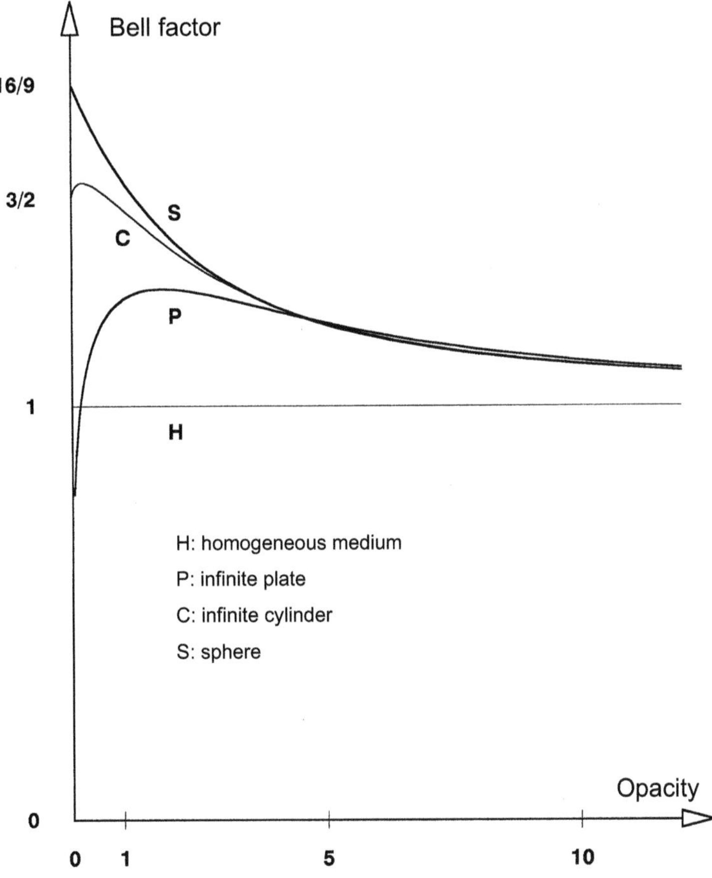

Figure 8.5. Bell function for a few geometries (opacity is the product of average chord by total cross-section).

we find the *same* fine structure equation with the *same* definition of equivalent section as before, i.e.:

$$\sigma_e = \frac{\sigma_{00}(1 - P_{00})}{P_{00}}. \tag{8.49}$$

With the Bell-Wigner approximation, this leads to:

$$\sigma_e \simeq \frac{b}{\ell N_0} + \frac{\Sigma'_f}{N_0}. \tag{8.50}$$

In this formula, the first term is the "heterogeneous term" mentioned earlier, and the second term is the "homogeneous term", i.e. (except for a detail of the notation) the term we had earlier for the homogeneous situation. We simply need to add these two terms.

8.3.4. The Dancoff effect

Figure 8.5, representing the Bell function, implicitly assumes that the fuel element (whether spherical, cylindrical, or a plate) is isolated: this means that a neutron leaving the fuel element without collision is certain to have its first collision with an atom of the moderator, not the fuel. In reality, in "tight" lattices such as water reactors or fast neutron reactors, a neutron leaving a fuel element can quite easily travel through the moderator, enter another fuel element, and undergo its first collision there... or even pass through this element, and then the moderator beyond it, and undergo its first collision in a third element...

If this effect exists, it is called the Dancoff effect, after the physicist who studied it. The resulting correction of P_{00} and σ_e is the Dancoff correction. After applying a few simple assumptions, it is expressed with the *Dancoff* C *factor*: the *probability* for a neutron leaving a fuel element of crossing the moderator without a collision, and therefore of entering another fuel element. To calculate this probability, we assume in general that neutrons leave the fuel element isotropically. In particular, this assumption was applied to the calculations concerning a regular square lattice of cylindrical elements in a homogeneous moderator, allowing the following curves to be plotted (Figure 8.6).

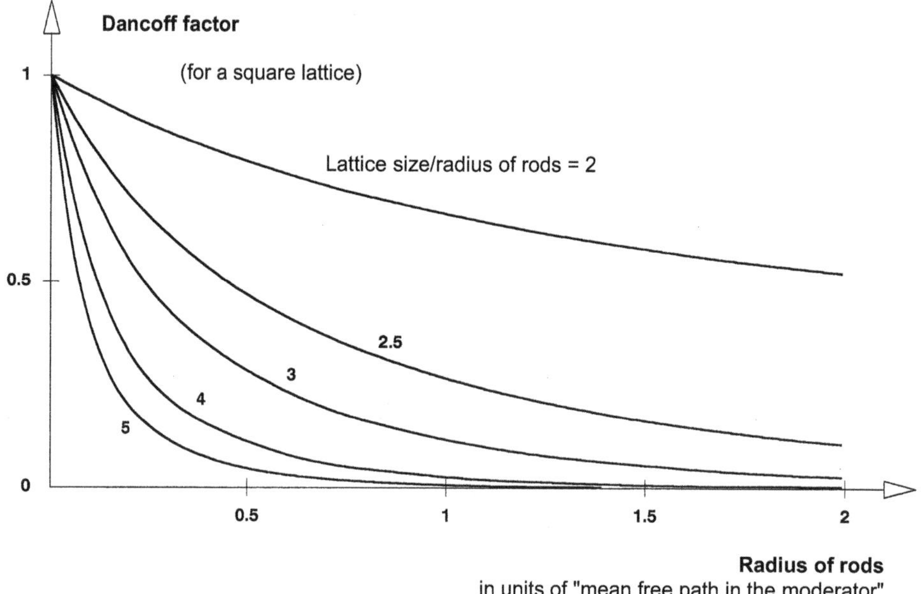

Figure 8.6. Dancoff factor for a square lattice of cylindrical fuel elements.

Note that the Dancoff factor for water reactors is of the order of 0.1 to 0.3.

Let P^+_{ff} denote the probability of a first collision in the fuel, without Dancoff effect, for a neutron emitted in the fuel. With the Dancoff effect, the series outlined above gives:

$$P_{ff} = P^+_{ff} + P^+_{fs}CP^+_{sf} + P^+_{fs}C(1 - P^+_{sf})CP^+_{sf} + \dots$$

$$P_{ff} = P^+_{ff} + \frac{P^+_{fs}CP^+_{sf}}{1 - C(1 - P^+_{sf})}, \qquad (8.51)$$

where $P^+_{fs} = 1 - P^+_{ff}$ is the probability of exit without collision from the fuel of origin, and P^+_{sf} is the probability of collision in the fuel for a neutron entering via its surface. We can show (*see* Chapter 14) that we have the following reciprocity relationship:

$$P^+_{sf} = \ell \Sigma_f P^+_{fs}, \tag{8.52}$$

such that all of the probabilities P^+ can be expressed according to P^+_{ff}.

Moving on to the equivalent cross-section, we can perform a somewhat tedious calculation which eventually becomes simpler, to obtain the following relationship:

$$\sigma_e \simeq \frac{b}{\ell N_0} + \frac{\Sigma'_f}{N_0}, \tag{8.53}$$

with:

$$b = \frac{(1 - C)b^+}{1 - C + Cb^+}, \tag{8.54}$$

where b^+ is calculated *without* the Dancoff effect (Figure 8.5).

8.3.5. Formula for the resonance escape probability in a heterogeneous situation

Let $Q(u)$ denote the slowing down current integrated over the entire volume of the cell: by definition, the resonance escape probability is the ratio of the value of this current at the end of the slowing down domain, after the resonances, to its initial value, just after the domain of emission by fission, and before the resonances.

1/ Because the lethargy integrals that define the current q concern the same intervals as those used to define the arrival density ρ, the same smoothing phenomenon occurs; this current is not related to the real "post-collision" flux Φ, but the macroscopic flux Ψ. Similarly, because we are reasoning based on the macroscopic behaviour, the Placzek transient can be neglected, i.e. the Fermi model can be used. Under these conditions, the current integrated over the cell is the product of the moderator power integrated over the cell and the macroscopic flux Ψ:

$$Q(u) \cong \left[(V\xi\Sigma_s)'_f + (V\xi\Sigma_s)_m\right] \Psi(u).$$

(The slowing-down power of the resonant material itself could also be added, but it is sure to be negligible compared to the others.)

2/ The derivative of this current with respect to lethargy is the product of the fuel volume and the average absorption rate per unit volume of fuel $N_0\sigma_{a,f}(u)\Phi_f(u)$. This rate has been rewritten in the form $N_0\sigma_{a,\mathrm{eff}}(u)\Psi(u)$ with Φ_f replaced by $\varphi\Psi$, and setting $\sigma_{a,\mathrm{eff}} = \sigma_{a,f}\varphi$:

$$\frac{dQ(u)}{du} = -V_c N_0 \sigma_{a,\mathrm{eff}}(u)\Psi(u). \tag{8.55}$$

3/ By eliminating Ψ between these two equations, we find the differential equation governing Q:

$$\frac{dQ(u)}{du} = -\frac{V_c N_0 \sigma_{a,\mathrm{eff}}(u)}{(V\xi\Sigma_s)'_f + (V\xi\Sigma_s)_m} Q(u). \tag{8.56}$$

We integrate to obtain the ratio $Q_{\text{final}}/Q_{\text{initial}}$, which is the resonance escape probability:

$$p = \exp\left[-\frac{V_c N_0 I_{\text{eff}}}{(V\xi\Sigma_s)'_f + (V\xi\Sigma_s)_m}\right], \tag{8.57}$$

setting:

$$I_{\text{eff}} = \int \sigma_{a,\text{eff}}(u)\, du. \tag{8.58}$$

This equation is the general version of the one we wrote for the homogeneous case (§ 8.1.4), and it is interpreted in the same way. Using the same approach, the following equation would be obtained for any geometry:

$$p = \exp\left[-\frac{V_c N_0 I_{\text{eff}}}{\sum_i (V\xi\Sigma_s)_i}\right]. \tag{8.59}$$

8.4. Doppler effect

8.4.1. Importance of the Doppler effect

In Chapter 13, we shall see that reactors are affected by a certain number of temperature effects; reactivities vary with temperature, creating counter-reactions and changing the kinetics. Here we wish to discuss one of these effects, because it is directly related to resonant absorption. It is called the Doppler Effect because it is related to a problem of the *relative velocity* of neutrons and resonant nuclei, and it is probably the most important one in terms of system stability and therefore safety[9].

As we shall see, this effect is characterised in practice by a *negative* coefficient: if the temperature increases, the reactivity decreases. This gives the reactor spontaneous stability, and guarantees a return to normal conditions in the event of an incident. For example, an unexpected power increase causes a temperature rise, and therefore a decrease in reactivity (if the coefficient is negative), and so the reactor, which was initially critical, becomes slightly sub-critical, and the power decreases. Similarly, an initial drop in power would be followed by an increase. In either situation, the counter-reaction cancels out the initial perturbation, and the system returns to its equilibrium power level (temperature giving zero reactivity).

8.4.2. Origin of the Doppler effect

In the previous chapter and the present chapter, we have always considered that the nucleus struck by a neutron was initially at rest. This appears to be a valid assumption, because the kinetic energy of a neutron slowing down — tens, hundreds, or thousands of electron volts — is far greater than the thermal agitation energy of the target nuclei (of the order of a few hundredths of an electron volt) related to their temperature in the reactor.

[9] The best-known case of the Doppler Effect in practice is the change in frequency of a wave if the emitting source is moving at a certain velocity with respect to the observer. For example, the sound of a fire engine's siren seems to change in frequency suddenly when the fire engine passes the observer, i.e. when the relative velocity of the source with respect to the observer changes sign.

With regard to scattering[10], the assumption can definitely be made, and so the arguments we have presented are not called into question. With regard to absorption, however, the effects of the thermal agitation of target nuclei are not negligible, and must be taken into account in the *effective resonance integral calculations* (the rest of the formalism remains unchanged).

The basic reason for this sensitivity is the *very fast change in cross-sections for materials with resonances,* which in practice means *uranium 238*. (Because this is the material concerned, the Doppler Effect is related to the temperature of the *fuel*; this is the main stabilising effect because changes in fuel temperature follow power variations almost instantaneously.)

The problem is as follows: by taking the (low) speed of the target nucleus at the moment of impact into account when considering the neutron-nucleus interaction, we very slightly modify the *relative velocity* of the neutron with respect to the nucleus, i.e. the velocity in the *centre of mass*. The cross-section, which is a function of the relative speed, is therefore changed. If the cross-section changes slowly, this change is small and probably negligible, but if the neighbourhood of a resonance peak, this small change in relative velocity can lead to a very significant change in cross-section.

The velocity of the target nucleus (considered in the laboratory system) varies in intensity and direction (in solids and liquids, the effect of temperature is a vibration of the atoms around a mean position). The correction can therefore vary in sign and absolute value. A complete calculation must be performed to find out what the overall effect will be. This calculation is a *convolution* (i.e. an integral) between the function representing the cross-section in the centre of mass system and the spectrum of speeds of the thermal agitation of the target nuclei, performed with the coordinate system change formulae taken into account. This calculation obviously assumes that the thermal agitation spectrum is known.

8.4.3. Doppler effect calculation

This spectrum is simple in gases: it is the *Maxwell spectrum* (the formulae are presented in the next chapter) giving the distribution of (scalar) speeds and the *isotropy* for the directions. In solid materials such as those found in nuclear fuels, the distributions are far more complicated and not well known. That is why a thermal agitation is often approximated according to a Maxwell spectrum (in practice, an attempt is made to correct the error by replacing the actual temperature of the fuel with an "effective temperature").

If we also use the Breit–Wigner relationship to represent the resonances in the centre of mass system, after performing all the calculations we can seen that the functions Ψ and χ in the formulae (§ 2.7.1) simply need to be replaced by integrals that have been tabulated for practical calculations:

$$\psi = \frac{1}{2\beta\sqrt{\pi}} \int_{-\infty}^{+\infty} \frac{\exp\{-(x-y)^2/(4\beta^2)\}}{1+y^2} dy, \tag{8.60}$$

and:

$$\chi = \frac{1}{2\beta\sqrt{\pi}} \int_{-\infty}^{+\infty} \frac{\exp\{-(x-y)^2/(4\beta^2)\}}{1+y^2} 2y\, dy, \tag{8.61}$$

10 Allowing for the Doppler Effect very slightly modifies the energy of the transfer nucleus during scattering, but the concrete consequences, which have been studied and estimated, are negligible.

where we set:

$$x = \frac{2(E - E_0)}{\Gamma} \qquad \Delta = \sqrt{4E_0kT/A} \qquad \beta = \Delta/\Gamma, \tag{8.62}$$

(Δ is the "Doppler width").

Figure 8.7 shows the impact of the Doppler Effect for the first and main resonance of uranium 238 (this figure shows the variations of the function Ψ with its arguments x (abscissa) and β, related to temperature. The function χ is antisymmetric, and tends to flatten and widen in a similar way when the temperature rises.)

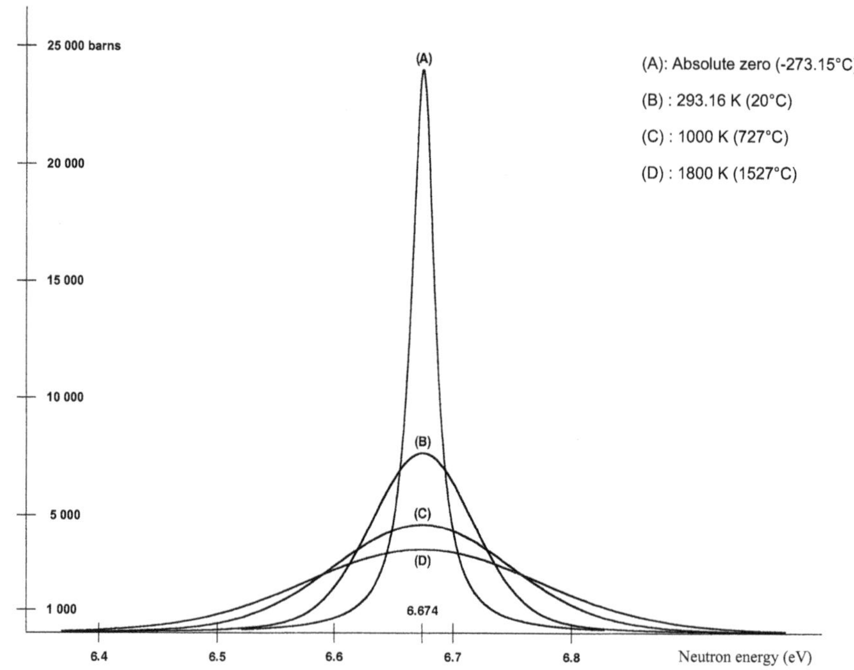

Figure 8.7. Doppler Effect widening of the major resonance of uranium 238.

Note two essential characteristics: the resonance widens, and the peak is lowered. It can be shown that this occurs with a *constant resonance integral* (area under the curve).

This might lead one to think that these two modifications with a constant integral have practically no effect on the neutron physics of the problem, but that is *wrong*; because of self-shielding, *the widening of the resonances has a much greater effect that the lowering of the peaks.* This is immediately apparent if we examine the formulae for the self-shielding factor and the effective resonance integral. In more basic terms, we can say that the Doppler Effect widens the resonance traps for the "kangaroos" whilst leaving them almost black in spite of the lowering of the cross-section curves.

This means that the Doppler Effect leads to *an increase in resonant absorption*; in particular, an increase in resonant capture by uranium 238 (fissionless capture) and therefore a reduction in the multiplication factor. That is why the Doppler Effect coefficient is negative in practice (roughly –2 to –3 pcm per degree Celsius).

In Figure 8.8, note that in the useful domain (equivalent cross-section in the neighbourhood of fifty barns), the *effective resonance integral for capture by uranium 238 varies approximately linearly with the square root of the dilution cross-section.* By performing calculations at various temperatures, we can also show that it varies approximately linearly with the *square root of the absolute temperature.*

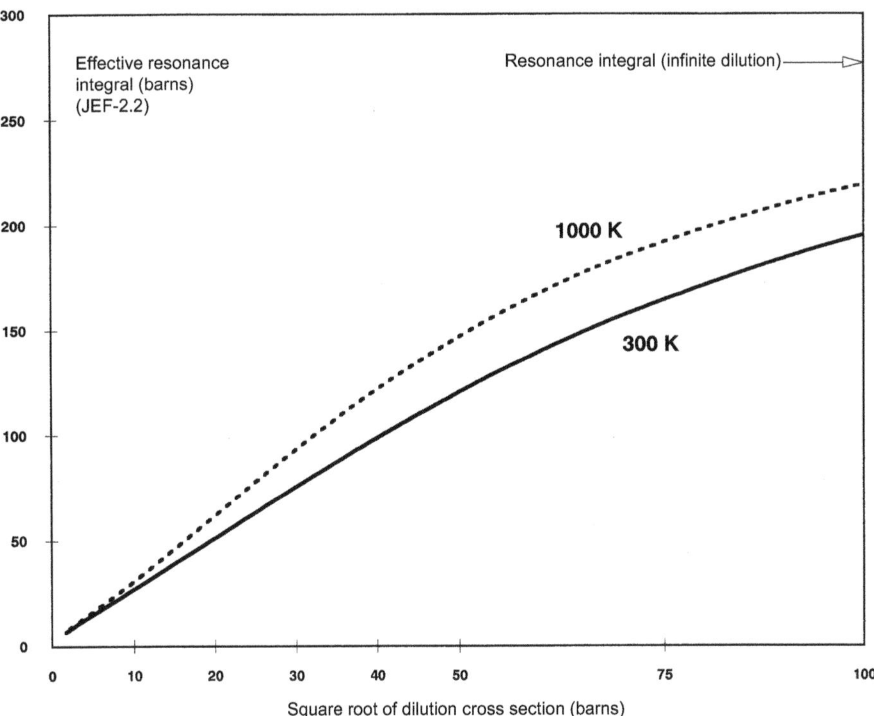

Figure 8.8. Effective resonance integral for capture by uranium 238.

In conclusion, note that the Doppler Effect makes it necessary to perform *tabulations of the effective resonance integral not only as a function of the dilution cross-section, but also as a function of temperature.*

8.5. Future prospects: problems that a resonant absorption theory will have to solve

To conclude this general chapter on resonant absorption, let us briefly outline why we cannot limit ourselves to using the formula for resonance escape probability p that we established, and why other developments for a theory of the resonant absorption of neutrons are necessary. (Some parts of the answer to this question will be given in Chapter 15).

8.5.1. Evaluating the Bell factor

As we have seen, the Bell factor, b, a constant replacing the Bell function, can be used to establish an equivalence between the real problem, which is heterogeneous, and a previously tabulated homogeneous problem. We must find the criterion for choosing this factor that will give the most accurate equivalence possible.

8.5.2. Continuous-multigroup equivalence

To find the equation for p, we had to assume that the macroscopic flux Ψ was "flat" in both lethargy and space. This is obviously an approximation.

A multigroup approach[11], if relatively fine (but not on the scale of the resonances), should eliminate the need for this assumption, but only if the multigroup cross-sections can be correctly defined and calculated. This problem will be handled by another equivalence procedure[12] called the "continuous-multigroup equivalence".

8.5.3. Allowing for complicated geometries

To handle the problems encountered in real reactors, it is necessary to deal with more complicated geometries than the simple two-zone cell, an element of a regular, infinite lattice, that we have considered here.

Two extensions are required:

1/ To be able to handle several non-resonant zones.

2/ To be able to handle several resonant zones.

We shall see that the first problem is a simple extension, but the second one requires far more difficult developments.

8.5.4. Cases with several resonant nuclei

Finally, a third extension will turn out to be essential: to be able to handle several resonant nuclei. This problem always arises in practice, partly because nuclear fuels contain several heavy nuclides (isotopes of uranium, plutonium, etc.) that all have resonances, and partly because a particular nuclide can be found in zones with different temperatures[13], and therefore with different deformations of the effective cross-section curves.

8.5.5. Definition and calculation of effective temperatures

This problem can be solved approximately by adopting a well-chosen average (or "effective") temperature. First, however, the selection criterion must be found and we must be able to implement it.

[11] The general principles of multigroup theory will be presented in Chapter 10.

[12] The general concept of equivalence, as used in neutron physics, will be presented in Chapter 17.

[13] For example, there can be a temperature difference of 1000 °C between the centre and the surface of fuel rods.

There is another problem of "effective" temperature that has already been mentioned: the need to rectify the error that is made in the Doppler Effect calculation due to equating the thermal agitation spectrum of the resonant nuclei with a Maxwell spectrum.

The resonant absorption theory, undoubtedly the most difficult theory in neutron physics, still presents a number of challenges that will need to be faced.

Exercises

Exercise 8.1: energy dependence of resonance widths

Assume that the resonances can be described by the single-level Breit-Wigner formula. In this exercise, the widths of the resonances will be compared with the lethargy gain by elastic scattering, at the resonance energy. In order to simplify the discussion, also assume that all the nuclear resonances have equal values for their widths Γ_n and Γ_γ.

a) *Nuclear resonance width:* this is given in terms of lethargy by $\gamma = \Gamma/E_0$, and lethargy gain due to scattering by its maximum value ε. How does the ratio γ/ε vary with E_0? Using Γ = 30 meV, determine the intervals where this ratio is greater than 5 (wide resonance), between 2 and 5 (rather wide resonance), between 1/2 and 2 (intermediate resonance), between 1/5 and 1/2 (rather narrow resonance) and less than 1/5 (narrow resonance).

b) *Practical nuclear resonance width:* this is defined as $\gamma_p = \Gamma_p/E_0$, where Γ_p is the energy interval where the resonant cross-section is greater than the potential cross-section. Repeat the previous exercise, using Γ = 30 meV, with a potential cross-section σ_p = 10 barns, and a resonant cross-section σ_r = 20 000 barns for a peak at 50 eV.

Exercise 8.2: window resonance

In order to simplify this exercise, the potential and resonant scattering cross-sections of the resonant nuclide are assumed to be zero. In the case of the potential cross-section, this is equivalent to including it in the dilution cross-section. The capture cross-section of the resonant nuclide is zero everywhere except over an energy interval $[E_1, E_2]$ where it is constant and equal to σ_r. This is the so-called 'window' approximation. The 'narrow resonance' hypothesis is assumed for all the nuclides included in the dilution cross-section σ_d.

a) Let $\gamma = \ln(E_1/E_2)$ be the lethargy width of the resonance. Show the expressions for the resonance integral I_{res}, the effective resonance integral I_{eff}, and the self-shielding factor $f = I_{eff}/I_{res}$.

b) Show the expressions for the resonance integral I_{res}, the effective resonance integral I_{eff}, and the self-shielding factor $f = I_{eff}/I_{res}$ for a narrow resonance without the interference term, which is described by the Breit-Wigner formula?

c) How should σ_r and γ be chosen in order to respect I_{res} and f in the 'window' approximation of the resonance?

d) In order to take account of Doppler broadening in the 'window' approximation, its width is increased and its height decreased without changing its integral. How does this change the effective integral? Compare the relative effects on I_{eff} and γ, of approximating all the uranium-238 nuclear resonances by replacing them with a single one, using I_{res} = 280 barns, I_{eff} = 20 barns, and σ_d = 60 barns.

Exercise 8.3: volume-surface and surface-volume probabilities

For a homogeneous, convex object of volume V and surface S, take the integral

$$I = \int_V d^3r \int_S d^2S \, \frac{e^{-\Sigma Y}}{Y^2} \cos\theta,$$

where $\vec{Y}$ is the vector joining a point inside the volume to a point on the surface, Y is its modulus, and θ is the angle between this vector and the external normal to the surface at each given point.

Show that the probabilities P_{VS} and P_{SV} can be expressed using I, and deduce the reciprocity relationship between both these probabilities.

Exercise 8.4: volume-volume probability for a small opacity

a) For a homogeneous, convex object of volume V and surface S, the probabilities P_{VV}, P_{VS}, P_{SV} and P_{SS} are applicable. Show that, when the opacity ω—which is product of the macroscopic total cross-section with the mean chord—is small, then P_{SS} can be approximated by an expansion in powers of ω,

$$P_{SS} = 1 - \omega + \frac{1}{2!}\frac{\langle X^2\rangle}{\langle X\rangle^2}\omega^2 - \cdots = 1 - \omega + \frac{Q}{2}\omega^2 - \cdots,$$

where $Q = \langle X^2\rangle/\langle X\rangle^2$.

b) When the expansion of P_{SS} is limited to these three terms, what is the expression for P_{VV}?

c) Supposing that V contains a resonant nuclide (and only this nuclide), find the first order approximation for the equivalent dilution cross-section σ_e.

d) Find Q for a sphere, and an infinitely long cylinder.

Exercise 8.5: effect of oxygen in the fuel

The presence of oxygen in the fuel modifies the resonance escape probability through two effects: it both contributes to the neutron slowing-down, and changes the dilution cross-section, thereby changing the effective integral. Using the classical formula for p, and neglecting uranium-235 and the cladding, evaluate both these effects for a PWR cell using the following data.

Data

- concentration of uranium-238 atoms in the fuel = 22×10^{27} m^{-3};
- concentration of oxygen atoms in the fuel = 45.5×10^{27} m^{-3};
- fuel radius = 4.1 mm;
- Bell's factor = 1.1;
- Dancoff's factor = 0.3;

- concentration of water molecules in the moderator = 24×10^{27} m^{-3};
- volume moderation ratio = 2;
- scattering cross-section of hydrogen[14] = 20.4 b;
- scattering cross-section of oxygen = 3.76 b;
- empirical formula for the effective integral (at 900 K):

$$I_{\text{eff}} = 2.36 + 2.80\sqrt{\sigma_e}$$

(I_{eff} and σ_e are given in barns).

Solutions

Exercise 8.1: resonance widths according to their energy

a) *Nuclear resonance width:* the ratio $r = \gamma/\varepsilon$ varies as $1/E_0$.

b) *Practical nuclear resonance width:* the practical width (in energy) is given by

$$\Gamma_p \simeq \Gamma\sqrt{\sigma_{\max}/\sigma_p},$$

where

$$\sigma_{\max} = 4\pi\lambda\!\!\!^{-2} g\frac{\Gamma_n}{\Gamma}.$$

This maximum cross-section varies as $1/E_0$ due to the contribution from $\lambda\!\!\!^{-2}$; therefore, the ratio $r = \gamma_p/\varepsilon$ varies as $1/E_0^{3/2}$.

Numerical values of E_0 are given in the table below in eV for the proposed limiting values of r, obtained using $\varepsilon = 0.0169$.

r	1/5	1/2	1	2	5
(nuclear)	8.8	3.5	1.8	0.9	0.4
(practical)	199	108	68	43	23

When the nuclear width is used, all the uranium-238 resonances appear to be narrow, except the first one at 6.7 eV. When the practical width is used, it is apparent that the low energy resonances become wide or rather wide. For the true resonances, a more precise examination must be made, taking into account the individual parameters of the resonances; nevertheless, the qualitative conclusion obtained here with the practical width is essentially correct.

---

14 *N.B:* this value is for the 'plateau' in the resonance range; it is smaller than the value given in appendix **B** which is relative to 2200 m s^{-1} neutrons.

Exercise 8.2: window resonance

a) Using $I_{res} = \gamma\sigma_r$ and, applying the 'narrow resonance' hypothesis

$$\varphi_{NR} = \frac{\sigma_d}{\sigma_d + \sigma_r}, \qquad I_{eff} = \frac{\gamma\,\sigma_d\,\sigma_r}{\sigma_d + \sigma_r};$$

therefore,

$$f = \varphi_{NR} = \frac{\sigma_d}{\sigma_d + \sigma_r}, \qquad f = \frac{\sigma_d}{\sigma_d + I_{res}/\gamma}.$$

b) If the potential cross-section and the resonant scattering are neglected, then the Breit-Wigner formula gives

$$I_{res} = \frac{\pi}{2}\frac{\sigma_{max}}{E_0}\Gamma, \qquad f = \frac{1}{\sqrt{1 + \sigma_{max}/\sigma_d}}.$$

c) The effective resonance integral and the self-shielding factor are properly taken into account when the 'window' is described by

$$\gamma = \frac{I_{res}}{\sigma_d\left(\sqrt{1 + \sigma_{max}/\sigma_d} - 1\right)}, \qquad \sigma_r = \sigma_d\left(\sqrt{1 + \sigma_{max}/\sigma_d} - 1\right).$$

When $\sigma_{max} \gg \sigma_d$, the first relationship simplifies to

$$\gamma E_0 \simeq \frac{\pi}{2}\Gamma\sqrt{\sigma_{max}/\sigma_d}.$$

Apart from the factor $\pi/2$, this is the same as the expression for the 'practical width' given in exercise **8.1**.

d) Using a logarithmic derivation gives

$$\frac{\Delta I_{eff}}{I_{eff}} = (1 - f)\frac{\Delta\gamma}{\gamma}, \qquad \frac{\Delta I_{eff}}{I_{eff}} \simeq \frac{\Delta\gamma}{\gamma}.$$

Indeed, this approximation is reasonable because $f = 1/14$. Thus, increasing the width has a more significant effect than reducing the height on the effectiveness of the 'trap'. When $\sigma_d = 60$ barns, $\sigma_r = 780$ barns.

Exercise 8.3: volume-surface and surface-volume probabilities

The probability P_{VS} is described by

$$P_{VS} = \frac{1}{4\pi V}\int_V d^3r \int_{(4\pi)} d^2\Omega\, e^{-\Sigma Y}.$$

Noticing that

$$d^2\Omega = \frac{d^2S}{Y^2}\cos\theta,$$

it can be seen

$$P_{VS} = \frac{1}{4\pi V}I.$$

The probability P_{SV} is described by

$$P_{SV} = \frac{1}{\pi S}\int_S d^2S\,\cos\theta \int_{(2\pi)} d^2\Omega\,\left(1 - e^{-\Sigma X}\right),$$

where X is the chord from the point on the surface where the neutron enters to the point on the other side where it exits. At the point of entry, the absolute value of the cosine must be taken. Noticing that

$$1 - e^{-\Sigma X} = \Sigma \int_0^X e^{-\Sigma Y}\,dY,$$

and

$$d^2\Omega\,dY = \frac{d^3r}{Y^2},$$

it can be seen

$$P_{SV} = \frac{\Sigma}{\pi S} I.$$

Consequently, the reciprocal relationship is

$$P_{SV} = \frac{4V\Sigma}{S} P_{VS}.$$

Exercise 8.4: volume-volume probability for a small opacity

a) Using the notation of the previous exercise, the probability P_{SS} is

$$P_{SS} = \frac{1}{\pi S}\int_S d^2S\,\cos\theta \int_{(2\pi)} d^2\Omega\,e^{-\Sigma X}.$$

When ω is small, the exponential can be expanded to give the approximate expression stated in the question.

b) Using the reciprocity and complementarity relationships,

$$P_{SV} = \frac{4V\Sigma}{S} P_{VS}, \qquad P_{VV} + P_{VS} = 1, \qquad P_{SV} + P_{SS} = 1,$$

gives

$$P_{VV} \simeq \frac{Q}{2}\omega.$$

c) If this volume contains only the resonant nuclide, then this formula can also be written

$$P_{00} \simeq \frac{Q}{2} N_0\,\sigma_0\,\ell.$$

Therefore,

$$\sigma_e \simeq \frac{2}{Q}\frac{1}{\ell N_0}.$$

The second factor is the asymptotic value for a large opacity. Recall that this result is similarly obtained from the expression $P_{SS} \simeq 1$.

d) For both geometries, any entry point can be chosen. For the sphere it is convenient to measure the polar angle from the normal at the point of entry. The calculation gives $Q = 9/8$. For the cylinder, it is convenient to measure the polar distance from the generatrix, and the azimuthal angle from the normal to the point of entry. The calculation gives $Q = 4/3$.

Exercise 8.5: effect of oxygen in the fuel

The data permit the value of p to be calculated without oxygen, giving $p = 0.797$.

To first order, the increase in slowing-down power affects p by an amount

$$\frac{\Delta p}{p} = \left[-\ln\frac{1}{p}\right]\left[\frac{\Delta\left(V\xi\Sigma_s\right)}{V\xi\Sigma_s}\right], \qquad \text{with} \qquad \frac{\Delta\left(V\xi\Sigma_s\right)}{V\xi\Sigma_s} = \frac{V_f(\xi\Sigma_s)_{O,f}}{V_m\left(\xi\Sigma_s\right)_m}.$$

In numerical terms, this represents about 1% greater slowing-down power, giving 235 pcm change in p.

The presence of oxygen mixed with uranium-238 reduces the effect of self-shielding. By scattering on oxygen, neutrons can acquire a resonance energy directly in the resonance region. The Σ_c'/N_0 term must be added to the equivalent dilution cross-section:

$$\sigma_e = \frac{1 - C}{1 - C + bC}\,\frac{b}{\ell N_0}.$$

Thus, it increases from 41.4 to 49.2 barns, and the effective integral is raised by +7.9% from 20.4 to 22.0 barns. However, the resonance escape probability falls by more than 1800 pcm. This second effect largely predominates.

Thermalisation of neutrons

Introduction

Neutrons will not slow down indefinitely in reactors, partly because they are sure to end up being absorbed[1], and partly because, even if not absorbed, they would end up reaching thermal equilibrium with the matter of which the system is made, and therefore a certain state of agitation characterised by a nonzero average kinetic energy. This average[2] is kT, where k is the Boltzmann constant, and T is the absolute temperature. For example, $kT = 0.0253$ eV, i.e. approximately one fortieth of an electron volt at normal temperature (20 °C), and approximately double that in an industrial water reactor, where that material (which has the function of the main "thermaliser") is at approximately 300 °C.

There is therefore a gradual transition between the "pure" slowing down we have studied and this asymptotic behaviour where thermal equilibrium is never quite reached. This transition is what we call "thermalisation". It begins to become apparent around a few tens times kT, i.e. a few electron volts.

In calculations, a cutoff energy is placed somewhat arbitrarily between the slowing down domain and the thermalisation domain. For example, 2.77 eV for the usual library of the APOLLO code. The energy domain we shall study in this chapter is therefore located below this cutoff.

9.1. Qualitative aspects of thermalisation

9.1.1. What distinguishes the thermal domain from the slowing-down domain?

When looking at neutron-nucleus interaction in the slowing-down domain, we have assumed the nucleus to be completely at rest before the interaction (only the absorption calculation in a material with resonances, i.e. very fast changes in cross-section, must

[1] This applies mainly to the domain we shall now study, but also in resonances in the case of a thermal neutron reactor; almost always at high energy in the case of a fast neutron reactor. In a fast neutron reactor, the flux of neutrons below about a hundred electron volts is generally negligible, and so the problem of thermalisation is even less liable to arise.

[2] Important note: as we shall see, there are different ways of defining this.

take the thermal agitation of target nuclei into account). By the very definition of the thermalisation domain, however, this assumption can no longer be made.

The essential consequence resulting from this is that neutrons can be *not only slowed down, but also accelerated* when scattering occurs. In a scenario where no absorption occurred, these two processes could even balance each other out, according to the definition of equilibrium.

We made another simplifying assumption (in addition to the *"target at rest"* assumption), but without giving the details: this was the *"free target"* assumption, meaning that the target was free of any restraint that might prevent its recoil on impact. In fact, however, this is not entirely true, because nuclei that are hit by neutrons are located at the centre of atoms, which are joined in structures (molecules or crystals) by chemical forces. As long as the energy of the incident neutrons is very much greater than the chemical bond energies, the bond energies can be neglected, as we did. For neutrons with an energy of the order of an electron volt or less, i.e. the order of magnitude of chemical bonds, this assumption is no longer acceptable[3].

These two aspects that must now be allowed for — thermal agitation and chemical bonds of the targets — will make neutron-matter interactions much more difficult to deal with. In the slowing-down domain we were able to handle this interaction using the phenomenological model of classical mechanics, but now we can no longer avoid using a quantum physics approach to describe scattering in the thermalisation domain. Like most neutron physicists working with reactors, we shall not enter into too many details of thermalisation theory; we simply choose to trust the work of the specialists who supply the cross-sections and transfer distributions to be used in codes. If the thermaliser were a monatomic gas (no chemical bonds), its thermal agitation spectrum would be a Maxwell spectrum (see below); a simple convolution with the velocity of the neutron would then give the cross-section and the transfer distribution (slowing down or acceleration) in the event of scattering.

In liquid or solid condensed matter, the agitation modes are far more complex[4]. In water, for example, in addition to their translations and rotations, the molecules also have internal vibration modes (Figure 9.1) according to the angle of the oxygen-hydrogen bonds or according to the axes of these bonds, in phase or in anti-phase.

As for the slowing down problem, we can distinguish *elastic scattering*, where the kinetic energy of the two "objects seen from the outside" is conserved, and *inelastic scattering*, where kinetic energy is not conserved because the internal energy of the object struck has been changed. In the slowing down domain, an inelastic collision corresponds to a change in the internal energy of the *nucleus*; in practice, this is an excitation from the fundamental to an excited level, and therefore a *loss* in terms of the energy of the neutron. In the thermal domain, and inelastic collision corresponds to a change in internal energy of a *molecule* (such as a water molecule) or a crystal; this than be a contribution or a withdrawal of internal energy, and therefore either a *loss* or a *gain* to the neutron energy (as for elastic collisions).

[3] Strictly speaking, the effect of the chemical bonds starts to be felt just before the end of the slowing down domain, for example below about forty electron volts in graphite. This is taken into account (in an indirect way, specialists speak of "linked slowing down".)

[4] These mechanical agitation modes, which are generally quantified, are called "phonons".

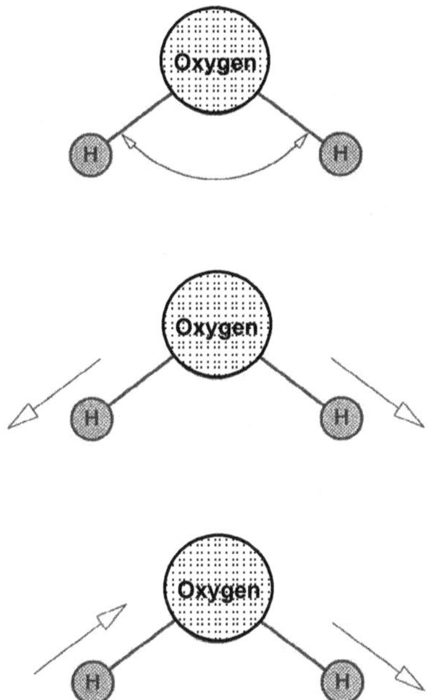

Figure 9.1. Internal vibration modes (phonons) of a water molecule.

Like any microscopic particle, the neutron not only has a material aspect, but also a wave aspect. Note that the wavelength of thermal neutrons is of the same order of magnitude as interatomic distances. That is why, in materials with a certain regularity, particularly crystals[5], as well as liquids such as heavy water[6], there can be interference effects between the waves scattered by each of the targets, leading to what is called *"coherent"* scattering. If this effect does not occur, then we have *"incoherent"* scattering.

9.1.2. Maxwell spectrum

For any thermaliser, the neutron spectrum (population comparable to an ideal gas) at equilibrium and in the absence of absorption would be a Maxwell spectrum. This spectrum has been well observed in materials with low capture (e.g. graphite, heavy water, etc.) when present in sufficient volume, is given by the following formula after normalising to one neutron:

$$n(E)\,dE = \frac{2\sqrt{E/kT}}{\sqrt{\pi}} \exp(-E/kT)\frac{dE}{kT}, \tag{9.1}$$

(proportion of neutrons with energy between E and $E + dE$). The most probable energy [at maximum $n(E)$] is $E_p = kT/2$; the average kinetic energy is $\bar{E} = 3kT/2$ (i.e. $kT/2$ per degree of freedom, with three degrees of freedom: translation along each direction of space).

5 Graphite is one example. The Bragg condition $2d\sin\theta = n\lambda$ (d: distance between atomic planes; θ: angle between neutron velocity and these planes; λ: neutron wavelength) cannot be satisfied for neutrons with energy below 0.004 eV; that is why the scattering cross-section almost vanishes above this threshold.

6 Interference between the waves scattered by each of the deuterons.

Taking speed from the equation for kinetic energy $E = mv^2/2$ and setting $mv_0^2/2 = kT$, the change of variable on this density gives:

$$n(v)\,dv = \frac{4}{\sqrt{\pi}}(v/v_0)^2 \exp\left[-(v/v_0)^2\right]\frac{dv}{v_0}, \tag{9.2}$$

(refer to Figure 9.2 for a graphical representation of this function). This formula gives a most probable speed v_p, which is equal to the reference speed v_0, and an average speed given by the following formula:

$$\bar{v} = \frac{2}{\sqrt{\pi}}v_0 = 1.128v_0. \tag{9.3}$$

Note that the most probable and average energies and speeds do not correspond to each other. If we use the flux nv for weighting, we obtain other coefficients.

The following values are associated with normal room temperature:

$$t = 20\ °\text{C}, \quad T = 293.15\ \text{K}, \quad kT = 0.0253\ \text{eV}, \quad v_0 = 2200\ \text{m/s}.$$

This speed, 2200 m/s, is often used as the reference speed in cross-section tables for the thermal domain.

9.1.3. Principle of microreversibility

As has been remarked, the theory of neutron-matter interactions in the thermal domain is likely to be complicated, and so we shall avoid discussing it in detail. Let us clarify just two points.

The first is the microreversibility principle, or *detailed balance*: in the absence of absorption and at equilibrium, when the neutron spectrum is the Maxwell spectrum, there is exactly the same number of transfers by diffusion from an energy interval dE to an energy interval dE' as transfers in the other direction, from dE' to dE. This leads to a mathematical constraint on the transfer relationship $P(E' \to E)$ that any model would have to observe:

$$E\exp(-E/kT)\,dE\,\Sigma_s(E)\,P(E \to E')\,dE' = E'\exp(-E'/kT)\,dE'\,\Sigma_s(E')\,P(E' \to E)\,dE. \tag{9.4}$$

(In each term we can identify the product of three terms: the Maxwell flux in the starting interval, the scattering cross-section, and the probability of transfer to another interval in the event of scattering.) This relationship can also be written by restricting the scattering events to those that cause the neutron to deviate by an angle ψ.

9.1.4. Scattering equations

The second point we shall raise is that the double differential (by energy and by angle or $\cos\psi = \mu$) scattering cross-section, $\Sigma_s(E')P(E' \to E)P(\mu)$, is a function of three variables E', E and μ which can in fact be expressed by S, a function of only two variables:

$$\Sigma_s(E', E, \mu) = \Sigma_s(E')\,P(E' \to E)\,P(\mu) = C^{te}\sqrt{E/E'}\exp[(E' - E)/(2kT)]S(\alpha, \beta), \tag{9.5}$$

where α and β characterise the momentum and energy transfers:

$$\alpha = \frac{E^2 + E'^2 - 2EE'\mu}{(kT)^2}, \qquad \beta = \frac{E' - E}{kT}. \tag{9.6}$$

This function S can be related to the *frequency spectrum*[7] $\rho(\omega)$, i.e. to the distribution of mechanical vibration modes (phonons) according to their frequency or their pulse ω.

9.1.5. Thermalisation equation

In formal terms, the thermalisation equation is just like the slowing-down equation, except for the fact that it has to be restricted to the thermal domain. Let us use it for a case that is independent of space and time. It can be written with either the lethargy, the speed, or the kinetic energy of the neutrons; for example, with kinetic energy:

$$\int_0^{E_{cutoff}} \Sigma_s(E')\,\Phi(E')\,dE'\,P(E' \to E) + S_{sl-d}(E) = \Sigma_t(E)\,\Phi(E). \tag{9.7}$$

This similarity reflects an analogous physical process — of energy changes by scattering — but hides a completely different mathematical aspect: whilst in the slowing down problem we always had E less than or equal to E' (and therefore E at the lower boundary of the integral), in the thermalisation problem, transfers can occur in both directions (the integral applies to the entire thermal domain). By discretising the equation, we obtain for the first case a *triangular* algebraic system that is solved explicitly from one term to the next starting from the highest energy. For the thermalisation problem, we obtain a *complete* algebraic system requiring a matrix inversion (generally performed by iteration, most notably in APOLLO). The upper boundary of the integral E_{cutoff} is the energy adopted to separate the thermalisation domain from the slowing down domain. The "source":

$$S_{sl-d}(E) = \int_{E_{cutoff}}^{\infty} \Sigma_s(E')\,\Phi(E')\,dE'\,P(E' \to E) \tag{9.8}$$

is no longer a true source in this case; it is a *density of arrival at energies below the cutoff energy* due to scattering events occurring in the last part of the slowing down domain and transferring the neutron beyond the cutoff energy, in the thermalisation domain. This term makes it possible to ensure flux continuity at the E_{cutoff} interface.

---

[7] This frequency spectrum not only affects the neutron physics properties, but also the heat transfer properties (specific heat) and optical properties (photon scattering) of the material concerned. Experimental data from these different branches of physics may be useful in the creation of models.

9.2. Appearance and characterisation of the thermal spectrum

9.2.1. Difference between the thermal neutron spectrum and the Maxwell spectrum

The difference between the real neutron spectrum and the Maxwell spectrum:

$$m(x) = \left(4/\sqrt{\pi}\right) x^2 \exp\left(-x^2\right) \tag{9.9}$$

is presented schematically as shown in Figure 9.2.

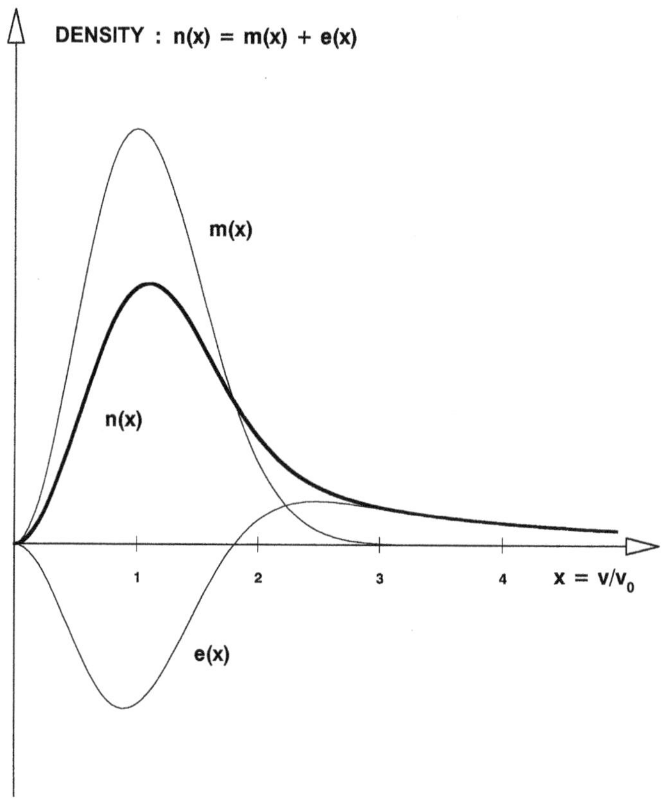

Figure 9.2. Difference between the real spectrum and the Maxwell spectrum.

This figure was plotted using the *normalised speed* $x = v/v_0$ where v_0 corresponds to kT, and the *density* is plotted as the ordinate. To break down the real density n into two components m (Maxwell) and e (difference), we normalised n and m to the same value; in other words, the integral of e is zero.

Using this convention, we observe that:

- in the truly Maxwellian part of the spectrum — approximately for speeds between zero and twice v_0 — the difference is *negative*. This reflects the *absorption* of neutrons that reduces their numbers;

- in the part corresponding to the higher speeds, the Maxwellian flattens out very quickly (it is practically zero above $x = 3$), whilst the real density falls slightly but remains significant. This is the "slowing down queue", where neutrons coming from high energies slow down and enter the thermal domain, compensating for the disappearances by absorption. As we have seen (§ 7.2.3), the slowing down flux is approximately constant in lethargy, and is given by the following equation:

$$\Phi(u) = \frac{q}{\xi\Sigma_s}, \tag{9.10}$$

where q is the slowing-down current in the neighbourhood of the cutoff energy, and $\xi\Sigma_s$ is the moderating power of the material. If we take the density and perform the change of variable, we find the following relationship for the behaviour of density in this range:

$$n_{as}(x) = \frac{2q}{\xi\Sigma_s v_0}\frac{1}{x^2}. \tag{9.11}$$

The current q is equal to the number of absorptions in the thermal domain, because it compensates for them, and so we can write:

$$q = \int \Sigma_a(v)\,\Phi(v)\,dv = \int \Sigma_a(x)\,n(x)x\,v_0\,dx, \tag{9.12}$$

and when we combine the two formulae, we obtain:

$$n_{as}(x) = N\frac{2\tilde{\Sigma}_a}{\xi\Sigma_s}\frac{1}{x^2}, \tag{9.13}$$

where:

$$N = \int n(x)\,dx, \tag{9.14}$$

is the total number of neutrons, and:

$$\tilde{\Sigma}_a = \frac{\int \Sigma_a(x)xn(x)\,dx}{N}, \tag{9.15}$$

is the absorption cross-section, traditionally called "effective". *If the cross-section is proportional to 1/v, then this effective cross-section is simply $\Sigma_a(v_0)$.*

If we then *normalise to one thermal neutron* ($N = 1$), the asymptotic density is:

$$n_{as} = 2r\frac{1}{x^2}, \tag{9.16}$$

where the following parameter:

$$r = \frac{\tilde{\Sigma}_a}{\xi\Sigma_s}, \tag{9.17}$$

characterises the *epithermal proportion* of the neutrons.

This coefficient can be introduced to renormalise the difference function presented above:

$$e(x) = 2r\varepsilon(x), \tag{9.18}$$

Note that, for a given thermaliser, the function $\varepsilon(x)$ (by definition asymptotically equivalent to $1/x^2$) has little dependence on the associated quantity of fuel, i.e. the absorption intensity. A good approximation of the thermal neutron spectrum is obtained by taking a "universal" function calculated once and for all.

9.2.2. Examples

Figure 9.2 showing the thermal spectrum can be compared to the spectra obtained by calculations using the APOLLO code (Figure 9.3) for a pressurised water reactor with fresh fuel (slightly enriched uranium on the one hand, and plutonium on a substrate of natural uranium on the other hand). The reader will have to become accustomed to the use of lethargy, which is the variable used by the code (although we have still oriented the abscissa axis in the direction of increasing energy). The spectra are directly those calculated by the code in multigroup theory, which gives the "staircase" shape of the curves. The calculation was performed on 99 energy groups, but only groups 52 to 99 (the last group of the slowing-down domain and the 47 groups of the thermal domain) are represented: this is the domain below 3.38 eV.

As in Figure 9.2, the curves concerning the uranium fuel spectrum and the Maxwell spectrum (this one is an analytical curve that can be plotted without the multigroup approximation) have been normalised to the *same number of neutrons* (same density integral over the domain considered). As in the previous figure, the slight difference between the maxima, the negative shift on the left (effect of absorption) and the positive shift on the right (slowing-down queue) can be observed. Apart from two small irregularities caused by small resonances of uranium 235, the constant-lethargy flux we would expect to find at the end of the slowing-down domain is indeed observed as far as 1 eV, and possibly even a little below.

We have chosen the normalisation used in APOLLO — a source of one neutron — to plot the spectrum for the fuel MOX. It is not surprising that both curves start from the same area at the far right, because the two situations are practically identical for the fast and epithermal domains: same quantity of moderator, approximately the same quantity of uranium 238, and the same cladding. On the other hand, the number of neutrons is far lower in the thermal domain; in this example, by approximately a factor of 4: this is because of the high absorption by MOX fuel of thermal neutrons, compared to the standard fuel. There are three combined reasons for this:

1/ the higher content of the plutonium than that of the uranium 235 it replaces, because the plutonium is only two-thirds fissile; this is plutonium obtained via the reprocessing of a standard irradiated PWR fuel, whose isotopic composition is approximately:

238/239/240/241/242 : 2/58/23/11/6 (in percentages),

(only odd-numbered isotopes 239 and 241 are fissile);

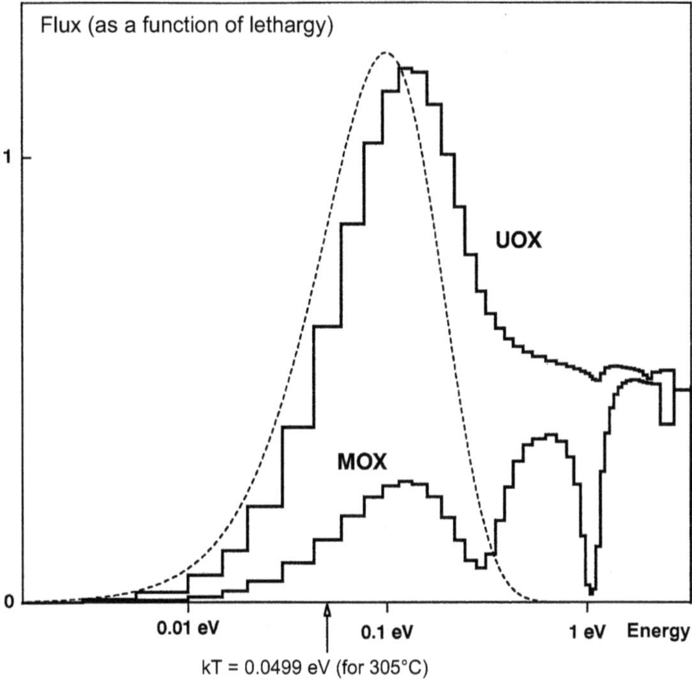

Figure 9.3. Thermal spectra in the fuel of a pressurised water reactor. 1) UOX: uranium, 3.7% of uranium 235. 2) MOX: mixed oxide at 6.5% plutonium. Dotted line: Maxwell spectrum normalised to the same number of thermal neutrons as the spectrum in the UOX case.

2/ the microscopic cross-sections at 2200 m/s that are higher for the fissile isotopes of plutonium than for uranium 235;

3/ the presence of very large resonances for four of these isotopes: around 0.3 eV for isotopes 239 and 241; at 1.06 eV for isotope 240 and around 2.5 eV for isotope 242 (refer to the cross-section curves of these nuclides presented in Figure 2.14).

The resonances at 0.3 and 1.06 eV create spectacular flux depressions, providing another illustration of the self-shielding phenomenon. The multigroup meshing has been refined in these regions so that the code will handle it reasonably well.

The resonance of plutonium 242, which is less abundant, is however poorly described[8]. With such strong absorption and such a complex structure, the fact of breaking down the spectrum into a Maxwellian component and a difference function obviously loses a lot of its interest.

[8] The most recent library with 172 groups further refines these resonances of plutonium, and makes it possible to handle self-shielding.

9.2.3. Average cross-sections

If we wish to characterise a cross-section $\sigma(E)$ with a *unique value* $\bar{\sigma}$ to perform a core calculation, for example, in the spirit of what was discussed in Chapter 6, an energy weighting rule must be adopted. There is no particular criterion that is imposed in an absolute way; the important thing is to ensure *coherence* between the convention used for cross-sections and the definition of the *total flux* $\bar{\Phi}$, in order to respect the *reaction rates*, the only truly measurable physical parameters:

$$\bar{\sigma}\bar{\Phi} = \int \sigma(E)\,\Phi(E)\,dE. \tag{9.19}$$

The *simplest* definition of total flux, and the one generally adopted[9], is the integral of the flux:

$$\bar{\Phi} = \int \Phi(E)\,dE. \tag{9.20}$$

The associated cross-sections, in this case called *"average" cross-sections*[10], must therefore be calculated by the following equation:

$$\bar{\sigma} = \frac{\int \sigma(E)\,\Phi(E)\,dE}{\int \Phi(E)\,dE}. \tag{9.21}$$

These formulae can be applied by integrating over all energies; this leads to a *one-group theory* (Chapter 6). More generally we isolate, the *group of thermal neutrons*[11]: the integrals are then taken between 0 and E_{cutoff}. For example, for a Maxwell spectrum *and* a 1/v cross-section:

$$\bar{\sigma} = \frac{\sqrt{\pi}}{2}\sigma(v_0) = \frac{\sqrt{\pi}}{2}\frac{\sqrt{293.15}}{\sqrt{T}}\sigma_{2200} = \frac{1}{1.128}\frac{\sqrt{293.15}}{\sqrt{T}}\sigma_{2200}. \tag{9.22}$$

(The factor $2/\sqrt{\pi} = 1.128$ is the average of $x = v/v_0$ on a Maxwell spectrum, and also the average of $1/x$.)

9.2.4. Processing a heterogeneous situation

In practice, the problem of thermalisation crops up in a heterogeneous structure such as a lattice of cells. To handle this problem, space and energy must be coupled. The handling of this problem in space will be described in detail in Chapter 14; but to give an overview, we take the example of the unit cell of a lattice assumed to be infinite and regular, consisting of a fuel rod surrounded by moderator (for simplicity, the cladding is ignored). The fuel is always fairly small, and it is no doubt acceptable to handle it by the

[9] Another possible definition: $v_0 N$, i.e. to within a factor, the *total number* of neutrons: this leads to the "effective" cross-sections. The expression for the parameter r reveals the macroscopic effective absorption cross-section.

[10] As opposed to "effective" cross-sections.

[11] This group will be, for example, the second of a two-group theory, where the first contains all fast and epithermal neutrons.

method of first collision probabilities assuming flat flux (as seen in the previous chapter, § 8.3.1, for the resonant absorption problem)[12].

If the moderator is also small in volume (e.g. water reactors), the same approximation can be applied. The equations to be solved then take the following form:

$$V_f R_f \Phi_f \times P_{ff} + V_m(R_m\Phi_m + S_{\text{sl-d}}) \times P_{mf} = V_f \Sigma_f \Phi_f,$$

$$V_f R_f \Phi_f \times P_{fm} + V_m(R_m\Phi_m + S_{\text{sl-d}}) \times P_{mm} = V_m \Sigma_m \Phi_m, \tag{9.23}$$

where the variable E has been omitted to simplify the expression, and where R now denotes the thermalisation operator; the slowing down source S_{ral} (per unit volume) has been assumed to appear in the moderator only.

If the moderator is large (for example, heavy water or graphite and natural uranium reactors), it must be divided into smaller volumes if a "transport" calculation is to be performed. For simplicity, it can also be handled using a diffusion approximation. For the energy aspect, we apply a multigroup process where the transfer probabilities are represented in matrix form[13].

9.3. Balance of thermal neutrons

9.3.1. General considerations

Assuming that all absorption cross-sections are proportional to 1/v and that all scattering cross-sections are constant, we would not be very far wrong if we took a monokinetic approach using the nuclear data taken at velocity v_0, i.e.:

$$2200 \text{ m/s} \times \sqrt{T/293.15}. \tag{9.24}$$

That is how Fermi and his associates proceeded[14]; the formulae they used, which we shall now introduce, allowed them to calculate the thermal part (factors f and η) of the multiplication factor (see the four-factor formula, § 3.3.2).

These days, calculations are performed numerically using codes such as APOLLO, with space and energy coupled. It is, however, still of interest to reproduce these factors in retrospect in order to determine the neutron balance. That is why we are presenting them.

We shall then examine three examples of how these basic formulae are used to analyse physical effects.

[12] Otherwise, the fuel must be cut up into smaller unit volumes.

[13] Just as slowing-down models have been developed (see § 7.2.5), differential thermalisation models have been constructed. The most elaborate one was that of Michel Cadilhac. These models have now fallen out of use.

[14] The piles of that period used uranium. For plutonium, whose cross-section is not proportional to 1/v, this approximation is more dubious.

9.3.2. Thermal utilisation factor

Treating the thermal neutron population as if it were monokinetic, then, the absorption rate in a region i is written $V_i\Sigma_{a,i}\Phi_i$. The thermal utilisation factor f which is, by definition, the proportion of thermal neutrons absorbed in the fuel, is therefore written as follows:

$$f = \frac{V_f\Sigma_{a,f}\Phi_f}{V_f\Sigma_{a,f}\Phi_f + V_m\Sigma_{a,m}\Phi_m + \dots}. \tag{9.25}$$

It is often easier to discuss in the following form:

$$\frac{1}{f} - 1 = \frac{V_m\Sigma_{a,m}\Phi_m}{V_f\Sigma_{a,f}\Phi_f}, \tag{9.26}$$

because this form isolates three ratios: the volume ratio known as the *"moderation ratio"*, the ratio of cross-sections, and the flux ratio known as the *"disadvantage factor"*, so called because it is greater than 1 (Figure 9.4), which is disadvantageous for the neutron balance (if its value were 1, the factor f would be better).

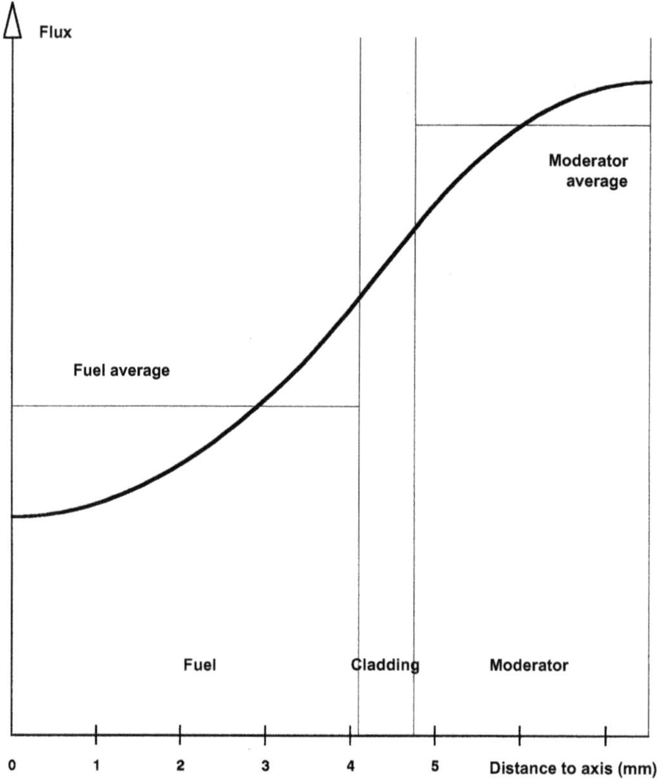

Figure 9.4. Appearance of the thermal flux in a cell. (The dimensions given here are relative to the unit cell of a pressurised water reactor, but the concavity is exaggerated. The disadvantage factor is of the order of 1.05 with a standard fuel and 1.1 with MOX.)

To calculate this disadvantage factor, the equations would have to be written and solved in space and in monokinetic theory. (In the formalism with first collision probabilities, these are the equations written above, with $R\Phi$ replaced by $\Sigma_s\Phi$.)

9.3.3. Reproduction factor

By definition, the reproduction factor is the number of neutrons produced by fissions induced by thermal neutrons, normalised to the number of absorptions of thermal neutrons in the fuel.

In this ratio, the product $V_f\Phi_f$ appears in the numerator and the denominator, such that the factor η is reduced to a simple ratio of macroscopic cross-sections:

$$\eta = \frac{\nu\Sigma_{f,f}}{\Sigma_{a,f}}. \tag{9.27}$$

In a uranium fuel, this ratio is expressed according to e, the isotope 235 content (in number of nuclei):

$$\eta = \frac{e\nu\sigma_{f,5}}{e\sigma_{a,5} + (1-e)\sigma_{a,8}}, \tag{9.28}$$

i.e.:

$$\eta = \eta_5\alpha_5, \quad \eta_5 = \frac{\nu\sigma_{f,5}}{\sigma_{a,5}}, \quad \alpha_5 = \frac{e\sigma_{a,5}}{e\sigma_{a,5} + (1-e)\sigma_{a,8}}, \tag{9.29}$$

where η_5 is the factor η for pure uranium 235, and α_5 is the probability that a neutron absorbed in the fuel will be absorbed by isotope 235.

Figures 9.5 and 9.6 show that the factor η, plotted as a function of the content e, increases very rapidly towards its asymptote.

Applying our reasoning in terms of reactivity only, it is not necessary to enrich a uranium fuel beyond a few percent. (More enrichment would, however, make the unit more compact. This aspect can be very important for a high-flux experimental reactor.)

9.3.4. Optimum moderation

The three examples of the application of the simple four factor formulae we shall now present will be illustrated numerically for the case of pressurised water reactors, but they have a more general scope.

Optimum moderation is the first problem: assuming that the fuel rod has been chosen (its geometry and composition defined), and we wonder what lattice size to choose in order to obtain the highest possible multiplication factor. In the context of neutron physics, this will involve finding the optimum value of the moderation ratio V_m/V_f.

In the basic formula[15], this parameter affects the factors p and f *only*.

If the moderation ratio increases from zero to infinity, p increases from 0 to 1 (without a moderator, no neutrons could be slowed down; and if the moderator is infinitely vast, all neutrons escape from the resonance traps); f, on the other hand, decreases from 1 to 0 (without a moderator, all neutrons are absorbed in the fuel, and, if the moderator is

[15] For simplicity, we shall not discuss the variations in the fast fission factor, which are small, and we reason based on the infinite multiplication factor, because leakage is minimal in power reactors.

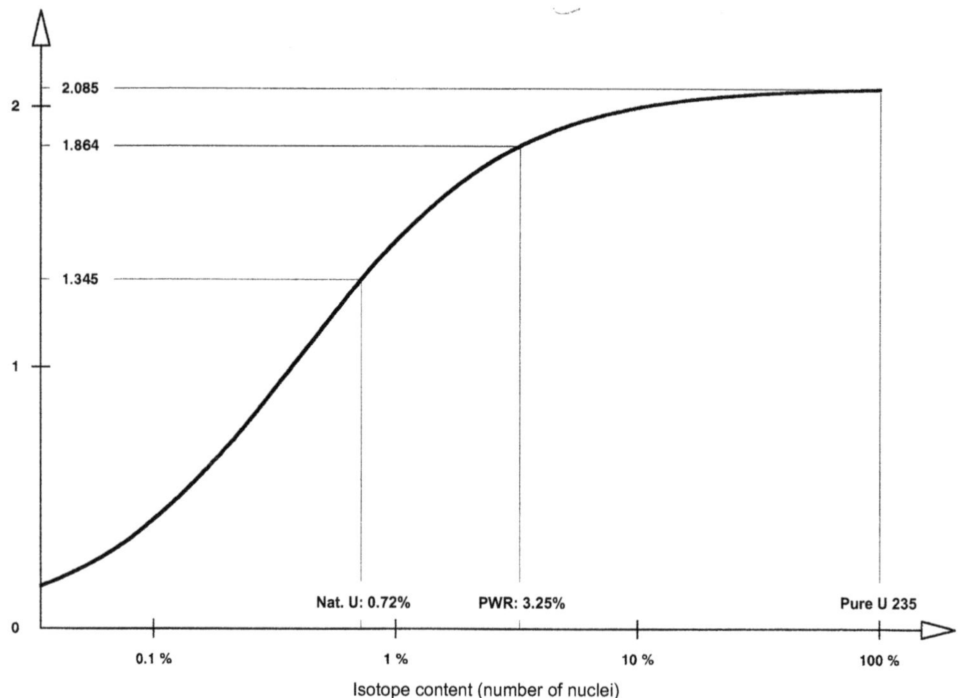

Figure 9.5. Reproduction factor η of uranium according to the isotope 235 content (Note: the abscissa is plotted on a log scale).

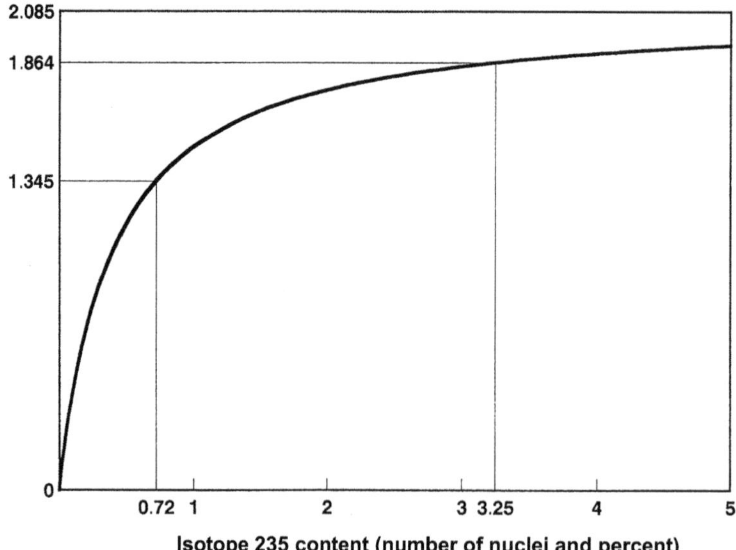

Figure 9.6. Reproduction factor η of uranium according to the isotope 235 content (zoom for low contents; linear scales).

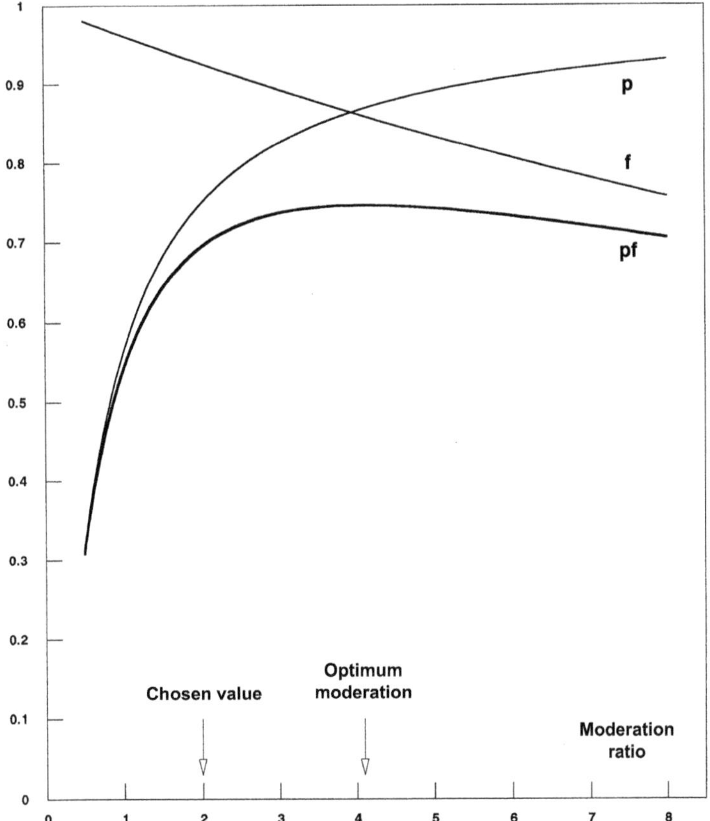

Figure 9.7. Optimum moderation (example of pressurised water reactors).

infinitely vast, all neutrons are lost in it). The analytical formulae confirm these qualitative comments. The product *pf*, and consequently k_∞, which are zero for both limits, must pass through an optimum value (see Figure 9.7)[16].

For pressurised water reactors, this optimum is located around 4 (the ratio by volume of water to uranium oxide). Note that the designers of this type of reactor chose a value around 2, which is very much under-moderated: a sacrifice of several thousands of pcm. There are two reasons for this:

1/ a more compact core;

2/ a water "loss" coefficient of the correct sign.

This latter reason is safety-related and is therefore obviously the essential reason. In the event of heating up, the expansion of the water or the formation of bubbles reduces the mass of water in the core. From a neutron physics point of view, this is equivalent to reducing the moderation ratio[17]; returning to the reasoning we applied concerning the

[16] Using the basic formulae, the position of this optimum is found via a second-order equation.

[17] The expressions for the factors *p* and *f* involve the product $(NV)_m$: this parameter is proportional to the mass of the moderator.

Doppler effect, if the system is under-moderated, this reduction in moderation leads to a lowering of the multiplication factor. As the reactor goes subcritical, the power, and then the temperature, become lower and counteract the initial perturbation.

9.3.5. Problem of using a boron solution in the primary circuit of PWRs

The designers of pressurised water reactors have also chosen to use boron in the form of a boric acid $\mathbf{B(OH)_3}$ solution in the water of the primary circuit to compensate for most spontaneous reactivity variations. (For example, as the fuel becomes worn out during an irradiation cycle, the multiplication factor can go down by more than ten thousand pcm. We therefore aim for a reactivity surplus of this amount at the beginning of the cycle, to be compensated for by a poison that will be removed gradually, in order to maintain the cycle.) The homogeneous distribution of the boric acid in the core makes it the ideal poison because it does not disturb the power distribution.

This poison has a major disadvantage, however: *it adds a positive component and therefore degrades the negative temperature coefficient of the core.* The boron solution's capture properties are added to those of the actual moderator in the numerator of the formula of $1/f - 1$ (for example, with 1000 ppm of boron in the core, the cross-section $\Sigma_{a,m}$ is approximately trebled with respect to that of water itself): the addition of boron lowers the factor f, and therefore the factor k_∞, which is the desired result (order of magnitude with a standard fuel: −10 pcm per ppm of boron). The presence of boron also deforms the curve of the factor f, however, and as it lowers this curve as well as the pf curve, *it moves the optimum moderation to the left,* as shown for 1000 ppm of boron on the optimum moderation diagram in Figure 9.8. The designers chose a fairly low moderation ratio to give themselves a bit of flexibility in the use of boron solution; there is *a limit to the allowable boron concentration,* corresponding to the concentration leading to the optimum moderation at the chosen moderation ratio. This limit must not be too restrictive.

The degradation of the temperature coefficient is due to the fact that the boron concerned, in aqueous solution, *expands like the water when the temperature rises.* If the limit is too low with respect to the reactivity compensation requirements, then it will be necessary to choose a different poison, one that does not expand (or not much), i.e. a solid poison. Because the effect to be compensated diminishes and finally disappears as the reaction progresses, the poison must also disappear. In practice, *burnable (consumable) poisons* are used for this purpose. These poisons are designed to be destroyed by neutron irradiation during the irradiation cycle. They cannot however exactly follow the multiplication factor curve, and so boron solution is still required, but in lesser concentration.

9.3.6. Problem of using plutonium in PWRs

The recycling of plutonium in water reactors also gives rise to problems because of the high absorption of this fuel for slow neutrons. We shall return to this problem in Chapter 18.

Note that the substitution of MOX for the standard fuel in this case increases not the numerator, but the denominator of the formula for $1/f - 1$. The effect of this is to move the optimum moderation *to the right* (see Figure 9.9).

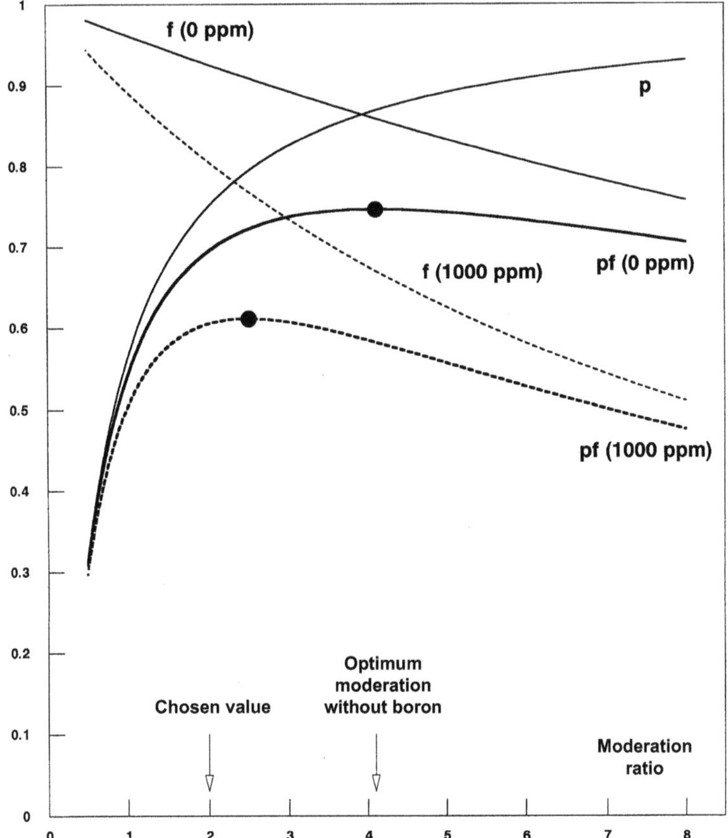

Figure 9.8. Optimum moderation: deformation of curves with the addition of boron in aqueous solution.

With the chosen value of moderation ratio, the system including MOX fuel becomes extremely under-moderated. This could also be seen in the thermal spectra we compared (see Figure 9.3) where the spectrum of the MOX case appears to be completely "crushed" by the high absorption of the fuel.

The direct consequence of this high absorption is the great loss of efficiency (by a factor in the region of 3) of the control absorbents, whether they are bundles or a boron solution. Either of two equivalent points of view can be used to explain this:

- the drop in thermal flux (compared to the standard case) leads to a drop in absorption rates in the bundles or the boron, which capture essentially in this energy domain, or
- the efficiency of an absorbent in terms of reactivity is proportional to the ratio of its absorption cross-section to that of the surrounding medium; if the latter is increased without any modification to the absorbent, then the efficiency decreases in inverse proportion.

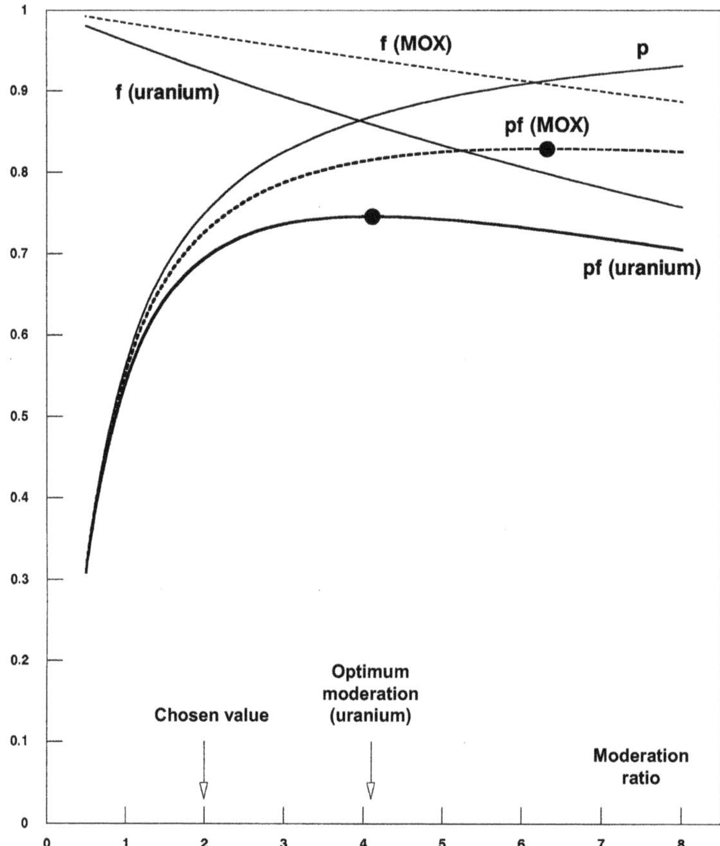

Figure 9.9. Optimum moderation: deformation of curves with the substitution UOX → MOX.

The loss of efficiency of boron is not very restrictive, because its concentration can simply be increased[18]. The loss of efficiency of bundles, however, has obliged engineers to limit the number of plutonium assemblies in a PWR core to one third in order to guarantee safety under any circumstances. This creates an inevitable juxtaposition of uranium and plutonium assemblies in a core, with considerable discontinuities in the neutron physics properties of thermal neutrons at the interfaces.

These discontinuities would cause unacceptable power peaks, and must therefore be attenuated. This is why MOX assemblies are "zoned": to soften the absorption cross-section gradients in the thermal domain, there are three levels of plutonium content, increasing from the edges towards the centre of the assembly.

[18] With MOX, there is an increased margin in terms of moderation ratio and therefore allowable concentration.

Exercises

Table 9.1. Typical data for a cell in a pressurized water reactor, operating with enriched uranium oxide fuel.

- *Nuclear data* (values for 2200 m s^{-1} neutrons)
  - σ_{a5} = 680 barns; σ_{f5} = 580 barns; ν_5 = 2.42
  - σ_{a8} = 2.7 barns
  - σ_{aH} = 0.322 barns
  - $\sigma_{aO} \simeq 0$
  - σ_{aB} = 760 barns (20% of 3800)
- *Nuclear concentrations in the fuel*
  - Uranium-238 atoms: 21.9×10^{27} m^{-3}
  - Uranium-235 atoms: 0.85×10^{27} m^{-3}
  - Oxygen atoms: 45.5×10^{27} m^{-3}
- *Concentrations in the moderator*
  - Water molecules: 24×10^{27} m^{-3}
  - Boron atoms: variable
- *Other data*
  - Cladding: not taken into account
  - Volume moderation ratio: 2
  - Disavantage factor: 1.1
  - Resonance escape probability: $p = 0.78$

Exercise 9.1: some average and most probable values

This exercise demonstrates that the notions of 'average' and of 'most probable' must be clearly specified!

For neutrons distributed according to a Maxwell spectrum, calculate the average value and the most probable value of the

a) velocity, weighted by the density,
b) energy, weighted by the density,
c) velocity, weighted by the flux,
d) energy, weighted by the flux,

and compare the results.

Exercise 9.2: average and effective cross-sections

a) In the theory of thermalisation, mean cross-sections are called *average cross-sections* when weighted by the *flux*:

$$\bar{\sigma} = \frac{\int \sigma(v)\,\Phi(v)\,dv}{\int \Phi(v)\,dv}.$$

Calculate the average value of a $1/v$ cross-section for a Maxwell flux.

b) Mean cross-sections are called *effective cross-sections* when weighted by the *density*. In order to have the reaction rate in the numerator and keep the dimensions right, a velocity is needed in the denominator. Two variants are possible. Either

$$\tilde{\sigma} = \frac{\int \sigma(v)\,v\,n(v)\,dv}{v_0 \int n(v)\,dv},$$

where v_0 is the velocity of neutrons with kinetic energy kT, or

$$\hat{\sigma} = \frac{\int \sigma(v)\,v\,n(v)\,dv}{v_{2200} \int n(v)\,dv},$$

where $v_{2200} = 2200\ \mathrm{m\,s^{-1}}$.

Give expressions for these effective values with a $1/v$ cross-section. What is the advantage of using effective cross-sections over using average cross-sections?

Exercise 9.3: sensitivity to the uranium-235 content

a) Using the data provided in Table 9.1, calculate the numerical values of the factors f and η for nominal conditions.

b) Show the logarithmic derivatives df/f and $d\eta/\eta$ for these factors with respect to the isotope-235 concentration $e = N_5/N_U$ of uranium, and calculate their values.

c) Estimate values for the factors f and η when $e = 5\%$, which is the maximum authorized at the EURODIF enrichment plant. Compare this result with the nominal values.

Exercise 9.4: optimum moderation ratio

a) Based on a single fuel element, calculate the optimum value of the moderation ratio, i.e. the value giving the peak infinite-multiplication factor k_∞. In this exercise, use the classical formulae for p and f, and neglect all the other factors, except the moderation ratio V_m/V_f. In particular, ignore variations of the Dancoff factor, the disavantage factor, and the average cross-sections.

b) Describe how p and f differ between the nominal and optimum cases.

c) Explain why PWRs are normally designed to have a moderation ratio $R \approx 2$?

Exercise 9.5: problem of using a boron solution in the primary circuit of the PWRs

In pressurized water reactors, boron, in the form of boric acid **B(OH)**$_3$, is dissolved in the primary circuit water, to compensate for the excess reactivity at the beginning of the cycle. The amount of boron is measured by its *concentration* C defined as the mass ratio of boron (N.B. not the acid) to water. Normally, this is expressed in parts per million (ppm). Natural boron consists of two isotopes, $^{10}\mathbf{B}$ and $^{11}\mathbf{B}$, in the ratio 1:4 by number. The $1/v$ capture cross-section of $^{10}\mathbf{B}$ at 2200 m s^{-1} is 3800 barns, while the commoner isotope has negligible capture cross-section, giving an average of 760 barns. At the concentrations used, any effect on the density of the water caused by the presence of boron is negligible.

a) Using the ratio of the macroscopic capture cross-sections of both components of the moderator $X = \Sigma_{a,B}/\Sigma_{a,water}$, determine the relationship between X and C, and calculate X for $C = 1000$ ppm.

b) Express f as a function of its value without boron f_0 and X. Calculate f for $C =$ 1000 ppm.

c) Calculate the boron efficiency, defined as the logarithmic derivative of f with respect to C, and express the result in the ratio pcm/ppm. Calculate this efficiency for $C = 0$ and $C = 1000$ ppm, and compare the values.

d) Calculate the temperature coefficient for k_∞ resulting from the thermal expansion of water containing dissolved boron, using the usual formulae for p and f, with the simplifications made in the previous exercise, for $C = 0$, $C = 1000$ ppm, and $C = 2000$ ppm. The relative variation of the density of water around nominal conditions is -250 pcm/K.

e) Determine the upper value of C not to be exceeded in order to maintain a negative temperature coefficient?

f) Describe how this limit changes when, 1) the moderation ratio is modified, 2) the uranium-235 content of the fuel is modified, and 3) plutonium is substituted for ^{235}U.

g) Explain how to compensate the reactivity excess when more poisoning is needed than is provided by boron while under this limit?

Exercise 9.6: overview of Westcott-Horowitz-Tretiakoff's formalism

The Canadian C. H. Westcott, and later French physicists Jules Horowitz and Oleg Tretiakoff, suggested that the spectrum of thermal neutrons in a homogeneous system is described well in terms of the reduced velocity $x = v/v_0$ by

$$n(x) \simeq C\left[m(x) + r\,e(x)\right].$$

Here, C is a normalization constant, and m the normalized Maxwell spectrum

$$m(x) = \frac{4}{\sqrt{\pi}}\,x^2\,e^{-x^2},$$

while e a universal 'difference' function with the property

$$\int_0^\infty e(x)\,dx = 0,$$

and equivalent to $1/x^2$ at higher energies. The coefficient r represents the proportion of epithermal neutrons relative to the number of thermal neutrons. The upper limit of the integral at $x = \infty$ represents the cut-off value of x separating the slowing-down range from the thermal regime.

a) When x increases towards the cut-off level between slowing-down and thermalisation, it is expected that the absorption cross-section is asymptotically negligible and the slowing-down power $\xi\Sigma_s$ becomes constant. Express $n(x)$ as a function of the slowing-down current Q, and the properties of the medium. Notice also that $m(x)$ is negligible in this regime; hence, deduce the relationship between Cr and Q.

b) It can be seen that the current Q of neutrons entering into the thermal range is equal to the total absorption rate A in the thermal range; consequently, show the relationship allowing r to be calculated, provided the function $e(x)$ is known.

c) In Westcott's notations, for a given cross-section, the following paramaters are defined.

$$g = \int_0^{\infty} \frac{\sigma(x)}{\sigma_{2200}}\, m(x)\, x\, dx, \qquad s = \int_0^{\infty} \frac{\sigma(x)}{\sigma_{2200}}\, e(x)\, x\, dx.$$

These parameters — which are temperature dependent — can be recorded in a table. Show that when this has been done, the calculation of r is explicit, and give the formula.

d) In this context, give the expression for the effective cross-sections (see exercise **9.2**).

Solutions

Exercise 9.1: some average and most probable values

Case	Variable	Weighting	Mean value	Modal value
a	Velocity	Density	$\frac{2}{\sqrt{\pi}} v_0$	v_0
b	Energy	Density	$\frac{3}{2} E_0$	$\frac{1}{2} E_0$
c	Velocity	Flux	$\frac{3\sqrt{\pi}}{4} v_0$	$\sqrt{\frac{3}{2}}\, v_0$
d	Energy	Flux	$2\, E_0$	E_0

All four of these criteria give different results.

Exercise 9.2: average and effective cross-sections

a) *If* the cross-section varies as $1/v$ *and* the spectrum follows a Maxwell distribution, *then* the average cross-section is proportional to the cross-section at 2200 m s^{-1}:

$$\bar{\sigma} = \frac{\sqrt{\pi}}{2}\sqrt{\frac{T_0}{T}}\,\sigma_{2200}.$$

b) For the effective cross-sections, a $1/v$ law alone is sufficient to give the required result:

$$\tilde{\sigma} = \sqrt{\frac{T_0}{T}}\,\sigma_{2200}, \qquad \hat{\sigma} = \sigma_{2200}.$$

The second definition avoids the need to include temperature. In order to obtain reaction rates from these formulae, an 'average' flux or an 'effective' flux must be provided. This is derived from the spectrum, of course.

Exercise 9.3: sensitivity to the uranium-235 content

a) $f = 0.94933$; $\eta = 1.87255$.

b) Using the probability for a neutron in the fuel to be absorbed by uranium-235 atoms,

$$\alpha = \frac{(\sigma_{a5} - \sigma_{a8})e}{(\sigma_{a5} - \sigma_{a8})e + \sigma_{a8}},$$

which gives

$$\frac{df}{f} = (1 - f)\,\alpha\,\frac{de}{e}; \qquad \frac{d\eta}{\eta} = (1 - \alpha)\,\frac{de}{e}.$$

These sensitivity coefficients are 0.0458 and 0.0964, and their sum is 0.1422.

c) Using the logarithmic formula, $f = 0.95900$, i.e. +1013 pcm; $\eta = 1.91932$, i.e. +2467 pcm; which in total is +3480 pcm. Extrapolating the sensitivity coefficients under nominal conditions, should yield the result $1335 + 2809 = 4143$ pcm.

Exercise 9.4: optimum of moderation

a) Define the moderation ratio as $R = V_m/V_f$. The formulae for p and f then are

$$p = \exp\left(-\frac{\lambda}{R}\right), \qquad f = \frac{1}{1 + \mu R},$$

where λ and μ are constants which do not depend on R. The optimum value of R is obtained by differentiating the expression for pf, then solving the resulting quadratic equation

$$\mu R^2 - \lambda\mu R - \lambda = 0.$$

Using $p = 0.78$ and $f = 0.94933$ obtained previously, gives $\lambda = 0.497$ and $\mu = 0.0267$; hence, $R_{\text{optimum}} = 4.57$.

b) $p_{\text{optimum}} = 0.89699$ (+13 975 pcm), $f_{\text{optimum}} = 0.89129$ (−6310 pcm), and k_∞ changes by +7665 pcm.

c) Under-moderation permits the core to be more compact and, above all, ensures that the moderator has a negative temperature coefficient, even when some boric acid is present in the primary circuit (see the following exercise).

Exercise 9.5: problem of using a boron solution in the primary circuit of the PWRs

a) The mass numbers A for water and natural boron are 18 and 10.8, respectively. Hence,

$$X = C\frac{(\sigma_a/A)_{\text{boron}}}{(\sigma_a/A)_{\text{water}}},$$

i.e. $X = 0.001967\,C \simeq 0.002\,C$, when C is expressed in ppm. When $C = 1000$ ppm, the effectiveness of the moderator at capturing neutrons is enhanced threefold in comparison with pure water.

b) The thermal utilisation factor f is given by the formula

$$\frac{1}{f} - 1 = \left(\frac{1}{f_0} - 1\right)\ (1 + X).$$

When $f_0 = 0.94933$ (see exercise **9.3**), and $C = 1000$ ppm of boron, the formula predicts $f = 0.86330$ (−9499 pcm). Therefore, the mean boron efficiency between zero and 1000 ppm is −9.5 pcm/ppm.

c) Using $\varepsilon = dX/dC$, gives

$$\frac{1}{f}\frac{\partial f}{\partial C} = -\frac{\varepsilon\,(1 - f_0)}{1 + \varepsilon C\,(1 - f_0)}.$$

Hence, the boron efficiency decreases very little from −9.97 pcm/ppm to −9.06 pcm/ppm when C increases from zero to 1000 ppm.

d) If ρ is the density of water, then

$$\frac{1}{p}\frac{\partial p}{\partial T} = \ln\frac{1}{p}\ \frac{1}{\rho}\frac{\partial \rho}{\partial T}, \qquad \frac{1}{f}\frac{\partial f}{\partial T} = (1 - f)\ \frac{1}{\rho}\frac{\partial \rho}{\partial T}.$$

Hence, the temperature coefficients are −62.1 pcm K^{-1} for p, and +12.7, +34.2, and +52.1 pcm K^{-1} for f when $C = 0$, 1000, and 2000 ppm, respectively.

e) Notice that the limit does not depend on the thermal expansion coefficient of water, due to both terms being proportional to it. The limit is given by

$$1 - f < \ln\frac{1}{p}; \qquad X < X_{\text{limit}} = \frac{\frac{\ln(1/p)}{1-f_0} - 1}{1 - \ln(1/p)}.$$

The result is 2641 ppm.

f) This value seems satisfactory in respect of reactivity compensation requirements, which are of order 1500 ppm at the beginning of the cycle. However, it would be lower in both cases for a greater moderation ratio and smaller uranium-235 content. For instance, for uranium enriched at 3.1% — needed to yield 33 000 MWd/t (2.85×10^{12} J kg^{-1}) in 1300 MWe PWRs — with the same values for the other parameters, yields 2156 ppm. Following the initial loading of a reactor, using an even less enriched fuel, the limit would be even lower. When the fuel contains plutonium, the discussion is more complex because its greater absorption cross-section simultaneously gives larger X and lower efficiency of the boron, or in other words more compensation is necessary to compensate for a given excess a reactivity. Notice also that the present calculation is very dependent on the value of p used. For instance, when $p = 0.85$, the limit falls to 1340 ppm for 3.7%-enriched uranium, and to only 1056 ppm for 3.1%-enriched uranium.

g) If the required reactivity compensation, while keeping below the limit, is greater than that which can be made by boron alone, burnable poisons must be used.

Exercise 9.6: overview of Westcott-Horowitz-Tretiakoff's formalism

a) The Placzek transitories are negligible near the cut-off; consequently, $\Phi(u) = Q/\xi\Sigma_s$, and

$$n(x) \simeq \frac{2Q}{v_0\xi\Sigma_s}\,\frac{1}{x^2}; \quad \text{therefore,} \quad Cr = \frac{2Q}{v_0\xi\Sigma_s}.$$

b) The neutron balance is

$$Q = A = \int_0^\infty \Sigma_a(x)\,\Phi(x)\,dx = \int_0^\infty \Sigma_a(x)\,v_0\,x\,n(x)\,dx.$$

Therefore,

$$r = \frac{2\int_0^\infty \Sigma_a(x)\,\left[m(x) + r\,e(x)\right]\,x\,dx}{\xi\Sigma_s}.$$

c) The definitions of the factors g and s lead to the expression

$$\int_0^\infty \sigma_a(x)\,\left[m(x) + r\,e(x)\right]\,x\,dx = \sigma_{a,2200}\,(g + rs).$$

Using this in the formula for r from part **b** gives

$$r = \frac{2\sum_k N_k\,\sigma_{a,k,2200}\,g_k}{\xi\Sigma_s - \sum_k N_k\,\sigma_{a,k,2200}\,s_k}.$$

d) The expression for the effective cross-sections is simply

$$\tilde{\sigma} = \sigma_{2200}\,\left(g + rs\right).$$

10 Multigroup theory

Introduction

In order to be dealt with accurately, the problems presented in the three preceding chapters require numerical processing of the neutron energy variable E, or a related variable, the speed v or lethargy u.

Virtually all neutron physicists use the "multigroup" approximation for this purpose. The principle of this approximation does not change from one application to the next, but the level of detail of the model, i.e. the number of groups, varies widely. In practice, anything from two groups up to several tens of thousands of groups can be used.

10.1. Principle of multigroup theory

Let E_0 be the maximum energy that neutrons can have. The principle of multigroup theory is to divide the overall energy interval $[E_0, 0]$ into sub-intervals numbered from 1 to N by positioning boundaries E_1, E_2, ..., E_{N-1}, $E_N = 0$, which in practice are numbered by decreasing energy. Neutron group g comprises all neutrons whose energy is in interval g, i.e. the interval $[E_{g-1}, E_g]$.

In each of the energy groups, neutron transport is treated as if neutrons were monokinetic.

The equations specific to each group are coupled, because there are not only real sources emitting in the group concerned, but also rates of transfer into this group from other groups; and in addition to the real absorptions in this group, there are also transfers to other groups.

10.2. Implementation of multigroup theory

It is also possible, although not obligatory, to approximate the transport operator (for example) by replacing it with the diffusion operator. As an abbreviation, we shall refer to *multigroup transport* and *multigroup diffusion* respectively. In the same way, the multigroup approach to a reactor situation can include (or not) a certain number of homogenisations.

The problem of defining and calculating the cross-sections arises in every case however, for every group and every reaction, and the probabilities of transfer from one group to another. Here we shall examine this problem, using the assumption that no other approximation has been introduced. (The problems of homogenisation and the use of the diffusion approximation will be discussed in Chapter 17.) For the sake of simplicity we shall take the steady-state case, but the introduction of time would not alter the discussion. Similarly, the variable $\vec{\Omega}$ is understood in what follows.

Firstly we must define *multigroup flux*; as one would expect in view of the very principle of this method, the *flux* $\Phi_g(\vec{r})$ *of group g at a point* $\vec{r}$ *must be defined as the integral over group g of the real flux* $\Phi(\vec{r}, E)$, i.e.:

$$\Phi_g(\vec{r}) = \int_{E_{g-1}}^{E_g} \Phi(\vec{r}, E)\, dE. \tag{10.1}$$

The *principle of conservation of reaction rates* then leads to defining the multigroup cross-sections as the averages at each point of the real cross-sections weighted by the real flux:

$$\sigma_g(\vec{r}) = \frac{\displaystyle\int_{E_{g-1}}^{E_g} \sigma(E)\, \Phi(\vec{r}, E)\, dE}{\displaystyle\int_{E_{g-1}}^{E_g} \Phi(\vec{r}, E)\, dE}. \tag{10.2}$$

It can be shown that, by calculating the fluxes using these cross-sections, we obtain the exact multigroup fluxes[1]. In addition to the fact that these multigroup cross-sections should be dependent on $\vec{r}$, even in a homogeneous medium, and possibly also on t and $\vec{\Omega}$ (not very convenient for calculation purposes), the problem is that they cannot be obtained by definition, because to calculate them it would be necessary to know the exact flux in space and energy—in other words, already to have solved the problem completely.

In order to apply the weighting that will enable the multigroup cross-sections to be found, we therefore replace the unknown exacted flux by a weighting flux $\varphi_g(E)$ chosen *a priori* which most closely resembles the exact flux:

$$\sigma_g = \frac{\displaystyle\int_{E_{g-1}}^{E_g} \sigma(E)\, \varphi_g(E)\, dE}{\displaystyle\int_{E_{g-1}}^{E_g} \varphi_g(E)\, dE}. \tag{10.3}$$

The multigroup approximation is applied at this level, and at this level only. This approximation obviously improves with finer partitioning, because the sensitivity of the multigroup cross-sections to the choice of weighting function diminishes with narrower groups.

The choice of weighting spectra depends on the type of problem. For example:

- to constitute a relatively fine library for cell or assembly calculations, a spectrum chosen once and for all is adopted. (For example, for APOLLO, we use a fission spectrum, a constant-lethargy flux, and a Maxwell spectrum respectively for the fast, epithermal, and thermal domains.)[2];

[1] The multigroup equations would simply be the integrals over each of the groups of the exact equation.

[2] For this type of calculation, the partitioning is performed with around a hundred groups. This is far too few to allow the resonances of heavy nuclei to be described, which is why a special resonance "self-shielding procedure" must also be applied (see Chapter 15).

– to constitute a library with less energy detail to be used for core calculations, we use the local spectrum (for each type of cell or assembly) obtained in the previous step[3]).

Note: The criterion of respecting the reaction rates leads to the definition of transfer matrices approximating the transfer probabilities using the following equation:

$$\sigma_{h\to g} = \sigma_h P_{h\to g} = \frac{\displaystyle\int_{E_{h-1}}^{E_h} \sigma_s(E')\,\varphi_h(E')\,dE' \int_{E_{g-1}}^{E_g} P(E' \to E)dE}{\displaystyle\int_{E_{h-1}}^{E_h} \varphi_h(E')\,dE'}. \qquad (10.4)$$

10.3. Examples of multigroup partitioning

Figure 10.1 compares a few multigroup partitions used in neutron physics calculations, presented according to the lethargy variable.

- The "universal" partition is the minimum partitioning required for accurate calculations without modelling the self-shielding (it turns out to be insufficiently detailed in the high-energy part of uranium 238 resonances, however). It gives an idea of how fine the partitioning needs to be in order to describe the cross-sections.

- The next four partitions are used by APOLLO and WIMS (the British equivalent of APOLLO). The 37-group partitioning was condensed from the 99-group partitioning using a characteristic PWR spectrum, and is hardly used today. The 172-group partitioning was constructed on the one hand to give details of certain resonances (e.g. that of plutonium 242 around 2.5 eV) and, on the other hand, to be compatible with the previous 99-group and 69-group partitions. This explains a few irregularities seen in the choice of group boundaries.

- The FNR partition is currently used for fast neutron reactor core calculations (a six-group partition is also used).

- The next four partitions were chosen by physicist in charge of interpreting criticality and neutron physics experiments in water system.

- The last two, finally, are those that are usually adopted for water reactor core calculations; for small cores (significant leakage), three groups are used in the domain of fast and epithermal neutrons[4] and one thermal group; for power reactors, *Électricité de France* generally limits itself to two-group calculations only.

[3] In addition to this *"energy condensation"*, i.e. going from a detailed multigroup structure to a more coarse structure by grouping "microgroups" into "macrogroups", homogenisations are also often performed (i.e. averages not over energy, but over space).

[4] Most leaks in water reactors are observed in the first two of these groups.

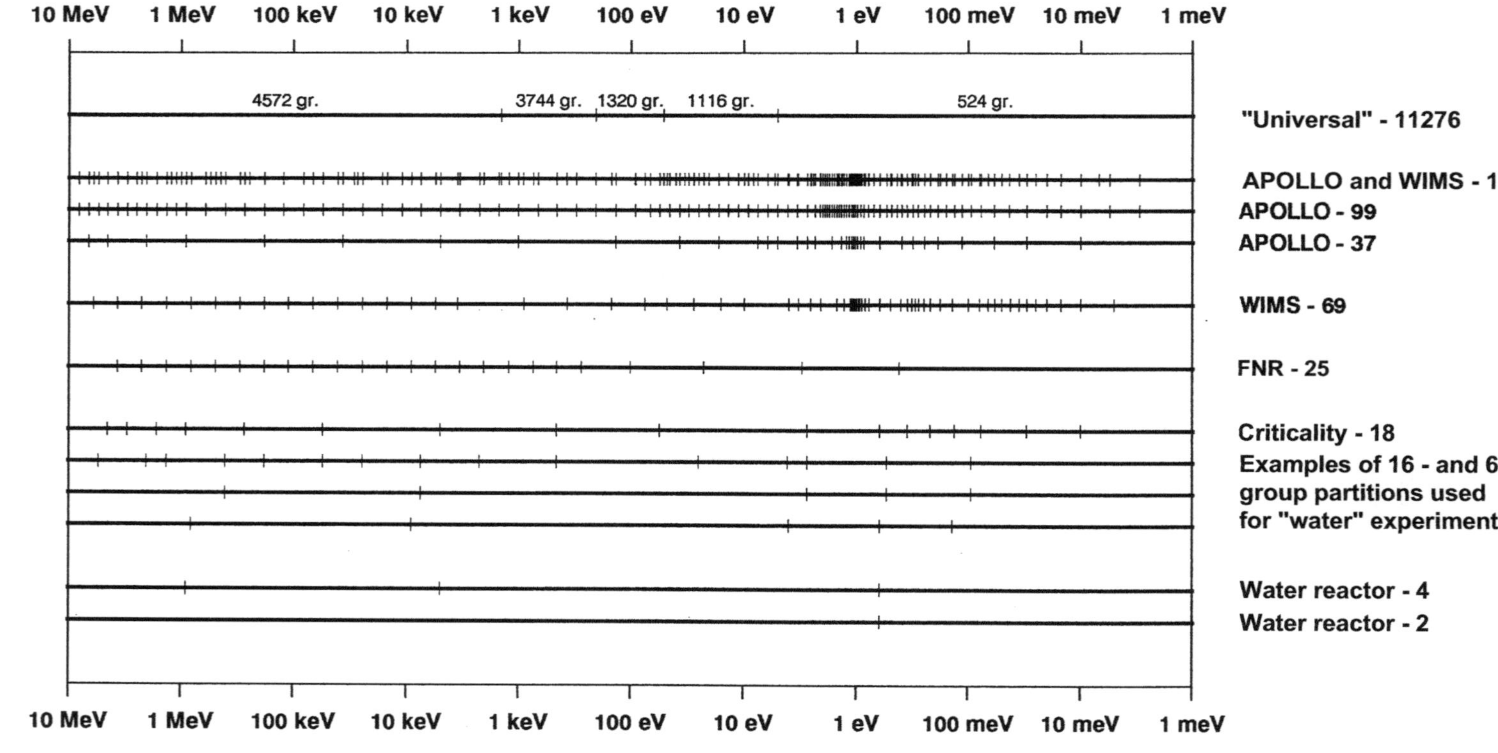

Figure 10.1. Comparison of a few multigroup partitions.

10.4. Multigroup diffusion theory

As an example, let us write out the multigroup equations in steady state using the diffusion approximation in each of the groups (the variable $\vec{r}$ is understood):

$$D_g \Delta \Phi_g - \Sigma_{a,g} \Phi_g - \sum_h \Sigma_{g \to h} \Phi_g + \sum_h \Sigma_{h \to g} \Phi_h + \chi_g \sum_h \nu \Sigma_{f,h} \Phi_h = 0,$$

$$(g = 1, 2, \ldots, N) \tag{10.5}$$

- the first term represents the leaks;
- the second term represents real absorptions;
- the third term represents disappearances from the group by transfer to another group;
- the fourth term represents contributions by transfer from another group;
- the last term represents contribution by fission (χ_g is the proportion of neutrons emitted by fission in group g, i.e. the integral of the fission spectrum over this group).

Note that transfers from the group into itself are eliminated between the third and fourth terms.

In the slowing-down domain, transfers occur only towards higher-numbered groups. If there is only one group to describe thermal neutrons, there is no up-scattering term, which simplifies the processing of the system.

It is important to note that there is not necessarily any advantage in increasing the number of groups in the slowing-down domain. A greater number of groups certainly improves the multigroup approximation, but also casts more doubt on the diffusion approximation. As we have seen (§ 5.1.7), this approximation is better at low absorption. The greater the number of epithermal groups, the narrower the groups, and therefore the greater the probability of escaping from them by scattering, which means that this "pseudo absorption" is greater. To obtain some idea of the order of magnitude, let us say that, if Δu is the width of the group in terms of lethargy, and if ξ is the average lethargy gain by scattering, a neutron (if not absorbed) will be scattered $n = \Delta u/\xi$ times before being transferred to the next group. The ratio of the "pseudo absorption" cross-section to the total cross-section (neglecting real absorption) is therefore in the region of $1/n = \xi/\Delta u$. If there are too many groups, then each one is narrow, and this ratio is high. (Note: The values of n for a few common nuclides and one epithermal group are given in Table 7.1, § 7.1.5.)

If the transport theory calculations are performed, however, it is desirable to have the largest possible number of groups, because there is no restriction on the absorption value.

10.5. Two-group theory calculation of a reflected cylindrical pile

As an example of the application of multigroup diffusion theory, let us take the simplest case, with two groups: one for fast and epithermal neutrons, and the other for thermal neutrons:

$$D_1\Delta\Phi_1 - \Sigma_1\Phi_1 + \nu\Sigma_{f1}\Phi_1 + \nu\Sigma_{f2}\Phi_2 = 0,$$
$$D_2\Delta\Phi_2 - \Sigma_2\Phi_2 + \Sigma_r\Phi_1 = 0. \quad (10.6)$$

To simplify the expressions, we have set:

- $\Sigma_r = \Sigma_{1\to 2}$: a single transfer section (*removal*);
- $\Sigma_1 = \Sigma_{a,1} + \Sigma_r$: disappearance cross-section of the first group;
- $\Sigma_2 = \Sigma_{a,2}$: disappearance cross-section of the second group.

Note that all of the neutrons emitted by fission appear in the first group.

Allowing for the fissions of group 1 with a factor ε, i.e. $\nu\Sigma_{f1}\Phi_1 + \nu\Sigma_{f2}\Phi_2 = \varepsilon\nu\Sigma_{f2}\Phi_2$, and if we note that the ratio Σ_r/Σ_1 can be interpreted as the resonance escape probability p and that the ratio $\nu\Sigma_{f2}/\Sigma_2$ can be taken as the product $f\eta$, we can re-write these equations in the following form:

$$D_1\Delta\Phi_1 - \Sigma_1\Phi_1 + \frac{k_\infty}{p}\Sigma_2\Phi_2 = 0,$$
$$D_2\Delta\Phi_2 - \Sigma_2\Phi_2 + p\Sigma_1\Phi_1 = 0, \quad (10.7)$$

by introducing the product k_∞ of the four factors.

To present the Bessel functions in a little more detail, we shall examine the solving of these equations on the example of a reflected cylindrical reactor of infinite height, where the core is homogeneous between radii 0 and a, and the reflector is homogeneous between radii a and b (extrapolation distance neglected). The reader can easily adapt this approach to other geometries defined by a single space variable. For this type of geometry, the diffusion equations are reduced to second-order differential equations. Here, with two groups, the general solution of the system is the linear combination of $2 \times 2 = 4$ particular solutions. To obtain these, we can look for fluxes that are proportional to the same eigenfunction of the Laplace operator, such as:

$$\Phi_2(\vec{r}) = s\Phi_1(\vec{r}), \quad (10.8)$$

with:

$$\Delta\Phi_1(\vec{r}) + \lambda\Phi_1(\vec{r}) = 0. \quad (10.9)$$

By substituting into the system of two-group equations, we find, on the one hand, the equation that the *eigenvalue* λ must satisfy:

$$(1 + L_1^2\lambda)(1 + L_2^2\lambda) = k_\infty, \quad (10.10)$$

(where we have set $L_i^2 = D_i/\Sigma_i$) and, on the other hand, the expression for the *coupling coefficient* s:

$$s = \frac{D_1\lambda + \Sigma_1}{k_\infty\Sigma_2/p} = \frac{p\Sigma_1}{D_2\lambda + \Sigma_2}. \quad (10.11)$$

The eigenvalue equation always has two real roots:

- if k_∞ is greater than 1, a positive root denoted μ^2 and a negative root denoted $-\nu^2$;
- if k_∞ is less than 1 (and, in particular, if it is zero), two negative roots denoted $-\mu^2$ and $-\nu^2$.

After calculating these roots, we determine the coupling coefficients.

For each eigenvalue, the equation (differential in this case):

$$\Delta\Phi_1(\vec{r}) + \lambda\Phi_1(\vec{r}) = 0,$$

has *two* linearly independent solutions. With the two eigenvalues, this gives the four particular solutions that were sought.

Solution of equations in the core

If the core is large, and therefore k_∞ is not much greater than 1, the first-order calculations give (setting $M^2 = L_1^2 + L_2^2$):

$$\mu^2 \simeq \frac{k_\infty - 1}{M^2}, \quad s_\mu \simeq p\frac{\Sigma_1}{\Sigma_2},$$

$$\nu^2 \simeq \frac{M^2}{L_1^2 L_2^2}, \quad s_\nu \simeq -p\frac{D_1}{D_2}, \tag{10.12}$$

We shall now examine the problem in space for the cylindrical reactor geometry. For $\lambda = +\mu^2$, the general solution of the eigenvalue/eigenfunction equation is the linear combination of the two zero-order Bessel functions of the first kind and the second kind:

$$\Phi_1(\rho) = AJ_0(\mu\rho) + A'Y_0(\mu\rho).$$

Because the function Y_0 is singular at the origin, in this case it is necessary to take A' zero for the flux to remain regular on the reactor axis; for $\lambda = -\nu^2$, the general solution of the eigenvalue/eigenfunction equation is the linear combination of the two zero-order modified Bessel functions of the first kind and the second kind:

$$\Phi_1(\rho) = BI_0(\nu\rho) + B'K_0(\nu\rho).$$

Because the function K_0 is singular at the origin, in this case it is necessary to take B' zero for the flux to remain regular on the reactor axis.

So, finally, the general solutions of the two-group equations in the core, observing regularity at the origin, are:

$$\Phi_1(\rho) = AJ_0(\mu\rho) + BI_0(\nu\rho),$$

$$\Phi_2(\rho) = As_\mu J_0(\mu\rho) + Bs_\nu I_0(\nu\rho). \tag{10.13}$$

Solution of equations in the reflector

For the reflector, where k_∞ is zero and where we suppose $\Sigma_{a_1} = 0$, we find (with the "reflector" subscript understood):

$$\mu^2 = \frac{1}{L_1^2}, \quad s_\mu = \frac{\Sigma_1}{\Sigma_2}\frac{L_1^2}{L_1^2 - L_2^2},$$

$$\nu^2 = \frac{1}{L_2^2}, \quad s_\nu = \infty. \tag{10.14}$$

(Because s_ν is infinite, the corresponding function should only be placed in the second group.)

Now, concerning the space problem, for the cylindrical reflector geometry:

- for $\lambda = -\mu^2$, the general solution of the eigenvalue/eigenfunction equation is the linear combination of the two zero-order modified Bessel functions of the first kind and the second kind:

  $$\Phi_1(\rho) = CI_0(\mu\rho) + C'K_0(\mu\rho).$$

  Because the axis is not included in the reflector, both functions must be kept, but the combination must be zero at $\rho = b$. This gives C' as a function of C:

  $$C' = -CI_0(\mu b)/K_0(\mu b);$$

- for $\lambda = -\nu^2$, the calculations are similar if we replace μ with ν, C with D, and C' with D'.

  Finally:

$$\Phi_1(\rho) = CI_0(\mu\rho) + C'K_0(\mu\rho),$$
$$\Phi_2(\rho) = s_\mu[CI_0(\mu\rho) + C'K_0(\mu\rho)] + DI_0(\nu\rho) + D'K_0(\nu\rho), \tag{10.15}$$

(expressions in which C' and D' must be replaced with their expressions).

Critical condition

The interface conditions at $\rho = a$ still need to be specified. There are four equations to write: continuity of the two fluxes, and continuity of the two currents. When these are written out, we obtain a homogeneous algebraic system of four equations in four unknowns A, B, C, and D.

Compatibility — the determinant of the system must be zero — gives the critical condition.

It is obviously a little bit tedious to write this out.

For the flux expressions, there is an arbitrary multiplication factor as for any critical problem, for example, the main constant A.

In conclusion, we can say that two-group theory (compared to one-group theory) gives an additional measure of freedom that allows us to describe *interface transients* (for the core, these are functions with the argument ν that decrease very quickly with distance from the interface). This is why at least two groups should be used in practice, because there are always interfaces in real reactors.

Exercises

Exercise 10.1: kernels in two-group theory

Recall that the kernel of the Boltzmann equation, in both exact and simplified forms, is the solution for a unit point, line, or plane source in an infinite, homogeneous medium without neutron multiplication. Here we are looking for the kernels of two-group diffusion theory, i.e. where the fluxes from two sources, group-1 and group-2, are mixed.

a) Let k be the kernel of the monokinetic diffusion equation (see chapter **5**). By expressing it as a linear combination of k_1 and k_2, and solving the equations, derive the kernel of two-group diffusion theory.

b) Find the mean squared length of the direct path between the point of emission of a neutron in group-1 and its point of absorption in group-2.

Exercise 10.2: critical condition in two-group theory

Supposing that fissions occur only with neutrons in group 2, compare the equations in the following two cases giving the critical condition of a bare, homogeneous pile, neglecting the extrapolation distance.

- 'Age + diffusion' theory (see exercise **7.9**),
- 'Two-group diffusion' theory.

N.B: assume that in both cases the migration areas for the two-group diffusion theory L_1^2 and L_2^2, and for the age + diffusion theory τ and L^2, respectively, are identical. Non-thermal neutrons includes both fast and epithermal ones.

Exercise 10.3: core-reflector interface in two-group theory

Let us consider the problem of a core-reflector interface, in two-group diffusion theory with the following assumptions.

- The interface is planar.
- The core is semi-infinite and occupies the region $x < 0$.
- The reflector is semi-infinite and occupies the region $x \geq 0$.
- The infinite multiplication factor is exactly equal to 1 in the core.
- No fission is induced by group 1 neutrons in the core.
- No absorption of group 1 neutrons occurs in the reflector.
- Both energy groups possess equal diffusion coefficients in both regions.
- Group 1 neutrons have equal macroscopic cross-sections in both regions.

a) Find the general solutions of the flux equations in each region. Show that the main mode in the core region takes the form $A(\delta - x)$, where A is a normalisation constant, and δ is the reflector saving.

b) What are the conditions at the interface? Deduce δ from these conditions.

Exercise 10.4: effect of a small, absorbing body in two-group diffusion theory

Apply two-group diffusion theory to the problem of a small, absorbing body located at the centre of a spherical, bare pile described in exercise **6.10**. To simplify the calculations, assume that fissions can be induced only by thermal neutrons, and that for fast and epithermal neutrons, the material is 'white' — i.e., has zero absorption cross-section. Also, use approximate expressions for the eigenvalues and coupling coefficients, with $k_\infty \approx 1$.

Solutions

Exercise 10.1: kernels in two-group theory

a) The equations to solve are

$$D_1\,\Delta\varphi_1 - \Sigma_1\,\varphi_1 + \delta = 0,$$

$$D_2\,\Delta\varphi_2 - \Sigma_2\,\varphi_2 + \Sigma_r\,\varphi_1 = 0.$$

Naturally, $\varphi_1 = k_1$. To find the solution, construct a linear combination of both kernels by substituting $\varphi_2 = \alpha k_1 + \beta k_2$ into the second equation, and replace $D_i \Delta k_i$ with $\Sigma_i k_i - \delta$ $(i = 1, 2)$. The result is

$$\alpha = \Sigma_r \frac{D_1}{D_1\,\Sigma_2 - D_2\,\Sigma_1}, \qquad \beta = \Sigma_r \frac{D_2}{D_2\,\Sigma_1 - D_1\,\Sigma_2}.$$

b) This expression is relative to Σ_r/Σ_1, i.e. those neutrons not absorbed into the first group. One sixth of the mean squared length of the direct path is $L_1^2 + L_2^2$. The migration areas must be added.

Exercise 10.2: critical condition in two-group theory

In the formalism of 'age + diffusion' theory for the migration areas, the critical conditions in each of the two cases are, respectively,

$$k_{a+d} = \frac{k_\infty e^{-\tau B^2}}{1 + L^2 B^2} = 1, \quad \text{and} \quad k_{2g-d} = \frac{k_\infty}{(1 + \tau B^2)(1 + L^2 B^2)} = 1.$$

Therefore,

$$\frac{k_{a+d}}{k_{2g-d}} \simeq 1 + \frac{(\tau B^2)^2}{2}.$$

The difference amounts to about 125 pcm when the leakage level of fast and epithermal neutrons is 5000 pcm.

Exercise 10.3: core-reflector interface in two-group theory

a) *In the core,* the main mode has a linear form $Ax + B$, which represents the general eigenfunction of the Laplace operator in a planar geometry possessing a null eigenvalue. Replacing B with $-A\delta$ and changing its sign to make the flux positive when A is positive, leaves $A(\delta - x)$. This function is zero at a distance δ beyond the interface, i.e. at the distance equal to the reflector saving. Hence, the general solution of the two-group diffusion equations is

$$\Phi_1 = A(\delta - x) + C\,e^{\nu x} + C'\,e^{-\nu x},$$

$$\Phi_2 = s\,A(\delta - x) + t\,(C\,e^{\nu x} + C'\,e^{-\nu x}),$$

where

$$\nu^2 \simeq \frac{L_1^2 + L_2^2}{L_1^2 L_2^2}, \qquad s \simeq p\frac{\Sigma_1}{\Sigma_2}, \qquad t \simeq -p\frac{D_1}{D_2}.$$

In the reflector,

$$\Phi_1 = E\,e^{-\alpha x} + E'\,e^{\alpha x},$$

$$\Phi_2 = u\,(E\,e^{-\alpha x} + E'\,e^{\alpha x}) + F\,e^{-\beta x} + F'\,e^{\beta x},$$

where,

$$\alpha = \frac{1}{L_1}, \qquad \beta = \frac{1}{L_2}, \qquad u = \frac{D_1\Sigma_1}{D_1\Sigma_2 - D_2\Sigma_1}.$$

b) The constants C', E', and F' must be zero to satisfy the conditions at infinite distance. The boundary conditions at the interface—continuity of the fluxes and their derivatives—give four equations for the four other constants:

$$A\,\delta + C = E,$$

$$-A + \nu\,C = -\alpha\,E,$$

$$s\,A\,\delta + t\,C = u\,E + F,$$

$$-s\,A + t\,\nu\,C = -u\,\alpha\,E - \beta\,F.$$

The determinant of this homogeneous system must be equal to zero; hence, the extrapolation distance is

$$\delta = \frac{u(\alpha - \beta) + t(\nu + \beta) - s(\nu + \alpha)}{-u\nu(\alpha - \beta) + t\alpha(\nu + \beta) - s\beta(\nu + \alpha)}.$$

Exercise 10.4: effect of a small absorbing body in two-group diffusion theory

The main eigenvalue of the equations

$$\mu^2 = \frac{k_\infty - 1}{L_1^2 + L_2^2},$$

is approximately the same as in one-group theory. The other eigenvalue $-\nu^2$, and the coupling coefficients s and t, are given in the previous exercise, core part.

Define the functions

$$f(r) = \frac{\sin \mu r}{r}, \qquad g(r) = \frac{\cos \mu r}{r},$$

$$h(r) = \frac{\sinh \nu r}{r}, \qquad k(r) = \frac{\cosh \nu r}{r}.$$

These enable the fluxes in each group to be re-expressed as

$$\Phi_1(r) = C\left[f(r) - \lambda g(r) - \alpha h(r) - \beta k(r)\right],$$

$$\Phi_2(r) = Cs\left[f(r) - \lambda g(r) - \alpha \frac{t}{s} h(r) - \beta \frac{t}{s} k(r)\right],$$

where the coefficients λ, α, and β–which are small–are due to the perturbation caused by the absorbing body.

The functions f and h are regular at the origin. The functions g and k are equivalent to $1/r$. Since the group-1 flux must be regular at the origin as well, it follows that $\beta = -\lambda$.

At the external radius R, the fluxes must be zero, i.e.

$$f(R) - \lambda g(R) - \alpha h(R) - \beta k(R) = 0,$$

$$f(R) - \lambda g(R) - \alpha \frac{t}{s} h(R) - \beta \frac{t}{s} k(R) = 0.$$

These equations give α and the negative reactivity ρ due to the absorber as functions of λ.

Replacing the previous expression for μ^2 with

$$\mu^2 \simeq \frac{k_\infty(1-\rho) - 1}{L_1^2 + L_2^2},$$

and writing $\mu R = \pi + u$, the result to first order is

$$u \simeq \lambda, \qquad \alpha \simeq \frac{\lambda}{\tanh(\pi\nu/\mu)}, \qquad \rho \simeq -\frac{2}{\pi}\frac{k_\infty - 1}{k_\infty}\lambda.$$

Notice that the expressions for u and ρ as functions of λ are *the same* as the ones obtained previously in exercise **6.10** for one-group theory.

Using

$$\frac{J_{+,2}(\varepsilon)}{J_{+,2}(\varepsilon)} = 1 - \gamma,$$

gives the result

$$\lambda \simeq \frac{\gamma\mu\varepsilon}{2D_2(2-\gamma)/\varepsilon + \gamma} w, \qquad \text{where} \qquad w = \frac{s}{s-t} \simeq \frac{L_2^2}{L_1^2 + L_2^2}.$$

If we ignore the factor w, then it is apparent that the expression for λ as a function of γ takes the same form as in one-group theory, except the diffusion coefficient now is D_2. The coefficient w means that the reactivity effect in two-group theory is less than in one-group theory by a factor equal to the ratio between the thermal migration area and the total migration area.

11 Poisoning by fission products

Introduction

As discussed in Chapter 2 (§ 2.10.2), the fission of heavy nuclei can produce a wide variety of *fission fragments* with very different *yields* (probability of production for one fission).

Fission is usually asymmetrical, with the result that there can be heavy fragments and light fragments (although there is no clearly-defined boundary between these two categories).

Because of the overall curvature of the valley of stability of nuclei, almost all nuclides obtained by fission are represented by points located *above* the valley of stability (assuming the number of protons, Z, is plotted along the abscissa axis, and the number of neutrons, N, along the ordinate axis). Many beta-minus decays will therefore be observed for these nuclides: around *four* decays on average before a stable nucleus is obtained. These decays are characterised by periods varying from a fraction of a second to millions of years. Almost all of these periods, however, are shorter than the 30-year period of caesium 137, one of the most abundant fission products[1].

As very often occurs with beta decay, many gamma emissions accompany this beta radioactivity.

Except in the event of an unforeseen incident, these products all remain confined within the nuclear fuel. The path of fission fragments does not exceed ten or so microns, and they are therefore unable to pass through the cladding. Volatile products and noble gases can, however, migrate within the fuel whilst remaining inside the cladding and, in particular, they can accumulate in the expansion chambers provided for this purpose on top of the cladding beyond the fuel rod.

Because they, like the fuel, are subjected to an intense flux, these nuclei are liable to capture neutrons. One of the consequences of this is to degrade the reactivity (this antireactivity due to the presence of fission products is called "poisoning"), and the other consequence is to change the concentrations (creation of a higher-rank isotope).

In general, the term *fission fragments* applies only to nuclides that come directly from fission reactions, and the term *fission products* refers to all nuclides obtained in reactors following fission, whether directly, by radioactive decay, or following capture. It is of course possible for the same nuclide to be obtained via more than one of these channels.

[1] Allowing 10 periods (reduction of the concentration by a factor of approximately a thousand) and taking this 30-year period as an "envelope", we consider that the radiotoxicity of these products becomes negligible after three centuries.

Approximately *eight hundred fission products*[2] have been catalogued. This illustrates how complicated the problem of neutron poisoning by fission products is, and why it must often be simplified.

This chapter is presented in three parts. The first part discusses fission products in general, and the second and third parts give details of the phenomena associated with two fission products that are especially significant in *thermal neutron reactors* because of their enormous cross-sections in this energy domain: xenon 135 and samarium 149.

11.1. Fission products

11.1.1. Equations governing fission products

In order to distinguish I fission products (possibly around 800), the I concentrations N_i (i from 1 to I), i.e. the number of atoms per unit volume, must be introduced as unknown functions of time. To simplify the problem, we often approach it as an average over the entire *fuel* volume, but to be strictly rigorous, these concentrations must be considered at each point $\vec{r}$ of the fuel: in this case, $\vec{r}$ is a parameter in the equations, and this is always taken to be the case in the discussions below.

The equation governing the temporal variation of N_i can involve *five terms* at most; in practice, many of these are zero or negligible, which simplifies the overall differential system:

$$\frac{dN_i}{dt} = +\gamma_i \Sigma_f \Phi - \lambda_i N_i - \sigma_i N_i \Phi + \lambda_j N_j + \sigma_k N_k \Phi, \tag{11.1}$$

- the first term on the right-hand side represents direct production by fission: $\Sigma_f \Phi$ is the number of fissions per unit volume and unit time; γ_i is the probability[3] of production (yield) of this product i: since there are several fissile nuclei in general, an average value weighted by fission rates must be used;

- the second term represents the disappearance of this nuclide i by radioactivity; λ is the radioactive decay constant;

- the third term represents the disappearance of this nuclide i by neutron capture; σ is the microscopic capture cross-section;

- the fourth term represents the production of the nuclide i by radioactive decay of another nuclide, which could be denoted j;

- the fifth term represents the production of the nuclide i by neutron capture by another nuclide, which could be denoted k.

[2] All of the nuclear properties required for these calculations (fission yields for the various fissile nuclei, details of the radioactive decays, etc.) are collected in databases for the use of specialists.

[3] Note, however, that the sum of all γ_i is equal to 2, since each fission produces *two* fragments.

Note: Flux and cross-sections are the "one-group" parameters: flux integrated over all energies; cross-sections averages over all energies by the flux. When the coefficients of the disappearance terms are high, the concentration quite rapidly tends towards an asymptote. When they are low, on the other hand, the concentration increases continuously and more or less linearly during irradiation. Every possible intermediate case can obviously also be encountered. In water reactors, for example, where the fuel is irradiated for a few years, the effect on the reactivity of capture by fission products lowers the multiplication factor by a few thousand pcm within a few days. The effect of poisoning[4] then continues to increase, but slowly. At the end of irradiation, it can reach fifteen to twenty thousand pcm. Because the cross-sections of fission products are lower, on average, in the fast domain than the thermal domain[5], the effect of poisoning by fission products are more modest (by about a factor of 10) in fast neutron reactors than thermal neutron reactors.

11.1.2. Fission pseudo-products

With the computing power now available, engineers could solve the full system of equations for fission products. This would make reactor calculations far more cumbersome, however, and above all it would be a waste of machine time in view of the level of accuracy required by engineers. Most fission products do not have much capture and therefore do not require a precise concentration calculation. The simplest way to identify the significant nuclides is to sort the fission products according to their one-group macroscopic cross-section, i.e., to within a factor, the contribution to the reactivity effect. The details obviously depend on the reactor (the average yields depend on the distribution of fissions according to fissile nuclei, and the one-group cross-sections depend on the spectrum, etc.) and the irradiation time (as has been stated, the concentrations do not change in proportion); nonetheless, for a given reactor type, the ranking depends little on this detail. As a guide, the following diagram presents this ranking based on calculations performed by processing all the equations and concerning a pressurised water reactor (Figure 11.1); the results were taken at 35 000 MWd/t[6], i.e. at the end of irradiation of a standard fuel.

The histogram gives the individual contributions in descending order of the first 50 products, and the curve gives the cumulative value of these contributions up to the nuclide concerned.

We can see, for example, that the first twelve fission products alone produce three quarters of the poisoning, and the first twenty-five produce 90% of the total.

That is why only a few dozen fission products will be dealt with in practice: the ones at the top of this list and, if necessary, those included in their chains and those, such as neodymium 148, that are often used in measurements to characterise the irradiation of a fuel.

The products that are not examined in detail cannot, however, be completely ignored; they are treated collectively as a *"pseudo fission product"*. The pseudo fission product is constructed once and for all from reference calculations like the one performed to create this diagram: the *average properties* of all of the fission products it represents are attributed to it.

[4] Refer to the exact definition of poisoning in § 11.1.3.

[5] In particular, strong poisoning by fission products with a resonance in the thermal domain, such as xenon 135 and samarium 149, does not exist in fast neutron reactors.

[6] This unit is defined in the next chapter.

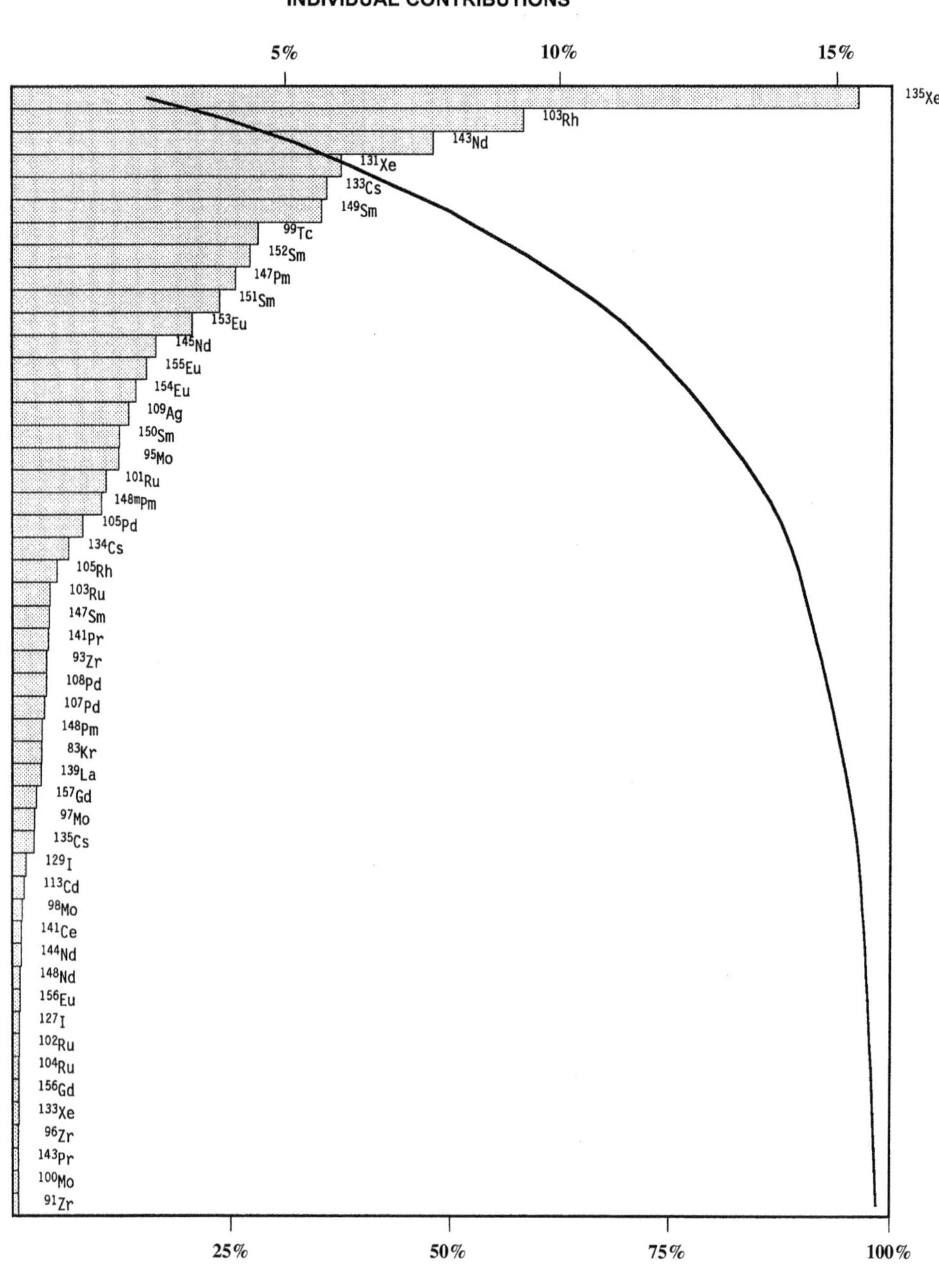

Figure 11.1. Ranking in descending order of the poisoning of the fifty main fission products (pressurised water reactor).

Note: Because the yield depends on the nucleus undergoing fission, in reality it is necessary to create and process a pseudo fission product for each of the main fissile nuclei (uranium 233, 235 and 238, plutonium 239 and 241, etc.).

11.1.3. Concept of poisoning

This term has already been used several times, but requires a more specific definition in the context of the four-factor formula for thermal neutron reactors. If we note that:

a/ the fission products appear and remain within the fuel, and that

b/ their capture, if it is to be taken into consideration, occurs essentially in the thermal domain[7],

then we can see that the fission products will affect the thermal utilisation factor slightly (*via* $\Sigma_{a,f}$ and very little *via* the disadvantage factor)[8], and will mainly affect the reproduction factor η because it affects it directly via this same cross-section[9] $\Sigma_{a,f}$.

By convention, poisoning is defined as the *effect on* η *only, expressed to the first order and as an absolute value*. It shall be written as π:

$$\pi = \left|\frac{\Delta\eta}{\eta}\right| = \frac{\Delta\Sigma_{a,f}}{\Sigma_{a,f}} = \frac{\Sigma_{a,FP}}{\Sigma_{a,f}}. \qquad (11.2)$$

Within this approximate expression for the effect on reactivity (except for the sign), the individual contribution of each fission product can easily be identified because the formula is additive:

$$\pi_i = \frac{\Sigma_{a,i}}{\Sigma_{a,f}} = \frac{N_i\sigma_{a,i}}{\Sigma_{a,f}}. \qquad (11.3)$$

11.2. The xenon effect

Xenon 135, an artificial nuclide since it is radioactive, is the nucleus with the largest known cross-section for thermal neutrons (approximately three million barns; see Figure 2.13). Added to the fact that a rather large amount of it is produced in reactors, this makes it the primary contributor to poisoning in thermal neutron power reactors.

11.2.1. The xenon 135 series

Figure 11.2 shows the two modes of production of this nuclide. Direct production by fission exists, but in a small quantity (yield γ_X of 0.1%). The main channel is *via* iodine 135.

[7] In the spirit of the four factors as defined by Fermi, we assume all absorptions to be "thermal" except fast fissions and resonant capture by uranium 238. In other words, the few epithermal absorptions by other materials are counted with the thermal utilisation and reproduction factors (§ 9.3.2 and 9.3.3). For water reactors, where the epithermal component of the spectrum is significant, this convention is debatable.

[8] This increase in thermal neutron absorption slightly reduces the diffusion area and therefore the leakage of thermal neutrons, but this can be neglected.

[9] The reader may wish to perform a comparison by calculating the logarithmic derivatives of both factors with respect to this cross-section.

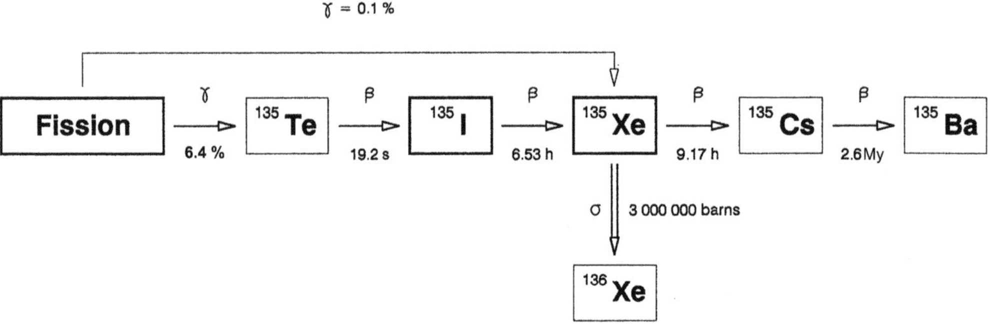

Figure 11.2. Chain of reactions that produce and destroy xenon 135 (the yield values are relative to fissions of uranium 235 induced by slow neutrons; the times indicated are the radioactive half-lives).

11.2.2. Simplified equations for the xenon 135 series

Because the lifetime of tellurium 135 is very short, it can be assumed that fission produces iodine 135 directly with a yield γ_I of 6.4%.

The nuclides situated beyond xenon can be ignored if it is xenon that interests us. Under these conditions, we can simplify by using I and X to denote the concentrations of iodine 135 and xenon 135, which are governed by the following equations (application of the general equations to this specific case):

$$\frac{dI}{dt} = \gamma_I \Sigma_f \Phi - \lambda_I I,$$
$$\frac{dX}{dt} = \gamma_X \Sigma_f \Phi + \lambda_I I - (\lambda_X + \sigma_X \Phi) X. \tag{11.4}$$

11.2.3. Reactor start-up, equilibrium, and shutdown

If the flux is kept constant, these equations can be integrated analytically. For example, the first curve in Figure 11.3 shows the evolution of xenon after the start-up of the reactor. We assume that there is initially no iodine 135 or xenon 135 because, even if the reactor has operated before, these two nuclides would have disappeared by radioactive decay within a few days. Along the abscissa axis, time is plotted as well as the dimensionless parameter λt where λ is the radioactive decay constant λ_I of iodine, whose value is 1/9.42 h^{-1} because the half-life is 6.53 hours. The ordinate axis shows not the concentration, but the associated poisoning. The value of 3000 pcm for the asymptote is the order of magnitude for pressurised water reactors.

Note that this asymptote is reached after a day or two. The equilibrium concentrations then observed are obtained by cancelling the time derivatives in the equations:

$$I_{eq} = \frac{\gamma_I \Sigma_f \Phi}{\lambda_I}, \qquad X_{eq} = \frac{(\gamma_I + \gamma_X)\Sigma_f \Phi}{\lambda_X + \sigma_X \Phi}, \tag{11.5}$$

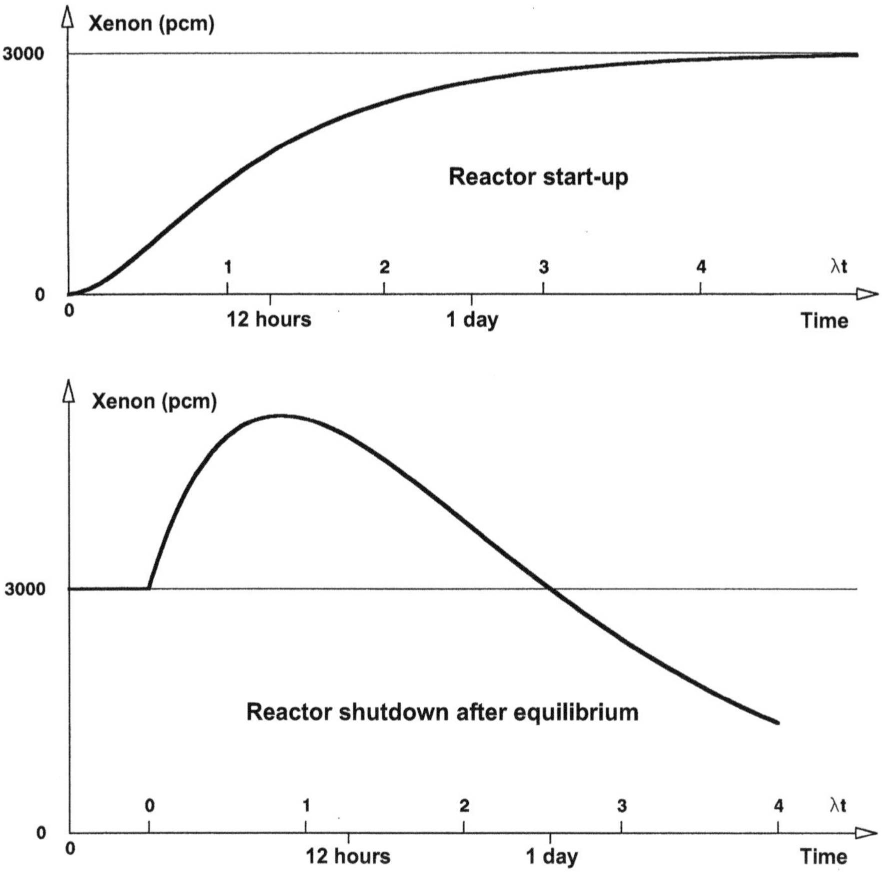

Figure 11.3. Evolution of xenon 135 poisoning.

which give the following poisoning:

$$\pi_{X,\text{eq}} = (\gamma_I + \gamma_X)\frac{\sigma_X\Phi}{\lambda_X + \sigma_X\Phi}\left[\frac{\Sigma_f}{\Sigma_a}\right]_f. \tag{11.6}$$

The first factor is the total yield, i.e. 6.5%; the second is the branching ratio of xenon 135 (neutron capture channel normalised to the whole of both channels, radioactive decay and neutron capture); the third is the fission/absorption ratio α of the fuel. For a PWR, this product is in the region of $6500 \times (2/3) \times 0.7 \simeq 3000$ pcm.

This equilibrium concentration therefore corresponds to significant poisoning in a power reactor. Note (second factor) that this poisoning decreases with decreasing reactor flux. For low flux, it is proportional to the flux, and for high flux, it tends towards an asymptotic value. Note also that (third factor) it increases slightly when the fuel is enriched.

The most spectacular and well-know phenomenon associated with xenon occurs when a reactor is shut down. Starting at that instant, the iodine 135 reservoir continues to empty into the xenon 135 reservoir, but the xenon 135 reservoir can no longer be emptied through capture; the xenon level therefore increases. This increase is not unlimited, because the iodine 135 reservoir is eventually depleted. The level of the xenon 135 reservoir therefore passes through a maximum value, and then decreases until it tends asymptotically to zero. According to calculations, this xenon peak lasts about a day, and its value is higher the more the iodine 135 reservoir was full, i.e. the higher the flux before shutdown (refer to the formula for I_{eq} above). The second curve presented (Figure 11.3) gives an order of magnitude for pressurised water reactors.

If it is significant — which it is for all thermal neutron power reactors — this xenon peak can give rise to a control constraint[10]: if the reactivity reserves (control rods or poisons that can be removed) are insufficient, the reactor cannot be restarted during this period of increased xenon poisoning, and it is necessary to wait until the level decreases by radioactive decay[11].

11.2.4. Spatial instabilities due to xenon 135

This peak associated with shutdown is a more general phenomenon; in the short term, any flux reduction leads to a rise in xenon concentration and, similarly, any flux increase leads to a xenon reduction. This is a counter-reaction mechanism that is naturally *unstable*. The following scenario could be imagined: a perturbation causes a slight reduction of the flux in one half of the core and a slight increase in the other half (caused for example by a change of operating speed that modifies the temperature distribution and therefore the distribution of neutron properties). In the first half, the drop in flux causes an increase in xenon, and therefore a lowering of the multiplication factor, which amplifies the drop in flux, etc. Similarly, in the other half, the flux increase can grow. If the decrease on the one hand and the increase on the other hand compensate for each other, then the total power will not change. If this total power is the only parameter being monitored, the operator is unaware of the growing tilt and the sudden change that could occur. In an extreme case, all the power could be released in one half of the core, and the resulting temperatures would almost certainly exceed the safety limits of the cladding.

Is this a realistic scenario? That would depend on the size of the core. In fact, there is another mechanism that counteracts the tilt created by the xenon effect: neutron migration, which tends to have the opposite effect because it evens out the flux among the various zones of the core. In a small core, migration operates very effectively and is the dominant effect, so that the tilt effect is spontaneously stifled. If the core is large, on the other hand, the xenon effect dominates, and the tilt can occur. In this case, the reactor is said to be *spatially unstable*.

This is why *pressurised water reactors* such as those used by *Électricité de France* — three-loop 900 MWe reactors and, even more so, four-loop 1300 and 1450 MWe reactors — *are axially unstable*. This means that a tilt in the power distribution can develop

[10] It is said that the operators of the first power reactor, at Hanford, were caught off guard by this temporary drop in reactivity. They had not expected it because the xenon effect was unknown.

[11] This period of unavailability is inconvenient in the operation of all industrial reactors. Sufficient reserves of reactivity must be provided in order to restart even during the xenon peak; this is important, for example, for the reactor in a nuclear submarine.

between the upper and lower parts of the core. The operators must therefore have a means of controlling this risk by detecting the nascent tilt and counteracting it.

To detect tilts, the operator monitors the axial offset, defined as:

$$AO = \frac{P_T - P_B}{P_T + P_B}, \tag{11.7}$$

where P_T and P_B are the power in the upper and lower halves of the core, respectively (the tilt can also be characterised by the product $AO \times P_{\text{relative}}$ where the relative power P_{relative} is the power normalised to the nominal power). In practice, this involves comparing the signals supplied by "external chambers" (neutron detectors placed outside the vessel) covering the upper and lower halves; these signals are practically proportional to the respective power levels.

To stifle an instability that could grow, a positive reactivity change must be applied to one half of the core, and a negative change to the other. In practice, this is done using control bundles and boron solution (the bundles modify only half of the core, and the boron modifies all of it) or with suitable displacement of at least two banks of control rod clusters.

Note that the time constant of these spatial instabilities is of the order of magnitude of the times seen in the xenon series, i.e. ten hours. This allows ample time in which to perform these procedures.

11.3. The samarium effect

It might seem surprising that the sixth-ranking poison in Figure 11.1 has been chosen as the second example, but there is a sound reason for this: just as xenon 135 is the main (but not the only) poison that gives rise to a *peak* after shutdown, samarium 149 is the main (but not the only) poison that gives rise to an *excess* after shutdown. Like the xenon effect, the samarium effect applies to thermal neutron reactors only (resonance in this energy domain).

11.3.1. Samarium 149 series

Figure 11.4 shows the samarium series. It is qualitatively similar to the xenon 135 series, with one important difference: samarium 149 is a *stable* nucleus, whilst xenon 135 is a *radioactive* nucleus. This means that only one channel is available for the elimination of samarium 149: destruction by neutron capture. Note also that there is no direct production by fission, and therefore only one channel for its formation.

11.3.2. Simplified equations in the samarium 149 series

If we neglect the neodymium step and use P and S to denote the concentrations of promethium and samarium 149, the equations can be written as:

$$\frac{dP}{dt} = \gamma\Sigma_f\Phi - \lambda P,$$
$$\frac{dS}{dt} = \lambda P - \sigma\Phi S. \tag{11.8}$$

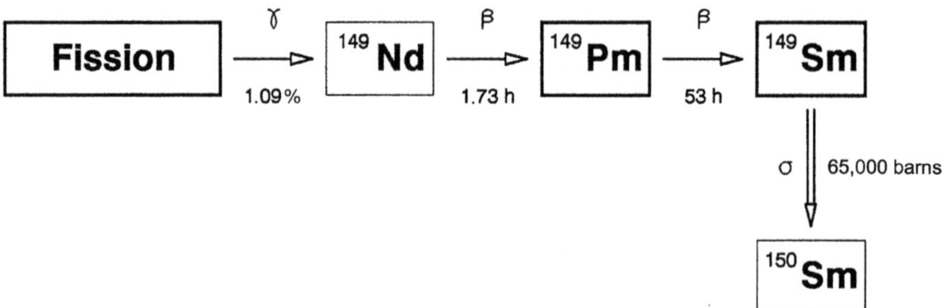

Figure 11.4. Series of reactions that produce and destroy samarium 149 (the yield value is relative to the fissions of uranium 235 induced by slow neutrons; the times given are the radioactive half-lives).

(There is no need to place the indices: $\gamma = 1.09\%$ is the yield for neodymium 149; $\lambda = 1/76\ \text{h}^{-1}$ is the radioactive decay constant of promethium 149; $\sigma = 65\,000$ b is the capture cross-section of samarium 149 for thermal neutrons.)

11.3.3. Reactor start-up, equilibrium and shutdown

Figure 11.5, which is presented in the same way as the diagram for xenon (Figure 11.3), shows the curve for the reaching of equilibrium on start-up without promethium or samarium (this therefore refers to the first start-up, since samarium is stable and some will remain until subsequent start-ups).

The equilibrium concentrations are as follows:

$$P_{eq} = \frac{\gamma \Sigma_f \Phi}{\lambda}, \qquad S_{eq} = \frac{\gamma \Sigma_f}{\sigma}, \tag{11.9}$$

giving a poisoning of:

$$\pi_{S,eq} = \gamma \left[\frac{\Sigma_f}{\Sigma_a} \right]_f . \tag{11.10}$$

The first factor is the yield, i.e. 1.09%; the second is the fission/absorption ratio α of the fuel. For a PWR, this product is in the region of 1090×0.7, or a little more than 700 pcm. Because there is only one exit channel, there is no longer a branching ratio as there was for the expression for xenon poisoning at equilibrium; the direct consequence of this is that *poisoning by samarium 149 at equilibrium is independent of flux.*

After reactor shutdown, the samarium concentration rises for the same reason as xenon 135: the "promethium reservoir" continues to empty out into the samarium 149 reservoir, which in turn is no longer being emptied at all. The final result will be that all of the promethium will have flowed into the "samarium 149 reservoir". This creates an excess that remains until the reactor is re-started. For the PWR example chosen for this diagram (Figure 11.5), the samarium excess is not huge: only 300 or 400 pcm added to the 700 present at equilibrium. If the designers allow for this excess in the sizing of the control elements, it will not create any insurmountable problems for the operators.

This does not hold true, however, for high flux reactors, where the equilibrium concentration of promethium, and therefore the samarium excess, are proportional to the flux

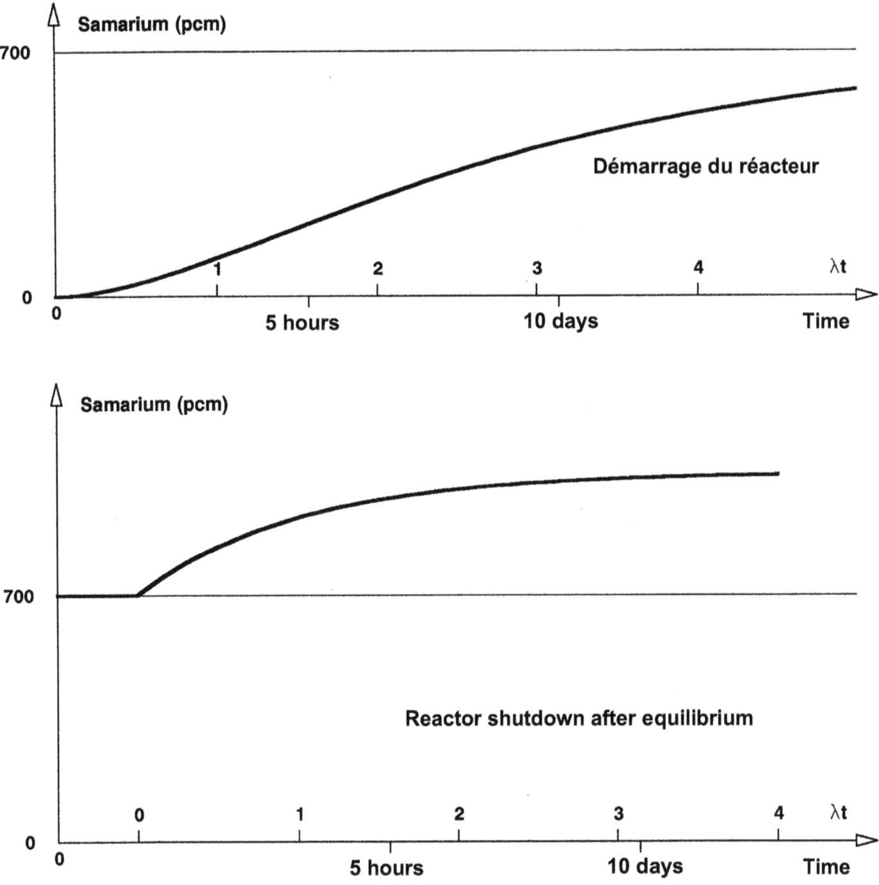

Figure 11.5. Evolution of poisoning by samarium 149.

level before shutdown. For example, a reactor operating at thirty times the flux of a pressurised water reactor, i.e. 10^{19} neutrons per m^2 per second instead of 3×10^{17} , which is achieved in certain experimental reactors such as the high-flux reactor at Grenoble, the excess will not be just 300 or 400 pcm, but thirty times higher, i.e. 10,000 pcm. This type of high-flux reactor must therefore not be shut down suddenly, at the risk of poisoning the core permanently, and possibly ruining any possibility of using it again. Instead, it is essential to lower the power level gradually according to carefully determined parameters, in order to burn off as much samarium 149 as possible before shutdown.

For exactly the same reasons as xenon 135, samarium 149 and other fission products could potentially lead to spatial instabilities. In practice, reactor cores are never big enough, by a long way, for instabilities other than those related to xenon 135 to arise. That is why only xenon 135 instabilities are taken into consideration.

Exercises

Exercise 11.1: canonical form of the equations describing the xenon effect

Using the reduced quantities θ, φ, i, and x defined by

$$t = \frac{\theta}{\lambda_I}, \qquad \Phi = \frac{\lambda_I}{\sigma_X}\varphi, \qquad I = \frac{\gamma_I \Sigma_f}{\sigma_X} i, \qquad X = \frac{\gamma_I \Sigma_f}{\sigma_X} x,$$

and the ratio $\alpha = \lambda_X/\lambda_I$, show the equations describing the xenon effect, based on the assumptions that iodine-135 is directly produced by fission, and that direct production by fission of xenon-135 can be ignored.

It is possible to use these canonical equations for the following exercises concerning the xenon effect. For a typical PWR, it is reasonable to assume that the quantity α can be rounded down to 0.7, and $\varphi = 3\alpha$.

Exercise 11.2: analytical solution to the equations describing the xenon effect

Assume that a reactor initially contains no iodine or xenon. Determine how their concentrations vary as functions of time t, with constant flux throughout,

a) from when the reactor starts at $t = 0$, and

b) after shutting it down from an equilibrium state.

Exercise 11.3: trajectories in the iodine-xenon plane

The variation of iodine and xenon concentrations can be plotted on a plane where the quantity of iodine is along the abscissa and xenon is along the ordinate.

a) What form does the curve take where the equilibrium points for a constant flux are situated?

b) Describe the path taken by the point (i, x) when the flux is constant, including the special case of zero flux.

Exercise 11.4: why does a peak in the concentration of xenon occur after a reactor is shut down?

A common misconception is that the peak which occurs in the concentration of xenon, following the shutdown of a reactor previously operating in an equilibrium state, is due to the lifetime of xenon being longer than iodine. Instead, the true cause is that the destruction of xenon by neutron capture ceases. To illustrate this, examine the effect on x of permuting the numerical values of λ_I and λ_X for a typical reactor in its equilibrium and post-shutdown states.

Exercise 11.5: analytical solution to the equations which describe the samarium effect

a) A reactor containing neither prometheum-149 nor samarium-149 is started with a constant flux. Using the data provided, estimate how long it takes for the samarium concentration to reach 90% of its equilibrium level.

b) Having achieved equilibrium, the reactor is then shut down instantaneously. How long now does it take for the excess samarium to reach 90% of its equilibrium level?

Data

- The half-life of prometheum-149 is 53 hours.
- The neutron capture microscopic cross-section of samarium-149 is 65 000 barns.
- The neutron flux is 2×10^{17} $m^{-2}s^{-1}$.

Exercise 11.6: shutting down a reactor with a constant samarium content

How would the flux level in a reactor vary with time, if it were shut down, while maintaining a constant samarium-149 concentration?

Exercise 11.7: spatial instabilities due to the samarium effect

It is well-known that spatial instabilities due to the xenon effect can occur in large reactors such as pressurized water reactors. Similar instablities could also originate from the samarium effect. In this exercise, a very simple model is developed and applied to the samarium case. The same model is also used in the following exercice for the xenon case.

In this model, the neutron flux is described by one-group diffusion theory using

$$M^2\,\Delta\Phi + (k_\infty - 1)\,\Phi - \beta\,S\,\Phi = \ell\,\frac{\partial\Phi}{\partial t},$$

The quantity ℓ is assumed to be zero, due to the neutron lifetime being much less than the time parameters related to poisoning by the fission products. The constant β represents the effect per unit concentration of samarium-149 on the infinite-multiplication factor. This concentration is calculated by the time-dependent prometheum-samarium equations for each point in the reactor. Apart from the latter, the reactor is assumed to be homogeneous with an ideal reflector, i.e. albedo equal to one.

a) Show the relationships linking the equilibrium values of Φ_0, P_0, and S_0. Demonstrate, for this model, that they are independent of *both* time and space.

b) Using $\Phi = \Phi_0 + \varphi$, and similarly for P and S, find the equations governing φ, p, and s. By cancelling the second order term—which is negligible provided φ, s, and p are small—give the linear approximation to these equations.

c) For this system of equations, which are linear and have constant coefficients, the solutions exhibit exponential behaviour with time. Consequently, they take the form

$$\varphi(\vec{r}, t) = \varphi(\vec{r})\,e^{\omega t}, \quad \text{etc.}$$

Derive the equations governing the functions φ, p, and s, purely in terms of $\vec{r}$.

d) Define $f_n(\vec{r})$ to be the eigenfunctions of the Laplace operator, and μ_n the associated eigenvalues, which are consistent with the boundary conditions for the model reactor. Show that $\mu_0 = 0$, f_0 is constant, and the flux and concentration projections on the main mode are the stationary components derived in part **a**. The transitory functions $\varphi(\vec{r})$, $p(\vec{r})$, and $s(\vec{r})$ may then be expanded as harmonic series taking the form

$$\varphi(\vec{r}) = \sum_{n=1}^{\infty} \varphi_n f_n(\vec{r}), \quad \text{etc.}$$

Find the linear relationships linking φ_n, p_n, and s_n.

e) Describe the compatibility condition that is necessary among the set of equations with index n. Deduce the equation giving ω for the nth mode.

f) Under what conditions does the real part of ω become negative, i.e. when do the functions decrease monotonically and provide stability?

g) Examine this condition for a cylindrical reactor, and its first radial and axial modes.

h) Based on the following data for a PWR, is there a risk it will suffer spatial instability due to the samarium effect? *Data:* radius = 1.6 m; height = 3.8 m; migration area $= 6 \times 10^{-3}$ m^2; equilibrium samarium poisoning = 700 pcm; see also exercise **11.5**.

Exercise 11.8: spatial instabilities due to the xenon effect

Apply the model developed in the previous exercise, and associated data, to xenon-135, assuming that the equilibrium xenon poisoning is 3000 pcm.

Exercise 11.9: natural recurrent perturbations of a reactor

Using a 'point reactor' model, the critical condition can be expressed as

$$k = k_0 - \alpha P - \beta X = 1,$$

where k_0 is the effective multiplication factor without the power effect and xenon poisoning, αP is the magnitude of the reactivity power effect (in particular the Doppler effect) for a given reactor power P (N.B. *not* Pm concentration), and βX is the magnitude of the reactivity effect due to xenon poisoning for a given concentration X.

Knowing the power P is proportional to the flux Φ, then replacing Φ and X with the reduced quantities φ and x (see exercise **11.1**), and substituting the numerical coefficients α and β by the appropriate coefficients a and b, permits the critical condition to be expressed as

$$k = k_0 - a\,\varphi - b\,x = 1.$$

a) Using this expression, and assuming that the iodine and xenon concentrations are at their equilibrium values for a constant flux, give the relationships for φ_0, i_0, and x_0 under steady-state conditions.

b) Now, suppose that around this stationary situation, a small perturbation occurs, and then evolves exponentially with time:

$$\varphi = \varphi_0 + u\,e^{\omega t},$$
$$i = i_0 + v\,e^{\omega t},$$
$$x = x_0 + w\,e^{\omega t}.$$

Find the system of equations giving the constants ω, u, v and w, assuming that u, v, and w are small, and consequently keep only first order terms.

c) From the compatibility condition for this system, deduce the equation giving ω.

d) Under what conditions is the real part of ω negative, i.e. when does the reactor remain stable with respect to small perturbations?

e) Apply the previous results to a typical PWR. In practice, negative feedback provided by the automatic control system suppresses instabilities; however, the natural reactors at Oklo may have experienced divergent modes.

*N.B: the formalism developed in exercises **11.7–9** is strictly applicable to small perturbations from equilibrium only; thus, it is purely indicative of a reactor's degree of stability under conditions close to steady-state operation. When divergent behaviour occurs, the previous approximations are invalid, and more rigorous treatment is required.*

Solutions

Exercise 11.1: canonical form of the equations describing the xenon effect

$$\frac{di}{d\theta} = \varphi - i, \qquad \frac{dx}{d\theta} = i - (\alpha + \varphi)\,x.$$

Exercise 11.2: analytical solution to the equations describing the xenon effect

The time-dependent general solution with constant flux is

$$i(\theta) = \varphi + \left[i(0) - \varphi\right] e^{-\theta},$$

$$x(\theta) = \frac{i(0) - \varphi}{\varphi + \alpha - 1}\, e^{-\theta} + \left[x(0) - \frac{i(0) - \varphi}{\varphi + \alpha - 1} - \frac{\varphi}{\varphi + \alpha}\right] e^{-(\varphi+\alpha)\theta}.$$

a) The initial state of the reactor is given by setting $i(0) = x(0) = 0$.

b) Using the equilibrium state for $t = 0$,

$$i(0) = \varphi, \qquad x(0) = \frac{\varphi}{\varphi + \alpha},$$

then setting $\varphi = 0$, gives the behaviour following shutdown of the reactor:

$$i(\theta) = \varphi\, e^{-\theta}, \qquad x(\theta) = \frac{\varphi(\varphi + 1)}{(1 - \alpha)(\varphi + \alpha)}\, e^{-\alpha\theta} - \frac{\varphi}{1 - \alpha}\, e^{-\theta}.$$

The peak xenon concentration occurs when

$$\theta = \frac{1}{1-\alpha} \ln \frac{\varphi + \alpha}{\alpha(\varphi + 1)},$$

and its magnitude is

$$x_{\text{max}} = \varphi \left[\alpha^{\alpha} \frac{\varphi + 1}{\varphi + \alpha} \right]^{1/(1-\alpha)}.$$

Exercise 11.3: trajectories in the iodine-xenon plane

a) The iodine-xenon concentration parameters for a reactor at equilibrium lie on the hyperbolic curve

$$x = \frac{i}{i + \alpha}.$$

Note that the abscissa i is also equal to the reduced flux φ.

b) The time-dependent curves for a reactor with a constant flux take a parametric form as given in exercise **11.1**. They can also be expressed in the form $x = f(i)$ by rearranging the first equation to make θ a function of i, then substituting it into the second.

For a given constant flux, the trajectories always move towards a point on the equilibrium curve. This is the origin when the flux is zero. Depending on the initial state, this occurs on both sides of the curve. Moreover, if the flux changes instantaneously, then from a starting point on the curve, the trajectory to the new equilibrium point lies above the curve for a lower flux, and below the curve for a higher flux.

Exercise 11.4: why does a peak in the concentration of xenon occur after a reactor is shut down?

The true reason for the peak in xenon concentratrion, which occurs after a reduction in flux, or reactor shutdown, is that xenon destruction by neutron capture decreases or ceases.

The xenon effect is illustrated by some examples in the following table, showing the equilibrium and peak values of the xenon concentration parameter x when a reactor is shutdown from steady-state operation, over a representative range of reduced flux parameter φ. The first pair of columns with $\alpha = 0.7$ show the standard result, while the middle pair of columns show what happens when λ_I and λ_{Xe} are permuted so that $\alpha = 1/0.7$. The peak in xenon concentration clearly still occurs in the second case; however, it underestimates the true flux which is proportional to φ/α, leading to a lower peak. To compensate for this, the third pair of columns shows the result when the flux parameter is adjusted by $\varphi/0.7$, making the true flux equal to the standard case. It can be seen that this correction nearly restores the magnitude of the peak to its original level in the standard case.

Reduced flux	$\alpha = 0.7$ Equil.	Peak	$\alpha = 1/0.7$ Equil.	Peak	$\alpha = 1/0.7$; $\varphi/0.7$ Equil.	Peak
$\varphi = 1$	0.588	0.748	0.415	0.479	0.500	0.636
$\varphi = 2$	0.741	1.236	0.583	0.832	0.667	1.113
$\varphi = 3$	0.811	1.693	0.677	1.159	0.750	1.566
$\varphi = 4$	0.851	2.139	0.737	1.476	0.800	2.011
$\varphi = 5$	0.877	2.581	0.788	1.789	0.833	2.452

Exercise 11.5: analytical solution to the equations which describe the samarium effect

a) This problem is similar to exercise **11.2.a** for xenon.

$$S(t) = \left[1 - e^{-\sigma\Phi t} + \frac{\sigma\Phi}{\lambda - \sigma\Phi}\left(e^{-\lambda t} - e^{-\sigma\Phi t}\right)\right]\frac{\gamma\Sigma_f}{\sigma}.$$

The samarium concentration reaches 90% of its equilibrium value after $t = 586$ hours.

b) The samarium excess is

$$\Delta S(t) = \left(1 - e^{-\lambda t}\right)\frac{\gamma\Sigma_f\Phi}{\lambda}.$$

It achieves 90% of its final value after 176 hours. This is a shorter time than in part **(a)** because there is only one step instead of two.

Exercise 11.6: shutting down a reactor with a constant samarium content

If the samarium concentration S is constant, then the samarium-effect equation gives $\Phi = \lambda P/(\sigma S)$. Substituting this into the prometheum equation, allows P and Φ to be found:

$$\Phi(t) = \frac{\lambda P_0}{\sigma S}\exp\left[\left(\frac{\gamma\Sigma_f}{\sigma S} - 1\right)\lambda t\right].$$

This flux decreases only when the samarium concentration is *greater* than the equilibrium concentration $\gamma\Sigma_f/\sigma$.

Exercise 11.7: spatial instabilities due to the samarium effect

a) In the equilibrium state, the equations for prometheum and samarium concentrations give

$$\gamma\Sigma_f\,\Phi_0 = \lambda\,P_0, \qquad \lambda\,P_0 = \sigma\,S_0\,\Phi_0.$$

Owing to the boundary condition, Φ_0, P_0 and S_0 are constant in space, and $\Delta\Phi_0$ is zero. Therefore,

$$k_\infty - 1 - \beta\,S_0 = 0.$$

This relationship sets the value of the infinite-multiplication factor, that normally is then to be determined by using the appropriate boron concentration.

b) Taking into account these relationships, replacing Φ with $\Phi_0 + \varphi$ etc., and neglecting the second order term $\sigma\varphi s$, gives

$$M^2\,\Delta\varphi - \beta\,\Phi_0\,s = 0,$$

$$\frac{dp}{dt} = \gamma\Sigma_f\,\varphi - \lambda\,p,$$

$$\frac{ds}{dt} = \lambda\,p - \sigma\,\Phi_0\,s - \gamma\Sigma_f\,\varphi.$$

c) Exponential time-dependence gives the result

$$M^2\,\Delta\varphi - \beta\,\Phi_0\,s = 0,$$

$$\omega p = \gamma\Sigma_f\,\varphi - \lambda\,p,$$

$$\omega s = \lambda\,p - \sigma\,\Phi_0\,s - \gamma\Sigma_f\,\varphi.$$

d) The first mode is the stationary state and has constant magnitude throughout space. The other modes describe the perturbation. It can be seen that they are independent from each other, within the first order approximation. For the nth mode

$$-M^2\,\mu_n\,\varphi_n - \beta\,\Phi_0\,s_n = 0,$$

$$\omega p_n = \gamma\Sigma_f\,\varphi_n - \lambda\,p_n,$$

$$\omega s_n = \lambda\,p_n - \sigma\,\Phi_0\,s_n - \gamma\Sigma_f\,\varphi_n.$$

e) This system is homogeneous: its determinant must be equal to zero. The rate ω must satisfy the equation

$$\omega^2 + \left(\lambda + \sigma\Phi_0 - \frac{\beta\,\Phi_0\,\gamma\Sigma_f}{M^2\,\mu_n}\right)\omega + \lambda\,\sigma\,\Phi_0 = 0.$$

f) The product of the roots is always positive; therefore, their real parts have the same sign. Hence, the reactor is stable with regard to the samarium effect when the sum is negative. Therefore, the *stability condition* is

$$M^2\,\mu_n > \frac{\beta\,\Phi_0\,\gamma\Sigma_f}{\lambda + \sigma\,\Phi_0}.$$

If the equilibrium poisoning is defined as $\pi_{eq} = \beta\,S_0 = \beta\,\gamma\Sigma_f/\sigma$, then the stability condition can be re-expressed as

$$M^2\,\mu_n > \frac{\sigma\,\Phi_0}{\lambda + \sigma\,\Phi_0}\,\pi_{eq}.$$

g) The modes of a cylindrical PWR are as follows.

- The first radial mode has a transverse gradient. Its eigenvalue is $\mu = (j_{11}/R)^2$ where $j_{11} = 1.84118$ is the first extremum of the Bessel function J_1.
- The first axial mode has a longitudinal gradient; $\mu = (\pi/H)^2$.

- The other radial and axial modes are more complex. They are characterized by higher values of μ.

h) The results of the stability calculation are as follows.

- $M^2\,\mu_{1,\mathrm{radial}} = 795$ pcm.
- $M^2\,\mu_{1,\mathrm{axial}} = 410$ pcm.
- Limit = 184 pcm.

This PWR is quite far from the limits of instability with respect to the samarium effect.

Exercise 11.8: spatial instabilities due to the xenon effect

Notice that under equilibrium conditions, provided k_∞ is constant, X_0—and consequently Φ_0—respond in such a way that

$$k_\infty - 1 - \beta X_0 = 0,$$

This differs from the samarium case in that the equilibrium concentration is independent of the flux level, and it is k_∞ that must be adjusted.

For the perturbation, the calculations are similar but a little more involved. Neglecting the direct production of xenon-135 by fission, the *stability condition* is

$$M^2\,\mu_n > \frac{\alpha\, u}{1-\alpha-u}\,\pi_{\mathrm{eq}}, \qquad u = \frac{\pi_{\mathrm{eq}}}{\pi_\infty} = \frac{\varphi}{\varphi+\alpha}.$$

When $\pi_{\mathrm{eq}} = 3000$ pcm and $\pi_\infty = 4000$ pcm (i.e. $\varphi = 3\alpha$), the limit is 1658 pcm. From this is might be concluded that the pressurized water reactor is *unstable* both radially and axially. Moreover, $M^2\,\mu_{1,\mathrm{axial}} < M^2\,\mu_{1,\mathrm{radial}}$. However, the present model is certainly inadequate. In particular it includes the assumption that perfect reflection occurs at the core boundaries. Nevertheless, this exercise demonstrates the well known axial instability of PWRs.

Exercise 11.9: natural recurrent perturbations of a reactor

The methodology of this exercise is similar to the approach used previously to examine spatial unstabilities, i.e. determine the response of the system to a small perturbation near the equilibrium state.

a) The equilibrium state is described by the relationships

$$k = k_0 - a\,\varphi_0 - b\,x_0 = 1, \qquad i_0 = \varphi_0, \qquad x_0 = \frac{\varphi_0}{\varphi_0+\alpha}.$$

b) The perturbation is described approximately to first order by

$$u = -\frac{b}{a}\,w, \qquad \omega\, v = u - v, \qquad \omega\, w = v - (\varphi_0+\alpha)\,w - u\,x_0.$$

c) The rate ω must satisfy the equation

$$\omega^2 + \left(1 + \alpha + \varphi_0 - \frac{b}{a} x_0\right) \omega + \alpha + \varphi_0 + \frac{b}{a}(1 - x_0) = 0.$$

d) The product of the roots is always positive because $x_0 < 1$. The *stability condition* is met when the sum of the roots is negative. This is

$$b\, x_0 < a\, \varphi_0 + a(1 + \alpha).$$

The left-hand term represents the effect of xenon poisoning on the reactivity under equilibrium conditions. The right-hand term is the sum of the reactivity effects resulting from the real flux and a reduced flux equal to $1 + \alpha$.

e) A typical PWR has

$$b\, x_0 = 3000 \text{ pcm}; \qquad x_0 = 0.75;$$

$$a\, \varphi_0 = 1200 \text{ pcm} \;\; (3 \text{ pcm/°C} \times 400\ \text{°C});$$

$$\varphi_0 = 3\alpha; \qquad \alpha = 0.7;$$

$$a\, \varphi_0 + a(1 + \alpha) = 2040 \text{ pcm}.$$

Hence, the reactor is unstable. The rate parameter is $\omega = \omega_1 \pm i\, \omega_2 = 0.725 \pm 2.01\, i$, with the unit $1/\lambda_I = 9.42$ hours.

Therefore, each cycle lasts $2\pi/\omega_2 = 3.13$ units $= 29.4$ hours, and doubles in intensity over $\ln(2)/\omega_1 = 0.96$ units $= 9.0$ hours.

12 Fuel evolution (heavy nuclei)

Introduction

Even though some nuclei can undergo fission after absorbing a neutron, we know that in many cases absorption is simply a sterile capture [(n,γ) radiative capture reaction]. An (n,2n) reaction is also sometimes observed. This illustrates the fact that it is not sufficient to consider only the transformation of fission products; the transformation of heavy nuclei by these reactions and by radioactive decay must also be examined.

The physics of these transformations and the associated equations are in fact similar in both cases. To follow the neutron physics convention, however, we must distinguish between "fission products" and "heavy nuclei", because these physical phenomena do not have the same consequences. In the first case, the materials concerned are all nuclear waste and, in some cases, poisons that degrade the reactivity and cause control problems. In the second case, some of the materials are energetic; they will contribute to fission as irradiation proceeds, and possibly, if the fuel is irradiated and reprocessed, end up being recycled in new fuel.

This chapter is devoted exclusively to the study of heavy nuclei; those that were introduced into the core, and those that will be formed by neutron irradiation if fission does not occur. There are relatively few of these nuclei. According to the desired accuracy of the neutron physics calculations, only a small number (from one to a few tens) of them need to be handled in detail (unless a very precise analysis of nuclear *waste* is to be performed). It is therefore unnecessary to introduce a model analogous to the pseudo fission product discussed in the previous chapter.

The first part of this chapter will be devoted to a physical analysis of the evolution: series, equations, and measurement of the evolution. The consequences of this evolution on the multiplication factor will be examined in the second part. The third part contains an analysis of the mechanisms for the conversion of fertile matter to fissile matter, and a discussion of the recycling that can be envisaged.

12.1. Evolution series and equations

12.1.1. Evolution series

The essential neutron physics aspects of the evolution of a uranium fuel are governed by the ten or so nuclides appearing in Figure 12.1. Note, however, that fission products

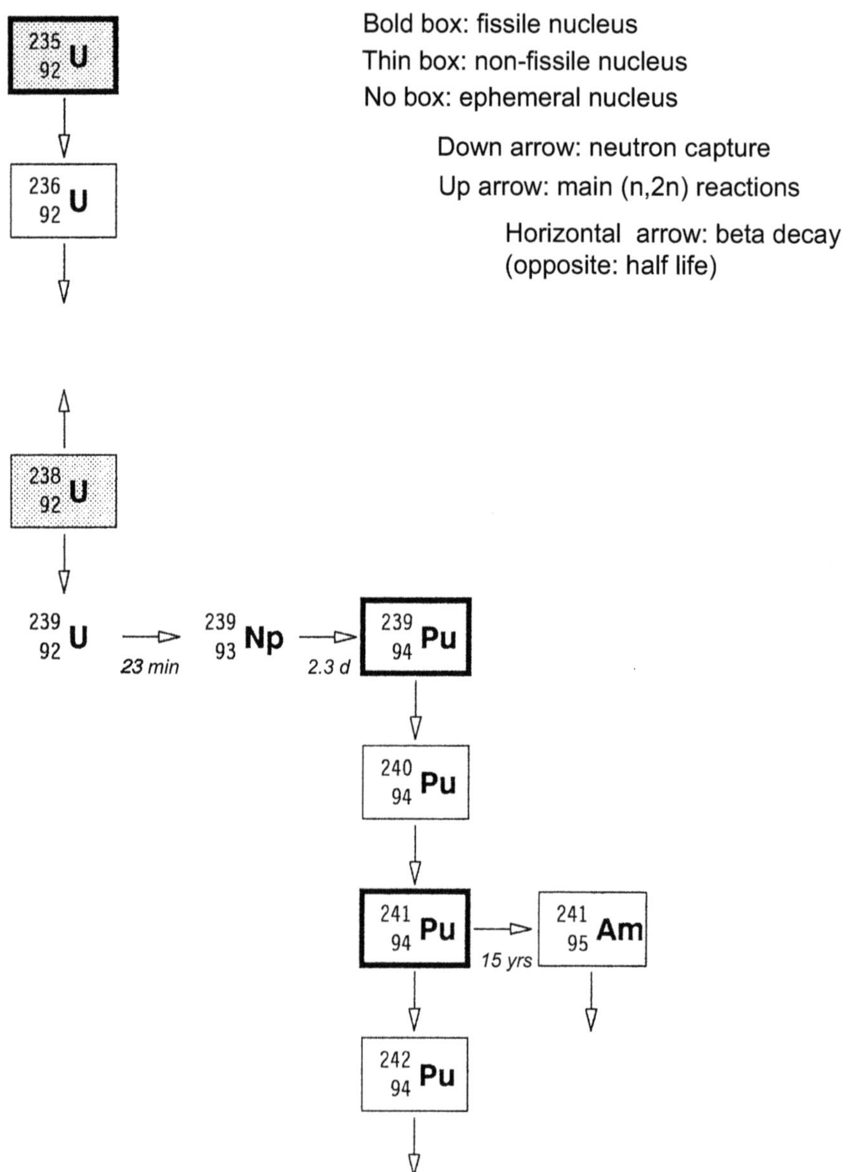

Figure 12.1. Simplified evolution series of uranium: main nuclei contributing to the reactivity effects.

are not included in this diagram in order to remain consistent with the distinction drawn between heavy nuclei and fission products. For all nuclei liable to undergo fission, the corresponding arrow must be added mentally; this will be important to write the evolution equations correctly.

Two essential aspects must be borne in mind:

1/ it is possible for uranium 235 not to undergo fission and to be transformed into uranium 236, which is a (mild) neutron poison because it is not fissile;

2/ if it absorbs a neutron, uranium 238 [except for the occasional rare fission and (n,2n) reaction] will be transformed into uranium 239. This very quickly (in a few days) is transformed into plutonium 239, the main *artificial* fissile nucleus (by neutrons of any energy[1]).

This transformation is called the *conversion* of a material known for this reason as "*fertile*", uranium 238, into a fissile material, plutonium 239.

As for uranium 235 (and even a little more frequently, for slow neutrons: one in four times instead of about one in seven), plutonium 239 has the possibility of not undergoing fission after absorbing a neutron; in this case, non-fissile plutonium 240 is then formed. If this material captures a neutron, it produces fissile plutonium 241. In the event of capture without fission, plutonium 241 produces plutonium 242. The plutonium isotope series ends there, because another capture would give plutonium 243 which very quickly decays to americium 243. Note that plutonium 241 can also disappear by beta-minus decay with a half-life of 15 years, which is neither very short nor very long (on the order of magnitude of the fuel cycle time constants).

Note also that *all* of the nuclei in this series (and in the ones to follow) are also alpha radioactive. All of the alpha decays that are not mentioned are long and can be neglected in reactor calculations: a good example is the 24,000-year half-life of plutonium 239[2].

If we wish to perform accurate neutron physics calculations and to evaluate the production of the main "minor actinides"[3], a slightly more detailed description of the nuclear reactions is required. Figure 12.2 indicates what should at least be added to the series we have just discussed.

Note the addition of neptunium 237 and plutonium 238 (the fifth and last of the plutonium isotopes formed in significant quantity in reactors), and a few other isotopes of americium and curium. Note also the two alpha decays (even-numbered isotopes of curium) that should be taken into account in view of the orders of magnitude of the half-lives.

Finally, the thorium 232 series is presented (Figure 12.3).

By neutron capture and a series very similar to the one leading to plutonium 239 from uranium 238, this nuclide (the only isotope of natural thorium) produces another artificial nuclide: uranium 233. Thorium, which is slightly more abundant on Earth than uranium,

1 In the rest of this chapter, the term "fissile" is reserved for nuclei that can be fissioned by neutrons of any energy. Even in fast neutron reactors, where the fission of nuclei that can only be fissioned by fast neutrons (nuclei with an even number of neutrons, such as uranium 238, plutonium 240, etc.) has more relative significance than in thermal neutron reactors, the fission of "fissile" nuclei (nuclei with an odd number of neutrons, such as uranium 235, plutonium 239, etc.) is still very much preponderant.

2 The plutonium isotope with the longest half-life is plutonium 244: 80 million years. This isotope is not produced in reactors, because plutonium 243 does not have the time to capture a neutron before disappearing by beta decay. With the exceptions of thorium 232, uranium 235, and uranium 238, these half-lives are all long in terms of the timescale of an engineer's concerns, but short compared to the age of the Earth.

3 The actinides are the elements of atomic number 89 and above. In dealing with the management of nuclear waste, a distinction is drawn between "major actinides", which are liable to produce energy (in practice, the fissile and fertile isotopes of uranium and plutonium, as well as thorium 232) and all others, which are called "minor actinides", which are currently considered as waste. This distinction is a little bit artificial, because these nuclei are always more or less susceptible to undergoing fission, either directly or after one or more neutron captures.

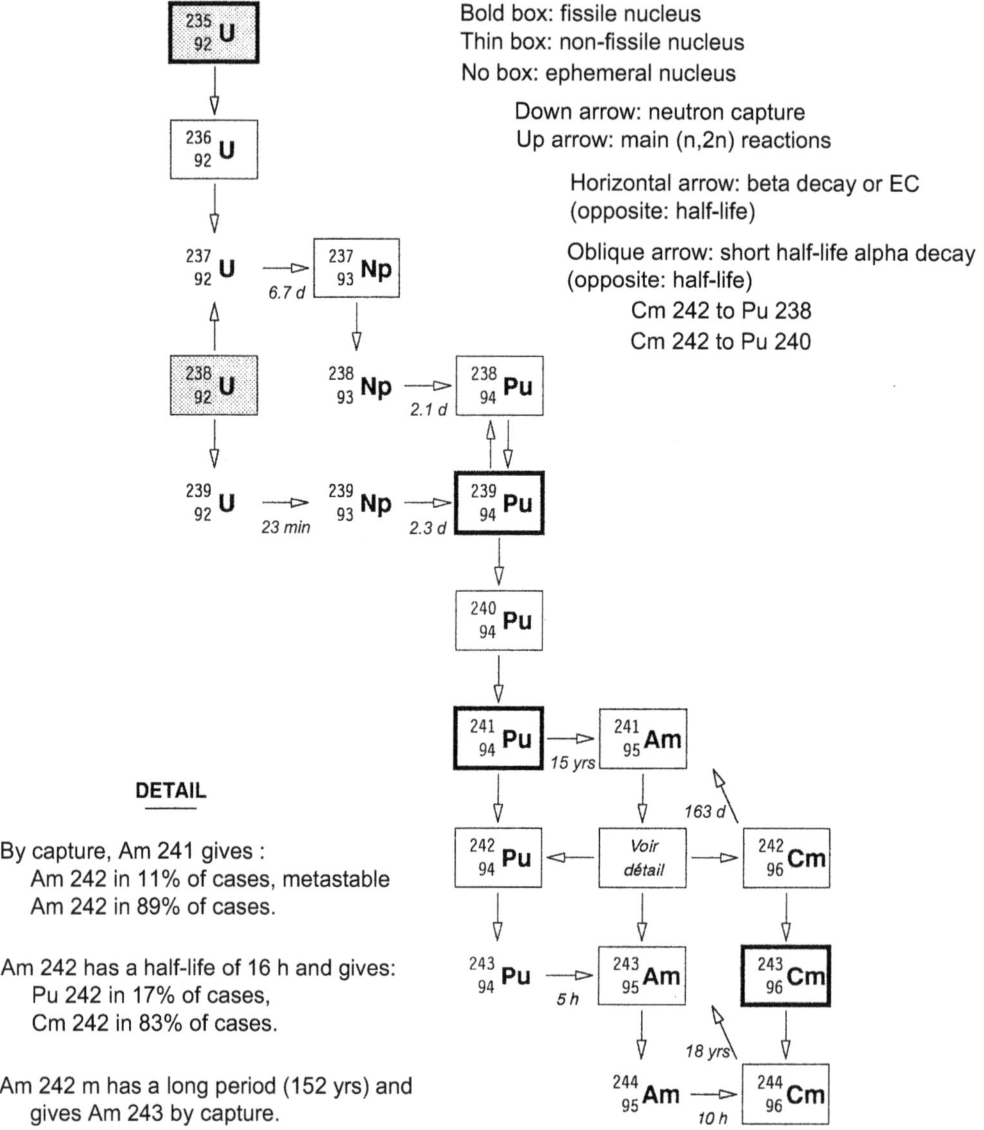

Figure 12.2. Simplified evolution series of uranium: the main nuclei contributing to the reactivity effects and the problems of alpha nuclear waste management.

is therefore a fertile material that could be used to supplement uranium 238. Because uranium 233 has good nuclear properties, it could be used with thorium to constitute the fuel for fast breeder reactors (for example, graphite-moderated thermal neutron reactors). The problem lies in constituting the initial stock of uranium 233, since thorium itself is not fissile and uranium 233 is not found in nature. This thorium-uranium 233 cycle could be initialised using another fissile material (uranium 235 or plutonium).

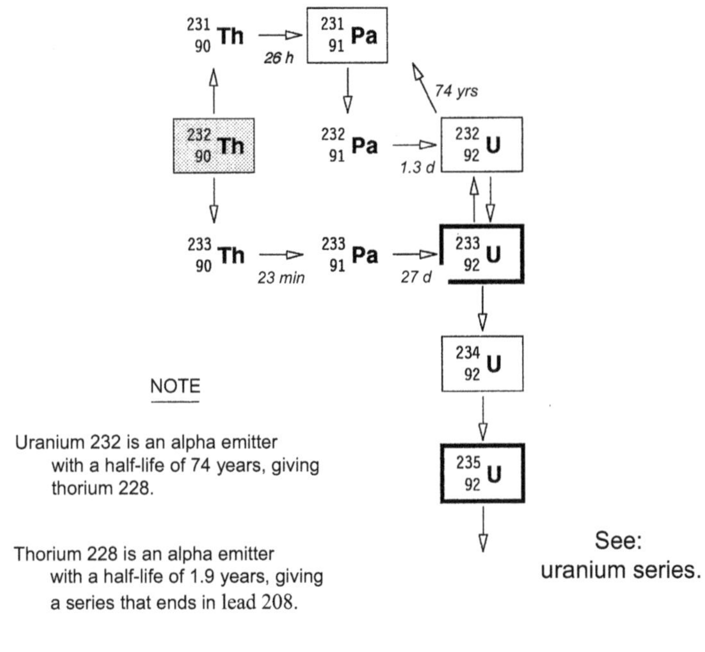

Figure 12.3. Simplified evolution series of thorium 232 (same conventions as for the previous figures).

12.1.2. Evolution equations

To identify the various actinide isotopes, it is convenient to use a double subscript *ij* where *i* is the last digit of the atomic number and *j* is the last digit of the mass number, for example: 25 for uranium 235 (atomic number 92: $i = 2$; mass number 235: $j = 5$).

The respective quantities of these various nuclides are characterised by their *volume concentrations* N, i.e. the number of atoms per unit volume (in practice: m^3). As with fission products, we can calculate average values over the entire volume of the *fuel*, or consider each point $\vec{r}$ or volume element V, in which case the parameter $\vec{r}$ or V is understood in the concentration calculations.

Because we are concerned with evolution, *the concentrations are functions of time t*.

The evolution equations describe the *variations* during a unit time interval: contribution by nuclear reaction [in general (n,γ) or (n,2n)]; elimination by neutron absorption [radiative capture, (n,2n) reaction or fission]. These reactions are quantified by their respective cross-sections σ, averaged over the entire spectrum and weighted by the flux[4]; associated with the flux Φ (integral over the entire spectrum), and they provide an expression for the reaction rates $\sigma N \Phi$. If appropriate, the rates of production or elimination λN

[4] According to the specific case, the average flux over the entire fuel space or the flux at the point or volume element concerned.

by radioactivity must be added. If we use the index a for absorption and the index c for radiative capture, some of these equations will be as follows[5]:

- *uranium 235:*

$$\frac{dN_{25}}{dt} = -\sigma_{a,25}N_{25}\Phi; \tag{12.1}$$

- *uranium 236:*

$$\frac{dN_{26}}{dt} = \sigma_{c,25}N_{25}\Phi - \sigma_{a,26}N_{26}\Phi; \tag{12.2}$$

- *uranium 238:*

$$\frac{dN_{28}}{dt} = -\sigma_{a,28}N_{28}\Phi; \tag{12.3}$$

- *plutonium 239:*

$$\frac{dN_{49}}{dt} = \sigma_{c,28}N_{28}\Phi - \sigma_{a,49}N_{49}\Phi; \tag{12.4}$$

(neglecting two short intermediate steps: the radioactive decay of uranium 239 and neptunium 239);

- *plutonium 240:*

$$\frac{dN_{40}}{dt} = \sigma_{c,49}N_{49}\Phi - \sigma_{a,40}N_{40}\Phi; \tag{12.5}$$

- *plutonium 241:*

$$\frac{dN_{41}}{dt} = \sigma_{c,40}N_{40}\Phi - \sigma_{a,41}N_{41}\Phi - \lambda_{41}N_{41}; \tag{12.6}$$

- *plutonium 242:*

$$\frac{dN_{42}}{dt} = \sigma_{c,41}N_{41}\Phi - \sigma_{a,42}N_{42}\Phi; \tag{12.7}$$

americium 241:

$$\frac{dN_{51}}{dt} = \lambda_{41}N_{41} - \sigma_{a,51}N_{51}\Phi. \tag{12.8}$$

12.1.3. Neutron fluence

Note that the flux Φ appears in almost all of the terms on the right-hand side, since the main rates are those of the neutron reactions, not the radioactive decays. That is why we usually *divide* all of these equations by Φ and introduce a new variable: *neutron fluence*, s, defined by:

$$ds = \Phi\, dt, \qquad s = \int_0^t \Phi(t')\, dt'. \tag{12.9}$$

By definition, fluence is a number of neutrons per surface unit. In practice, it is often expressed in *neutrons per kilobarn*:

$$1 \text{ n/kb} = 10^{25} \text{ neutrons per m}^2.$$

[5] Nuclei shown in boxes on the simplified diagram in Figure 12.1; (n,2n) reactions being neglected.

For example, the fuel in pressurised water reactors is irradiated with a flux on the order of 3×10^{17} neutrons per m^2 per second for approximately 3 years: the fluence on discharge of the fuel is in the region of 3 n/kb.

Following the introduction of this variable, the equations take the following form:

- *uranium 235:*

$$\frac{dN_{25}}{ds} = -\sigma_{a,25}N_{25}; \tag{12.10}$$

- *uranium 236:*

$$\frac{dN_{26}}{ds} = \sigma_{c,25}N_{25} - \sigma_{a,26}N_{26}; \tag{12.11}$$

- *etc.;*
- *americium 241:*

$$\frac{dN_{51}}{ds} = \frac{\lambda_{41}N_{41}}{\Phi} - \sigma_{a,51}N_{51}. \tag{12.12}$$

In addition to simplifying the equations, this has the great advantage of automatically allowing for the variations in flux over time. In particular, for a reactor performing load monitoring, the flux varies daily, and the introduction of fluence removes the complicated "noise" from the curve $\Phi(t)$. The constants λ, on the other hand, are replaced by $\lambda/\Phi(t)$; but it is acceptable to replace these expressions by their average values because, as has been remarked, the radioactivity terms are not very significant in the heavy nucleus evolution equations.

12.1.4. Variation of cross-sections

It is important to note that, in spite of the introduction of fluence, the system of evolution equations *is not a constant-coefficient system*. Not only do the coefficients of the radioactivity terms vary if the flux varies, but the cross-sections, above all, are *implicitly* dependent on time. They are in fact averages weighted by the neutron spectrum $\Phi(E)$, and the spectrum in a reactor depends on the composition of the matter inside it, i.e. on the concentrations N. Because the concentrations vary with time (or fluence) according to the evolution equations, the flux $\Phi(E)$ varies, and so do the cross-sections σ.

For example, in a thermal neutron and uranium reactor, the spectrum is initially more or less Maxwellian, and is gradually deformed with the formation of plutonium because of the resonances at 0.3 eV of plutonium 239 and at 1 eV of plutonium 240, and the average cross-sections of these isotopes in particular, but also of the other nuclides, change during irradiation.

These variations, however, are slow and regular. To account for them, it is necessary to integrate the evolution equations numerically[6], not analytically[7], and to recalculate the neutron spectrum regularly during irradiation with the concentrations at the fluence under consideration.

---

[6] A Runge-Kutta method is generally used.

[7] If alpha decay and (n,2n) reactions are ignored, the system of equations is triangular and can easily be solved analytically if the coefficients are constant (combinations of exponentials).

Note: In fast neutron reactors, the spectrum variations are relatively small (no effect of the large resonances of plutonium in the thermal domain) and the constant cross-section approximation can reasonably be applied.

12.1.5. Burn-up and combustion rate

Fluence might be a useful variable for physicists, but it does not mean much to an engineer. Engineers prefer to talk about kilogrammes or tonnes of fuel placed in the core, in megawatts of power output, and days of irradiation.

The combination of these parameters leads to the concept of *burn-up* or *specific burn-up*: this is the (thermal) energy produced normalised to the mass of fuel placed in the core[8]. If SI units were to be used, burn-up would be expressed in joules per kilogramme (J/kg) or, given the orders of magnitude concerned, terajoules per kilogramme (1 TJ/kg = 10^{12} J/kg).

In practice, the unit *megawatt-day per tonne* (MWd/t, 1 megawatt-day = 1 megawatt over the course of one day, i.e. $8.64.10^{10}$ joules), or sometimes *gigawatt-day per tonne* (GWd/t) or *megawatt-day per kilogramme* (MWd/kg):

$$1000 \text{ MWd/t} = 1 \text{ GWd/t} = 1 \text{ MWd/kg} = 0.0864 \text{ TJ/kg},$$
$$1 \text{ TJ/kg} = 11\,574 \text{ MWd/t}.$$

Order of magnitude on discharge of a pressurised water reactor fuel: 30 000 to 45 000 MWd/t, i.e. 3 to 4 TJ/kg.

Physicists also use the *burn-up fraction* or *fission burn-up*: the number of fissions normalised to the initial number of heavy nuclei, i.e. the proportion of heavy nuclei placed in the core that have undergone fission either directly or after conversion. It would be possible to verify that:

$$1\% \text{ in fission burn-up} \cong 10\,000 \text{ MWd/t}.$$

Table 12.1 gives a few approximate values for some reactor types. Note that, because of conversion, the fission burn-up can reach or exceed the initial fissile material content even though the irradiated fuel still contains some: it is in fact necessary for the reactor to remain critical throughout irradiation, particularly at the end.

Table 12.1. Orders of magnitude of burn-up for a few reactor types.

Type	UNGG	CANDU	PWR	FNR
Initial content (%)	0.7	0.7	4	15
Fission burn-up (%)	0.4	1	4	10
Burn-up (GWd/t; TJ/kg)	4; 0.35	10; 0.9	40; 3.5	100; 9

[8] Initial mass of heavy nuclei only. "Initial" must be specified because this mass then decreases due to fission. If the fuel is an oxide, for example, the mass of oxygen is not counted.

12.1.6. Example of heavy nucleus balance (pressurised water reactor)

As an example, Figure 12.4 gives the evolution curves for the main isotopes of uranium and plutonium as a function of burn-up. The points give an indication of the time discretisation that is performed in practice (it must be finer at the beginning because of the fission products that reach saturation fairly quickly, such as samarium 149)[9].

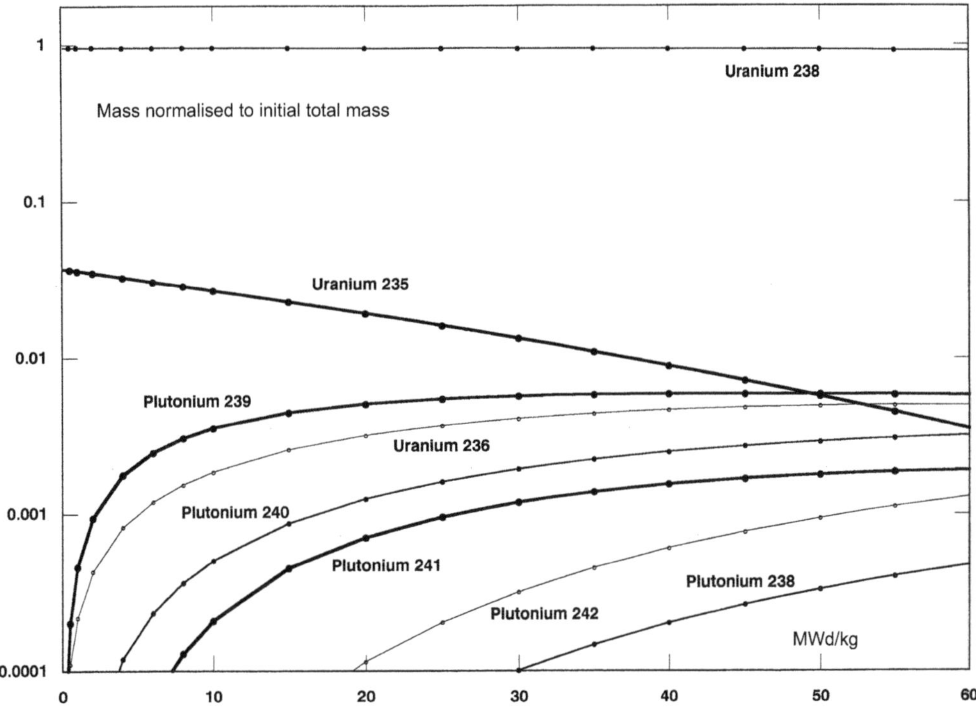

Figure 12.4. Evolution of mass the main uranium and plutonium isotopes for a pressurised water reactor with respect to burn-up (the ordinate axis is on a log scale).

Figure 12.5 very approximately represents the balance for a classic management of three one-year cycles (in its reactors, *Électricité de France* now performs slightly longer irradiations). Note the following key points:

- approximately 3% of the heavy nuclei are fissioned, i.e. 1% per year (10,000 MWd/t; almost 1 TJ/kg);
- two thirds of these fissions come directly from 235, and the other third from uranium 238 after conversion to plutonium (for the longer irradiations now performed, the contribution of plutonium goes up to about 40%);

[9] In evolution calculations, xenon is immediately taken to be at equilibrium.

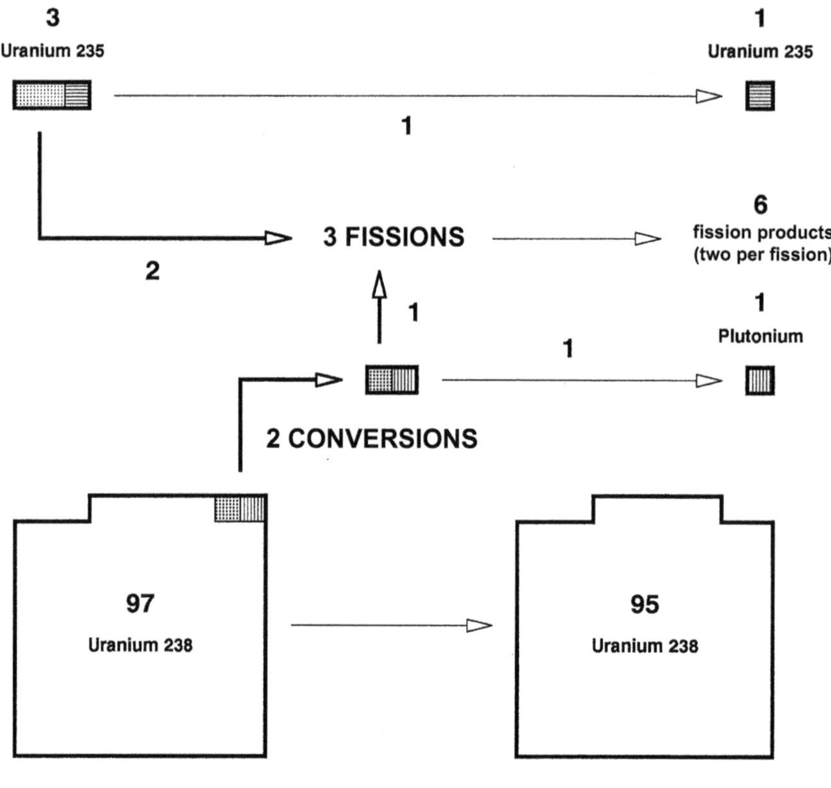

Figure 12.5. Approximate balance of heavy nuclei between the start and end of irradiation for a pressurised water reactor (for 100 heavy nuclei loaded in the core).

- the irradiated fuel contains approximately 1% plutonium and uranium with 1% uranium 235 (roughly equivalent to natural uranium, allowing for poisoning by uranium 236). It can be worthwhile to recycle these two energetic materials.

12.2. Evolution of the multiplication factor

12.2.1. Evolution of the infinite multiplication factor

The initial variation of the infinite multiplication factor depends on the abundance of fissile material in the fuel. For a *natural uranium reactor*, the initial conversion factor — the ratio of the quantity of plutonium 239 produced to the quantity of uranium 235 destroyed — is relatively high because the concentration of uranium 238 is quite high compared to that of uranium 235; for example, this factor is approximately 0.8 for graphite and natural uranium reactors. Although the production of plutonium 239 does not fully compensate

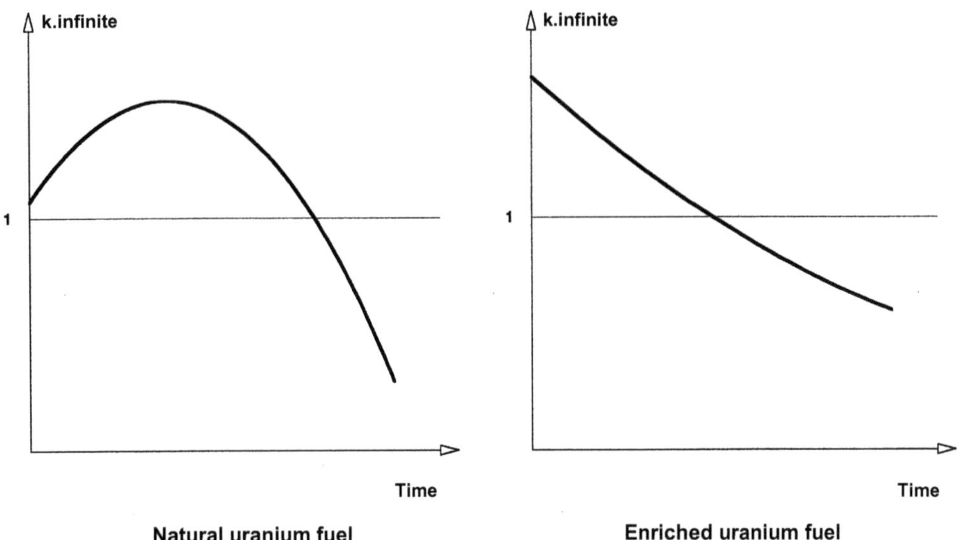

Figure 12.6. Evolution of the infinite multiplication factor (on these approximate curves, we have assumed that fission products that reach saturation quickly, such as xenon 135, samarium 149, had their equilibrium concentration values from the start).

for the disappearance of uranium 235, the reactivity improves at the beginning because the fission cross-section of plutonium 239 is significantly higher than that of uranium 235 (Figs. 2.13 and 2.14).

After a certain amount of irradiation, however, the plutonium itself starts to be consumed to a significant degree, and the factor k_∞, after passing through a maximum, ends up collapsing (refer to the diagram on the left in Figure 12.6: no scales are indicated because the exact values depend on the reactor, but note that, for a graphite and natural uranium reactor, the initial value of k_∞ is restored after 4000 MWd/t, after it had gained 1000 to 2000 pcm with respect to the initial value).

For a *uranium 235- or plutonium-enriched fuel reactor,* the conversion factor is lower (the ratio of concentrations between fertile nuclei and fissile nuclei is lower), for example in the region of 0.6 for water reactors, and the factor k_∞ decreases from the start of irradiation (see the right-hand diagram in Figure 12.6: the two diagrams have different scales; the important point to remember is that the variation is approximately linear for a pressurised water reactor, and has a value of roughly −1 pcm per MWd/t, i.e. −30,000 pcm for a combustion of 30,000 MWd/t over approximately three years; for a fast neutron reactor, the slope of pcm to MWd/t is approximately ten times lower).

12.2.2. Multiple-frequency core management

On some parts of the curves in Figure 12.6, the value of k_∞ is less than 1, and one might think that this extrapolation serves no purpose, since a reactor must be critical to operate, and must therefore have an infinite multiplication factor greater than 1 to compensate for leakage. In fact, however, this section of the curves is useful because, in a working core, there is always a *juxtaposition of fuels at different irradiations*.

The simplest case to imagine would be the situation where the core is initially homogeneous with fresh fuel everywhere (if the reactor is supercritical, this can be compensated by a poison). After operating for a while, the core contains fuels at different irradiations because the distribution of flux, and therefore also fluence, is never spatially uniform (see Chapter 6). After a certain time, there is a likelihood of finding some highly irradiated fuel with k_∞ less than1 and, to compensate, some less irradiated fuel with a k_∞ greater than 1. (Note that the concept of k_∞ is *local*, whilst the concept of k_{eff} is *global*, i.e. characteristic of the entire system.)

In practice, it is desirable to mix fuels of different irradiations as much as possible. The ideal situation would be to have a complete and uniform mixture of fuels of every irradiation from zero to maximum at each volume element.

In reactors where fuel is renewed during operation (UNGG, CANDU), and therefore almost infinitesimally, there is an attempt to approach this *ideal management* as closely as possible.

In reactors where fuel renewal occurs during shutdown periods (water reactors, fast neutron reactors), this ideal inevitably remains further out of reach. In practice, at each scheduled shutdown, a fraction $1/n$ of the fuel assemblies is replaced. For example, for water reactors, the fraction is 1/3 or 1/4 approximately each year.

The following formula is simple and useful for optimising the management of water reactors. Let $1/n$ be the fraction of the core that is renewed at each scheduled shutdown, and let T be the burn-up increment of the core at each cycle. We assume that the flux distribution is more or less uniform and therefore that each assembly and each of the n batches of fuel undergoes the same irradiation at each cycle; the combustion at fuel discharge is therefore $\tau_{\max} = nT$. We also assume that the multiplication factor of the core is given by the usual formula:

$$k_{\text{eff}} = \frac{\langle k_\infty \rangle}{1 + M^2B^2}, \tag{12.13}$$

where the denominator is independent of time and the numerator is simply the arithmetic mean of the k_∞ values of all batches; if the k_{eff} given by this formula is greater than 1, the excess is counterbalanced by a poison (for example, boron solution). Finally, we assume that the k_∞ of each assembly, initially equal to k_0, then decreases linearly with the burn-up τ:

$$k_\infty = k_0 - \alpha\tau. \tag{12.14}$$

Based on these simple assumptions, the following reasoning is applied:

- At the beginning of the cycle, the batches have the following burn-up values:

  $$0, T, 2T, \ldots (n-1)T,$$

  and, at the end of the cycle:

  $$T, 2T, 3T, \ldots nT;$$

- The average of these values is $(n+1)T/2$. The average multiplication factor of the core is the infinite multiplication factor corresponding to this average, since the relationship between k_∞ and τ is assumed to be linear:

  $$\langle k_\infty \rangle = k_0 - \alpha(n+1)T/2; \tag{12.15}$$

– The effective multiplication factor (without poison) is just equal to 1 at the end of the cycle; this is the criterion that defines the cycle period T:

$$k_{eff} = \frac{k_0 - \alpha(n+1)T/2}{1 + M^2B^2}; \tag{12.16}$$

– This gives T, from which we obtain the burn-up at discharge:

$$\tau_{max} = nT = \frac{2n}{n+1}\frac{k_0 - 1 - M^2B^2}{\alpha}. \tag{12.17}$$

We can therefore see that, for a given fuel (k_0, α) and a given core (M^2B^2), the energy supplied by a fuel increases as $2n/(n+1)$ as the number of batches in the core is increased. To illustrate this comment, Table 12.2 gives the burn-ups obtained with a uranium oxide fuel enriched to 3.25% in a 900 MWe pressurised water reactor core (given the value for $n = 3$).

Table 12.2. Maximum burn-up according to number of batches in core *n*.

n	$2n/(n+1)$	nT
1	1	22,222
2	4/3	29,630
3	3/2	33,333
4	8/5	35,556
5	5/3	37,037
6	12/7	38,095
∞	2	44,444

Between the extreme cases, a factor of 2 is observed on the energy produced. The $n = 3$ case is just half-way between these extreme cases. This obviously results from a compromise: an increased n increases the energy output, but reduces the availability of the installation, because it is shut down more frequently. A the infinite n limit, 44,444 MWd/t would be obtained with a reactor that was permanently shut down. Changing from $n = 3$ to $n = 4$ or $n = 5$ provides an increase of 7 and 11% respectively: this is the evolution currently observed in the management of *Électricité de France* cores (at the time of writing, in 2008, 900 MWe reactors are managed according to four fuel batches). To avoid cycles with too short a period, EdF simultaneously increases the initial abundance of the fuel (if k_0 increases, then T increases).

12.2.3. Other core management problems (pressurised water reactors)

Whenever a pressurised water reactor is shut down to renew a fraction of the core, the shutdown is used as an opportunity to rearrange the partially irradiated assemblies in order to obtain the best possible power distribution, i.e. the smallest possible shape factor $F = P_{max}/P_{av}$ that is compatible with the reactivity constraint (a sufficient multiplication

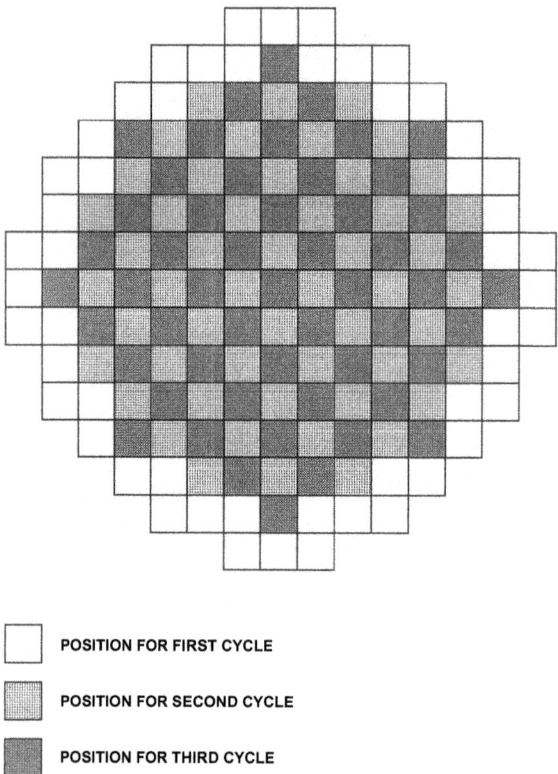

Figure 12.7. Overall rearrangement diagram for third-core management of a 900 MWe pressurised water reactor. The fresh assemblies are placed at the edge. Assemblies that have already done one or two cycles are placed in a chessboard pattern in the central area.

factor to last for the expected cycle period) and any other possible constraints[10]. The drawing up of a rearrangement plan is a difficult optimisation problem because the number of possible combinations is huge[11], even if symmetries are observed and we restrict ourselves to a general plan (e.g. Figures 12.7 and 12.8, showing management by thirds and by quarters[12] of a core), since the number of combinations remains for each batch. In practice, the combustion undergone by the assemblies of a single batch is different in each case.

Engineers must draw up *reloading plans* for each reactor at each shutdown, because normal operating fluctuations make it impossible to predict with any accuracy the state of

[10] *Électricité de France* is now looking for "low vessel fluence" management methods. The fast neutron fluence received by the vessel causes a weakening in the long term; it would therefore be desirable to reduce it, because this problem could be the limiting factor of the power station's working life. In practice, this means avoiding placing very reactive assemblies, liable to have a high flux level, in the positions nearest the vessel.

[11] There are $N!$ ways of placing N assemblies in N positions.

[12] The diagram for quarter-core management cannot be equally simple, but it follows the "in/out" principle: new assemblies, which are the most reactive, are preferentially placed at the edges in order to raise the power level in areas where it tends to drop because of leakage. In subsequent cycles, they are brought in towards the central areas. The vessel fluence constraint makes it necessary to review this simple principle.

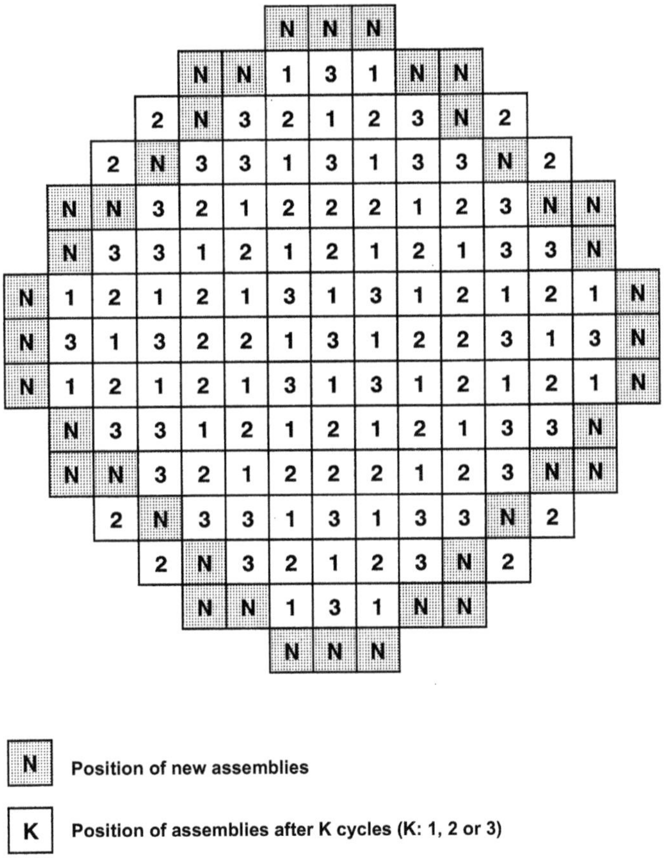

						N	N	N						
				N	N	1	3	1	N	N				
			2	N	3	2	1	2	3	N	2			
		2	N	3	3	1	3	1	3	3	N	2		
	N	N	3	2	1	2	2	2	1	2	3	N	N	
	N	3	3	1	2	1	2	1	2	1	3	3	N	
N	1	2	1	2	1	3	1	3	1	2	1	2	1	N
N	3	1	3	2	2	1	3	1	2	2	3	1	3	N
N	1	2	1	2	1	3	1	3	1	2	1	2	1	N
	N	3	3	1	2	1	2	1	2	1	3	3	N	
	N	N	3	2	1	2	2	2	1	2	3	N	N	
		2	N	3	3	1	3	1	3	3	N	2		
			2	N	3	2	1	2	3	N	2			
				N	N	1	3	1	N	N				
						N	N	N						

N Position of new assemblies

K Position of assemblies after K cycles (K: 1, 2 or 3)

Figure 12.8. Rearrangement diagram for quarter-core management of a 900 MWe pressurised water reactor. Note the second-cycle assemblies placed at the edges to limit the fluence received by the vessel.

the core at a given instant. The output of a set of reactors is in fact subject to economic and climatic fluctuations as well as technical ones. In particular, to harmonise power production by all of the reactors, the possibility of *early cycle shutdown* provides a certain amount of flexibility. Any loss will be partially made up at the next cycle, which can be longer because it will begin with a less irradiated fuel. The possibility of a *stretch-out*, which, like the early shutdown, can represent a shift of up to a few weeks, also provides some flexibility. It is possible to extend the cycle thanks to the strongly negative temperature coefficient at the end of the cycle, as we shall see in the next chapter. By reducing the water temperature by a few degrees, which slightly degrades the thermodynamic efficiency of the installation, we gain the reactivity required to continue the irradiation.

To draw up a loading plan, it is impossible to explore every case and select the best[13]; nowadays, the engineers in charge of this type of work have codes based on heuristic

[13] In any case, one would have to define exactly what constitutes the "best" configuration. In general, the shape factor of the power is minimised with constraints on the reactivity and fluence of the vessel.

methods, but they still rely a great deal on common sense and experience to find, if not the best, then at least a suitable loading plan.

It is also worth mentioning the problem of reaching the equilibrium cycle. Because they remain in the core for a shorter time, the assemblies of the n-1 first fuel loads, intended to perform only $1, 2, \dots n-1$ cycles, are made with less enriched uranium than those of the subsequent reloads.

Example for management by thirds of a 900 MWe reactor: 2.1%, 2.6%, 3.1%, and then 3.25% (equilibrium).

Note also that the problem of boron and consumable poisons and the problem of plutonium if recycled both add to the management difficulties (see § 9.3.5 and 9.3.6).

12.3. Conversion and recycling of materials

12.3.1. Fissile, fertile, and sterile nuclei

There is a broader problem of the management of natural and artificial nuclear material. For ease of discussion, these materials can be divided into three categories:

1/ *Fissile* materials consist of heavy nuclei (actinide isotopes) that can undergo fission by absorption of neutrons of any energy, including thermal neutrons. Even if they are not the only ones to undergo fission in reactors, since other nuclei can be fissioned by fast neutrons only, these fissile materials constitute the essential contribution to the production of nuclear energy, even in fast neutron reactors. Note that, in practice, the fissile actinide isotopes are those with an odd number of neutrons[14]. Main examples: uranium 233, uranium 235, plutonium 239, and plutonium 241.

2/ *Fertile* materials consist of heavy nuclei that are transformed into fissile nuclei after absorbing a neutron, either directly or after fast radioactive decay. Main examples: thorium 232, uranium 238, plutonium 238, and plutonium 240.

3/ All other materials are *sterile*. The narrow sense of the term would be limited to the other actinide isotopes, noting that they are always more or less susceptible to undergo fission induced by fast neutrons and that they can lead to a fissile nucleus after not one, but several neutron absorptions. Main examples: other isotopes of thorium, uranium, and plutonium; the isotopes with an even number of neutrons of protactinium, neptunium, americium, and curium. The wider sense of the term would also include the other radioactive nuclei produced in reactors and fission and activation products[15], which are currently all considered as waste.

The only nuclear materials in nature are uranium (0.7% uranium 235, 99.3% uranium 238, and negligible traces of uranium 234) and thorium (isotope 232 only).

The management of these materials concerns all of the problems involved in their use to produce energy by fission, either directly or indirectly after the *conversion* of fertile material into fissile material; in particular, problems related to reprocessing, reconditioning,

[14] Strictly speaking, even nuclei with an even number of neutrons can sometimes undergo slow-neutron fission caused by the "tunnel effect". In general, this can be neglected. More specifically, we could define as "fissile" any nuclei with a large fission cross-section for slow neutrons.

[15] Activation: formation of a radioactive nucleus by neutron capture.

and recycling. An entire book could usefully be written about these problems. Here we shall discuss only a few essential points.

12.3.2. Conversion factor and regeneration gain

The concept of the *conversion factor* (§ 12.2.1) has already been introduced: this is the ratio of the number of fissile nuclei produced to the number of fissile nuclei destroyed.

These numbers can be considered either at a given instant (for example, at the beginning of irradiation), or integrated over a certain time (for example, over the entire combustion period). Note that fissile nuclei can be "destroyed" by fission, which is the main channel, by the definition of fissile materials, as well as by sterile capture or by (n,2n) reaction.

The *regeneration gain* is the *net* number of fissile nuclei produced (production minus consumption) normalised to the number of *fissions*. As for the conversion factor, these numbers can be evaluated at a given instant or integrated over a certain time. If we use P, A, and F to denote the numbers of fissile nuclei produced, fissile nuclei destroyed (by neutron absorption), and fissions, we can write the following relationships; for the conversion factor:

$$C = \frac{P}{A}, \tag{12.18}$$

and for the regeneration gain:

$$G = \frac{P - A}{F} = (C - 1)\frac{A}{F}. \tag{12.19}$$

(In practice, A/F is close to 1: the numerator contains the fissions of fissile nuclei plus the captures by these nuclei, and the denominator contains the fissions of fissile nuclei plus the fissions by other nuclei.)

The disadvantage of these definitions is that they place all fissile nuclei in the same role. That is why a *weight* is sometimes assigned to each nuclide for the purpose of calculating these balances. For example, reactor physicists sometimes introduce the concept of a *plutonium 239 equivalent*. (The definition of this equivalence presents a problem in and of itself: under given conditions, should we reason in terms of reactivity, energy output, or some other parameter?)

Note (§ 1.9) that certain reactors can become *breeders*: this means that C is greater than 1 or that G is positive, i.e. that the system produces by conversion more fissile material than it consumes. Plutonium fast neutron reactors, in particular, can become breeders *if (and only if) a blanket is placed on it*, i.e. a layer of fertile matter is arranged around the core. Leaks in fast neutron reactors are very significant because the core is small (high power density) and the migration area is large (small fast neutron cross-sections); with a uranium 238 blanket, neutrons that escape from the core are used to convert this uranium to plutonium 239. This conversion is added to the conversion that already exists in the core itself, consisting of mixed uranium (natural or impoverished) and plutonium (approximately 15%) oxide fuel assemblies. For these reactors, it is useful to distinguish the following parameters in the numerator of the formula for, G: the net production of plutonium in the core (always negative) and the net production of plutonium in the blankets (positive). These are known as internal and external regeneration gains, and the overall regeneration gain is written as follows:

$$G \text{ (or } ORG) = IRG + ERG.$$

Table 12.3 gives a few orders of magnitude of the conversion factor for a few reactor types, in particular for fast neutron reactors without a blanket, with axial blankets (AB) only, and with radial (RB) and axial blankets. (To find the orders of magnitude of the regeneration gain, we can settle for the approximation $G \cong C - 1$.)

Table 12.3. Order of magnitude of the conversion factor for a few reactor types.

Type	UNGG	CANDU	PWR	FNR (bare)	FNR + AB	FNR + AB + RB
C	0.8	0.9	0.6	0.8	1.0	1.2

One might wonder why fast neutron reactors can be breeders, and not the other types. The answer lies in the high value of the η factor of plutonium in these conditions. Note that this factor is the number of neutrons emitted per neutron absorbed for the material concerned. In a critical chain reaction, of the η neutrons produced each time a neutron is absorbed by the fissile material, one neutron must be reserved for the chain reaction (next absorption in the fissile material) and the complement $\eta - 1$ must be consumed by fissionless captures (in this simple reasoning, we neglect fissions other than those in the fissile material). Some of these captures are inevitable, purely sterile captures in the various materials of the reactor, and can also be fertile captures giving a new fissile nucleus. We also see that the conversion factor must be less than $\eta - 1$; or, in other words, overbreeding is possible only if η is greater than 2, and becomes easier as this factor increases. Table 12.4 gives the orders of magnitude of the factor η for the three main fissile nuclei and the fast and thermal spectra.

Table 12.4. Order of magnitude of the reproduction factor η for the main fissile nuclei, in the fast spectrum and thermal spectrum.

Fissile nucleus	Fast spectrum	Thermal spectrum
Uranium 233	2.3	2.3
Uranium 235	1.9	2.1
Plutonium 239	2.3	2.1

It is clearly preferable to use uranium 235 in thermal neutron reactors and plutonium 239 in fast neutron reactors (in thermal neutrons, however, it is equivalent to uranium 235); for uranium 233, the two modes are equivalent. Breeding appears to be possible using plutonium in the fast spectrum and with uranium 233, but it appears to be virtually impossible with uranium 235 in view of the small margin of $\eta - 1$ with respect to 1.

12.3.3. Recycling plutonium

This basic considerations have led to the development of a strategy that has been applied since the 1940s to promoters of nuclear energy for the optimum use of natural uranium, which is a limited resource (see § 1.9): the first step is to create thermal neutron and uranium 235 reactors and reprocess the irradiated fuel in order to build up a stock of plutonium. The second step is to build fast neutron reactors to use this plutonium and reconstitute the stock, or even increase it, thanks to breeding. Ultimately, when all reactors are fast neutron reactors, the regeneration gain can be brought back down to zero (unless the production of plutonium is required to help start nuclear energy production in other countries), and all natural uranium can be used to produce energy by fission.

This is a long-term scenario, because breeding takes a long time to implement. It requires time to produce plutonium in the core and the blankets, to allow the radioactivity to decrease slightly, to reprocess the materials and recover the plutonium to constitute fuel elements, and then to load them into the core: this amounts to a total of almost twenty years from one recycling to the next. Finally, note that many cycles are necessary in order to burn all of the uranium (the 235 directly and the 238 after conversion to plutonium); this could take centuries.

With respect to current thermal neutron reactor types, such as pressurised water reactors, the energy that can be extracted from uranium is multiplied by a factor greater than 50 if this type of cycle is used. In this case, the reserves would be able to cover energy requirements not only for a few decades, but for millennia. These two aspects of the problem are therefore consistent, and both represent a complete shift of the horizon with respect to the usual industrial prospects.

France, in particular, promoted this plutonium strategy very vigorously, which influenced the following series of industrial decisions: *choice of natural uranium-graphite gas* (UNGG)[16] reactor type with the G1, G2, and G3 reactors at Marcoule, and then the six *Électricité de France* (EdF) reactors on the Loire and at Bugey, adoption of a "closed" fuel cycle, and therefore *development of reprocessing technology*, and a concentration of R & D efforts *on fast neutron reactors* (FNR) with Rapsodie at Cadarache, Phenix at Marcoule, and Superphenix[17] at Creys-Malvile.

The scenario did not turn out as expected: FNRs were not developed as quickly as expected, and, towards the end of the 1960s, France abandoned UNGG reactors, and chose instead to proceed with pressurised water reactors (PWRs). Framatome, assisted by the CEA (French atomic energy commission), acquired a licence from Westinghouse, "Frenchified" PWR technology, and finally freed itself from the licence. Between 1977 and 1999, 58 PWR plants (three-loop 900 MWe models and 1300 and 1450 MWe four-loop models) were put in service. These plants now account for three-quarters of French electricity production. The reprocessing policy, implemented for UNGG reactors, was pursued with the extension of the plants at La Hague for PWR reactors. It soon became apparent that the production of plutonium by the reprocessing of irradiated PWR fuel (a dozen tonnes per year) far exceeded the requirements of the existing FNRs. That is why the recycling of plutonium in the PWRs themselves was studied and then implemented (in 1987) in some of the 900 MWe reactors. This created a few problems, which will be mentioned again in Chapter 18, and is certainly a less attractive proposition than FNR

[16] There was obviously also the intention of someday acquiring nuclear weapons. At that time, France did not have uranium enrichment technology, which ruled out any water reactor types.

[17] The 1997 decision to shut down Superphenix was not taken purely for technical reasons (a few difficulties were encountered, which is normal for a prototype) or economic reasons (there was every reason to think that the plant would become profitable through its electricity production; note that the core in place was not irradiated very much and that a new core had been supplied), but also for vote-garnering reasons.

recycling, but is still better than storing plutonium whilst waiting for some hypothetical future use[18].

The plutonium from PWRs is currently recycled only once in a PWR, which does not really solve the problem. The studies that have been carried out to try to improve on this situation are mentioned in Chapter 18.

12.3.4. Thorium-uranium 233 cycle

Thorium 232 is the other natural fertile nucleus. It leads to uranium 233, which has good neutron physics properties (low capture/absorption ratio). At equilibrium, a thorium 232-uranium 233 cycle can be self-sustaining, or even breeding (for example, in high-temperature graphite "HTR" reactors, using the fuel in the form of small graphite-coated particles). Because thorium is a little more abundant than uranium, this represents a potentially significant resource.

This cycle has been studied, but so far not put into practice very much. There are a few important points to note:

- protactinium 233 (the precursor of uranium 233) has a relatively long half-life (27 days, compared to 2.3 days for neptunium 239 giving plutonium 239); moreover, protactinium has a significant cross-section (40 barns) for thermal neutrons: consequently, if the flux is too high[19], the yield of the conversion of thorium 232 into uranium 233 is lowered because of the neutron captures by protactinium 233;
- as has already been pointed out, the thorium 232-uranium 233 cycle must be primed with another fissile material;
- it would be possible to reprocess the thorium 232-uranium 233 fuel, but this would require technological developments;
- in this cycle, a small quantity of uranium 232 is formed: its radioactive descendants include very high-energy gamma photon emitters, such as thallium 208 in particular, that would raise some tricky radiation protection problems when reprocessing and recycling materials.

12.3.5. Incineration of nuclear waste in reactors

To conclude this chapter, let us briefly mention the problems of incinerating nuclear waste in reactors.

The waste produced by the nuclear industry (energy and other applications) includes radioactive products with short half-lives (up to thirty years) which cause no long-term problem, since they will have completely disappeared spontaneously after a few centuries, and long half-life products that create both technical[20] and ethical problems (the

[18] In addition to the expense of this type of storage, it is important to note that it causes the loss of the (fissile) plutonium 241 part, which transforms itself by radioactivity into (non fissile) americium 241 with a half-life of 15 years.

[19] The problem arises if the flux significantly exceeds that of a PWR.

[20] In discussing nuclear waste problems, it is important to bear in mind the fact that the activity is inversely proportional to the half-life. In particular, waste with a long half-life is low-activity waste.

problem of the legacy to future generations). These latter products are essentially minor actinides (neptunium, americium, curium) that are long half-life alpha emitters. There can also be a few beta-emitting long-life fission products (LLFP): selenium 79, zirconium 93, technetium 99, tin 126, iodine 129, and caesium 135.

One avenue of research to solve this problem is the incineration of these products, i.e. irradiation by neutrons until fission or the formation of a stable or short half-life nucleus. This incineration could be carried out in dedicated machines (critical cores or hybrid systems[21]) or in industrial fast-neutron or thermal neutron reactors (this would be a sensible means other than conversion to use the $\eta - 1$ neutrons left available by the chain reaction). Theoretical calculations and irradiations are underway to compare the possible procedures: reactor type, homogeneous recycling (the material to be incinerated is mixed with the fuel in the pellets) or heterogeneous recycling (the material to be incinerated is placed in separate capsules), etc. Incineration is unlikely to provide a complete solution to the problem of waste, but it might provide a considerable reduction in the mass that needs to be managed (for example, to be deposited in an underground storage facility).

[21] See Chapter 18.

Exercises

Exercise 12.1: formation of plutonium

Show that for a uranium fuel, and for small fluences, the concentrations of plutonium isotopes in it are approximately proportional to the

- fluence for plutonium-239,
- fluence-squared for plutonium-240,
- fluence-cubed for plutonium-241,

and give the coefficients of proportionality.

Exercise 12.2: branching in the uranium-233 chain

Calculate the branching ratio due to radioactive decay and neutron capture for species produced after proactinium-233 in thorium-232 fuel, assuming that the concentration of thorium-232 is constant, and in equilibrium with thorium-233 and protactinium-233, using the following data provided.

Data (thermal neutrons)

- Microscopic absorption cross-section of protactinium-233: 43 barns.
- Radioactive half-life of protactinium-233: 27 days.
- Flux: 10^{16}; 10^{17}; 10^{18}; then 10^{19} m^{-2} s^{-1}.

Exercise 12.3: effect of the excess of uranium-233

a) A reactor operating in the equilibrium state described in the previous exercise is shut down instantaneously. Using the data provided below, calculate how the uranium-233 concentration subsequently changes with time. How much does the reactivity increase due to this excess? To simplify the calculation, consider only the variation of the factor η.

Data (thermal neutrons)

- When the shutdown occurs, the fuel is a mixture composed of 3% uranium-233 and 97% thorium-232.
- Uranium-233: $\sigma_a = 580$ barns; $\eta = 2.3$.
- Thorium-232: $\sigma_a = 7.4$ barns.

b) Compare the previous result with the reactivity excess due to plutonium-239 that occurs in a uranium reactor.

Data (thermal neutrons)

- When the reactor is shut down, the fuel is a mixture of 3% uranium-235, 97% uranium-238, and negligible plutonium-239.

- Uranium-235: $\sigma_a = 680$ barns; $\eta = 2.1$.
- Uranium-238: $\sigma_a = 2.7$ barns.
- Plutonium-239: $\sigma_a = 1010$ barns; $\eta = 2.1$.

c) Compare approximately the previous two reactivity effects and their time constants with the effect of samarium excess after shutdown (see exercise **11.5**).

Exercise 12.4: equilibrium compositions

a) In fuel containing uranium-238 and plutonium isotopes 239, 240, 241, and 242, show that equilibrium can be achieved when the relative concentrations of these nuclides are independent of the fluence, neglecting consumption of uranium-238.

b) Estimate the proportion of the fuel that is plutonium $N_{Pu}/(N_U + N_{Pu})$ at equilibrium, and the isotopic composition of the plutonium.

c) Calculate the infinite multiplication factor of the fuel in part **(b)**, assuming that 5% of absorbed neutrons are captured outside the fuel.

d) Is it possible to envisage a uranium reactor that would have constant fuel composition?

e) Describe the outcome when the fuel consists of a mixture of thorium-232 and uranium-233.

Data (one group averages)

The indices have their usual meaning.

- *Uranium-238-plutonium cycle*
  - $\sigma_{a9}/\sigma_{a8} = 200/3$; $\sigma_{a0}/\sigma_{a9} = 1/4$; $\sigma_{a1}/\sigma_{a0} = 5$; $\sigma_{a2}/\sigma_{a1} = 1/9$.
  - Uranium-238: $\sigma_{c8}/\sigma_{a8} = 9/10$; $\eta_8 = 0.27$.
  - Plutonium-239: $\sigma_{c9}/\sigma_{a9} = 1/4$; $\eta_9 = 2.1$.
  - Plutonium-240: $\sigma_{c0}/\sigma_{a0} = 1$.
  - Plutonium-241: $\sigma_{c1}/\sigma_{a1} = 1/4$; $\eta_1 = 2.1$; radioactive decay neglected.
  - Plutonium-242: $\sigma_{c2}/\sigma_{a2} = 1$.
- *Thorium-232-uranium-233 cycle*
  - $\sigma_{a3}/\sigma_{a2} = 30$.
  - Thorium-232: $\sigma_{c2}/\sigma_{a2} = 1$.
  - Uranium-233: $\eta_3 = 2.3$.

Exercise 12.5: derivative at the origin of the reproduction factor

Find the derivative at the origin of the reproduction factor η with respect to the fluence s for a uranium fuel, assuming that all neutrons have thermal energies. At what value of the conversion factor C does the derivative become positive? Note that C takes into account resonant captures.

Compare a natural uranium-gas-graphite core, where $N_5/[N_5 + N_8] = 0.72\%$ and $C = 0.85$, with a PWR core, where $N_5/[N_5 + N_8] = 3.7\%$ and $C = 0.55$.

Thermal neutron data

- Uranium-235: σ_a = 680 barns; η = 2.1.
- Uranium-238: σ_a = 2.7 barns.
- Plutonium-239: σ_a = 1010 barns; η = 2.1.

Exercise 12.6: doubling time for a group of generating stations powered by breeder reactors

Consider a group of generating stations powered by breeder reactors with a total power output $P(t)$ as a function of time t. Their reactors contains a mass m of fuel per unit of power, and consume a mass c per unit of power and unit of time. These reactors achive a *positive* global regeneration gain G, by virtue of their breeder design.

a) Assume that each *available* atom of fuel—i.e. produced by conversion beyond what is consumed—is immediately recycled to increase the power of the station.
Also assume that P is a continuous variable. Derive and integrate the time-dependent differential equation for P. How long is the ideal doubling time D_0 of the station, i.e. the time to double the power output as a consequence of the gain G, based on the present assumptions?

b) In reality, additional time d is taken to recycle used fuel in the reprocessing plants, including operations such as transportation and storage.

Show the equation governing P when the time d is taken into account, integrate it, and compare the true doubling time D with D_0.

Data (optimized fast breeder reactors)

- m/c = 10 years.
- G = 0.2.
- d = 5 years.

Exercise 12.7: use of fissile material

Assume that the reactors at the generating stations use a fuel consisting of a mixture of a fissile and fertile materials, where q is the fraction of fissile material, and the remainder is fertile material. During irradiation in the reactors, a proportion p of the initial fissile material is consumed and Cp of new fissile material is created by transmutation. All the irradiated fuel is reprocessed and recycled, i.e. it is a closed-cycle system. There are assumed to be no obstacles to achieving unlimited recycling. Neglecting the losses at the reprocessing plants, and assuming all fissile nuclides (e.g. uranium-235 and plutonium-239) can be classified as usable fuel, calculate the amount of energy generated by the plants with a closed-cycle system, and compare it with the amount of energy generated by the plants with a open-cycle system—where no recycling occurs—for the following cases, using the information provided in Table 12.1.

a) Reactors where the conversion ratio $C < 1$, i.e. (*i*) water reactors ($C = 0.55$) starting from unenriched stock; (*ii*) natural uranium-graphite reactors ($C = 0.85$).

b) Breeder reactors where the conversion ratio $C \geq 1$, starting from (*i*) natural uranium; (*ii*) plutonium only. Specifically, calculate the time taken to consume the entire fuel stock, assuming that recycling is repeated at a time interval $t_r = 10$ years—which includes irradiation, cooling, reprocessing, storage, etc.—when $C = 1$

Table 12.1. Plant information.

Fuel stock	p	q
Natural uranium	2/3	0.72%
Plutonium	2/3	15%

Solutions

Exercise 12.1: formation of plutonium

The concentrations are initially zero, and are small for small fluences. Consequently, the destruction terms in the equations can be neglected, and only the creation terms retained.

The creation term in the equation for plutonium-239 is constant; therefore, N_9 is proportional to the fluence.

The creation term in the equation for plutonium-240 is proportional to the fluence; therefore, N_0 is proportional to the fluence-squared.

Similarly, the concentration of plutonium-241 is proportional to the fluence-cubed.

The formulae are

$$N_9(s) \simeq \sigma_{c8}\, N_8\, s,$$

$$N_0(s) \simeq \sigma_{c9}\, \sigma_{c8}\, N_8\, \frac{s^2}{2},$$

$$N_1(s) \simeq \sigma_{c0}\, \sigma_{c9}\, \sigma_{c8}\, N_8\, \frac{s^3}{6}.$$

Exercise 12.2: branching in the uranium-233 chain

At equilibrium, the rates of destruction for the three nuclides, thorium-232, thorium-233 and protactinium-233, are equal:

$$\sigma_{c,02}\, \Phi\, N_{02} = \lambda_{03}\, N_{03} = (\lambda_{13} + \sigma_{c,13}\, \Phi)\, N_{13}.$$

The branching ratio—which is the same whether the equilibrium is reached or not—is the ratio between the $\sigma_{c,13}\,\Phi$ term for a species and the sum of the rates. For instance, the branching ratio towards the formation of protactinium-234, i.e. the relative loss of production of uranium-233, is $b = \sigma_{c,13}\,\Phi/(\lambda_{13} + \sigma_{c,13}\,\Phi)$. Examples are as follows.

- $\Phi = 10^{16}$ m^{-2} s^{-1} $\Longrightarrow$ $b = 0.01\%$.
- $\Phi = 10^{17}$ m^{-2} s^{-1} $\Longrightarrow$ $b = 0.14\%$.

– $\Phi = 10^{18}$ m^{-2} s^{-1} $\Longrightarrow$ $b = 1.43\%$.

– $\Phi = 10^{19}$ m^{-2} s^{-1} $\Longrightarrow$ $b = 12.64\%$.

For neutron fluxes $\gtrsim 10^{19}$ m^{-2} s^{-1}, the loss of uranium-233 production is noticeable.

Exercise 12.3: effect of the excess of uranium-233

a) At equilibrium (see previous exercise),

$$N_{03} = \frac{\sigma_{c,02}\, \Phi\, N_{02}}{\lambda_{03}}, \qquad N_{13} = \frac{\sigma_{c,02}\, \Phi\, N_{02}}{\lambda_{13} + \sigma_{c,13}\, \Phi}.$$

Following shutdown of the reactor, both these nuclides will decay into uranium-233: $\Delta N_{23} = N_{03} + N_{13}$. In practice, note that the thorium-233 concentration is negligible in comparison with the protactinium-233 concentration.

The formula for the reproduction factor is

$$\eta = \frac{\eta_{23}\, \sigma_{a,23}\, N_{23}}{\sigma_{a,23}\, N_{23} + \sigma_{a,02}\, N_{02}},$$

Its derivative is

$$\frac{\Delta\eta}{\eta} = \frac{\sigma_{a,02}\, N_{02}}{\sigma_{a,23}\, N_{23} + \sigma_{a,02}\, N_{02}}\, \frac{\Delta N_{23}}{N_{23}}.$$

Results

$\eta = 1.628$, and

– $\Phi = 10^{16}$ m^{-2} s^{-1} $\Longrightarrow$ $\Delta N_{23}/N_{23} = 0.08\%$; $\Delta\eta/\eta = 24$ pcm;

– $\Phi = 10^{17}$ m^{-2} s^{-1} $\Longrightarrow$ $\Delta N_{23}/N_{23} = 0.80\%$; $\Delta\eta/\eta = 235$ pcm;

– $\Phi = 10^{18}$ m^{-2} s^{-1} $\Longrightarrow$ $\Delta N_{23}/N_{23} = 7.94\%$; $\Delta\eta/\eta = 2318$ pcm;

– $\Phi = 10^{19}$ m^{-2} s^{-1} $\Longrightarrow$ $\Delta N_{23}/N_{23} = 70.3\%$; $\Delta\eta/\eta = 20\,544$ pcm.

For the fourth result, a first order calculation is, of course, not adequate. Notice also that η increases monotonically, due to neutron capture by protactinium-233 becoming rarer as its concentration diminishes:

$$\frac{\Delta'\eta}{\eta} = \frac{\sigma_{c,13}\, N_{13}}{\sigma_{a,23}\, N_{23} + \sigma_{a,02}\, N_{02}}, \qquad \frac{\Delta'\eta/\eta}{\Delta\eta/\eta} = \frac{\sigma_{c,13}\, N_{23}}{\sigma_{a,02}\, N_{02}}.$$

This second effect is equal to 18% of the first one; hence, the reactivity increases of the preceding examples must be multiplied by 1.18.

b) For the uranium-plutonium cycle, the reactivity effect of the excess is

$$\frac{\Delta\eta}{\eta} = \frac{\sigma_{a,49}\, \Delta N_{49}}{\sigma_{a,25}\, N_{25}} \left[\frac{\eta_{49}}{\eta_{25}} - \frac{\sigma_{a,25}\, N_{25}}{\sigma_{a,25}\, N_{25} + \sigma_{a,28}\, N_{28}} \right].$$

Results
$\eta = 1.861$ and

- $\Phi = 10^{16}\ \text{m}^{-2}\,\text{s}^{-1} \quad \Longrightarrow \quad \Delta\eta/\eta = 0.4$ pcm;
- $\Phi = 10^{17}\ \text{m}^{-2}\,\text{s}^{-1} \quad \Longrightarrow \quad \Delta\eta/\eta = 3.9$ pcm;
- $\Phi = 10^{18}\ \text{m}^{-2}\,\text{s}^{-1} \quad \Longrightarrow \quad \Delta\eta/\eta = 39$ pcm;
- $\Phi = 10^{19}\ \text{m}^{-2}\,\text{s}^{-1} \quad \Longrightarrow \quad \Delta\eta/\eta = 386$ pcm.

At normal flux levels, this effect is negligible.

c) In a PWR, the flux $\Phi \approx 2 \times 10^{17}\ \text{m}^{-2}\,\text{s}^{-1}$. Following shutdown, the samarium effect is about −250 pcm. Its kinetics is similar to that for plutonium-239. The kinetics of uranium-233, however, is more than ten times longer.

- Uranium-233 effect (27 days): $1.18 \times 469 = 553$ pcm;
- Plutonium-239 effect (2.1 days): 77 pcm;
- Samarium effect (2.2 days): −250 pcm.

Exercise 12.4: equilibrium compositions

a) The equilibrium compositions are obtained by cancellation of either the time- (t) or fluence-dependent (s) derivatives.

b) The result is as follows.

$$\begin{aligned}
N_9 &= \frac{\sigma_{c8}}{\sigma_{a9}} N_8, \\
N_0 &= \frac{\sigma_{c9}\,\sigma_{c8}}{\sigma_{a0}\,\sigma_{a9}} N_8, \\
N_1 &= \frac{\sigma_{c0}\,\sigma_{c9}\,\sigma_{c8}}{\sigma_{a1}\,\sigma_{a0}\,\sigma_{a9}} N_8, \\
N_2 &= \frac{\sigma_{c1}\,\sigma_{c0}\,\sigma_{c9}\,\sigma_{c8}}{\sigma_{a2}\,\sigma_{a1}\,\sigma_{a0}\,\sigma_{a9}} N_8.
\end{aligned}$$

At equilibrium the plutonium fraction in the fuel is $N_{Pu}/(N_U + N_{Pu}) = 5.1\%$. Its isotopic composition is

- Plutonium-239: 25%;
- Plutonium-240: 25%;
- Plutonium-241: 5%;
- Plutonium-242: 45%.

c) The infinite-multiplication factor is simply expressed using the fuel-related terms $\eta = \nu\sigma_f/\sigma_a$ and $\gamma = \sigma_c/\sigma_a$, together with a factor to account for neutron capture by other materials $f = 1/1.05$:

$$k_\infty = \frac{\eta_8 + \eta_9 + \gamma_9\gamma_0\eta_1}{1/\gamma_8 + 1 + \gamma_9 + \gamma_9\gamma_0 + \gamma_9\gamma_0\gamma_1} f.$$

The result is $k_\infty = 1.031$.

d) Unfortunately, k_∞ is too small to compensate for the neutron leakage from the core (about 3000 pcm), and poisoning by the fission products (typically 10 000 pcm at the end of the cycle). Therefore, a constant fuel composition will not work. Furthermore, a full assessment requires the temperature-dependence of fuel containing a significant amount of even-numbered plutonium isotopes to be taken into account.

e) For the thorium-232-uranium-233 cycle, with similar hypotheses, and neglecting species created via uranium-233 neutron capture, the result is

$$N_3 = \frac{\sigma_{c2}}{\sigma_{a3}} N_2,$$

and

$$k_\infty = \frac{\eta_3}{1/\gamma_2 + 1} f.$$

At equilibrium the uranium fraction in the fuel is $N_U/(N_{Th} + N_U) = 3.2\%$, and $k_\infty = 1.095$. This latter value is greater than it is for the uranium-plutonium cycle, and nearly sufficient to compensate for leakage and fission product poisoning. Thus, a smaller loss of reactivity with irradiation is expected in a thorium-232-uranium-233 reactor than in a uranium-plutonium reactor, and lies behind the idea of an energy amplifier described by Carlo Rubbia[22].

Exercise 12.5: derivative at the origin of the reproduction factor

The derivative of the expression for η at the origin, without plutonium, is

$$\frac{\Delta\eta}{\eta} = \sigma_{a5}\, s \left[\left(\frac{\eta_9}{\eta_5} - \alpha_5\right) \frac{\sigma_{a9}}{\sigma_{a5}} C - (1 - \alpha_5)\right],$$

where

$$\alpha_5 = \frac{\sigma_{a5}\, N_5}{\sigma_{a5}\, N_5 + \sigma_{a8}\, N_8}.$$

This can be simplified by assuming that $\eta_9 \simeq \eta_5$; hence,

$$\frac{\Delta\eta}{\eta} = \sigma_{a5}\, s\, (1 - \alpha_5) \left(\frac{\sigma_{a9}}{\sigma_{a5}} C - 1\right).$$

The microscopic absorption cross-section of plutonium-239 is greater than the uranium-235 one; therefore, this derivative is positive when C exceeds $\sigma_{a5}/\sigma_{a9} \simeq 2/3$. This is true even in a non-breeding system. For example, in a UNGG-type reactor $C \approx 0.85$, and this gives $1/(\sigma_{a5}\, s)\,(\Delta\eta/\eta) = 0.093$.

In water reactors, C is below this limit, and the derivative is negative. For example, $1/(\sigma_{a5}\, s)\,(\Delta\eta/\eta) = -0.017$ when $C = 0.55$.

Notice that if the preceding expressions are multiplied by the fraction e of isotope-235 in uranium, then, except for a common factor, the derivative is with respect to the mass burn-up, rather than the fluence s. Hence, the corresponding results are 0.00067 and −0.00063, respectively, while the absolute values are almost the same.

---

[22] See for example, Rubbia *et al.*, *Conceptual Design of a Fast Neutron Operated High Power Energy Amplifier*, CERN/AT/95-44 (ET) (1995).

Note also that, when the derivative of the infinite-multiplication factor k_∞ is required, proper account must be taken of the variation in the thermal utilisation factor f. Moreover, it is particularly important to include the effect of poisoning by the fission products on η.

Exercise 12.6: doubling time for a group of a generating stations powered by breeder reactors

a) The ideal system is described by

$$m\,dP = c\,G\,P\,dt.$$

Therefore,

$$P(t) = P(0)\,\exp\left(\frac{cG}{m}t\right).$$

Hence, the doubling time is

$$D_0 = \frac{m}{cG}\ln 2.$$

b) Including an additional time d in the model, simply changes the differential equation to

$$m\,dP = c\,G\,P(t-d)\,dt.$$

Its solution is still an exponential function,

$$P(t) = P(0)\,\exp(\alpha t),$$

where

$$\alpha\exp(\alpha d) = \alpha_0 = \frac{cG}{m}.$$

The doubling time is now

$$D = \frac{\ln 2}{\alpha}.$$

The results of the calculations are $1/\alpha_0 = 50$ years and $D_0 = 34.7$ years; $1/\alpha = 54.8$ years and $D = 38.0$ years.

Exercise 12.7: use of fissile material

a) For each unit mass of stock, the amount consumed by fission is qp, leaving $q(1-p+Cp)$ of fissile material and $1-q-qCp$ of fertile material at the end of irradiation.

When fissile material is recycled in full, x new fuel elements per old element are produced by the reprocessing facility, where x is defined by

$$xq = q(1-p+Cp),$$

which represents the net balance of fissile material. In making x fuel elements, the quantity of fertile material mixed with reprocessed fuel is $x(1-q)$. Hence, the unused remainder is $p(1-q-C)$. This is positive unless if $C \approx 1$ because, in practice, q is small. Essentially, q represents the natural abundance of isotope-235 in uranium.

If the fuel elements are repeatedly recycled until all the fuel is spent, then the total that can be made is

$$g = 1 + x + x^2 + x^3 + \cdots = \frac{1}{1-x} = \frac{1}{p(1-C)}.$$

Hence, in a closed-cycle system, the net proportion of the original stock utilized by fission is

$$\frac{qp}{1-x} = \frac{q}{1-C},$$

while in an open-cycle system it is qp.

When $C < 1$, the results of the calculations are as follows.

- Water reactors utilize 0.48% and 1.6% of the stock in open- and closed-cycle plants, respectively ($g = 3.3$).

- Graphite-natural uranium reactors utilize 0.48% and 4.8% of the stock in open- and closed-cycle plants, respectively ($g = 10$).

b) Irradiation of the fuel in breeder reactors provides more fissile material during each cycle than will be consumed in the next one, until the original stock is exhausted. The excess $q(1 - p + Cp) - q = q(C - 1)p$ can be 'sold' to an external customer, or used to increase the capacity of the facility (see the previous exercise). When the fertile material is recycled to the maximum extent possible, each irradiated fuel element yields x new ones, where x is chosen in such a way that any remaining fertile material in irradiated elements always is used fully in the manufacture of new ones. Therefore,

$$x(1-q) = 1 - q - qCp.$$

Thus, in a closed-cycle system, for each of the fuel elements made from the original stock, the total number of fuel elements that are made is

$$g = 1 + x + x^2 + x^3 + \cdots = \frac{1}{1-x} = \frac{1-q}{qCp}.$$

The results of the calculations are as follows.

- When the stock is natural uranium, $x = 0.9952$, and $g = 207$.

- When the stock is plutonium only, $x = 0.8824$, and $g = 8.5$.

In order to estimate the time it takes to consume all the original stock of fuel, assume that the quantity qp undergoes fission during each cycle. Therefore, the number of cycles is $n \approx 1/(qp)$ and the total time is $D \approx nt_r$.

- Starting from natural uranium, $n = 208$, and $D = 2080$ years.

- Starting from plutonium only, $n = 10$, and $D = 100$ years.

13 Temperature effects

Introduction

We have repeatedly underlined (§ 8.4.1 and 9.3.5) the importance of having negative temperature coefficients in nuclear reactors, for the sake of ease of control (stable power levels) and a certain measure of safety (a fast counter-reaction that limits the power excursion in the event of a reactivity accident). It is therefore fundamental for an engineer to have a clear understanding of these effects and to ensure that they are correctly modelled. That is why we intend to review and supplement what has already been outlined in a structured presentation of this problem.

Following a few general points and a physical analysis of the phenomena that can affect the neutron physics characteristics of a core, we provide details of the effects of temperature on the example of pressurised water reactors, the most common reactor type in France. The chapter ends with a few comments about the effects of temperature in other reactor types.

13.1. Counter-reaction loop

In general terms, the state of a reactor at a given instant can be characterised by three parameters:

- reactivity ρ of the core;
- its power output P;
- its temperature T.

The *derivative* of the power with respect to time is related to the reactivity by the kinetics equations. The temperature is related to the power, possibly with a slight time lag. There are two mechanisms that can make the reactivity vary[1]: deliberate actions by the operator (in general, absorbent movements), and the effects of temperature that we shall study in this chapter. All of this is summarised in Figure 13.1.

[1] This reasoning applies to a short term, and so the reactivity fluctuations associated with the fission products and the changes in concentrations of heavy nuclei can be ignored.

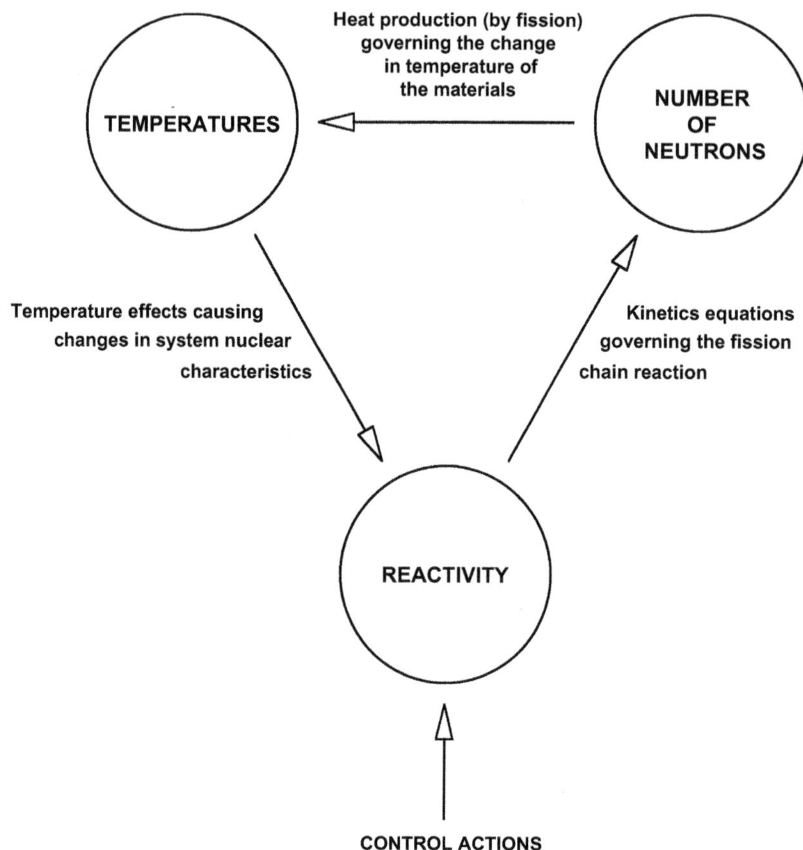

Figure 13.1. Temperature counter-reaction loop in a reactor.

A study of the consequences of changes shows that *the system is stable if the temperature coefficient is negative* (i.e. if a temperature change causes a change of the opposite sign in the reactivity), and unstable otherwise. In a stable system, control actions are necessary only to change the power level and to compensate for gradual changes in reactivity (fission products and evolution of heavy nuclei).

13.2. Definition of temperature coefficients

In general, the temperature coefficient α is defined as the logarithmic derivative of the effective multiplication factor k with respect to the reactor temperature T:

$$\alpha = \frac{1}{k}\frac{dk}{dT}. \tag{13.1}$$

The (usual) derivative of k with respect to T and the derivative of the reactivity $\rho = (k-1)/k$ with respect to T are practically equivalent definitions, since the system is always close to criticality.

In practice, α is expressed in *pcm per degree Celsius*, i.e. 10^{-5} per degree Celsius[2].

The benefit of using a log derivative is that it allows us to evaluate α by summing the coefficients for each of the factors if k is expressed in factorised form. For example, using Fermi's equation:

$$k = \frac{\varepsilon p f \eta}{1 + M^2 B^2}, \tag{13.2}$$

we write:

$$\alpha = \frac{1}{k}\frac{dk}{dT} = \frac{1}{\varepsilon}\frac{d\varepsilon}{dT} + \frac{1}{p}\frac{dp}{dT} + \frac{1}{f}\frac{df}{dT} + \frac{1}{\eta}\frac{d\eta}{dT} - \frac{M^2B^2}{1+M^2B^2}\left[\frac{1}{M^2}\frac{dM^2}{dT} + \frac{1}{B^2}\frac{dB^2}{dT}\right]. \tag{13.3}$$

The second reason to talk about coefficients in the plural is the practical need to distinguish between the materials. In particular, it is important to distinguish the fuel from the moderator both in terms of neutron physics parameters and the dynamics of temperature variations.

13.3. Physical effects that contribute to temperature coefficients

The multiplication factor of a reactor is a *reaction rate ratio.* Any rate of reaction is the integral (in space and energy) of the product of three terms: a microscopic cross-section, a concentration, and a flux. Each of these terms can vary with temperature, and so there are three fundamental types of temperature effect: one associated with each of these variations.

1/ The *Doppler Effect,* already described (§ 8.4), is related to a change in microscopic cross-sections caused by the thermal agitation of atoms, acting essentially in the form of resonances. The chief practical consequence of this effect is the *increase in resonant absorption of the heavy nuclei in the fuel,* particularly *resonant capture by uranium 238.* This effect is therefore associated with the *temperature of the fuel,* and is characterised by a negative coefficient (order of magnitude: -1 to -3 pcm/°C). Because it is also an effect that occurs quickly, since the fuel temperature changes almost instantaneously if the power varies, the Doppler Effect is the main phenomenon that ensures power stability.

In the event of a disturbance, the power returns to the equilibrium value so that at the associated temperature the multiplication factor is exactly equal to one. This return to equilibrium is achieved with a time constant on the order of magnitude of the time constants of delayed neutrons, i.e. a few tens of seconds.

2/ The concentrations vary with temperature because of the effects of thermal expansion and possibly boiling. Expansion affects all materials, but liquids more than solids. The associated coefficients vary a great deal according to material and reactor type. One of the most notable effects of expansion is the one associated with water in pressurised water reactors (see below).

[2] Be aware of coefficients expressed per degree Fahrenheit, often found in American publications.

3/ The *spectrum effects* associated with flux variations are a little more difficult to analyse: because the multiplication factor (or one of the factors expressing it) is a reaction rate ratio. The flux level has no effect; only the *distribution* of the neutrons in energy and space. This can affect the multiplication factor when, and only when, the cross-sections in the numerator and denominator do not vary proportionally, i.e. if the *average* cross-sections do not vary in the same way.

The spectrum can vary because of the effects mentioned above, and particularly because of expansion effects. Note that these spectrum changes result not only from absorption variations, but also from diffusion and energy transfer variations.

There is another spectrum effect in thermal neutron reactors that is nuclear in origin, and could therefore be placed in the first category: the modification of transfers in the thermalisation domain if the temperature of the thermalising materials (essentially the moderator) varies. If the temperature increases, this results in the *hardening of the thermal neutron spectrum* (moving towards higher energies) associated with the shift of the Maxwell spectrum.

Like other spectrum effects, this does not affect the reactivity directly, because it acts upon scattering but not absorption. It can, however, affect it via the average cross-sections. The following figure shows that effects can be observed on the factors f and η because the cross-sections of the main fissile nuclei do not really obey the 1/v relationship, and the fission and capture cross-sections do not vary proportionally.

The effect on the factor η, in the case of uranium fuel, is easier to analyse if this factor is written in the following form:

$$\eta = \nu_5 \frac{\bar{v}\bar{\sigma}_{f5}}{\bar{v}\bar{\sigma}_{a5}} \frac{1}{1+\zeta}, \qquad \zeta = \frac{N_8 \bar{v}\bar{\sigma}_{a8}}{N_5 \bar{v}\bar{\sigma}_{a5}}. \tag{13.4}$$

These equations show that this factor does not vary to the first order because the fission and absorption cross-sections are approximately proportional to 1/v and the product of average speed and the average cross-section does not change in these conditions if the spectrum moves. The effect of temperature on this factor is a second-order effect caused by *deviations* of the cross-section curves with respect to the 1/v relationship. If the temperature of the *thermalising material* increases, the average speed increases, which means a shift to the right in Figure 13.2.

- The factor ν_5 does not vary in the thermal domain (the variation is on the order of 0.14 per MeV).
- The second factor tends to decrease slightly (see Figure 13.2).
- The third factor also (Figure 13.2: ζ increases).

The associated temperature coefficient is therefore negative and, in practice, on the order of a few pcm per °C (absolute value).

For a plutonium fuel, the third factor, on the contrary, increases.

In practice, this third term is the one that dominates, and this coefficient is slightly positive.

Regarding the thermal utilisation factor, the analysis is simpler if it is written in the following form:

$$\frac{1}{f} - 1 = \frac{V_m \bar{v}\bar{\Sigma}_{am} \Phi_m}{V_f \bar{v}\bar{\Sigma}_{af} \Phi_f}. \tag{13.5}$$

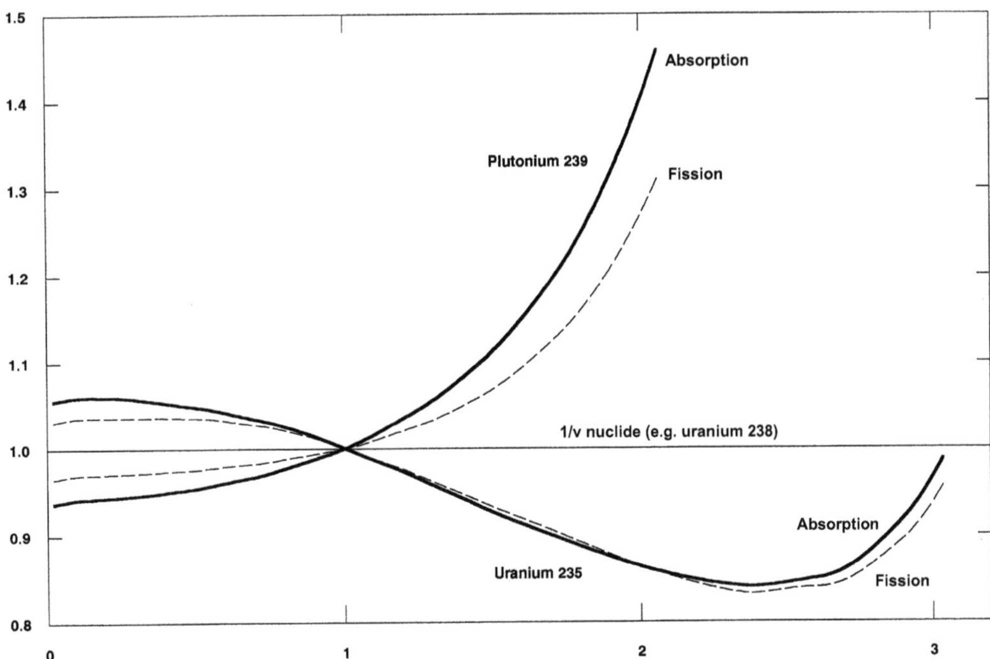

Figure 13.2. Cross-section × velocity products (normalised to the values at 2200 m/s) as a function of velocity (normalised to 2200 m/s) for uranium 235 and plutonium 239 in the thermal domain.

If the temperature of the *thermalising material* increases:

- the ratio of cross-sections increases (uranium) and *f* decreases (and inversely in the case of plutonium);
- the disadvantage factor decreases (in both cases) because absorption in the fuel is less intense (1/v relationship giving the first-order behaviour) and *f* increases[3].

In practice, both of these effects are weak.

13.4. Effects of temperature on reactivity in pressurised water reactors

in water reactors, as in all reactors, the stabilising Doppler Effect will be observed (on the order of −2 pcm/°C in this case). And, as in all thermal neutron reactors, there will be the effects of a few pcm/°C associated with ηf (negative coefficient in the case of uranium and positive coefficient in the case of MOX). Even though the Doppler Effect remains significant, although its absolute value is low, because it is related to the temperature of the *fuel* and is therefore the fastest, the spectrum effects on ηf related to the *water*

[3] This is an example of a "space spectrum", rather than energy spectrum, effect.

temperature are almost negligible in practice compared to a much more significant effect also related to the *water* temperature: the expansion effect[4].

The density ρ of liquid water passes through a maximum at 4 °C. At higher temperatures, this density decreases: gradually to begin with, and then at an accelerating rate as boiling point is approached (345 °C at a pressure of 155 bars). At the nominal operating conditions of a PWR, around 310 °C, the density (specific mass normalised to the cold specific mass) is only about 0.7, and decreases by approximately a quarter of a percent (250 pcm) per degree Celsius.

By applying a very similar argument to the one used for optimum moderation (§ 9.3.4), we can see that this expansion will essentially affect the factors p and f (neglecting the effect on η induced by the thermal spectrum change due to expansion).

If the water temperature increases, and therefore if the water expands, the resonance escape probability decreases because the neutrons are less effectively slowed down; the associated temperature coefficient is *negative*. This is confirmed by examining the usual formula for p (§ 8.3.5)[5]:

$$p = \exp\left[-\frac{V_f N_f I_{\text{eff}}}{V_m N_m (\xi\sigma_s)_m}\right], \tag{13.6}$$

in which N_m must be reduced in order to analyse the effect of an increase in water temperature. Approximate value of this coefficient: −50 pcm/°C.

The thermal utilisation factor, on the other hand, improves because there will be less moderator to capture neutrons in a sterile way. This is confirmed by the formula for the factor f (§ 9.3.2):

$$\frac{1}{f} - 1 = \frac{V_m N_m \sigma_{am} \Phi_m}{V_f N_f \sigma_{af} \Phi_f} \tag{13.7}$$

(if N_m decreases, f increases). Neglecting variations in the disadvantage factor, we differentiate the formula and note that N_m is proportional to ρ, and we find:

$$\frac{1}{f}\frac{df}{dT} = -(1-f)\frac{1}{\rho}\frac{d\rho}{dT}, \tag{13.8}$$

(which is positive, since f is less than 1 and ρ decreases with temperature). Note that this coefficient is proportional to $1 - f$, i.e. approximately proportional to Σ_{am}, the macroscopic cross-section of capture by the moderator. If an aqueous boron solution is added, this cross-section increases, and so does this positive coefficient. Orders of magnitude: +15 pcm/°C without boron; +45 pcm/°C with 1000 ppm boron. This considerable effect obviously comes from the high capture ability of boron, and the fact that, if used in the form of an aqueous boric acid solution, it expands as water does, which reduces its concentration and therefore its macroscopic cross-section.

We therefore see that, in the absence of boron (i.e. at the end of a cycle), the overall coefficient of expansion of water is a large negative value:

$$-50 + 15 = -35 \text{ pcm/°C},$$

[4] There are, of course, effects related to the expansion of solids, but in practice they are weak and almost negligible. We therefore base our reasoning on the assumption of constant volume of the assemblies. The expansion of water (density reduction) therefore results in the expulsion of a certain mass of water from the core.

[5] To simplify, we use the formula for a cell consisting only of a fuel and a moderator, neglecting slowing down by the fuel.

whilst with 1000 ppm of boron it practically vanishes (with the chosen values):

$$-50 + 45 = -5 \text{ pcm/°C}.$$

With a slightly higher boron concentration, this overall coefficient would become positive. Because this is unacceptable for safety reasons (§ 9.3.5), a *limit is imposed on the allowable boron concentration.* (On § 9.3.5, we saw that this limit corresponds to the concentration for which the moderation ratio adopted is the optimum.)

Note that, if the compensation for excess reactivity at the beginning of the cycle exceeds the possibilities of boric acid, an additional process must be used: *burnable poisons*. In practice, *Électricité de France* uses two types of burnable poison in its pressurised water reactors:

- For the first core (less enriched uranium than in the cycle at equilibrium), there are rods made of borate Pyrex inserted in all or some of the twenty-four guide tubes, with a boron concentration calculated to more or less vanish before the end of the cycle. At the end of the first cycle, these rods are removed.

- For the core at equilibrium, gadolinium is used in the form of an oxide mixed in appropriate proportions with the uranium oxide[6] in a certain number of rods (Figure 13.3). The gadolinium remains throughout the three or four cycles, leaving a small residual penalty after the first cycle (the consumable isotopes, characterised by a large cross-section, are 155 and 157; after neutron capture, they produce isotopes 156 and 158, which have capture that is low but not completely negligible).

Finally, we note (§ 9.2.2) that the absorption of thermal neutrons in the MOX assemblies of pressurised water reactors is much higher than in standard uranium assemblies. We have seen that the efficiency of the control absorbents is very much reduced by this (by a factor of approximately 3). (To compensate for this, it is possible to replace the SIC [silver-indium-cadmium] rods of the control bundles with boron 10-enriched $\mathbf{B_4C}$ boron carbide rods; similarly, it might be possible to enrich the boron of the boric acid with the boron 10 isotope.) Another consequence is a positive temperature coefficient on a lower f —since the absorption of the fuel is higher, $1 - f$ is lower— and therefore an overall temperature coefficient related to the expansion of water with a greater absolute value than in the standard case.

13.5. Overview of counter-reaction effects in pressurised water reactors

In reactors, the term "counter-reactions" is used to describe the couplings between neutron physics and thermal hydraulics, affecting not only the reactivity, but also the power distribution.

The neutron physics and thermal hydraulics calculations are coupled because:

- the temperature distribution depends on the heat sources, i.e. the fission distribution given by the neutron physics calculation;

[6] This uranium is less enriched than that of standard rods in order to avoid "hot spots" after the disappearance of the gadolinium.

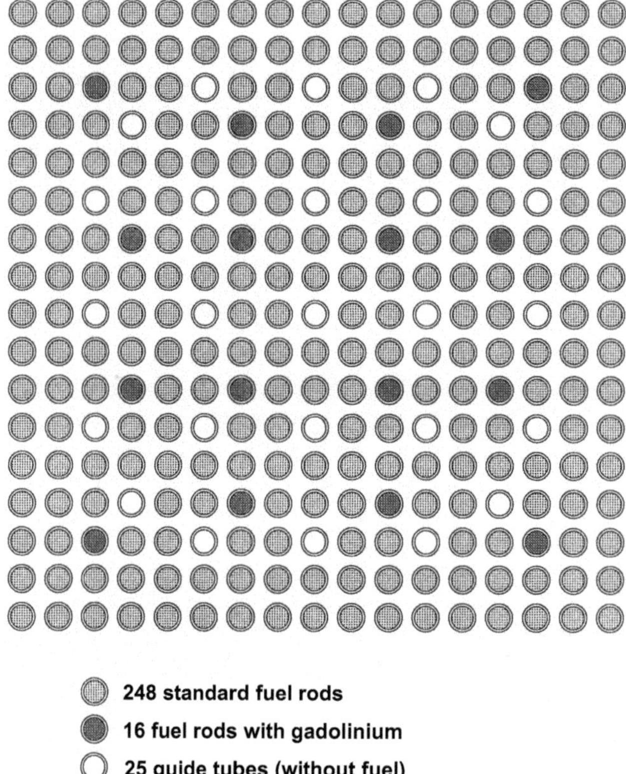

Figure 13.3. Cross-section of a pressurised water reactor assembly with 16 gadolinium rods.

- the macroscopic cross-sections involved in the neutron physics calculation depend on the temperatures via the various mechanisms already described.

In practice, this coupling is allowed for by incorporating the neutron physics and thermal hydraulic codes into a single computer system, and performing iterations by alternating the two calculations until they converge. In general, convergence is achieved after a few iterations.

Figures 13.4, 13.5, and 13.6 do not, strictly speaking, illustrate this coupling, but the need to take it into account point by point rather than as an average. These figures compare the power distributions obtained in a pressurised water reactor, taking the counter-reactions point by point on the one hand[7], and as an average only on the other hand.

In all of the results presented, the power is renormalised to the average power over the core. The axial distributions show the integrals over each of the "slices" of the finite elements calculation. The radial distributions show the integrals over the entire height for each fuel assembly.

[7] In this case the calculation was performed by the finite element method, and the counter-reaction was calculated in each element (a parallelepiped measuring approximately 10 cm along its side).

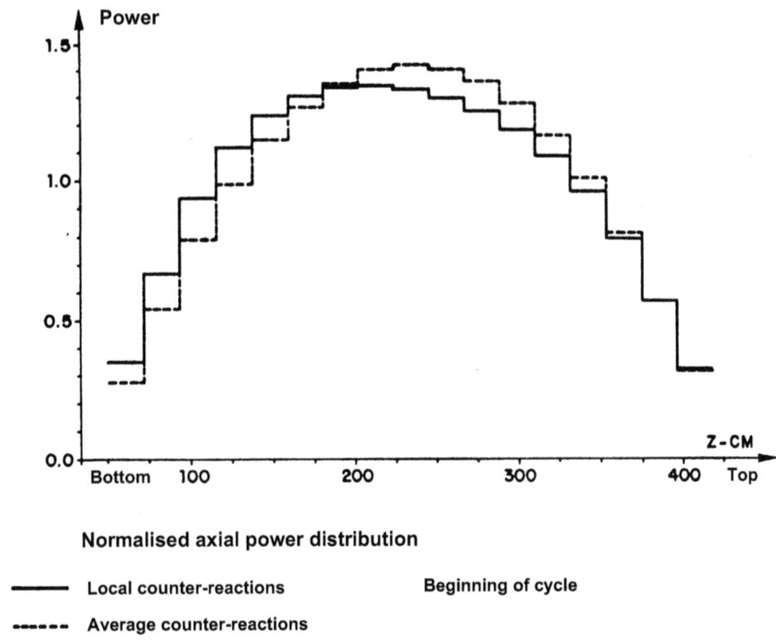

Figure 13.4. Axial power distribution at start of cycle.

- The first figure (Figure 13.4) shows the axial power distribution at the beginning of the cycle. Taking the counter-reactions as an average gives a symmetric power curve similar to the expected cosine. The counter-reactions cause a downwards axial offset to where the water temperature is lower and the multiplication factor is higher than in the average calculation.

- The second figure (Figure 13.5) shows the same axial power distribution at the end of the cycle: the same axial offset appears, but is attenuated by evolution. A far more spectacular effect is the deformation of the power distribution, which has gone from a cosine shape to a "camel-hump" shape.

This effect is not due to counter-reactions, but evolution: in the median part of the core, the flux was high, particularly at the beginning of the cycle, and there was high burn-up, such that the infinite multiplication factor fell below one. This is shown by the fact that the curvature was inverted. In the upper and lower parts, on the other hand, the burn-up has been less intense, and the disappearance of the boron solution has led to an increase in k_∞ revealed by an accentuated curvature.

- The third figure (Figure 13.6) shows the same effects in a radial presentation (an eighth of a core assumed to be symmetric). At the beginning of the cycle, we observe a flattening of the power distribution for the local counter-reaction calculation. In high-power assemblies, the temperature is higher and the reduction in k_∞ is greater than average, giving a power drop in the central part of the core. The opposite effect

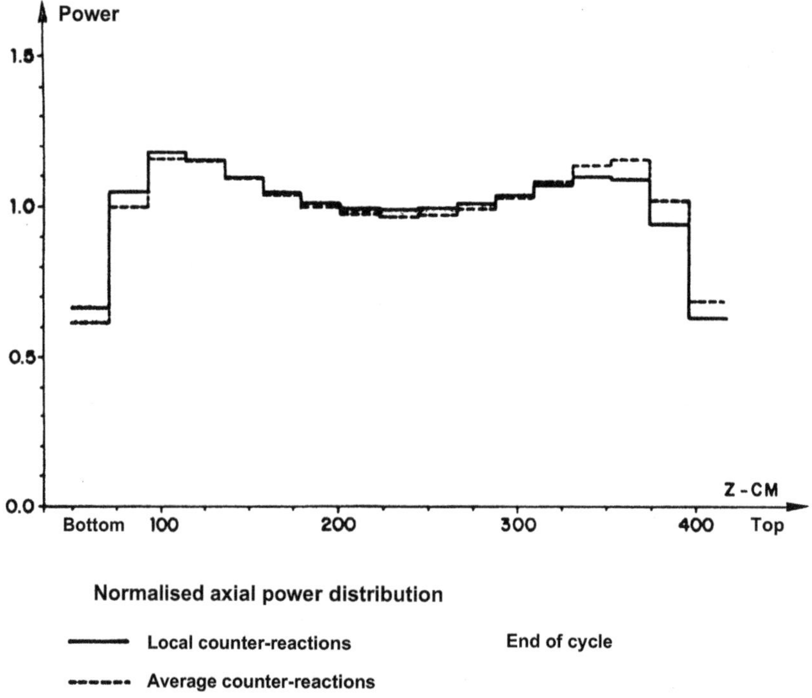

Figure 13.5. Axial power distribution at end of cycle.

is seen at the edges, where the power is lower. Note that, at the end of the cycle, these effects can practically no longer be seen, as in the axial distribution.

Generally speaking, the evolution tends to flatten the power distribution, and therefore to improve the shape factor. At a peak, the combustion (and the degradation of k_∞) is faster, which will tend to level down the peak; and the inverse effect will apply to a trough.

13.6. Temperature effects in other reactors

The effects described above will be observed to varying degrees in all reactors; here is an overview:

a) Boiling water reactors

The vacuum effects caused by the boiling of water in the core are similar to the expansion effects seen in pressurised reactors, but are accentuated because of higher average density variations (the vacuum level is zero at the bottom of the core and high at the top; the average density of the water/vapour mixture goes roughly from 0.7 to 0.3).

Under certain conditions, these strong counter-reaction effects can give rise to unstable transients.

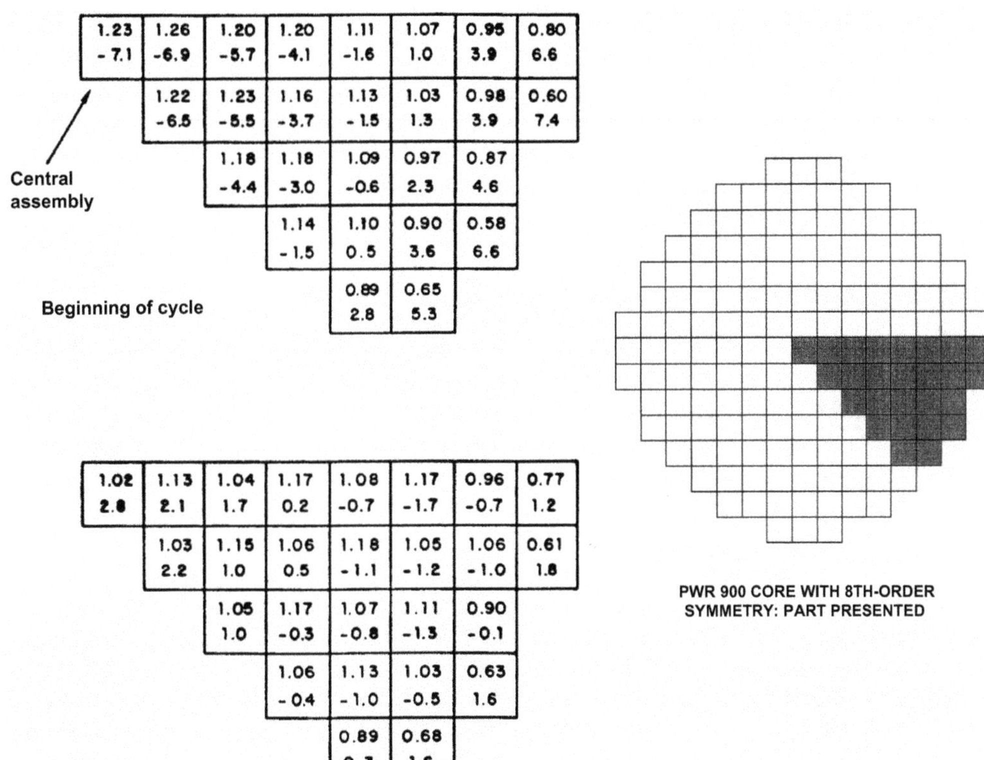

1.23 -7.1	1.26 -6.9	1.20 -5.7	1.20 -4.1	1.11 -1.6	1.07 1.0	0.95 3.9	0.80 6.6
	1.22 -6.5	1.23 -5.5	1.16 -3.7	1.13 -1.5	1.03 1.3	0.98 3.9	0.60 7.4
		1.18 -4.4	1.18 -3.0	1.09 -0.6	0.97 2.3	0.87 4.6	
			1.14 -1.5	1.10 0.5	0.90 3.6	0.58 6.6	
				0.89 2.8	0.65 5.3		

1.02 2.8	1.13 2.1	1.04 1.7	1.17 0.2	1.08 -0.7	1.17 -1.7	0.96 -0.7	0.77 1.2
	1.03 2.2	1.15 1.0	1.06 0.5	1.18 -1.1	1.05 -1.2	1.06 -1.0	0.61 1.8
		1.05 1.0	1.17 -0.3	1.07 -0.8	1.11 -1.3	0.90 -0.1	
			1.06 -0.4	1.13 -1.0	1.03 -0.5	0.63 1.6	
				0.89 0.3	0.68 1.6		

Figure 13.6. Radial power distribution at start and end of cycle.

b) CANDU-type heavy water reactors

The density effects are more modest in this case, because most of the heavy water is outside the pressure tubes and remains at low temperature. Only the pressurised heavy water, which circulates in the pressure tubes to cool the fuel rod bundles, leads to expansion effects similar to those in ordinary pressurised water reactors; but, because there is no boron solution, the (negative) effect on the resonance escape probability dominates.

c) RBMK reactors

As in the case of CANDU reactors, the expansion effect of the essential part of the moderator (here, graphite between the pressure tubes) is low. On the other hand, the effect of boiling water circulating in the pressure tubes can be significant. Even though there is no boron, the effect on the thermal utilisation factor (effect of capture by water) wins out over the effect on the resonance escape probability caused by the slightest slowing, at least in the standard concept with uranium enriched to 1.8%: that is why the abundance of uranium was raised to 2.4% in RBMK reactors (refer to the formula for the expansion coefficient on *f*, § 13.4: for more absorbent fuel, the factor *f* is higher and the positive coefficient on this factor is lower).

d) Graphite and high temperature reactors (HTR)

Because all of the materials are solid (except helium, which has practically no neutron physics function), the expansion effects are low. The thermal spectrum effects are therefore important to take into account. In particular, there are complex effects that are difficult to model related to graphite reflectors, because the spectrum is more Maxwellian there than in the core, giving rise to interface transients.

e) Fast neutron reactors

The Doppler effect even affects fast neutron reactors[8]; its absolute value is lower, however, than in thermal neutron reactors, partly because there are fewer neutrons in the domain of uranium 238 resonances (especially the large, low-energy resonances), and because there is a small positive Doppler effect related to the plutonium, for which fission dominates over capture. Although small, the overall Doppler effect guarantees the stability of the reactor and the counter-reaction in the event of a reactivity accident.

The other effects observed in these reactors are related to expansion: non-negligible effects arise from the expansion of the fuel elements and control rods, but the main effect is related to the expansion of sodium. This has three consequences:

- less capture by the sodium (positive effect similar to that on the factor *f* in thermal neutron reactors);
- hardening of the spectrum (less slowing), which generally has a positive effect, because the factor η tends to increase with average neutron energy;
- increased leakage (because the medium becomes more neutron-transparent), which is a negative effect.

In practice, the total effect is positive for a central assembly (not very sensitive to leaks) and negative for an edge assembly (more affected by leakage); on average it is positive for a large core[9] and negative for a small core.

[8] Unless the quantity of uranium 238 is reduced too much, as was recommended at the time of the initial studies of the CAPRA concept (increased plutonium consumption in fast neutron reactors).

[9] This remains acceptable from a safety point of view, particularly because the (negative) Doppler effect comes into action first.

Exercises

Exercise 13.1: temperature-dependence of the Doppler coefficient

Calculations and measurements of the effective integral show that it varies approximately linearly as a function of the square root of the absolute temperature T:

$$I_{eff}(T) \simeq I_{eff}(T_0)\left[1+\beta\left(\sqrt{T}-\sqrt{T_0}\right)\right].$$

How does the temperature coefficient vary according to this model? *Application:* compare the values in cold (293 K), unpowered hot (573 K), and nominal running (900 K) conditions, using -2.5 pcm K^{-1} for the temperature coefficient in the third case.

Exercise 13.2: stability of a reactor

To treat the time-dependence of power for a reactor, taking into account the feedback effects due to temperature (e.g. the Doppler effect), the following approximations can be made.

- point kinetics with only one group of delayed neutrons,
- zero lifetime for prompt neutrons (see Chapter **4**),
- power P proportional to the number N of neutrons,
- temperature T proportional to the power,
- linear variation of the reactivity ρ with temperature T (this is correct for small variations).

a) Show that, according to this model, the reactivity is

$$\rho(t)=\alpha\,[N_0-N(t)],$$

where α is a coefficient (assumed to be positive) characterizing the temperature feedback, and N_0 is the number of neutrons N for the power level that yields zero reactivity.

b) Show the equation governing $N(t)$.

c) Integrate this equation (note that the variables are separated), where $N(0)$ is defined as the initial value of N at $t = 0$.

d) In pratice, the absolute value of the reactivity ρ is always small in comparison with the fraction β of delayed neutrons. Based on this, simplify and integrate the differential equation. *Hint:* use $1/N$ as the unknown function.

e) When a reactor is approaching equilibrium, $N(t)$ varies at a characteristic rate ω_0. Estimate ω_0 using the following data.

- feedback effect between the unpowered hot temperature and the working temperature = -800 pcm.
- $\beta = 650$ pcm } delayed neutrons.
- $1/\lambda = 13$ s

Are the assumptions and approximations made in this exercise reasonable?

Exercise 13.3: reactivity event

A potentially hazardous situation occurs when there is a sudden increase in reactivity, for example, due to rapid removal of a control rod. Subsequently, a large, transient pulse of energy is released, possibly followed by a slow rise in power output.

To examine what happens during the initial stages of such an event, assume that at time t_i, an instantaneous increase in reactivity occurs that is larger than the fraction β of delayed neutrons. Then, apply a point kinetic model together with several further simplifications as follows.

- The initial power P_i is very small before the event.
- Neglect delayed neutrons—very few are present before the event, and very few are generated in the period after, because it is short.
- Heating of the fuel is adiabatic, i.e. there is no transfer of heat outside the fuel elements.
- The heat capacity of the fuel is independent of its temperature.
- The Doppler coefficient is independent of the fuel temperature.

Data (900 MWe PWR)

- initial power: $P_i = 1$ MW;
- mass of fuel: $m = 82$ tonnes;
- heat capacity of the fuel: $C_p = 300\ \mathrm{J\,kg^{-1}\,K^{-1}}$;
- Doppler coefficient (absolute value): $\alpha = 9\ \mathrm{pcm\,K^{-1}}$;
- neutron lifetime: $\ell = 25\ \mu\mathrm{s}$;
- initiating reactivity increase: $\rho_i = 1.5\,\beta$ with $\beta = 650$ pcm.

Remark: the paramater α represents an 'effective' value that compensates for perturbations to the power-density distribution.

a) According to the present model, show that the power P is governed by the differential equation

$$\frac{dP}{dt} = \frac{\rho - \beta}{\ell} P.$$

b) Define $\theta = T - T_i$ to be the rise in temperature of the fuel above its initial temperature T_i. Show the relationship linking θ with P, within the adiabatic approximation.

c) Define r to be the excess of the initial reactivity ρ_i above β. Noticing that $\rho - \beta = r - \alpha\theta$, deduce the differential equation governing θ.
Rewrite this equation with the function h defined by

$$\theta = \frac{r}{\alpha} + h.$$

d) Integrate the previous result once, and make the constant term $\alpha h_\infty^2/(2\ell)$.

e) Integrate again (note that the variables are separated). The second integration constant is an initial time t_0.

f) Noticing that

$$\frac{dh}{dt} = \frac{P}{mC_p},$$

find the power as a function of the time $P(t)$, and the temperature change θ. *Hint:* express the integration constant h_∞^2 as a function of the power level P_0 at time t_0.

g) Find the two remaining constants P_0 and t_0 by solving the previous expressions for $P(t)$ and θ at $t = t_i$ when $P(t) = P_i$, and $\theta = 0$. Note that P_0 is easily obtained using the identity

$$1 - \tanh^2 x = \frac{1}{\cosh^2 x}.$$

It is possible to simplify the expression for t_0 by assuming that $P_i \ll P_0$.

h) Describe the general behaviour of P and θ versus t. What physical interpretation can be placed on the parameters t_0 and P_0?

i) Estimate numerically P_0, $t_0 - t_i$ and the peak temperature. How long is the characteristic period for the event?

Exercise 13.4: condition for a negative water expansion coefficient

Using the classical formulae for the factors p and f, and neglecting variations in the other terms that describe the neutron balance, find the condition for a negative expansion coefficient of the water in a pressurized water reactor.

a) Verify that this condition is satisfied for a reactor without boron in solution ($p = 0.78, f = 0.95$).

b) Above which boron concentration is the condition no longer satisfied? Assume that the boron efficiency is independent of its concentration and is −10 pcm/ppm.)

N.B: see also **Exercise 9.5**.

Exercise 13.5: effect of the thermal expansion of water on the leakage

Using the usual formulae

$$k = \frac{k_\infty}{1 + M^2 B^2}, \qquad M^2 = \frac{D}{\Sigma_a}, \qquad D = \frac{1}{3\Sigma_t},$$

and the data provided, calculate the effect on the temperature-dependence of the migration area due to the thermal expansion of water. Assume that the water—which expands—accounts for two thirds of the homogenized values of Σ_a and Σ_t, and the fuel and cladding—which do not expand—for one third.

Data

a) Power reactor, water temperature $\approx 300\,°C$, leakage rate $M^2B^2 = 3000$ pcm, and expansion coefficient = -250 pcm K^{-1}.

b) Critical facility, water temperature $\approx 50\,°C$, leakage rate $M^2B^2 = 30\,000$ pcm, and expansion coefficient = -50 pcm K^{-1}.

Solutions

Exercise 13.1: temperature-dependence of the Doppler coefficient

The temperature coefficient α is inversely proportional to the square root of the absolute temperature; hence,

- $T = 293$ K $\Longrightarrow$ $\alpha = -4.4$ pcm K^{-1};
- $T = 573$ K $\Longrightarrow$ $\alpha = -3.1$ pcm K^{-1};
- $T = 900$ K $\Longrightarrow$ $\alpha = -2.5$ pcm K^{-1}.

Exercise 13.2: stability of a reactor

a) The formula for the reactivity can be derived simply from the basic assumptions of the model.

b) Starting from the slow kinetic equation,

$$\frac{d}{dt}\left[(\beta-\rho)\,N\right] = \lambda\rho N,$$

and substituting ρ by the reactivity formula gives

$$\frac{d}{dt}\left[(\beta-\alpha N_0+\alpha N)\,N\right] = \lambda\alpha\,(N_0-N)\,.$$

c) After integration,

$$\lambda t = \frac{\beta+\alpha N_0}{\alpha N_0}\ln\frac{N_0-N(0)}{N_0-N(t)} + \frac{\beta-\alpha N_0}{\alpha N_0}\ln\frac{N(t)}{N(0)}.$$

Solving this expression for $N(0)$ at $t = 0$ yields the time-dependent behaviour $N(t)$. For all $N(0)$, this asymptotically goes to N_0, which corresponds to the power level where $\rho = 0$.

d) Since $\beta \gg \rho$,

$$\beta\frac{dN}{dt} = \lambda\rho N, \qquad \beta\frac{dN}{dt} = \alpha\lambda\rho(N_0-N)N.$$

Therefore,

$$N(t) = \frac{N_0}{1+\left[N_0/N(0)-1\right]\,\exp(-\omega_0 t)},$$

where

$$\omega_0 = \frac{\alpha N_0 \lambda}{\beta}.$$

e) Hence, the result, $\omega_0 = 0.094\ s^{-1}$, demonstrates that the slow kinetic approximation is valid.

Exercise 13.3: reactivity event

a) Within the point kinetic model, and when delayed neutrons are neglected, the equation for the neutrons reduces to

$$\frac{dN}{dt} = \frac{\rho - \beta}{\ell} N.$$

P can simply replace N provided they are assumed to be proportional.

b) The adiabatic approximation means all the energy generated during the event heats the fuel; hence,

$$mC_p\,\theta(t) = \int_0^t P(t')\,dt', \qquad mC_p\,\frac{d\theta}{dt} = P.$$

c) Substitute the second equation in part **(b)** into the equation for the power P found in part **(a)**, then replace the term $\rho - \beta$ with $r - \alpha\theta$, to obtain

$$\frac{d^2\theta}{dt^2} = \frac{r - \alpha\theta}{\ell}\,\frac{d\theta}{dt}.$$

Thus, when the unknown function θ is replaced by h, the result is

$$\frac{d^2 h}{dt^2} = -\frac{\alpha}{\ell} h \frac{dh}{dt}.$$

d) The first integration is straightforward, giving

$$\frac{dh}{dt} = -\frac{\alpha h^2}{2\ell} + C^t.$$

Replacing the constant of integration with $\alpha h_\infty^2/(2\ell)$ gives

$$\frac{dh}{dt} = \frac{\alpha}{2\ell}(h_\infty^2 - h^2).$$

e) The variables are then separated, which permits integration for a second time:

$$\frac{1}{h_\infty}\,\arg\tanh\frac{h}{h_\infty} = \frac{\alpha}{2\ell}(t - t_0), \qquad h = h_\infty \tanh\left[\frac{\alpha h_\infty (t - t_0)}{2\ell}\right].$$

f) Owing to $dh/dt = P/(mC_p)$, it can be seen that the power is

$$P = \frac{\alpha m C_p}{2\ell}\,\frac{h_\infty^2}{\cosh^2\left[\alpha h_\infty (t - t_0)/(2\ell)\right]}.$$

At time $t = t_0$, this gives

$$h_\infty^2 = \frac{2\ell P_0}{\alpha m C_p}.$$

Combining the above results gives

$$P = \frac{P_0}{\cosh^2\left[\sqrt{(\alpha P_0)/(2\ell m C_p)}(t - t_0)\right]}.$$

Then, using again $\theta = r/\alpha + h$, taking the expression for h from the result in part **(e)**, and combining it with the square-root of the formula above for h_∞^2, gives

$$\theta = \frac{r}{\alpha} + \sqrt{(2\ell P_0)/(\alpha m C_p)}\, \tanh\left[\sqrt{(\alpha P_0)/(2\ell m C_p)}(t - t_0)\right].$$

g) When $t = t_i$, $\theta = 0$. Using this in the previous expression for θ, and combining it with the previous but one expression for P, together with the hyperbolic-function identity given in the question, gives the result

$$P_0 = P_i + \frac{m C_p r^2}{2\ell\alpha}.$$

Therefore,

$$t_0 - t_i = \sqrt{(2\ell m C_p)/(\alpha P_0)}\, \operatorname{arg\,cosh} \sqrt{P_0/P_i}.$$

Recall it can be assumed that $P_i \ll P_0$; hence, $P_0 \simeq (m C_p r^2)/(2\ell\alpha)$, and $\operatorname{arg\,cosh} x \simeq \ln 2x$. Thus,

$$t_0 - t_i \simeq \frac{2\ell}{r} \ln\left(2\sqrt{P_0/P_i}\right).$$

h) Starting at time t_i, the power rises rapidly until time t_0, whereupon it reaches its peak value P_0, and the reactivity due to prompt neutrons vanishes. The fall in power after the peak mirrors its rise, symmetrically. Consequently, the energy released during the event generates a rise in temperature that takes an antisymmetric form, centred on time t_0. Thus, provided the initial power is small, $P(t)$ takes the form essentially of a sharp pulse, and the total rise in temperature is divided nearly equally between the periods before and after t_0.

i) Assuming, as before, that the initial power is small, it can be seen from the previous results that

$$P_0 \simeq \frac{m C_p r^2}{2\ell\alpha}, \qquad \theta_0 \simeq \frac{r}{\alpha}, \qquad \theta_\infty \simeq 2\theta_0 \simeq 2\frac{r}{\alpha}.$$

The smaller the initial power P_i, the longer the interval is from t_i to t_0. More significant is the time constant appearing in the arguments of the hyperbolic functions,

$$t_p \simeq \frac{2\ell}{r},$$

which represents the duration of the power pulse. Using the data provided gives the following results.

- $P_0 = 58$ GW;
- $\theta_0 = 36\,°C$; $\theta_\infty = 72\,°C$;

- $t_p = 15$ ms;
- $t_0 - t_i = 6.2\, t_p = 95$ ms.

The peak power is huge; however, since its duration is short, the temperature increase is modest, and the consequences of the event are relatively minor, at least in this example.

Exercise 13.4: condition for a negative water expansion coefficient

The temperature coefficients for p and f due to the thermal expansion of water are

$$\frac{1}{p}\frac{\partial p}{\partial T} = \ln\frac{1}{p}\,\frac{1}{\rho}\frac{\partial \rho}{\partial T} \qquad \frac{1}{f}\frac{\partial f}{\partial T} = -(1-f)\,\frac{1}{\rho}\frac{\partial \rho}{\partial T},$$

respectively, where ρ represents the density of water. The derivative of ρ is negative; therefore, the condition is satisfied when

$$\ln\frac{1}{p} > 1 - f.$$

The absolute value of the derivative cancels because it appears in each component.

a) $\ln(1/0.78) = 0.2485$, and $1 - 0.95 = 0.05$.

b) The presence of boron modifies f. The maximum value ($1 - 0.2485 = 0.7515$) is exceeded when the concentration of boron is more than 2300 ppm.

Exercise 13.5: effect of the thermal expansion of water on the leakage

Differentiation yields

$$\frac{dM^2}{M^2} = \frac{dD}{D} - \frac{d\Sigma_a}{\Sigma_a}, \qquad \frac{dD}{D} = -\frac{d\Sigma_t}{\Sigma_t}.$$

In the present model, the thermal expansion of water accounts for two-thirds of each term; therefore,

$$\frac{dM^2}{M^2} = -\frac{4}{3}\frac{d\rho}{\rho}.$$

Hence,

$$\frac{1}{T}\frac{dk}{k} = \frac{4}{3}\,\frac{M^2B^2}{1+M^2B^2}\,\frac{1}{T}\frac{d\rho}{\rho}.$$

The results of the calculations are as follows.

a) Power reactor: $-1.33 \times 0.029 \times 250 = -9.7$ pcm K^{-1}.

b) Critical facility: $-1.33 \times 0.231 \times 50 = -15.4$ pcm K^{-1}.

Part II

ELEMENTS OF NEUTRON PHYSICS CALCULATIONS

14 Boltzmann equation

Introduction

The Boltzmann equation was briefly introduced in Chapter 4, with an outline of the principles for solving it (§ 3.2). This topic will now be discussed in greater detail. Firstly we review the operators involved in this equation: the collision operator and the transport operator, and then we present the principles of numerical processing for each of them. For the transport operator there are two variants: the integral approach and the differential approach, which lead to completely different numerical methods. We then examine the behaviour of the solution to this equation in fundamental mode, which is simpler than the general situation, and is often taken as a reference case. Finally, we present the probabilistic "Monte Carlo" method.

This book is intended for engineers who are likely to use these calculation codes or to work with results thus obtained, not for specialists in the development of these codes. The discussion is therefore often limited to presenting the principles of the numerical methods, without providing too many details or mathematical proofs[1].

Because numerical neutron kinetics calculations[2] are not very original compared to what is done in other fields[3] and because, except for the study of accident scenarios, all reactor design and monitoring calculations are performed in a steady-state situation, the variable *time* (t) is not used in this chapter[4].

According to convention, we use the lethargy u to describe the neutron spectrum. The other two categories of variable to be discussed are the usual space variables $\vec{r}$ (three coordinates) and phase variables $\vec{\Omega}$ (two coordinates).

---

1 Moreover, a separate book would be required to provide a reasonably complete description of the methods for solving the Boltzmann equation.

2 Kinetics as defined in Chapter 4, i.e. fast kinetics with not change in the atomic concentrations by irradiation.

3 It is, however, appropriate to point out the special nature in neutron physics of two very different timescales: the one related to prompt neutrons, and the one related to delayed neutrons. Certain kinetics codes attempt to deal with phenomena separately according to each of these scales.

4 The problems related to allowing for evolution and counter-reactions, where time is not really an argument of the flux governed by the Boltzmann equation, but simply a parameter handled by other evolution equations, will be discussed in Chapter 17.

14.1. The two forms of the Boltzmann equation

14.1.1. Incoming density, outgoing density, and neutron flux

The neutron population can be represented equally well by three density functions:

- the *density* n and the *neutron flux* $\Phi = nv$ characterise neutrons that "travel", i.e. that are followed from their starting point to their next collision;

- the *emission density*, also called *outgoing density*, here denoted by Q—often written as χ — (these letters suggest the neutrons that *quit* a point), represents the neutrons emitted by fission sources or other sources and the neutrons re-emitted by scattering;

- the *collision rate*, also called *incoming density* (coming into collision) is the product $\Psi = \Sigma \times \Phi$ of the total cross-section of the material by the neutron flux.

These three functions are connected by the two operators in the Boltzmann equation:

- The *collision operator* C is used to express the outgoing density as a function of the incoming density (if there are any independent sources S_a, they should be added to the outgoing density):
$$Q = C\Psi + S_a;$$

- The *transport operator* T is used to express the flux and therefore, after multiplication by Σ, the incoming density based on the outgoing density:
$$\Psi = TQ.$$

The Boltzmann equation is obtained by eliminating one of the densities, generally Q:

$$\Psi = T(C\Psi + S_a),$$

or, if it is preferable to work with the flux:

$$\Sigma\Phi = T[C(\Sigma\Phi) + S_a].$$

For most practical applications, the independent source can be neglected. Under these conditions, the Boltzmann equation is homogeneous, which has two consequences:

- the (non-trivial) solution cannot be obtained unless a critical condition is fulfilled (physically, this condition expresses the exact equality between the number of neutrons disappearing per unit time and the number of neutrons produced during the same unit time: a necessary equality for the steady-state regime to exist);

- if this condition is satisfied, the solution is obtained only *to within a factor* (physically, the equilibrium corresponding to steady-state can be created at any level).

14.1.2. Collision operator

The collision operator expresses the number of neutrons leaving collisions as a function of the number of neutrons entering collisions. In practice, a distinction is made between the physical processes of scattering and fission:

$$Q = D + S_f + S_a.$$

For the case of fission, if emission is assumed to be isotropic (always an allowable assumption) and that the fission spectrum[5] χ_f is independent of the energy of the neutron that caused the fission and the nucleus that underwent fission (non-mandatory assumptions), this gives:

$$S_f(\vec{r}, u, \vec{\Omega}) = \frac{1}{4\pi}\chi_f(u) \int_0^{\infty} du' \int_{(4\pi)} d^2\Omega' \nu\Sigma_f(\vec{r}, u')\Phi(\vec{r}, u', \vec{\Omega}'), \tag{14.1}$$

where the integral expresses the total number of neutrons produced by fission at a point $\vec{r}$. (All neutrons are assumed to have positive lethargy.)

The second part of the collision operator [including, where applicable, (n,2n) reactions], is written as:

$$D(\vec{r}, u, \vec{\Omega}) = \int_0^{\infty} du' \int_{(4\pi)} d^2\Omega' \Sigma_s[\vec{r}, (u', \vec{\Omega}') \to (u, \vec{\Omega})]\Phi(\vec{r}, u', \vec{\Omega}'), \tag{14.2}$$

where the integral expresses all of the transfers from any lethargy u' and direction $\vec{\Omega}'$ liable to take the neutron to the lethargy u and direction $\vec{\Omega}'$ considered on the left-hand side.

14.1.3. Transport operator (integral form)

Neutrons of lethargy u travelling in direction $\vec{\Omega}$ seen by an imaginary observer at a point O are neutrons that have left (after emission or scattering) one of the points M located a distance s upstream of the observer on the line of direction $\vec{\Omega}$ passing through O (see Figure 14.1), provided that such neutrons have not undergone any collision on the path MO.

The probability of no collisions occurring on the path MO is $\exp(-\tau)$, where τ is the integral along the line segment MO of the total cross-section of the material for neutrons of lethargy u:

$$\tau = \int_0^{s} \Sigma(\vec{r} - s'\vec{\Omega}, u)\, ds', \tag{14.3}$$

an expression where $\vec{r}$ denotes the position of the observer O. (This parameter τ is called the "optical path".)

By summing over all points M, the flux counted by the observer is found to be:

$$\Phi(\vec{r}, u, \vec{\Omega}) = \int_0^{\infty} ds \exp(-\tau) Q(\vec{r} - s\vec{\Omega}, u, \vec{\Omega}). \tag{14.4}$$

[5] The notation χ for the fission spectrum is standard: do not confuse χ_f with the outgoing density χ.

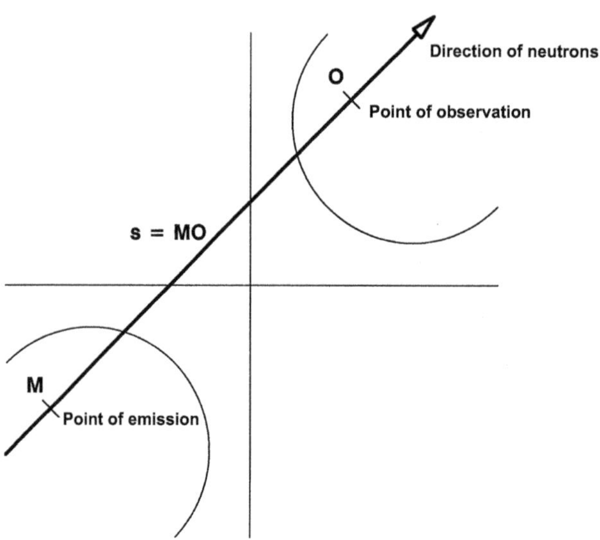

Figure 14.1. Path of neutrons of direction $\vec{\Omega}$ in a structure that can be heterogeneous.

14.1.4. Transport operator (differential form)

The differential form of the transport operator can be obtained *via* physical arguments based on an analysis of the neutron balance, like the reasoning used to obtain the diffusion equation (§ 5.1.1). Let us apply this reasoning again here in the context of a steady-state regime.

Let us consider any domain in space, D, and look at the number X of neutrons of lethargy u and direction $\vec{\Omega}$ it contains (for the sake of simplicity, this will always be understood to mean "per unit lethargy and per unit solid angle"). If a steady-state regime is established, this number does not change with time; its variation over a unit time interval is zero:

$$dX = 0.$$

This means that the algebraic sum of all the variations that can modify X is zero. Three mechanisms can affect this number X:

1/ ***Emission and re-emission*** of neutrons at lethargy u and in the direction $\vec{\Omega}$, which contribute to *increasing* X by the following amount during the unit time interval:

$$d_{(1)}X = \int_D Q(\vec{r}, u, \vec{\Omega})\, d^3r; \tag{14.5}$$

2/ ***Collisions*** between neutrons of lethargy u and direction $\vec{\Omega}$, which contribute to *decreasing* X by the following amount during the unit time interval:

$$d_{(2)}X = \int_D \Sigma(\vec{r}, u)\Phi(\vec{r}, u, \vec{\Omega})\, d^3r, \tag{14.6}$$

(absorptions eliminate neutrons, and scattering events make them go to a different lethargy and send them in a different direction);

3/ *Inputs and outputs:* Inputs contribute to *increasing* X and outputs contribute to *decreasing* it. We have seen (§ 3.1.4) that the vector $\vec{J}(\vec{r}, u, \vec{\Omega})$ is used to express this *current*: $\vec{J}(\vec{r}, u, \vec{\Omega}).\vec{N}dS$ is the *algebraic* number of crossings of a surface element dS oriented by its normal vector $\vec{N}$ (it is positive if the vectors $\vec{\Omega}$ and $\vec{N}$ form an acute angle, and therefore if crossings occur in the direction of the normal, and negative if the vectors $\vec{\Omega}$ and $\vec{N}$ form an obtuse angle, therefore if crossings occur in the opposite direction). If the normal is directed towards the outside of the domain D and if we integrate over its entire surface S, we can express the third variation of X (decreasing if positive, increasing if negative):

$$d_{(3)}X = \int_S \vec{J}(\vec{r}, u, \vec{\Omega}).\vec{N}\, dS. \tag{14.7}$$

To write this integral in an analogous form to the previous ones, i.e. in the form of a volume integral, we use the *divergence theorem:*

$$d_{(3)}X = \int_D \operatorname{div}[\vec{J}(\vec{r}, u, \vec{\Omega})]\, d^3r. \tag{14.8}$$

Finally:

$$dX = d_{(1)}X - d_{(2)}X - d_{(3)}X = 0,$$

expresses the balance in D during the unit time.

Because D can be any domain, it can be reduced to the volume element d^3r by removing the summation signs and then simplifying by d^3r :

$$Q(\vec{r}, u, \vec{\Omega}) - \Sigma(\vec{r}, u)\Phi(\vec{r}, u, \vec{\Omega}) - \operatorname{div}[\vec{J}(\vec{r}, u, \vec{\Omega})] = 0. \tag{14.9}$$

Finally, note (§ 3.1.4), that the phase current is simply the product of the phase flux by $\vec{\Omega}$; the transport operator relating Φ to Q is therefore written as follows, with the terms rearranged:

$$\operatorname{div}[\vec{\Omega}\Phi(\vec{r}, u, \vec{\Omega})] + \Sigma(\vec{r}, u)\Phi(\vec{r}, u, \vec{\Omega}) = Q(\vec{r}, u, \vec{\Omega}). \tag{14.10}$$

Note that, unlike the transport operator in integral form, the differential form of the transport operator is not written in the form: $\Psi = TQ$ (where $\Psi = \Sigma\Phi$ and where T is the transport operator), but in the following form:

$$T^{-1}\Psi = Q.$$

Note: The operator $\operatorname{div}(\vec{\Omega}\cdot)$ is called the *"streaming operator"*.

14.1.5. Equivalence between the two forms of the transport operator

It can be shown, either by differentiating the integral form or by integrating the differential form, that these two expressions for the transport operator are strictly equivalent from a mathematical point of view[6]. (Compare this to the lethargy reasoning in § 7.1.10.)

[6] If working on a finite geometry, there are a few precautions to be taken (we shall not insist on them here) when defining the boundary conditions.

Physically, this equivalence expresses the comparison of the counts performed by the observer O of Figure 14.1 and an observer O′ shifted from O by an infinitesimal distance *ds* in direction $\vec{\Omega}$: the difference in the counts, i.e. the derivative of the flux, $\mathrm{div}(\vec{\Omega}\Phi)$, along direction $\vec{\Omega}$, comes from:

a/ Neutrons emitted between O and O′, i.e. Q, seen by O′ and not by O;

b/ Neutrons undergoing a collision between O and O′, i.e. $\Sigma\Phi$, seen by O and not by O′;

which gives:

$$\mathrm{div}(\vec{\Omega}\Phi) = Q - \Sigma\Phi.$$

14.1.6. The two deterministic approaches to the Boltzmann equation

Even if the two forms of the transport operator are equivalent from a mathematical point of view, they are not equivalent from a practical point of view, i.e. in terms of the solutions, whether analytical or, more often, numerical, of the Boltzmann equation.

Each form of the equation has been the subject of many studies by mathematicians, computational scientists, and physicists. As we have mentioned, it is not appropriate to attempt an exhaustive review here[7]. We shall limit ourselves to presenting the methods that have turned out to be the most effective cost/accuracy tradeoffs[8]:

- Concerning the integral form of the transport operator, the first collision probabilities method;
- Concerning the differential form of the transport operator, the method of spherical harmonics and the method of discrete ordinates.

In both cases, the collision operator must be processed in parallel. The technique used is the multigroup approximation that was presented in Chapter 10 and which is briefly reviewed below.

14.1.7. Probabilistic approach to the Boltzmann equation

In addition to these deterministic treatments of the Boltzmann equation, probabilistic calculations can be performed, and are often valued by neutron physicists: this is the Monte Carlo approach.

One of the immediate advantages of this approach is that it eliminates the need to write out the Boltzmann equation explicitly[9].

It can also produce results that provide a reference to validate the deterministic methods, which are generally less costly to run.

[7] For more information, please use the bibliography as a guide to further reading.
[8] It has been common to refer to the "cost" of a calculation ever since the days when computer-based calculations were subcontracted out to a service provider. Today, most calculations are performed by the engineers concerned at their own workstations, and so it is more appropriate to think of the "cost" as "machine time".
[9] This remains necessary, however, if "biasing" is to be performed.

Note that this method consists in simulating the neutron paths as closely as possible and, after many simulations, in performing a statistical analysis of the results. For certain problems, it can be useful, or even essential, to "bias" the phenomena to favour the events of interest, thus improving the statistical accuracy. This probabilistic approach will be presented at the end of this chapter.

14.2. Processing the collision operator

Collisions can change both the energy and the direction of neutrons. The directional aspect is closely related to the spatial aspect, i.e. the transport, since the direction of a neutron affects the points it will be able to reach, and will be examined when transport is calculated.

In deterministic solutions of the Boltzmann equation, the energetic aspect is taken into account by a multigroup process (see Chapter 10): the flux $\Phi(u)$ and the cross-sections $\sigma(u)$, as lethargy functions, are replaced by "vectors" Φ_g and σ_g, where the group number is a whole number from 1 to N. Note that Φ_g must be interpreted as the *integral* of flux on the group g, i.e. over the interval $[u_{g-1}, u_g]$, and σ_g as the *flux-weighted average* of the cross-section in this interval. (The multigroup theory approximation relies on the fact that this average is not calculated, in practice, with strictly the exact flux.)

In this formalism, the collision operators take the following form:

– concerning the *fission operator*:

$$S_{f,g}(\vec{r}, \vec{\Omega}) = \frac{1}{4\pi}\chi_{f,g} \sum_{h=1}^{N} \nu\Sigma_{f,h}(\vec{r}) \int_{(4\pi)} \Phi_h(\vec{r}, \vec{\Omega}')d^2\Omega'; \qquad (14.11)$$

– concerning the *diffusion operator*:

$$D_g(\vec{r}, \vec{\Omega}) = \sum_{h=1}^{N} \int_{(4\pi)} \Sigma_{s,h\to g}(\vec{r}, \vec{\Omega}' \to \vec{\Omega})\Phi_h(\vec{r}, \vec{\Omega}')d^2\Omega'. \qquad (14.12)$$

(Note that these relationships would be absolutely rigorous if the multigroup cross-sections had been obtained by weighting by the exact flux.)

With regard to Monte Carlo solutions of the Boltzmann equation, the general recommendation in the simulation of collisions is a continuous, and therefore exact, treatment of the lethargy variable. A multigroup treatment can also be introduced, however, either to simplify the simulation calculations[10], or to validate a deterministic calculation that is itself multigroup.

[10] For example, a deterministic calculation by the APOLLO code can give more or less homogeneous structure multigroup cross-sections, which are then input into a Monte Carlo code.

14.3. Treatment of the integral form of the transport operator

14.3.1. Isotropic collision assumption

The transport operator T, in the integral form just written, is more difficult to handle than the inverse operator T^{-1}, which is differential. That is why the integral problem is never considered in this form in calculation codes.

The advantage of the integral approach is that, applying an assumption that is not too drastic, *the integral form lets us eliminate the directional variable $\vec{\Omega}$ from the equations,* which is not possible with the differential form. The required assumption is that of *isotropic scattering and sources*, or "isotropic collision".

This is a reasonable assumption because neutrons are emitted by fission in an effectively isotropic manner, and the anisotropy of scattering is not very pronounced, particularly for nuclei that are not too light (§ 7.1.3). Moreover, as we shall see below, most of the error caused by this approximation is easy to correct using transport cross-sections[11].

The isotropic collision assumption means that $\Sigma_{s,h\to g}(\vec{r'},\vec{\Omega}' \to \vec{\Omega})$, and therefore the density $D_g(\vec{r},\vec{\Omega})$, is independent of $\vec{\Omega}$ (note that, as written here, $S_{f,g}(\vec{r},\vec{\Omega})$ is also independent of $\vec{\Omega}$). The sum Q of these densities is therefore also independent of $\vec{\Omega}$.

Returning to the integral form of the transport operator under these conditions, we have:

$$\Phi(\vec{r},u,\vec{\Omega}) = \int_0^\infty ds \exp(-\tau) Q(\vec{r} - s\vec{\Omega}, u, \vec{\Omega}). \tag{14.13}$$

We can:

- Not write $\vec{\Omega}$ as the last variable of Q.
- Set $\vec{r'} = \vec{r} - s\vec{\Omega}$.
- Integrate both sides of the equation over $\vec{\Omega}$.
- Note that, except for a factor, $ds\, d^2\Omega$ is the volume element about the point $\vec{r'}$:

$$d^3r' = 4\pi s^2\, ds\, d^2\Omega = 4\pi R^2\, ds\, d^2\Omega.$$

In this context, the distance from $\vec{r}$ to $\vec{r'}$ is written as R rather than s:

$$R = |\vec{r'} - \vec{r}|.$$

After performing the calculations, and allowing that the functions that do not contain $\vec{\Omega}$ among their arguments are the *integrals* over $\vec{\Omega}$ of the corresponding phase functions, we obtain the following sets of equations:

[11] Note that it is also possible to construct a nearly exact integral transport theory with an assumption of "linearly anisotropic collision", i.e. with a scattering law that is linearly dependent on the cosine of the scattering angle ψ in the laboratory system.

a/ Continuous form:

$$\begin{aligned}
Q &= D + S_f + S_a, \\
S_f(\vec{r}, u) &= \chi_f(u) \int_0^\infty \nu\Sigma_f(\vec{r}, u')\Phi(\vec{r}, u')du', \\
D(\vec{r}, u) &= \int_0^\infty \Sigma_s(\vec{r}, u' \to u)\Phi(\vec{r}, u')du', \\
\Phi(\vec{r}, u) &= \int_{(\infty)} \frac{e^{-\tau}}{4\pi R^2} Q(\vec{r'}, u)d^3r',
\end{aligned} \tag{14.14}$$

(τ: optical path from $\vec{r}$ to $\vec{r'}$).

b/ Multigroup form:

$$\begin{aligned}
Q_g &= D_g + S_{f,g} + S_{a,g}, \\
S_{f,g}(\vec{r}) &= \chi_{f,g} \sum_{h=1}^{N} \nu\Sigma_{f,h}(\vec{r})\Phi_h(\vec{r}), \\
D_g(\vec{r}) &= \sum_{h=1}^{N} \Sigma_{s,h\to g}(\vec{r})\Phi_h(\vec{r}), \\
\Phi_g(\vec{r}) &= \int_{(\infty)} \frac{e^{-\tau_g}}{4\pi R^2} Q_g(\vec{r'})d^3r',
\end{aligned} \tag{14.15}$$

Again we find the simple form of the transport operator that was obtained using the same assumptions in § 3.1.6. This is known as the *Peierls operator*.

Note that these relationships make *no approximation* concerning the phase distribution of travelling neutrons (flux).

14.3.2. Transport correction

The parameters that affect the multiplication factor of a reactor and the power distribution within it are, on the one hand, the number of neutrons regenerated when a neutron is absorbed and, on the other hand, the distance separating one fission from the next. The first aspect is characterised by the ratio $\nu\bar{\Sigma}_f/\bar{\Sigma}_a$ of the average production and absorption cross-sections. The second aspect is described by the Green's function of the migration and, more specifically, as is easily seen by analysing the critical condition of the bare homogeneous pile using one-group theory (Chapter 6), its second-order moment, i.e. the mean square of the crow-fly distance travelled by the neutron from emission to absorption. Allowing for a factor of 1/6, this is called the *migration area* M^2. It is therefore important when carrying out modelling to respect the three synthetic parameters $\nu\bar{\Sigma}_f$, $\bar{\Sigma}_a$, and M^2 as much as possible.

The migration area involves three aspects:

1/ The elementary path of the neutrons between the point of emission or re-emission and the point of the first subsequent collision, governed by the total cross-section Σ.

2/ The number of elementary paths travelled by the neutron during its migration, governed by the ratio Σ_s/Σ.

3/ The scattering deflection angle ψ and particularly the mean value μ of its cosine. For example, if μ is positive, which is the most frequent case, i.e. if scattering tends to occur in a forward direction, then the neutrons tend to travel farther on average than if scattering were isotropic.

Even if it is obvious that the isotropic collision assumption we have introduced does not affect the average cross-sections, it does lead us to replace μ by 0. This assumption therefore does not respect the migration area.

The transport correction is applied to restore the correct migration area value.

In practice, to apply this correction, we use the equation that gives the migration area in monokinetic theory for an infinite homogeneous medium[12]:

$$M^2 = \frac{1}{3\Sigma_a\Sigma_{tr}}, \tag{14.16}$$

where the so-called *"transport" cross-section* is defined by:

$$\Sigma_{tr} = \Sigma - \mu\Sigma_s. \tag{14.17}$$

To avoid modifying Σ_a, the total cross-section Σ is replaced by the transport cross-section Σ_{tr}. In order to respect M^2, we also abandon the constraint of respecting the elementary mean free paths, which is less of a problem than not respecting the migration area.

In a spectrum theory, this correction is applied to each lethargy or in each group. It is not strictly equivalent to respecting M^2, but it can be shown, for the overall migration area, that it is a good approximation.

14.3.3. First collision probabilities

To present the formalism for first collision probabilities, we return to the expression for the isotropic collision transport operator, where the lethargy variable u or the group index g is understood, as well as the transport correction:

$$\Phi(\vec{r}) = \int_{(\infty)} \frac{e^{-\tau}}{4\pi R^2} Q(\vec{r'}) d^3r'. \tag{14.18}$$

To comply with the usual practice, although this is not essential[13], we go from the flux Φ to the *collision density* (or incoming density) $\Sigma\Phi$ by multiplying both sides of the equation by the cross-section at point $\vec{r}$:

$$\Sigma(\vec{r})\Phi(\vec{r}) = \int_{(\infty)} \Sigma(\vec{r}) \frac{e^{-\tau}}{4\pi R^2} Q(\vec{r'}) d^3r'. \tag{14.19}$$

The kernel of the operator is now no longer interpreted as the flux at the point $\vec{r}$, but as the collision density at this point for a neutron emitted isotropically at the point $\vec{r}'$.

The "first collision probabilities" method consists of the following steps:

1/ Cutting up the object to be processed into small volumes V_i, with i from 1 to M;

2/ Approximating the emission density $Q(\vec{r'})$ in each of the little volumes V_j by its average value Q_j, i.e.:

$$Q_j = \frac{1}{V_j} \int_{(j)} Q(\vec{r'}) d^3r'. \tag{14.20}$$

---

[12] The proof of this formula involves calculating the mean of the square of the vector sum of the elementary paths.

[13] In the APOLLO code, for example, flux is used rather than collision density, so that calculations can be performed even for a vacuum.

In practice, the volumes are homogeneous for practical reasons (for example, one or several volumes for the fuel, one or several volumes for the moderator, etc.), and this is what we assume here. Obviously the choice of smaller volumes will improve the accuracy of the approximation, but at the cost of longer calculation times.

Let Φ_i be the average value of flux in the volume V_i, i.e.:

$$\Phi_i = \frac{1}{V_i} \int_{(i)} \Phi(\vec{r}) d^3 r. \tag{14.21}$$

This average is calculated by integrating the equation over the volume V_i. Moreover, if the integral on the right-hand side is broken down into M integrals on the volumes V_j, we immediately see that:

$$V_i \Sigma_i \Phi_i = \sum_{j=1}^{M} V_j Q_j \tilde{P}_{ji}, \tag{14.22}$$

with:

$$\tilde{P}_{ji} = \frac{1}{V_j Q_j} \int_{(i)} d^3 r \int_{(j)} \Sigma(\vec{r}) \frac{e^{-\tau}}{4\pi R^2} Q(\vec{r'}) d^3 r'. \tag{14.23}$$

This quantity is interpreted as the probability for a neutron emitted isotropically in V_j and according to density $Q(\vec{r'})$ to undergo its first collision in V_i.

This expression is rigorous but impossible to use, because the distribution $Q(\vec{r'})$ of emissions in each volume is unknown. That is why $Q(\vec{r'})$ is replaced by the constant Q_j (which is eliminated between the numerator and the denominator), i.e. the exact probabilities $\tilde{P}_{ji}$ by the probabilities P_{ji} corresponding to uniform emission:

$$P_{ji} = \frac{1}{V_j} \int_{(i)} d^3 r \int_{(j)} \Sigma(\vec{r}) \frac{e^{-\tau}}{4\pi R^2} d^3 r'. \tag{14.24}$$

This quantity is interpreted as the *probability for a neutron emitted* ***uniformly*** *and* ***isotropically*** *in V_j to undergo its first collision in V_i.*

This "flat emission" approximation in terms of space and angle is the only approximation applied by this method. In particular, no space or angle approximation is made concerning the flux. The first collision probabilities method therefore consists of the following steps:

1/ Calculating the double integrals numerically:

$$P_{ji} = \frac{\Sigma_i}{V_j} \int_{(i)} d^3 r \int_{(j)} d^3 r' \frac{e^{-\tau}}{4\pi R^2}, \tag{14.25}$$

by suitable quadrature formulae (the volume elements are assumed to be homogeneous, and the function $\Sigma(\vec{r})$ to be taken in V_i can be replaced by the constant Σ_i that comes out of the double integral);

2/ And then calculating the flux Φ as a function of the emission density Q by simply multiplying a vector by a matrix:

$$V_i \Sigma_i \Phi_i = \sum_{j=1}^{M} V_j Q_j P_{ji}. \tag{14.26}$$

In practice, this must be done for each of the energy groups with which the cross-sections and the spectrum are processed:

1/ Calculating the first collision probabilities $P_{ji,g}$ for each of the groups with the corresponding total cross-sections (or transport cross-sections if appropriate);

2/ And then calculating the flux values of this group according to the emission density in the group:

$$V_i \Sigma_{i,g} \Phi_{i,g} = \sum_{j=1}^{M} V_j Q_{j,g} P_{ji,g}. \tag{14.27}$$

Naturally, Q must also be expressed as a function of the flux values Φ by writing out the collision operators, which can be done in each of the volumes using the multigroup formalism, which can be reduced to multiplications of a flux vector by fission and scattering matrices:

$$\begin{aligned} Q_{j,g} &= D_{j,g} + S_{f,j,g} + S_{a,j,g}, \\ S_{f,j,g} &= \chi_{f,j,g} \sum_{h=1}^{N} \nu \Sigma_{f,j,h} \Phi_{j,h}, \\ D_{j,g} &= \sum_{h=1}^{N} \Sigma_{s,j,h \to g} \Phi_{j,h}. \end{aligned} \tag{14.28}$$

The looping between these two types of formula is usually performed by iteration. To accelerate convergence, scattering is generally isolated in the group:

$$Q_{j,g} = \Sigma_{s,j,g \to g} \Phi_{j,g} + Q'_{j,g},$$

and we work with the vectors Φ and Q'.

14.3.4. Reciprocity and complementarity relationships between the first collision probabilities

Note that the double integral in the formula defining P_{ji} is symmetric; this implies an obvious reciprocity relationship[14]:

$$V_i \Sigma_i P_{ij} = V_j \Sigma_j P_{ji}. \tag{14.29}$$

(This relationship reflects the fact that the probability of a neutron travelling from one point to another without collision does not change if the direction of travel is inverted.)

Moreover, if there is no leakage, every neutron undergoes a collision in the system; the sum of the probabilities on the target volumes for any emission volume is therefore equal to 1:

$$\sum_{i=1}^{M} P_{ji} = 1. \tag{14.30}$$

(These are "complementarity" or "neutron conservation" relationships.)

The number of double integrals to be calculated can be almost halved by applying the reciprocity and complementarity relationships.

[14] In the following equations, the group subscript is understood.

14.3.5. Probabilities involving a surface

In practice, it is often necessary to deal with a *finite* "object" bounded by a surface S, for example an elementary cell or a reactor assembly. This leads to considering the neutrons leaving the "object" on the one hand, and the future of the neutrons entering the neighbouring "object" on the other hand. That is why we must consider not only volume probabilities P_{ji}, but also probabilities P_{jS} for a neutron emitted in V_j to exit *via* the surface S, and first collision probabilities P_{Si} and re-exit probabilities P_{SS} for a neutron entering *via* the surface S.

If the neutrons can exit, the above complementarity relationships no longer apply, because the probability of the "exit" event would need to be added:

$$\sum_{i=1}^{M} P_{ji} + P_{jS} = 1. \tag{14.31}$$

In practice, this formula makes it possible to obtain the probabilities P_{jS} once the probabilities P_{ji} have been calculated.

Just as it was necessary when defining P_{ji} to specify how neutrons were emitted in V_j, likewise it is necessary to specify how the neutrons enter *via* the surface S in order to define P_{Si} and P_{SS} unambiguously. Here again, ***uniformity*** *and* ***isotropy***[15] will be assumed. In this case:

- *Uniformity* means that the same number of neutrons enters *via* any unit surface element; in other words, the probability of the neutron entering *via* an element dS is dS/S;
- *Isotropy* concerns the *incoming* phase *flux* of neutrons; in other words, the number of neutrons entering *via* a solid angle element $\sin\theta\, d\varphi\, d\theta$ defined by $d\varphi$ and $d\theta$ (counting θ from the incoming normal) is $\cos\theta \sin\theta\, d\varphi\, d\theta/\pi$ so that a total of one neutron is concerned for the 2π steradians of input directions.

14.3.6. Reciprocity and complementarity relationships between probabilities involving a surface

Under these conditions, if we write out the integrals we can show that there is a reciprocity relationship between the mixed surface/volume probabilities:

$$P_{Si} = \frac{4V_i\Sigma_i}{S} P_{iS}. \tag{14.32}$$

This means that the P_{Si} probabilities can be deduced from the P_{iS} probabilities.

By listing all the possibilities, we can write the complementarity relationship:

$$\sum_{i=1}^{M} P_{Si} + P_{SS} = 1, \tag{14.33}$$

which then allows us to obtain the probability P_{SS}.

[15] Note that an assumption must be made about the *incoming* neutrons, but no assumption is made concerning the distribution of the *outgoing* neutrons.

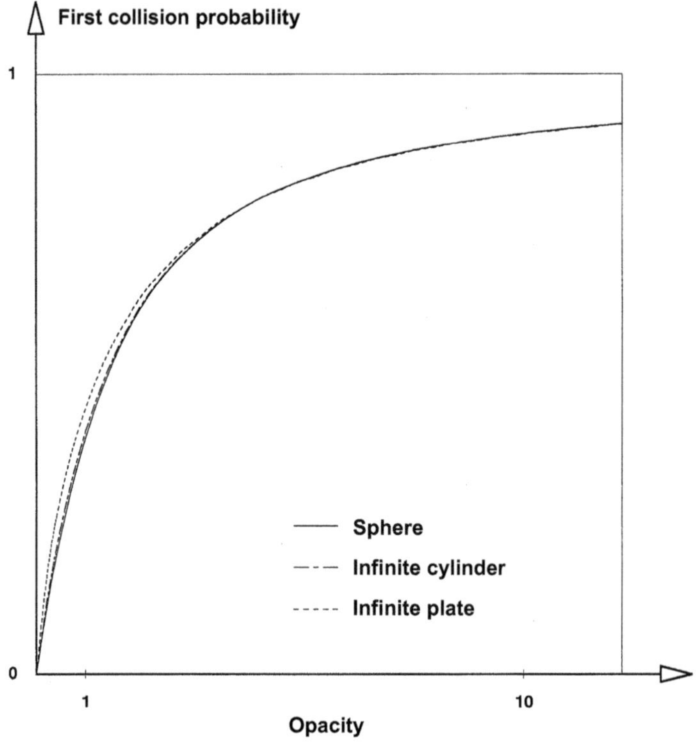

Figure 14.2. Probability P_{VV} for three simple geometries (in schematic form, these are the geometries generally used for reactor fuel elements).

Note that, with the assumptions made, only the $M(M+1)/2$ P_{ji} probabilities with (for example) $j \leqslant i$ need to be calculated explicitly by quadrature; all others can be deduced using reciprocity and complementarity.

14.3.7. First collision probabilities for a homogeneous convex body

The simplest case is that of a single volume V bounded by its surface S, i.e. the case $M = 1$. There are then four probabilities to consider: P_{VV}, P_{VS}, P_{SV}, and P_{SS}, but only one to be calculated, thanks to the reciprocity and conservation relationships. (Note in the examples below that P_{SS} is the easiest probability to express and calculate.)

The first collision probability P_{VV} in this body for a neutron emitted uniformly and isotropically in the volume V is plotted in Figure 14.2 for three simple geometries: a sphere, a cylinder of infinite height, and a plate of infinite length and width. The variable plotted along the abscissa is not the radius or thickness, but the *opacity* ω: the dimensionless product $\omega = \Sigma\bar{X}$ of the total cross-section Σ by the average chord[16] $\bar{X} = 4V/S$ (Cauchy's theorem)[17].

A series expansion of the probability P_{SS} allows the expansion of P_{VV} to be deduced:

- for *low values of opacity*:

$$P_{VV} = \frac{Q\omega}{2} + \ldots, \quad Q = \frac{\langle X^2 \rangle}{(\langle X \rangle)^2}. \tag{14.34}$$

The coefficient Q characterises the slope of the curves at the origin. Its value is 9/8 for the sphere, 4/3 for the cylinder, and infinity for the plate;

- for *high values of opacity*:

$$P_{VV} = 1 - \frac{1}{\omega} + \ldots, \quad P_{VV} \simeq \frac{\omega}{1 + \omega}. \tag{14.35}$$

(This equation is the Wigner approximation.)

The asymptotic behaviour of the curves is therefore the same for all geometries.

(Note that this observation is the basis for the possible equivalence between the heterogeneous case and a homogeneous case in resonant absorption theory: see § 8.3.2.)

14.3.8. Calculation of collision probabilities in x Geometry and $x - y$ Geometry

In these three examples, probability calculations are possible thanks to geometric symmetries. In general, after a few legitimate simplifications, we often find problems that are independent of the variable z, or independent of y or the azimuth φ. The following equations are useful in these cases because they simplify the general form of the Peierls operator:

$$\Phi(\vec{r}) = \Phi(x, y, z) = \int_{-\infty}^{+\infty} \int_{-\infty}^{+\infty} \int_{-\infty}^{+\infty} \frac{e^{-\tau}}{4\pi R^2} Q(x', y', z') dx'\, dy'\, dz'; \tag{14.36}$$

- for a *planar geometry* problem, meaning one involving the variable x only, it is possible to integrate over y and z to obtain:

$$\Phi(x) = \int_{-\infty}^{+\infty} \frac{1}{2} E_1(\tau_x) Q(x') dx', \tag{14.37}$$

and the resulting formulae for the first collision probabilities. The argument τ_x is the projection of the optical path $\vec{\tau}$ on the $\vec{x}$ axis, and the functions E_n are the *integral exponentials*[18]:

$$E_n(u) = \int_0^1 \exp\left(-\frac{u}{\mu}\right) \mu^{n-2} d\mu; \tag{14.38}$$

---

16 To define the average chord, it is necessary to specify the probability distribution with which the chord was chosen. Cauchy selected the chord's point of origin according to a probability that is uniform on the surface, and an incoming direction according to an isotropic distribution. These are the assumptions we have adopted for the distribution of incoming neutrons.

17 Note that this opacity is also the coefficient that relates the two surface/volume probabilities in this case.

18 See appendices.

– for a problem involving only the variables x and y, it is possible to integrate over z to obtain:

$$\Phi(\vec{\rho}) = \Phi(x, y) = \int_{-\infty}^{+\infty} \int_{-\infty}^{+\infty} \frac{Ki_1(\tau_{xy})}{2\pi R_{xy}} Q(x', y') dx'\, dy', \tag{14.39}$$

and the resulting formulae for the first collision probabilities. The argument τ_{xy} is the projection of the optical path $\vec{\tau}$ on the $x - y$ plane, and R_{xy} is the projection of the real path $\vec{R}$ of the neutron on the $x - y$ plane; Ki_n are the *Bickley functions*[19]:

$$Ki_n(u) = \int_0^{\pi/2} \exp\left(-\frac{u}{\sin\theta}\right) \sin^{n-1}\theta\, d\theta. \tag{14.40}$$

For problems in the cylindrical geometry (infinite height), this formula is used, with the revolution symmetry taken into account[20].

For general two-dimensional $x - y$ problems, the first collision probabilities are calculated using the variables R, Φ, t and t' as defined in the diagram below.

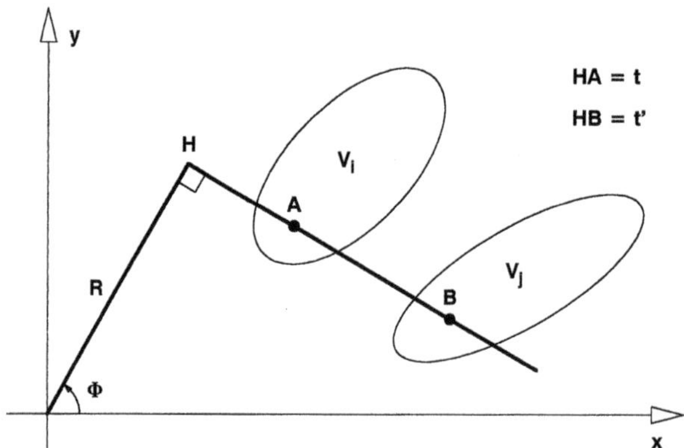

Figure 14.3. Variables used to calculate first collision probabilities.

Under these conditions, the first collision probability is written as follows:

$$P_{ji} = \frac{\Sigma_i}{V_j} \iiiint \frac{Ki_1(\tau_{xy})}{2\pi} dR\, d\Phi\, dt\, dt', \tag{14.41}$$

where the variables describe the volumes V_i and V_j.

---

19 See appendices.

20 The formulae can also be simplified in problems with spherical symmetry.

14.3.9. Calculation of probabilities on an infinite lattice of identical cells

The structure used in nuclear reactor cores is often relatively regular, so that each fuel element and its cladding can be associated with a part of the volume of the coolant and the moderator (in the case of a thermal neutron reactor) to form the mesh element of this structure, which neutron physicists call a "cell". By juxtaposing these cells, we obtain what neutron physicists call a *"lattice"*.

These lattices are not always very regular. Four types of cell can be identified in the example presented in Figure 13.3, concerning a pressurised water reactor assembly[21] with consumable poison rods:

- 248 standard fuel cells[22];
- 16 cells with a gadolinium-poisoned fuel;
- 24 cells able to accommodate the control bundle consisting of 24 absorbent rods;
- The central cell, reserved for instrumentation[23].

To simplify the calculations, the lattice is extended to infinity by assuming that the geometry and therefore the neutron flux are periodic; this creates the *regular infinite lattice*, which has become a paradigm in neutron physics, as it is used as a reference between the two steps of the calculation[24]: the fine calculation (on the scale of a cell), and the macroscopic calculation (calculation of the whole core).

Lattices can be:

- One-dimensional, i.e. consisting of flat cells made up of infinite plates,
- Two-dimensional, i.e. consisting of prismatic cells of infinite height,
- Three-dimensional, i.e. consisting of box-shaped cells.

All three cases are found in practice; the two-dimensional case is the most common, and this is the case we shall consider here.

The simplest lattice consists of cells that are all identical; for example, the one represented in Figure 14.4 (imagine that it continues *ad infinitum*), a fictitious lattice made up of fuel cells in a pressurised water reactor. We begin by considering this case, and then we shall look at "multi-cell" lattices.

Whether the lattice is rectangular, hexagonal, or triangular, the planes containing the interfaces between cells are all symmetry planes. When a neutron travelling in a given cell passes through one of these interfaces, the part of the path located beyond it can be replaced by its symmetric counterpart[25]; in other words, without changing the physics, we can reason on this one cell, assuming that the neutrons reaching the surface are sent back to the inside as if by a perfect mirror. This boundary condition is introduced in the processing of the Boltzmann equation, and is called a *specular reflection condition*.

---

[21] Note that, in water reactors, this fluid acts as both a coolant and a moderator.

[22] In practice, the peripheral cells must also be distinguished; they include the half-water gap between assemblies and the corner cells with two half-water gaps.

[23] On the simplified diagram (Figure 13.3), it is represented like the previous 24 cells.

[24] The link between these two steps (equivalence) will be specified in Chapter 17.

[25] In fact, this symmetry means that the neutron under consideration is replaced by an equivalent symmetric neutron.

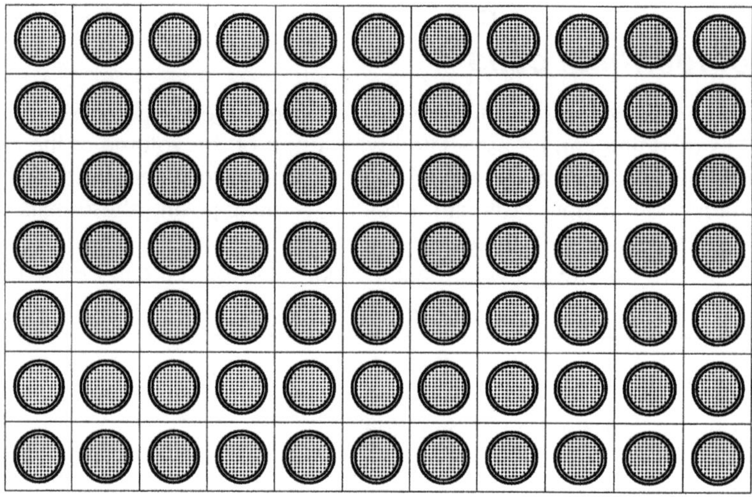

Figure 14.4. The lattice: paradigmatic neutron situation of reactors (here: pressurised water reactor lattice). Note that the contours of the squares correspond to purely fictitious limits.

It is obvious that the use of the first collision probabilities method in this type of problem will involve a two-dimensional $x-y$ calculation, either in the lattice itself by identifying the similar volumes of all the cells[26], or in the isolated cell with the specular reflection conditions[27]. In codes such as APOLLO, this type of calculation is possible, but is still costly. That is why an interface current uniformity and isotropy approximation is often made. This consists of replacing the correct boundary condition (specular reflection) by an approximate condition called *white reflection*, where every neutron reaching the surface of the cell is assumed to "forget" its state (position and direction), and is sent according to a distribution that is uniform along the surface and isotropic towards the inside. Note that this makes it possible to use the surface probabilities introduced above, which do not need to be calculated if the volume-volume probabilities have already been obtained.

The following equations are to be used in the context of this simplifying assumption. Let:

- P^+_{ji} be the probability for a neutron emitted in the volume V_j in one of the cells of undergoing its first collision in the volume V_i of the *same* cell;

- P^+_{jS} be the probability for a neutron emitted in the volume V_j of one of the cells of leaving this cell;

- P^+_{Si} be the probability for a neutron entering one of the cells of undergoing its first collision in the volume V_i of *that* cell.

[26] For P_{ji} to be calculated, the neutrons must be emitted in one of the volumes V_j and the impacts in *all* volumes V_i must be counted.

[27] The path of neutrons reaching the surface must be continued by reflection until the neutrons have had their first collision.

The probability P_{ji} for a neutron emitted in the volume V_j of one of the cells undergoing its first collision in the volume V_i of the same cell or any other cell[28] is calculated by adding the probabilities of events with 0, 1, 2, 3 ... exits:

$$P_{ji} = P^+_{ji} + P^+_{jS}P^+_{Si} + P^+_{jS}P^+_{SS}P^+_{Si} + P^+_{jS}P^+_{SS}P^+_{SS}P^+_{Si} + \cdots$$

A geometric series is obtained:

$$P_{ji} = P^+_{ji} + \frac{P^+_{jS}P^+_{Si}}{1 - P^+_{SS}}. \tag{14.42}$$

The "no exit" probabilities P^+ are those that were introduced above (§ 14.3.3 and 14.3.5); we noted that all probabilities involving the surface are deduced from the volume-volume probabilities. This means that only the latter need to be calculated (taking the reciprocity relations into account): there are far fewer of them[29] and they are far more simple to calculate than the true probabilities, because they relate to a single cell which we imagine to be isolated. The true probabilities, on the other hand, involve calculating and summing, for a fixed V_j, the probabilities for the volumes V_i of all the cells[30] and require the handling of complicated optical paths for the outside cells.

When it is acceptable, which it is for most calculations, this simplifying assumption considerably reduces the volume of calculations to be performed.

14.3.10. Cylindrisation of cells

To further simplify the calculations, another (non-mandatory) approximation is often introduced: cell cylindrisation[31].

This idea arises from the observation that, in many reactor concepts, (UNGG, PWR and BWR, FNR, etc.), the unit cell consists of a fuel and a cladding, i.e. a channel with rotational symmetry, where only the outside contour of the cell breaks the symmetry. The cylindrisation approximation (see Figure 14.5) consists of replacing the prismatic contour of the cell with a cylindrical contour[32] in order to obtain complete rotational symmetry, and therefore simplify the calculation of the probabilities P^+_{ji}[33].

Conservation of matter is obviously essential in this operation, but this leaves an open choice of radius R. It might be difficult to choose between "Askew cylindrisation", which conserves the exchange surface with the neighbouring cells and therefore the *external perimeter* (with reduction of the moderator density to conserve its mass), and "Wigner cylindrisation", which conserves the *cross-section* (and the density of the moderator). For cells whose cross-section is shaped like a regular polygon (hexagon, square, or triangle), Wigner cylindrisation is usually chosen. For cells with a rectangular cross-section[34] with

28 Or, in the context of the problem of the unique cell which we imagine to be isolated, in the volume V_i of this cell after any number of reflections on the surface.

29 For example, six probabilities if three volumes are placed in the cell: the fuel, the cladding, and the moderator.

30 In practice, for this type of calculation, the infinite series is truncated, neglecting the cells beyond a certain "optical" distance.

31 Cylindrisation of the cells whilst conserving specular reflection would not make sense, because the probability calculations would not be simplified much; moreover, it has been shown that the errors due to both approximations (white reflection and cylindrisation) partially cancel each other out.

32 Cylindrisation can only be envisaged in the image of the cell that we imagine to be isolated, because space can obviously not be paved with cylinders.

33 Rotational symmetry simplifies the integrals expressing these probabilities.

34 For example, the cells at the edge of a PWR assembly, including the water gap on one of their faces.

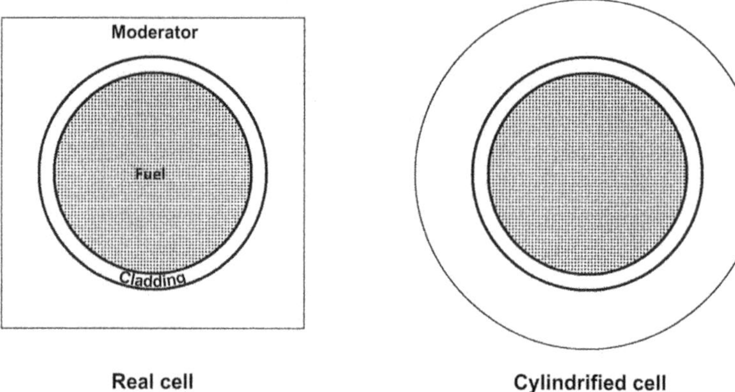

Figure 14.5. Cylindrisation of a cell (here, a pressurised water; real size of the square cell: 12.6 mm).

sides a and b, it is recommended to use "Lefebvre cylindrisation" (with a density reduction), which is better suited to "multicell" calculations[35] (see next section), and is reduced to Wigner cylindrisation if the rectangle is a square:

$$R = \frac{a+b}{2\sqrt{\pi}}. \tag{14.43}$$

14.3.11. Principle of "multicell" geometry calculations

We have seen that, in practice, the objects dealt with by a neutron physicist are often assemblies of unit cells that are not always identical.

The method presented above can be generalised (with exactly the same assumptions) to the regular infinite lattice whose elementary pattern comprises a set of cells of different types. This is the "multicell" formalism.

The pattern can contain several cells that are identical in geometry and composition. Theoretically, they cannot be considered as being of the same type (from a neutron physics point of view) unless they also have an identical environment (for example, being symmetrical with respect to a general plane of symmetry for the pattern), but in practice, they can be declared to be of the same type even if this condition is not fulfilled. This reduces the number of types, and therefore the calculation cost, but is obviously an additional approximation whose validity needs to be verified.

Let I, J or K be the index used to identify the type, and α_I, α_J or α_K be the proportion in the lattice (or pattern) of cells of type I, J or K, and let S_{JI} be the probability that a neutron leaving a type J enters a type I cell. To remain consistent with the assumption of uniformity of the currents leaving the cells, S_{JI} must be the proportion of the side surface of all cells J of the pattern in contact with a type I cell. It is therefore a strictly geometric parameter that characterises the connections between the various types of cell. Because

[35] Because it satisfies the reciprocity relationships on the "contact probabilities" S_{JI}.

S_{JI} are probabilities (for fixed J), the following complementarity relationships exist:

$$\sum_I S_{JI} = 1. \tag{14.44}$$

In addition, because the surfaces of the type J cells in contact with type I cells are obviously the same as the surfaces of the type I cells in contact with type J cells, the S_{JI} values must satisfy the following reciprocity conditions:

$$\alpha_I p_I S_{IJ} = \alpha_J p_J S_{JI}, \tag{14.45}$$

where p_I is the surface of a type I cell, i.e. its perimeter. (These relationships are useful on the one hand to check that no errors were made in calculating the S_{JI} of a pattern, and on the other hand, as we have just seen, to obtain a cylindrisation rule if we also wish to introduce this approximation.)

Using the same type of argument as the one used to express the probabilities P of a lattice as a function of the probabilities P^+ of a cell, the "multicell"[36] formalism provides all the first collision probabilities P_{ji} in a zone i of a type I cell for a neutron emitted in a zone j of a type J cell based on the cell-specific probability sets P^+ — the only ones, finally, that need to be calculated:

1/ Using δ_{JI} to denote the Kronecker symbol and Q_{JI} to denote the number of neutrons entering a type I cell, without an intermediate collision, for a neutron leaving a type J cell, we have:

$$P_{ji} = P^+_{ji}\delta_{JI} + P^+_{jS} Q_{JI} P^+_{Si}; \tag{14.46}$$

2/ By writing out the definition of these probabilities, we obtain the equation used to calculate Q_{JI} :

$$Q_{JI} = S_{JI} + \sum_K S_{JK} P^+_{SS,K} Q_{KI}. \tag{14.47}$$

In a lattice consisting of only one type of cell, there is obviously no need to distinguish between the faces because they are equivalent. In a multicell lattice, however, this assumption of complete uniformity, named after Roth (the author who suggested it), is very much debatable.

For example, in the pattern in Figure 14.6, it is obvious that a neutron leaving a fuel cell adjacent to the absorbent cell is not likely to have the same outcome as if it emerged facing the absorbent or *via* the opposite face. That is why it is helpful to *distinguish the faces* of cells, i.e. to have uniformity of interface currents not over the entire surface of cells, but face by face.

The simplest improvement of the Roth assumption, known as ROTH-4 for square cells and ROTH-6 for hexagonal cells, consists of preventing a neutron that enters a cell and passes through it without collision from exiting *via* the face of entry. It can, however, leave with equal probability *via* the three (or five) other faces. This approximation does not lead to new P^+ probability calculations, but makes it necessary to generalise the matrix Q (distinction of faces).

[36] It can be verified that it gives the formulae for the lattice when there is only one type of cell.

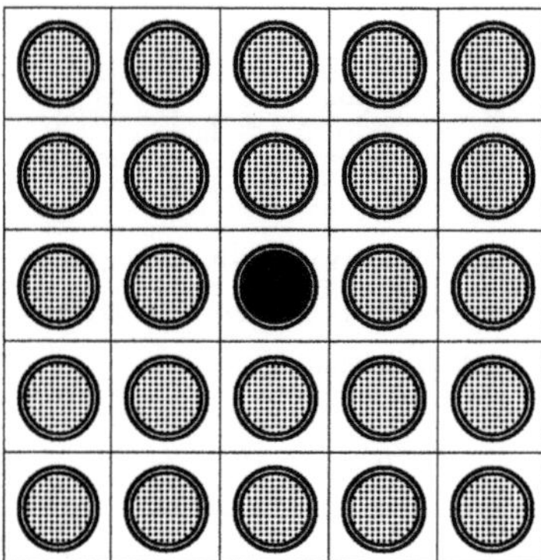

Figure 14.6. Example of heterogeneity in a PWR lattice: an absorbent rod in the centre of a 5×5 pattern, where the other twenty-four cells contain fuel (imagine that the edges of the pattern are planes of symmetry).

A more accurate method can also be envisaged: calculating the exact probabilities of crossing a cell from one face to the other. In this case, in addition to the generalisation of the matrix Q, some surface probabilities P^+ must be calculated, because the complementarity and reciprocity relationships do not distinguish between the faces in the surface probabilities that are deduced from the P^+_{ji}.

Note also that the treatment of interface currents can be improved not only with respect to uniformity, but also with respect to isotropy. The assumption of isotropy (over 2π) of the flux of neutrons passing through the interfaces can be abandoned in favour of a *linear anisotropy* assumption, i.e. a flux that is linearly dependent on $\cos\theta$. Here again, on the one hand, certain components must be added to the interface currents and the matrix Q must be generalised. On the other hand, a few additional probabilities P^+ must be calculated.

The multicell method can be generalised to handle any structure that has been imaginarily cut into sub-structures that exchange neutrons *via* their interfaces. The uniformity and isotropy (or linear anisotropy) approximation at each interface can reduce the number of probabilities and considerably simplify the calculation, because only P^+ type probabilities are involved.

14.4. Handling the differential form of the transport operator

14.4.1. Handling the diffusion operator

We have seen that the isotropic collision approximation (possibly with a transport correction) considerably simplifies the integral form of the Boltzmann equation, because the variable $\vec{\Omega}$ can be made to disappear simply by integrating over the phases. The transport operator, on the other hand, can hardly be simplified if written in differential form, because in that case it is impossible[37] to return to a problem without $\vec{\Omega}$ even if the collisions are isotropic. In contrast, taking the exact angular scattering distribution into account does not make it more difficult to handle the integral/differential equation[38].

In its continuous form, the scattering operator is written as follows:

$$D(\vec{r}, u, \vec{\Omega}) = \int_0^\infty du' \int_{(4\pi)} d^2\Omega' \Sigma_s \left[\vec{r}, (u', \vec{\Omega}') \to (u, \vec{\Omega})\right] \Phi(\vec{r}, u', \vec{\Omega}'). \tag{14.48}$$

After multigroup discretisation, this takes the following form:

$$D_g(\vec{r}, \vec{\Omega}) = \sum_{h=1}^{N} \int_{(4\pi)} \Sigma_{s,h\to g}(\vec{r}, \vec{\Omega}' \to \vec{\Omega}) \Phi_h(\vec{r}, \vec{\Omega}') d^2\Omega'. \tag{14.49}$$

In practice, materials placed in reactors can *always* be considered to be *isotropic*; as we have seen, this leads us to separate v (scalar speed, which can be replaced by u) and $\vec{\Omega}$ (direction of speed) since the isotropy of matter means that the cross-sections do not depend on $\vec{\Omega}$. This applies to all reactions, and scattering in particular. This illustrates the fact that writing the differential scattering cross-section in these integrals is not a good idea, because it does not reveal it. With regard to the angular aspect, the differential scattering cross-section is not dependent on four variables (θ', φ', θ and φ), but only one: the scattering angle or, which amounts to the same thing, its cosine, written as μ or $\vec{\Omega}' \cdot \vec{\Omega}$, i.e. the scalar product of two vectors. (Moreover, there can be a partial or total correlation between this angle and the group or lethargy change; this aspect is mentioned here as a reminder only.) These operators are then rewritten as:

$$D(\vec{r}, u, \vec{\Omega}) = \int_0^\infty du' \int_{(4\pi)} d^2\Omega' \Sigma_s \left(\vec{r}, \vec{\Omega}' \cdot \vec{\Omega}, u' \to u\right) \Phi(\vec{r}, u', \vec{\Omega}'), \tag{14.50}$$

and:

$$D_g(\vec{r}, \vec{\Omega}) = \sum_{h=1}^{N} \int_{(4\pi)} \Sigma_{s,h\to g}(\vec{r}, \vec{\Omega}' \cdot \vec{\Omega}) \Phi_h(\vec{r}, \vec{\Omega}') d^2\Omega'. \tag{14.51}$$

We allow $\vec{r}$, u' and u (or h and g) to be understood, and concentrate on the integral operator:

$$D(\vec{\Omega}) = \int_{(4\pi)} \Sigma_s(\vec{\Omega}' \cdot \vec{\Omega}) \Phi(\vec{\Omega}') d^2\Omega', \tag{14.52}$$

with kernel $\Sigma_s(\vec{\Omega}' \cdot \vec{\Omega})$.

As a function of $\mu = \vec{\Omega}' \cdot \vec{\Omega}$, this kernel can be broken down into Legendre polynomials[39]:

$$\Sigma_s(\mu) = \sum_{n=0}^{\infty} \frac{2n+1}{2} \Sigma_{s,n} P_n(\mu), \tag{14.53}$$

with:

$$\Sigma_{s,n} = \int_{-1}^{+1} \Sigma_s(\mu) P_n(\mu) d\mu. \tag{14.54}$$

---

[37] Unless an approximation such as the diffusion approximation is applied.

[38] Note again that emission by fission is isotropic.

[39] The definitions and main properties of Legendre polynomials are reviewed in an Appendix.

In particular, $\Sigma_{s,0}$ is the usual scattering cross-section Σ_s (including all angles), and $\Sigma_{s,1}$ is the product $\bar{\mu}\Sigma_s$ of this cross-section by the average cosine of the deflection angle of the neutron when scattered.

The following theorem describes the advantage of this breakdown: *the scattering operator, whose kernel depends only on $\vec{\Omega}' \cdot \vec{\Omega}$, is rotationally invariant; its eigenfunctions are the spherical harmonics*[40] $Y_n^m(\vec{\Omega})$; the associated eigenvalues are the coefficients $\Sigma_{s,n}$:

$$\int_{(4\pi)} \Sigma_s(\vec{\Omega}' \cdot \vec{\Omega}) Y_n^m(\vec{\Omega}') d^2\Omega' = \Sigma_{s,n} Y_n^m(\vec{\Omega}). \tag{14.55}$$

14.4.2. Spherical harmonic method

The spherical harmonics $Y_n^m(\vec{\Omega})$ are the equivalent for three-dimensional space of the trigonometric functions used to perform a Fourier expansion of a periodic function; they are functions of $\vec{\Omega}$ (or of the angles θ and φ), i.e. of a point on the unit sphere. Just as trigonometric functions return to the same value after one full revolution, i.e. when the argument changes by 2π, spherical harmonics are invariant after one full rotation around the unit sphere along, for example, a large circle or a parallel circle. These functions can be numbered with two subscripts: a main subscript n with values from zero to infinity, and a secondary subscript m with values (for fixed n) from $-n$ to $+n$.

The functions of $\vec{\Omega}$ that can be encountered in physics, for example a neutron phase flux, can be expanded using spherical harmonics[41]:

$$\Phi(\vec{r}, u, \vec{\Omega}) = \sum_{n=0}^{\infty} \sum_{m=-n}^{n} \Phi_n^m(\vec{r}, u) Y_n^m(\vec{\Omega}). \tag{14.56}$$

Because spherical harmonics are normed, by definition the coefficients of the expansion are as follows[42]:

$$\Phi_n^m(\vec{r}, u) = \int_{(4\pi)} \Phi(\vec{r}, u, \vec{\Omega}) Y_n^{*m}(\vec{\Omega}) d^2\Omega, \tag{14.57}$$

where the asterisk denotes the complex conjugate function.

The formulae are analogous in the multigroup approximation.

By writing this expansion into the integral/differential Boltzmann equation and then:

a/ using the property that spherical harmonics are eigenfunctions of the diffusion operator, and

b/ expressing the products $\Omega_k Y_n^m(\vec{\Omega})$ (with $k = x$, y or z) as a function of the neighbouring spherical harmonics using the recurrence relationships between these functions,

we reach an infinite system of equations numbered with the subscripts n and m *where the variable $\vec{\Omega}$ no longer appears.*

---

[40] The definitions and main properties of spherical harmonics are reviewed in an Appendix. Legendre polynomials are the spherical harmonics that dependent on the angle θ (or its cosine μ) only.

[41] This is the same idea as expanding the cross-section, except that for the cross-section the invariance along φ leads to an expansion along n only, i.e. in Legendre polynomials.

[42] A formula comparable to the one defining $\Sigma_{s,n}$.

In practice, we truncate this system by eliminating all terms of rank n greater than a given value N, and then process it according to $\vec{r}$ and u using the usual numerical methods. This approach is traditionally called the $\mathbf{P_N}$ *approximation.*

The number of coefficients $\Phi_n^m(\vec{r}, u)$ to be calculated is:

$$1 + 3 + 5 + 7 + \cdots + (2N + 1) = (N + 1)^2.$$

To illustrate this general principle on a simple example, let us examine a monokinetic problem in planar geometry. In monokinetic theory, the equation to be solved has the following form[43]:

$$\text{div}\left[\vec{\Omega}\Phi(\vec{r}, \vec{\Omega})\right] + \Sigma(\vec{r})\Phi(\vec{r}, \vec{\Omega}) = Q(\vec{r}, \vec{\Omega}) + \int_{(4\pi)} d^2\Omega'\Sigma_s(\vec{r}, \vec{\Omega}' \to \vec{\Omega})\Phi(\vec{r}, \vec{\Omega}'), \tag{14.58}$$

where $Q(\vec{r}, \vec{\Omega})$ is assumed to be known, at least provisionally.

"In planar geometry" means that, in terms of space, the problem involves the variable x only (the system is assumed to consist of a series of infinite plates along y and z). In this case, concerning $\vec{\Omega}$, it is advisable to identify the colatitude from the direction of the $\vec{x}$ axis, since the longitude will not be involved for reasons of symmetry. If we set $\mu = \cos\theta$, the equation is reduced to:

$$\mu\frac{\partial\Phi(x, \mu)}{\partial x} + \Sigma(x)\Phi(x, \mu) = Q(x, \mu) + \int_{-1}^{+1} \Sigma_s(x, \mu' \to \mu)\Phi(x, \mu')d\mu'. \tag{14.59}$$

Under these conditions, the Legendre polynomials (except for a factor, the φ-independent spherical harmonics) are sufficient to expand the phase flux. We therefore set:

$$\Phi(x, \mu) = \sum_{n=0}^{\infty} \Phi_n(x)P_n(\mu), \tag{14.60}$$

where the coefficients are interpreted like the integrals[44]:

$$\Phi_n(x) = \frac{2n + 1}{2}\int_{-1}^{+1} \Phi(x, \mu)P_n(\mu)d\mu. \tag{14.61}$$

In the first instance, we replace the flux appearing under the "summation" sign by its expansion, bearing in mind that Legendre polynomials are eigenfunctions of the diffusion operator. The equation then takes on the following form:

$$\mu\frac{\partial\Phi(x, \mu)}{\partial x} + \Sigma(x)\Phi(x, \mu) = Q(x, \mu) + \sum_{k=0}^{\infty} \frac{2k + 1}{2}\Sigma_{s,k}P_k(\mu)\int_{-1}^{+1} \Phi(x, \mu')P_k(\mu')d\mu', \tag{14.62}$$

[43] In multigroup theory, the equations would have the same form in each group, because, in order to improve the convergence of the iteration between Q and Φ, it is advisable to place the scattering within the group itself on the right-hand side. In this case, in addition to the absorptions, the term $\Sigma\Phi$ on the left-hand side includes transfers to other groups, and the term Q on the right-hand side represents the transfers from other groups (as well as fissions).

[44] Unlike spherical harmonics, Legendre polynomials are not normed. The coefficient $2/(2n+1)$ representing the square of the norm of polynomial P_n is therefore placed in an arbitrary position, which different authors choose differently. Here, a different convention was used for the diffusion cross-section and the flux, in order to simplify the formulae.

where the integrals that appear implicitly represent the coefficients of the Legendre expansion of the flux[45].

This form suggests handling the equation by "internal" iteration (as opposed to the "external" iteration between Q and Φ): based on an initial estimate of the last term, Φ is calculated by inverting the transport operator on the left-hand side[46]; after obtaining Φ, we re-evaluate the integrals on the right-hand side; we then re-calculate Φ; and so on until convergence.

Strictly speaking, the method of spherical harmonics consists of introducing on the left-hand side also the expansion according to these base functions to invert the advection operator.

In the current example, where only Legendre polynomials appear, the recurrence formula for these polynomials can be used:

$$nP_{n-1}(\mu) - (2n+1)\mu P_n(\mu) + (n+1)P_{n+1}(\mu) = 0, \tag{14.63}$$

to transform the terms of the μP_n form. We now expand the "source" Q like the flux:

$$Q(x,\mu) = \sum_{n=0}^{\infty} Q_n(x)P_n(\mu), \tag{14.64}$$

with:

$$Q_n(x) = \frac{2n+1}{2}\int_{-1}^{+1} \Phi(x,\mu)P_n(\mu)d\mu. \tag{14.65}$$

By stating that the algebraic sum of the coefficients in front of each of the polynomials is identical to zero, we finally obtain an infinite system of differential equations governing the functions Φ_n of the space variable x. In practice, this system will be truncated at the *Nth* order:

$$\begin{aligned}
&-\tfrac{1}{3}\Phi_1' - \Sigma\Phi_0 + \Sigma_{s,0}\Phi_0 + Q_0 = 0,\\
&-\Phi_0' - \tfrac{2}{5}\Phi_2' - \Sigma\Phi_1 + \Sigma_{s,1}\Phi_1 + Q_1 = 0,\\
&-\tfrac{n}{2n-1}\Phi_{n-1}' - \tfrac{n+1}{2n+3}\Phi_{n+1}' - \Sigma\Phi_n + \Sigma_{s,n}\Phi_n + Q_n = 0,\\
&-\tfrac{N}{2N-1}\Phi_{N-1}' - \Sigma\Phi_N + \Sigma_{s,N}\Phi_N + Q_N = 0.
\end{aligned} \tag{14.66}$$

This system of $N+1$ equations governs $N+1$ functions. Note that the equations of this system are relatively uncoupled because each equation only involves three successive unknown functions. Even by combinations, however, it is not possible in the general case to obtain a system of decoupled equations.

For reasons that will be explained below, related to an equivalence between the $\mathbf{P_N}$ and $\mathbf{S_{N+1}}$ approximations, an odd value of N is usually chosen.

14.4.3. Diffusion approximation and transport correction

The $\mathbf{P_0}$ approximation would lead us to assume the flux to be everywhere isotropic, which would eliminate any migration. The "minimum" approximation is therefore $\mathbf{P_1}$:

$$\begin{aligned}
&-\tfrac{1}{3}\Phi_1' - \Sigma\Phi_0 + \Sigma_{s,0}\Phi_0 + Q_0 = 0,\\
&-\Phi_0' - \Sigma\Phi_1 + \Sigma_{s,1}\Phi_1 + Q_1 = 0.
\end{aligned} \tag{14.67}$$

45 In the general case, these would be the coefficients of the spherical harmonic expansion.

46 The tricky part is the "advection operator".

If we also assume that the "sources" are isotropic, i.e. taking Q_1 to be zero and noting that:

- Φ_0 is $\Phi/2$ where Φ is the usual flux integrated over the phases, because $P_0 = 1$;
- similarly, Q_0 is $Q/2$;
- Φ_1 is $3J/2$ where J is the usual current integrated over the phases, because $P_1 = \mu$;
- $\Sigma - \Sigma_{s,0} = \Sigma - \Sigma_s$ is the absorption section Σ_a;
- $\Sigma - \Sigma_{s,1} = \Sigma - \bar{\mu}\Sigma_s$ is the transport section Σ_{tr},

we see that this system can be rewritten in the following form:

$$-J' - \Sigma_a\Phi + Q = 0,$$
$$-\Phi' - 3\Sigma_{tr}J \quad = 0.$$

The first equation sets out the neutron balance per unit volume (including all neutron directions); the second is the approximate expression for the current, known as *"Fick's Law"*:

$$\vec{J} = -D\,\overrightarrow{\text{grad}}\,\Phi \tag{14.68}$$

(§ 5.1.2), with the *transport correction* (§ 5.1.8) for the expression of the diffusion coefficient $D = 1/3\Sigma_{tr}$.

More generally, the approximation $\mathbf{P_1}$ that consists of representing the phase flux by the zeroth-order spherical harmonic (a constant) and the three first-order spherical harmonics (linear combinations of the three components of the vector $\vec{\Omega}$) amounts to approximating the phase flux at each point by an expression of the following type:

$$\Phi(\vec{r}, \vec{\Omega}) \simeq A + \vec{B}\cdot\vec{\Omega}.$$

By identification, we can see that, except for a factor, the scalar constant A is the integrated flux and the vector constant $\vec{B}$ is the integrated current:

$$\Phi(\vec{r}, \vec{\Omega}) \simeq \frac{1}{4\pi}\Phi(\vec{r}) + \frac{3}{4\pi}\vec{\Omega}\cdot\vec{J}(\vec{r}). \tag{14.69}$$

This approach to the diffusion approximation as a $\mathbf{P_1}$ approximation is the mathematical justification for the physical and intuitive approach suggested in Chapter 5.

Note: In monokinetic theory, the transport correction is equivalent to approximating linearly anisotropic diffusion (probability distribution for the deflection of the neutron during a linear collision at $\cos\theta$); in multigroup theory, the transport correction is "concentrated" on the initial group, when it should be "broken down" in the arrival groups. The practical effects of this second approximation turn out not to be very major.

14.4.4. Method of simplified spherical harmonics

The spherical harmonics method quickly produces a large number of unknown functions to be calculated if a somewhat high order N is used. On the other hand, limiting the calculation to $N = 1$, i.e. the diffusion approximation, can sometimes turn out to be insufficient[47]. The method of "simplified" spherical harmonics can be a good compromise between the cost and the precision of the calculation.

The idea is to identify the direction of the *current* $\vec{J}$ at each point of the reactor, described in two- or three-dimensional geometry; this means describing the axis along which migration mainly occurs and assuming that, according to this axis, the *local* phase flux has *rotational symmetry*. This assumption allows it to be represented by a Legendre expansion only (taking a local reference with its axis along $\vec{J}$ to measure θ) and therefore without the φ-dependent harmonics. Under these conditions, at the Nth order, this "$\mathbf{SP_N}$" approximation involves $N+3$ unknown functions instead of $(N+1)^2$ for the standard $\mathbf{P_N}$ approximation; for example, 12 instead of 100 for $N = 9$.

The advantage of the $\mathbf{SP_N}$ approximation is its ability to improve the diffusion approximation at little additional cost, by taking, for example, $N = 3$ or 5. The disadvantage is that the solution does not converge towards the exact solution when N tends to infinity: adopting very high values of N brings only an illusory improvement, and does not allow the error to be evaluated (to do this, it is necessary to perform an "exact" calculation, for example a complete $\mathbf{P_N}$ approximation).

14.4.5. Method of discrete ordinates

As we saw in § 6.2.1, there are two main types of method used to represent functions by a finite number of numerical values: discretisation, and series representation. Where functions of the variable $\vec{\Omega}$ are concerned, the method of spherical harmonics illustrates the series representation; the method of "discrete ordinates" illustrates the concept of discretisation (here, "ordinates" refers to the variable $\vec{\Omega}$).

The difficulty in discretising the variable $\vec{\Omega}$, i.e. the point on a unit sphere, is obviously caused by the curvature: a finite number of points and associated area elements on the sphere must be distributed as uniformly as possible. The most frequently used technique is illustrated on the diagrams in Figure 14.7. This technique can be improved, as is discussed below. To construct these "$\mathbf{S_N}$" diagrams, we start by cutting up the sphere into N bands delineated by circles that are parallel to the equator at regularly spaced colatitudes, i.e. multiples of π/N. Then, from the north pole down to the equator, these bands are cut up, from the meridian of origin, by meridian segments into 4, 8, 12, etc. identical trapezoidal elements; the southern hemisphere is cut up symmetrically from the south pole. The "discrete ordinates" are the directions $\vec{\Omega}_n$ associated with the points placed at the centres of the mesh elements; the corresponding weights are the mesh element areas normalised to 4π.

Discrete ordinates $\vec{\Omega}_n$ on the equator should be avoided, because the value $\mu = \cos\theta = 0$ could cause some difficulties for the numerical processing; that is why an even value of N is always used.

---

[47] Example: processing of core-reflector or standard assembly-plutonium assembly interface transients.

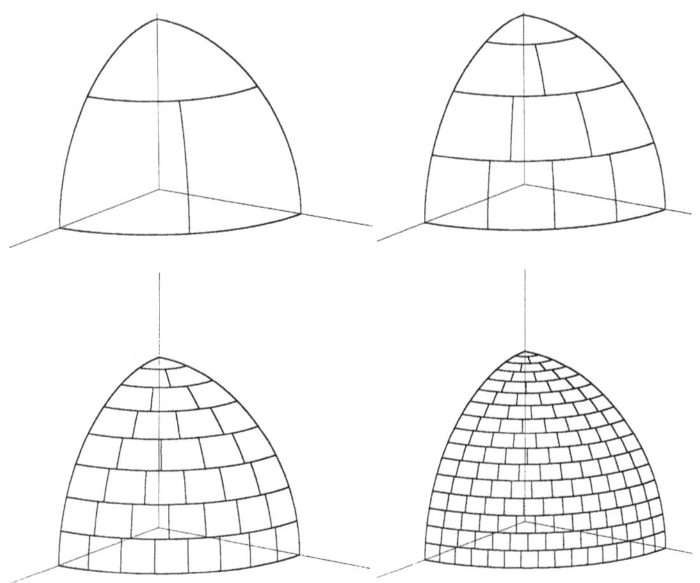

Figure 14.7. Representation on an octant of meshes S_4, S_8, S_{16} and S_{32} (24, 80, 288 and 1088 mesh elements respectively, therefore discrete ordinates).

A first possible way to improve this meshing consists of replacing the regular latitude divisions by a discretisation of the variable μ along the Gauss points: for a given N and with the associated Gauss weights, these values lead to the best possible quadrature formula in μ. The longitudinal discretisation is performed in the same way. It can be shown that, for a problem that is dependent on μ only, the $\mathbf{S}_N$-Gauss and $\mathbf{P}_{N-1}$ approximations are rigorously equivalent.

Another possible improvement, but one which is incompatible with the previous one, consists of seeking a partitioning that is symmetric with respect to the three Cartesian axes[48], which is not the case for the previous partitionings because they give a special role to the $\vec{z}$ axis. A symmetric meshing can be of interest if there is no direction that plays a special part in the problem, but this is not usually the case in neutron physics, where the $\vec{z}$ axis along which the coolant flows[49] is different from the perpendicular axes.

14.4.6. Ray effects

An artefact related to the method of discrete ordinates, known as the "ray effect", is illustrated below (Figure 14.8) for an extreme example.

For a two-dimensional case, which is easier to represent, we have considered the problem of a point source in a purely absorbent medium. The arrows represent the directions of the discrete ordinates, and the squares represent the meshing of the spatial discretisation. Because neutrons are obliged to travel along the discrete directions, we note that only the

[48] This is shown to be possible; there is even a degree of freedom.

[49] Generally vertical, but sometimes horizontal as in *CANDU* reactors.

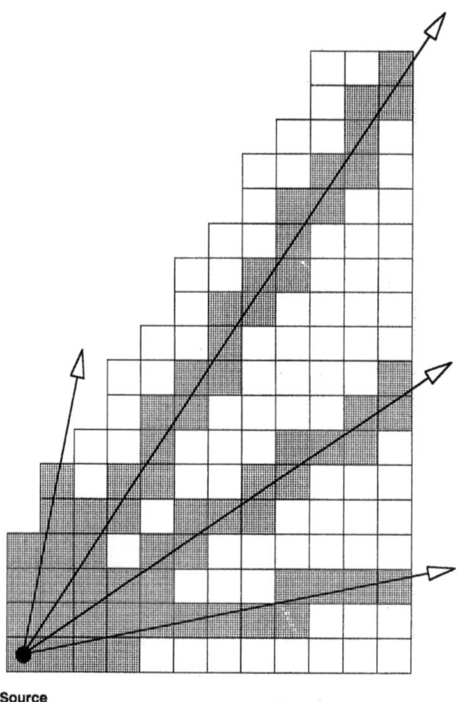

Figure 14.8. Example of the ray effect.

shaded mesh elements will "see" a certain neutron flux, since the others cannot receive any neutrons. The resulting lines are clearly visible on the diagram.

In practical cases, the artefact is never so clear, since the sources are spread out and there is scattering, but it can still be identified. The only way to improve this situation is to refine the angular meshing.

14.4.7. Handling the space variable

In any phase flux calculation, the handling of the space variable $\vec{r}$ is always associated with the handling of the angular variable $\vec{\Omega}$. If it was decided to handle the angular variable with discrete ordinates, there are many possible variants for processing the space variable. The most classic method is known as the "diamond scheme". More recently, several other methods have been suggested.

The diamond scheme[50] is illustrated in Figure 14.9 for a two-variable case only: a space variable x or r, and an angular variable μ[51].

[50] The term "$\mathbf{S_N}$ method" is often associated with "discrete ordinates + diamond scheme"; here we shall use the term "$\mathbf{S_N}$ method" to refer solely to the processing of angular variables by "discrete ordinates".
[51] Note that problems with a cylindrical symmetry, and of course problems with two or three space variables, must involve **both** angular variables.

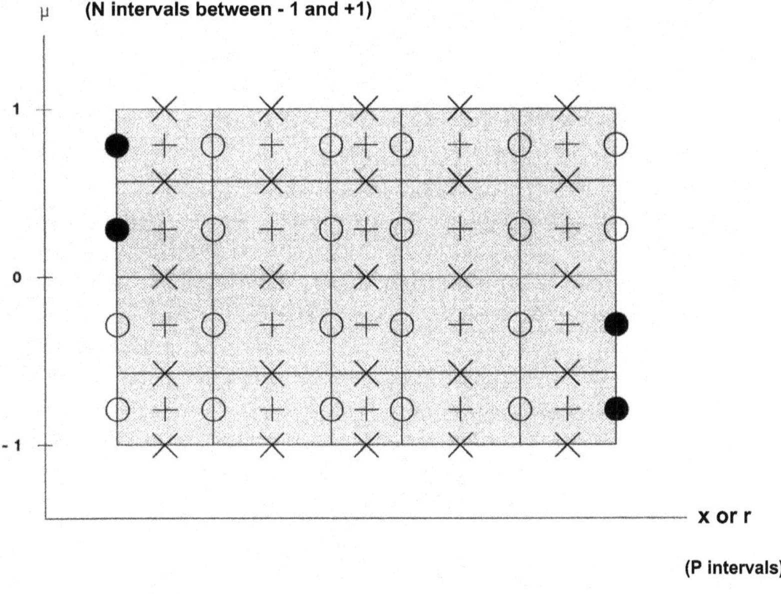

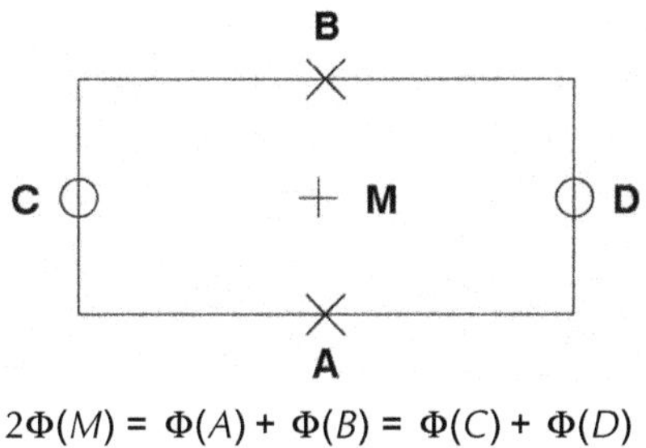

$$2\Phi(M) = \Phi(A) + \Phi(B) = \Phi(C) + \Phi(D)$$

Figure 14.9. Identification of the unknowns in the S_N + finite difference method.

During the iterative process, the equation to be processed at each step of the "internal" iteration is, for example for a problem that depends on x and μ:

$$\mu\frac{\partial\Phi(x,\mu)}{\partial x} + \Sigma(x)\Phi(x,\mu) = E(x,\mu), \tag{14.70}$$

where E is known (emissions calculated using the flux estimated at the previous iteration).

The angular variable μ was discretised according to N values μ_n (4 on the diagram, shown by horizontal lines of "circles"). In an analogous way, the space variable is discretised in P intervals (5 on the diagram).

The above neutron physics equation is written for each "box" centre identified by the + sign and the letter M; the derivative with respect to x or r is replaced by the finite difference quotient:

$$\frac{\Phi(D) - \Phi(C)}{x_D - x_C}, \quad \text{or:} \quad \frac{\Phi(D) - \Phi(C)}{r_D - r_C}, \tag{14.71}$$

and, similarly, the derivative with respect to μ (which does not appear in the above equation, but would enter into the similar equation with r because of the curvature) is approximated by: $[\Phi(B) - \Phi(A)]/(\mu_B - \mu_A)$.

Because there are more unknowns — all the circles, + signs and × symbols — than equations — one per + sign — the "diamond" equations are used in addition (refer to the bottom of Figure 14.9), which is equivalent to stating that the flux variation in a "box" is *assumed to be linear with respect to each of the variables.*

Under these conditions, the equation can be solved very quickly (inversion of the advection operator giving Φ if E is known) because the unknowns can be calculated from one term to the next, working line by line, i.e. with fixed μ, from the point of entry of the neutrons where the phase flux is known. That is the boundary condition of the problem (this input flux is often assumed to be zero)[52]. Note that these known values (black circles on Figure 14.9) are located to the left if μ is positive, and to the right if μ is negative. The line by line process is therefore carried out from left to right if μ is positive, and from right to left if μ is negative[53].

The method can be extended to problems with a greater number of variables, where the rectangular "boxes" become 3, 4, or 5-dimensional parallelepiped boxes, and the parallelograms *ABCD* become true "diamonds", hence the name of the method.

The disadvantage of the diamond scheme is that it can lead to negative flux values. This problem is solved by replacing an equation of the following type:

$$\Phi(M) = \frac{1}{2}\Phi(C) + \frac{1}{2}\Phi(D),$$

with:

$$\Phi(M) = \alpha\Phi(C) + (1 - \alpha)\Phi(D), \tag{14.72}$$

and by choosing α suitably to eliminate the problem. This inevitably leads to the iterations converging more slowly.

It is worth mentioning some of the other methods that use discrete ordinates:

- *Finite element*-type methods (the principle is explained in Chapter 6);
- "*Nodal*" methods. Their principle is as follows for the example of the planar geometry problem and a linear approximation in x:

$$\mu\frac{\partial\Phi(x,\mu)}{\partial x} + \Sigma(x)\Phi(x,\mu) = E(x,\mu). \tag{14.73}$$

  - The equation is written for a direction μ_n and a homogeneous spatial mesh element, assuming that $E(x, \mu_n)$ was previously approximated by an expression that is linear in x:

$$\mu\frac{\partial\Phi(x,\mu_n)}{\partial x} + \Sigma\Phi(x,\mu_n) = E_0 + E_1 x. \tag{14.74}$$

[52] The boundary condition is often also a reflection condition, in which case an iteration is necessary.

[53] To be precise, it should be noted that there are only $3PN$ equations for $3PN + P$ unknowns; an additional assumption must therefore be made to obtain the P missing equations.

  - This equation is integrated analytically by x.
  - The integration constant is calculated from the incoming flux, which is known because it is the flux leaving the previous mesh element (or the boundary condition).
  - The outgoing flux is deduced from this.
  - By projection on the polynomials 1 and x, we approximate the flux that was calculated in the mesh element with a linear expression $\Phi_0 + \Phi_1 x$.
  - This linear approximation is used to calculate the linear approximation $E_0 + E_1 x$ of $E(x, \mu)$, which will be used for the next iteration;

- The so-called *"characteristics"* methods. These are similar in principle to nodal methods, but instead of integrating the neutron physics equation along the directions of the coordinate axes, we integrate along the direction $\vec{\Omega}_n$ of neutrons travelling parallel to the discrete ordinate under consideration. These methods are of particular interest in dealing with complicated geometries.

  The calculation is performed by iteration, as with all numerical methods for solving the Boltzmann equation:

  - In a given spatial mesh element, where the right-hand side E is assumed to be known, the flux Φ is calculated analytically along $\vec{\Omega}_n$ according to the abscissa s on the characteristic using:

$$\operatorname{div}\left[\vec{\Omega}_n \Phi(\vec{r}, \vec{\Omega}_n)\right] + \Sigma\Phi(\vec{r}, \vec{\Omega}_n) = E(\vec{r}, \vec{\Omega}_n). \tag{14.75}$$

    The integration constant is determined from the current entering the mesh element along direction Ω_n.
  - All other spatial mesh elements are handled in a similar way; the calculation is explicit if the mesh elements are processed according to the path of the neutrons for the direction concerned.
  - Using the flux thus obtained, E is re-evaluated. The integrals on the direction are evaluated using the quadrature formula:

$$I = \int_{(4\pi)} F(\vec{\Omega}) d^2\Omega \simeq \sum_n w_n F(\vec{\Omega}_n). \tag{14.76}$$

    With respect to space, the function E is approximated by a constant in each mesh element, obtained by taking the average over space of the mesh elements of the functions considered (where appropriate, for Cartesian geometries, a polynomial approximation, first-order at the most, can be performed). The currents leaving the faces of the mesh element are also evaluated in this way (currents entering the next mesh element).
  - We recalculate Φ, and then E, and then Φ, and then E, and so on until convergence. In practice, the iterations are separated into internal iterations (handling of scattering in the group) and external iterations (handling of inter-group transfers and fissions).

Finally, note that it is possible to construct a simplified $\mathbf{S_N}$ method, called $\mathbf{SS_N}$, which is similar to the simplified spherical harmonic method; by assuming rotational symmetry of the local phase flux around the direction of the overall current, we can greatly reduce the number of discrete directions to be handled because the azimuth is no longer involved.

14.5. Concept of fundamental mode

14.5.1. Why is the fundamental mode of interest?

The "fundamental mode" is the name given to the neutron physics situation observed in an infinite homogeneous medium or an infinite regular lattice. The symmetry of the system leads to a solution of the Boltzmann equation that is simpler than the solution of a general case. This alone makes it of interest to study the problem, in order to reveal physical or mathematical aspects that can be more or less generalised. More specifically, the fundamental mode will be introduced very naturally when seeking to simplify the problem of a two-stage reactor calculation: the "mesh", handled as finely as possible, and the whole core, handled with a simpler model that is adjusted according to the fine mesh calculation.

For the elementary mesh calculation, the boundary conditions must be specified: the simplest way, even if it is not exactly realistic, is to use a *zero flux derivative*; in other words, as we have seen (§ 14.3.9), we imagine inserting this mesh element in an infinite, regular lattice of identical mesh elements, i.e. we place it in fundamental mode.

In this section, we begin by examining a few simple solutions of the Boltzmann equation in an infinite, homogeneous medium and in monokinetic theory: these will reveal the mathematical nature of the solutions in fundamental mode. We shall then generalise to spectrum theory, and then to the case of the infinite regular lattice.

14.5.2. A few analytical solutions of the Boltzmann equation in monokinetic theory

a) No-absorption case

In monokinetic theory, we consider an infinite, homogeneous medium that is purely scattering, with a given angular scattering distribution. In the absence of a source, we shall look for a solution that depends on space by x only, and therefore depends on the phase by μ only. If we seek this solution in the form of a Legendre expansion (see the equations in § 14.4.2), we note that only the coefficients Φ_0 and Φ_1 are nonzero; the phase flux has the following form:

$$\Phi(x,\mu) = A\left(x - \alpha - \frac{\mu}{\Sigma_{tr}}\right), \tag{14.77}$$

where A and α are constants. This solution is obviously physically acceptable only in the area of space where the flux is positive. It assumes sources at infinity, either to the right or to the left according to the sign of A.

It is remarkable that the Legendre expansion of the flux involves only two terms, even if the scattering cross-section involves all of the terms.

We also note that the integrated flux is:

$$\Phi(x) = 2A(x - \alpha), \tag{14.78}$$

and the integrated current is (directed along the $\vec{x}$ axis):

$$J(x) = -\frac{2A}{3\Sigma_{tr}}, \tag{14.79}$$

and therefore that these parameters are related by Fick's law with the following scattering coefficient:

$$D = \frac{1}{3\Sigma_{tr}}. \tag{14.80}$$

b) Linearly anisotropic scattering distribution; absorbent medium

Still in monokinetic theory, we now assume the homogeneous material to be scattering and absorbent and, as before, we look for a solution without a source at finite distance that is dependent only on x and μ.

By direct examination of either the Boltzmann equation for this case or the system of equations $\mathbf{P_N}$, we see that the flux can only depend on x by an exponential distribution, either increasing or decreasing, according to whether the sources are at infinity to the right or to the left. Let us take the latter case as an example:

$$\Phi(x, \mu) = \varphi(\mu)e^{-\kappa x}.$$

The constant κ and the phase distribution $\varphi(\mu)$ must be determined by the Boltzmann equation.

The equations $\mathbf{P_N}$ could be used, but this would require manipulating an algebraic system of infinite dimension (whatever the scattering distribution). If the scattering distribution is represented by a finite number of terms (expansion to order K), it is simpler to work directly on the integral/differential Boltzmann equation in the form shown in § 14.4.2, eq. (14.62), which is written out with a finite sum. After substituting in the factorised form of the phase flux, and then simplifying by $e^{-\kappa x}$, it gives:

$$-\kappa\mu\varphi(\mu) + \Sigma\varphi(\mu) = \sum_{k=0}^{k} \frac{2k+1}{2}\Sigma_{s,k}P_k(\mu)\int_{-1}^{+1} \varphi(\mu')P_k(\mu')d\mu'. \tag{14.81}$$

If we divide this equation by $\Sigma - \kappa\mu$, multiply it by $P_l(\mu)$ (for values of l from 0 to K) and integrate from -1 to $+1$, we obtain a linear system of $K + 1$ equations giving the $K + 1$ unknown integrals:

$$\varphi_k = \frac{2k+1}{2}\int_{-1}^{+1} \varphi(\mu')P_k(\mu')d\mu', \tag{14.82}$$

appearing on the right-hand side. This system involves the following coefficients:

$$G_{kl} = \int_{-1}^{+1} \frac{P_k(\mu)P_l(\mu)}{\Sigma - \kappa\mu}d\mu, \tag{14.83}$$

which are calculated analytically:

$$G_{00} = \frac{1}{\kappa}\ln\frac{\Sigma+\kappa}{\Sigma-\kappa}, \quad G_{01} = G_{10} = \frac{1}{\kappa}G_{00} - \frac{2}{\kappa}, \quad G_{11} = \frac{\Sigma^2}{\kappa^2}G_{00} - \frac{2\Sigma}{\kappa^2}, \tag{14.84}$$

and so on. For example, for $K = 1$, i.e. the so-called $\mathbf{B_1}$ approximation, the system is written as:

$$\begin{aligned} 2\varphi_0 &= \Sigma_{s,0}G_{00}\varphi_0 + \Sigma_{s,1}G_{01}\varphi_1, \\ \tfrac{2}{3}\varphi_1 &= \Sigma_{s,0}G_{10}\varphi_0 + \Sigma_{s,1}G_{11}\varphi_1. \end{aligned} \tag{14.85}$$

This system is homogeneous; the compatibility condition (zero determinant) gives the value of the constant κ; this is the solution of the following equation:

$$\frac{\kappa[\kappa^2 + 3\bar{\mu}c(1-c)\Sigma^2]}{c\Sigma[\kappa^2 + 3\bar{\mu}(1-c)\Sigma^2]} = \frac{1}{2}\ln\frac{\Sigma+\kappa}{\Sigma-\kappa} = \text{argth}\frac{\kappa}{\Sigma}, \tag{14.86}$$

with $c = \Sigma_s/\Sigma$, called the *"number of secondaries per collision"*. The constant κ is called the *"relaxation constant"* and its inverse $1/\kappa$ is the *relaxation length*.

c) Linearly anisotropic scattering distribution; multiplying medium

If the medium is multiplying, but remaining in monokinetic theory (assuming therefore that neutrons emitted by fission are at the same energy as that where neutrons are scattered and absorbed), a similar approach can be used by adding the cross-section of production by fission, $\nu\Sigma_f$, to the scattering cross-section Σ_s. We now set:

$$c = \frac{\nu\Sigma_f + \Sigma_s}{\Sigma}. \tag{14.87}$$

- If c is less than 1, all the above formulae apply exactly as they are[54].
- If c is greater than 1, i.e. if production wins out over absorption (k_∞ greater than 1), the constant κ becomes purely imaginary. We therefore set $\kappa = i\chi$ and look for a flux of the following form:

  $$\Phi(x,\mu) = \varphi(\mu)e^{-i\chi x}.$$

  The calculations are similar if we set:

  $$G_{kl} = \int_{-1}^{+1}\frac{P_k(\mu)P_l(\mu)}{\Sigma - i\chi\mu}d\mu, \tag{14.88}$$

  i.e.:

  $$\begin{aligned} &G_{00} = \tfrac{2\xi}{\chi}, \quad G_{01} = G_{10} = \tfrac{2i(1-\xi)}{\chi}, \quad G_{11} = \tfrac{2\Sigma(1-\xi)}{\chi^2}, \\ &\text{with} \quad \xi = \tfrac{\Sigma}{\chi}\text{Arctg}\tfrac{\chi}{\Sigma}. \end{aligned} \tag{14.89}$$

  For example, for $K = 1$ (linearly anisotropic scattering), the relaxation constant is given by the following equation:

  $$\frac{\chi[\chi^2 + 3\bar{\mu}c(c-1)\Sigma^2]}{c\Sigma[\chi^2 + 3\bar{\mu}(c-1)\Sigma^2]} = \text{Arctg}\frac{\chi}{\Sigma}. \tag{14.90}$$

---

[54] In the formulae, $\bar{\mu}$ is now the average cosine of the deflection angle of **all** re-emitted neutrons, whether re-emitted by scattering or by fission.

d) Linear combinations of the above solutions

If the medium in which the neutrons scatter is isotropic, which it is in most media, the $\vec{x}$ axis that we chose plays no special role.

In the above formulae, we can therefore replace:

$$\Phi(x,\mu) = \varphi(\mu)e^{-\kappa x} \quad \text{or} \quad \Phi(x,\mu) = \varphi(\mu)e^{-i\chi x},$$

by:

$$\Phi(\vec{r},\vec{\Omega}) = \varphi(\mu)e^{-\kappa\vec{u}\cdot\vec{r}} \quad \text{or} \quad \Phi(\vec{r},\vec{\Omega}) = \varphi(\mu)e^{-i\chi\vec{\mu}\cdot\vec{r}}, \tag{14.91}$$

where $\vec{u}$ is any unit vector in space, and μ is the scalar product $\vec{u}\cdot\vec{\Omega}$.

Moreover, because neutron physics is linear, any linear combination of the above functions with a weight $\Delta(\vec{u})$:

$$\Phi(\vec{r},\vec{\Omega}) = \int_{(4\pi)} \Delta(\vec{u})\varphi(\mu)e^{-\kappa\vec{u}\cdot\vec{r}}d^2u, \tag{14.92}$$

or:

$$\Phi(\vec{r},\vec{\Omega}) = \int_{(4\pi)} \Delta(\vec{u})\varphi(\mu)e^{-i\chi\vec{u}\cdot\vec{r}}d^2u, \tag{14.93}$$

is also a solution of the Boltzmann equation.

14.5.3. Concept of fundamental mode in a homogeneous medium in monokinetic theory

This solution, which appears in the infinite homogeneous and, strictly speaking, multiplying (k_∞ greater than 1) medium, is called the "fundamental mode". The equations we have just seen show that it can also be considered in a submultiplying medium or non-multiplying medium provided that sources are placed at infinity to "feed" the exponential.

This fundamental mode can be generalised:

- to exponential functions $e^{-i\vec{b}\cdot\vec{r}}$ that are space-dependent *via* any vector $\vec{b}$; in practice, the real values of this vector are the most interesting ones to consider, but the extension to complex vectors is possible[55];
- to the general Boltzmann equation, i.e. to a spectrum theory;
- to the case of the infinite regular lattice.

We shall examine each of these three points in turn.

In monokinetic theory and in an infinite and homogeneous medium, let the source be isotropic and of the following form:

$$S(\vec{r}) = se^{-i\vec{b}\cdot\vec{r}}. \tag{14.94}$$

[55] The notation b is generally chosen for the argument of Fourier transforms. It is used in discussions of the fundamental mode because there is a close connection between this mode and the Fourier transform of the kernel of the Boltzmann equation, i.e. its solution for a point source in an infinite homogeneous medium.

By substituting it into the Boltzmann equation:

$$\operatorname{div}\left[\vec{\Omega}\Phi(\vec{r},\vec{\Omega})\right] + \Sigma\Phi(\vec{r},\vec{\Omega}) = \int_{(4\pi)} \Sigma_s(\vec{\Omega}' \to \vec{\Omega})\Phi(\vec{r},\vec{\Omega}')d^2\Omega' + S(\vec{r}), \tag{14.95}$$

we note that the flux must be of the form

$$\Phi(\vec{r},\mu) = \varphi(\mu)e^{-i\vec{b}\cdot\vec{r}}, \tag{14.96}$$

with $\mu = (\vec{b}/b)\cdot\vec{\Omega}$ and:

$$-ib\mu\varphi(\mu) + \Sigma\varphi(\mu) = \int_{-1}^{+1} \Sigma_s(\mu' \to \mu)\varphi(\mu')d\mu' + s, \tag{14.97}$$

If the source comes from fission, it is expressed as a function of the flux:

$$S(\vec{r}) = \nu\Sigma_f\Phi(\vec{r}) \quad \text{with:} \quad \Phi(\vec{r}) = \int_{(4\pi)} \Phi(\vec{r},\vec{\Omega})d^2\Omega. \tag{14.98}$$

Therefore:

$$s = \nu\Sigma_f\varphi \quad \text{with:} \quad \varphi = \int_{-1}^{+1} \varphi(\mu)d\mu. \tag{14.99}$$

It is convenient to study this mode in the context of the so-called $\mathbf{B_K}$ approximation, which involves expanding the scattering angular cross-section to the order K in Legendre polynomials (bearing in mind that no additional approximation concerning the phase flux is applied). For example, approximation $\mathbf{B_1}$, the flux and the source are related by the following equations:

$$\begin{aligned} -2\varphi_0 + \Sigma_{s,0}G_{00}\varphi_0 + \Sigma_{s,1}G_{01}\varphi_1 + G_{00}\tfrac{s}{2} &= 0, \\ -\tfrac{2}{3}\varphi_1 + \Sigma_{s,0}G_{10}\varphi_0 + \Sigma_{s,1}G_{11}\varphi_1 + G_{10}\tfrac{s}{2} &= 0, \end{aligned} \tag{14.100}$$

with:

$$s = 2\nu\Sigma_f\varphi_0 \quad \text{since:} \quad \varphi = 2\varphi_0.$$

The coefficients G_{kl} are those written above in Paragraph **c**, replacing χ with b.

We have mentioned that, *a priori*, $\vec{b}$ is any vector; however, if we substitute the expression for the source s into the flux equations, we see that the system becomes homogeneous and that, therefore, there is no non-trivial solution unless its determinant is zero, i.e. unless *the modulus b of the vector $\vec{b}$ is equal to the relaxation constant* χ which is a solution of the equation written in Paragraph **c** above. Physically, this expresses the concept of a *critical condition*.

Linear combinations of solutions of this type with vectors $\vec{b}$ having the same modulus b are still solutions of the Boltzmann equation, and can therefore still be considered as "fundamental mode":

$$\begin{aligned} S(\vec{r}) &= \textstyle\int_{(4\pi)} \Delta(\vec{u})se^{-ib\vec{u}\cdot\vec{r}}d^2u, \\ \Phi(\vec{r},\vec{\Omega}) &= \textstyle\int_{(4\pi)} \Delta(\vec{u})\varphi(\mu)e^{-ib\vec{u}\cdot\vec{r}}d^2u \quad \text{with:} \quad \mu = \vec{u}\cdot\vec{\Omega}. \end{aligned} \tag{14.101}$$

They correspond to a critical situation if and only if b is equal to χ.

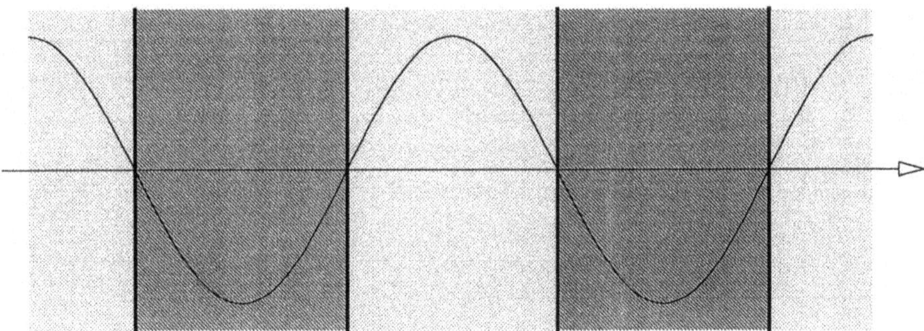

Figure 14.10. Image piles of a bare homogeneous pile, here in the form of an infinite plate. The curve represents the flux as a function of the abscissa *x*.

14.5.4. Physical interpretation of the fundamental mode

These equations might satisfy mathematicians because they satisfy the Boltzmann equation, but they can cause puzzlement to the poor physicist who is looking for a flux, which is by its nature both real and positive. They can represent a physical situation:

a/ If we select linear combinations that lead to real functions,

b/ If we restrict them to a region where they remain positive or zero.

For example, by combining the exponential functions e^{-ibx} and e^{+ibx} with the same weight, 1/2, we obtain $\cos(bx)$. If considered in the interval $-\pi/2b < x < +\pi/2b$, $\cos(bx)$ represents the flux that would be obtained in a homogeneous reactor in the shape of a plate bounded by these two planes, and therefore of thickness π/b, including the extrapolation distance. By other linear combinations it would be possible to find the flux for other pile shapes; for example: parallelepiped, cylinder, sphere, etc.

The fundamental mode therefore generalises, by analytical extension of the functions over all space, neutron physics situations encountered in bare homogeneous piles (including the extrapolation distance in the dimensions).

If the medium is multiplying, the curvature induced by a nonzero value of b simulates the leaks in an actual pile. For example, with the function $\cos(bx)$, fictitious (negative) piles can be seen to appear on either side of the actual pile, in plates where the cosine is negative. These negative piles are adjacent to positive piles, etc. (Figure 14.10). "Negative" neutrons emitted by the "negative" image piles can migrate to the "positive" piles and annihilate the "positive" neutrons emitted by the positive piles, and vice-versa. This mechanism provides a strictly equivalent replacement for leaks *via* the surface of a real pile of finite dimensions.

14.5.5. Existence and calculation of the leakage coefficient

The fundamental mode is characterised by a very interesting property: proportionality at every point between the current (integrated over the phases) and the flux gradient (integrated over the phases). In other words, Fick's law, which we proposed and attempted to justify in Chapter 5, to simplify the Boltzmann equation:

$$\vec{J}(\vec{r}) = -D\,\overrightarrow{grad}\,\Phi(\vec{r}), \tag{14.102}$$

becomes rigorously satisfied for this fundamental mode. In the first instance, we shall demonstrate the existence of this law. In the second instance, we shall examine the calculation of the coefficient of proportionality, D.

Assume the phase flux has the following form:

$$\Phi(\vec{r}, \vec{\Omega}) = \varphi(\mu)e^{-ib\vec{u}\cdot\vec{r}}, \tag{14.103}$$

with $\mu = \vec{u} \cdot \vec{\Omega}$. On the one hand, we conclude that:

$$\Phi(\vec{r}) = e^{-ib\vec{u}\cdot\vec{r}} \int_{-1}^{+1} \varphi(\mu)d\mu, \tag{14.104}$$

$$\overrightarrow{grad}\Phi(\vec{r}) = -ib\vec{u}e^{-ib\vec{u}\cdot\vec{r}} \int_{-1}^{+1} \varphi(\mu)d\mu, \tag{14.105}$$

and, on the other hand, since the phase current is the product of phase flux by $\vec{\Omega}$, that:

$$\vec{J}(\vec{r}) = \int_{(4\pi)} \vec{\Omega}\Phi(\vec{r}, \vec{\Omega})d^2\Omega = \vec{u}e^{-ib\vec{u}\cdot\vec{r}} \int_{-1}^{+1} \mu\varphi(\mu)d\mu, \tag{14.106}$$

(Only the integral along the $\vec{u}$ axis is nonzero, for symmetry reasons.) Note that these two vectors are in fact proportional at every point.

This remains true for the general fundamental mode, i.e. if we take any combination of functions of this type with various vectors $\vec{u}$ and weights $\Delta(\vec{u})$, but the same modulus b. Indeed, $\varphi(\mu)$ depends on the modulus of $\vec{b}$ *via* the coefficient $ib\mu$ appearing in the equation that governs this function and the coefficients G_{kl} that are introduced in the handling of $\mathbf{B_K}$, but not its direction[56].

These formulae also show that the diffusion coefficient is given by the following formula:

$$D = \frac{\int_{-1}^{+1} \mu\varphi(\mu)d\mu}{ib\int_{-1}^{+1} \varphi(\mu)d\mu}. \tag{14.107}$$

If we break down $\varphi(\mu)$ into Legendre polynomials:

$$\varphi(\mu) = \sum_{k=0}^{\infty} \varphi_k P_k(\mu) \quad \text{with:} \quad \varphi_k = \frac{2k+1}{2} \int_{-1}^{+1} \varphi(\mu')P_k(\mu')d\mu', \tag{14.108}$$

and if we note that $P_0(\mu) = 1$ and that $P_1(\mu) = \mu$, we can also write:

$$D = \frac{\varphi_1}{3ib\varphi_0}. \tag{14.109}$$

[56] If the medium were anisotropic, $\varphi(\mu)$ would also depend on $\vec{u}$ and Fick's law would no longer be exactly satisfied for linear combinations of exponential modes.

Note that the diffusion coefficient defined in this way in fundamental mode is *dependent*[57] on the value of the parameter b. That is why this coefficient $D(b)$ appearing in the fundamental mode is sometimes called the *leakage coefficient*[58] to distinguish it from the usual diffusion coefficient of Fick's phenomenological law.

For example, in the $\mathbf{B_1}$ approximation, the system of two equations (14.100) gives the two coefficients φ_0 and φ_1 (both proportional to s), giving D. After performing all necessary calculations, we obtain:

$$D = \frac{1}{3(\gamma\Sigma - \Sigma_{s,1})}, \tag{14.110}$$

with:

$$\gamma = \frac{1}{3}\frac{\frac{b}{\Sigma}\mathrm{Arctg}\frac{b}{\Sigma}}{1 - \frac{\Sigma}{b}\mathrm{Arctg}\frac{b}{\Sigma}} = 1 + \frac{4}{15}\left(\frac{b}{\Sigma}\right)^2 + \cdots \tag{14.111}$$

Note that, at the limit $b \to 0$, i.e. a pile whose dimensions tend to infinity, the coefficient γ tends to 1 and the diffusion coefficient tends to the usual value of $1/3\Sigma_{tr}$.

14.5.6. Balance in fundamental mode

Returning to the first of the $\mathbf{B_K}$ equations, the one expressing the neutron balance for all phases[59]:

$$-2\varphi_0 + \Sigma_{s,0}G_{00}\varphi_0 + \Sigma_{s,1}G_{01}\varphi_1 + G_{00}\frac{s}{2} = 0. \tag{14.112}$$

If we substitute in $\varphi_1 = 3ib\varphi_0 D$ and then perform certain algebraic manipulations, we obtain:

$$-Db^2\varphi - \Sigma_a\varphi + s = 0.$$

Here we have the equation expressing the neutron balance of a bare homogeneous pile dealt with using one-group theory (see Chapter 6). This observation again shows the complete equivalence between the bare homogeneous pile problem (with the neglected extrapolation distance) and the fundamental mode problem. The only detail that this latter approach changes, is to replace the constant D with a function $D(b)$ that depends on the modulus of the vector $\vec{b}$.

Even if $\vec{b}$ can be any vector *a priori*, in practice we generally take the value that lets us "loop" the neutron balance in a chain, i.e. such as:

$$s = \nu\Sigma_f\varphi.$$

This value is therefore given by the following equation:

$$Db^2\varphi + \Sigma_a\varphi = \nu\Sigma_f\varphi,$$

i.e.:

$$Db^2 + \Sigma_a = \nu\Sigma_f,$$

[57] This is a weak dependence in practice.

[58] It refers to leaks in the sense of "image piles".

[59] This equation is the projection on P_0 of the equation giving the phase flux, and thus the integral of this equation over all neutron directions.

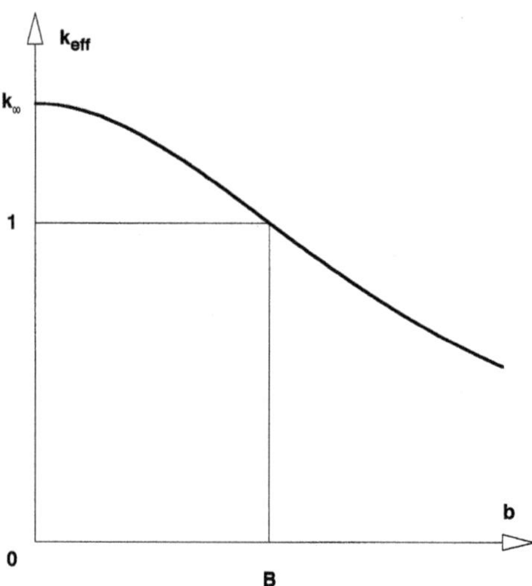

Figure 14.11. Critical value *B* of the parameter *b*, square root of the material buckling.

(where the diffusion coefficient depends on b). This *critical condition* expresses the equality between production and elimination (leakage and absorption) that ensures the existence of a *steady-state mode*. This is usually written in the following form:

$$k_{\text{eff}} = \frac{\nu\Sigma_f}{\Sigma_a + Db^2} = 1, \tag{14.113}$$

or

$$k_{\text{eff}} = \frac{k^*}{1 + M^{*2}b^2} = 1, \tag{14.114}$$

by setting:

$$k^* = \frac{\nu\Sigma_f}{\Sigma_a}, \quad \text{and:} \quad M^{*2} = \frac{D}{\Sigma_a}. \tag{14.115}$$

These expressions are reminiscent of those we write in "one group-diffusion theory". The factor k^* is here identified with the usual factor k_∞ (production by absorption in the absence of leakage), but this will no longer hold true, as we shall see, in spectrum theory, where k^* becomes a function of b; the area M^{*2} — which depends on b even in monokinetic theory — is similar to the migration area.

The value of b that achieves criticality (see Figure 14.11) is generally written as B (in monokinetic theory it is the relaxation constant χ for the multiplying medium that we introduced and calculated below, § 14.5.2, in theory $\mathbf{B_1}$).

The square B^2 of this parameter generalises the concept of "material buckling" (§ 6.1.2). Note that b^2 is a "geometric buckling", since:

$$\Delta e^{-i\vec{b}\cdot\vec{r}} = -b^2 e^{-i\vec{b}\cdot\vec{r}}.$$

The critical condition in fundamental mode can therefore be written, as for a bare homogeneous pile, in the following form:

material buckling = geometric buckling

14.5.7. Generalisation to the spectrum case

All of the results seen above and the proofs used to establish them remain practically unchanged when going from a monokinetic theory to a spectrum theory. The essential points can be summarised using, for example, the variable v to describe the spectrum:

- if the neutron source is an exponential:

$$S(\vec{r}) = se^{-i\vec{b}\cdot\vec{r}}, \tag{14.116}$$

the same applies to the flux, which thus appears in *factorised* form:

$$\Phi(\vec{r}, v, \mu) = \varphi(v, \mu)e^{-i\vec{b}\cdot\vec{r}}, \tag{14.117}$$

with $\mu = \vec{b} \cdot \vec{\Omega}$;

- the spectral and angular part of the flux is obtained by substituting these expressions into the Boltzmann equation; the exponentials representing the behaviour in space are simplified:

$$-ib\mu\varphi(v, \mu) + \Sigma(v)\varphi(v, \mu) = \int_0^\infty \int_{-1}^{+1} \Sigma_s[(v', \mu') \to (v, \mu)]\varphi(v', \mu')dv'\, d\mu' + s\chi(v); \tag{14.118}$$

- if the source is from fissions, it can be expressed as a function of the flux:

$$S(\vec{r}) = \int_0^\infty \nu\Sigma_f(v)\Phi(\vec{r}, v)dv, \tag{14.119}$$

with:

$$\Phi(\vec{r}, v) = \int_{(4\pi)} \Phi(\vec{r}, v, \vec{\Omega})d^2\Omega. \tag{14.120}$$

And therefore:

$$s = \int_0^\infty \nu\Sigma_f(v)\varphi(v)dv \quad \text{with:} \quad \varphi(v) = \int_{-1}^{+1} \varphi(v, \mu)d\mu; \tag{14.121}$$

- *Fick's law* is strictly satisfied with a diffusion coefficient that depends not only on v, but also on b. If the medium is isotropic, this result remains valid for combinations of modes with $\vec{b}$ vectors with the same modulus;
- the *flux* integrated over the phases is a solution of the following equation:

$$-D(v)b^2\varphi(v) - \Sigma(v)\varphi(v) + \int_0^\infty \Sigma_s(v' \to v)\varphi(v')dv' + s\chi(v) = 0; \tag{14.122}$$

- it can be established that in the $\mathbf{B_1}$ approximation, the *diffusion coefficient* is itself given by an integral equation:

$$D(v) = \frac{1}{3\gamma(v)\Sigma(v)}\left[1 + 3\int_0^\infty \Sigma_{s,1}(v' \to v)\frac{\varphi(v')}{\varphi(v)}D(v')dv'\right]. \tag{14.123}$$

This integral equation can be treated as a multigroup approximation, like the one giving the flux. Iterations between the two equations (flux and diffusion coefficient) must be performed;

- by integrating the flux equation over all speeds, scattering is simplified between the second and third terms; there remains an equation expressing the overall evaluation in any finite or infinitesimal volume:

$$\int_0^\infty D(v)b^2\varphi(v)dv + \int_0^\infty \Sigma_a(v)\varphi(v)dv = \int_0^\infty \nu\Sigma_f(v)\varphi(v)dv, \tag{14.124}$$

(total production equal to the total number of eliminations by absorption and leakage);

- by setting:

$$k^* = \frac{\int_0^\infty \nu\Sigma_f(v)\varphi(v)dv}{\int_0^\infty \Sigma_a(v)\varphi(v)dv} \quad \text{and:} \quad M^{*2} = \frac{\int_0^\infty D(v)\varphi(v)dv}{\int_0^\infty \Sigma_a(v)\varphi(v)dv}, \tag{14.125}$$

it is possible to write this balance equation in an identical form to the one obtained in monokinetic theory:

$$k_{\text{eff}} = \frac{k^*}{1 + M^{*2}b^2} = 1. \tag{14.126}$$

Note that, now, not only M^{*2} but also k^* are dependent on b, since φ and D are dependent on it:

- we can show that, *when b tends to zero:* **a/** *k^* tends to the multiplication factor in the absence of leakage k_∞;* **b/** *M^{*2} tends to the migration area M^2*, i.e. a sixth of the mean square of the crow-fly distance of neutrons from birth to absorption in the infinite homogeneous medium;

- the concept of material buckling B^2 as a value of b^2 reaching criticality ($k_{\text{eff}} = 1$) remains unchanged.

14.5.8. Concept of fundamental mode in a regular lattice

A fundamental mode does not appear only in infinite homogeneous media, but also in infinite regular lattices. The theory is similar, but more difficult. The essential aspects are as follows:

- for a source of the form $S(\vec{r}) = s(\vec{r})e^{-i\vec{b}\cdot\vec{r}}$ the flux is *factorised* and has the form $\Phi(\vec{r}, v, \vec{\Omega}) = \varphi(\vec{r}, v, \mu)e^{-i\vec{b}\cdot\vec{r}}$; in these expressions, s and φ are *functions with lattice periodicity* with respect to space, meaning that they return to the same value at analogous points of the various mesh elements;

- the functions s and φ have complex values; the real parts of $s(\vec{r})e^{-i\vec{b}\cdot\vec{r}}$ and $\varphi(\vec{r}, v, \mu)e^{-i\vec{b}\cdot\vec{r}}$, when positive, simulate the situation in a pile made up of this lattice. If we set $\varphi = f - ig$, where f and g are real and periodic, the flux appears as the sum of two terms: in each mesh element (assumed to be symmetric), it is the sum of a *symmetric* part $f \cos\vec{b}\cdot\vec{r}$ and an *antisymmetric* part $g \sin\vec{b}\cdot\vec{r}$;

- to calculate the functions f and g, which are also space-dependent, a transport calculation on the mesh scale must be performed. In the APOLLO code, for example, this is done using the method of collision probabilities (only the symmetric term is calculated in this code);

- on a macroscopic scale, a lattice is more or less anisotropic. The same applies to the diffusion coefficient, which is now a *tensor*, not a scalar. A theory $\mathbf{B_K}$ of this tensor is possible, but it would become complicated because transport problems would arise in addition to the aspects already discussed. (In the APOLLO code, only reasonably simplified calculation options are offered.) The material buckling concept also becomes anisotropic.

14.6. Use of Monte Carlo techniques in neutron physics

14.6.1. Outline of the Monte Carlo method

The Monte Carlo method is often used in physics; increasingly so as computing power grows. Methods like Monte Carlo are found to be useful because they can handle problems with few approximations. Their disadvantage is that they require many simulations and therefore use a great deal of machine time: especially when great accuracy is required, because the statistical uncertainty of a result obtained by the Monte Carlo method decreases with the number N of simulations only as an inverse square root (for example, a ten-times increase in accuracy requires 100 times more simulations, and therefore that much more computer time).

The Monte Carlo method takes its name from the fact that it is based on the random selection of random events — an activity for which this city in Monaco is renowned.

The following definition of the Monte Carlo method was stated by Iván Lux and László Koblinger (refer to bibliography):

In all applications of the Monte Carlo Method a stochastic model is constructed in which the expected value of a certain random variable is equivalent to the value of a physical quantity to be determined. This expected value is then estimated by the average of several independent samples representing the random variable introduced above. For the construction of the series of independent samples, random numbers following the distributions of the variable to be estimated are used.

14.6.2. Analogous simulation and non-analogous simulation

We generally draw a distinction between:

- *analogous* simulations, when the stochastic model *copies* the physical phenomenon concerned as closely as possible (e.g. path of a neutron), and

- *non-analogous* simulations otherwise (e.g. calculation of an integral);

- and *intermediate* cases (e.g., calculation of a biological protection with a particle path biasing technique).

The stochastic model is called a *game*. This game is a random process which, when carried out, attributes a value X to a random variable. The value obtained is called the *score*.

If we are examining a physical problem whose solution is characterised by a numerical value Y and we wish to use the Monte Carlo method to solve the problem, we must choose a game and a score such that:

$$E(X) = Y, \tag{14.127}$$

where $E(X)$ denotes the mathematical expectation of the random variable. The value of Y will be estimated by playing many times and taking the arithmetic mean of the scores X obtained. In accordance with the law of large numbers, its value will tend towards the mathematical expectation of this variable.

Buffon's needle is a simple, classic example of a non-analogous simulation. The purpose of this game is to estimate π. It consists of throwing a needle of length b on a parquet floor with slats of width a, and scoring as follows:

- $X = 1$ if the needle cuts across two slats;
- $X = 0$ if the needle rests on only one slat,

and then evaluating the mean value of X after a large number of throws.

It can easily be shown that this stochastic model satisfies the objective; for example, if $a = b$:

$$E(X) = 2/\pi.$$

Although Buffon's game might be very entertaining, it is clearly not very effective[60] at evaluating π.

Another application of the Monte Carlo method is to evaluate an integral:

$$S = \int_{(D)} f(p)dp, \tag{14.128}$$

where D is a multi-dimensional domain, and p is the "point" running in this domain[61]. If we choose a point P at random in D with a normed probability distribution $g(p)\,dp$ (for example, a uniform distribution: $g(p)dp = dp/V$, where V is the volume of D) and if we calculate the random variable $X = f(P)/g(P)$, we immediately see that the mathematical expectation of X is the sought-after integral S:

$$E(X) = \int_{(D)} Xg(p)dp = \int_{(D)} \frac{f(p)}{g(p)} g(p)dp = \int_{(D)} f(p)dp = S. \tag{14.129}$$

Note: Calculation of the variance under the same conditions shows that the *ideal game* is obtained by choosing $g(p)$ equal to $f(p)/S$, because the score is then equal to S irrespective

[60] A well-chosen series expansion, for example, performs far better.

[61] For example, the Monte Carlo method can be used to evaluate the no-collision flux and the associated dose in a given location for a source of ionising radiation; p is then the set of variables "emission point + emission direction + emission energy". Note that, in this context, the simulation is analogous, but this is not necessarily true in general for problems in which the Monte Carlo method is used to calculate integrals.

of the P that is drawn and the variance is zero; but this ideal case is theoretical because it assumes that S is known, i.e. the problem has been solved. This does however illustrate the fact that it is preferable to choose an (integrable) function g that is "similar" to f to reduce the statistical error of the Monte Carlo calculation.

14.6.3. Overview of sampling problems

Whatever the problem being examined, the use of Monte Carlo involves the drawing of one or more random variables according to probability distributions that are given. We use a capital letter, e.g. X, to denote the random variable concerned or the value obtained by drawing. The probability distribution governing this variable will be characterised by the *probability density* $f(x)$ or the *distribution function* $F(x)$:

$$\begin{aligned} &Pr\{x < X \leqslant x + dx\} = f(x)dx, \\ &Pr\{X \leqslant x\} = F(x), \end{aligned} \tag{14.130}$$

where $Pr\{\ldots\}$ is the probability of the event shown between curly brackets. We can see that the distribution function is the integral of the probability density:

$$\begin{aligned} &F(x) = 0 && (x < a), \\ &F(x) = \int_a^x f(x')dx' && (a \leqslant x \leqslant b), \\ &F(x) = \int_a^b f(x')dx' = 1 && (x > b), \end{aligned} \tag{14.131}$$

where a and b denote the limits between which X can be selected.

In practice, whether a table, a calculator, or a computer is used, it is possible to generate a random variable[62] Ξ that is uniformly distributed[63] between 0 and 1:

$$\begin{aligned} &f(\xi) = 0, \; F(\xi) = 0, \; (\xi < 0), \\ &f(\xi) = 1, \; F(\xi) = \xi, \; (0 \leqslant \xi < 1), \\ &f(\xi) = 0, \; F(\xi) = 1, \; (\xi \geqslant 1). \end{aligned} \tag{14.132}$$

To go from the selection of Ξ to that of a random variable X of density $g(x)$, *we identify the probabilities of events* $\Xi \leqslant \xi$ and $X \leqslant x$, i.e. *the values of the distribution functions*:

$$F(\xi) = \xi = G(x). \tag{14.133}$$

In other words, if the value Ξ was chosen for the variable equidistributed between 0 and 1, we deduce X from this by writing:

$$X = G^{-1}(\Xi), \tag{14.134}$$

where G^{-1} is the inverse function of G.

If this inverse function is simple, this calculation can be performed directly.

62 This is actually a pseudo-random variable, i.e. one obtained by a perfectly deterministic process but generating numbers with all the properties of a random variable. For example: the decimals of the number π. In practice, we use the remainder of a whole-number division by a carefully-chosen large prime divisor.

63 We shall assume that 0 can be obtained but that 1 cannot. If necessary, replace ξ by $1 - \xi$ if taking the logarithm, for example; this avoids possibly having to perform computer troubleshooting later on.

- *Example 1:* selection of a longitude Φ distributed isotropically between $-\pi$ and $+\pi$: $\Phi = \pi(2\Xi - 1)$.
- *Example 2 :* selection of an optical path Ω of probability density $e^{-\omega}$ between zero and infinity: $\Omega = -\ln(1 - \Xi)$.
- *Example 3:* selection of a discrete variable k, e.g. of the k *type* of a nuclear reaction, knowing that it took place:

$$\sum_{i=1}^{k-1} \frac{\sigma_i}{\sigma_t} < \Xi \leqslant \sum_{i=1}^{k} \frac{\sigma_i}{\sigma_t}. \tag{14.135}$$

  The probability of reaction i is the quotient of the cross-section σ_i of this reaction divided by the total cross-section σ_t; the distribution function of the discrete variable k rises in steps as a function of k. Another, similar, example: determining the element struck by a neutron if the collision took place in a mixture.

If the function G is not a simple analytical function, but has been tabulated, we can return to the previous case in each interval if we assume that the variation is linear between the tabulated points. *Example:* selecting the deflection angle Θ of a scattered neutron from a table of $G(\theta)$ (we generally tabulate the $n + 1 \cos\theta_i$ values leading to intervals of equal probability $1/n$).

For certain probability distributions, methods that are more economical in terms of calculation time than the direct determination of G^{-1} have been developed. *Example:* for the variable X governed by the power law:

$$g(x) = (n + 1)x^n, \quad G(x) = x^{n+1},$$

generating Ξ from $n + 1$ draws, and then finding the maximum:

$$X = \max(\Xi_1, \Xi_2, \ldots, \Xi_{n+1}), \tag{14.136}$$

turns out to be more economical than calculating an $(n + 1)^{th}$ root:

$$X = \Xi^{1/(n+1)}.$$

(*Proof:* the probability that X is less than x is the probability that Ξ_1 is less than x, multiplied by the probability that Ξ_2 is less than x,..., multiplied by the probability that Ξ_{n+1} is less than x, i.e., since each of these independent probabilities is equal to x: $x \times x \times \cdots \times x = x^{n+1}$.)

Another example: to select a point in a circle uniformly, it can be more economical to uniformly choose the Cartesian coordinates in the circumscribed square and to reject any points that are outside the circle, rather than selecting the azimuth and the distance to the centre, r, according to a distribution function in r^2.

An analogous principle is seen in the *rejection method,* which can usually be used when g is calculable: Ξ_1 is drawn uniformly between a and b, and then Ξ_2 uniformly between 0 and the value g_{max} that g can reach in the interval $[a, b]$. The selection is accepted if $\Xi_2 < g(\Xi_1)$, and we take $X = \Xi_1$; otherwise, this dual draw is repeated.

(*Proof:* a/ The probability of obtaining Ξ_1 in the interval $]x, x + dx]$ is $dx/(b - a)$; b/ The probability of obtaining X_2 less than $g(x)$ is $g(x)/g_{max}$; c/ By taking the product of these two probabilities and renormalising the distribution thus obtained to 1 to take the success rate into account, we do indeed find the desired distribution $g(x)\,dx$ for the probability of obtaining X between x and $x + dx$.)

14.6.4. Analogous simulation of a neutron path

The path of a neutron from emission to disappearance by absorption in the system or outside (leakage) is a series of *independent random events,* which are either *transport*-type events (travelling through space in a straight line) or *collisions* with an atomic nucleus. The process is *Markovian*[64], which means, in this case, that the event about to occur is dependent only on the current state of the neutron (its position and velocity), and is independent of the details of what occurred before[65].

Figure 14.12, shows how the simulated "life story" of a neutron might look if plotted on a flowchart. In practice, the selection of random variables and the calculations are programmed in a computer code so that many stories — typically, anywhere from a few thousand to a few million — can be simulated before the statistical processing of the results.

In this diagram, the story starts at the "emission" box and continues until the "end of story" box, taking various paths through the chart. A shaded rectangle represents the selection of a random variable. A circle represents a direction imposed according to the result of the selection. A white rectangle represents a calculation.

The source is assumed to be distributed in space, energy, and angle according to given distributions: the flowchart begins with the corresponding selections. The diagram is centred on the selection of the optical path. We examine whether the neutron remains in or leaves the homogeneous medium. If it remains, we deal with the interaction; if it leaves, we see whether the interface it crossed is at the surface of the domain under consideration, or whether it is an internal interface. If the neutron has reached the limit, we evaluate the leak or handle the reflection, according to the chosen distribution, if applicable. If the neutron crosses an internal interface in the domain concerned, we must calculate the trajectory beyond and repeat the test; rather than working with the remainder of the optical path beyond the interface reached, it is simpler (and strictly equivalent given the Markovian nature of the process) to reposition the neutron at the interface with its velocity, and repeat the selection of an optical path as if the point were an emission point. For a collision, the element concerned, and then the reaction concerned, must be defined by random selection. In the case of absorption, the story is brought to an end. In the event of scattering, we must define whether it is elastic or inelastic (and, in this case, choose the excitation energy), randomly select both scattering angles and, finally, calculation the post-collision energy using the laws of conservation of momentum and energy.

14.6.5. Estimating the multiplication factor

In problems where neutrons are emitted by fission, the source is unknown but results from the flux. Moreover, if the calculation is performed at steady state, a critical parameter must be introduced. In general, the effective multiplication factor k_{eff} is used. This is defined as the number by which the fission products must be divided to achieve system criticality.

These two aspects require an iterative calculation:

$$S^{(0)} \rightarrow \Phi^{(0)} \rightarrow S^{(1)} \rightarrow \Phi^{(1)} \rightarrow S^{(2)} \rightarrow \Phi^{(2)} \rightarrow \cdots$$

[64] Andrei Andreyevich Markov, Russian mathematician (1856–1922).

[65] In more descriptive terms: as the neutron goes about living its life, it forgets its past and is only aware of its present state.

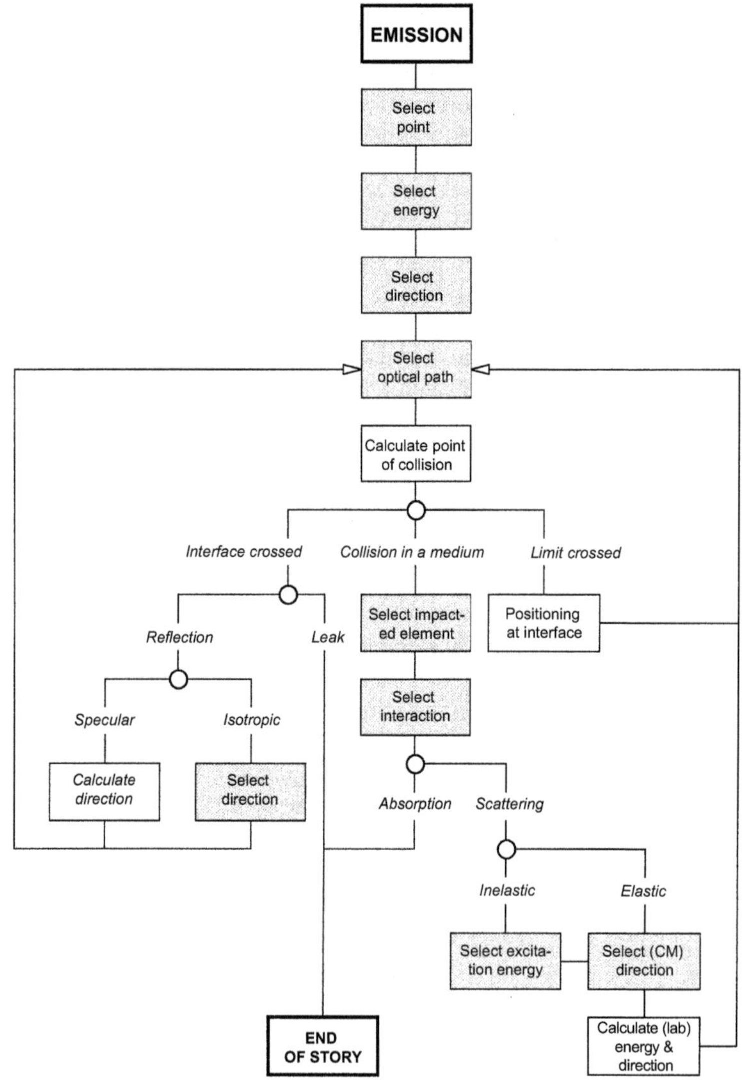

Figure 14.12. Simulation of the life story of a neutron using the Monte Carlo method.

It is standard practice to simulate successive batches of neutrons. The first batch is emitted according to $S^{(0)}$, giving $\Phi^{(0)}$ and $S^{(1)}$; the second is emitted according to $S^{(1)}$, giving $\Phi^{(1)}$ and $S^{(2)}$; and so on. The multiplication factor $k^{(n)}$ is calculated at each stage as the ratio of the number of neutrons obtained by fission to the number of neutrons emitted in the batch (the same number of neutrons is generally emitted in each batch).

The multiplication factor k_{eff} is finally obtained by taking the arithmetic mean of all[66] $k^{(n)}$. Note: To calculate this mean, and to avoid any bias, it is necessary to eliminate the *first*

[66] Weighted by the number of neutrons in each batch if the batches are not equal.

generations[67], for which the sources do not have their energy and space equilibrium distribution, and for which the number of descendants has not reached its asymptotic value.

Note: Working on successive batches also has the advantage of providing statistical distributions for the parameters of interest (reaction rate, multiplication factor, etc.) and therefore, in addition to an estimate of the parameter, an estimate of the standard deviation with which it is obtained. After dividing by the square root of the number N of batches used, we then have the estimate of the standard deviation on the average of the N batches.

14.6.6. Semi-analogous simulation of neutron paths

In reactor core neutron physics problems, flux levels are relatively uniform and analogous simulations are sufficient. In protection calculations, however, we often have to deal with problems where the flux varies by several decades between the level observed in the neighbourhood of sources (e.g. in the core) and the level of the room in which the dose of ionising radiation must be evaluated. In this type of problem[68], an analogous simulation would perform very badly. For example, if the protection attenuates the flux by a factor of 10^6, then 1,000,000 particles must be simulated, of which 999,999 unnecessarily, to obtain the one particle that provides the pertinent information (or: a thousand million to obtain a thousand useful ones with the desired information). Clearly, under these conditions, the Monte Carlo approach would be useless. That is why *biased simulation* (or semi-analogous) methods were developed.

References to many biasing techniques are given in the bibliography, and so we shall not discuss them in detail. Two ideas should however be mentioned:

- the idea of giving a weight to the particles, in proportion to their likely contribution to the result being sought. This should be associated with the Russian roulette consisting of "killing" particles whose weight is too low, with a probability that is proportional to the weight, and the *"duplication"* or *"splitting"* of particles that are too heavy. It can also sometimes be of interest to replace absorptions (capture and fission) by weight changes, without stopping the particles;

- the idea being to change the cross-sections to favour trajectories in the desired direction, by replacing the section Σ with:

$$\Sigma^* = \Sigma(1 - a\cos\theta),$$

  where a is a biasing coefficient between 0 and 1, and θ is the angle between the direction of the neutron and the direction of interest. This obviously leads to the true flux Φ being replaced by a biased flux Φ^*, but when we substitute into the Boltzmann equation, we can see that the biasing factor Φ^*/Φ is fairly simple to evaluate.

[67] To give a rough idea, let us say that this means about ten generations.

[68] The problem arises in a similar way (the same Boltzmann equation) for the transport of neutrons and gamma photons.

We should also mention *correlated* Monte Carlo simulations, where a simulation is repeated with the same selection of pseudo-random numbers in two very slightly different situations. This provides an evaluation of the sensitivity coefficients (reactivity effects, sensitivity to cross-sections or composition, etc.) without having the effect of the great statistical error on the difference between two neighbouring results that would exist if they had been obtained by independent simulations.

Along the same lines, a Monte Carlo calculation can also be used to evaluate not only a given physical parameter, but also its derivatives with respect to a given parameter, so that the variations can then be evaluated using a Taylor series expansion.

Exercises

A. Study of the Boltzmann equation

Exercise 14.1: equivalence of the two forms of the transport operator

Let $Q(\vec{r}, v, \vec{\Omega}, t)$ and $\Phi(\vec{r}, v, \vec{\Omega}, t)$ be the emission density and the resulting flux, respectively. Recall that the transport operator links Φ to Q.

a) Show this operator in its integral form.

b) Apply a small shift of magnitude ds, to the location of a notional detector, along $\vec{\Omega}$, with the operator.

c) Derive the differential form of the transport operator from the effect on it of the shift $\vec{\Omega}ds$.

d) Give a physical interpretation of the difference between observations made by two notional detectors separated by $\vec{\Omega}ds$.

Exercise 14.2: exact solution without absorption

In a homogeneous, non-absorbent material, without neutron sources at finite distances, show the general solution of the monokinetic, stationary, one-dimensional Boltzmann equation.

Exercise 14.3: relaxation length

a) In a homogeneous, diffusive, absorbing material, without neutron sources, show that the solution of the monokinetic, stationary, one-dimensional Boltzmann equation, is in the form $\Phi(x, \mu) = \varphi(\mu)e^{-\kappa x}$. Recall that the quantity $L^* = 1/\kappa$ is called the 'relaxation length' (see exercise **5.18**).

b) Using the Boltzmann equation for this case in its differential form, and assuming that the scattering is isotropic in the laboratory system, derive the equation giving κ.

c) Based on the same assumptions as used previously, derive the equation giving κ, from the Boltzmann equation in its integral form.

d) Compare the relaxation length with the diffusion length (see exercise **3.2**) for several example values of the number of secondaries per collision $c = \Sigma_s/\Sigma_t$.

e) Repeat part **b** for the linearly anisotropic collision hypothesis.

Exercise 14.4: diffusion length

The aim of this exercise is to find the general expressions describing the diffusion length L for a homogeneous, infinite medium, as in Exercise **3.2**, except no longer assuming that scattering is isotropic.

a) Begin by considering the neutrons performing exactly n elementary paths $\vec{\rho}_i$ ($i = 1, \cdots, n$), and define

$$\vec{R}_n = \vec{\rho}_1 + \vec{\rho}_2 + \cdots + \vec{\rho}_n;$$

thereby taking the average,

$$\langle \vec{R}_n^{\,2} \rangle = \langle \vec{\rho}_1^{\,2} \rangle + \langle \vec{\rho}_2^{\,2} \rangle + \cdots + \langle \vec{\rho}_n^{\,2} \rangle + 2 \langle \rho_1 \cdot \rho_2 \cdot \cos\theta_{12} \rangle + \cdots,$$

where θ_{ij} is the angle between the vectors $\vec{\rho}_i$, and $\vec{\rho}_j$. Write out the sum in full. Then show that the variables ρ and θ are independent, and hence simplify the result.

b) Define $\bar{\mu}$ to be the average cosine of the scattering angle. This angle, $\theta_{i,i+1}$ is a random variable, independent of i. Employ recursion and spherical trigonometry to show that $\langle \cos\theta_{i,i+k} \rangle = \bar{\mu}^k$. A spherical triangle drawn on the surface of a sphere is described by either its three dihedral *face-angles* A, B, C measured at the centre of the sphere, or the three angles a, b, c of its *sides* on the surface of the sphere. They are related by the formula

$$\cos a = (\cos b)(\cos c) + (\sin b)(\sin c)(\cos A).$$

c) Find the expression for $\langle \vec{R}_n^{\,2} \rangle$.

d) Weighting the $\langle \vec{R}_n^{\,2} \rangle$ by the probability p_n that a neutron carries out exactly n paths (see exercise **3.1**), find $\langle \vec{R}^2 \rangle$.

e) Show that the diffusion area $L^2 = \langle \vec{R}^2 \rangle / 6$ can be written in the form $L^2 = D/\Sigma_a$ with $D = 1/(3\Sigma_{tr})$ and $\Sigma_{tr} = \Sigma_t - \bar{\mu}\Sigma_s$. Notice that this diffusion area — calculated exactly here — is expressed using only the first moment, $\bar{\mu}$, of the collision law.

Exercise 14.5: integral kernel in one- and two-dimension geometries

Recall that the point kernel of the integral transport operator is $e^{-\tau}/(4\pi R^2)$, provided emission occurs isotropically.

a) Show by taking the integral along y and z that for a system with planar geometry, i.e. depending on x only, the kernel becomes $E_1(\tau_x)/2$. Substitution by other variables is necessary. Define τ_x to be the projection of the optical path $\vec{\tau}$ on the $\vec{x}$ axis.
Recall of the definition of the integral exponentials,

$$E_n(u) = \int_0^1 \exp\left(-\frac{u}{\mu}\right) \mu^{n-2}\, d\mu.$$

b) Show by taking the integral along z only that the kernel becomes $Ki_1(\tau_{xy})/(2\pi\rho)$ when the geometry of the system depends on x and y only. Again, it is necessary to make a suitable substitution. Define ρ to be the projection on the x-y plane of the true neutron path $\vec{R}$, and τ_{xy} the projection of the optical path $\vec{\tau}$ on this plane.
Recall of the definition of the Bickley functions,

$$\mathrm{Ki}_n(u) = \int_0^{\pi/2} \exp\left(-\frac{u}{\sin\theta}\right) \sin^{n-1}\theta\, d\theta.$$

B. Treatment of energy and time

Exercise 14.6: multigroup theory and treatment of the time derivative

a) In multigroup theory, any function $f(u)$ of the lethargy u is replaced by a set of discrete group values f_g, which can be interpretated as averaged values, while the lethargy integrals are replaced by corresponding sums over the groups.
For example, in exercise **7.8**, the integral equation

$$\int_{u-\varepsilon}^{u} f(u')\,P(u-u')\,du' = (1+a)\,f(u),$$

is replaced by the algebraic expression

$$\sum_h f_h\,P_{g-h} = (1+a)\,f_g,$$

where h is taken for the groups in the interval $[u-\varepsilon, u]$. Also assume that the groups are divided up over intervals with equal lethargy width $\Delta = \varepsilon/n$, where n is integer.
The solutions to these equations take the form

$$f(u) = C^t \exp(-mu)\,,$$

and

$$f_g = C^t \exp\left(-\tilde{m}\,\Delta g\right).$$

To simplify matters, assume that the probability $P = 1/\varepsilon$, and is constant over the interval $[u-\varepsilon, u]$. This approximation is accurate for a heavy nucleus. Compare m and $\tilde{m}$ according to a and n.

b) Consider the differential equation

$$\frac{df(t)}{dt} = f(t),$$

together with the discrete forms divided into equal time intervals Δ, which are either

$$\frac{f_{i+1} - f_i}{\Delta} = f_i,$$

or

$$\frac{f_{i+1} - f_{i-1}}{2\Delta} = f_i.$$

Compare the exact and approximate solutions — notice that they are exponential functions. Apart from its symmetry, what other advantage does the second discrete form have?

C. Collision probabilities

Exercise 14.7: Cauchy's theorem

Consider any convex body of volume V and surface S. The mean chord X, averaged isotropically over all directions is $\langle X \rangle = 4V/S$.

This theorem is due to Augustin Cauchy (1789-1857). It is important in neutron physics because the mean chord multiplied by the macroscopic cross-section represents in effect the 'opacity' of a volume 'seen' by the neutrons.

Previously, in chapter **3**, Exercise **7**, this theorem was demonstrated using a model with a physical basis, i.e. monokinetic particles, such as neutrons, flowing uniformly and isotropically throughout all space, and passing through a body. This exercise takes a more formal mathematical approach to the problem.

a) Define AB to be a chord and X its length. The point A on the surface is chosen at random with equal probability, i.e. proportional to the surface element d^2S. The direction of entry $\vec{\Omega}$ of the chord is made proportional to $d^2\Omega \cos\theta$, where $\theta < \pi/2$ is the angle between $\vec{\Omega}$ and the normal at A, meaning that the angular distribution is isotropic. This determines where B is. Give the expression for the probability density governing chords.

b) Show the double integral giving the mean chord $\langle X \rangle$.

c) Show that the integral with d^2S is equal to the volume. Then deduce the Cauchy theorem.

Exercise 14.8: reciprocity theorem (and absorption probabilities)

In Exercise **8.3** a direct demonstration of the reciprocity relationship between the probabilities P_{VS} and P_{SV} was proposed. This relationship can also be obtained using a notional physical model similar to the one in Exercise **3.7**.

The volume V — which is assumed to contain homogeneous, diffusive, absorbing material — with surface S, is divided and distributed evenly, thereby transfoming it into an infinite, homogeneous medium. A uniform, isotropic neutron source with intensity s is located in this infinite medium; consequently, a flux Φ is generated. Assume also that the neutrons are monokinetic with a velocity v.

a) Express, as functions of s, and the cross-sections of the medium, the flux Φ, the number of neutrons J_- crossing the surface per unit time and area, in the direction opposite to the normal, and the emission density $Q = s + \Sigma_s\Phi$.

b) How many collisions are observed per unit time in volume V?

c) By considering that these collisions represent both, the first ones occuring to neutrons produced or scattered in V before they exit, and the collisions of neutrons entering through S and suffering their first collision in V, show the relationship linking P_{VV} with P_{SV}.

d) Deduce the reciprocal relationship connecting P_{VS} with P_{SV}.

e) By following a similar argument, next establish the relationship between the *absorption* probabilities P^*_{VV} and P^*_{SV} in volume V for a neutron emitted uniformly and isotropically in V, and a neutron entering uniformly and isotropically through S, respectively.

Exercise 14.9: moments of the chord

Show the integrals giving the moments $\langle X^n \rangle$ of the chord X, with isotropic entry. Then find expressions for the first values of n for,

a) an infinite slab of thickness $2a$;

b) an infinite cylinder of radius R;

c) a sphere of radius R.

Exercise 14.10: first collision probability for a sphere

a) Find the probability P_{SS} for a homogeneous sphere of radius R.

b) Deduce the probabilities P_{SV}, P_{VS} and P_{VV}.

Exercise 14.11: first collision probabilities in a checkerboard pattern

Consider an 'infinite checkerboard' consisting of two types of alternating cells A and B. Define i or j to be regions in type A cells, and k or ℓ are regions in type B cells.

Find expressions giving the first collision probabilities P_{ij}, P_{ik}, P_{ki}, and $P_{k\ell}$, assuming that neutrons crossing the interfaces between the two types of cells exhibit isotropic, uniform behaviour. Why is this problem interesting?

Exercise 14.12: calculation of the disavantage factor

Consider a cell containing two regions labelled 1 and 2, representing fuel and moderator, respectively. The objective is to find the thermal utilisation factor f, which is the probability that a neutron emitted uniformly and isotropically in the moderator is eventually absorbed in the fuel.

a) Assuming that the neutrons are monokinetic, and considering the total reaction rates in each zone, show the equations giving the fluxes Φ_1 and Φ_2 as functions of the cross-sections, of the volumes, and of the first collision probabilities P_{ij} $(i, j = 1 \text{ or } 2)$.

b) From these equations, find the disavantage factor Φ_2/Φ_1, and the thermal utilisation factor f. *Hint:* use the reciprocity and conservation relationships in order to keep only the probability P_{11}.

c) Let P_{11}^{+} be the probability for a neutron generated uniformly and isotropically in a fuel element to suffer its first collision in this fuel element, without exit; and the Dancoff factor C, which is the probability that a neutron leaving uniformly and isotropically a fuel element enters another fuel element without an intermediate collision in the moderator. Find the probability P_{11} as a function of both these probabilities, and the physical and geometrical characteristics on the cell. Define S to be the area of the fuel-moderator interface, and assume that the neutrons cross the moderator-fuel interface uniformly and isotropically.

d) According to tabulated values in the literature, the Dancoff factor for square cells is $C_{\text{exact}} = 0.179$, while the corresponding figure for cylindrical cells is $C_{\text{cylinder}} = 0.158$. Given that $f \approx 0.94$ and $4V_2\Sigma_{a2}/S \approx 0.015$, estimate the error in f when a cylindrical geometry is adopted. Is this amount significant?

Exercise 14.13: numerical calculation of the collision probabilities

The first collision probability method introduces two approximations which have to be distinguished:

- the so-called 'flat flux' approximation which assumes that the flux, and the related quantities, reaction rates, sources, etc., are constant relative to the space variable $\vec{r}$ in each elementary volume. This constant can be interpretated as the mean value over the volume;
- the use of approximate numerical quadrature formulae for the calculation of the first collision probabilities.

Both approximations are illustrated here for a fictitious, elementary example: one-dimensional neutron migration. Imagine a 'string' made from a homogeneous medium. Scattering in the string is assumed to be isotropic, meaning that when it occurs there is an equal probability for a scattered neutron to go in either direction. It is also assumed that the neutrons are monokinetic. In these circumstance the Boltzmann equation reduces to

$$\Phi(x) = \int \frac{1}{2} e^{-|x-x'|} \left[S(x') + c\,\Phi(x') \right] dx',$$

where $c = \Sigma_s/\Sigma_t$, and where $1/\Sigma_t$ is taken as unit of length. It can be proved that, in this case, the diffusion equation becomes rigourous; however, we wish to examine an integral treament of the first collision probability type.

a) Firstly, consider the relaxation mode $\Phi(x) = A\,e^{-\kappa x}$ without a source, then find the exact value of the parameter κ.

b) The $\vec{x}$ axis is divided into finite intervals of length h labelled i or j, and the discretised expression of the previous equation is then

$$\Phi_i = \sum_{j=-\infty}^{+\infty} P_{ji} \left[S_j + c\,\Phi_j \right].$$

Calculate the exact values of the first collision probabilities P_{ji}. Verify that their sum over $i = 1$, exactly. Show that the solution of the discretised equation is in the form $\Phi_i = A\,e^{-\tilde{\kappa}x}$. Show the equation giving $\tilde{\kappa}$, and show that

$$\frac{\tilde{\kappa}}{\kappa} = 1 - \frac{ch^2}{6} + \cdots$$

c) Now, the exact calculation of the probabilities is replaced with

$$P_{ji} \simeq \frac{h}{2} e^{-h|i-j|} \quad (i \neq j); \qquad P_{jj} = 1 - \sum_{i \neq j} P_{ji}.$$

Repeat the previous problem, and show that, owing to a fortuitous compensation, the error on κ is smaller than in the preceding case.

Exercise 14.14: Wigner, Bell-Wigner, and Carlvik approximations

Apart from in a few cases such as the sphere (Exercise **14.10**), the formulae for the first collision probabilities are complicated, and have no analytical solution. For this reason approximate formulae sometimes are employed. Three examples are given below. These are based on a (infinite) cylinder, the most usual geometry of the fuel elements.

a) Recall that the probability P_{VV} is a function, for a given form, of only one parameter, namely the opacity $\omega = \Sigma\langle X\rangle$, which is the product of the macroscopic total cross-section and the mean chord. How does this function behave for the small and the large opacities? *Hints:* for small opacities, see exercise **8.4**; large opacities, start from $P_{SS} \simeq 0$, then use the complementarity and reciprocity relationships.

b) *Wigner's approximation:*

$$P_{VV} \simeq \frac{\omega}{1+\omega}.$$

Show this is consistent with the asymptotic behaviour.

c) *Bell-Wigner's approximation:* this is a generalization of the previous approximation by virtue of an adjustable coefficient b, defined by

$$P_{VV} \simeq \frac{\omega}{b+\omega}.$$

Which value of b is consistent with the behaviour at the origin?

d) *Carlvik's approximation:* it is a generalization of the previous approximation, with two rational fractions instead of one:

$$P_{VV} \simeq \omega\left[\frac{\alpha}{\beta+\omega} - \frac{\gamma}{\delta+\omega}\right].$$

Show that the set $\alpha = 2$; $\beta = 2$; $\gamma = 1$; $\delta = 3$ is consistent with both the asymptotic behaviour and the behaviour at the origin.

e) Compare numerically these three approximations with the exact values given in the table below.

Table showing the first collision probability for an infinite cylinder

ω	P	ω	P	ω	P
0.04	0.02561	0.6	0.28351	3	0.69843
0.08	0.04967	0.8	0.34838	4	0.76355
0.2	0.11498	1.2	0.45225	5	0.80677
0.4	0.20697	2	0.59285	10	0.90077

D. Treatment of the integral and differential forms

Exercise 14.15: unknowns of the 'diamond scheme' calculation

For geometries depending only on the variable x (planar symmetry) or the variable r (spherical symmetry), the monokinetic transport equation contains only one angular variable, the angle θ, or of the cosine of this angle, $\mu = \cos\theta$, between the direction of the neutron and either the axis $\vec{x}$ or the direction $\vec{r}$. In a planar geometry, the only derivative is the one with respect to x; however, in spherical geometry, there is not only a derivative with respect to r but also with respect to μ, owing to the curvature of the coordinate system.

When the numerical treatment employs an angular 'discrete ordinate' and a spatial 'finite difference' approach, which unknowns are required in each case?

Exercise 14.16: maximum of the space mesh

In a planar geometry and for neutron directions characterized by a given value of μ, the transport equation is

$$\mu\frac{\partial\Phi(x,\mu)}{\partial x} + \Sigma(x)\,\Phi(x,\mu) = \left[\text{Second member}\right],$$

where Σ is the total macroscopic cross-section.

Consider now the general solution of the equation without the second member, in a homogeneous zone, and in a given direction $\Phi(x,\mu) = C^t\, e^{\kappa x}$, with $\kappa = \Sigma/\mu$.

When using a finite difference treatment, this equation without the second member is replaced by

$$\mu\frac{\Phi_i - \Phi_{i-1}}{h} + \Sigma\frac{\Phi_i + \Phi_{i-1}}{2} = 0,$$

where $h = x_i - x_{i-1}$ and $\Phi_i = \Phi(x_i, \mu)$.

Assuming the mesh h is regular, show that the solution remains of the same exponential type, $\tilde{\Phi}_i = C^t\, e^{\tilde{\kappa} ih}$, with $\tilde{\kappa}$ instead of κ. Compare the two constants $\tilde{\kappa}$ and κ. In what way is h constrained?

When applying one-group theory to a homogeneous, spherical system, the value $h = \varepsilon\mu_1/\Sigma$ is chosen such that μ_1 is the smallest absolute value of μ in the 'discrete ordinates', and $\varepsilon = 0.1$. How many spatial meshes are necessary?

Numerical example: Find the number of meshes needed for a 50-litre sphere containing an aqueous solution, where $\Sigma = 300\ \text{m}^{-1}$, when using $\mathbf{S_4}$, $\mathbf{S_8}$, and $\mathbf{S_{16}}$ schemes.

Exercise 14.17: Milne's problem

Consider the Milne problem for the simplest case:

- planar interface between a homogeneous medium and a vacuum;
- monokinetic neutrons;
- no absorption;
- isotropic scattering;
- sources situated deep within the homogeneous medium.

a) Write the system of $\mathbf{P_N}$ equations for this problem.

b) Examine the $\mathbf{P_1}$ approximation, then compare two possibilities for taking into account the boundary condition $\Phi(0, \mu) = 0$ for μ negative, which are either

$$\int_{-1}^{0} \Phi(0, \mu)\, d\mu = 0,$$

or

$$\int_{-1}^{0} \Phi(0, \mu)\, \mu\, d\mu = 0,$$

and calculate the extrapolation distance in each case.

c) Apply the $\mathbf{P_3}$ approximation to the problem.

Exercise 14.18: even-odd formulation of the Boltzmann equation

a) Write the Boltzmann equation in its integral/differential form for the following conditions.

- Stationary situation.
- Monokinetic neutrons.
- Isotropic sources.
- Streaming operator given by $\vec{\Omega} \cdot \overrightarrow{\text{grad}}$.
- Linearly anisotropic scattering, i.e.

$$\Sigma_s(\vec{r}, \vec{\Omega}) = \frac{1}{4\pi}\Sigma_{s0}(\vec{r}) + \frac{3}{4\pi}\Sigma_{s1}(\vec{r})\, \vec{\Omega}' \cdot \vec{\Omega},$$

where Σ_{s0} is the scattering cross-section integrated over all directions, and Σ_{s1} is the mean cosine of the deflecting angle multiplied by Σ_{s0}.

b) The phase flux $\Phi(\vec{r}, \vec{\Omega})$ is split into the sum of its even part ψ and its odd part χ with respect to $\vec{\Omega}$:

$$\Phi(\vec{r}, \vec{\Omega}) = \psi(\vec{r}, \vec{\Omega}) + \chi(\vec{r}, \vec{\Omega}),$$

where

$$\psi(\vec{r}, \vec{\Omega}) = \frac{\Phi(\vec{r}, \vec{\Omega}) + \Phi(\vec{r}, -\vec{\Omega})}{2},$$

$$\chi(\vec{r}, \vec{\Omega}) = \frac{\Phi(\vec{r}, \vec{\Omega}) - \Phi(\vec{r}, -\vec{\Omega})}{2}.$$

The Greek letters ψ and χ are used to indicate the symmetry or antisymmetry of the state; the notations Φ^+ and Φ^- are also used.
By substitution into the Boltzmann equation, find the system of two equations (even and odd parts) coupling the even and odd parts with respect to $\vec{\Omega}$.

c) Assume temporarily that the scattering is isotropic, i.e. that Σ_{s1} is equal to zero, then express χ from the second equation and, by substituting the result into the first equation, find the equation governing only ψ.

d) For a given fixed direction $\vec{\Omega}$ (for instance, during a treatment by discrete ordinates), compare the previous equation with the diffusion equation.

e) To treat the second equation when $\Sigma_{s1} \neq 0$, a '$\mathbf{P_1}$' hypothesis is introduced, meaning it is assumed that $\chi(\vec{r}, \vec{\Omega})$ is given approximately by

$$\chi(\vec{r}, \vec{\Omega}) \simeq \vec{u}(\vec{r}) \cdot \vec{\Omega}.$$

Calculate the vector $\vec{u}(\vec{r})$ using the second equation, then write the equation governing ψ by substituting χ into the first equation.

f) Show that the '$\mathbf{P_1}$' hypothesis made in the previous question for Σ_s, and for χ can, in fact, be made only for Σ_s, and avoided for χ. In other words, a '$\mathbf{B_1}$' instead of a '$\mathbf{P_1}$' treatment can be made for the second equation. Find the equation for ψ with this '$\mathbf{B_1}$' treatment.

E. Fundamental mode

Exercise 14.19: elementary eigenfunctions of the Laplace operator

Consider a linear combination with the same weight of $e^{-i\vec{b}.\vec{r}}$ functions with vectors $\vec{b}$ whose extremity is situated on,

a) a point of the sphere of radius b and its seven symmetrical points with respect of the coordinate planes,

b) two parallel circles symmetrical with respect to the equator,

c) the whole sphere.

Show this yields the eigenfunctions of the Laplace operator which, respectively, go to zero on,

a) the faces of a rectangle parallelepiped,

b) the surface of a cylinder,

c) the surface of a sphere.

Recall:

$$\frac{1}{2\pi} \int_{(2\pi)} \exp(-i\,u\,\cos\varphi)\,d\varphi = J_0(u).$$

Exercise 14.20: diffusion coefficient in fundamental mode

For the fundamental mode in a homogeneous medium — meaning the situation where the flux and the reaction rates varies in space as $e^{-i\vec{b}.\vec{r}}$ — it can be observed that Fick's law is rigorous. Thus, there is proportionality between the current $\vec{J}(\vec{r})$ and the gradient of the flux $\Phi(\vec{r})$, with a proportionality coefficient which can depend on the neutron velocity, but not on the position in space. This coefficient (after changing the sign) is called 'diffusion coefficient' or 'leakage coefficient' according to its association with the neutron migration, or with the neutron balance. This exercise examines the calculation of this coefficient, employing monokinetic theory in order to simplify matters. An extension to the polykinetic case is possible.

a) Since the medium can be assumed to be isotropic, the vector $\vec{b}$ can take any direction, e.g. $\vec{x}$. Then, the source and the flux are

$$S(x,\mu) = \frac{1}{2} s\, e^{-ibx}; \qquad \Phi(x,\mu) = \varphi(\mu)\, e^{-ibx},$$

respectively. Find the equation governing φ and the integral giving s from $\varphi(\mu)$.

b) The phase distribution $\varphi(\mu)$ can be expanded in Legendre polynomials:

$$\varphi(\mu) = \sum_{n=0}^{\infty} \varphi_n P_n(\mu).$$

Show that the integrated flux is $\Phi(x) = 2\,\varphi_0\, e^{-ibx}$; therefore, that the component of its gradient along the axis $\vec{x}$ is $-2ib\,\varphi_0\, e^{-ibx}$. Then show that the component along $\vec{x}$ of the integrated current is $J_x(x) = (2/3)\,\varphi_1\, e^{-ibx}$, and consequently that Fick's law is consistent with the diffusion coefficient

$$D = \frac{\varphi_1}{3ib\varphi_0}.$$

c) To find φ_0 and φ_1, express the equation governing $\varphi(\mu)$ in terms of an expansion in Legendre polynomials. Since the Legendre polynomials are eigenfunctions of the scattering operator, notice that

$$\int_{-1}^{+1} \Sigma_s(\mu' \to \mu)\, P_n(\mu')\, d\mu' = \Sigma_{s,n}\, P_n(\mu),$$

Thus, the eigenvalues $\Sigma_{s,n}$ are the moments of the differential scattering cross-section.

d) Divide the previously obtained equation by $\Sigma - ib\mu$. Next, multiply by one of the Legendre polynomials, $P_k(\mu)$, and integrate over μ in order to obtain its projection on this polynomial. Introduce the coefficients

$$G_{nk} = \int_{-1}^{+1} \frac{P_k(\mu)\, P_n(\mu)}{\Sigma - ib\mu}\, d\mu,$$

and calculate G_{00}, $G_{01} = G_{10}$, and G_{11}.

e) Assuming that only the first moment $\Sigma_{s0} \equiv \Sigma_s$ (scattering cross-section integrated over the angles) is not zero, find φ_0 and φ_1 from the two first equations, and deduce D.

f) Supposing now that only the two first moments Σ_{s0} and $\Sigma_{s1} \equiv \bar{\mu}\,\Sigma_s$ — where $\bar{\mu}$ is the mean cosine of the deflecting angle — are not zero, find D from the two first equations, and compare with the previous result.

g) Finally, express D in a general form. In pratice, such as in the APOLLO code, $\mathbf{B_1}$ theory is employed to estimate the diffusion coefficient in a manner similar to **f** with Σ_{s0} and Σ_{s1}; however, multigroup theory is used instead of the monokinetic assumption.

Exercise 14.21: Behrens's correction

In a heterogeneous medium, such as a reactor lattice, the theory of the fundamental mode and of the diffusion coefficient is much more difficult. It appears that, owing to streaming effects, the use of simple averages of cross-sections or their inverses (mean free paths) is not sufficient to take into account the heterogenity effect on the diffusion coefficient[69].

Behrens studied the regular, infinite lattice constituted by a homogeneous moderator, and empty cavities. This is similar to a natural uranium-gas-graphite UNGG system, when the fuel element is neglected because its volume in the gas duct is relatively small. The (thermal) neutrons — which are assumed to be monokinetic — are emitted by a uniform source in the moderator. This represents neutrons near the end slowing down. Consequently, the flux everywhere — in the moderator as well as in the cavities — is uniform and isotropic.

As the flux is constant, the averaged macroscopic absorption cross-section can be defined by the homogenization formula:

$$\bar{\Sigma}_a = \frac{V_m}{V_m + V_c}\Sigma_{am}.$$

Then, conservation of the diffusion area[70] L^2 implies D. This criterion is correct at the limit $b \longrightarrow 0$, leading to

$$L^2 = \frac{D}{\bar{\Sigma}_a}, \qquad D = L^2\,\bar{\Sigma}_a.$$

Therefore, the diffusion area must be obtained from its definition,

$$L^2 = \frac{1}{6}\langle \vec{R}^2\rangle.$$

Behrens performed this calculation of $\langle \vec{R}^2\rangle$ by generalizing the argument made in exercises **3.1** and **3.2**. He also assumed that isotropic scattering occurs; hence,

$$\langle \vec{R}^2\rangle = \bar{n}\,\langle \vec{\rho}^{\,2}\rangle, \qquad \bar{n} = \frac{\Sigma_m}{\Sigma_{am}}.$$

To calculate $\langle \vec{\rho}^{\,2}\rangle$, he distinguished the neutron paths crossing a cavity or not, and neglected paths crossing more than one cavity. This is a reasonable assumption for a UNGG system of this type. The following exercise demonstrates his reasoning.

a) By counting, the numbers of collisions and neutrons entering a cavity per unit of time, calculate the proportion γ of neutron paths crossing a cavity.

b) Calculate $\langle \rho_m\rangle$ and $\langle \rho_m^2\rangle$ for a path which does not cross a cavity.

c) The vector representing a neutron path crossing a cavity can be given as the sum $\vec{\rho}_{m1} + \vec{\rho}_c + \vec{\rho}_{m2}$, where the terms are the parts of the path before, within, and after the cavity,

[69] D. J. Behrens Proc. Phys. Soc. A, **62**(10) 607–616 (1949), and P. Benoist, thesis, University of Paris (1964); CEA-R-2278.

[70] The diffusion area is one sixth of the averaged square direct distance between the creation of the neutron and its absorption. In diffusion theory, this area is equal to the diffusion coefficient divided by the macroscopic absorption cross-section. Here, the diffusion area will be calculated, and the diffusion coefficient defined in order to obtain this equality.

respectively. The averages $\langle\rho_{m2}\rangle$ and $\langle\rho^2_{m2}\rangle$ are equal to the averages $\langle\rho_m\rangle$ and $\langle\rho^2_m\rangle$ calculated in part **b**. Symmetry in time dictates that the averages $\langle\rho_{m1}\rangle$ and $\langle\rho^2_{m1}\rangle$ are equal to the averages $\langle\rho_{m2}\rangle$ and $\langle\rho^2_{m2}\rangle$. Find $\langle\rho_c\rangle$ and $\langle\rho^2_c\rangle$ as functions of the parameters S (surface), V_c (volume) and Q (form parameter — see exercise **8.4**) of the cavities. Deduce the average of the square of the length of a path crossing a cavity.

d) Using these results, give the formula for the diffusion coefficient. What size is the ratio between this coefficient and that for the moderator? Show that this ratio is not the expression of a simple homogenization.

e) *Numerical example:* apply the previous results to a lattice containing cylindrical cavities 100 mm in diameter, embedded inside square-section 200 × 200 mm graphite blocks, with infinite length in the z-direction. Assume that the neutron mean free path in graphite is 26 mm.

F. Monte Carlo method

N.B: in the following exercises, ξ is the random variable governed by a uniform law in the interval $[0, 1[$. A pseudo-random algorithm is employed to generate ξ, in practice.

Exercise 14.22: power law probability distribution

Assume that n random values of ξ are taken, and that the largest one is selected. Show that this maximum is the random variable X governed by the probability law $P(x)\,dx = n\,x^{n-1}\,dx$ in the interval $[0, 1[$. How can this variable be obtained in a single step? What is the best way to generate it?

Exercise 14.23: random point inside a circle

Consider the following methods for picking a random point inside a circle, where ξ_1 and ξ_2 are two random numbers used to generate the coordinates.

a) Why is $\rho = \xi_1$ and $\varphi = 2\pi\xi_2$ unsatisfactory?

b) How can the previous method be modified to yield a better outcome?

c) An alternative might be to generate Cartesian coordinates using

$$x = 2\xi_1 - 1, \qquad y = 2\xi_2 - 1,$$

and discard points outside the circle when $x^2 + y^2 \geq 1$. Does it produce a satisfactory result? What proportion of the points are discarded?

d) Which of these methods is best?

Exercise 14.24: Buffon's needle

Recall that the Buffon experiment consists of throwing a needle of length b on a plane divided into parallel strips of width a, and observing how often the needle lies across a boundary between two strips. The result is related to the constant π.

a) Taking $a = b$, give the probability that the needle lies across a boundary.

b) How do the variance and standard deviation in the estimate for π depend on the number of trials n?

Exercise 14.25: evaluation of a resonance escape probality

a) Assume that neutron slowing down occurs in a monoatomic material by elastic, isotropic scattering in the centre of mass system. Devise a 'Monte Carlo' algorithm to evaluate the probability that a neutron emitted at high energy escapes a resonance simulated by a unique 'black trap' (infinite absorption cross-section in the trap and no absorption outside the trap).

b) Perform some numerical simulations, and compare the outcome with the result given by the exact analytical expression for the resonance escape probability.

Solutions

A. Study of the Boltzmann equation

Exercise 14.1: equivalence of the two forms of the transport operator

a) The integral form of the transport operator is

$$\Phi\left(\vec{r}, v, \vec{\Omega}, t\right) = \int_0^\infty e^{-\tau} Q\left(\vec{r} - s\vec{\Omega}, v, \vec{\Omega}, t - s/v\right) ds,$$

where

$$\tau = \int_0^s \Sigma\left(\vec{r} - s'\vec{\Omega}, v\right) ds'.$$

The integral expressing Φ makes explicit the neutron paths without collision over a distance s, at the velocity v, and in the direction $\vec{\Omega}$ from the emission point, to the point where the flux is evaluated. Q is the emission density; $e^{-\tau}$ is the probability that the neutron performs the path without collision; the integral sums all the possible contributions.

b) In order to express a shift in the direction $\vec{\Omega}$ of the point where the neutron flux is observed, it is convenient to *fix* a point on the path taken by the neutrons to define the origin of the abscissa s — which can be different from the observation point — and to rewrite the equation giving the flux when the observation point is located at the abscissa s as follows.

$$\Phi\left(\vec{r} - s\vec{\Omega}, v, \vec{\Omega}, t - s/v\right) = \int_s^\infty e^{-\tau} Q\left(\vec{r} - s'\vec{\Omega}, v, \vec{\Omega}, t - s'/v\right) ds',$$

where

$$\tau = \int_s^{s'} \Sigma\left(\vec{r} - s''\vec{\Omega}, v\right) ds''.$$

For the sake of clarity, s in the integral expressing the transport operator is replaced with s', and s' in the expression of the optical path with s''.

c) All that remains is to differentiate with respect to s everywhere it appears, as follows.

$$-ds\,\vec{\Omega}.\overrightarrow{\text{grad}}\,\Phi\left(\vec{r}-s\vec{\Omega},v,\vec{\Omega},t-s/v\right)$$

$$-\frac{ds}{v}\frac{\partial}{\partial t}\left[\Phi\left(\vec{r}-s\vec{\Omega},v,\vec{\Omega},t-s/v\right)\right]$$

$$=-ds\,Q\left(\vec{r}-s\vec{\Omega},v,\vec{\Omega},t-s/v\right)$$

$$+\int_s^{\infty} ds\,\Sigma\left(\vec{r}-s\vec{\Omega},v\right)\,e^{-\tau}\,Q\left(\vec{r}-s'\vec{\Omega},v,\vec{\Omega},t-s'/v\right)\,ds'.$$

In the third term, the exponential for $\tau = 0$ occurs; hence, it is omitted. For the last term, $ds\,\Sigma$ can be outside the integral. The remaining integral is the flux Φ. Finally, by simplifying ds, changing the sign, and setting $s = 0$ gives

$$\vec{\Omega}.\overrightarrow{\text{grad}}\,\Phi\left(\vec{r},v,\vec{\Omega},t\right)+\frac{1}{v}\frac{\partial}{\partial t}\Phi\left(\vec{r},v,\vec{\Omega},t\right)$$

$$=Q\left(\vec{r},v,\vec{\Omega},t\right)-\Sigma\left(\vec{r},v\right)\,\Phi\left(\vec{r},v,\vec{\Omega},t\right).$$

This can be recognized as being the differential form of the transport operator. *Recall,*

$$\vec{\Omega}.\overrightarrow{\text{grad}}\,\Phi\left(\vec{r},v,\vec{\Omega},t\right)\equiv\text{div}\left[\vec{\Omega}\,\Phi(\vec{r},v,\vec{\Omega},t)\right].$$

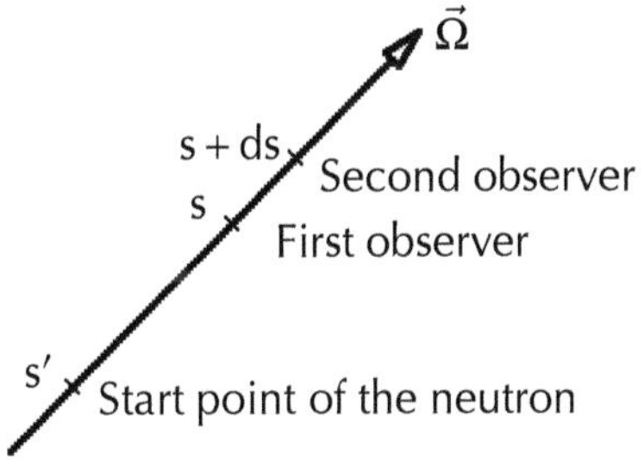

Figure 3

d) The left hand side describes the difference between the observations made in two points separated by $\vec{\Omega}\,ds$ (see figure **3**). These observations — which compare two instants shifted by $dt = ds/v$ — concern almost the same neutrons. The only differences — described by the right hand side — are as follows.

- The neutrons emitted along the line segment joining both the points, seen by the 'observer' located downstream, but not by the 'observer' upstream.

- The neutrons undergoing a collision on this segment, seen by the upstream 'observer' but not by the downstream 'observer'.

Exercise 14.2: exact solution without absorption

The equation to be solved is

$$-\mu \frac{\partial \Phi (x, \mu)}{\partial x} - \Sigma \Phi (x, \mu) + \int_{-1}^{+1} \Sigma_s (\mu' \longrightarrow \mu) \, \Phi (x, \mu') \, d\mu' = 0.$$

This has a solution linear with x and with μ,

$$\Phi (x, \mu) = A \left(x - a - \frac{\mu}{\Sigma_{tr}} \right),$$

where A and a are any constants, and where $\Sigma_{tr} = \Sigma - \bar{\mu}\Sigma_s$ is the transport cross-section (here $\Sigma_s = \Sigma$). It is noteworthy that only the moments of order 0 and 1 of the scattering law appear.

Exercise 14.3: relaxation length

a) The equation to solve is the same as the previous one, except now with $\Sigma_s \neq \Sigma$. A factorized solution appears, $\Phi(x, \mu) = \varphi(\mu)\, e^{-\kappa x}$, where

$$\kappa\mu \, \varphi (\mu) - \Sigma \varphi (\mu) + \int_{-1}^{+1} \Sigma_s (\mu' \longrightarrow \mu) \, \varphi (\mu') \, d\mu' = 0.$$

b) Provided the scattering is isotropic,

$$\kappa\mu \, \varphi(\mu) - \Sigma \varphi (\mu) + \frac{\Sigma_s}{2} \int_{-1}^{+1} \varphi (\mu') \, d\mu' = 0.$$

Dividing by $\Sigma - \kappa\mu$, then integrating over μ, yields

$$I = \frac{1}{\kappa} \ln \left(\frac{\Sigma + \kappa}{\Sigma - \kappa} \right) \frac{\Sigma_s I}{2},$$

where I is the integral of $\varphi(\mu)$. Following simplification by I, the equation giving κ is obtained:

$$\ln \frac{\Sigma + \kappa}{\Sigma - \kappa} = \frac{2\kappa}{\Sigma_s}.$$

c) The equation for this problem, in the integral form, is

$$\Phi(x) = \int_{-\infty}^{+\infty} \frac{1}{2} E_1 \left(\Sigma \,|\, x - x' \,| \right) \Sigma_s \, \Phi(x') \, dx',$$

where

$$E_1 (u) = \int_0^1 \exp \left(-\frac{\mu}{u} \right) \frac{d\mu}{\mu}.$$

See exercise **14.5** for a demonstration of this problem in one dimension x geometry.

Replacing $\Phi(x)$ with $A\, e^{-\kappa x}$, and simplifying by $A\, e^{-\kappa x}$, the equation giving κ is obtained. The double integral can be calculated by integrating firstly over x' and then over μ. Thus, we obtain again the equation of the paragraph **b**.

d) When the collisions are isotropic, the diffusion length L is given by the formulae

$$L^2 = \frac{D}{\Sigma_a}, \qquad D = \frac{1}{3\Sigma}.$$

Some numerical values for various values of $c = \Sigma_s/\Sigma$ are provided in the following table, where the unit of length is $1/\Sigma$.

c	κ	$1/L$	Difference (%)
0.99	0.17251	0.17321	0.40
0.9	0.52543	0.54772	4.24
0.8	0.71041	0.77460	9.03
0.5	0.95750	1.22474	27.91
0.2	0.99991	1.54919	54.93
0.1	1.00000	1.64317	64.32

e) Expanding $\varphi(\mu)$ into Legendre polynomials, dividing by $\Sigma - \kappa\mu$, and projecting out the two first polynomials, i.e. 1 and μ, yields a homogeneous system governing the two integrals I_0 and I_1, similar to I. The equivalence of the system gives the equation for κ:

$$\ln \frac{\Sigma + \kappa}{\Sigma - \kappa} = \frac{2\kappa}{\Sigma_s} \frac{\kappa^2 + 3\bar{\mu}\Sigma_s\Sigma_a}{\kappa^2 + 3\bar{\mu}\Sigma\Sigma_a}.$$

Exercise 14.4: diffusion length

a) For the neutrons performing exactly n paths, the sum to be calculated is

$$\langle \vec{R}_n^2 \rangle = \langle \vec{\rho}_1^{\,2} \rangle + \langle \vec{\rho}_2^{\,2} \rangle + \cdots + \langle \vec{\rho}_n^{\,2} \rangle + 2 \sum_{i=1}^{n-1} \sum_{j=i+1}^{n} \langle \rho_i \rho_j \cos\theta_{ij} \rangle.$$

The various random variables are independent, owing to the neutron migration being a Markovian process; hence,

$$\langle \rho \rangle = \frac{1}{\Sigma}, \qquad \langle \rho^2 \rangle = \frac{2}{\Sigma^2}, \qquad \langle \cos\theta_{i,i+1} \rangle = \bar{\mu}.$$

b) Assuming $\vec{\rho}_{i,i+1}$ is located at the vertex A, and defining φ to be the dihedral angle at this vertex, using spherical trigonometry gives

$$\cos\theta_{i,i+2} = \cos\theta_{i,i+1} \cos\theta_{i+1,i+2} + \sin\theta_{i,i+1} \sin\theta_{i+1,i+2} \cos\varphi.$$

The average value of $\cos\varphi$ is zero owing to symmetry; hence,

$$\langle \cos\theta_{i,i+2} \rangle = \bar{\mu}^2.$$

Repeating the reasoning, it can be seen that

$$\langle \cos\theta_{i,j} \rangle = \bar{\mu}^{j-i}.$$

c) This allows the expression

$$\langle \vec{R}_n^2 \rangle = \left[n - \frac{\bar{\mu}\,(1-\bar{\mu}^n)}{1-\bar{\mu}} \right] \frac{2}{(1-\bar{\mu})\,\Sigma^2}$$

to be evaluated.

d) Weighting by the probabilities p_n, gives

$$\langle \vec{R}^2 \rangle = \frac{2}{\Sigma_a\,(\Sigma - \bar{\mu}\Sigma_s)}.$$

e) Hence, this allows $D = 1/(3\Sigma_{tr})$ to be found.

Exercise 14.5: integral kernel in one- and two-dimension geometries

a) Use $\mu = \cos\theta$, where θ is the angle between the axis $\vec{x}$ and the vector $\vec{R}$ representing the neutron path.

b) Express the integral as a function of the angle θ between the axis $\vec{z}$ and the vector $\vec{R}$ representing the neutron path.

B. Treatment of energy and time

Exercise 14.6: multigroup theory and treatment of the time derivative

a) The exact value of m is given by the equation

$$\frac{e^{m\varepsilon} - 1}{m\varepsilon} = 1 + a,$$

or, with $x = m\varepsilon$,

$$\frac{e^{x} - 1}{x} = 1 + a.$$

When the parameter a characterizing the intensity of the absorption ($a = \Sigma_a/\Sigma_s$) is small, this allows an expansion in powers of a:

$$x = 2a\left(1 - \frac{2a}{3} + \cdots\right).$$

If a regular dicretization is made, with n groups in the interval ε, then $n+1$ probabilities appear in the sum on the left hand side. Integrating $P = 1/\varepsilon$ over the departure and arrival groups, yields

$$P_{gg} = P_{g-n,g} = \frac{1}{2n}; \qquad P_{g-k,g} = \frac{1}{n} \qquad (1 \le k \le n-1).$$

It can be seen by algebraic substitution that it is possible to replace m with $\tilde{m}$ or x with $\tilde{x}$ such that

$$\frac{e^{\tilde{x}} - 1}{2n} \coth \frac{\tilde{x}}{2n} = 1 + a.$$

This formula can be applied even for $n = 1$ and $n = 2$, but then can be simplified:

$$\tilde{x}_{(n=1)} = \ln(1 + 2a); \qquad \tilde{x}_{(n=2)} = 2 \ln\left(2\sqrt{1 + a} - 1\right).$$

If the parameter a is small, then a series expansion can be employed:

$$x = 2a \left[1 - \frac{2a}{3}\left(1 + \frac{1}{2n^2}\right) + \cdots\right].$$

The term $1/2n^2$ represents, approximately, the error due to the use of the multigroup theory.

When a is not sufficiently small to do this expansion, only numerical solution of the equations giving x and $\tilde{x}$ permits the error to be estimated. For example, for $a = 1/2$ and some values of n, the results are as follows.

n	x	Error (%)
1	0.693147	−9.72
2	0.742423	−2.66
3	0.753381	−1.22
5	0.759279	−0.45
10	0.761830	−0.11
∞	0.762688	0

b) The solutions are exponential functions. Without the normalisation factor, they are $f = e^t$, $\check{f} = e^{i\Delta\alpha}$, and $\hat{f} = e^{i\Delta\beta}$, respectively. The quantities α and β describe the errors due to the numerical calculations of the derivatives by quotients of finite differences. Without the error, these numbers would be equal to one. The equations giving α and β can be found by substitution:

$$e^{\Delta\alpha} = 1 + \Delta; \qquad \sinh \Delta\beta = \Delta.$$

The expansions in powers of Δ,

$$\alpha = 1 - \frac{\Delta}{2} + \frac{\Delta^2}{3} + \cdots, \qquad \beta = 1 - \frac{\Delta^2}{6} + \frac{\Delta^4}{12} + \cdots,$$

show that the error in the first formula is of order Δ, while it is of order Δ^2 in the second. This result also demonstrates that the tangent at a point on a curve is better approximated by a line joining two points at either side of the point than by a line from the point to a point on one side of it.

C. Collision probabilities

Exercise 14.7: Cauchy's theorem

a) After normalisation, the probability law is $d^2S\, d^2\Omega \cos\theta/(\pi S)$.

b) The mean chord is given by the double integral

$$\langle X\rangle = \frac{1}{\pi S}\int_{(S)} d^2S \int_{(4\pi)} d^2\Omega \cos\theta\, X\, \Upsilon(\cos\theta),$$

where $\Upsilon(\cos\theta) = 1$ when the cosine is positive, otherwise is zero.

c) Notice that $d^2S \cos\theta\, X$ is the element of volume and that the integral of this term — when taking the Heaviside function is into account — is the total volume:

$$\int_{(S)} d^2S \cos\theta\, X\, \Upsilon(\cos\theta) = V.$$

Therefore, the double integral is

$$\int_{(4\pi)} d^2\Omega\, V = 4\pi V,$$

i.e. the Cauchy theorem.

Exercise 14.8: reciprocity theorem (and absorption probabilities)

a) The neutron densities are described by

$$\Phi = \frac{s}{\Sigma_a}, \qquad J_- = \frac{s}{4\Sigma_a}, \qquad Q = \frac{s\,\Sigma_t}{\Sigma_a}.$$

b) There are $V\Sigma_t\Phi$ collisions in the volume V per unit of time.

c) Distinguishing both these categories of neutrons, gives

$$V\Sigma_t\Phi = V Q P_{VV} + S J_- P_{SV}.$$

d) Therefore, owing to $1 - P_{VV} = P_{VS}$,

$$P_{SV} = \frac{4V\Sigma_t}{S} P_{VS}.$$

e) Similarly,

$$P^*_{SV} = \frac{4V\Sigma_a}{S} P^*_{VS} = \frac{4V\Sigma_a}{S}\left(1 - P^*_{VV}\right).$$

Exercise 14.9: moments of the chord

a) *Slab:* the $\vec{z}$ axis is normal to the surface at the point of entry; hence,

$$\langle X^n \rangle = 2 \int_0^1 \left(\frac{2a}{\mu}\right)^n \mu \, d\mu.$$

Only the first moment, $\langle X \rangle = 4a$, is finite.

b) *Cylinder:* the $\vec{x}$ axis is normal to the surface at the point of entry, and the axis $\vec{z}$ is along the generatrix; hence,

$$\langle X^n \rangle = \frac{1}{\pi} \int_{-\pi/2}^{+\pi/2} d\varphi \int_0^{\pi} \sin\theta \, d\theta \left(\frac{2R\cos\varphi}{\cos\theta}\right)^n \sin\theta \cos\varphi.$$

In particular,

$$\langle X \rangle = 2R, \qquad \langle X^2 \rangle = \frac{16\,R^2}{3}, \qquad Q = \frac{4}{3}.$$

c) *Sphere:* the $\vec{z}$ axis is normal to the surface at the point of entry; hence

$$\langle X^n \rangle = 2 \int_0^1 (2R\mu)^n \, \mu \, d\mu.$$

Particularly:

$$\langle X \rangle = \frac{4}{3}R, \qquad \langle X^2 \rangle = 2\,R^2, \qquad Q = \frac{9}{8}.$$

Exercise 14.10: first collision probabilities for a sphere

When $u = R\Sigma$, the formula is

$$P_{SS} = \frac{1 - (1 + 2u)\,e^{-2u}}{2u^2}; \qquad P_{VV} = 1 - \frac{3\left[2u^2 - 1 + (1 + 2u)\,e^{-2u}\right]}{8u^3}.$$

Exercise 14.11: first collision probabilities in a cherckerboard pattern

Owing to the hypothesis the calculation is reduced to the probabilities without exit for each type of cell. For two zones in the same cell,

$$P_{ij} = P_{ij}^{+} + P_{iS}^{+} P_{SS,B}^{+} P_{Sj}^{+} + P_{iS}^{+} P_{SS,B}^{+} P_{SS,A}^{+} P_{SS,B}^{+} P_{Sj}^{+}$$
$$+ P_{iS}^{+} P_{SS,B}^{+} P_{SS,A}^{+} P_{SS,B}^{+} P_{SS,A}^{+} P_{SS,B}^{+} P_{Sj}^{+} + \cdots;$$

$$P_{ij} = P_{ij}^{+} + \frac{P_{iS}^{+} P_{SS,B}^{+} P_{Sj}^{+}}{1 - P_{SS,A}^{+} P_{SS,B}^{+}}.$$

For two zones in different cells,

$$P_{ik} = P_{iS}^{+} P_{Sk}^{+} + P_{iS}^{+} P_{SS,B}^{+} P_{SS,A}^{+} P_{Sk}^{+} + P_{iS}^{+} P_{SS,B}^{+} P_{SS,A}^{+} P_{SS,B}^{+} P_{SS,A}^{+} P_{Sk}^{+} + \cdots;$$

$$P_{ik} = \frac{P_{iS}^{+} P_{Sk}^{+}}{1 - P_{SS,A}^{+} P_{SS,B}^{+}};$$

and similar formulae for P_{ki} and $P_{k\ell}$.

Exercise 14.12: calculation of the disavantage factor

a) The numbers of collisions in each zone are

$$V_1\,\Sigma_{t1}\,\Phi_1 = V_1\,\Sigma_{s1}\,\Phi_1\,P_{11} + (\,V_2\,\Sigma_{s2}\,\Phi_2 + 1\,)\,P_{21},$$
$$V_2\,\Sigma_{t2}\,\Phi_2 = V_1\,\Sigma_{s1}\,\Phi_1\,P_{12} + (\,V_2\,\Sigma_{s2}\,\Phi_2 + 1\,)\,P_{22}.$$

b) Owing to these equations, both fluxes Φ_1 and Φ_2, their ratio Φ_2/Φ_1 (disavantage factor), and the thermal utilisation factor f can be calculated:

$$\frac{1}{f} - 1 = \frac{\Sigma_{a2}}{\Sigma_{t2}}\left[\frac{V_2\,\Sigma_{t2}}{V_1\,\Sigma_{t1}}\left(\frac{P_{11}}{1-P_{11}} + \frac{\Sigma_{t1}}{\Sigma_{a1}}\right) - 1\right].$$

c) Expressing the series of all the possible events as

$$P_{11} = P_{11}^+ + P_{1S}^+\,C\,P_{S1}^+ + P_{1S}^+\,C\,P_{SS}^+\,C\,P_{S1}^+ + P_{1S}^+\,C\,P_{SS}^+\,C\,P_{SS}^+\,C\,P_{S1}^+ + \cdots,$$

$$P_{11} = P_{11}^+ + \frac{P_{1S}^+\,C\,P_{S1}^+}{1 - C\,P_{SS}^+},$$

with

$$P_{1S}^+ = 1 - P_{11}^+, \qquad P_{S1}^+ = \frac{4V_1\Sigma_{t1}}{S}P_{1S}^+, \qquad P_{SS}^+ = 1 - P_{S1}^+,$$

gives

$$\frac{P_{11}}{1-P_{11}} = \frac{P_{11}^+}{1-P_{11}^+} + \frac{C}{1-C}\frac{4V_1\Sigma_{t1}}{S}.$$

The value of f is, therefore, obtained when this expression is substituted into the equation in part **b**.

d) Differentiating the formulae gives

$$\frac{\Delta f}{f} = -f\frac{4V_2\Sigma_{a2}}{S}\frac{\Delta C}{(1-C)^2}.$$

Example: $\Delta f/f = 44$ pcm; the error in f due to the cylindrical approximation is not important. Notice that, for a similar reason, there is also an error of opposite sign in p.

Exercise 14.13: numerical calculation of the collision probabilities

a) The relaxation parameter is obtained by substitution of the exponential solution into the equation without source:

$$\kappa = \sqrt{1-c}.$$

b) The first collision probabilities in the given 'string' can be evaluated exactly using

$$P_{ji} = \frac{(e^h - 1)(1 - e^{-h})}{2h}e^{-|i-j|h}(j \neq i);$$

$$P_{jj} = \frac{h - e^{-h} - 1}{h}.$$

It can be verified that the sum over the second index is one.

The relaxation constant $\tilde{\kappa}$ is solution of the equation

$$\frac{h}{c} = h + e^{-h} - 1 + \frac{(e^h - 1)(1 - e^{-h})}{2} E,$$

where

$$E = \frac{e^{-(1+\tilde{\kappa})h}}{1 - e^{-(1+\tilde{\kappa})h}} + \frac{e^{-(1-\tilde{\kappa})h}}{1 - e^{-(1-\tilde{\kappa})h}}.$$

Limited expansion gives the result

$$\frac{\tilde{\kappa}}{\kappa} = 1 - \frac{ch^2}{6} + \cdots$$

c) Using

$$P_{ji} \simeq \frac{h}{2} e^{-h|i-j|} \quad (j \neq i),$$

and P_{jj}, owing to complementarity

$$P_{jj} = 1 - \sum_{i \neq j} P_{ji} = 1 - \frac{he^{-h}}{1 - e^{-h}}.$$

Hence, $\tilde{\kappa}$ is given by

$$\frac{1}{c} = 1 - \frac{he^{-h}}{1 - e^{-h}} + \frac{h}{2} E,$$

with the same formula for E. Then, by limited expansions

$$\frac{\tilde{\kappa}}{\kappa} = 1 - \frac{c^2 h^4}{120} + \cdots$$

Owing to a fortuitous cancellation of errors, the terms of order h^2 vanish!

Exercise 14.14: Wigner, Bell-Wigner, and Carlvik appproximations

a) The boundary conditions are

$$\omega \longrightarrow 0 : \; P_{VV} \simeq \frac{Q}{2}\omega; \qquad \omega \longrightarrow \infty : \; P_{VV} \simeq 1 - \frac{1}{\omega};$$

where $Q = \langle X^2 \rangle / \langle X \rangle^2$.

b) The Wigner approximation obeys the asymptotic behaviour

$$\omega \longrightarrow 0 : \; P_{VV,W} \simeq \omega; \qquad \omega \longrightarrow \infty : \; P_{VV,W} \simeq 1 - \frac{1}{\omega}.$$

c) For the Bell-Wigner approximation,

$$\omega \longrightarrow 0 : \; P_{VV,B-W} \simeq \frac{\omega}{b}; \qquad \omega \longrightarrow \infty : \; P_{VV,B-W} \simeq 1 - \frac{b}{\omega},$$

it obeys the behaviour at the origin — but not at infinity — provided that $b = 2/Q$. For example, $b = 3/2$ for a cylinder. For intermediate ω, intermediate values of b can be used.

d) The Carlvick approximation gives

$$\omega \longrightarrow 0: \; P_{VV,C} \simeq \omega(\frac{\alpha}{\beta} - \frac{\gamma}{\delta}); \qquad \omega \longrightarrow \infty: \; P_{VV,C} \simeq \alpha - \gamma - \frac{\alpha\beta - \gamma\delta}{\omega}.$$

This suggests that

- *Origin:* $\alpha/\beta - \gamma/\delta = Q/2$;
- *Infinity:* $\alpha - \gamma = 1 \qquad \alpha\beta - \gamma\delta = 1$.

However, there are only three equations with four unknowns; the proposed choice is the simplest one.

e) The following table gives some numerical examples.

ω	Exact	Wigner	(%)	Bell-W.	(%)	Carlvik	(%)
0.04	0.02561	0.03846	50.18	0.02597	1.42	0.02606	1.75
0.08	0.04967	0.07407	49.13	0.05063	1.94	0.05095	2.58
0.2	0.11494	0.16666	44.95	0.11765	2.32	0.11932	3.77
0.4	0.20697	0.28571	38.05	0.21053	1.72	0.21579	4.21
0.6	0.28351	0.37500	32.27	0.28571	0.78	0.29487	4.01
0.8	0.34838	0.44444	27.57	0.34783	−0.16	0.36090	3.59
1.2	0.45225	0.54545	20.61	0.44444	−1.73	0.46429	2.66
2	0.59285	0.66667	12.45	0.57143	−3.61	0.60000	1.21
3	0.69843	0.75000	7.38	0.66667	−4.55	0.70000	0.22
4	0.76355	0.80000	4.77	0.72727	−4.75	0.76190	−0.22
5	0.80677	0.83333	3.29	0.76923	−4.64	0.80357	−0.40
10	0.90077	0.90909	0.92	0.86957	−3.46	0.89744	−0.37

Notice that the Wigner approximation always overestimates the probability P_{VV}, and by a large relative amount for the small values of the opacity ω. The Bell-Wigner approximation improves the situation for the small opacities, but greatly underestimates the exact values of P_{VV} for the large values of ω. The Carlvik approximation satisfies both for the small and large values of ω, with good accuracy for the intermediate opacities. Furthermore, within the Bell-Wigner approximation, the sign of the error changes according to ω, and some compensation may occur; for example, in resonant-absorption calculations.

D. Treatment of the integral and differential forms

Exercise 14.15: unknows of the 'diamond scheme' calculation

The treatment of the Boltzmann equation must be performed in a rectangle limited horizontally by $x_{min} \leq x \leq x_{max}$, or $r_{min} \leq r \leq r_{max}$, and vertically by $-1 \leq \mu \leq +1$. This rectangle is itself divided into rectangular meshes. This introduces two kinds of unknowns: the fluxes at the intersections of the boundaries of the rectangles, and at their centres. In the 'diamond scheme' the unknown fluxes at the centres of the edges of the rectangles are also introduced.

In a planar geometry, the transport operator contains only a derivative with respect of x. To invert this operator, it is possible to keep only the unknowns at the centres of the meshes, and at the centres of the vertical edges. Indeed, when a mesh is treated,

- the source and the collision rates are considered at the mesh centres; and
- the derivative with respect of x is evaluated from the difference between two fluxes at neighbouring vertical edge centres.

The equations can be treated sequentially along x for the various discrete values of μ, following the direction of the neutrons, i.e.

- from the left and the entering flux (generally zero) when μ is positive; and
- from the right and the entering flux (also generally zero) when μ is negative.

N.B: an even value of N is always chosen for discrete ordinate calculations, to avoid the centres of the mesh from falling on the equator ($\mu = 0$).

In spherical geometry, the transport operator contains derivatives with respect to both r and μ. This is because some coordinate lines are curved while the neutrons travel in straight lines. To invert the transport operator, implies unknowns are needed at the mesh centres. These express the sources and the collision rates. Similarly, unknowns are needed at the vertical edge centres (to express the derivatives with respect of r), and at the horizontal edge centres (to express the derivatives with respect of μ). When a mesh is treated, the neutron balance equation has to be completed with two 'diamond' equations which permit these supplementary unknowns to be found, i.e.

- the half-sum of the fluxes on the horizontal edge centres = the flux at the mesh centre; and
- the half-sum of the fluxes on the vertical edge centres = the flux at the mesh centre.

As before, the equations can be treated sequentially along r, for each discrete value of μ, in the direction of the neutrons.

Note: in the 'diamond scheme', the fluxes at the mesh vertices are not introduced.

Exercise 14.16: maximum of the space mesh

For the discretised solution,

$$\Phi_i = \Phi_{i-1} \frac{1 - (h\Sigma)/(2\mu)}{1 + (h\Sigma)/(2\mu)} = \Phi_{i-1}\, e^{-\tilde{\kappa}h}.$$

If h is small, then $\tilde{\kappa} \to \kappa$. Clearly, h must not be greater than $2\mu/\Sigma$, otherwise some fluxes would be negative. This condition is the most restrictive for $\mu = \mu_1$. Using $h = \varepsilon\mu_1/\Sigma$, ε must be less than (preferably much less than) 2.

Example: the radius of the sphere: 228.5 mm. Using regular meshes according to the latitude, $\mu_1 = \sin \pi/(2N)$.

- $\mathbf{S_4}$: $\mu_1 = 0.38268$; $h = 0.012756$; 1 792 meshes;
- $\mathbf{S_8}$: $\mu_1 = 0.19509$; $h = 0.006503$; 3 514 meshes;
- $\mathbf{S_{16}}$: $\mu_1 = 0.09802$; $h = 0.003267$; 6 995 meshes.

Exercise 14.17: Milne's problem

a) In $\mathbf{P_1}$ and $\mathbf{P_3}$ approximations, the equations are

$$-\frac{1}{3}\varphi_1' = 0,$$

$$-\varphi_0' - \Sigma\varphi_1 = 0;$$

and

$$-\frac{1}{3}\varphi_1' = 0,$$

$$-\varphi_0' - \frac{2}{5}\varphi_2' - \Sigma\varphi_1 = 0,$$

$$-\frac{2}{3}\varphi_1' - \frac{3}{7}\varphi_3' - \Sigma\varphi_2 = 0,$$

$$-\frac{3}{5}\varphi_2' - \Sigma\varphi_3 = 0,$$

respectively.

b) In $\mathbf{P_1}$ approximation, the solution takes the form

$$\varphi_0 = a + bx, \qquad \varphi_1 = -\frac{b}{\Sigma},$$

$$\Phi(x,\mu) = a + bx - \frac{b\mu}{\Sigma}, \qquad \Phi(x) = 2(a + bx),$$

where a and b are constants.

The last relationship shows that the scalar flux goes to zero at the abscissa $x = d = -a/b$, known as the extrapolation distance.

The phase flux at the interface is

$$\Phi(0,\mu) = a - \frac{b\mu}{\Sigma}.$$

When the integral of this *phase flux* over the negative values of μ is zero, $b = -2a\Sigma$; therefore, the reduced extrapolation distance is $\Sigma d = 1/2$. When the integral over the negative values of μ of the *phase current*, $\mu\,\Phi(0,\mu)$ is zero, $b = -3a\Sigma/2$; therefore, the reduced extrapolation distance is $\Sigma d = 2/3$. This last value — which corresponds to the diffusion approximation — is based on a more physical criterion, namely a null total entering current. Hence, this implies that it is nearer to the exact value, 0.710446.

c) In $\mathbf{P_3}$ approximation, a fourth (and transitory) term is added to the previous expression of the phase flux:

$$c\left[1 - \frac{5}{2}P_2(\mu) + \frac{\sqrt{35}}{2}P_3(\mu)\right]\exp\left(\frac{\sqrt{35}}{3}\Sigma x\right).$$

When the moments of orders 0 and 1 of the flux, or the moments of orders 1 and 2 of the current, are zero,

$$\Sigma d = \frac{160 - 3\sqrt{35}}{195} = 0.7295 \qquad \text{and} \qquad \Sigma d = \frac{11}{3} - \frac{\sqrt{35}}{2} = 0.7086,$$

respectively. The second approximation is better: the error is about –0.26% instead of +2.68%.

Exercise 14.18: even-odd formulation of the Boltzmann equation

a) With the hypotheses, the equation to be solved is

$$\vec{\Omega}.\overrightarrow{\text{grad}}\Phi(\vec{r},\vec{\Omega}) + \Sigma_t(\vec{r})\,\Phi(\vec{r},\vec{\Omega}) =$$

$$\frac{\nu\Sigma_f(\vec{r})}{4\pi}\int_{(4\pi)}\Phi(\vec{r},\vec{\Omega}')\,d^2\Omega' + \int_{(4\pi)}\frac{\Sigma_{s0}(\vec{r}) + 3\Sigma_{s1}(\vec{r})\vec{\Omega}'\cdot\vec{\Omega}}{4\pi}\Phi(\vec{r},\vec{\Omega}')\,d^2\Omega'.$$

b) Substitute $\Phi = \psi + \chi$ into the equation, and express the even and odd parts as follows.

$$\vec{\Omega}\cdot\overrightarrow{\text{grad}}\chi(\vec{r},\vec{\Omega}) + \Sigma_t(\vec{r})\,\psi(\vec{r},\vec{\Omega}) =$$

$$\frac{\nu\Sigma_f(\vec{r})}{4\pi}\int_{(4\pi)}\psi(\vec{r},\vec{\Omega}')\,d^2\Omega' + \frac{\Sigma_{s0}(\vec{r})}{4\pi}\int_{(4\pi)}\psi(\vec{r},\vec{\Omega}')\,d^2\Omega';$$

$$\vec{\Omega}\cdot\overrightarrow{\text{grad}}\psi(\vec{r},\vec{\Omega}) + \Sigma_t(\vec{r})\,\chi(\vec{r},\vec{\Omega}) = \frac{3\Sigma_{s1}(\vec{r})}{4\pi}\vec{\Omega}\cdot\left[\int_{(4\pi)}\vec{\Omega}'\,\chi(\vec{r},\vec{\Omega}')\,d^2\Omega'\right].$$

c) When the scattering is isotropic, i.e. Σ_{s1} is null, the right hand side of the second equation is zero, and this equation gives explicitly

$$\chi(\vec{r},\vec{\Omega}) = -\frac{1}{\Sigma_t(\vec{r})}\vec{\Omega}\cdot\overrightarrow{\text{grad}}\psi(\vec{r},\vec{\Omega}).$$

Substituting this into the first equation, yields an equation governing only the unknown function ψ:

$$-\vec{\Omega}\cdot\overrightarrow{\text{grad}}\left[\frac{1}{\Sigma_t(\vec{r})}\vec{\Omega}\cdot\overrightarrow{\text{grad}}\psi(\vec{r},\vec{\Omega})\right] + \Sigma_t(\vec{r})\,\psi(\vec{r},\vec{\Omega}) =$$

$$\frac{\nu\Sigma_f(\vec{r})}{4\pi}\int_{(4\pi)}\psi(\vec{r},\vec{\Omega}')\,d^2\Omega' + \frac{\Sigma_{s0}(\vec{r})}{4\pi}\int_{(4\pi)}\psi(\vec{r},\vec{\Omega}')\,d^2\Omega'.$$

Notice that only the even flux need be calculated when only the reaction rates or associated quantities are wanted (power distribution, multiplication factor, etc.) This explains the purpose of eliminating χ.

The elimination of χ is similar to the elimination, in the diffusion approximation, of the current $\vec{J}$ between the balance equation and Fick's law, in order to get an equation containing only the flux Φ.

d) For a given and fixed value of $\vec{\Omega}$, the first term is a combination of second derivatives of ψ; hence, may be expressed as

$$-\sum_k \Omega_k \frac{\partial}{\partial k}\left[\frac{1}{\Sigma_t}\sum_\ell \Omega_\ell \frac{\partial \psi}{\partial \ell}\right],$$

where k and ℓ represent x, y, and z. This operator is very similar to the streaming operator in diffusion approximation:

$$-\text{div}\left[\frac{1}{3\Sigma_{tr}}\overrightarrow{\text{grad}}\,\Phi\right] = -\sum_k \frac{\partial}{\partial k}\left[\frac{1}{3\Sigma_{tr}}\frac{\partial \Phi}{\partial k}\right].$$

Except for the replacement of the scalar flux with the integral of the even phase flux, the three other terms are the same as the ones which appear in diffusion theory.

Finally, it appears that the algorithm needed to solve the equation for the even flux can be a simple adaptation of an existing code solving the diffusion equation. All the modules concerning the multigroup treatment, the iterations, the feedbacks, the evolution, etc. are indeed the same. Only the treatment of the phase variable $\vec{\Omega}$ must be added, and the treatment of the streaming operator slightly modified. The phase treatment can be made by one of the usual techniques: spherical harmonics, complete or simplified, discrete ordinates, etc. The following demonstrates that this conclusion remains true even when the anisotropy of the scattering is taken into account.

e) Compare the even-odd splitting to the spherical harmonic expansion, and notice that ψ is the sum of the harmonic terms of even orders n and χ, the sum of the harmonic terms of odd orders n. Therefore, the proposed approximation consists of keeping its main term $n = 1$ for χ only. Recall that the three first-order spherical harmonics can be replaced with the three components of the vector $\vec{\Omega}$; hence, the proposed expression.

With this hypothesis, the term between brackets in the second equation of part **b** is a vector with components

$$\int_{(4\pi)} \Omega'_k \sum_\ell u_\ell \Omega'_\ell \, d^2\Omega' = \frac{4\pi}{3} u_k,$$

since the integrals of the rectangle terms are equal to zero. The right hand side of this equation is, therefore,

$$\Sigma_{s1}\,\vec{\Omega}\cdot\vec{u} = \Sigma_{s1}\,\chi.$$

Notice again that χ can be expressed explicitly by simply replacing Σ_t with the transport cross-section:

$$\Sigma_{tr} = \Sigma_t - \Sigma_{s1},$$

taken from part **c**.

f) Consider again the equation of χ obtained in part **b**, then use

$$\vec{v}(\vec{r}) = \int_{(4\pi)} \vec{\Omega}'\,\chi(\vec{r},\vec{\Omega}')\,d^2\Omega'.$$

Hence,

$$\vec{\Omega}\cdot\overrightarrow{\text{grad}}\psi(\vec{r},\vec{\Omega}) + \Sigma_t(\vec{r})\,\chi(\vec{r},\vec{\Omega}) = \frac{3\Sigma_{s1}(\vec{r})}{4\pi}\,\vec{\Omega}\cdot\vec{v}(\vec{r}).$$

Multiply this by $\vec{\Omega}$ and integrate over all the directions $\vec{\Omega}$ in order to get $\vec{v}$ in the second term of the left hand side. This gives

$$\int_{(4\pi)} \vec{\Omega}\left[\vec{\Omega}\cdot\overrightarrow{\text{grad}}\psi(\vec{r},\vec{\Omega})\right] d^2\Omega + \Sigma_t(\vec{r})\,\vec{v}(\vec{r}) = \Sigma_{s1}(\vec{r})\,\vec{v}(\vec{r}).$$

The term on the right hand side of the equation is solved in a similar manner to before. It then allows $\vec{v}$ to be found. Hence, when substituted into the equation for χ, this gives $\chi(\vec{r},\vec{\Omega})$. Finally, this leaves an equation for ψ only, when the expression for χ found in part **b** is substituted into the equation. Notice that this equation for ψ is exact when the scattering is assumed to be linearly anisotropic.

E. Fundamental mode

Exercise 14.19: elementary eigenfunctions of the Laplace operator

a) For a rectangular parallelepiped of edges equal to π/u, π/v and π/w, where u, v and w are the components of the vector $\vec{b}$,

$$f(x,y,z) = \sum_{1}^{8} \exp\left[-i(\pm ux \pm vy \pm wz)\right] = 8\,\cos ux \cos vy \cos wz.$$

b) For a cylinder of radius $j_{01}/\sqrt{u^2+v^2}$ and of height π/w,

$$f(\rho,z) = \sum_{1}^{2} \int_{(2\pi)} \exp\left[-i(\beta\rho\cos\varphi \pm wz)\right] d\varphi = 2\pi\,J_0(\beta\rho)\cos wz$$

$$(\beta^2 = u^2 + v^2).$$

c) For a sphere of radius π/b,

$$f(r) = \int_{(4\pi)} \exp(-ibr\cos\theta)\,\sin\theta\,d\theta\,d\varphi = 4\pi\frac{\sin br}{br}.$$

Exercise 14.20: diffusion coefficient in fundamental mode

a) The equations giving φ and s are

$$ib\mu\,\varphi(\mu) - \Sigma\,\varphi(\mu) + \int_{-1}^{+1} \Sigma_s(\mu' \longrightarrow \mu)\,\varphi(\mu')\,d\mu' + \frac{s}{2} = 0;$$

$$s = \nu\Sigma_f \int_{-1}^{+1} \varphi(\mu)\,d\mu.$$

b) Now notice that

$$\Phi(x) = \int_{-1}^{+1} \Phi(x,\mu)\,d\mu, \qquad J_x(x) = \int_{-1}^{+1} \Phi(x,\mu)\,\mu\,d\mu,$$

and that $P_0(\mu) = 1$ and $P_1(\mu) = \mu$; then, take into account the orthogonality and normalisation relationships of the Legendre polynomials.

c) Substituting the expansion into Legendre polynomials, gives

$$-(\Sigma - ib\mu)\sum_{n=0}^{\infty} \varphi_n P_n(\mu) + \sum_{n=0}^{\infty} \Sigma_{s,n}\, \varphi_n P_n(\mu) + \frac{s}{2} P_0(\mu) = 0.$$

d) After division by $\Sigma - ib\mu$, and projection on to each Legendre polynomial, the result is

$$-\frac{2}{2k+1}\varphi_k + \sum_{n=0}^{\infty} G_{kn}\, \Sigma_{s,n}\, \varphi_n + \frac{s}{2} G_{k0} = 0.$$

It more judicious to project to the Legendre polynomials after having divided by $\Sigma - ib\mu$ because it is possible to restrict the expansion of the differential scattering cross-section to only few terms, while the projection of the initial equation would give a $\mathbf{P_N}$ type system of equations which would require a very many terms for the flux expansion.

The G coefficients can finally be calculated, where the first ones are

$$G_{00} = \frac{2\xi}{\Sigma}; \qquad G_{01} = G_{10} = \frac{2i\,(1-\xi)}{b}; \qquad G_{11} = \frac{2\Sigma\,(1-\xi)}{b^2};$$

with

$$\xi = \frac{\Sigma}{b} \arctan \frac{b}{\Sigma}.$$

e) Noticing that $s = 2\nu\Sigma_f\, \varphi_0$ and using the first and the second equations to get φ_0 and φ_1, gives an equation expressing the neutron balance (production = absorption + leakage) and a formula giving the diffusion coefficient

$$D = \frac{1}{3\gamma\Sigma},$$

where

$$\gamma = \frac{\frac{b}{3\Sigma} \arctan \frac{b}{\Sigma}}{1 - \frac{\Sigma}{b} \arctan \frac{b}{\Sigma}} \simeq 1 + \frac{4}{15}\left(\frac{b}{\Sigma}\right)^2.$$

f) Provided the scattering is assumed to be linearly anisotropic — i.e. only the two first moments $\Sigma_{s,0}$ and $\Sigma_{s,1} \equiv \bar{\mu}\,\Sigma_{s,0}$ of the scattering law are retained—then it is still possible to deduce the neutron balance and the diffusion coefficient from the two first equations. For this last one, the expression which is obtained—beyond the 'buckling correction' by the coefficient γ—shows also the transport correction:

$$D = \frac{1}{3\left(\gamma\Sigma - \bar{\mu}\Sigma_s\right)}.$$

At the limit $\gamma = 1$, the usual expression $1/3\Sigma_{tr}$ for the diffusion coefficient is obtained.

g) More generally, the so-called $\mathbf{B_N}$ approximation consists of keeping the first moments of the scattering law up to order N. The $N+1$ first equations constitute a system giving the $N+1$ first unknown coefficients $\varphi_0, \varphi_1, \cdots, \varphi_N$, and particularly the diffusion coefficient $D = \varphi_1/(3ib\varphi_0)$ for the fundamental mode.

Exercise 14.21: Behrens's correction

a) Per unit of time, the number of collisions (therefore of paths) is $V_m\Sigma_m\Phi$, and the number of entrances into a cavity is SJ with $J = \Phi/4$. The ratio of these rates is the fraction γ of the paths crossing a cavity. N.B. this 'γ' has no relation with the 'γ' of the previous exercise.

$$\gamma = \frac{S}{4V_m\Sigma_m}.$$

b) It was shown previously in exercise **3.2** that

$$\langle\rho_m\rangle = \frac{1}{\Sigma_m}; \qquad \langle\rho_m^2\rangle = \frac{2}{\Sigma_m^2}.$$

c) The Cauchy theorem gives

$$\langle\rho_c\rangle = \frac{4V_c}{S}.$$

Then from the definition of the factor Q,

$$\langle\rho_c^2\rangle = Q\,\langle\rho_c\rangle^2.$$

Therefore,

$$\langle(\vec{\rho}_{m1} + \vec{\rho}_c + \vec{\rho}_{m2})^2\rangle = \frac{6}{\Sigma_m^2} + 4\frac{4V_c}{S\Sigma_m} + Q(\frac{4V_c}{S})^2.$$

d) The two last terms of the right hand side represent the *increase* of the mean square of the paths crossing a cavity. When it is multiplied by γ, the mean increase of the squared paths, either crossing or not crossing a cavity, is obtained. Finally, multiplying by the mean number of paths, $\langle n\rangle = \Sigma_m/\Sigma_{am}$, and dividing by six, gives the increase of the diffusion area:

$$\Delta L^2 = \frac{2V_c}{V_m}\left(1 + Q\frac{V_c\Sigma_m}{S}\right)\frac{1}{3\Sigma_m\Sigma_{am}} = \frac{2V_c}{V_m}\left(1 + Q\frac{V_c\Sigma_m}{S}\right)L_m^2.$$

By combining with the homogenisation formula for the absorption cross-section, the result for the diffusion coefficient is

$$D = (1 + h)\,D_m,$$

where D_m is the moderator diffusion coefficient, and where

$$h = \frac{V_c}{V_c + V_m}\left(1 + 2Q\frac{V_c\Sigma_m}{S}\right)$$

is the heterogeneity correction. This last correction is *not* a simple homogenisation formula, due to it being an average of *squared* quantities.

e) For a cylinder, $Q = 4/3$, the result is $h = 0.585$.

F. Monte Carlo method

Exercise 14.22: power law probability distribution

Since the n events are independent, the distribution function $F(x)$ of the maximum of the random values is x^n, i.e. the probability that the first selection gives a value smaller than x, multiplied by the probability that the second selection gives a value smaller than x, ..., multiplied by the probability that the nth selection gives a value smaller than x. In other words, the appropriate probability law is $P(x)\,dx = dF = n\,x^{n-1}\,dx$, in the interval $[0, 1[$.

It is also possible to make the selection in one step by applying directly the function $y = x^n$, and then taking its nth root; however, this is expected to require a greater amount of computer time.

Exercise 14.23: random point inside a circle

a) This method of selection is biased toward a greater probability of giving a point near the centre than near the periphery.

b) The distribution is uniform provided that $\rho^2 = \xi_1$, and $\varphi = 2\pi\xi_2$. For the radius, the calculation of the square root can be replaced with a double selection and subsequent search for the maximum—see the previous exercise.

c) This method gives a uniform distribution; however, $1 - \pi/4 = 21.5\%$ of the double selections will give a point outside the circle, which will be rejected.

Exercise 14.24: Buffon's needle problem

a) This probability is $2/\pi$.

b) Let X be the random variable which value is zero when the needle rests on one strip, and one when it lies across two slats. The mathematical expectation of X is $2/\pi$, and the standard deviation is

$$\sigma = \sqrt{2/\pi - 4/\pi^2} = 0.481.$$

After n trials, this standard deviation is reduced by a factor $\sqrt{n}$; for example, when $n = 10\,000$, $\sigma = 0.5\%$.

Exercise 14.25: evaluation of a resonance escape probability

A suitable algorithm is as follows.

- `repeat;`
  - `select a neutron possessing lethargy` u' `chosen uniformly at random within a length` ε `before the trap;`
  - `increment the lethargy by a random amount chosen according to the formula`

$$P(u)\,du = \frac{e^{-(u-u')}}{1-\alpha}\,du, \qquad 0 < u - u' < \varepsilon;$$

```
    – if u is beyond the trap, then increment the counter
      'surviving-neutrons';
    – if u is inside the trap, then increment the counter
      'captured-neutrons';

      /* else u is still before the trap, so continue; */
– until sufficient trials have been conducted;
– calculate statistical parameters;
– output the results;
– stop.
```

Recall that the exact expression of the probability to escape the trap is

$$p = 1 - \frac{1 - e^{-\gamma} - \alpha\gamma}{\xi(1 - \alpha)}.$$

15 Theory of Resonant Absorption of Neutrons

Introduction

Chapter 8 ended with a discussion of the physical aspects of the resonant absorption of neutrons and listed some calculation and modelling problems, which were deliberately put aside for later. We now return to the subject of resonant absorption of neutrons in order to clarify these theoretical aspects (although we do not claim to be giving a detailed analysis of this difficult problem of neutron physics) to give the reader an overview of the types of problem that can be solved by codes such as APOLLO-2, and of the points that still create some difficulties.

The main assumption made from the beginning, which was already applied in Chapter 8 (§ 8.1) was that the resonant material was purely absorbent. By applying the approach used by Michel Livolant and his doctoral student Françoise Jeanpierre in the late 1960s, we saw how to allow for scattering and slowing down by the resonant material, which affect the aspects that we introduced (self-shielding factors, heterogeneous-homogeneous equivalence, Dancoff effect, Doppler effect). This theory was introduced at that time in the first version of APOLLO. With the new developments introduced in Version 2 of this code, there was renewed interest in the theory of resonant absorption of neutrons, because the inadequacies of Livolant and Jeanpierre's original theory had been clearly identified. We shall now present these developments.

15.1. Energy scales of different neutron physics problems

Broadly speaking, neutron physics problems need to be solved at three levels of precision with respect to the energy variable:

1/ In the epithermal domain, the many resonances of heavy nuclei such as uranium 238 typically require *a few tens of thousands of energy groups* in order to properly describe each of the resonances. It is not essential to handle the heterogeneities very precisely at this level; in fact, as we have seen, by using an equivalence we can relate the real geometry to a *homogeneous geometry*.

2/ Handling the spectrum requires about *a hundred energy groups*; this can be done at the assembly constituting the elementary "mesh" of the core of a nuclear reactor, but it requires "microscopic" heterogeneity to be taken into account; this means on the scale of the fuel element or assembly.

3/ In practice, the multiplication factor of a core and the power distribution in it can be calculated to a few energy groups only, e.g. two groups for the usual calculations for pressurised water reactor cores. This calculation takes "macroscopic" heterogeneity into account, i.e. the differences between assemblies and axial variations.

These three types of calculation must be performed one after the other: when calculations have been carried out at a given level, the mean values in space (homogenisation) and energy (condensation) must be taken in order to prepare for the calculation at the next level[1].

Chapter 17 mentions the passage from level **2** to level **3**. Here we shall concentrate on the passage from the first to the second level.

15.2. The heterogeneous-homogeneous equivalence: choice of Bell factor

15.2.1. Principle of pre-tabulations (review)

In Chapter 8, three key points of the Livolant–Jeanpierre theory were presented:

1/ The passage from an actual situation to a "fine structure" situation characterised by a macroscopic flux that is constant in lethargy and uniform in space.

2/ A heterogeneous-homogeneous equivalence, which we are about to describe in greater detail.

3/ A continuous-multigroup equivalence, which has not yet been mentioned but will be presented in the next section.

The heterogeneous-homogeneous equivalence leads to the concept of pre-tabulation in a homogeneous medium which, when the resonant nucleus and its temperature have been defined, turns out to be characterised by a single parameter: the dilution cross-section σ_d. It can therefore easily be explored and tabulated once and for all.

This tabulation can concern the effective resonance integral I_{eff}, which characterises all resonances. With a view to the assembly calculation (Level **2**) that follows, using the APOLLO code, for example, it is preferable to tabulate by group. In this case, we refer to *"effective reaction rates"* rather than an *"effective integral"*, but the idea remains the same, except for the limits of the integral used in the expression:

$$\int \sigma_{a,\text{eff}}(u)du \quad \text{with:} \quad \sigma_{a,\text{eff}}(u) = \varphi(u)\sigma_{a,0}(u). \tag{15.1}$$

We must now specify how the equivalence between the heterogeneous situation under consideration and a homogeneous situation can be established.

[1] More precisely, as we shall see, these are "equivalences" that aim to match the reaction rates as closely as possible, these being the only truly relevant physical parameters in this case because they can be measured.

15.2.2. Principle of heterogeneous-homogeneous equivalence

As we have seen (Figure 8.5, § 8.3.2), the equivalent section $\sigma_e(u)$ of the heterogeneous case is almost a constant. In practice, *a Bell factor b* must be chosen.

This is the ratio $\bar{\sigma}_e/\sigma_{e,\infty}$ between the unique value $\bar{\sigma}_e$ that will be adopted and the asymptotic value $\sigma_{e,\infty}$ of the equivalent section for an infinite cross-section of the resonant material.

The principle applied is very simple: the essential parameter of the problem must be observed. This is the effective integral:

$$I_{\text{eff, heterogeneous}} = I_{\text{eff, homogeneous}}. \tag{15.2}$$

The left-hand side has a given value, and the right-hand side is a function of the dilution cross-section σ_d which characterises the "equivalent" homogeneous medium. The equality (theoretically) defines the value of σ_d, and therefore this equivalent homogeneous medium. We say "theoretically" because the left-hand side is unknown; in fact, it is what we are looking for!

That is why this equation will be re-written in a simplified context: on the one hand, the exact equation for slowing down by the resonant material will be replaced by a simplified model, and on the other hand, the real problem will be replaced by another simplified problem, the **P** problem where the macroscopic flux Ψ is "flat" in lethargy and space. The equivalent cross-section $\bar{\sigma}_e$ or the associated Bell factor b is therefore defined by the following equation:

$$I_{\text{eff, heterogeneous, model, "P" problem}} = I_{\text{eff, homogeneous, model, "P" problem}}. \tag{15.3}$$

Note: This heterogeneous-homogeneous equivalence can apply to the entire resonant domain or to several macrogroups or all resonant groups in the intermediate (Level **2**) multigroup mesh.

15.2.3. Definition of the simplified problem

This "**P**" problem, where the macroscopic function Ψ is "flat" in space and lethargy, gives rise to the fine structure equation (§ 8.2.1 and 8.3.1):

$$r_0\varphi + \sigma_e = (\sigma_0 + \sigma_e)\varphi. \tag{15.4}$$

Its solution φ is interpreted physically as the flux that would exist if the non-resonant materials (moderator etc.) were replaced by purely absorbent materials without changing their total cross-section, and if a neutron source were added with intensity equal to the cross-section of these non-resonant materials.

15.2.4. Implementation with the "narrow resonance" model

If slowing down by resonant nuclei is dealt with using the "narrow resonance" assumption (§ 8.2.2):

$$r_0\varphi \simeq \sigma_{p,0}, \quad \varphi(u) \simeq \varphi_{NR}(u) = \frac{\sigma_{p,0} + \sigma_e(u)}{\sigma_0(u) + \sigma_e(u)}, \tag{15.5}$$

[$\sigma_e(u)$ is replaced by σ_d for the homogeneous case], the heterogeneous-homogeneous equivalence equation is written as:

$$\int \frac{\sigma_{p,0} + \sigma_e(u)}{\sigma_0(u) + \sigma_e(u)} \sigma_{a,0}(u)du = \int \frac{\sigma_{p,0} + \sigma_d}{\sigma_0(u) + \sigma_d} \sigma_{a,0}(u)du, \tag{15.6}$$

where σ_d is the unknown value to be determined, and the equivalent cross-section of the heterogeneous case must be calculated by its definition formula (§ 8.3.1):

$$\sigma_e(u) = \frac{\sigma_0(u)[1 - P_{00}(u)]}{P_{00}(u)}. \tag{15.7}$$

Note that P_{00} depends on the lethargy u because it is a function of σ_0 which, in turn, is dependent on u; similarly, σ_e is actually a function of σ_0. To calculate these integrals, it is therefore simpler to avoid using the lethargy u and instead to use the total cross-section σ_0 of the resonant nucleus as the integration variable[2]. The equivalence equation is written as:

$$\int \frac{\sigma_{p,0} + \sigma_e(\sigma_0)}{\sigma_0 + \sigma_e(\sigma_0)} \pi(\sigma_0)d\sigma_0 = \int \frac{\sigma_{p,0} + \sigma_d}{\sigma_0 + \sigma_d} \pi(\sigma_0)d\sigma_0. \tag{15.8}$$

The boundaries of the integrals are now the extreme values of the total cross-section σ_0 of the resonant nucleus; the equivalent cross-section σ_e is considered as a function of σ_0; the "weight function" π is defined as follows: $\pi(\sigma_0)d\sigma_0$ is the sum of all the elements $\sigma_{a,0}(u)du$ for all lethargy intervals du where the total cross-section of the resonant nucleus is between σ_0 and $\sigma_0 + d\sigma_0$. This density π is obviously rather difficult to calculate, because to do this would require listing all the resonances, but this calculation can be performed once and for all like the effective integral tabulations. This second form of the equivalence equation is far simpler to implement than the first, because some very complicated functions of the lethargy u are replaced by far more regular functions of the cross-section σ_0. In practice, it suffices to tabulate π for a small number of values of σ_0, and then to calculate P_{00} and σ_e for the σ_0 values of the table for each case and to evaluate the integrals using a quadrature formula.

15.2.5. Implementation with the "wide resonance" model

If we choose to use the "wide resonance" approximation (§ 8.2.2):

$$r_0\varphi \simeq \sigma_{s,0}\varphi, \quad \varphi(u) \simeq \varphi_{WR}(u) = \frac{\sigma_e(u)}{\sigma_{a,0}(u) + \sigma_e(u)}, \tag{15.9}$$

a similar equivalence procedure can be implemented. It is, however, slightly more complicated because a weight must be tabulated as a function of both variables: σ_0 and $\sigma_{a,0}$.

[2] This type of change of variable where the second variable is not a monotonic function of the first is called the "Lebesgue integration technique".

15.2.6. Examining the width of resonances: example of uranium 238

To decide which model is better in practice, we need to compare the width of the resonance at the energy interval that can be reached when scattering occurs. This interval can be estimated (in terms of energy) by the product $\varepsilon_0 \times E_0$ where E_0 is the energy at the resonance peak, i.e. approximately the energy of the neutrons concerned and where ε_0, the maximum lethargy gain during scattering by a resonant nucleus, is approximately the maximum relative change in energy. The nuclear width Γ is not the relevant parameter for evaluating the width of the resonance. Using the Breit-Wigner formula, we can see that it is the width at half the height of the peak, but half the cross-section at the peak is still practically infinite. The relevant width is therefore likely to be significantly greater than Γ. In general, we use the "practical width", defined as the energy interval where the resonant cross-section exceeds the potential cross-section, i.e. where the total cross-section is more than double the asymptotic value. Using the Breit-Wigner formula (§ 2.7.1), neglecting the interference term, and noting that the cross-section at the peak, σ_{max}, is much larger than the potential cross-section, σ_p, the following expression is found for this practical width (where the index "0" is understood):

$$\Gamma_p \simeq \Gamma\sqrt{\sigma_{max}/\sigma_p}. \tag{15.10}$$

By way of example, if we take the resonances of uranium 238 that we listed (Table 2.4), we can create Table 15.1, where energies are expressed in eV and the last column is $\Gamma_p/(\varepsilon_0 E_0)$.

Ignoring any irregularities, it seems that the resonances are wider at low energies. A simple empirical model that is sometimes used involves handling all resonances above 50 eV with the "narrow resonance" model, and all resonances below 50 eV with the "wide resonance" model.

15.2.7. Macrogroup-by-macrogroup equivalence

If we wish to examine more than these two energy domains, an equivalence, and therefore a determination of the Bell factor, could be performed for each of the groups used for the spectrum calculation, in the APOLLO calculation groups for example. The model for the equivalence can be chosen differently for each group. In any case, because the equivalence error is second-order with respect to the error of the model itself, the choice of best model is not very important.

15.3. Continuous-multigroup equivalence

15.3.1. Why calculate the real flux rather than the macroscopic flux?

After introducing the factorisation $\Phi = \varphi\Psi$ and applying the principle of pre-tabulating the microscopic aspect φ, it would seem logical to try to calculate the macroscopic flux Ψ. But this was not the recommendation of the authors of the theory, who suggested the opposite

Table 15.1. Estimation of the width of the first resonances of uranium 238 (σ_p = 8.9 barns; ε_0 = 0.0170).

Energy of peak	Nuclear width	Practical width	Slowing down interval	Ratio
6.674	0.024	1.27	0.11	11.3
20.871	0.033	2.19	0.35	6.3
36.682	0.057	3.96	0.62	6.4
66.032	0.048	2.30	1.11	2.1
80.749	0.025	0.41	1.36	0.30
102.56	0.095	4.43	1.72	2.6
116.902	0.048	1.77	1.97	0.90
145.66	0.024	0.20	2.45	0.08
165.29	0.026	0.40	2.78	0.14
189.67	0.196	7.26	3.19	2.3
208.51	0.075	2.33	3.50	0.66
237.38	0.052	1.32	3.99	0.33
273.66	0.048	1.15	4.60	0.25
291	0.039	0.82	4.89	0.17

approach of using the real flux Φ to avoid the approximation that was used to go from the real problem to the simplified "**P**" problem, i.e. the "flat macroscopic flux" approximation. (As seen in § 8.2.1, this approximation is expressed by: $R_0(\varphi\Psi) \cong \Psi R_0(\varphi)$, where R_0 is the operator for slowing down by the resonant material.)

We will obviously not try to calculate the real flux Φ in the ultra-fine lethargy mesh used to calculate all resonances (requiring around 20,000 groups to process the resonant domain), but in the usual multigroup mesh for the calculation of cells or assemblies (about twenty groups in this energy domain), which is far too coarse to describe the resonances. In fact, the whole purpose of resonant absorption theory is to avoid always having to perform the fine calculation. The remaining problem is therefore to be able to calculate the real flux Φ correctly on a coarse mesh.

15.3.2. Principle of continuous-multigroup equivalence

When we say "correctly" in neutron physics, we mean that the reaction rates —the only parameters that are actually measurable— are correct. It does not matter if the flux itself —a non-measurable parameter— is not accurate, as long as it leads to correct reaction rates. This assumes that the flux values are associated with cross-sections that lead to the right reaction rates.

Note also that, in neutron physics, flux values depend on cross-sections because they are obtained by solving the Boltzmann equation containing the cross-sections. In practice, version of this equation that has been simplified to some extent is always used. For example, for the present problem of calculating Φ, a wide multigroup approximation is used to handle a resonance problem that requires a practically continuous approach. The procedure applied to overcome this difficulty is known as "equivalence". This requires the approximate theory used (in this case, wide multigroup theory) to observe the reaction rates. In this case, *multigroup cross-sections* will be sought that, in association with the multigroup flux values, will lead to correct reaction rates in each group.

Because there are obviously as many reaction rates as cross-sections[3], this criterion leads to an equal number of equations and unknowns (the multigroup cross-sections). Multigroup fluxes depend, via the Boltzmann equation, on the multigroup cross-sections to be determined, and so the equations defining these cross-sections are implicit. They must be solved by iteration[4].

This approach might appear reasonable, but unfortunately it is impossible to apply rigorously. Because it is based on the idea of observing the exact reaction rates, it assumes the rates to be known, which means assuming that we know the solution to the problem we are attempting to solve. That is why it will not be applied to the real problem, which has not yet been solved, by definition, but to a simplified problem. If the simplified problem is close to the real problem, we might hope that the multigroup cross-sections determined on the simplified problem and then used for the real problem will give, if not exact results to the real problem, then at least results that are very close to being exact.

In our example, we obviously choose the "**P**" problem as the simplified problem; on the one hand it is close to the real problem[5], and on the other hand, the reaction rates of this problem are known because we obtained them by interpolation in the effective reaction rate tables.

After being obtained by "continuous-multigroup equivalence" on the "**P**" problem, the multigroup cross-sections will be used to calculate the multigroup flux $\mathbf{\Phi}$ of the real problem, and then the reaction rates.

We can conclude that the complete calculation of resonant absorption in a heterogeneous problem (cell, assembly, etc.) involves *two equivalences:*

- the "heterogeneous-homogeneous" equivalence, which is used to define the homogeneous situation that represents the heterogeneous situation under consideration, and then to determine the effective reaction rates by interpolation in the tables previously established for the homogeneous case;

- the "continuous-multigroup" equivalence, intended to establish the correct multigroup "averages" of the cross-sections which, when associated with the multigroup flux calculated with these cross-sections, will lead to the reaction rates just obtained as a result of the "heterogeneous-homogeneous" equivalence.

These two equivalences are applied to the same "**P**" problem (macroscopic flux that is absolutely "flat" in terms of space and lethargy), simplified with respect to the real problem, but representative of the self-shielding situation. With regard to the "lethargy" aspect, the first equivalence is also based on a model of slowing down by the resonant material.

Finally, with practically negligible errors, "Livolant-Jeanpierre" theory reduces the number of groups to be used by a factor of about 1000 (from 20,000 to 20), which greatly reduces the number of calculations to be performed.

[3] Capture, fission, scattering, etc. on the resonant material and non-resonant materials for each group.
[4] Note that the multigroup cross-sections thus determined are the "reasonable" averages of the true cross-sections; in particular, for a constant-lethargy cross-section, the multigroup cross-section is equal to this constant.
[5] The only difference between these two problems is the macroscopic function Ψ which varies slowly in the real problem, and is rigorously constant in the simplified problem.

15.4. "One-by-one" handling of situations with several resonant nuclei

15.4.1. The need to handle situations with several resonant nuclei

The theory discussed above, which was introduced into APOLLO code as soon as it had been put forward, applies to problems involving one and only one resonant material, since the other nuclei are considered to be non-resonant. In fact, we are always dealing with a mixture and/or juxtaposition of different types of resonant material:

- Initially, the fuel has at least two types of resonant nucleus, for example uranium isotopes 238 and 235; after irradiation, other resonant nuclei appear (isotopes of neptunium, plutonium, americium, etc.);
- Around the fuel there can be other materials with resonances (zirconium, hafnium, etc.);
- A single material, such as uranium, can be at different temperatures in different regions of space, and can therefore have resonances that are differently widened by the Doppler effect (for example, the uranium oxide in the fuel pellets of water reactors varies from a little over 300 °C at the surface to well over 1000 °C at the centre);
- A single resonant material can appear in elements with different characteristics (for example, the plutonium assemblies of water reactors have zones of fuel rods with different plutonium content), etc.

15.4.2. Principle of "one-by-one" handling

The only approach that is directly compatible with the assumptions sub-tending the Livolant-Jeanpierre theory is the "one-by-one" process: this is what the APOLLO code does. For the first resonant nucleus, all other nuclei, whether resonant or not, are considered to be non-resonant and are handled as such for the calculation of the equivalent dilution cross-section of the nucleus declared resonant. For the second resonant nucleus, the self-shielding calculation is repeated, with the assumption that all other nuclei, including the first resonant nucleus handled, are non-resonant. This proceeds until the last resonant nucleus.

In practice, for nuclei that are effectively resonant but considered as non-resonant at the stage of calculation reached, we adopt:

- Non-self-shielding average multigroup cross-sections (those found in the library) if the nucleus has not yet been processed.
- Self-shielding multigroup cross-sections (the ones from the continuous-multigroup equivalence) if the nucleus has been processed.

To reduce error as much as possible, the nuclei should be processed in decreasing order of resonant absorption rate (for example, for natural or slightly enriched uranium reactors: uranium 238 before uranium 235).

It would be possible to iterate this procedure in order to replace the cross-sections of all the nuclei that are actually resonant but considered to be non-resonant with self-shielding multigroup cross-sections, but this leads to results that are approximate in any case, and not necessarily better. In practice, this iteration is not performed, which saves on calculation time.

15.5. Extensions of the Livolant-Jeanpierre theory

With Version 2 of the APOLLO code, theoretical developments were carried out to try to extend the scope and accuracy of the self-shielding theory. An overview is presented below.

15.5.1. Allowing for capture in the moderator

In the energy domain of the resonances of uranium, capture by the true moderators is totally negligible (1/v rule for capture cross-sections): that is why its authors created the resonant absorption theory with the assumption that non-resonant nuclei had a purely scattering effect. Structural materials, such as the zirconium in the cladding of fuel rods in water reactors, are obviously chosen with their low capture in mind (amongst other criteria). To consider them as purely scattering does not therefore lead to a great error in the self-shielding calculation.

We have seen, however, that in the "one-by-one" procedure, at certain stages in the calculation resonant materials could be treated as moderator materials. Obviously, however, these materials cannot be treated as purely scattering materials.

That is why the formalism set out by the authors had to be modified to take into account any capture by materials that act as moderators. Using the arguments that led to the fine structure equation[6], we see that to take this capture into account, it is necessary to replace the fine structure equation previously obtained:

$$r_0\varphi + \sigma_e = (\sigma_0 + \sigma_e)\varphi,$$

with:

$$r_0\varphi + \gamma\sigma_e = (\sigma_0 + \sigma_e)\varphi, \tag{15.11}$$

where $\gamma = \Sigma_{s1}/\Sigma_{t1}$ is the probability that an impact on the moderator[7] is a scattering event. It then becomes evident[8] that the solution φ_γ with capture is simply the product $\gamma \times \varphi_1$ (the factor γ times the solution φ_1 without capture). To take the capture into account, we therefore merely need to multiply by γ the effective reaction rates obtained by interpolation in the tables drawn up without capture. No additional tabulation or calculation is required.

[6] In the context we studied, either homogeneous (§ 8.2.1) or heterogeneous with only one fuel zone and a moderator zone (§ 8.3.1); but, as we shall see, it is possible to generalise.

[7] For the energy group in which the self-shielding calculation concerned is performed.

[8] The neutron source in the "**P**" problem was multiplied by γ.

15.5.2. Self-shielding in the thermal domain

The creators of the APOLLO code chose a two-part multigroup subdivision; one for the fast and epithermal domain, and the other for the thermal domain, with the dividing line at an energy E_c (for example: 52 + 47 = 99 groups with the dividing line at 2.7 eV).

The subdivision in the epithermal domain clearly cannot "follow" the resonances. On the other hand, the subdivision in the thermal domain was constructed so that the main resonances, such as those of plutonium 239 and plutonium 241 at 0.3 eV and that of plutonium 240 at 1 eV, would be accurately described. This subdivision can however be a little bit "borderline" and, in particular, can be poorly suited to other resonances, such as that of plutonium 242, located slightly below the cutoff energy. That is why the self-shielding formalism and the associated tables were extended to the thermal domain in Version 2 of the code.

15.5.3. Other slowing-down models

We have seen (§ 15.2.2) that the heterogeneous-homogeneous equivalence could in practice be applied only in the context of a model of slowing-down by the resonant nucleus in order to explicitly obtain the solution φ of the fine structure equation.

The accuracy of this equivalence is clearly likely to increase with increasing accuracy of the slowing down model.

We have already examined two classic models: the NR (narrow resonance) model and the WR (wide resonance) model. There is another model of this type: the IR (intermediate resonance) model. This consists of using an intermediate formula between the first two, with a coefficient λ between 0 and 1 that weights them in a way[9]:

$$\varphi_{NR} = \frac{\sigma_{p0} + \sigma_d}{\sigma_0 + \sigma_d}, \quad \varphi_{WR} = \frac{\sigma_d}{\sigma_{a0} + \sigma_d}, \quad \varphi_{IR} = \frac{\lambda\sigma_{p0} + \sigma_d}{\sigma_{a0} + \lambda\sigma_{s0} + \sigma_d}. \tag{15.12}$$

This model involves a difficulty because it is purely empirical and it is difficult to know how to choose λ, and it has therefore fallen into disuse.

In the context of the developments associated with APOLLO 2, two new models were proposed:

- The *statistical model* ST is a generalisation of the NR model. It consists of approximating the slowing down $r_0\varphi$ with the resonant nucleus instead of with σ_{p0}, which assumes that there is no other resonance above the one being dealt with (therefore $\varphi = 1$), but with the lethargy average of the scattering rate $\sigma_{s0}\varphi$ in the current domain or group. If we substitute this approximation into the fine structure equation, we can see that this model only requires the tabulation of one additional parameter: the average by group or lethargy domain of $\sigma_{s0}/(\sigma_0 + \sigma_d)$;
- The *all-resonance model (*AR) is a generalisation of the previous model (ST). It consists of approximating the diffusion rate in each of the groups by its average value in the group and calculating the transfers $r_0\varphi$ using this assumption; for a given group, these transfers are then the average diffusion rates in this group and the few preceding groups weighted by transfer coefficients expressed with simple analytical formulae for elastic and isotropic slowing down in the centre of mass.

[9] Variant: $\varphi_{IR} = \lambda\varphi_{NR} + (1 - \lambda)\varphi_{WR}$. Here we present the formulae for the homogeneous case.

The tests performed tend to suggest that other models should be replaced by this AR model for all resonant nuclei and all resonances (hence its name "all resonance").

15.5.4. Handling geometries with several moderator zones

Following the study of the homogeneous case (§ 8.2.1), we presented the Livolant–Jeanpierre theory for the heterogeneous case for the simple example of a two-zone geometry, a fuel containing the resonant nucleus and a moderator (§ 8.3.1).

If we assume that there is always only one resonant zone but any number of non-resonant but moderator zones, the calculation will be almost as simple[10] and we end up with the fine structure equation with the same expression for the equivalent cross-section, i.e.:

$$\sigma_e = \frac{\sigma_0(1 - P_{00})}{P_{00}}. \tag{15.13}$$

With the Bell-Wigner approximation, we thus generalise the breakdown into a homogeneous term and a heterogeneous term (§ 8.3.3).

15.5.5. Handling of cases with several resonant zones

On the other hand, the treatment of cases with several resonant zones leads to additional problems. Such situations will be encountered whenever we wish to give details of the distribution of resonant absorptions within the zone concerned. For example, the very large cross-sections in the resonances mean that the captures of neutrons from the moderator by uranium 238 largely occurs in the immediate neighbourhood of the surface of the fuel and not very deep; the formation of plutonium is therefore fairly heterogeneous and it is necessary to divide the fuel into "rings" to take this into account (see Figure 15.1).

More heterogeneous cases can also be encountered: irregular lattices, fuel irradiated during dissolving for reprocessing[11], etc.

If the resonant zones are numbered by an index α or β and the concentration of resonant material in zone α is denoted $N_{0\alpha}$, the equations for the "**P**" problem can be established using the general assumptions of Livolant and Jeanpierre (macroscopic flux Ψ that is "flat" in space and lethargy):

$$\varphi_\alpha = \sum_\beta \frac{P_{\alpha\beta}}{\Sigma_\beta} N_{0\alpha} r_0 \varphi_\beta + \sum_\beta \frac{P_{\alpha\beta}}{\Sigma_\beta} \sum_{i\in\beta} N_{i\beta}\gamma_i\sigma_i. \tag{15.14}$$

There is now one fine structure function φ_α per resonant zone α. The index i denotes the non-resonant nuclei, and the $N_{i\beta}$ values denote the concentrations of the non-resonant nuclei mixed with the resonant nuclei in resonant zones. All other notation has the same meaning as before.

10 The sum must be performed on the index m, applying the reciprocity and complementarity relationships to the first collision probabilities.

11 The first stage of reprocessing fuel elements of water reactors is to cut them into segments a few centimetres long, and then to dissolve them in nitric acid (the fuel only, not the cladding, goes into solution). There is a risk of criticality during this dissolution. It is therefore necessary to be able to form an accurate evaluation of this neutron physics situation where some of the resonant nuclei are in solution in the liquid and others are in solid form.

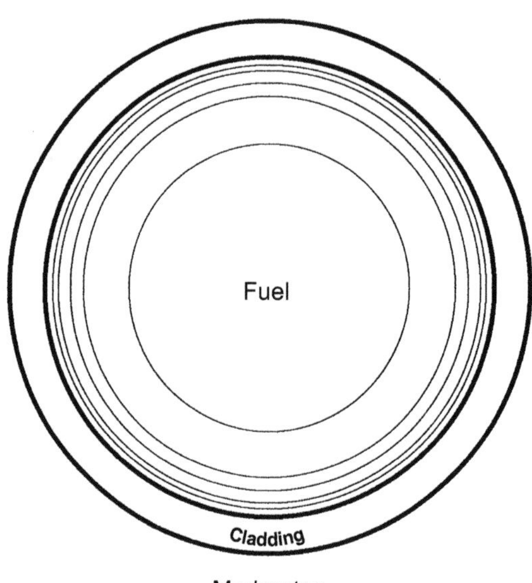

Figure 15.1. Example of the division of a water reactor fuel pellet into concentric crowns to take the "rim effect" into account in the formation of plutonium (from the centre to the edge, the ring volumes are 40%, 30%, 10%, 10%, 5% and 5% of the volume of the pellet).

Livolant and Jeanpierre suggested an additional approximation to simply decouple the equations of this system. This "Pic" approximation (name given by the authors to the probability here written as $P_{\alpha 0}$) consists of assuming that the incoming densities by slowing down on a resonant nucleus are the same in all resonant zones:

$$r_0\varphi_\alpha = r_0\varphi_\beta. \tag{15.15}$$

By replacing $r_0\varphi_\beta$ on the right-hand side with $r_0\varphi_\alpha$, we obtain independent fine structure equations that can be handled by the usual approach. This approximation is not completely arbitrary; it is correct in particular for the NR assumption, since in that case $r_0\varphi_\beta = \sigma_{p0}$ for any zone.

Livolant and Jeanpierre did not see that this additional approximation was not essential. In fact, it can be shown that the system of equations can be diagonalised, which means that it is possible to find a set of linear combinations $\psi_1, \psi_2 \ldots$ of the unknown functions φ_1, $\varphi_2 \ldots$ leading to independent equations for the new unknowns ψ_α. This was programmed in APOLLO 2 under the name of *dilution matrix method* (this matrix generalises the nearly-constant dilution cross-section).

In fact, it is simpler to implement the *"direct method"*, which consists of inverting the order of the two calculation steps: 1/ diagonalising; 2/ using a slowing-down model. (We show that in the WR model, matrix inversions are necessary to handle the coupling in space, but with NR, ST, and AR models, there are only matrix × vector products to calculate.)

15.5.6. Treatment of the general case

If we remain in the context of "flat" macroscopic flux, it is practically as easy to write the equations of the "**P**" problem in the general case with several resonant zones and several resonant materials (index w instead of 0 to designate the various resonant materials):

$$\varphi_\alpha = \sum_\beta \frac{P_{\alpha\beta}}{\Sigma_\beta} \sum_{w \in \beta} N_{w\alpha} r_w \varphi_\beta + \sum_\beta \frac{P_{\alpha\beta}}{\Sigma_\beta} \sum_{i \in \beta} N_{i\beta} \gamma_i \sigma_i. \tag{15.16}$$

It is no longer possible to diagonalise the equations except in the case where the resonant compositions are the same in all zones[12], and so the "dilution matrix method" is no longer applicable. The "direct method" can however be applied if it is acceptable to use the approximation of handling the resonant materials "one by one". Note that this "one by one" treatment can be done by taking the problem globally or group by group. In the context of the AR model, modelling inter-group transfers by scattering on the resonant nuclei, the group-by-group option is preferable.

15.5.7. Problem of interference between resonances

Finally, this problem of interference between the resonances of different materials is the only one that has not been fully solved. In the context of the Livolant-Jeanpierre method of effective reaction rates, the only way to take this interference into account properly would be to establish and use tabulations as a function not of one dilution cross-section, but of N dilution cross-sections, corresponding to the number of resonant nuclei (without mentioning the temperatures): clearly the complexity of the calculations with a half-dozen or more resonant nuclei interfering with each other would make this approach unrealistic. We must emphasise, however, that the one-by-one procedure does not lead to unacceptable errors in practice. There are three possible ways of improving the situation without the need for terribly difficult calculations:

1/ *Tabulating mixtures*: if the composition of the resonant material does not change, or changes little, the mixture can be directly tabulated once and for all. For example, natural hafnium, which contains several resonant isotopes, can be tabulated as such if we assume that the isotopic proportions change little with irradiation in the reactor. There are a few counterexamples where this approximation is not possible, including: gadolinium used as a burnable poison and, obviously, actinides (uranium, plutonium, etc.);

2/ *Method of probability tables*: this approach is presented below;

3/ *Approximately allowing for the interference using a few carefully-chosen coefficients:* this third method has not yet been explored.

[12] In other words, if the *relative* proportions of the resonant nuclei are the same. We could imagine this to be the case, for example, in the dissolution problem where the mixture of resonant nuclei is the same in the solid and dissolved phases, but obviously with very different concentrations.

15.6. Method of probability tables

15.6.1. Introduction

The "sub-group" method is even older than the effective reaction rate method. It was most notably used in the development of calculation codes for fast neutron reactors. Over the last two decades, this approach has attracted renewed interest. Under the name "method of probability tables", it has been given a mathematical foundation. Like the Livolant-Jeanpierre effective reaction rate method, this method is based on the principle of tabulation that is accurate, preliminary, and established once and for all. It is not just an alternative; it is in fact a complementary approach. As we shall see, each method has a preferred domain of application: the high-energy domain for the probability tables (narrow, numerous resonances) and the lower energy domain for effective reaction rates (wider resonances, but fewer of them).

15.6.2. Principle of a probability table

Figure 15.2 (adapted from the publications of Pierre Ribon and Alain Hébert), shows the principle of a probability table:

- On the left is a curve representing a cross-section σ as a function of energy or lethargy in a group normalised to the interval [0,1] by a change of variable. In this example, four resonances can be seen in this group.
- In the middle, shown horizontally, is the probability density $\pi(\sigma)$ as a function of σ (vertically): by definition, $\pi(\sigma)d\sigma$ is the probability, in this group, of the cross-section having a value between σ and $\sigma + d\sigma$;
- On the right, an approximation of this probability density by three Dirac distributions is proposed.

The advantage of this approach can be explained as follows: imagine that we wish to calculate an integral of the following form in the energy group:

$$I = \int F[\sigma(E)]dE, \tag{15.17}$$

where F is a function that is either written out in analytical form or that can be calculated by a calculation code or subroutine. In this form, this integral obviously requires discretisation with many points in the energy group because the function $\sigma(E)$ is complicated there. Since we assume that the function F is actually a function of σ (and therefore of E, but only indirectly), it would seem more sensible to use σ rather than E as the integration variable. This so-called Lebesgue integration technique consists of rewriting the integral I in the following form:

$$I = \int F[\sigma(E)]dE = \int F(\sigma)\pi(\sigma)d\sigma. \tag{15.18}$$

(The limits of the second integral are the extreme values between which the cross-section varies in the group.) The middle diagram (Figure 15.2) appears to suggest that the probability density π is rather complicated, which casts doubt on the benefits of this change of

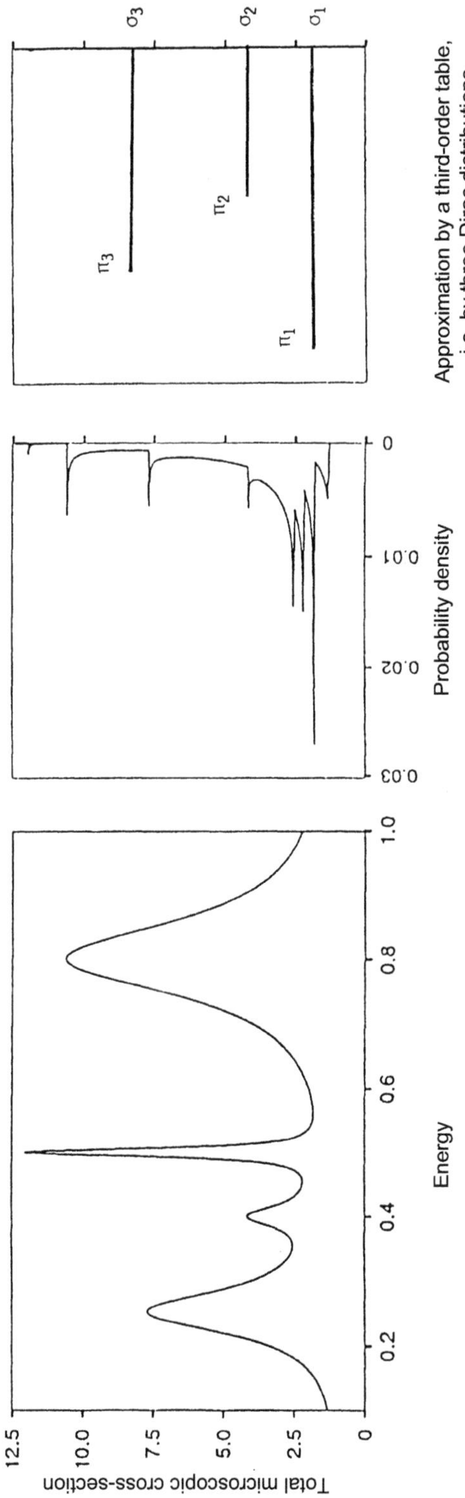

Figure 15.2. Probability density and table of a cross-section.

variable. It turns out, however, that for neutron physics problems where functions F exhibit simple, regular behaviour, the details of π have little importance. In other words, this density can be approximated as long as its essential properties are preserved. The simplest approximation is to replace π by Dirac distributions centred on discrete values σ_i and with weight π_i. This is equivalent to approximating the integral I with a discrete sum:

$$I = \int F[\sigma(E)]dE = \int F(\sigma)\pi(\sigma)d\sigma$$

$$\simeq \sum F(\sigma_i)\pi_i. \tag{15.19}$$

The approximation of an integral using a discrete sum is called a "quadrature formula"[13]. The accuracy of this method obviously depends on the number and choice of elements σ_i and π_i in the "probability table" (third-order table in Figure 15.2).

For an *Nth*-order table, $2N$ elements must therefore be chosen. As a general rule, the choice suggested by Gauss is the best possible selection. It consists of ensuring that the quadrature formula is exact for $2N$ particular functions F. In general, we use the functions $F(\sigma) = \sigma^n$ for n from 0 to $2N-1$: under these conditions, the quadrature formula is exact for all polynomials of an order equal to or less than $2N-1$ (there are tables of coefficients to be used in making this selection). In view of the types of function F appearing in neutron physics (often in hyperbolic form), it is preferable to observe both the positive and negative "moments", i.e. to use values of n to either side of zero; e.g. from $-N+1$ to N.

In practice, we calculate (once and for all) the moments by integrating the cross-sections as they are given, i.e. in terms of energy. We then determine the elements of the probability table by writing out the expression for conservation of momentum:

$$M_n = \int \sigma^n(E)dE, \quad \sum_{i=1}^{N} \pi_i\sigma_i^n = M_n, \quad (-N+1 \leqslant n \leqslant N). \tag{15.20}$$

15.6.3. Table of partial cross-sections

The integrals found in the resonant absorption problem (e.g. on § 15.2.2) involve not one, but two cross-sections of the resonant nucleus; e.g. the total cross-section and the absorption cross-section. We often find, as in this example, that the second cross-section comes into play linearly in the expression for the integral, which has the following form:

$$J = \int \sigma_r(E)F[\sigma(E)]dE, \tag{15.21}$$

where σ is the total cross-section and σ_r is one of the partial cross-sections (capture, fission, scattering, etc.).

We can show that, to handle this type of problem, it is very easy to generalise the $[\sigma_i, \pi_i]$ table by adding the partial cross-sections: $[\sigma_i, \pi_i, \sigma_{c,i}, \sigma_{f,i}, \sigma_{s,i} \ldots]$. These partial cross-sections are obtained by adding the conservation of the appropriate number of linear

[13] Physicists have introduced the same formula under the name of *"sub-groups method"*, with the idea of breaking down the group concerned, for the resonant nucleus, into N sub-groups characterised by the cross-sections σ_i.

co-moments for the partial cross-sections, i.e. integrals in the group of expressions of the type $\sigma_r\sigma^n$. Defined in this way, they are consistent; i.e. the sum of the partial cross-sections is equal to the total cross-section for any i. They are used to calculate type J integrals using the quadrature formula:

$$J = \int \sigma_r(E)F[\sigma(E)]dE \simeq \sum \sigma_{r,i}F(\sigma_i)\pi_i. \tag{15.22}$$

To deal with the more general problem of an integral involving two resonant cross-sections in any way:

$$K = \int F[\sigma_1(E),\ \sigma_2(E)]dE, \tag{15.23}$$

calculation using a probability table becomes more complicated. A table of second-order tensors must be introduced: $[\sigma_{1,i}, \sigma_{2,j}, \pi_{i,j}]$ where i goes from 1 to N and j goes from 1 to M. Just as, for the first-order table, the weight π_i can be interpreted as the probability that σ is approximately equal to σ_i, for a second-order table the weight $\pi_{i,j}$ can be interpreted as the probability that σ_1 is approximately equal to $\sigma_{1,i}$ and that σ_2 is approximately equal to $\sigma_{2,j}$. The table is drawn up by writing out the conservation of the appropriate number of moments and co-moments, i.e. integrals in the group of expressions of the types σ_1^n, σ_2^m and $\sigma_1^n\sigma_2^m$. The table is used to express the integral K with a second-order tensor quadrature formula:

$$K = \int F[\sigma_1(E),\ \sigma_2(E)]dE = \iint F(\sigma_1, \sigma_2)\pi(\sigma_1, \sigma_2)d\sigma_1\, d\sigma_2$$

$$\simeq \sum_{i=1}^{N}\sum_{j=1}^{M} F(\sigma_{1,i}, \sigma_{2,j})\pi_{i,j}. \tag{15.24}$$

This method can be tedious to apply unless the table can be simplified, either because there is little correlation between the cross-sections or because they are very strongly correlated[14].

15.6.4. Mixtures

Mixtures of resonant nuclei can in theory be dealt with by the probability table method by generalising the above method to any number of cross-sections characterising the various resonant nuclei and their reactions. Obviously, however, the complexity increases exponentially and, here again, this approach is not useful unless it can be simplified. For example, the correlation between two nuclides is weak and could no doubt be taken into account by a very small number of parameters; perhaps just one. Like partial cross-sections of the same nuclide, cross-sections at different temperatures are very strongly correlated, which could also allow it to be simplified. For the time being, however, these are merely avenues of research that remain to be explored.

[14] This would certainly apply to two partial cross-sections of the same nuclide, because they will have resonances at the same energies.

15.6.5. Conclusion

This approach is very advantageous for problems that can be handled with one-dimensional probability tables, because it gives rise to low-cost calculations. It is also very well suited to Monte Carlo calculations, because it immediately suggests a selection with the probabilities π_i of the cross-section from the values σ_i. Problems of this type have no correlation between the slowing down and the absorption by resonant nuclei, i.e. for high-energy resonances that are both narrow and statistically dispersed[15]. They can also be problems of mixtures of resonant nuclei taken under these conditions because "one by one" treatment is acceptable in this case.

For problems involving large resonances at lower energies, the slowing down-absorption correlations are stronger[16]; the approach using the effective reaction rate method, which takes a snapshot of this correlation at the instant when the tables are compiled, is probably more suitable. Like the method of probability tables, however, it would assume the use of "cross-referenced" tabulations to allow for the correlations between the resonances of different nuclides in the event that they more or less overlap.

As predicted, the two methods of handling resonant absorption —tabulation of effective reaction rates and creation of probability tables— are complementary, rather than competing, methods.

15.7. Treatment of the Doppler effect

As we have seen (§ 8.4.2), the Doppler effect due to thermal agitation of resonant nuclei leads to a widening of resonances and a lowering of peaks. Because of self-shielding, these two effects do not cancel each other out; since the widening effect dominates, there is an increase in resonant absorption rates as the temperature of the resonant nuclei increases. We shall now present some of the difficulties involved in calculating the Doppler effect.

15.7.1. Calculating the Doppler widening of resonances

In Chapter 8 we also pointed out that the calculation of the deformation of cross-section curves gives rise to a dual problem: 1/ Knowledge of the vibration spectrum of the resonant target nuclei; 2/ The convolution of this spectrum with the cross-sections.

The second point does not lead to insurmountable problems; if the conditions for an analytical calculation (the functions ψ and χ) are not fulfilled, it is still possible to perform the convolution numerically using a sufficiently fine mesh to describe the resonances and the thermal agitation spectrum.

The first point is far trickier, however, because it requires the use of an atomic vibration model in more or less crystalline structures; i.e. it takes us into a complex and poorly understood realm of physics. Research is currently being carried out on such problems. Meanwhile, neutron physicists mainly rely on Doppler effect measurements that are generally expressed in terms of "effective temperature", defined as the temperature that would

[15] Note that only probabilistic treatment is possible for resonances in the statistical domain that are not discriminated by measurement.

[16] Note that the NR assumption is equivalent to decorrelating these two aspects, whilst the WR assumption is equivalent to assuming complete correlation.

give the correct resonant absorption rate if the Doppler widening calculation were performed with a Maxwell spectrum at that temperature. The effective temperature is higher than the true temperature; all the more so at low temperatures. For uranium oxide, the difference is a few tens of degrees Celsius at ambient temperature.

15.7.2. Allowing for temperature gradients

Another problem involved in Doppler effect calculations is the need to take into account the different temperatures of resonant nuclei, such as those of uranium, according to whether they are located near the centre or the edge of fuel elements.

To take this gradient completely into account, we would need to cut up the fuel (for example, as in Figure 15.1) and introduce a different temperature in each zone. This leads to a problem with several resonant nuclei having cross-sections that are somewhat, but not fully, correlated — which, as we have seen, can create difficulties[17]. That is why we generally work with a single temperature throughout the fuel, also known as the "effective temperature", and not to be confused with the previous use of this term.

A more or less empirical rule, attributed to Rowlands, is based on an evaluation of the average temperature experienced by neutrons passing through the fuel. This rule gives satisfactory results in practice:

$$T_{\text{eff}} = aT_{\text{surface}} + (1 - a)T_{\text{centre}},$$

with a equal to 1/3 for a plate, 5/9 for a cylinder, and 2/3 for a sphere.

15.8. Prospects

15.8.1. Validation of self-shielding calculations

Calculating the resonant absorption of neutrons has undoubtedly been the toughest challenge faced by neutron physicists. In the pre-War years, Fermi looked at the physical aspects and expressed them using the concept of the effective integral. The developments that were to follow, and which are still in progress, have served to refine to a high degree of detail both our physical understanding of self-shielding phenomena and their modelling using calculation codes. Today there is still room for improvement in some areas, but the overall situation is satisfactory with regard to the calculations that engineers need to perform.

We now have precise knowledge about resonant cross-sections thanks to the unceasing efforts of experimentalists and analysts. These efforts are still a long way from reaching their conclusion; even for nuclides that we might consider to be very well known, integral experiments occasionally reveal errors[18]; new measurement requirements can also appear[19]. It is therefore necessary to preserve equipment for the measurement of nuclear data and to renew the associated teams of experimentalists and analysts.

[17] A simplified procedure is proposed in APOLLO 2.
[18] A recent example: epithermal capture by uranium 235.
[19] An example: the cross-section of long half-life fission products and of minor actinides liable to be incinerated.

Moreover, the models used in calculation codes have been refined but, likewise, this does not mean that efforts in this area have reached their conclusion.

With regard to modelling, we should point out that reference codes have been developed in parallel with codes intended for standard calculations. These are tools, generally very costly ones in terms of computing time, for the purpose of validating the standard tools. The purpose of validation is to obtain results that are practically free of errors due to the numerical approximations used, in order to support the results obtained with standard tools and to estimate the amount of error resulting from the use of these standard tools[20].

Regarding neutron physics, and the resonant absorption problem in particular, there are two distinct types of reference code: on the one hand, Monte Carlo-type codes that can provide an exact simulation of nuclear reactions, and in which the exact geometry and concentrations can be entered[21]; on the other hand, ultra-fine deterministic codes, able to describe both the geometry and the cross-sections in as much detail as necessary; with this in mind, for example, a 11 276-group library has been created for APOLLO 2, which is almost sufficient to handle resonant absorption without modelling[22].

15.8.2. Problems still pending

In conclusion, we should mention a few problems related to resonant absorption that are still waiting to be solved.

- The role of crystalline effects on thermal agitation and therefore on the Doppler effect in neutron physics is not well known and is difficult to model.

- The Doppler effect on scattering and on transfers is obviously less important that its direct effect on absorption; the few studies that have been carried out reveal weak but non-negligible effects. This subject deserves further investigation.

- The statistical domain, not well known by definition, could undoubtedly be examined more accurately by a probabilistic approach.

- The main problem that is poorly handled by the "one-by-one" approach is that of mixtures (or juxtaposition) of different types of resonant nuclei. This is undoubtedly the most crucial problem at the moment. In her thesis (2006, ref. R-10), Mireille Coste gives up the idea of doing a preliminary tabulation for the mixture: after the heterogeneous-homogeneous equivalence, she performs an "on line" calculation of the effective reaction rates for the homogeneous medium containing this mixture. On the other hand, Noureddine Hfaiedh and Alain Santamarina (ref. R-11) propose to extend the fine multigroup calculation up to about 20 eV in order to treat without any selfshielding modelling the main resonances and the associated interference effects.

[20] Note that we can never be completely free of errors concerning nuclear data and the data for a particular case (dimensions, densities, chemical and isotope composition, etc.). It is therefore essential for validation to be performed with exactly the same sets of data: both the nuclear data and data concerning geometry and concentrations.

[21] This can take a very long time because of these very precise descriptions, and especially because a very large number of particles must be simulated to reduce the statistical uncertainty of the results analysed.

[22] "Sub-group"-type modelling is still necessary in the upper resonance zone.

Exercises

Exercise 15.1: Bell's function in Calvik's approximation

Recall from exercise **14.14** that Carlvik proposed to approximate the first collision probability for an isolated cylinder by the expression

$$P(\omega) = \omega \left(\frac{2}{2+\omega} - \frac{1}{3+\omega} \right),$$

where ω is the opacity, which is the product of the macroscopic total cross-section by the mean chord (in this case, the diameter). What is the Bell function,

$$b(\omega) = \frac{\omega [1 - P(\omega)]}{P(\omega)},$$

when this approximation is used? Compare with the exact function—see exercise **14.14** for a table of the exact first collision probability.

Exercise 15.2: study of the equivalent dilution cross-section

Consider next the self-shielding of uranium in a PWR-UOX type cell whose characteristics are as follows.

- Diameter of the fuel pellet = 8.2 mm.
- Uranium-235 concentration = 0.75×10^{27} nuclei per m^3.
- Uranium-238 concentration = 22×10^{27} nuclei per m^3.
- Oxygen concentration in the fuel = 45.5×10^{27} nuclei per m^3.
- Potential cross-sections (in barns):
    - uranium-235: 13.8
    - uranium-238: 8.9
    - oxygen: 3.76
- Dancoff's factor = 0.1.

a) In order to simplify matters, assume that the resonant cross-section of uranium-238 is 200 barns at resonance, and zero elsewhere. Calculate the opacity of the fuel at a uranium-238 resonance, using the potential cross-sections for uranium-235 and oxygen. Deduce the Bell factor without Dancoff's effect, b^+, based on the table generated in the previous exercise; then the Bell factor with Dancoff's effect, b.

b) Calculate the homogeneous and the heterogeneous terms of the equivalent dilution cross-section. For the heterogeneous term, compare the values without and with the Dancoff effect.

c) Repeat this exercise for uranium-235 with a resonant cross-section equal to 100 barns, and comment the differences.

Exercise 15.3: statistical slowing-down model

The statistical model consists of approximating the slowing operator appearing in the fine structure equation,

$$r_0\varphi + \sigma_d = (\sigma_0 + \sigma_d)\,\varphi,$$

by the average value of the scattering rate,

$$r_0\varphi \simeq \langle \sigma_{s0}\varphi \rangle,$$

which is assumed to be independent of the lethargy in the group or energy range.

Using the expression for φ obtained from the fine structure equation to express the average scattering rate, show that this rate can be calculated and permanently recorded for this group or this energy range.

Exercise 15.4: self-shielding and continuous-multigroup equivalence

Consider the neutron slowing-down process in a homogeneous mixture comprising a purely scattering material, characterized by constant values of ξ and Σ_s, and a purely absorbing material. For the latter material, the cross-section is assumed to be zero everywhere except over an interval of lethargy of width equal to γ, where $\Sigma_a = I/\gamma$, I being the integral of this "window" resonance.

a) Using the "narrow resonance" assumption, i.e. where γ is assumed to be much smaller than ξ, calculate the resonance escape probability.

b) Examine how this probability varies when γ increases without variation of I, thereby simulating of the Doppler effect.

c) Assume that this resonance is in a lethargy group of width Δ greater than γ. For a multigroup treatment, this group is characterized by the cross-sections and moderation power $\tilde{\Sigma}_a$, $\tilde{\Sigma}_s$, and $\tilde{\xi}\tilde{\Sigma}_s$. Since these quantities are constant in lethargy, it is logical to choose $\tilde{\Sigma}_s = \Sigma_s$ and $\tilde{\xi} = \xi$. Which value must be taken for $\tilde{\Sigma}_a$ in order to maintain the resonance escape probability in the multigroup treatment? The ratio $\tilde{\Sigma}_a/\Sigma_a$ can be expressed as the product of γ/Δ—which describes the averaging in the group—by a factor e giving the continuous-multigroup equivalence. Examine this last factor.

Exercise 15.5: interference effect between resonant nuclei

Consider a mixture of two types of resonant nuclei, which are characterized by the indices s and t, and of non-resonant nuclei, characterized by the index 1. Compare the expressions for the effective reaction rates obtained with the "narrow resonance" assumption in each of the following cases.

a) Using an exact calculation.

b) Using a stepwise treatment of the self-shielding effects, without iteration, first for the nuclei of type s, then for the nuclei of type t.

c) Using stepwise treatment of the self-shielding effects, with iteration.
Compare the last two cases with the exact calculation.

Exercise 15.6: probability density for a Gaussian cross-section

Consider an isolated resonance whose cross-section is given by the formula

$$\sigma = e^{-x^2},$$

where x is the reduced energy measured from the resonance peak. With an adequate choice of the units, this expression could represent the cross-section for a resonance widely broadened by the Doppler effect. This resonance is a the centre of a group limited by $-X < x < +X$. Calculate the probability density of σ in the group.

Numerical example: tabulate this density for $X = 2$. Calculate the positive moments of this cross-section.

Exercise 15.7: representation of effective integrals by a probability table

In terms of functions of the dilution cross-section σ_d, how should the effective integrals be expressed, calculated with the "narrow resonance" assumption in a energy group, where the cross-sections of a resonant nuclide are represented by probabilities tabulated according to $[p_i, \sigma_i, \sigma_{ai}\,;\ i = 1 \text{ to } N]$?

Solutions

Exercise 15.1: Bell's function in Carlvik's approximation

The result is

$$b(\omega) = \frac{6+\omega}{4+\omega}.$$

This approximation has, by construction, the correct behaviour at the origin ($b = 1.5$) and at infinity ($b = 1$). The following table facilitates comparison of the intermediate values.

ω	$P(\omega)$	$b(\omega)$	Approx.	Error (%)
0.04	0.02561	1.5219	1.4950	−1.76
0.08	0.04967	1.5306	1.4902	−2.64
0.2	0.11494	1.5394	1.4762	−4.11
0.4	0.20697	1.5326	1.4545	−5.10
0.6	0.28351	1.5163	1.4348	−5.38
0.8	0.34838	1.4963	1.4167	−5.32
1.2	0.45225	1.4534	1.3846	−4.73
2	0.59285	1.3735	1.3333	−2.93
3	0.69843	1.2953	1.2857	−0.74
4	0.76355	1.2387	1.2500	+0.91
5	0.80677	1.1976	1.2222	+2.06
10	0.90077	1.1016	1.1429	+3.74

Exercise 15.2: study of the equivalent dilution cross-section

a) The opacity of the fuel,

$$\omega = 0.82 \times (0.022 \times 200 + 0.00075 \times 13.8 + 0.0455 \times 3.76)$$

$$\omega = 3.61 + 0.01 + 0.14 = 3.76,$$

is dominated by the resonant nuclei. According to the preceding table, $b^{+} = 1.25$; hence, using

$$b = \frac{(1 - C)b^{+}}{1 - C + Cb^{+}}$$

gives $b = 1.10$.

b) The result are $\sigma_e = 8.2 + 60.9 = 69.2$ barns, and 69.4 barns for the heterogeneous term without Dancoff effect.

c) The opacity of the fuel

$$\omega = 0.82 \times (0.00075 \times 100 + 0.022 \times 8.9 + 0.0455 \times 3.76$$

$$\omega = 0.06 + 0.16 + 0.14 = 0.36,$$

is dominated by the *non*-resonant nuclei. The results are $b^{+} = 1.53$, and $b = 1.31$. Therefore, $\sigma_e = 489 + 2131 = 2620$ barns, or 2494 barns for the heterogeneous term without Dancoff effect.

The Bell factor is greater for isotope 235 than for isotope 238 due to its smaller resonant cross-section. Moreover, the concentration is much smaller as well. Consequently, a factor ≈ 10 is observed on the opacities. The equivalent cross-section for isotope 235 is much greater owing to the small concentration of this isotope.

The two self-shielding factors—approximately $(\sigma_{p0} + \sigma_e)/(\sigma_0 + \sigma_e)$—with the chosen model, for uranium-238 and uranium-235 are about 0.28 and 0.96, respectively.

Exercise 15.3: statistical slowing-down model

The equations can be rearranged to give

$$\langle \sigma_{s0}\varphi \rangle = \frac{m\,\sigma_d}{1 - m}, \qquad \text{where} \qquad m = \langle \frac{\sigma_{s0}}{\sigma_0 + \sigma_d} \rangle,$$

which can simply be tabulated as a function of σ_d.

Exercise 15.4: self-shielding and continuous-multigroup equivalence

a) The number of neutrons falling into the trap is γ/ξ, and the probability of absorption for each of these neutrons is $\Sigma_a/(\Sigma_s + \Sigma_a)$. Therefore, the probability of escape from the trap is

$$p = 1 - \frac{\gamma}{\xi}\frac{\Sigma_s}{\Sigma_s + \Sigma_a}, \qquad p = 1 - \frac{\gamma}{\xi}\frac{I}{\gamma\Sigma_s + I}.$$

b) The second expression shows that p decreases when γ increases.

c) Rewriting p with the multigroup parameters, gives

$$\frac{\tilde{\Sigma}_a}{\Sigma_a} = \frac{\gamma}{\Delta} e; \qquad e = \frac{\Sigma_s}{(1 - \frac{\gamma}{\Delta})\Sigma_a + \Sigma_s}.$$

This last factor expresses the continuous-multigroup equivalence for the absorption rate with a modified self-shielding factor.

Exercise 15.5: interference effect between resonant nuclei

a) Starting with the fine structure equation for the mixture,

$$R_s\varphi + R_t\varphi + \Sigma_1 = (N_s\sigma_s + N_t\sigma_t + \Sigma_1)\varphi,$$

and using the "narrow resonance" asumption for the heavy nuclides,

$$R\varphi \simeq N\sigma_p,$$

the fine structure φ can be found explicitly; hence, the effective absorption rate is

$$a = \int (N_s\,\sigma_{as} + N_t\,\sigma_{at}) \frac{N_s\,\sigma_{ps} + N_t\,\sigma_{pt} + \Sigma_1}{N_s\,\sigma_s + N_t\,\sigma_t + \Sigma_1}\, du.$$

Note that σ_s, σ_t, σ_{as} and σ_{at} are functions of lethargy with resonances; the other quantities are constants.

b) Within this treatment,

$$a_s = \int N_s\,\sigma_{as}\, \frac{N_s\,\sigma_{ps} + N_t\,\sigma_{pt} + \Sigma_1}{N_s\,\sigma_s + N_t\,\bar{\sigma}_t + \Sigma_1}\, du,$$

and

$$a_t = \int N_t\,\sigma_{at}\, \frac{N_s\,\sigma_{ps} + N_t\,\sigma_{pt} + \Sigma_1}{N_s\,\tilde{\sigma}_s + N_t\,\sigma_t + \Sigma_1}\, du.$$

Here, $\bar{\sigma}$ is the average value of the unshielded cross-section (the multigroup value in the library), and $\tilde{\sigma}$ is the multigroup value calculated by the continuous-multigroup equivalence that matches the effective absorption rate. For the calculation of the nuclide s, the nuclide t is unknown, and the unshielded cross-sections of the library are used. For the calculation of the nuclide t, the nuclide s is known, and the multigroup self-shielded cross-sections can be used.

c) The self-consistent solution yields both the effective reaction rates using the shielded multigroup cross-sections $\tilde{\sigma}$ for *both* nuclides, in the denominator of the formulae. Owing to the stepwise approximation, the latter calculation is different from the exact calculation in part **a**, and is not necessarily better than the non-iterated calculation in part **b**. It is generally recommended not to perform the iteration and to proceed to the stepwise calculation in order of *decreasing* absorption rates.

Exercise 15.6: probability density for a Gaussian cross-section

To obtain the probability density of the cross-section σ, use

$$p(\sigma)\,d\sigma = \frac{|\,dx\,|}{X}, \qquad \text{where} \qquad \sigma = e^{-x^2}.$$

This is

$$p(\sigma) = \frac{1}{2X\sigma\sqrt{\ln 1/\sigma}} = \frac{1}{2\sigma\sqrt{\ln(1/\Sigma)\,\ln(1/\sigma)}}.$$

with $\Sigma = e^{-X^2}$.

The following table gives some numerical values using $X = 2$, i.e. $\Sigma = 0.0183$. The curve representing this probability, shown in the accompanying graph, has a vertical asymptote for $\sigma = 1$, associated with the horizontal tangent of the curve representing $\sigma(x)$ at its maximum for $x = 0$. The probability p is minimum for $\sigma = 1/\sqrt{e}$, then increases when σ is small, i.e. in the wings of the resonance.

σ	$p(\sigma)$
1	∞
0.999	7.912
0.99	2.519
0.95	1.162
0.9	0.693
0.8	0.662
0.7	0.598
0.6	0.583
0.5	0.601
0.4	0.653
0.3	0.759
0.2	0.985
0.1	1.648
0.05	2.889
0.0183	6.825

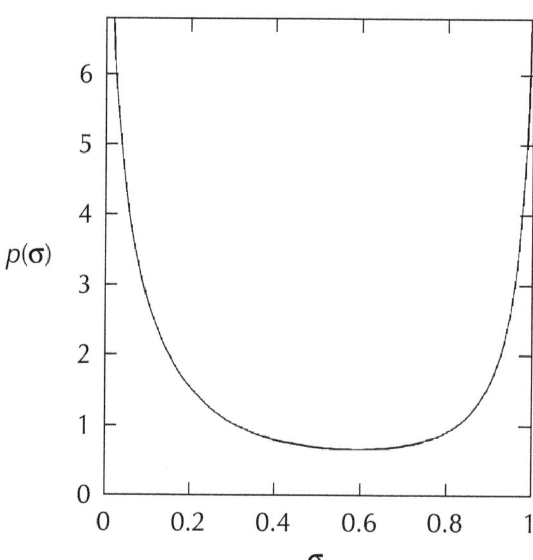

By defining the kth moment of σ as the integral over the given interval of the cross-section to the power k, then

$$m_k = \int_{-X}^{+X} \sigma^k(x)\,dx = X\int_{\Sigma}^{1} \sigma^k\,p(\sigma)\,d\sigma = \sqrt{\frac{\pi}{k}}\,\mathrm{erf}\,(\sqrt{k}X).$$

Exercise 15.7: representation of effective integrals by a probability table

The effective integral is

$$I_{\text{eff}} = \int \sigma_a(u)\,\varphi(u)\,du,$$

where

$$\varphi(u) = \frac{\sigma_p + \sigma_d}{\sigma(u) + \sigma_d},$$

when using the "narrow resonance" hypothesis.

Using a probability table to evaluate the integral, yields a sum of rational fractions:

$$I_{\text{eff}} = \sum_{i=1}^{N} p_i \, \sigma_{ai} \frac{\sigma_p + \sigma_d}{\sigma_i + \sigma_d}.$$

16 Perturbation theory

Introduction

Perturbation theory is not specific to neutron physics; it is frequently used in physics as a whole. Its purpose is to evaluate as accurately as possible the changes in parameters of interest that result from a change applied to a system. In particular, it aims to avoid the effects of differences in the estimation of the consequences of a small perturbation[1].

For example, because astronomers do not know how to solve the N-body problem analytically for N greater than 2, they use perturbation theory to estimate the change in orbit of a planet in the solar system caused by the gravitational perturbations due to the presence of the other planets.

In neutron physics, there is a variety of parameters of interest. The most important of these is undoubtedly the multiplication factor, and the neutron physics version of perturbation theory was created primarily to evaluate the changes in this factor. There are, however, other parameters that an engineer needs to be aware of: power distribution, change of composition by irradiation, etc. That is why generalised perturbation theories have been developed to evaluate the variations in these parameters.

Here we intend to provide only an introduction to this theory, and so we present only the standard perturbation theory as it applies to the multiplication factor. The generalisations will be discussed briefly at the end of the chapter.

16.1. Concept of adjoint flux

16.1.1. Importance in neutron physics

The perturbation formula involves an "adjoint flux" and, even though a purely mathematical definition would suffice in this context, it is of interest to introduce this function based on physical concepts.

From a physical point of view, adjoint flux quantifies the concept of "neutron importance". Intuitively, it is easy to understand that a neutron placed at the centre of a reactor, with a good chance of causing fission, would be more "important" than a neutron placed at the surface with a high probability of escape. Similarly, a neutron placed in the thermal

[1] If, for example, we wish to evaluate an effect of a few tens of pcm, it will obviously not be possible to do so by taking the difference between two calculation results obtained to within 100 pcm.

domain is likely to be more important than a neutron placed at a resonance energy of uranium 238. To quantify this importance, it is easiest to count the descendants of the neutron concerned (children, grandchildren, etc.), obviously applying a reasoning based on averages. This might be possible for a subcritical system, but a difficulty arises for a critical system and, even more so, for a supercritical system, because in that case the average descendancy becomes infinite on average[2]. We will therefore have to reject the concept of adjoint flux for a supercritical system, and consider that the adjoint flux of a critical case is the limit of the adjoint flux of the subcritical case when its (negative) reactivity tends to zero. In a subcritical system, all neutrons have finite descendancy. In a critical system, an infinitesimal fraction of neutrons has infinite descendancy. The evaluation of the limit is therefore a mathematical problem of the type: "$0 \times \infty$"[3].

16.1.2. Mathematical definition of adjoint flux

In mathematics, the scalar product $\langle \Psi, \Phi \rangle$ of two functions Φ and Ψ is defined as the integral over the entire system of their product[4]. For example, for a neutron system described by the space variables and the neutron speed and direction variables:

$$\langle \Psi, \Phi \rangle = \iiint \Psi(\vec{r}, v, \vec{\Omega})\Phi(\vec{r}, v, \vec{\Omega}) d^3r\, dv\, d^2\Omega. \tag{16.1}$$

An operator A is a mathematical entity that associates a function $g = Af$ to any function f (taken within a certain set). This operator is said to be "linear" if it satisfies linear combinations.

The adjoint operator A^+ of any linear operator A is defined by the following property:

$$\langle \Psi, A\Phi \rangle = \langle A^+\Psi, \Phi \rangle, \tag{16.2}$$

whatever the functions Φ and Ψ.

For example, for an operator in integral form:

$$A\Phi = \int k(X, X')\Phi(X')dX', \tag{16.3}$$

(where X represents all of the variables), we can simply permute the two arguments of the kernel k to obtain the adjoint operator:

$$A^+\Psi = \int k(X', X)\Psi(X')dX', \tag{16.4}$$

(the two double integrals expressing the scalar products are identical).

For the neutrons, this is equivalent to reversing the direction of time, i.e. travelling the opposite path to true neutrons, for example by rising from the thermal domain to the fast domain.

---

2 Certain neutrons obviously have finite descendancy irrespective of the multiplication factor.

3 In a supercritical system, the fraction of neutrons with infinite descendancy is finite and grows as the system becomes increasingly supercritical.

4 If the functions have a complex value, the function Ψ must be replaced by its complex conjugate.

The flux (referred to as "direct" as opposed to adjoint) is the solution of an equation of the following type:

$$H\Phi = \frac{1}{v}\frac{\partial \Phi}{\partial t}, \tag{16.5}$$

where H is the (time-independent) Boltzmann operator describing transport and collisions. If the reactor is critical, then the stationary equation:

$$H\Phi = 0, \tag{16.6}$$

has a non-trivial solution (defined to within a factor). We can say that this solution is the eigenfunction of H:

$$H\Phi = \mu\Phi, \tag{16.7}$$

associated with the eigenvalue $\mu = 0$. (Criticality can be defined as the situation characterised by a zero eigenvalue.) We can show that an operator and its adjoint have the same eigenvalues; H^+ therefore also has an eigenvalue of zero. Adjoint flux is therefore defined as the solution (also defined to within a factor) of:

$$H^+\Phi^+ = 0. \tag{16.8}$$

16.1.3. Examples

The equation for one-group diffusion theory, in a reactor assumed to be critical, is:

$$D\Delta\Phi - \Sigma_a\Phi + \nu\Sigma_f\Phi = 0. \tag{16.9}$$

The Laplace operator and multiplication by a function are self-adjoint operators. In this theory, therefore, the direct flux and adjoint flux are identical (to within a factor).

This result, which is true in monokinetic theory even with the exact transport operator, reflects the fact that the physics is not changed by the fact of reversing the direction of the neutron path (there is no way of knowing whether the "film was shown forwards or backwards").

This does not hold true in spectrum theory (for example, we cannot reverse neutron slowing-down). In "two-group diffusion theory", for example:

$$\begin{aligned} D_1\Delta\Phi_1 - \Sigma_1\Phi_1 + \tfrac{k_\infty}{p}\Sigma_2\Phi_2 &= 0, \\ D_2\Delta\Phi_2 - \Sigma_2\Phi_2 + p\Sigma_1\Phi_1 &= 0, \end{aligned} \tag{16.10}$$

the 2×2 matrix of operators must be transposed to write the adjoint equations:

$$\begin{aligned} D_1\Delta\Phi_1^+ - \Sigma_1\Phi_1^+ + p\Sigma_1\Phi_2^+ &= 0, \\ D_2\Delta\Phi_2^+ - \Sigma_2\Phi_2^+ + \tfrac{k_\infty}{p}\Sigma_2\Phi_1^+ &= 0. \end{aligned} \tag{16.11}$$

16.1.4. Physical definition of adjoint flux

In a critical system, the adjoint flux, like the direct flux, is defined only to within a multiplication factor. It is convenient to choose both normalisation constants such that:

$$\iiint \Phi^+(\vec{r}, v, \vec{\Omega})\Phi(\vec{r}, v, \vec{\Omega})d^3r\,dv\,d^2\Omega = 1. \tag{16.12}$$

Under these conditions the following theorem can be demonstrated: in a critical system free of all neutrons let us place a neutron at point $\vec{r}_0$, at speed v_0 and in direction $\vec{\Omega}_0$; after extinction of the transients, the flux observed asymptotically will be:

$$\Phi^+(\vec{r}_0, v_0, \vec{\Omega}_0)\Phi(\vec{r}, v, \vec{\Omega}). \qquad (16.13)$$

This definition is obviously consistent with the intuitive concept of importance introduced above, and solves the problem of normalisation that is created by the evaluation of the limit of the type "$0 \times \infty$".

16.2. Perturbation formulae

16.2.1. Associated critical reactor

Neutron physics calculations are usually performed at steady state. If all of the neutrons are emitted by fission, which we assume here, this calculation requires the reactor to be exactly critical. Otherwise, the flux is bound to vary with time. But naturally a given reactor *a priori* (whether physically existing or described by numerical data in a calculation code) is never exactly critical. This leads us to search for a (fictitious) critical reactor associated with the real reactor.

In practice, this critical reactor is obtained by modifying any state parameter: the boundary between the core and the reflector, the penetration of a control element, concentration of an absorbent or a fissile material, etc. In most cases, as in this case, criticality is obtained in the neutron physics calculations by changing *the production operator by an appropriate factor* λ. This is very easy to do: simply replace the number ν of neutrons emitted by fission by $\lambda\nu$ each time this parameter appears in a term of an equation. This factor λ has a simple physical interpretation: its inverse $1/\lambda$ is the effective multiplication factor k of the initial reactor. Note that the multiplication factor k is the number of neutrons re-emitted at the time of absorption of the initial neutron, for each neutron emitted by fission and on average. The number of neutrons re-emitted is normalised to 1 if we divide by k the number ν of neutrons emitted when the absorption is a fission. In other words, if the system is supercritical, it is made critical by dividing productions by the factor k greater than 1, and conversely, if it is subcritical, we can make it critical by dividing productions by the factor k less than 1.

More formally, we can separate the Boltzmann operator H introduced above into two components:

$$H = P - K, \qquad (16.14)$$

where P is the production operator (all of the terms proportional to ν) and $-K$ represents all the other operators.

With the real reactor we can associate a family of fictitious reactors that are dependent on a parameter λ and characterised by:

$$H(\lambda) = \lambda P - K, \qquad (16.15)$$

and we select the critical one from this family[5].

[5] To be strictly rigorous, we must point out that several values of λ can satisfy this problem, but only one gives a flux and adjoint flux that are everywhere positive, and that is obviously the one we shall use. (This difficulty and this conclusion are easy to see and to analyse on the bare homogeneous pile problem handled by the one-group theory seen in Chapter 6.)

It is possible to calculate a stationary flux for this critical reactor by solving the following equation:

$$(\lambda P - K)\Phi = 0, \tag{16.16}$$

and, if this is also desired, an adjoint flux can be calculated by solving the adjoint equation:

$$(\lambda P^{+} - K^{+})\Phi^{+} = 0. \tag{16.17}$$

16.2.2. Exact perturbation formula

We now consider an "initial" reactor (denoted with the subscript 1) and a "perturbed" reactor (with the subscript 2).

The flux of each of these reactors will be given by the following equations:

$$(\lambda_1 P_1 - K_1)\Phi_1 = 0, \quad (\lambda_2 P_2 - K_2)\Phi_2 = 0, \tag{16.18}$$

where the critical values of parameter λ are chosen so that these equations will have a solution.

If we take the difference between these two equations and group the terms, we can write:

$$(\lambda_2 - \lambda_1)P_2\Phi_2 + (\lambda_1 P_1 - K_1)(\Phi_2 - \Phi_1) + [(\lambda_1(P_2 - P_1) - (K_2 - K_1)]\Phi_2 = 0.$$

This equation governing functions can be replaced by a scalar equation if we perform scalar multiplication by any function. In this case, it is a good idea to multiply it by Φ_1^{+} to eliminate, thanks to the definition of the adjoint operator, the intermediate term containing a flux *difference* and therefore a risk of numerical inaccuracies on the parameters calculated:

$$\langle\Phi_1^{+}, (\lambda_1 P_1 - K_1)(\Phi_2 - \Phi_1)\rangle = \langle(\lambda_1 P_1^{+} - K_1^{+})\Phi_1^{+}, (\Phi_2 - \Phi_1)\rangle, \tag{16.19}$$

this expression is zero because the first factor of the second scalar product is zero by definition of Φ_1^{+}.

The remaining expression leads to:

$$\Delta\rho = \lambda_1 - \lambda_2 = \frac{\langle\Phi_1^{+}, [\lambda_1(P_2 - P_1) - (K_2 - K_1)]\Phi_2\rangle}{\langle\Phi_1^{+}, P_2\Phi_2\rangle}, \tag{16.20}$$

i.e. a formula that lets us calculate the difference between the values of the critical parameter without any risk of error due to difference effects on calculated results (the remaining differences in the expression concern the operators only, i.e. calculation *data*, and are therefore not subject to this problem). The difference between the values of λ can be written (except for its sign) as a reactivity difference $\Delta\rho = \rho_2 - \rho_1$, (since reactivity is defined by the following equation (§ 4.1.3):

$$\rho = \frac{k-1}{k} = 1 - \lambda. \tag{16.21}$$

If the initial reactor is critical, this perturbation equation can be rewritten in its simplest form:

$$\Delta\rho = \frac{\langle\Phi_1^{+}, \Delta H\Phi_2\rangle}{\langle\Phi_1^{+}, P_2\Phi_2\rangle}, \tag{16.22}$$

with:

$$\Delta H = H_2 - H_1.$$

Note that *two* calculations are required to apply this formula: the adjoint calculation for the first case, and the direct calculation for the second case. In more general terms, to study n perturbations, n direct calculations must be performed in addition to the initial adjoint calculation. Note also that no assumption about the size of the perturbation was made and no approximation was introduced.

16.2.3. First-order perturbation equation

The perturbations studied are often *small*. In this case, the higher-order terms can be neglected in favour of the main terms in the equation. In the numerator, if we set $\Phi_2 = \Phi_1 + \Delta\Phi$, the $\Delta\Phi$ term can be omitted, leaving Φ_1, and likewise in the denominator, if we set $P_2 = P_1 + \Delta P$, the $\Delta P \Phi_1$ and $\Delta P \Delta\Phi$ terms can be neglected. We thus obtain the following first-order equation:

$$\Delta\rho \simeq \frac{\langle \Phi_1^+, \Delta H \Phi_1 \rangle}{\langle \Phi_1^+, P_1 \Phi_1 \rangle}, \tag{16.23}$$

or, if the index 1 is implied:

$$\Delta\rho \simeq \frac{\langle \Phi^+, \Delta H \Phi \rangle}{\langle \Phi^+, P \Phi \rangle}. \tag{16.24}$$

The advantage of this equation is that *it does not require the flux of the perturbed state to be calculated*; only the flux and adjoint flux of the initial case need to be calculated. In particular, if there are several (small) perturbations to be evaluated, no new flux or adjoint flux calculation is required.

16.3. Application examples

Perturbation theory has a very wide variety of applications. We shall present just three examples to illustrate this diversity.

16.3.1. Plutonium 239 equivalent

The fuel loaded into a fast neutron reactor can have a relatively varied composition because this type of reactor allows a great deal of flexibility in this respect. In particular, the plutonium can come from different sources (UNGG, PWR, FNR, etc.) and its composition can be quite different, depending on the case. It is therefore useful to construct equivalence coefficients for the various heavy nuclei. These coefficients let us define the proportions to be used when manufacturing the various heavy nuclei in order to obtain the desired performance characteristics, in particular the correct cycle duration.

As a first approximation, if we assume that cycle duration is affected by the initial reactivity of the fuel, we can then define the equivalences using a reactivity criterion. Two fuel compositions will be considered as equivalent if they give the same initial multiplication factor.

Table 16.1. Approximate reactivity weights in a fast neutron reactor.

Nuclide	Weight
Uranium 235	0.8
Uranium 238	0
Plutonium 239	1
Plutonium 240	0.1
Plutonium 241	1.5
Plutonium 242	0

Under these conditions, it is convenient to normalise the composition of the fuel used, a mixture of uranium 238 with abundance t'_8 and other materials (various plutonium isotopes, uranium 235, etc.) with abundance values t'_i, to an equivalent composition containing only uranium 238 and plutonium 239 in abundance $t_8 = 1 - t$ and $t_9 = t$ respectively. We can assume that only the absorption and production cross-sections in the fuel are different, and that the other terms (transfers, leakage, etc.) are the same in both situations. Under these circumstances, the numerator of the first-order perturbation equation, which must be zero by the definition of equivalence, has the following form:

$$\int \left[\sum_i t'_i \sigma_i^+ - (1-t)\sigma_8^+ - t\sigma_9^+ \right] \Phi^+ \Phi du = 0, \tag{16.25}$$

with $\sigma^+ = \nu\sigma_f - \sigma_a$ (the flux and adjoint flux values in the fuel must be considered and, like the cross-sections, they are functions of the lethargy u). Allowing for the fact that the sum of all t'_i is equal to 1, we see that the plutonium 239 equivalent content t is given by:

$$\sum_i t'_i(\bar{\sigma}_i^+ - \bar{\sigma}_8^+) = t(\bar{\sigma}_9^+ - \bar{\sigma}_8^+), \tag{16.26}$$

$$t = \sum_i W_i t'_i, \quad \text{with:} \quad W_i = \frac{\bar{\sigma}_i^+ - \bar{\sigma}_8^+}{\bar{\sigma}_9^+ - \bar{\sigma}_8^+}, \tag{16.27}$$

where $\bar{\sigma}^+$ denotes the average of σ^+ weighted by $\Phi^+\Phi$.

In a sense, we have placed each nuclide on a scale where uranium 238 is at zero and plutonium 239 has the value 1.

Note that, unlike the flux, the adjoint lethargy flux is relatively "flat" in all reactors[6]. Figure 16.1 shows this for an example concerning a water reactor lattice[7]. Note the strong effect of leakage on the importance of the fast neutrons.

[6] Also note that flux is a density function, whilst adjoint flux is a "true" function. If the adjoint flux is "flat" in terms of lethargy, it is also "flat" in terms of energy.

[7] 172-group calculations using APOLLO-2 code, courtesy of Olivier Litaize.

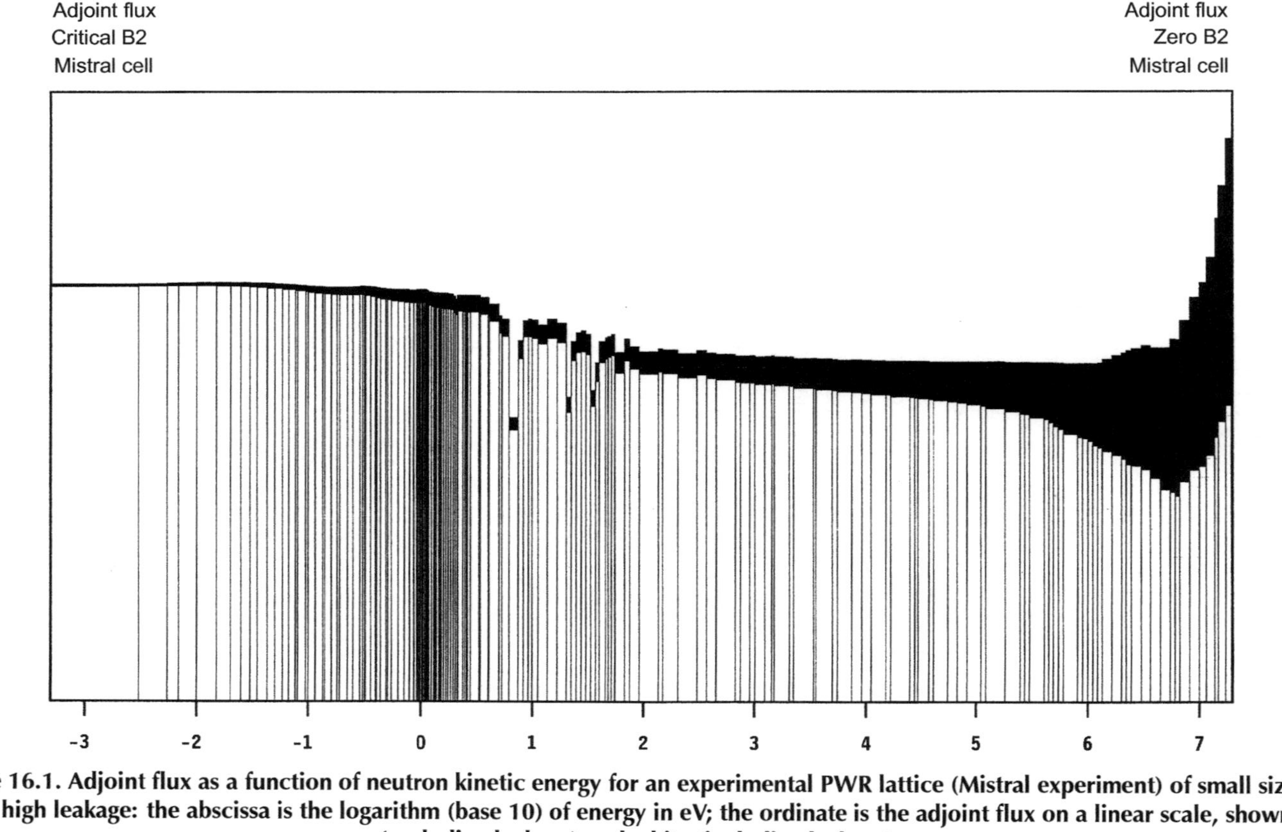

Figure 16.1. Adjoint flux as a function of neutron kinetic energy for an experimental PWR lattice (Mistral experiment) of small size and therefore high leakage: the abscissa is the logarithm (base 10) of energy in eV; the ordinate is the adjoint flux on a linear scale, shown in black (excluding leakage) and white (including leakage).

16.3.2. Differential and integral efficiency of a control rod

In most reactors, control rods or bundles are used to regulate the reactivity. They penetrate via one of the faces of the core, and can be inserted partially or fully until they reach the opposite face. It is obviously crucial to know the efficiency of these rods, i.e. the change in reactivity caused by its insertion.

We examine this problem using one-group theory and the first-order equation (noting however that a first-order calculation can be only approximate at high efficiency). We assume that the core is a cylinder or parallelepiped and that the bar is inserted vertically.

If the rod is laterally small, only the z dimension running along the rod will be considered. For an insertion e from the entry face, under these conditions the efficiency E is given by:

$$E(e) = C^{te} \int_0^e \Phi^+(z)\Phi(z)dz, \tag{16.28}$$

where the constant allows for the absorption difference ΔH between zones with and without rods and the normalisation (in particular, the denominator of the perturbation equation).

In one-group theory, the flux and adjoint flux are identical, such that the function within the integral can be replaced by the square of the flux[8].

Moreover, if we liken the flux $\Phi(z)$ to the axial flux in a bare homogeneous pile, we obtain (still to within a coefficient):

$$\Phi(z) = \sin \pi \frac{z}{h}, \tag{16.29}$$

where h is the height of the core. We then find:

$$E(e) = C^{te} \left(\frac{e}{h} - \frac{1}{2\pi} \sin \frac{2\pi e}{h} \right). \tag{16.30}$$

Graphically (Figure 16.2), this gives an increasing "S" curve. Also note that the *differential efficiency dE/de*, i.e. the efficiency corresponding to an elementary insertion step, is proportional to the square of the flux and reaches its maximum for insertion to mid-core, with a value of zero for zero or full insertion. That is why rods that are partially inserted to a greater or lesser degree are always used for control.

16.3.3. Error due to nuclear and technological uncertainties

In physics, it is very important to estimate the uncertainties associated with calculation results.

In general, there are four different types of error:

- Error due to a simplified physical model (for example, in the field of neutron physics, replacing the transport operator by the diffusion approximation);

[8] In multigroup theory, the flux and adjoint flux keep a similar shape if the curves are plotted as a function of space.

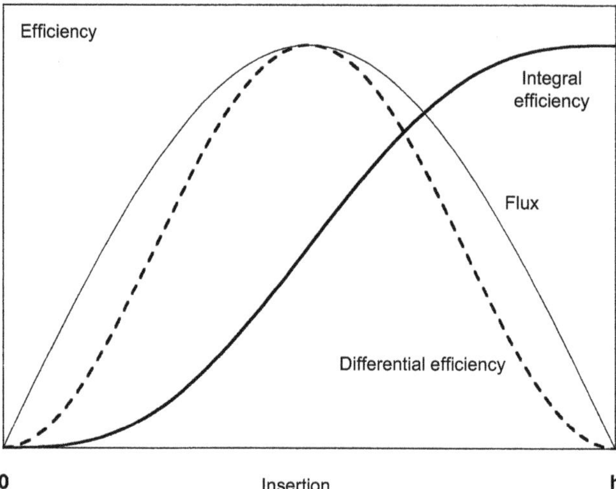

Figure 16.2. Schematic differential and integral efficiency curves as a function of control rod insertion.

- Error due to imperfect numerical schemes used to solve equations (multigroup approximation of a spectrum, discretisations introduced for calculations in space, etc.),
- Error in numerical values of basic calculation data; nuclear data in our case;
- Errors in the description of the system (dimensions, densities, isotopic composition, etc.).

The neutron physicist is fortunate to have an equation that can be described as exact — the Boltzmann equation — and to have developed reference codes, which are costly to implement, but which can provide very accurate results (Monte Carlo codes using a continuous energy representation, deterministic codes with a very fine multigroup mesh). These tools allow a comparison within the same problem and using the same data of reference results to the results given by the codes used in the normal way. This comparison lets us *validate* the results obtained via the codes, i.e. to give an accurate estimate of the first two types of error, and to show that they are not prohibitive.

Errors concerning the nuclear data and technological errors (involving the system, essentially arising from manufacturing tolerances) are obviously unknown. The best we can do is to estimate the standard deviations and covariances (correlations between the possible errors). Under these conditions, the only possible analysis is to evaluate the consequences on the parameters of interest, such as the multiplication factor. The perturbation equations are very useful in estimating these consequences.

In practice, these uncertainties are sufficiently small to allow the first-order equations to be used in this context. They give *sensitivity coefficients,* which are the partial derivatives of the integral parameters with respect to the nuclear or technological data (for practical reasons, the logarithmic derivative is generally used; i.e. the relative change in the integral parameter concerned with respect to a relative change in the datum). Note that all sensitivity coefficients for a given integral parameter are calculated using the same flux and adjoint flux; however, the definition of the adjoint flux and therefore the equation giving it depend on the integral parameter, and Φ^+ must be calculated for each.

Research carried out on this theme has shown that the consequences of uncertainties, concerning nuclear data and technological data, are far from negligible. They can affect the multiplication factor by at least several hundreds of pcm, and the local power with respect to average power (notably the hot spot factor) by several percent. This leads to two conclusions:

- It is important to continue performing nuclear physics measurements, not only for the materials that have not received much attention to date, but also for standard nuclear materials. Likewise, careful manufacture is very important;
- There is no use refining the calculations performed by the usual codes to a level of error that is far smaller than the inevitable error introduced by uncertainties in the data.

16.4. Generalised perturbation theory

We have just mentioned the possibility of using perturbation theory to examine integral parameters other than the multiplication factor, e.g. a local power level. This would allow us to generalise the concept of adjoint fluxes (plural) and perturbation formulae (also plural).

Moreover, in each case, it is possible to work to the first order or on an exact formula. It is also possible to expand intermediate approximations to higher orders: second, third, etc.

Exercises

Exercise 16.1: reactivity weight of a nuclide

The reactivity weight of a nuclide can be defined as the change in reactivity resulting from the insertion of *one atom* of this nuclide at a given point in a system. A relative scale can also be used when this reactivity effect is divided by the reactivity effect of a reference atom inserted at the same point.

a) Assuming only the absorption owing to fission and capture is modified, and the adjoint flux is constant in lethargy, express the relative weight of the given nuclide in comparison with the reference nuclide at the same point.

b) When using one-group theory, how does the weight vary as a function of the point in the system?

Exercise 16.2: axial weighting of the xenon poisoning

Notation: see exercise **11.1**.

Assume that the reduced flux φ in the reactor is radially uniform and varies axially according to the formula

$$\varphi(z) = \beta \cos z,$$

where β is a normalisation constant, and where the reduced height varies from $-\pi/2$ to $+\pi/2$. The axial reflectors are neglected.

a) When the xenon is assumed to be at equilibrium, what form does its axial distribution take along z?

b) Calculate its average concentration. Compare with the concentration calculated at equilibrium with the averaged flux $\bar{\varphi}$.

c) If the concentration is weighted by the *square* of the flux, as suggested by the perturbation theory, then how much is the "effective" concentration?

d) *Numerical example:* compare these values for $\bar{\varphi} = 3\alpha$.

Exercise 16.3: "clear water" event

This hypothetical event in a pressurized water reactor would result from the insertion of reactivity due to the injection of some water without boron in a core containing boron, such as at the beginning of a cycle.

Assume here that borated water is replaced with unborated water for eight fuel assemblies in a total of 157. The initial concentration of boron in the core is 1000 ppm. The boron efficiency is −10 pcm/ppm.

a) Using the elementary perturbation formula in neutron one-group theory, and assuming that before the event the flux in the reactor is perfectly uniform, evaluate the reactivity effect of this clear water injection.

b) How is the result modified if this injection is made into a zone where the ratio of the flux to the core average flux is 1.25?

c) Compare this reactivity effect with the proportion β of the delayed neutrons.

Solutions

Exercise 16.1: reactivity weight of a nuclide

a) The reactivity effect is proportional to $\langle \Phi^+, \Delta H\Phi \rangle$. Therefore, with the proposed hypotheses, it is proportional to

$$\int \left[\nu\sigma_f(u) - \sigma_a(u)\right]\ \Phi(u)\, du,$$

or $\bar{\nu}\bar{\sigma}_f - \bar{\sigma}_a$, where the bars mean averages over the lethargy spectrum. The ratio of such an expression to the reference nuclide permits the weight w_k of the given nuclide k to be defined as

$$w_k = \frac{(\bar{\nu}\bar{\sigma}_f - \bar{\sigma}_a)_k}{(\bar{\nu}\bar{\sigma}_f - \bar{\sigma}_a)_{\text{ref}}}.$$

b) In one-group theory, the weight varies in space as the square of the flux.

Exercise 16.2: axial weighting of the xenon poisoning

a) When equilibrium is reached, the xenon-135 concentration x is linked to the flux by the relationship

$$x = \frac{\varphi}{\varphi + \alpha}.$$

b) When $\varphi(z) = \beta \cos z$, the average of this concentration is

$$\bar{x} = 1 - \frac{2\alpha}{\pi s} \ln \frac{u+v}{u-v},$$

where

$$u = \sqrt{\beta + \alpha}, \qquad v = \sqrt{\beta - \alpha}, \qquad s = \sqrt{\beta^2 - \alpha^2}.$$

This average is different from the value of x corresponding to $\bar{\varphi} = 2\beta/\pi$:

$$x(\bar{\varphi}) = \frac{2\beta}{2\beta + \pi\alpha}.$$

c) Weighting by the square φ^2 of the flux, yields an effective concentration:

$$\hat{x} = 1 - \frac{4\alpha}{\pi\beta} + \frac{2\alpha^2}{\beta^2} - \frac{4\alpha^3}{\pi\beta^2 s} \ln \frac{u+v}{u-v}.$$

d) When $\beta = 3\pi\alpha/2$,

- $x(\bar{\varphi}) = 0.75$;
- $\bar{x} = 0.6915$;
- $\hat{x} = 0.7921$.

Exercise 16.3: "clear water" event

a) Weighting only by the volumes (8/157) the local reactivity effect of $1000 \times 10 = 10\,000$ pcm, yields 510 pcm.

b) Weighting also by the square of the local relative flux (1.25^2), increases the result to 796 pcm.

c) In the first case, the reactor remains undercritical by prompt neutrons; however, in the second case, the reactivity insertion due to the clear water injection is greater than the proportion β of the delayed neutrons.

17 Overview of the "Calculation Scheme"

Introduction

It would be nice to be able to simply enter the details of a problem into a calculation code, press a button, and obtain the result of a complete and exact simulation. Unfortunately, things are not quite that simple. A reactor is such a complex object in terms of geometry, composition, and the nuclear data involved, that an all-numerical approach will probably always be just a pipe dream.

To give some idea, let us imagine calculating a pressurised water reactor core without modelling. The core contains approximately 200 assemblies plus the reflector. Each assembly comprises 289 cells. Each cell has at least ten radial zones and around fifty axial zones. To identify the axes along which neutrons travel, about a hundred angular directions are required. The very complicated structure of the cross-section curves requires an energy mesh of at least 20 000 points. An evolution calculation (i.e. the normal change over a few years, or a short accidental sequence), taking counter-reactions into account, involves at least fifty or so calculations. If we multiply all these considerations together, we find that approximately 3×10^{15} flux values need to be determined. This is clearly far beyond the capabilities of even the most powerful machine.

This demonstrates how essential modelling is if we wish to perform calculations in practice. The talent of a neutron physicist lies in the ability to choose the best model to achieve two apparently antagonistic goals: obtaining very precise results, whilst keeping the volume of numerical calculations down to a "reasonable" level. The choices made will constitute the neutron physicist's "calculation scheme".

The creation of the calculation scheme clearly depends on the "object" to be calculated, the required accuracy, and the definition of what is a "reasonable" calculation time. According to context, these choices can be quite different; we must then refer to "calculation schemes" in the plural. Overall, however, the calculation objectives for different engineers and physicists tend to be fairly similar, and the calculation options tend to remain relatively close, this middle path is what we shall attempt to describe in this chapter. We shall use the example of calculations on pressurised water reactors, but the general principles should also apply to calculations for other reactor types.

Obviously, before performing any neutron physics calculation, it is necessary to gather the necessary nuclear data. Before discussing the calculation scheme, we shall review a few points about nuclear data that were introduced in Chapter 2 (part E); we shall then see that this scheme is based on the simple observation that there are three distinct levels of detail in the neutron physics of a reactor:

- on a macroscopic level, a core has a heterogeneous structure because of the differences between the assemblies placed in it;
- at an intermediate level, we see a heterogeneous structure within each assembly: fuel, cladding, moderator, etc.;
- on a finer scale, we observe complex self-shielding phenomena due to the presence of many resonances of heavy nuclei.

These problems must clearly be addressed in reverse order; dealing with the fine detail first, and then performing the essential averaging before processing the next level. The relevance of the calculation scheme will be related to the careful choice of boundary between these levels and the averaging formulae to be introduced. Its quality will also depend on the accuracy of the models and the numerical calculations used at each stage.

17.1. Nuclear data

In Chapter 2 we saw that the nuclear data emerging from the huge effort to gather nuclear measurements in collected in databases after being "evaluated". We also saw that these data are not generally used for direct input into neutron physics codes.

Interface codes therefore need to be developed and implemented to develop the libraries required for neutron physics calculations.

17.2. Tabulation for the processing of resonance self-shielding

The two main operations to be performed are as follows:

- grouping for the transport calculations (typically about a hundred or a few hundred groups);
- tabulation of the parameters characterising the resonant absorption problem (effective reaction rates, probability tables, etc.).

We can consider that this latter operation constitutes the first stage of the neutron physics calculations; the one that accounts for the resonant structure of the cross-sections in all the necessary detail.

Without repeating what was done in Chapter 15, note that this stage is, in principle, carried out once and for all[1], for each of the nuclides of interest with resonances.

The relevant parameters can, for several temperatures if necessary, either be tabulated in the form of carefully-chosen functions of a parameter (effective reaction rates as a function of the dilution cross-section) or summarised in the form of quadrature parameters (probability tables).

[1] In practice, new evaluations are proposed periodically, and these groupings and tabulations must be repeated each time. Of course, we might also wish to process the evaluations performed by different bodies for the purpose of comparison.

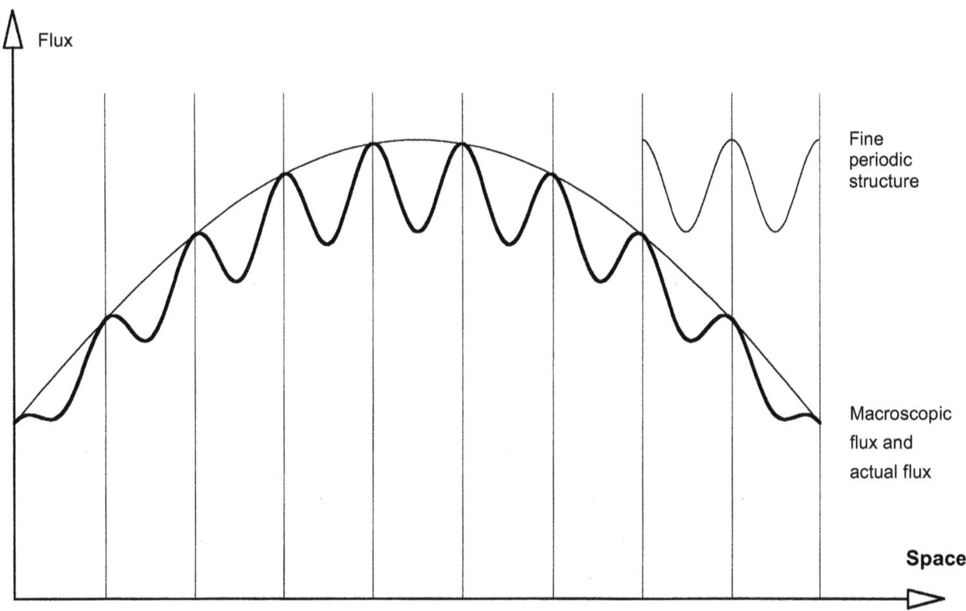

Figure 17.1. Factorisation of flux in an infinite regular lattice (here: schematic representation of the thermal flux in a lattice of cells consisting of fuel plates where the flux is depressed with respect to the flux in the adjacent moderator). In bold: true flux; in plain characters: fine structure and macroscopic flux.

17.3. Assembly calculations

The second (assembly calculations) and third (core calculations) stages must, however, be repeated for each new problem.

An assembly represents only a very limited part of a core, whose boundary conditions should be specified. The simplest boundary conditions are chosen: a "mirror" condition on each face of the boundary; this is equivalent to the situation in the infinite, regular lattice obtained by the infinite juxtaposition of the images of the assembly concerned.

As we have seen (§ 14.5), in this type of infinite regular lattice, the flux is established according to a fundamental mode factorised into the product of a macroscopic function of the form $exp(-i\vec{B}\cdot\vec{r})$ and a fine structure φ with the periodicity of the lattice (Figure 17.1) or according to a linear combination of such modes with vectors $\vec{B}$ all ensuring criticality.

In practice, only the fine structure is calculated, either by adding a "leakage cross-section" DB^2 playing the same role as a capture, or using equivalent surface leakage. Heterogeneous $\mathbf{B_K}$ theory is usually replaced by a homogeneous $\mathbf{B_K}$ calculation (in practice, $\mathbf{B_1}$) giving the diffusion coefficient of the previously homogenised medium. One of the advantages of this approximation is that it leads to a *real* fine structure.

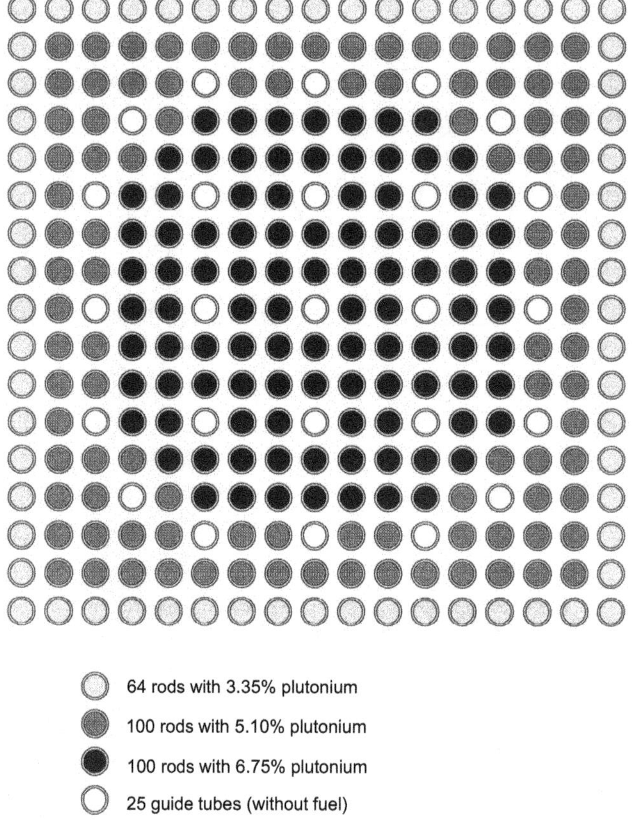

Figure 17.2. Cross-section of an MOX assembly for a pressurised water reactor.

In most calculations, the mesh element of this infinite, regular lattice is an assembly (narrow pitch reactors such as water reactors or fast neutron reactors) or a cell (wide pitch reactors). This mesh element is assumed to be axially infinite[2], which reduces it to a two-dimensional $x - y$ problem. The transport calculation is generally more or less simplified. Let us take the example of MOX assemblies (MOX, or "mixed oxide", is a fuel consisting of a mixture of plutonium and uranium oxides) for a pressurised water reactor[3] and the calculation scheme for such an object using the APOLLO code.

Figure 17.2 presents the 2-D structure of a cross-section of the assembly.

The transport calculation can be simplified by introducing an *imaginary* cell structure (Figure 17.3) separating the water mass into parcels associated with each of the 289 fuel tubes and other tubes[4].

[2] In most reactors, fuel assemblies are prismatic; the height of the mesh elements is far greater than their width. Exception: high-temperature pebble bed reactors.

[3] As we shall see in the next chapter, the rods must be grouped into zones of increasing abundance from the edge to the centre. There are zoning variants; the one presented here was used in the first PWR loads using MOX.

[4] Note that the edge cells incorporate the water gap that remains between the assemblies once they have been placed in the core.

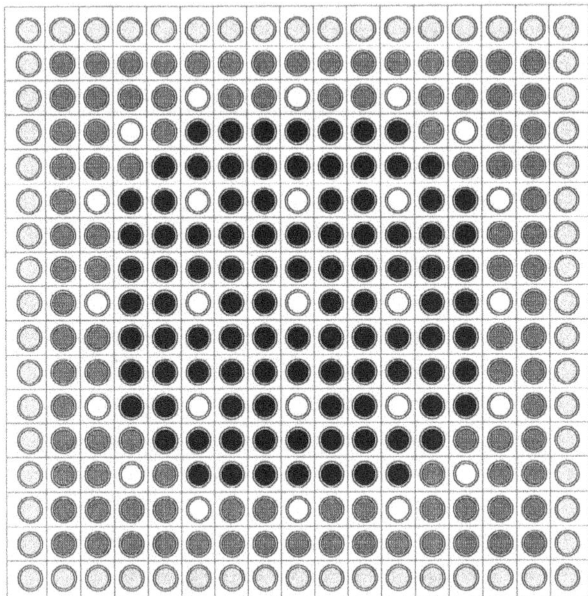

Figure 17.3. Multicell representation.

We have seen (§ 14.3.11) that, if we introduce a few assumptions about the treatment of interface currents between cells, the treatment of the transport operator in its integral form can be considerably simplified.

The main simplifying assumption is the uniformity and isotropy of the interface currents[5]: uniformity can be assumed on all four faces of the cells (moreover, as discussed on § 14.3.10, we can cylindrise the cells to simplify the first collision probability calculation even further) but it is preferable to distinguish the faces and restrict the uniformity approximation at each face; the isotropy approximation, $\mathbf{P}_0$[6], can be replaced, at the cost of a significantly higher volume of calculations, by an approximation called $\mathbf{P}_1$, of a linearly anisotropic current.

It is also often possible to reduce the number of cells to be processed not only by allowing for symmetries (in our example, this results in 45 cells instead of 289), but also by taking as identical cells with a similar position and environment in the assembly[7] (Figure 17.4).

Note that the division of the cells into volume elements must be chosen. The minimum is three volumes: fuel, cladding, and moderator[8]. For more precise calculations, we might wish to divide the fuel (Figure 15.1) and the moderator more finely.

Note also that the self-shielding effects of the resonances must be taken into consideration for the main heavy nuclides and for the zirconium of the cladding.

[5] "Isotropy" of the current is the traditional term but, strictly speaking, it is the phase flux passing through the interface in a given direction that is assumed to be isotropic.

[6] A zeroth-order approximation in a Legendre expansion.

[7] It is important to distinguish the neighbouring cells from "water holes", which are not subject to the same Dancoff effect.

[8] In practice, grids are homogenised with the moderator.

28	27	26	25	25	25	25	25	25	25	25	25	25	25	26	27	28
27	24	23	23	23	21	22	22	21	22	22	21	23	23	23	24	27
26	23	20	19	19	3	18	18	3	18	18	3	19	19	20	23	26
25	23	19	4	17	16	15	14	13	14	15	16	17	4	19	23	25
25	23	19	17	12	11	10	10	9	10	10	11	12	17	19	23	25
25	21	3	16	11	2	9	9	2	9	9	2	11	16	3	21	25
25	22	18	15	10	9	8	8	7	8	8	9	10	15	18	22	25
25	22	18	14	10	9	8	6	5	6	8	9	10	14	18	22	25
25	21	3	13	9	2	7	5	1	5	7	2	9	13	3	21	25
25	22	18	14	10	9	8	6	5	6	8	9	10	14	18	22	25
25	22	18	15	10	9	8	8	7	8	8	9	10	15	18	22	25
25	21	3	16	11	2	9	9	2	9	9	2	11	16	3	21	25
25	23	19	17	12	11	10	10	9	10	10	11	12	17	19	23	25
25	23	19	4	17	16	15	14	13	14	15	16	17	4	19	23	25
26	23	20	19	19	3	18	18	3	18	18	3	19	19	20	23	26
27	24	23	23	23	21	22	22	21	22	22	21	23	23	23	24	27
28	27	26	25	25	25	25	25	25	25	25	25	25	25	26	27	28

Figure 17.4. Example of cell grouping (here: 28 cells instead of 45).

17.4. Reflector calculation

A reactor core contains a certain number of fuel assemblies and a certain number of peripheral structures, most notably a reflector. Before running the core calculation, it is therefore necessary to consider not only the different types of assembly to be placed there, but also these other structures.

In rigorous terms, this problem is complex because the core–reflector interface is a heterogeneous structure (water and steel plates) with offsets and therefore typically two-dimensional. In practice, tests have shown that it was sufficient to run a planar model of the interface and to use the "reflector constants" deduced from the planar calculation in the two- or three-dimensional calculations.

The relevant parameters for the characterisation of reflectors in one-dimensional geometry are albedos β_{hg}, the number of neutrons sent back into the core in group g for a neutron entering the reflector in group h. These are the parameters that must be determined in the planar geometry transport calculation and used subsequently. Because they will be used in another geometry, however, and in general in a "multigroup diffusion" approximation, we tend to go via the parameters of the equations for "multigroup diffusion" theory in an equivalent reflector, known as the "reflector constants". The *equivalent reflector* must, by definition, have the same *matrix of albedos* as the real reflector. We select the simplest possible version, i.e. homogeneous and infinite. Note that, in diffusion theory

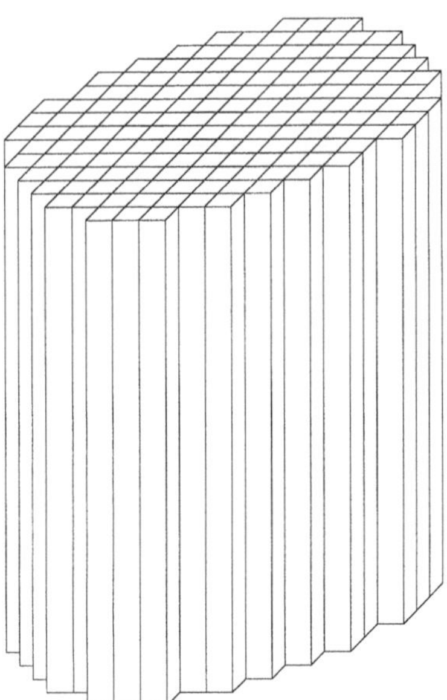

Figure 17.5. Arrangement of the 157 fuel assemblies of a 900 MWe pressurised water reactor.

with G groups without up-scattering there are $G(G+1)/2+G$ parameters to be determined from $G(G+1)/2$ albedos. In general, the equations are supplemented using an additional criterion to select the diffusion coefficients[9].

17.5. Core calculation

For the core calculation, we shall use a far simpler neutron physics model than the one used for the assembly and reflector calculations. For water reactors, for example, a diffusion approximation with only two energy groups is sufficient for the usual calculations. If high precision is required, a few more groups can be used, and/or a simplified transport calculation $\mathbf{SP_N}$ or $\mathbf{SS_N}$ can be performed instead of the diffusion (§ 14.4.4 and 14.4.7).

To prepare the data for this type of calculation, energy averages — the *condensation* of a fine multigroup mesh into a wide multigroup mesh — and space averages — *homogenisation* — are required. The averages are taken by weighting, respectively, by the lethargy intervals and flux values, or by the volumes and flux values.

Where necessary, these averages are corrected by "equivalence" as we shall see later.

Concerning homogenisation, we might wonder at what level it should be performed. For water reactors, for example, we might hesitate between cell-by-cell homogenisation — Figure 17.6 — and complete homogenisation of the assemblies[10] — Figure 17.7 — (and the analogous case for the reflector).

[9] Several reflector calculation methods have been suggested; some using a heterogeneous representation of the reflector, and others using a homogeneous representation. From a physical aspect, respecting the albedos would seem to be the only truly relevant criterion. It turns out that, in practice, probably through the effect of error compensation, other approaches sometimes lead to better results.

[10] For wide pitch reactors (UNGG, RBMK, CANDU, etc.) only cell-by-cell homogenisation could be considered.

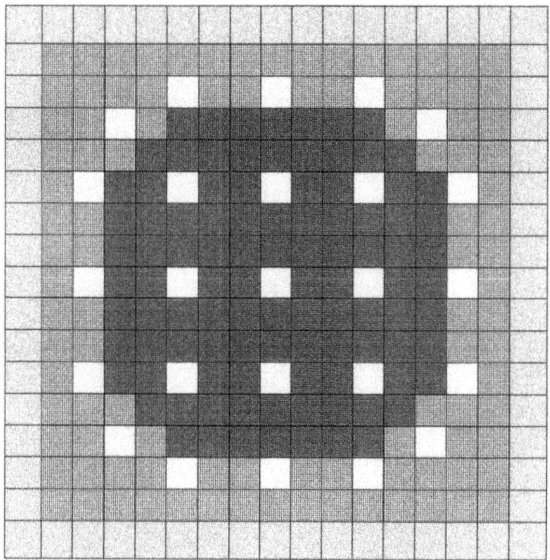

Figure 17.6. Cell-by-cell homogenisation.

Figure 17.7. Complete homogenisation of the assembly.

Cell-by-cell homogenisation provides the fine rod-by-rod power distribution that is a fundamental piece of information[11].

Under these conditions, the diffusion calculation is performed by the finite difference method (§ 6.2.3) with (in the plane of the figure) 1 mesh element or 2 × 2 mesh elements per cell.

A *complete homogenisation of the assembly* obviously does not make it possible to obtain the fine rod-by-rod power distribution; this must be approximately reconstituted *a posteriori* by placing the fine structure obtained from the assembly calculation on top of the macroscopic power layer.

The advantage of this larger-scale homogenisation is that we can then adopt a numerical method that can make do with large mesh elements — finite elements or nodal method (§ 6.2.4 and 6.2.5) — which, as we have seen, is far more economical in terms of calculation time. In particular, this approach makes it easier (than with the finite difference method) to allow for the axial aspects related to evolution and for counter-reactions, which turn out to be significant in water reactors.

Note: Fast neutron reactor assemblies have dimensions that are comparable to those of water reactor assemblies, but the effects of heterogeneities are far less pronounced because the mean free paths of their neutrons are far longer; cell-by-cell homogenisation is not necessary. Finite difference calculations[12] with relatively large mesh elements on homogenised assemblies turn out to be sufficient in practice.

17.6. Problem of homogenisation and condensation

Equivalence is the term adopted by neutron physicists to designate the procedure intended to correct artefacts introduced during operations such as condensation, homogenisation and, more generally, the transition from a reference theory to a simplified theory. We shall present this for the example of homogenisation, bearing in mind that the approach can be generalised. As a concrete example, consider the cell-by-cell homogenisation of a pressurised water reactor assembly (Figure 17.6).

Homogenisation consists of replacing a structure described by a set of small zones (index m) with a structure formed by a smaller number of larger zones (index M) obtained by grouping a certain number of small zones to form a larger one.

Volumes are obviously conserved in this operation:

$$V_M = \sum_{m \in M} V_m. \tag{17.1}$$

[11] For safety reasons, a limit is imposed on the power "peak" (hot spot). This limit must not be exceeded, at the risk of damaging the cladding. To improve the performance of the reactor, it is obviously advantageous to reduce the margin with respect to this limit as much as possible, which assumes that the hot spot has been correctly located and correctly calculated.

[12] The discretisation formulae must be adapted to suit the hexagonal geometry adopted in these reactors.

It is logical to conserve the numbers of neutrons also, and therefore the flux values, by defining the average flux per macrozone with the following formula:

$$V_M \Phi_M = \sum_{m \in M} V_m \Phi_m. \tag{17.2}$$

Likewise, conservation of reaction rates for all reactions (reaction index implied) leads us to adopt the following rule:

$$V_M \Phi_M \Sigma_M = \sum_{m \in M} V_m \Phi_m \Sigma_m, \tag{17.3}$$

i.e.:

$$\Sigma_M = \frac{\sum_{m \in M} V_m \Phi_m \Sigma_m}{\sum_{m \in M} V_m \Phi_m}. \tag{17.4}$$

Unfortunately, the conservation of reaction rates that this weighting rule appears to ensure is only an illusion, because if the neutron physics calculation is repeated in the macrozone structure with these cross-sections, the expected reaction rates will not be obtained. That is because this calculation does not restore the average flux per macrozone. In other words, the homogenisation problem *is not linear*; the average of a nonlinear function (here, flux) is not the function taken for the average arguments (cross-sections in this case).

The solution to this problem consists of setting the problem in terms of equivalence: determine the Σ_M values such that the neutron physics calculation performed with these cross-sections in the macrozone structure gives the correct reaction rates. In this equivalence problem, the true unknowns are no longer the flux and reaction rates, but the cross-sections introduced in calculations.

In practice, this equivalence is applied in the context of a multigroup theory (group index:g); it is required to conserve all of the reaction rates (reaction index:α) of the reference calculation for the case concerned. The equivalence equations are therefore:

$$V_M \Phi_{M,g} \Sigma_{\alpha,M,g} = T_{\alpha,M,g}, \tag{17.5}$$

where Σ (cross-sections) are the unknowns, Φ (flux) are calculated from the cross-section by solving the neutron physics equations, and T (reaction rates given by the reference calculation) are the data for the equivalence calculation. Note that, in a macrozone and a given group, since the cross-section values $\Sigma_{\alpha,M,g}$ satisfying these equations are proportional to the reaction rates $T_{\alpha,M,g}$; the actual unknowns are therefore the cross-sections concerning one of the reactions, for example the total cross-sections $\Sigma_{M,g}$. In reality, we tend to use so-called *SPH* (for "superhomogenisation") factors. The SPH factor $\mu_{M,g}$ is defined as the ratio between the cross-sections $\Sigma_{\alpha,M,g}$ satisfying the equivalence equations and the "average" cross-sections $\bar{\Sigma}_{\alpha,M,g}$, obtained by the rule for weighting by flux, volume, and lethargy interval[13]:

$$\mu_{M,g} = \frac{\Sigma_{\alpha,M,g}}{\bar{\Sigma}_{\alpha,M,g}} = \frac{\Sigma_{M,g}}{\bar{\Sigma}_{M,g}}. \tag{17.6}$$

These factors are obtained by iterating the equivalence equations. The iteration is started by taking 1.

[13] They too are proportional to the reaction rates for fixed M and g.

Note: In a given energy group, it is possible to multiply all the cross-sections (and therefore all the SPH factors) by one factor k_g and to divide the fluxes by this same factor. This does not affect the reaction rates and, therefore, is compatible with the equivalence equations. This arbitrary coefficient must be supported by a criterion other than conservation of reaction rates, such as:

- Conservation of flux average over the entire volume of the pattern concerned,
- Conservation of flux average over the entire length of the surface,
- Conservation of the average current leaving over the entire length of the surface,
- etc.

17.7. Transport-diffusion equivalence

The equivalence operation can be used to try to correct other approximations of the calculation model that will be applied to the macroscopic problem, the core calculation. In particular:

- Replacement of the transport operator used in the reference calculation by the diffusion operator, in practice used for the core calculation[14];
- The approximations of the numerical scheme to be implemented (finite differences, finite elements, etc.).

In this type of context, we speak of a "transport–diffusion equivalence", which implies four simplifications that we rectify: condensation, homogenisation, diffusion approximation, and numerical processing approximation.

17.8. Generalisation: the concept of equivalence in neutron physics

The concept of equivalence in the sense that has just been specified is in fact very general in neutron physics. The term itself, in this sense[15], seems to have been introduced by the authors of the effective reaction rate theory: M. Livolant and F. Jeanpierre, who used two equivalences:

- The heterogeneous–homogeneous *equivalence* (§ 8.3.2 and 15.2),
- The continuous–multigroup equivalence (§ 8.5.2 and 15.3).

Here are a few other examples:

- The six *equivalent* groups of delayed neutrons, replacing about a hundred processes (§ 2.10.1 and 4.2.1);

[14] The diffusion coefficient is, in this case, handled by equivalence like the macroscopic cross-sections.

[15] Note that this word is sometimes used with a different meaning; for example, "plutonium 239 equivalent" (§ 16.3.1).

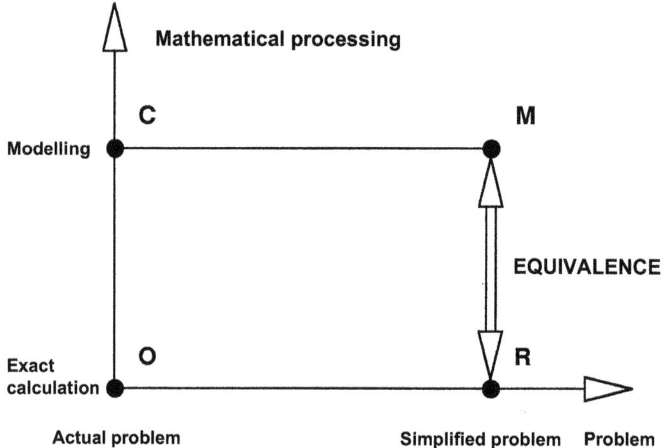

Figure 17.8. General principle of equivalence in neutron physics (O, R, M, C: see text).

- The multiplication factor and lifetime of the point kinetic model *equivalent* to spatial kinetics[16];
- The pseudo-fission product that is *equivalent* to the products neglected in the calculations (§ 11.1.2);
- The two *equivalent* effective temperatures for the Doppler effect calculation (§ 15.7.1 and 15.7.2);
- The transport correction, i.e. the isotropic treatment that is *equivalent* to a linearly anisotropic treatment (§ 5.1.8 and 14.4.3);
- Cancellation of the flux on the surface extrapolated by the black body extrapolation distance *equivalent* to the zero input current condition (§ 5.1.5);
- The diffusion coefficient leading to a diffusion equation *equivalent* to the transport equation in fundamental mode (§ 14.5.7);
- The infinite homogeneous reflector *equivalent* to the real reflector (§ 17.4);
- etc.

The equivalence principle can be summarised by the diagram above (Figure 17.8).

Point **O** (objective) represents the calculation we would like to perform: the actual, complex object handled with the exact theory; but this objective cannot be attained at a reasonable calculation cost.

This calculation will be replaced by calculation **C** using a simpler theory with equivalence corrections.

To apply this correction, we replace the real problem with a problem that is close but sufficiently simplified that the exact theory can be applied to it. We therefore perform this reference calculation **R** on this problem.

[16] The definition of the equations for this equivalence is another application of perturbation theory.

We then continue with this simplified problem using the simplified theory or **M** model; we assume that this model contains a certain number of numerical parameters, and we choose their values carefully so that in this simplified problem the model preserves the results we deem to be essential. This is the *equivalence*.

Finally, by applying the model "adjusted" in this way (calculation **C**) to the real problem, we can hope that the results will be very close to the exact results, or at least better than the results we would have obtained without the equivalence.

It is obviously impossible to estimate the residual error of calculation **C**. We can only assume that it will decrease as:

- The simplified problem approaches the real problem,
- The simplified model approaches the exact model,
- The equivalence criteria are increasingly relevant[17].

Quantifying these three aspects is not a simple matter; a great deal of physical intuition will be required in order to apply the best tradeoffs.

17.9. Evolution and counter-reactions

The calculation scheme we have described, and equivalence in particular, were presented for a given reactor state. In reality, however, this state is unknown and must be calculated, because it depends on counter-reactions and changes over time.

In § 13.5, we saw the importance of these effects for the example of pressurised water reactors. To take these effects into account, we must tabulate the parameters of the model used for the core calculation, such as the two-energy group diffusion equations, according to the main parameters needed to describe the state of each assembly. To pressurised water reactors, at least the following must be introduced:

- Average *burn-up* of the assembly (§ 12.1.5);
- The *fuel temperature* which governs the Doppler effect (§ 8.4 and 13.3);
- The *moderator temperature* or its density, which governs spectrum effects and, in particular, expansion (§ 13.4).

During the "assembly" step of the calculation, we obviously do not know how these three parameters will change over time. We are therefore obliged to create tables in which the calculation code for the core can interpolate at each instant according to the local values. In practice, these multi-parameter tables are created in a simplified manner:

- *The evolution of the assembly* concerned occurs *in the nominal conditions* concerning power and temperatures[18,19];

[17] The number of parameters that can be preserved by equivalence can obviously not exceed the number of parameters available in the model.

[18] The time steps must be sufficiently short at the beginning to correctly calculate the bodies that are quickly at saturation or quickly consumed; after that, they can be longer. Order of magnitude: a total of about twenty time steps.

[19] Note also that we do not know what the concentration of the boric acid will be at the instant corresponding to the burn-up concerned. That is why assembly evolution calculations are generally performed at a constant boron concentration (average value).

- For a certain number of evolution steps (for example, every 10 000 MWd/t), the evolution calculation is temporarily suspended to *perform new calculations* with the concentrations of heavy nuclei and fission products at this instant, the code repeats some calculations with a change in fuel temperature and/or moderator temperature in order to construct an element of the tabulation[20]; the calculation of evolution under the nominal conditions is then resumed and continues to the next step.

In the core calculation, the neutron flux and compositions must be calculated according to calendar time:

- Fluence, and therefore local burn-up, are determined by integrating the local flux over time;

- Flux distribution at a given instant is calculated by taking counter-reactions into account, i.e., concretely, by coupling the core neutron physics and thermal hydraulics codes and by performing iterations until convergence is achieved.

With regard to neutron physics, these calculations will look in the specific tables for each type of assembly to find the cross-sections for the burn-up and the temperatures at the point and the instant concerned.

17.10. Fast kinetics

In core evolution calculations for which the time constants are of the order of a year, quickly saturated bodies such as xenon 135 are taken at equilibrium.

Conversely, if we wish to study the evolution of such bodies — for example, to monitor a spatial instability due to xenon 135 for which the time constant is of the order of a day — the core calculations can neglect long-term evolution. For the study of accident scenarios (e.g. pressurised water reactors: unplanned ejection of a control bundle, steam pipe failure, etc.), the time constants are even shorter: of the order of a second (bundle ejection) or a minute (steam pipe failure). For these studies, even the evolution of fission products can be neglected. On the other hand, the assumption of an evolution with reactivity remaining at zero can then no longer be made. This means that the equations for the concentrations of the precursors of delayed neutrons must be written down and then processed[21].

Clearly, in any case, the thermal hydraulics counter-reactions must be taken into account. The general calculation scheme therefore remains similar to the one described: iterations between the neutron physics and thermal hydraulics codes and interpolation of the parameters of the neutron physics equations in tables according to thermal hydraulic conditions. For accident scenarios, it is also generally necessary to couple the system thermal hydraulic code (processing of all circuits and organs of the plant) and the thermal code of the fuel (handling of heat transfer and any overheating or degradation).

[20] Orders of magnitude: five values for each of the two temperatures.

[21] For this type of problem, it is of interest to optimise the numerical schemes by distinguishing the two timescales: that of prompt neutrons (of the order of a millisecond), and that of precursors (a few seconds).

17.11. Summary of the main approximations of the calculation scheme

In conclusion, we can see that a reactor core neutron physics calculation is divided into three consecutive steps intended to handle the resonant structure of cross-sections, fine heterogeneity in the cells and assemblies and, finally, the core with its macroscopic heterogeneity and the couplings between the three dimensions of space.

The approximations of this scheme appear, on the one hand, at these boundaries between the three stages and, on the other hand, within each of these stages. We shall now review the main approximations using the example of pressurised water reactor calculations, although most of the following comments are generally applicable.

a) Approximations due to the separation in three stages

These approximations come from the fact that the problem handled at step $n-1$ is simpler than the problem encountered at step n; more specifically:

- The effective reaction rate tabulations at Step 0 are done:
  - by taking the resonant material in a *homogeneous* situation, not in its real situation,
  - by assuming that there is *only one resonant nuclide*,
  - by assuming, for the *Doppler effect* calculation, that the thermal agitation spectrum of the resonant nuclei is a *Maxwell spectrum*;

  (For the probability tables approach, the first point must be replaced by: the sources of slowing down are assumed to be lethargy-constant and therefore not correlated to the resonant structure).
- The assembly calculations at Step 1 will be performed in fundamental mode, i.e. by surrounding the assembly concerned with identical assemblies, not different assemblies, as found in reality;
- At this stage also, the evolution calculations are not performed at real temperature and boron concentration conditions, but in nominal conditions: we do not know in advance what the exact conditions will be because they depend on everything else in the core and, for a given assembly, on the height. (Moreover, the nominal relationship for the change in concentration of boric acid is usually replaced by an average constant value, because it has been shown that the resulting error is quite small.)

b) Step 0 Approximations (handling of resonances)

Leaving aside the fact that our knowledge of resonance parameters is necessarily approximate, particularly in the high-energy domain, in principle there is no approximation other than a numerical one; fine (but not infinitely fine) multigroup treatment of resonances, tabulation of effective reaction rates in a (relatively) wide meshing, or finite-order probability tables.

c) Step 1 Approximations (handling of assemblies)

In simplified terms, we can say that this step deals with three problems: resonant absorption, fine transport, and evolution and counter-reactions.

For the **calculation of resonant absorption:**

- Resonant nuclides are generally handled one by one, *disregarding the effects of interference between resonances* on the self-shielding factors;
- *The heterogeneous-homogeneous equivalence* is applied using an *approximate slowing-down model* for slowing down by the resonant material, and replacing the Bell function by a constant Bell factor;
- The *continuous-multigroup equivalence* is applied to the **P** *problem*, which represents the real case only approximately;
- To calculate the *Doppler effect*, the thermal agitation spectrum is likened to a *Maxwell spectrum at an "effective" temperature*;
- Furthermore, the spatial distribution of temperature in the fuel is not taken into account; the *Doppler effect* calculation is carried out at a *uniform temperature*.

Concerning the **transport calculation**:

- Space is discretised into *finite volumes*, and energy into *groups*;
- The calculation is performed by the method of collision probabilities, i.e. from the integral form of the transport operator. This almost always requires the use of an *isotropic diffusion assumption*, which is in fact applied; but a *transport correction* corrects much of the error that was introduced in this way;
- The transport calculation for the assembly is usually carried out with *multicell approximations* which can be more or less accurate depending on the purpose of the calculation;
- The *reflector calculation* is almost always performed in *planar geometry*.

Concerning the **evolution calculation**:

- Evolution is handled by *numerical integration of the* (nonlinear) *evolution equations* (with a step that can vary according to the physical phenomena to be considered; for example, the depletion of a burnable gadolinium poison requires a fairly fine temporal discretisation at the beginning);
- The *boron solution* is usually taken at a *constant concentration*;
- Evolution is dealt with under nominal conditions, not the actual temperature conditions, and the difference is taken into account by coefficients describing the *counter-reaction effects* evaluated by *repeated calculations* during the nominal evolution of the assembly.

d) Step 2 Approximations (handling of core)

- The *numerical processing of neutron physics equations* requires *numerical discretisations* that are more or less fine depending on the purpose of the calculation.
- *Hot spot calculations* are often performed in an approximate manner, either by *factorising a fine assembly structure on a macroscopic distribution* calculated with homogeneous assemblies, or by *cell-by-cell calculations, but in two dimensions only*.
- *Condensation and homogenisation* are performed, which gives rise to error even if an equivalence procedure is applied to try to minimise it.
- Moreover, the transport operator is usually replaced by the *diffusion approximation*.
- The *evolution calculation* uses *tables or physical models* that are inevitably *simplified* to a greater or lesser extent.

17.12. Validation of calculation schemes

Obviously, all of these approximations have been carefully compared and optimised in order to reach the best possible compromise between the volume of calculations to be performed and the accuracy of the results. Determining the best compromise involves a very large number of tests. These tests are carried out by physicists either, if possible, using the same codes as engineers, but selecting more precise options in order to estimate error, or using *reference codes* specifically intended for this purpose of validating calculation scheme options.

As has been mentioned, engineers almost always use deterministic codes because of the shorter calculation times. For the purpose of validation, we can also take deterministic codes (the same ones or different ones), but we often prefer to perform the reference calculations using a Monte Carlo code, because the constraint on computing time is less strict for validation procedures that will be performed only a small number of times.

For the conclusions drawn from these comparisons to be relevant, it is very important to ensure that the calculation data — concerning the specific case as well as nuclear data — are strictly the same in the routine codes and the reference codes. This consistency is easier to ensure if the same code is used in both cases. Note also that the complete calculation scheme is so complex that each step will need to be validated separately, and that an overall validation is not really conceivable (a complete core calculation, even a reference one, is totally unrealistic). In any case, to define the best options at each stage, a specific validation for each stage is required. To evaluate the quality of the whole calculation scheme, only a comparison with actual experience will provide a relevant indicator.

17.13. Qualification of calculation schemes

The term "validation" is normally used exclusively to describe comparisons of one calculation to another, as we have just described. The term "qualification" refers to comparisons of calculations to experimental data. Experiments are obviously the ultimate test of a code's ability to describe reality.

The qualification approach is more complex than the validation approach, because it involves not only calculation approximations, but also uncertainties concerning the data for these calculations.

Qualification can be performed either using results measured on experimental installations created specifically for the purpose, or by analysing measurements performed on industrial installations.

Measurements on industrial installations allow the overall qualification of the entire calculation scheme, whereas experimental installations are intended mainly to test certain points.

The design of an experiment can aim at two different goals: either to qualify the calculation scheme or to test nuclear data. In practice, it is often a case of seeking answers to both types of question at the same time.

Mock-up experiments to test the calculation scheme against an assembly or core configuration (absorbents, burnable poisons, uranium-plutonium interface, abundance zoning, etc.) are rarely performed nowadays because they are expensive and can often be replaced by reference calculations.

Nonetheless, experiments are, and always will be, essential in order to test nuclear data. That is why it is very important to maintain experimental installations (critical experiments, irradiation reactors, etc.) to perform qualification.

17.14. Tendency search

As opposed to mock-up experiments, measurements on a critical experiment or small reactor, intended for the qualification of nuclear data, are sometimes known as "clean" experiments. This adjective does not denote any kind of value judgement; it is merely used to indicate that, since the purpose is to test *nuclear data,* we try to simplify the configuration as much as possible so that calculation errors will be considered as negligible in our comparison between the measurement results and the corresponding calculations. In this way, the differences will reflect errors in the nuclear data only.

This information clearly has an integral character — hence the term *integral experiments*[22] used to qualify this source of information — because the measured parameters (reaction rate or combination of reaction rates) are expressed with integrals on the cross-sections. For example:

- Measurement of reactivity by divergence or antireactivity by source ejection or rod drop (using the kinetics equations to process the measurements of flux variation with respect to time);
- Measurement of critical size by adjustment of a water height or number of fuel rods loaded;

[22] As opposed to direct nuclear data measurements, known as "differential experiments".

- Measurement of the critical concentration of a soluble absorbent (generally boric acid);
- Measurement of the spatial distribution of the power in regular or heterogeneous configurations[23];
- Measurement of material buckling by adjustment of cosine or J_0 functions on the spatial distribution of power in a regular configuration;
- Measurement of relative reaction rates, e.g. the plutonium 239/uranium 235 spectrum index[24]:

$$I = \frac{[\text{Pu 239 Fission/U 235 Fission}]\text{lattice spectrum}}{[\text{Pu 239 Fission/U 235 Fission}]\text{reference spectrum}}, \tag{17.7}$$

  where the reference spectrum is usually a Maxwell spectrum created at the centre of a "thermal column" (generally a block of graphite);
- Oscillation of samples at the centre of an experimental lattice giving a reactivity effect that can be calibrated by the reference samples;
- "Neutron noise" measurement (low flux level fluctuations) which, after statistical analysis, gives information about the kinetic parameters;
- Chemical and isotopic analyses of fuels irradiated in a power, experimental, or industrial reactor,
- etc.

The rest of this section gives more technical details about the processing of this integral information, which we call "tendency search".

As we have said, the nuclear data supplied by evaluators are, in principle, *qualified*, i.e. they have a guarantee of a certain level of quality. The reactor physicist, however, cannot be satisfied with this assurance, because it is always advisable to confirm an estimate, and because the engineers who use the codes insist on being shown more direct evidence of how representative the calculations are.

This proof comes from the comparison of experimental information to the calculated parameters that interest the engineer directly: reaction rate, multiplication factor, composition of an irradiated material, etc. These are the "integral" parameters.

If the analysts have any such integral information, they take it into account when making their choices and proposals. The use of integral data, however, was mainly developed by reactor physicists. The comparison of experiment to calculation on these parameters allows a choice, for each nuclide, between the evaluations. If none is truly satisfactory, it suggests modifications of the libraries.

These modifications used to be performed ad hoc because they were *ad hoc* adjustments enabling the calculated results to be properly centred on the measured results. As

[23] It is obviously not the power that is being measured, but, for example, gamma activity of the rods after a few minutes of irradiation at a power of a few tens of watts.

[24] The "ratio of ratios" liberates us from normalisation problems related to the characteristics of the detectors, the irradiation mode, and the metering device.

everyone knows, however, using adjustments outside the narrow range of situations just studied can lead to disaster. That is why more physical, and therefore safer, approaches were developed. Several variants of these can be seen; here we shall summarise the method developed at the CEA (French Atomic Energy Commission) under the name of *tendency search*.

A tendency search begins with the idea that integral measurements give a "boost" to differential measurements, at least in simple geometries where the errors related to the numerical approximations in the calculations are small compared to the uncertainties caused by inaccurate knowledge of the nuclear data[25]. In fact, these integral parameters are often measured with greater precision than this uncertainty arising from the nuclear data, and therefore provide *relevant* information about the nuclear data. This information is very *indirect,* however. The differences between experimental and calculated data can in fact be attributed to each of the many nuclear data concerning the many nuclei involved, and this information must be decoded in some way; this will clearly not be possible unless we have *several* pieces of integral information and are able to *correlate* these differences. In any event, we can only reach a tentative conclusion about the "guilty parties" and the orders of magnitude, which is why we speak of "tendencies". The better these tentative conclusions are supported, the more reliable the extrapolations will be; unlike simple adjustments. In particular, that is why the proponents of this method have always emphasised the advantages of using measurements from different reactor types even if only one type is actually of interest at the time. These ideas are applied via a mathematical analysis that is rigorous in terms of probability distributions[26].

The measured parameters Y_i — differential *and* integral — are random Gaussian variables of which experimentalists have estimated the variances E_i^2 . If F_i is the exact value, then the probability of having measured Y_i to within dY_i is:

$$\frac{dY_i}{E_i\sqrt{2\pi}} \cdot \exp\left(-\frac{(Y_i - F_i)^2}{2E_i^2}\right). \tag{17.8}$$

The overall probability — the product of M expressions of this type — is proportional to $\exp(-S)$ with:

$$S = \sum_{i=1}^{M} \frac{(Y_i - F_i)^2}{2E_i^2}. \tag{17.9}$$

Let us now assume that the theoretical values F_i are obtained as functions of a certain number of parameters $P_1, P_2, \ldots P_K$ depending on experimental conditions $C_1, C_2, \ldots C_L$:

$$F_i = F(C_{1,i}, C_{2,i}, \ldots, C_{L,i}; P_1, P_2, \ldots, P_K). \tag{17.10}$$

For example, F_i is a multiplication factor, the parameters P are the nuclear data, and C are the physical and geometric data of the configuration whose multiplication factor has been measured; the function F, which gives the value of this factor, can be a code such as APOLLO.

In a tendency search, we shall consider that the specific data C are known, but that the basic data P are (at least to some extent) unknown. The reasoning applied in order

[25] i.e. for experiments we would call "clean".

[26] As presented here, the correlations between measurements are neglected.

to deduce these values from the Y_i measurements is based on the *principle of maximum likelihood*. This is a simple, common-sense principle stating that it is more likely that most of the measurements actually obtained had a high prior probability of occurring, rather than a low probability. In other words, the sum S is more likely to be small than large. By taking this observation to its limit, we end up looking for the minimum of S. Because this quantity takes the form of a sum of squares, we often call this approach the *method of least squares*. Concretely, its application involves the cancellation of K partial derivatives:

$$\frac{\partial S}{\partial P_k} = 0, \quad (k = 1, 2, \ldots, K) \tag{17.11}$$

which gives an equal number of equations and unknowns P, and therefore a solution. This says nothing about its relevance, however.

A tendency search attempts to answer this question about relevance. Before discussing this point, however, a few comments about implementation are required:

1) The nuclear data are obviously not totally unknown; in the equations we do not use P_k, but the *differences* X_k between P_k and the value $P_{k,0}$ used in the calculation code. We also take these differences to be small, and assume that a first-order expansion of the functions F is sufficient. Under these conditions, the system of equations becomes linear and is written as follows:

$$\sum_{l=1}^{K} X_l \sum_{i=1}^{M} \frac{S_{ik}S_{il}}{E_i^2} = \sum_{i=1}^{M} \frac{S_{ik}(Y_i - F_{i,0})}{E_i^2}, \quad (k = 1, 2, \ldots, K), \tag{17.12}$$

where $F_{i,0}$ is the value currently calculated by the calculation code, and $S_{ik} = \partial F_i/\partial P_k$ is the coefficient of sensitivity of this value to the kth nuclear datum.

2) The differential (nuclear) parameters, like the integral parameters, come from measurements and are therefore also included in the equations amongst the M parameters processed (in this case, we have simply $F = P_k$). We therefore always have more measurement results than parameters to be determined.

3) Even though, mathematically, the linear system has a solution, we might decide that it is nonsensical to try to determine thousands of nuclear data from a few tens or, at best, a few hundreds of integral measurements. To restrict the scope of these problems and, above all, to improve the relevance of the conclusions, we *reduce* thousands of nuclear data to a small number (a few tens) of more compact parameters: average value in an energy domain, effective resonance integral, normalisation to 2200 m/s, migration area, etc.

Despite these simple precautions, there is no guarantee that the "tendencies" obtained are significant, i.e. that the values X_k correspond, even roughly, to errors actually committed on the nuclear data. A more in-depth analysis has therefore been suggested for the purpose of tendency searches. Please note the following points:

1) By inversion of the linear system, the unknowns X_k are obtained in a perfectly deterministic fashion as a function of Y_i; because these are random variables, the same applies to X_k. It is therefore possible to determine the probability distributions governing X_k knowing the (Gaussian) probability distributions governing Y_i. In particular, the *variances* of X_k give an idea of the significant nature of the tendencies;

2) This is not sufficient, however, because the X_k obtained are *correlated* with each other[27]; in other words, *compensation* effects can occur between different parameters (for example, based on multiplication factor measurements, it is difficult to distinguish between the fission cross-section σ_f and the number of neutrons emitted by fission ν). By minimising with combinations of a reduced number of parameters X_k from one term to the next, it is possible to gain a better grasp of the relevance of each tendency;

3) By examining the values of the sum S and its "differential" and "integral" components before and after minimising, it is possible to evaluate whether the error bars adopted beforehand on the Y_i values were, on the whole, correctly estimated. It is also possible to quantify the role of unidentified parameters: nuclear data removed from the analysis following the reduction and the calculation errors (numerical discretisation, etc.) not taken into consideration. These unidentified parameters can be partly compensated for by (non-physical) changes to the parameters identified and partly uncompensated, thereby increasing the final dispersion of the difference between experimental and calculated values.

This brings us to the limits of the tendency search, which can provide *much* information about nuclear data, but not *complete* information. We conclude by noting that the quality of the integral information in terms of the accuracy and variety of the parameters considered (several types of parameter and different reactor types) is more important than the quantity in identifying the significant tendencies.

17.15. Conclusions

In this book, we have attempted to give a glimpse of many facets of reactor physics and, in particular, we have illustrated calculation methods in this chapter by analysing the calculation scheme, all of which goes to show that the developments in this field call on a rather wide variety of specialities or trades. Thus, we can quite easily pick out the following areas:

- *Physics*, which lets us identify and describe phenomena;
- *Modelling*, translates the physical parameters into mathematical entities and establishes relationships (or equations) between them;
- *Mathematics* used to solve these equations, sometimes analytically, but more often numerically by computer; this is *numerical analysis*, i.e. selecting the most effective techniques to obtain numerical solutions to the equations[28];

[27] Note that the Y_i values can also be correlated with each other to a certain extent; the most detailed studies take this into account not only via variances, but also via covariances.

[28] With regard to the Boltzmann equation, we have seen that very different methods can be considered: collision probabilities, discrete ordinates, Monte Carlo techniques, etc. They are different not only in the numerical method itself (finite differences, finite elements, etc.), but also in the mathematical form of the equation being handled.

- *Programming* which, these days, goes well beyond simply transposing mathematical formulae into a programming language (such as FORTRAN): modularity (and therefore "macro languages"), dynamic memory management, programming techniques ensuring the best possible quality, etc.;

- *Creation and validation of calculation schemes* to seek a tradeoff between the best possible accuracy and the budget constraints (availability of computers and qualified personnel);

- *Physical qualification* using a "tendency search" to check the quality of the basic data (cross-sections, etc.) and, if possible, to improve them;

- *Global and final qualification* on industrial objects providing proof that the calculations are representative, and therefore giving us the confidence to run applied studies.

In the golden age of reactor physics, these different tasks could be performed by one person, but nowadays *teamwork* is essential. To develop and qualify an APOLLO code, for example, must take hundreds of man-years. There is no absolute rule, but these tasks are generally shared amongst the team members more or less according to these categories, and we can therefore speak of different "trades".

There must be a great deal of synergy between the various specialists contributing to the developments and with the "customers" who will use the codes for engineering research (the requirements at this level provide the orientation for development work to some extent)[29].

Finally, the calculation code (the program itself, its documentation, its qualification and the studies applied) that provides cohesion to the project; all the knowledge and know-how acquired over the years are collected there.

The final and crucial question is: *how far should these developments be pursued?*

Firstly, we should note that any science that does not progress will die out sooner or later, because knowledge is never acquired permanently and must be nurtured.

We should also note that constantly developing the accuracy of neutron physics calculations is both illusory and useless;

- *useless* because the engineer and the safety analyst do not require infinite precision to perform a good technical and cost optimisation and to demonstrate that the design is safe;

- *illusory* because some uncertainties will always remain in the data, and increasing the accuracy of the calculations beyond the order of magnitude of these uncertainties will not improve the overall accuracy. It has been shown that these inaccuracies in the data are far from negligible; and this applies to both nuclear data and technological data (dimensions, specific mass, isotopic composition, etc.). In both cases, the resulting uncertainty is of the order of a few hundred pcm on the multiplication factor and a few percent on the power peak.

A healthy combination of physical models and numerical approaches is — and in our opinion will always be — essential.

[29] There are often heated debates about the extent to which code users should orient development work.

Reactor concepts and the requirements of engineers are evolving, new concepts are emerging (examples are given in the next chapter), computing power is increasing, and numerical analysis is progressing. The compromises discussed are therefore not permanent; on the contrary, they are fleeting. That is why a hard core of talented neutron physicists must always be on hand to track developments and to keep this branch of physics alive.

Exercises

Exercise 17.1: evaluation of the numbers of unknowns

This exercise will show that a core calculation in a single step, even simplified, would be extremely expensive in computer running time and memory, while a calculation in two steps, is significantly less demanding, yet retains satisfactory accuracy. In order to simplify matters, here only the numbers of unknowns are considered. It would be necessary to analyse also the structures of the matrices to invert, in order to obtain a complete evaluation.

For this exercise, consider a 900 MWe PWR core comprising 157 fuel assemblies with 17×17 cells per assembly. The unknowns related to the treatment of the reflectors are neglected. Then, for the next step, the numbers of values of flux to determine for different types of static flux calculations will be evaluated within the one- and two-step formulations in order that they may be compared.

a) *One step calculation, simplified model*

— 3 zones per cell,

— 50 axial slices,

— 100 groups.

b) *Assembly simulation, simplified model*

— *x*-*y* cross-section calculation only,

— 5 assembly types,

— 3 zones per cell,

— 10 cell types per assembly,

— 100 groups.

c) *Assembly simulation, precise model*

— *x*-*y* cross-section calculation only,

— 5 assembly types,

— 6 + 1 + 3 zones per cell,

— 45 cell types per assembly (order 8 symmetry),

— 100 groups.

d) *Finite difference core simulation, simplified model*

- Radial calculation only:

— 1 mesh per cell,

— 2 groups.

- Three dimensional calculation:

— as before with

— 50 axial slices.

e) *Finite difference core simulation, precise model*

- Radial calculation only:
  - — 4 meshes per cell,
  - — 4 groups.
- Three dimensional calculation:
  - — as before with
  - — 200 axial slices.

f) *Finite element core simulation, simpified model*

- Radial calculation only:
  - — 4 elements per assembly (parabolic expansion),
  - — 2 groups.
- Three dimensional calculation:
  - — as before with
  - — 15 axial slices (linear expansion).

g) *Finite element core simulation, precise model (example 1)*

- Radial calculation only:
  - — 4 elements per assembly (cubic expansion),
  - — 4 groups.
- Three dimensional calculation:
  - — as before with
  - — 15 axial slices (parabolic expansion).

h) *Finite element core simulation, precise model (example 2)*

- Radial calculation only:
  - — 16 elements per assembly (parabolic expansion),
  - — 4 groups.
- Three dimensional calculation:
  - — as before with
  - — 30 axial slices (linear expansion).

Exercise 17.2: error due to the interface current isotropy hypothesis

Consider the problem of an infinite slab of thickness $2a$ containing a medium labelled "0", and all the remaining space containing a medium labelled "1" (figure **17.9**).

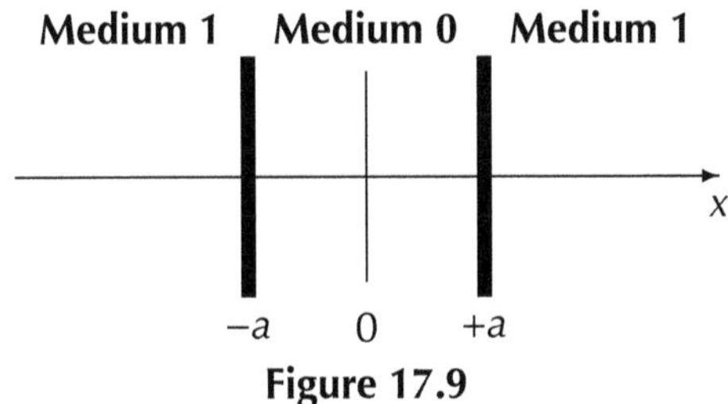

Figure 17.9

These media are purely absorbent. There is a uniform source emitting one neutron per unit of time and per unit of volume in medium "1", and no source in medium "0", thereby providing a simulation of a narrow neutron group at a resonance energy. The neutrons arrive in this group only in the moderator and approximately uniformly. Immediately upon suffering a collision, they are lost from the group. When this occurs in the fuel, it means they are absorbed; when this occurs in the moderator, it means they are lost by virtue of possessing less energy.

a) Recall from exercise **14.5.a** that the kernel of the transport operator in a planar geometry is $E_1(\tau_x)/2$. Show the flux as a function of x in medium "0" and the number α of neutrons absorbed in this medium.

b) Next, the notion of isotropy is introduced for the currents of neutrons crossing the planes $x = -a$ and $x = +a$ in each direction. Let T be the probability for a neutron entering isotropically in medium "0" to cross it without collision. Find this probability and these currents. Deduce the value $\tilde{\alpha}$ of the number of neutrons absorbed in medium "0" when the present idea is applied. Explain why $\tilde{\alpha} = \alpha$.

c) The notion of isotropy of the currents crossing the symmetry plane $x = 0$ in each direction can be added to the previous formulation. In other words, the problem is limited to the part $x > 0$ and an isotropic reflection condition is introduced at $x = 0$. Find the probability T' of crossing the half-slab $0 < x < a$ without collision, together with the associated currents. Deduce the value $\hat{\alpha}$ of the number of neutrons absorbed in medium "0" within this model. Compare $\hat{\alpha}$ and α. The function E_3 is tabulated below to provide the necessary data.

u	$E_3(u)$	u	$E_3(u)$	u	$E_3(u)$
0.01	0.490277	0.1	0.416291	1	0.109692
0.02	0.480968	0.2	0.351945	2	0.030133
0.04	0.463324	0.4	0.257286	4	0.002761
0.05	0.454919	0.5	0.221604	5	0.000878

Exercise 17.3: parameters for an equivalent group of delayed neutrons

a) Show the Nordheim equation for a point kinetic theory with G groups of delayed neutrons.

b) Give the approximate expressions of the dominant solution ω_0 for each of the following cases.

1) reactivity ρ very near zero,

2) reactivity ρ significantly more than the total proportion β of the delayed neutrons.

c) The theory may be simplified by replacing the G groups of delayed neutrons with only one group. Then, owing to the "equivalence procedure", the parameters $\bar{\beta}$ and $\bar{\tau}$ of this unique group can be determined. What are the appropriate values to use as equivalence criteria that are consistent with both situations considered in part **b**?

Exercise 17.4: equivalent reflector

Consider a reflector comprising a set of slabs infinite along directions y and z, numbered from 1 — the interface with the core — to n. "One-group diffusion" theory is employed to calculate the neutron flux.

a) The slab number i can be characterized by its transmission probabilities α_i, representing the number of neutrons emerging from the opposite face for one neutron entering through a face, and β_i, representing the number of neutrons emerging from the same face for a neutron entering through a face. These quantities can be calculated as in the exercise **6.9**. Number the interfaces from 0 — the interface with the core — to n. Give the recurrence relationships linking the currents $J_{+,i}$ and $J_{-,i}$.

b) Show that the general solution of the system of equations is the linear combination of two particular solutions; for instance, the solution obtained from $J_{+,0} = 1$ and $J_{+,1} = 0$, and from $J_{+,0} = 0$ and $J_{+,1} = 1$.

c) Which conditions must be used in order to calculate the albedo of this multi-layer reflector? How can the coefficients λ and μ of the linear combination be deduced?

d) Define "reflector equivalent to the real multi-layer reflector" to be the homogeneous, infinite reflector having the *same albedo*. This equivalent reflector is characterized by two parameters $\hat{D}$ and $\hat{\Sigma}_a$. However, the conservation of albedo gives only one equation. Assuming that $\hat{D}$ is known, write the equation giving $\hat{\Sigma}_a$ due to this equivalence. Here, $\hat{D}$ might represent the diffusion coefficient of the first layer, or any other conventional value.

Exercise 17.5: homogenisation in transport theory

In monokinetic theory, or for a given energy group, a previous calculation for a heterogeneous object of volume V and surface S, gave the probability π for a neutron generated in this object to be absorbed in this volume, and the probability τ for a neutron entering uniformly and isotropically through the surface to re-emerge without absorption.

When this object is replaced with a homogeneous object of same volume and surface, how are the macroscopic absorption cross-section $\bar{\Sigma}_a$ and the macroscopic total cross-section $\bar{\Sigma}_t = \bar{\Sigma}_s + \bar{\Sigma}_a$ of the homogeneous object related, such that they are consistent with π and τ?

Exercise 17.6: homogenisation and transport-diffusion equivalence

Following a fine transport calculation of an elementary "mesh", cell, or assembly, the problem of transport-diffusion equivalence arises. This is subsequently homogenised it in order to perform a more macroscopic diffusion calculation, where this mesh appears among others. For the group collapsing from a fine multigroup description to a coarser description a similar problem arises. Here, this problem is applied to an elementary example: the repeated two-zone pattern generated according to an "infinite sandwich" construction. This will next be examined using monokinetic theory.

The following describes the elementary pattern.

For the *numerical model*, use two typical examples of thermal neutron range occuring in pressurized water reactors.

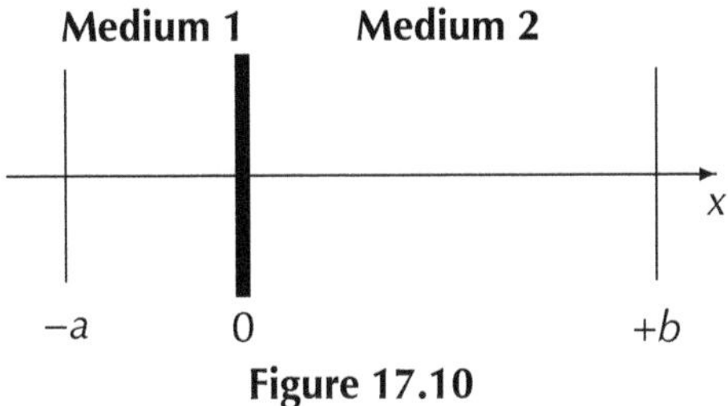

Figure 17.10

1) a pattern representative, in plane geometry, of an elementary cell (the cladding is neglected):

- *Zone 1*
  - — half-thickness = 2 mm,
  - — macroscopic absorption cross-section = 40 m^{-1},
  - — macroscopic scattering cross-section = 40 m^{-1};
- *Zone 2*
  - — half-thickness = 4 mm,
  - — macroscopic absorption cross-section = 1 m^{-1},
  - — macroscopic scattering cross-section = 159 m^{-1};

2) a pattern representative of a heterogeneity (absorbent) in a fuel assembly:

- *Zone 1*
  - — half-thickness = 6 mm,
  - — macroscopic absorption cross-section = 45 m^{-1},
  - — macroscopic scattering cross-section = 120 m^{-1};
- *Zone 2*
  - — half-thickness = 18 mm,
  - — macroscopic absorption cross-section = 15 m^{-1},
  - — macroscopic scattering cross-section = 120 m^{-1}.

The function E_3 takes the values given in the following table.

u	$E_3(u)$	u	$E_3(u)$
0.16	0.375938	0.99	0.111188
0.64	0.180857	2.43	0.017744

a) Find the cross-sections of the homogenised medium, weighted by their volumes.

b) Let medium "2" contain a uniform and isotropic source, normalised to one neutron per unit of time. Use the first collision probability method to find the fluxes, assuming reflection occurs isotropically at $x = -a$ and $x = +b$. Next, give absorption rate f in medium "1". See exercise **14.12**. Here, it is recommended to keep the probability P_{21}. Deduce the cross-sections of the homogenised medium with a flux × volume weighting.

c) Calculate the probability f, and the cross-sections of the homogenised medium using diffusion theory, and compare the result with the values obtained in part **b**.

d) Which cross-section(s) should be modified, and in which direction, in order that diffusion theory gives the same value of f as obtained using transport theory?
Is the transport calculation performed here a good reference for the transport-diffusion equivalence?

Exercise 17.7: axial effect of feedbacks in a core

When using one-group diffusion theory, provided that radial flux variations are neglected, the equation giving the flux Φ in a cylindrical reactor (e.g. a PWR) reduces to

$$\frac{d^2\Phi(z)}{dz^2} + \chi^2(z)\,\Phi(z) = 0,$$

where χ^2 — which can depend on z — is the local "material buckling".

In the following, the axial refelectors are neglected; hence, the boundary condition $\Phi(\pm H/2) = 0$ is applicable, where z lies on the core median plane and H is the core height.

a) Find, with an arbitrary normalisation factor A, the solution Φ_0, and the critical condition for $\chi^2(z) = m^2 = C^t$.

b) Next, axial feedbacks are taken into account using

$$\chi^2(z) = m^2 + \Delta\chi^2_{\text{Boron}} + \Delta\chi^2_{\text{Doppler}} + \Delta\chi^2_{\text{Moderator}}.$$

- $\Delta\chi^2_{\text{Doppler}}$ is the effect on χ^2 of the temperature of the fuel, i.e. the Doppler effect.
- $\Delta\chi^2_{\text{Moderator}}$ is the effect on χ^2 of the temperature of the moderator, mostly due to the water expansion.
- $\Delta\chi^2_{\text{Boron}}$ is the effect on χ^2 of the variation of boron concentration which is necessary to maintain criticality when the previous feedbacks occur.

In order to simplify the calculations, assume the following. The first corrective term is proportional to the local flux — which determines the local fuel temperature. The second term is proportional to the integral of the flux over the whole height — which determines the water temperature increase. The third term is independent of z. Let $-\beta$, $-\gamma$, and α be the coefficients describing these effects. The equation to be solved then becomes

$$\frac{d^2\Phi(z)}{dz^2} + \left[m^2 + \alpha - \beta\,\Phi(z) - \gamma \int_{-H/2}^{z} \Phi(z')\,dz'\right] \Phi(z) = 0.$$

This equation does not seem to have an obvious analytical solution. Hence, to make the problem tractable,

- replace Φ with Φ_0 in the expressions for the feedback effects,
- use $\Phi = \Phi_0 + \varphi$ and $\varphi = f\Phi_0$,
- neglect the second order terms ($\alpha\varphi$, $\beta\varphi$, and $\gamma\varphi$).

Now give the equation governing f.
After muliplication by Φ_0, show that f can be obtained by integrating twice.

c) Show that α and the two integration constants are determined jointly by the boundary conditions, and the flux normalisation. Apply the normalisation condition, integral over the core height of the flux Φ equal to the integral over the same height of Φ_0, to find the result.

Exercise 17.8: axial effect of fuel consumption in a core

A similar formalism can be employed to treat the axial effect of fuel consumption. The effects of feedback are ignored to simplify matters. A term $\Delta\chi^2_{\text{Evolution}}$ represents the effect on the material buckling factor. This is proportional to the integral of the local flux over time from when irradiation begins up to a given time. The equation to solve — which now governs a flux depending not only on z, but also of t — is

$$\frac{d^2\Phi\,(z,t)}{dz^2} + \left[m^2 + \alpha(t) - \varepsilon \int_0^t \Phi\left(z,t'\right)\,dt'\right] \Phi\,(z,t) = 0.$$

This can be solved by using an expansion in powers of t:

$$\Phi(z,t) = \sum_{n=0}^{\infty} \varphi_n(z)\,t^n; \qquad \alpha(t) = \sum_{n=0}^{\infty} \alpha_n\,t^n.$$

Show the first equations and find the first coefficients of these expansions.

Solutions

Exercise 17.1: evaluation of the number of unknowns

a) *One step calculation, simplified model*

$$157 \times 17 \times 17 \times 3 \times 50 \times 100 = 680\,000\,000.$$

b) *Assembly simulation, simplified model*

$$5 \times 10 \times 3 \times 100 = 15\,000.$$

c) *Assembly simulation, precise model*

$$5 \times 45 \times 10 \times 100 = 225\,000.$$

d) *Finite difference core simulation, simplified model*

- Radial calculation only:

$$157 \times 17 \times 17 \times 2 = 91\,000.$$

- Three dimensional calculation:

$$157 \times 17 \times 17 \times 50 \times 2 = 4\,500\,000.$$

e) *Finite difference core simulation, precise model*

- Radial calculation only:

$$157 \times 17 \times 17 \times 4 \times 4 = 730\,000.$$

- Three dimensional calculation:

$$157 \times 17 \times 17 \times 4 \times 200 \times 4 = 150\,000\,000.$$

f) *Finite element core simulation, simpified model*

- Radial calculation only:

$$157 \times 4 \times 4 \times 2 = 5000.$$

- Three dimensional calculation:

$$157 \times 4 \times 4 \times 15 \times 2 = 75\,000.$$

1D finite elements

Linear
(2 × 1/2 = 1 node)

Parabolic
(2 × 1/2 + 1 = 2 nodes)

Figure 17.11a

2D finite elements (squares and rectangles)

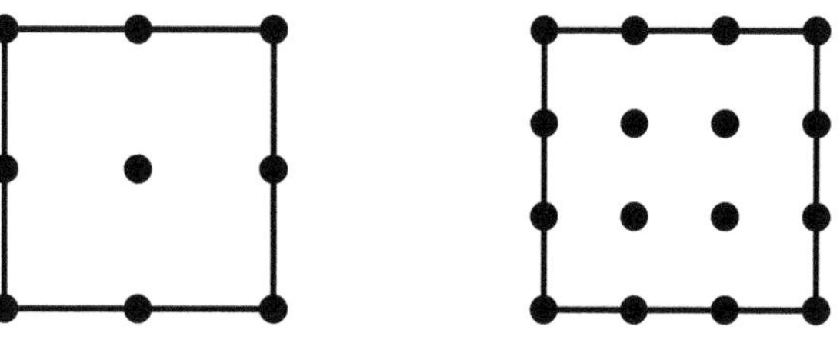

Parabolic
(4 × 1/4 + 4 × 1/2 + 1 = 4 nodes)

Cubic
(4 × 1/4 + 8 × 1/2 + 4 = 9 nodes)

Figure 17.11b

g) *Finite element core simulation, precise model (example 1)*

- Radial calculation only:

$$157 \times 4 \times 9 \times 4 = 23\,000.$$

- Three dimensional calculation:

$$157 \times 4 \times 9 \times 15 \times 2 \times 4 = 680\,000.$$

h) *Finite element core simulation, precise model (example 2)*

- Radial calculation only:

$$157 \times 16 \times 4 \times 4 = 40\,000.$$

- Three dimensional calculation:

$$157 \times 16 \times 4 \times 30 \times 4 = 1\,200\,000.$$

Exercise 17.2: error due to the interface current isotropy hypothesis

a) Integrating the transport kernel gives

$$\Phi(x) = \frac{1}{2\Sigma_1}\left\{ \mathrm{E}_2\left[\Sigma_0(x+a)\right] + \mathrm{E}_2\left[\Sigma_0(a-x)\right] \right\} \qquad (-a < x < +a).$$

Integrating for a second time gives

$$\alpha = \frac{1}{\Sigma_1}\left[\frac{1}{2} - \mathrm{E}_3\,(2a\Sigma_0)\right].$$

b) Let J_{in} and J_{out} be the currents through each interface of the neutrons entering medium "0" and leaving it, and let T be the probability that a neutron entering isotropically into medium "0" crosses it without collision. Hence,

$$J_{out} = J_{in}\,T; \qquad \alpha = 2\,J_{in}\,(1-T).$$

Thus, J_{in} is

$$J_{in} = \int_0^\infty \frac{1}{2}\mathrm{E}_2(\xi\Sigma_1)\,d\xi = \frac{1}{2\Sigma_1}\mathrm{E}_3(0) = \frac{1}{4\Sigma_1},$$

since $\mathrm{E}_2(\tau_x)/2$ integrated between τ_x and infinity of the collision rate is the probability that a neutron emitted at the optical distance τ_x from the interface suffers its first collision beyond the interface. Also,

$$T = \int_0^{\pi/2} 2\cos\theta\sin\theta\,d\theta\,\exp\left(-\frac{2a\Sigma_0}{\cos\theta}\right) = 2\mathrm{E}_3\,(2a\Sigma_0).$$

Therefore,

$$\tilde{\alpha} = \alpha.$$

This result is exact for the following reasons.

- Here, the neutrons enter medium "0" isotropically, meaning that the expression for T contains no approximations.
- the neutrons do *not* leave medium "0" isotropically. Instead, they are all absorbed whatever their (outgoing) direction. Consequently, they have no effect on the expected outcome.

c) In addition to the previous currents, now include the current J_{refl} representing neutrons that reach the plane $x = 0$ from one side and are (isotropically) reflected, together with the probability T' that a neutron isotropically entering a half-slab of medium "0" crosses it. The current J_{in} is the same as previously, with the following modification:

$$J_{\text{refl}} = J_{\text{in}}\, T'; \qquad J_{\text{out}} = J_{\text{refl}}\, T'.$$

A calculation similar to the calculation for T, except now only for the half-slab gives

$$T' = 2\mathrm{E}_3(a\Sigma_0).$$

Then,

$$\hat{\alpha} = \frac{1}{2\Sigma_1}\left\{1 - 4\,[\mathrm{E}_3\,(a\Sigma_0)]^2\right\}.$$

Some numerical results for various values of $u = a\Sigma_0$ are shown in the following table.

u	α	$\hat{\alpha}$	**Difference**
0.01	0.0381	0.0385	1.2%
0.02	0.0734	0.0747	1.8%
0.05	0.1674	0.1722	2.9%
0.1	0.2961	0.3068	3.6%
0.2	0.4854	0.5045	3.9%
0.5	0.7806	0.8036	2.9%
1	0.9397	0.9519	1.3%
2	0.9945	0.9964	0.2%
5	1.0000	1.0000	0.0%

Exercise 17.3: parameters for an equivalent group of delayed neutrons

a) The Nordheim equation is

$$\rho = \omega\left[\ell + \sum_{i=1}^{G}\frac{\beta_i}{\lambda_i + \omega}\right].$$

b) When ω is near zero, the denominators can be replaced with λ_i; therefore,

$$\rho \simeq \omega\left[\ell + \sum_{i=1}^{G}\frac{\beta_i}{\lambda_i}\right].$$

When ω goes to infinity, the denominators can be replaced with ω; therefore,

$$\rho \simeq \omega \left[\ell + \frac{1}{\omega} \sum_{i=1}^{G} \beta_i \right].$$

c) Both these limits are consistent for a single "equivalent" group when its parameters are defined by

$$\bar{\beta} = \sum_{i=1}^{G} \beta_i; \qquad \frac{\bar{\beta}}{\bar{\lambda}} = \sum_{i=1}^{G} \frac{\beta_i}{\lambda_i}.$$

In other words, the total proportion of the delayed neutrons, and the average lifetime of the precursors are mutually consistent.

Exercise 17.4: equivalent reflector

a) The recurrence relationships between the currents are

$$J_{+,i} = \alpha_i J_{+,i-1} + \beta_i J_{-,i}; \qquad J_{-,i} = \alpha_{i+1} J_{-,i+1} + \beta_{i+1} J_{+,i}.$$

Normalising to one neutron entering into the reflector ($J_{+,0} = 1$), and taking into account the boundary condition ($J_{-,n} = 0$), yields as many equations as unknowns. The albedo of the reflector is $\beta = J_{-,0}$.

b) When the second relationship $J_{-,i}$ is substituted into the first relationship, this yields a recurrence formula linking three successive currents J_+, thereby giving all the currents from the first and the second ones. When different sets for both these ones are used, two different particular solutions can be constructed, and subsequently combined linearly to give the general solution.

c) The boundary conditions then allow both coefficients of the combination to be determined. Thus, knowing the currents J_+, permits the currents J_- to be deduced, owing to the second recurrence relationship, and particularly the albedo.

d) Using a "hat" to denote the parameters of the equivalent, infinite and homogeneous reflector,

$$\hat{\beta} = \frac{1 - 2\hat{\kappa}\hat{D}}{1 + 2\hat{\kappa}\hat{D}}, \qquad \text{where} \qquad \hat{\kappa}^2 = \frac{\hat{\Sigma}_a}{\hat{D}}.$$

The equivalence equation $\hat{\beta} = \beta$ gives $\hat{\Sigma}_a$ when $\hat{D}$ is chosen.

Exercise 17.5: homogenisation in transport theory

For the homogenised object, let P be the first collision probability for a neutron emittted uniformly and isotropically in this object, and α be the ratio $\bar{\Sigma}_a/\bar{\Sigma}_t$. The probabilities π and τ are then

$$\pi = \frac{\alpha P}{1 - (1 - \alpha)P}; \qquad \tau = T + \frac{(1 - \alpha)(1 - P)(1 - T)}{1 - (1 - \alpha)P};$$

where $T = 1 - \omega(1 - P)$ is the probability of crossing without collision for a neutron entering uniformly and isotropically, and where the opacity ω is the product of the mean chord $\bar{\ell}$ by the total cross-section.

When these relationships are combined, the result is

$$\frac{\omega(1-P)}{P}=\frac{1-\tau}{\pi}; \qquad \alpha\omega=\Sigma_a\bar{\ell}=\frac{1-\tau}{1-\pi}.$$

The first expression is an implicit equation giving ω. Consequently, since the volume and the surface are known, $\bar{\Sigma}_t$ can be found. The second formula then gives $\bar{\Sigma}_a$.

Exercise 17.6: homogenisation and transport-diffusion equivalence

a) For the different cross-sections, the elementary homogenisation formulae are

$$\bar{\Sigma}=\frac{V_1\Sigma_1+V_2\Sigma_2}{V_1+V_2}.$$

b) Let $c=\Sigma_s/\Sigma$ be the ratio between the scattering cross-section and the total cross-section. Next, the flux and the thermal utilisation factor can be found by using the first collision probability P_{ij} method. Three of these probabilities can be expressed as functions of the last one. For example, using P_{21} yields

$$\frac{\Phi_2}{\Phi_1}=(1-c_1)\frac{V_1\Sigma_1}{V_2\Sigma_2}\left(\frac{1}{P_{21}}-1\right)+c_1,$$

and

$$\frac{1}{f}-1=(1-c_2)\left[\frac{1}{P_{21}}-1+\frac{V_1\Sigma_1}{V_2\Sigma_2}\frac{c_1}{1-c_1}\right].$$

When all the possibilities are listed, and using the complementarity and reciprocity relationships among the "without exit" probabilities P^+, gives

$$P_{21}=\frac{1}{4\tau_2}\frac{\left(1-T_1^2\right)\left(1-T_2^2\right)}{1-T_1^2\,T_2^2},$$

where

$$\tau_1=a\Sigma_1, \qquad \tau_2=b\Sigma_2,$$

and

$$T_i=P^+_{ss,i}=2E_3(\tau_i).$$

Next, the homogenised cross-sections can be deduced from the flux × volume weighting:

$$\bar{\Sigma}=\frac{V_1\Sigma_1\Phi_1+V_2\Sigma_2\Phi_2}{V_1\Phi_1+V_2\Phi_2}.$$

c) In diffusion theory, the formulae are

$$\frac{1}{f}=\frac{b\Sigma_{a2}}{a\Sigma_{a1}}\frac{\kappa_1 a}{\tanh\kappa_1 a}+\frac{\kappa_2 b}{\tanh\kappa_2 b}, \qquad \text{where} \qquad \kappa_i=\sqrt{3\Sigma_{ai}\Sigma_i},$$

and

$$\frac{\Phi_2}{\Phi_1}=\left(\frac{1}{f}-1\right)\frac{V_1\Sigma_{a1}}{V_2\Sigma_{a2}}.$$

Numerical example

The following table gives the values of the cross-sections homogenised by the volumes ($\bar{\Sigma}$), by the volumes and the fluxes calculated by the P_{ij} method ($\tilde{\Sigma}$), and by the volumes and the fluxes calculated by the diffusion theory ($\hat{\Sigma}$).

Quantity	Case 1	Case 2
$\bar{\Sigma}_a$	14.000	22.500
$\bar{\Sigma}_s$	119.333	120.000
$\bar{\Sigma}$	133.333	142.500
$\tilde{\Sigma}_a$	12.752	19.792
$\tilde{\Sigma}_s$	123.142	120.000
$\tilde{\Sigma}$	135.894	139.792
$f_{\text{transport}}$	0.94521	0.36315
$\hat{\Sigma}_a$	13.469	19.607
$\hat{\Sigma}_s$	120.955	120.000
$\hat{\Sigma}$	134.424	139.607
$f_{\text{diffusion}}$	0.94949	0.35242

d) Here, the equivalence can be obtained thanks to the modification of only one cross-section — since the equivalence criterion is unique — namely the conservation of f. The most natural choice is a modification of Σ_{a1}:

- decreasing in case 1,
- increasing in case 2.

The transport calculation with the "flat flux" approximation (only one volume per physical zone) is probably sufficient in case 1; however, it is not in case 2, where the optical thicknesses is greater.

When compared with a fine transport calculation, it can be seen that the diffusion approximation underestimates the disavantage factor; therefore, the thermal utilisation factor is overestimated. The absorption cross-section of medium 1 (the more absorbent) must be reduced in order to maintain consistent absorption rates.

Exercise 17.7: axial effect of feedbacks in a core

a) The flux and the critical condition for the case without feedback are

$$\Phi(z) = A \cos \frac{\pi z}{H}; \qquad m = \frac{\pi}{H}.$$

b) The simplifications and changes of functions give the equation

$$f'' \, \Phi_0 + 2 f' \, \Phi_0' + \left(\alpha - \beta \, \Phi_0 - \gamma \int_{-H/2}^{z} \Phi_0 \, dz' \right) \Phi_0 = 0.$$

After multiplication by Φ_0, the derivative of $f'\,\Phi_0^2$ appears:

$$(f'\,\Phi_0^2)' = \left(-\alpha + \beta\,\Phi_0 + \gamma \int_{-H/2}^{z} \Phi_0\,dz'\right)\Phi_0^2.$$

Integrating once, yields f'; integrating again, gives f, and hence φ. It is simpler to integrate from $z = 0$, which only modifies the integration constants D and E. Using $u = \pi z/H$, yields

$$\varphi = \Phi_0 \frac{\pi}{H} \int_0^{u} \left\{ \frac{\pi}{H} \int_0^{u'} \left(-\alpha + \beta\,\Phi_0 + \frac{\pi\gamma}{H} \int_{-H/2}^{u''} \Phi_0\,du^{(3)}\right) \Phi_0^2\,du'' + D \right\} \frac{du'}{\Phi_0^2} + E\,\Phi_0.$$

These integrals have analytical solutions:

$$\varphi = EA\cos u + \frac{\pi D}{HA}\sin u - \frac{\pi^2 A}{2H^2}\beta\,u\sin u + \frac{\pi^2 A^2}{3H^2}\beta\,(2 - \cos^2 u - \cos u)$$

$$+\frac{\pi^3 A^2}{6H^3}\gamma\,(3u\sin u + 2\sin u - u\sin u\cos u).$$

c) The boundary conditions (zero flux for $u = \pm\pi/2$) give

$$\alpha = \frac{8A\beta}{3\pi} + \frac{A\gamma}{H}; \qquad D = -\frac{\pi^3 A^3 \gamma}{3H^2}.$$

Finally, setting the integral of φ equal to zero yields the constant E; hence,

$$\varphi = \frac{\pi^2 A^2}{3H^2}\left\{\beta\left[2 - \frac{3\pi^2 - 16}{4\pi}\cos u - \cos^2 u - \frac{4}{\pi}u\sin u\right]\right.$$

$$\left.+\frac{\pi\gamma}{H}\left[\frac{3\pi - 2}{2\pi}(u\sin u - \cos u) - \sin u\cos u\right]\right\}.$$

Exercise 17.8: axial effect of fuel consumption in a core

The equation to solve is

$$\sum_{n=0}^{\infty} \varphi_n''\,t^n + \left(m^2 + \sum_{n=0}^{\infty} \alpha_n\,t^n - \varepsilon \sum_{n=0}^{\infty} \varphi_n \frac{t^{n+1}}{n+1}\right) \sum_{n=0}^{\infty} \varphi_n\,t^n = 0.$$

The zero-order terms give

$$\varphi_0'' + (m^2 + \alpha_0)\,\varphi_0 = 0.$$

Therefore,

$$\varphi_0 = A\cos\frac{\pi z}{H}; \qquad m = \frac{\pi}{H}; \qquad \alpha_0 = 0.$$

The first-order terms give

$$\varphi_1'' + m^2\varphi_1 + (\alpha_1 - \varepsilon\varphi_0)\,\varphi_0 = 0.$$

By replacing φ with φ_1, α with α_1, β with ε, and γ with 0, the equation becomes the same as the one solved in the previous exercise. In particular,

$$\alpha_1 = \frac{8A\varepsilon}{3\pi}.$$

The second-order terms give

$$\varphi_2'' + m^2\varphi_2 + (\alpha_1 - \varepsilon\varphi_0)\,\varphi_1 + \left(\alpha_2 - \frac{\varepsilon\varphi_1}{2}\right)\varphi_0 = 0.$$

Given φ_0 and φ_1, it is still possible to obtain an analytical solution, although more tedious. Similarly, the remaining terms may be calculated.

18 Overview of core design problems

Introduction

To conclude this book, we offer a few thoughts about some problems that go beyond the realm of neutron physics strictly speaking, but where knowledge of neutron physics is required. To provide a solution to these problems, or even to discuss them in great detail, is beyond the scope of this book[1]; we merely wish to give the reader a few pointers to other technical aspects of nuclear energy, and an outline of the prospects as they appear today.

In the first and second parts of this chapter, we shall use and expand on several elements touched upon in the previous chapters in order to provide a quick summary of core design and management problems. Part Three gives a brief presentation of a few current research topics in the laboratories concerned, particularly in France, and mentions a few short-, medium-, and long-term prospects.

18.1. General elements of core design

Reactor core design poses many problems involving different branches of physics and engineering. These problems interact and, in practice, design work must be performed iteratively between the specialists involved.

18.1.1. Reactor target

The first stage in design is to assign a target to the reactor. If the reactor is designed for the purpose of *energy production* (electronuclear reactor, naval propulsion reactor, nuclear heat reactor, desalination reactor, reactor for a space station, etc., or a mixed-purpose reactor), it is necessary to specify the power requirement, the operating mode (basic, occasional or back-up), etc. For a *research reactor*, the type of experiment must be specified: technological irradiation, production of radionuclides or doped silicon, neutron flux for imaging or structural research on matter, neutron qualification, teaching, etc. We might also wish to design a reactor with enough flexibility to be used for several of these applications.

[1] Many additional elements can be found in other books: reactor types, water reactors, thermal hydraulics, nuclear safety, nuclear economy, etc.

Table 18.1. Main electronuclear reactor types.

TYPE	MODERATOR	COOLANT	FUEL	CLADDING
PWR and VVER	Water	Pressurised water	Uranium oxide 3–4% or mixed oxide U+Pu (MOX)	Zr
BWR	Water	Pressurised water	Uranium oxide 2–3% or mixed oxide U+Pu (MOX)	Zr
CANDU	Heavy water	Heavy water (pressure tubes)	Uranium oxide, natural or very slightly enriched	Zr
UNGG and Magnox	Graphite	Carbon dioxide	Natural metallic uranium	Mg
AGR	Graphite	Carbon dioxide	Uranium oxide 2%	Stainless steel
HTR	Graphite	Helium	Uranium oxide 5–10%	Graphite
RBMK	Graphite	Boiling water (pressure tubes)	Uranium oxide 1.8–2.4%	Zr-Nb
FNR	(N/A)	Sodium	Mixed oxide U+Pu (MOX) 15–20% Pu	Stainless steel

18.1.2. Choice of reactor type

Once these broad categories have been determined, the overall system must be defined (moderator, fuel, coolant, etc.), i.e., the reactor type must be chosen. With all the different possible combinations of *fuel* (type of fissile and fertile material, abundance, chemical form, geometric arrangement, etc.) with the fuel's *cladding* (steel, magnesium, zirconium, graphite, etc.), as well as the *moderator* (graphite, ordinary water, heavy water, beryllium, beryllia, etc. or absence of moderator) and the *coolant* (liquid: water, heavy water, organic liquid, sodium, lead, etc., or gas: carbon dioxide, helium, etc.), thousands of different reactor types could be created.

The pioneers of the nuclear age more or less considered all of these possible types, ruled out many of them[2], and studied a few dozen of them to a greater or lesser extent. Finally, a very small number of reactor types was developed to the manufacturing stage (see Table 18.1 concerning electronuclear reactor types. We included the HTR, even though this type is almost non-existent today, because interest in it seems to have been revived).

The second table (Table 18.2) gives the same type of information as Table 1.1, but by reactor type instead of by country. Note that water reactors are by far the most common (two thirds in pressurised technology and one third in boiling technology). Most of the rest are CANDU and RBMK; this latter type, as well as Magnox and AGR, are reaching the end of their life cycle. Fast neutron reactors and the other types only represent a tiny proportion these days.

[2] Some criteria might change over half a century of research; concepts that were rejected in the past can sometimes re-emerge.

Table 18.2. Nuclear Electric Power Plant as at 31/12/2006: breackdown by reactor types. (Source: ELECNUC, 2007 Edition)

TYPE	Installed power (GWe)	Number of units
Magnox and AGR	9.0	18
RBMK	11.4	16
Heavy water	21.5	42
Water (PWR and VVER)	242.3	264
Water (BWR)	83.9	93
Fast neutrons	0.7	2
TOTAL	368.9	435

18.1.3. Elements required for lattice sizing

Once these choices have been made, the neutron lattice must be sized. For the sake of simplicity, let us say that three parameters must be defined:

1/ Diameter of the fuel elements,

2/ Abundance of fissile material in the fuel,

3/ Lattice step.

The first parameter is not decisive in neutron physics if we apply a reasoning by fixed moderation ratio. In the context of the four-factor formula, and if we neglect variations in the fast fission factor ε and in the disadvantage factor that enters into the thermal utilisation factor f, we note that only the resonance escape probability p is affected by this choice. Increasing the diameter of the fuel elements leads to greater self-shielding (§ 8.3.2), which is favourable in terms of reactivity, and leads to a reduction in the Dancoff effect (§ 8.3.4) because of the greater thickness of moderator between the fuel elements — this works in the opposite direction. In practice, thermal considerations will have the most influence on this decision; in view of the specific power objectives, a certain fuel temperature limit must not be exceeded at the core of the element or pellet. For a thermal neutron reactor, the choice of lattice step affects the moderation ratio when the diameter of the fuel elements has been fixed. In Chapter 9 (§ 9.3.4), we saw that there is an optimum moderation ratio corresponding to the best tradeoff between minimising resonant capture in the fuel and minimising thermal capture in the moderator. It might seem logical to take this optimum value but, as we saw in the example of pressurised water reactors, there might be other factors to consider, particularly any arguments related to control and safety (which in this case lead us to adopt an under-moderated situation).

Finally, if the abundance of fissile material in the fuel can be chosen (which it can for pressurised water reactors, but not for a natural uranium reactor), it is defined with a view to the cycle duration aimed for in the core management mode to be adopted (§ 12.2.2).

The example of pressurised water reactors shows very clearly how retroactive considerations must be introduced to determine the lattice sizing; in this case, safety considerations and the choice of multiple frequency core management mode.

As a general rule, it is clear that an analysis of the reactivity coefficients can lead to drastic restrictions of the range of possible choices. We shall see other examples, concerning the multiple recycling of plutonium in water reactors and CAPRA reactors.

Table 18.3. Approximate specific power for the main electronuclear reactor types.

TYPE	Specific power ($MWth/m^3$)
UNGG and Magnox	1
AGR	2
HTR	8
CANDU	12
PWR and VVER	100
BWR	50
FNR	500

18.1.4. Elements for core sizing and choice of reflector

The choice of reflector is related to the characteristics of leaks, which involve the product of two factors: B^2 (geometric buckling), which is inversely proportional to the square of a characteristic dimension of the core, and M^2 (migration area), which characterises the average displacement of neutrons in the lattice.

The dimension of the core depends on the total power output, which can vary within a relatively wide range for different examples, and on the specific power, which is a characteristic of the reactor type and is more or less independent of power (Table 18.3).

For a given reactor type, the migration area depends relatively little on the details of lattice sizing. Orders of magnitude for a few examples are given in Table 18.4.

Note that:

- The slowing-down area of a lattice is a little larger than that of the pure moderator because slowing-down is negligible in the volume added for the fuel. On the other hand, the diffusion area of the lattice is far smaller than that of the moderator because of the great amount of absorption added by the presence of the fuel.

- In graphite and heavy water, the diffusion area is far greater than the slowing-down area but, in the corresponding lattices, the two terms have the same order of magnitude;

- In water reactors, the migration area is very small and most of the leaks involve fast neutrons because the diffusion area is close to zero (a thermalised neutron in a water reactor is practically absorbed on the spot).

If there are *significant leaks*, the choice of reflector is crucial; we can try to use the escaping neutrons. A typical example is provided by fast neutron reactors, which have a large migration area because cross-sections are smaller for fast neutrons than slow neutrons, and which are characterised by the very high power density allowed by the thermal qualities of sodium, leading to a small core volume. By adding a uranium 238 cover, we obtain the albedo of a reflector and improve the conversion factor thanks to the captures in this material[3].

[3] Note that a fast neutron reactor cannot become a breeder unless surrounded by radial *and* axial blankets.

Table 18.4. Approximate migration area (m^2) for the main moderators and a few electronuclear reactor types (for thermal neutron reactor types, the migration area is broken down into slowing-down area and diffusion area).

Moderator	Migration area	Reactor type	Migration area
Graphite	$3.5 + 25 \simeq 30$	UNGG	$4 + 4 = 8$
Heavy water	$1.3 + 100 \simeq 100$	CANDU	$1.5 + 1.5 = 3$
Water at 20 °C	$0.27 + 0.08 = 0.35$	***	***
Water at 300 °C	$0.50 + 0.25 = 0.75$	PWR	$0.5 + 0.06 = 0.56$
***	***	FNR	2.5

If there is *little leakage*, the role of the reflector is more to contribute to the flattening of the power distribution than to improve the neutron balance. Water reactors provide a typical example, as they are characterised by a very small migration area and often have a high power rating; thus they are often large despite having high power density. Because leaks often concern fast neutrons, the steel surrounding the core (a bad reflector for thermal neutrons because of high capture, but a good reflector for fast and epithermal neutrons) has about as much effect as water on the reflection of neutrons[4].

We have highlighted the importance in core design of analysing the reactivity coefficients. Clearly, the effect of the reflector on these coefficients should also be examined. For example, the study of HTRs with internal and external reflectors has revealed complex mixing phenomena between the lattice spectrum and the Maxwell spectrum of graphite — which are difficult to model with calculations — affecting the temperature coefficient of the moderator.

18.2. General remarks about core control and management

A discussion of core control and management problems is outside the remit of this book, but we wish to give this theme a passing mention because it draws a great deal on neutron physics. We shall merely state the example of pressurised water reactors[5]; the general principles are the same for other reactor types, although the details of implementation might be different.

[4] Placing blankets in a water reactor would improve the conversion factor only very slightly — at the cost of making the design and management of the core far more complicated and expensive.

[5] Further details can be found in other neutron physics books dealing with these reactors and with safety.

18.2.1. Control concepts

"Reactor control" refers to all the operations carried out to start up, shut down, regulate or change the power level of a reactor. To initiate the chain reaction, the few neutrons from the reactions induced by cosmic rays or spontaneous fission of uranium 238[6] are theoretically sufficient. The flux in the initial reactor, subcritical and characterised by its multiplication factor k, has the following form:

$$\Phi \simeq \frac{\Phi_s}{1-k}, \quad \text{with:} \quad \Phi_s = \frac{S}{\Sigma_a} \tag{18.1}$$

where S is the source, Σ_a is the average absorption cross-section, and Φ_s is the neutron flux without multiplication; power is governed by a similar expression. By raising the value of k until it approaches criticality, any level of flux can theoretically be reached.

Concretely, we observe that, even taking into account the amplification factor[7] $1/(1-k)$, with the reactor not operating, the flux due to spontaneous sources only is extremely low compared to nominal. No instrumentation would be able to monitor the twelve or fifteen decades separating these two levels. That is why the initial level must be raised thanks to a much more intense additional source, to avoid blindly diverging with the risk of reaching an excessive level of reactivity or even exceeding the prompt neutron criticality threshold.

In practice, we generally use a source using (α, n) or (γ, n) reactions on beryllium, created by mixing an α emitter (radium, polonium, etc.) or a γ emitter (antimony 124, sodium 24, etc.) with this material (see § 2.8.2).

In principle, there is no problem shutting down a reactor; simply insert the absorbent rods intended for this purpose. Note, however, that even in the event of an emergency, shutdown is never instantaneous, no matter how much antireactivity is introduced. After a fast drop in power (prompt jump if the antireactivity is introduced almost instantaneously), the neutron population only decreases at the rate of the disintegration of the delayed neutron precursors and, in particular, of those with the longest half-life (around 55 seconds).

Moreover, the chain reaction can of course be stopped, but the radioactivity of the fission products and actinides that have accumulated in the fuel cannot be stopped. This *residual power* P_r is in the region of 6% of the initial power P_0 just after shutdown, and then decreases very gradually and after a few seconds or a few tens of seconds it exceeds the neutron power. A precise evaluation of this power is essential for accident studies[8] and studies on the management of irradiated fuel. This requires processing all the radioactive series, and assumes reasonable knowledge of the nuclides involved (decay half-lives, modes, and energies). If we are only looking for orders of magnitude, however, we can simply use the relationship suggested by K. Way and E. Wigner in 1948:

$$P_r = 5.9 \cdot 10^{-3} P_0 [t^{-0.2} - (t+T)^{-0.2}], \tag{18.2}$$

where t is the time in days after shutdown (the formula is not suitable for values of t below 10 seconds) and T is the time in days during which the reactor operated at power P_0.

---

[6] Approximately 26 spontaneous fissions are observed per hour and per gramme of uranium 238. The process exists for all heavy nuclei, especially those with an even number of neutrons.

[7] For example, to give a rough idea of the gain, let us say a factor of 100.

[8] It is absolutely essential to cool the core, even a very long time after normal or accidental shutdown. The fusion of a large part of the core, observed when the containment vessel was opened from the ill-fated Three Mile Island reactor, is an unfortunate illustration of this.

						52			51					
		53	SA		SB 43		R 42		SB		SA			
				G2		N2	56	N2	44	G2	50		41	
	SA 34	45	N1		SC		G1 32		SC		N1		SA	
		G2		R 54				57		R 49		G2 31		
46	SB	55	SC		SD 48		N1 36		SD		SC		SB 37	
		N2	47			33			39			N2 58		
27	R1	26	G1	35	N1	28	R		N1 38		G1 24	22	R1 40	
	23	N2						13		2		N2		3
	SB		SC	14	SD	25	N1		SD		SC 21		SB	
6		G2		R 30			4			R 15		G2		29
	SA		N1		SC 12		G1	18	SC		N1 1		SA	
		16		G2 7		N2	5	N2		G2			11	
		8	SA		SB	19	R		SB 20		SA 10			
				17			9							

Figure 18.1. Chart of control bundles and assemblies wired up for flux measurement (1300 MWe pressurised water reactor; mode "G").

With regard to normal operation, note that, thanks to the Doppler effect, reactor power is stable. Operator intervention is required only to modify the power level if necessary and to compensate for spontaneous changes in reactivity (accumulation of fission products and changes in heavy nuclei).

As a general rule, it is preferable to operate reactors at constant power, i.e. to use them in "basic" mode. Nuclear power plants are not as flexible as classic power plants in adjusting to fluctuating power demand. In particular, thermal cycling of the fuel induced by power jumps must be avoided. If the power needs to be changed, this should be done (except obviously for an emergency shutdown) in sufficiently gradual ramps (typical value for the power increase of a pressurised water reactor: 5% of nominal power per minute). When nuclear power stations constitute a large proportion of the power generating plant, however, as in France where approximately 75% of electricity is nuclear, the nuclear power plants themselves must perform *load monitoring*. Thus, after control in "mode A" which is well suited to basic operation, Framatome and Électricité de France developed a control mode "G", which is far more sophisticated and combines the usual SIC (silver indium cadmium) "black" bundles with "grey" bundles made of steel that are less absorbent and therefore disturb the power distribution less (Figure 18.1).

Obviously it does not make sense to have control without monitoring. There are different ways of monitoring the situation in a reactor core either continuously or discontinuously. That is why, for pressurised water reactors, for example, in addition to constantly measuring a certain number of temperatures at the inlet and outlet of the core (to give a continuous measurement of the power), two types of neutron physics measurement are performed:

Table 18.5. Reactivity balance example.

Reactivities in PCM	Beginning of cycle	End of cycle
Antireactivity of bundles		
Hot operation, zero power, 48 bundles	9720	9270
Most reactive stuck bundle	2050	1550
	---	---
Net	7670	7730
10% margin	770	770
Insertion of power bundles	500	500
	---	---
(1) Antireactivity of 47 bundles	6400	6460
Insertion of reactivity due to the passage from rated power to zero power		
Doppler effect	1350	1200
Temperature effect of moderator	40	1030
Vacuum effect	50	50
Redistribution effect	200	850
	---	---
(2) Total reactivity insertions	1640	3130
(1) – (2) : Antireactivity margin when shut down	4760	3330
Required antireactivity margin when shut down	1000	1770

- Measuring the *axial offset*, from chambers external to the containment vessel[9], in order to monitor the axial power distribution continuously (but only approximately) and, in particular, to detect an oscillation due to xenon 135 (§ 11.2.4);
- Measuring the *axial and radial neutron flux* in the core *via* miniature fission chambers inserted into the instrumentation tube[10] from below using a flexible guide. Only about fifty assemblies can be explored in this way. The measurements are performed periodically (e.g. monthly)[11].

18.2.2. Reactivity balance

In Figure 18.1, note that some of the bundles (regulating bundles, grey bundles and black bundles) are used for control, and the others ("S" bundles, where S stands for *shutdown* or *safety*) are exclusively for the purpose of shutting down the reactor. It is important to ensure, with safety margins as imposed by specifications, that the available antireactivity is sufficient to halt the chain reaction under any circumstances. The designer must therefore draw up a balance of the effects to be compensated and the efficiencies available in order to show that they are sufficient. An example of such a balance is given in Table 18.5, for the nominal situation at the beginning and the end of a cycle (source: J. Bussac, P. Reuss, *Traité de neutronique*, Hermann, 1985; for a 900 MWe reactor).

[9] Originally there were two chambers (top and bottom), but now there are six chambers.

[10] In a central position in the assembly; see for example Figure 17.2, § 17.3 (the other 24 guide tubes are intended to accommodate a control bundle).

[11] Continuous "in-core" measurements by collectrons are currently under research and development.

The purpose of the balance is to show that the antireactivity margin when shut down, i.e. the difference between the antireactivity of the bundles and the effects to be compensated, is greater than the required margin. The antireactivity of the bundles is determined whilst allowing for small insertion during normal operation, assuming that a bundle (the most efficient one) does not fall, and taking a margin of 10% on the efficiency calculations.

Concerning the effects to be compensated, other than the Doppler effect and the moderator effect described in Chapter 13, we take into account the small vacuum effect due to a few bubbles that appear at full power and in particular the redistribution effect: when the temperature gradients (essentially axial) are cancelled on shutdown, the spatial distribution of the flux tilts towards the top of the core; this modifies the average poisoning due to xenon. In practice, the effect is unfavourable for the balance, as it can be seen.

18.2.3. Core management

We shall not go into detail about this subject, which was presented in Chapter 12 (§ 12.2.2 and 12.2.3). Let us review the essential points related to neutron physics:

- The variation in multiplication factor as a function of *average* core burnup becomes slower as the renewal of the fuel becomes more fractionated; in particular, there is approximately a factor of 2 between the energy produced by a given fuel between a block loading-unloading management and an ideal continuous loading-unloading management. In reactors where fuel renewal is performed during operation (UNGG, RBMK, CANDU), there is an attempt to approach this ideal case; in reactors where this procedure is performed with the reactor shut down (PWR, BWR, FNR), a compromise between these two extreme cases is sought;
- In these searches for an optimum solution, other constraints must be introduced. This is evident, for example, in the drawing up of reloading and repositioning plans, which are drawn up for each shutdown of the pressurised water reactors; the main ones are:
  - Checking that the reactivity is sufficient for the cycle duration envisaged,
  - The minimisation as far as possible of the form factor, i.e. the power peak normalised to the average power,
  - The minimisation also as far as possible of the fluence that the containment vessel will receive in fast neutrons;

for the management of an entire power plant network, such as all French pressurised water reactors, a certain amount of flexibility is required to harmonise the shutdown dates of each unit and to cope with unplanned events. The possibility of early shutdown and of irradiation campaign extension, allowed by the highly negative moderator temperature coefficient at the end of a cycle, provides this flexibility.

18.2.4. Plutonium recycling

In the introductory chapter (§ 1.7) and Chapter 12 (§ 12.3.3), we saw that it can be of interest to recycle the plutonium produced from neutron capture by uranium 238 in reactors but not burnt up *in situ*. This involves reprocessing the irradiated fuel and manufacturing a plutonium fuel.

We also saw (see analysis of η factors, § 12.3.2) that the plutonium is used most efficiently in fast neutron reactors, where the excellent neutron balance leaves enough neutrons available for good conversion, breeding, or any other application, such as the incineration of waste.

We shall not go into details about the history of plutonium recycling, but it is of interest to mention a few milestones to illustrate the connection between technical — particularly neutron physics — considerations and economic and political decisions. Let us refer to the French example.

The potential interest of fast neutron reactors was revealed in the early days of nuclear energy research. It appears as a recurrent theme in speeches of the 1950s: let us build a few thermal neutron reactors[12] in order to constitute an initial stock of plutonium; this will allow us to start up a fast neutron reactor; because this will be a breeder reactor, the number of fast neutron reactors will be able gradually to grow...until most energy production requirements are satisfied in this way.

The decision to develop reprocessing technology and to build the Marcoule and La Hague plants was a response to this objective.

In the 1970s, however, the masses of plutonium thus obtained by reprocessing far exceeded the quantity that could be absorbed in the short term by the few existing or planned fast neutron reactors. The long-term prospect might remain, but it appears that an intermediate stage will be possible: a first recycling in thermal neutron reactors will allow this plutonium to be used with barely any reduction of the stock whilst waiting for the arrival of fast neutron reactors. In particular, a major R&D programme on this theme was promoted by the European Commission. Initially reluctant, France finally rallied round and made a substantial contribution to the work devoted to the study of plutonium recycling in pressurised water reactors[13].

As we have seen (§ 9.2.2), by replacing the usual uranium oxide fuel enriched by a few percent with MOX fuel with a slightly higher plutonium content (equivalence of maximum burnup), we considerably change the neutron physics situation in the thermal domain; overall, the macroscopic absorption cross-section in the thermal domain is three times greater and the flux is three times lower. This leads to the control means (boron solution and bundle) being three times less efficient because their absorption takes place essentially in the thermal domain.

This does not completely rule out boron; boron can be enriched[14] to isotope 10. It turns out, however, in view of the reactivity balances similar to those presented in Table 18.5, that the number of available control bundles would be insufficient to ensure safety in a core completely loaded with MOX fuel. A core loaded to 50% with MOX would, however be possible. For the sake of caution, MOX loading is authorised for one third of reload assemblies only. In France, this has been applied since 1987 in one of the 900 MWe reactors at Saint-Laurent-des-Eaux, and has since been put into wider practice by successive administrative authorisations, now covering most PWR-900 reactors.

The mixed loading of a core with standard and MOX assemblies must naturally create interfaces between the two types of lattice. This hardly disturbs the distribution of fast and epithermal neutrons, since loading plans are designed to give approximately the same

[12] In that era, UNGG reactors were the favoured type.

[13] Meanwhile, the UNGG programme was abandoned and the programme to build pressurised water reactors was launched.

[14] There is 20% boron 10 — absorbent — in natural boron.

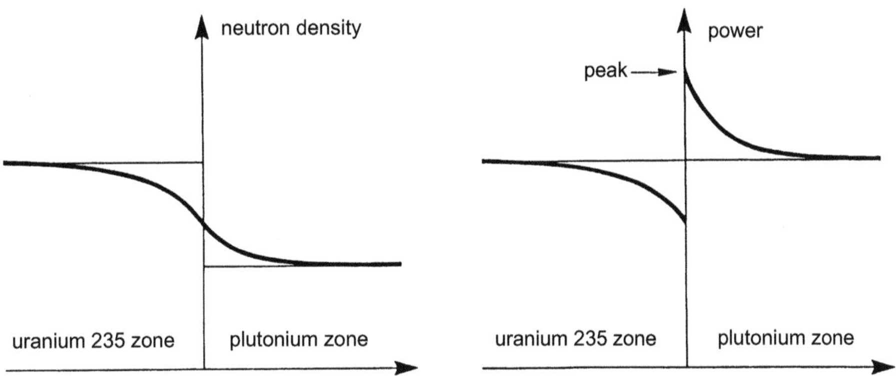

Figure 18.2. Density (or flux) distribution curve and power curve near an interface between uranium and plutonium lattices.

fission rate in both types of assembly; but the great difference in thermal absorption cross-section gives rise to the risk of a local power peak.

The diagram (Figure 18.2) shows the thermal flux curve on the left. Because the thermal neutron "sources" are similar on either side of the interface, we would expect to see flux levels inversely proportional to the macroscopic absorption cross-sections (factor of 2 to 4), and this is so in the asymptotic case. A flux must be continuous, however; and so there is a gradual transition between these levels in the neighbourhood of the interface. To find the power, we must multiply by a macroscopic fission cross-section. The factor between these fission cross-sections is roughly the same as the factor between the absorption cross-sections. Under these conditions, the power distribution resembles the curve shown on the right-hand side in Figure 18.2: note that the asymptotic levels are equal, in accordance with our initial assumption. Note also the strong perturbation near the interface: a "peak" on the plutonium side and a "hole" on the uranium side. The hole has no disastrous consequences, but the peak on the plutonium side is unacceptable[15].

The only way to reduce this peak is to lower the cross-section, i.e., in practice, the plutonium content, in the area concerned. These interfaces therefore create the need to "zone" the MOX assemblies placed in water reactors[16]. There are different designs (see Figure 17.2 and Figure 18.3); three zones with three different abundances turns out to be the number required to "crush" the power peak.

Such zoning is obviously an additional obstacle to the economical recycling of plutonium in water reactors; a process already burdened by the cost of reprocessing[17] and the additional manufacturing cost arising from the need to work in glove boxes for MOX.

[15] Note that there is a safety-related limit on the power peak: the higher the peak with respect to average, the more the average core power must be lowered.

[16] Zoning is not applied to uranium assemblies, partly because the "holes" do not create the same constraints, and partly because there would be no need for zoning between two uranium assemblies. Note that there is never an interface between MOX assemblies in PWR loading plans.

[17] It is very difficult to estimate this cost because the plants have been partially amortised and because the improved packaging of waste is also a justification for reprocessing.

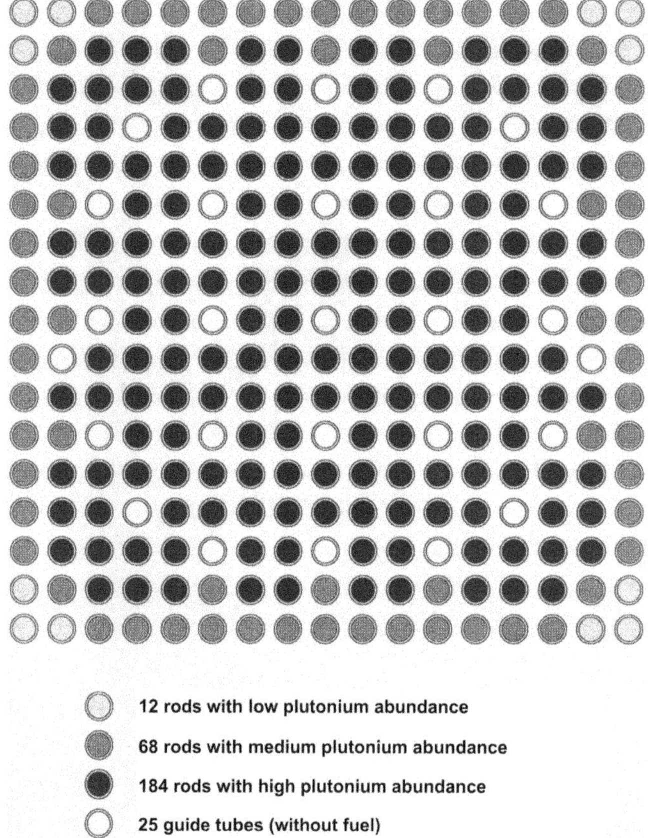

Figure 18.3. Zoned MOX assembly for pressurised water reactor.

At present, plutonium multirecycling is not carried out, partly for economic reasons[18], and partly for reasons related to neutron physics. As it is currently practised, the recycling of first-generation plutonium gives second-generation plutonium[19] — the type found in irradiated MOX assemblies — which is highly charged with higher isotopes, particularly plutonium 242. This would not rule out the use of fast neutrons. A very high abundance of 10 to 15% would however need to be achieved for a second recycling in a water reactor. In addition to the power distribution problems this would cause, that abundance level would result in a positive moderator temperature coefficient, which would be unacceptable.

[18] The reprocessing of MOX assemblies would involve the handling of masses and concentrations of plutonium far above those involved in standard fuel reprocessing. Currently in France, of the 1050 tonnes of irradiated fuel discharged from reactors each year, approximately 850 tonnes are reprocessed. Irradiated MOX assemblies are not reprocessed.

[19] See Table 18.6.

Table 18.6. Abundance and isotopic composition of plutonium in MOX fuel before and after irradiation in a PWR (approximate values in number of nuclei for management over four cycles).

Ratio (%)	Fresh fuel	Irradiated fuel
Pu/(U + Pu)	6.4	4.5
^{238}Pu/Pu	2	2
^{239}Pu/Pu	58	39
^{240}Pu/Pu	23	29
^{241}Pu/Pu	11	19
^{242}Pu/Pu	6	11

18.3. Prospects for nuclear energy

18.3.1. Strategic considerations

Any strategy for the development (or abandonment) of nuclear energy cannot ignore public discussion about the future of this resource. Much debate about this topic is heard in the media, and is oriented along two axes:

1/ The concern for sustainable development, i.e. taking care of the limited resources our planet has to offer;

2/ The concern for protecting the environment and preserving the quality of our habitat.

To confine our discussion to strictly technical aspects, however, we shall give a few approximate values concerning the use of nuclear fuel in reactors, and supplement the discussion of plutonium recycling. We shall then conclude by presenting a few current avenues of research that address these concerns.

Firstly, however, it should be noted that the main advantage of nuclear energy in environmental terms is that it emits almost no greenhouse gases, most notably carbon dioxide.

18.3.2. Use of nuclear fuel

Table 18.7 recalls Table 12.1 and gives approximate values for the energy produced in a few nuclear reactor types and, for the purpose of comparison, shows the approximate initial content of fissile matter in the heavy nuclei constituting the fuel. Note that a one percent fission burn-up rate (number of fissions over the irradiation duration normalised to the initial number of heavy nuclei) approximately corresponds to a burn-up of 10 000 MWd/t.

Table 18.7. Approximate burn-up rates in a few reactor types.

Type	Initial abundance (%)	Burn-up rate in fission (%)	Burn-up (MWd/t)
UNGG	0.7	0.4	4000
CANDU	0.7	1	10 000
PWR	4	4	40 000
FNR	15	10	100 000

Note that only heavy water, with its favourable conversion factor (Table 12.3), makes it possible to burn more fissile nuclei than there are in the initial fuel. In other reactor types, the best that can be achieved is equality between these two terms thanks to the contribution of plutonium[20]. For UNGG reactors, this rate is not achieved because of insufficient reactivity. For FNRs, the burn-up rate could probably be higher if the mechanical strength of the cladding, which is subjected to very high fluence, would allow it. Considering, however, that about 8 kg of natural uranium is required to obtain 1 kg of 4% enriched uranium, it turns out that the performance of the PWR in terms of energy produced per kg of natural uranium is similar to the performance of the UNGG type. In this respect, the performance of BWR and RBMK reactors is also roughly the same.

Plutonium recycling in a thermal neutron type reactor, such as water reactors, increases the quantity of energy per kg of natural uranium by about 20%, as can be seen by looking back at the approximate values given for PWRs.

A further gain of about ten percent could be made in this reactor type by *recycling the reprocessed uranium*. This uranium still contains about 1% of uranium 235, which makes it about equivalent to natural uranium, taking into account the penalty for uranium 236. After re-enrichment[21], this uranium could be used to manufacture a new fuel[22].

To take the use of natural uranium in water reactors even further, conversion would have to be pursued as far as possible (in particular, by a sharp reduction of the moderation ratio), and irradiated (uranium and plutonium) fuel would have to be systematically reprocessed.

Studies carried out about twenty years ago[23] gave some hope of gaining a factor of about 4 compared to standard PWRs, but this idea was not developed to the production stage.

The only method that provides a really big jump is breeding; it can give a possible gain of more than 50 times compared to the energy obtained in water reactors per kg of natural uranium. Note that this gain can be achieved only with the use of axial and radial blankets and by systematically reprocessing the core assemblies and blankets. Because in practice the recycling of plutonium requires about fifteen years, multirecycling would only provide this gain factor of 50 after several centuries. But this timescale is compatible with the prospect, offered by breeder reactors, of an energy source available for several millennia. Note (§ 12.3.4) that the thorium 232-uranium 233 cycle could also breed, either in fast or thermal neutron reactors, obviously with the same timescales. It is estimated that thorium in mines is approximately four times more abundant than uranium from mines, but, unlike uranium (see § 1.9), there is little thorium in sea water.

18.3.3. Multirecycling of plutonium in reactors

Today's policymakers seem to have a vision in which uranium resources are abundant, and breeding no longer seems to be a concern.

---

[20] Approximately 40% of fissions for PWRs in four-cycle management.

[21] This could be done by ultracentrifuging in special installations to avoid polluting the enrichment plant by gaseous diffusion with minor uranium isotopes such as 232, 236, etc.

[22] Uranium recovered from reprocessing currently constitutes a 'strategic stockpile' in the event of a crisis in the uranium market.

[23] *Cf.* the Framatome "RCVS" concept: a convertible reactor (uranium or plutonium) using spectrum variation (increasing the moderation ratio by withdrawal of fertile rods, in order to gain, at the end of irradiation, the reactivity required to continue irradiating).

This means that plutonium can now be seen from a dual perspective: improved utilisation of the natural uranium energy resource, as well as reduction of the potential long-term radiotoxicity of the materials discharged from nuclear plant cores (an environmental concern rather than an energy-saving concern). Approximately 90% of this radiotoxicity comes from plutonium, which means that burning plutonium can be seen as the first action to apply in order to reduce the quantities of HALL (high activity long life)[24] waste to be managed in the long term. That is why, in just a few years, the status of plutonium has changed from that of an energetic material to that of waste to be eliminated as a priority.

We have seen that the use of MOX in water reactors reduces the mass of plutonium fairly little[25]. More radical solutions will have to be considered if the incineration of the plutonium is our objective. With that in mind, here are three paths that have been explored in France. In all three cases, a multirecycling liable to completely eliminate the plutonium produced in standard reactors is recommended.

a) First example: the MIX concept

As we have seen, the isotopic composition of the MOX irradiated in a PWR does not allow a second recycling under similar conditions. To make recycling (preferably unlimited) possible, the formation of higher isotopes must be minimised and "dirty" plutonium (recycled and therefore highly irradiated) must be mixed with "clean" plutonium (first generation).

The MIX (as in "mixture") concept involves distributing the plutonium among *all the assemblies* of a core, not just a few special assemblies. If the total amount of plutonium loaded in a core remains approximately the same, or is reduced if the operation involves a larger number of reactors than MOX operation, this means that the plutonium content of the mixed oxide pellets is largely insufficient to ensure criticality. The plutonium must therefore be mixed not with natural or depleted uranium, but *enriched uranium*.

Calculations show that, under these conditions, unlimited recycling is possible and the mass of plutonium can be gradually reduced.

This would, however, involve considerable cost: firstly, the extra manufacturing costs associated with handling plutonium would affect all of the assemblies rather than just some of them, and secondly, cycle studies have shown that the uranium 235 content of the MIX would be barely less than that of standard assemblies. In other words, under these conditions, the use of plutonium would only provide a slight saving of natural uranium and enrichment work. The use of MIX would, however, prevent the interface problems encountered with MOX.

b) Second example: the APA concept

The APA concept (advanced plutonium assembly, Figure 18.4), like the previous concept, was designed to be used in the whole core of an existing PWR. Once again we find the idea of placing enriched uranium and plutonium together, but this time not as a homogeneous

[24] Note that the concepts "high activity" and "long life" are antinomic by virtue of the law of radioactivity (activity is inversely proportional to the half-life); we can talk about "HALL" because nuclides of very different half-lives (isotopes of neptunium, plutonium, americium, curium, etc., as well as some long half-life fission products) are treated more or less together.

[25] This is logical in the context of the initial prospect of a transitory stage before the arrival of fast neutron reactors.

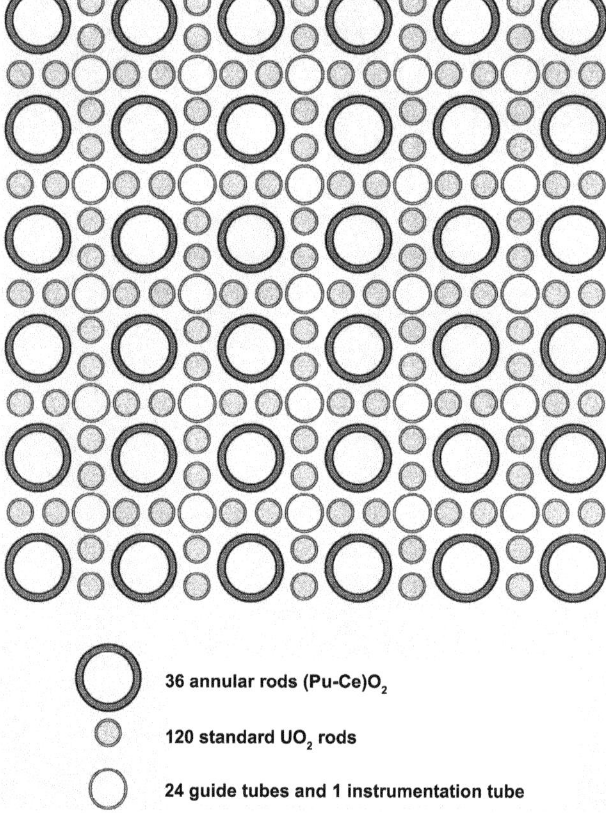

Figure 18.4. APA assembly for a pressurised water reactor.

mixture; they are placed in a heterogeneous structure created by alternating two types of fuel element within the assembly:

- The uranium rods would be identical to the $\mathbf{UO_2}$ rods of present-day assemblies;
- The plutonium elements, however, would be annular and in a cladding in order to be cooled internally and externally; the dimensions are chosen so that this type of element replaces four standard rods; for manufacturing reasons, sintered oxide pellets would be replaced by a ceramic such as a mixed plutonium and cerium oxide.

The overall design of the assembly keeps the central instrumentation tube and the 24 guide tubes, but the arrangement of the guide tubes cannot be the same as for standard assemblies. To adopt this concept in existing reactors, the control bundles would have to be replaced.

The tubular shape used for plutonium fuel elements gives a local increase in the moderation ratio, which allows better use of the plutonium than in the very under-moderated MOX situation (see Figure 9.9).

c) Third example: the CAPRA concept

The CAPRA concept (whose acronym stands for "increased plutonium burn-up in fast reactors" in French) was designed for fast neutron reactors. (This research programme was launched before the decision to shut down *Superphénix,* which was to be used to carry out irradiation testing to qualify the concept.) Whereas a standard fast neutron reactor design aims to achieve the highest possible regeneration gain[26], CAPRA does the opposite by trying to reduce conversion as much as possible, in order to burn up the plutonium and produce as little of it as possible.

The first thing to do is obviously to remove the fertile blankets and replace them with inert materials such as steel, or possibly even waste transmutation targets.

The second thing to do is to reduce internal conversion. The only radical solution is to eliminate uranium 238 also from the core. This leads to a major difficulty, however: disappearance of the Doppler effect on the capture resonances of this material, and therefore the reactor stabilising coefficient[27]. We might imagine replacing the uranium 238 with a non-fertile material having capture resonances, such as tungsten or technetium 99 (a fission product that could be partially incinerated at the same time), but in practice, the promoters of this concept sought a tradeoff between reducing the Doppler effect and reducing conversion. This compromise lies somewhere around a composition of 50% plutonium and 50% uranium 238, instead of the usual 15% and 85% in the standard concept.

18.3.4. Next-Generation reactor types

The MIX and APA concepts could be implemented in the relatively near future, because current reactors could accommodate them either directly or following fairly minor modifications to the control elements[28]. The same thing could have been said of the CAPRA concept if the FNR programme had not been discontinued.

Physicists and engineers are therefore working on more long-term prospects, i.e. on reactors that could replace the current PWRs when the time comes[29] or their successors.

It is obviously impossible to summarise in just a few lines all the ideas being considered by the entire community of specialists, so let us simply mention two projects of interest.

a) First example: HTR

We have already mentioned the potential interest of graphite and high-temperature reactors.

This reactor type has given rise to a few prototypes in the past, and more recently the concept has been taken up by a consortium of nuclear companies (most notably Framatome) in a project known as GT-MHR (*gas turbine - modular helium reactor*).

26 See approximate values in Table 12.3.

27 Note that the Doppler effect for plutonium is slightly positive because the effect due to fission dominates the effect due to capture.

28 For APA, it would be necessary to conduct an R&D programme concerning the manufacture of the plutonium fuel.

29 It seems likely that, if the current reactors are replaced by nuclear reactors when they reach the end of their life cycle (in France, this would be in 2010, 2020 or 2030, according to the lifetime that can be achieved for the plants in operation), they would be replaced with reactors using currently technology or a slightly improved version of it, such as the EPR (*Evolutionary Pressurised Reactor*) by Areva. In that case, the concepts discussed here would be applicable to the subsequent generation.

The fuel in these reactors comes in small particles with a diameter of a few hundreds of micrometres containing oxides of uranium, plutonium and/or thorium, coated in several layers of graphite to contain the fission products. These particles are compacted with graphite, also in fuel elements, which are cylindrical in this case. These cylinders are placed in graphite bricks which have cylindrical cavities — some to accommodate the fuel elements, and some to allow the circulation of helium coolant.

In the GT-MHR concept, the actual core has an annular shape, with internal and external graphite reflectors.

The system uses a direct cycle, i.e. without steam generators, because the helium is sent directly to the turbine.

The concept is intended for reactors with a relatively modest power not exceeding a few hundred MWe: a size suited to the export market for developing countries as well as for industrialised countries, since several reactors can be coupled on a single site (hence the word "modular" in the name).

b) Second example: fast neutron gas reactors

In the current climate of the late 1990s, it seemed unlikely that sodium-cooled fast neutron reactors could return to centre stage. But in the long term, breeder reactors must become the accepted choice, unless another energy source is discovered in the meantime.

Of course, HTRs can become breeders if the thorium-uranium 233 cycle is used, but their performance in this respect will never match that of FNRs.

Moreover, the high temperature reached thanks to helium, allowing improved thermodynamic efficiency and the prospect of energetic uses other than electricity production, such as chemical applications, is an important objective. That is why the "fast gas" (helium) concept is currently the subject of active research.

It could turn out to be the reactor that fulfils a technician's dream: producing energy and incinerating waste at the same time.

18.3.5. Hybrid reactors

As we have mentioned, all possible reactor types were more or less considered right from the beginning of nuclear energy research. Abandoned concepts have sometimes been revived a few decades later in a changed context. Thus, the idea of a hybrid reactor has existed for a very long time.

a) Fusion–fission hybrid concept

The first type of hybrid concept combines fusion with fission. We recall that the deuterium + tritium fusion reaction gives an alpha particle and a 14 MeV neutron. It might be a good idea to use this neutron.

The method most often considered is to use a tritium breeding blanket surrounding the fusion machine, in practice in the shape of a torus (Tokamak). This favours the multiplication of neutrons by (n,2n) reaction with the aim of producing, *via* neutron capture by the lithium, at least as much tritium as the fusion burns up.

Another possible method is to use a blanket made of heavy materials in which the neutrons produced by fusion provoke fission, which in turn causes further fissions thanks to the neutrons emitted[30]; this supplements the energy supply.

The balance between fusion power and fission power depends on the overall design of the machine.

Along the same lines, instead of favouring fission we could seek to favour the conversion of fertile matter into fissile matter[31].

These concepts will continue to seem rather futuristic until someone manages to demonstrate the industrial feasibility of fusion.

b) Spallation–fission hybrid concept

Spallation is a nuclear reaction that is very familiar to astrophysicists. It was rediscovered a few decades ago by nuclear reactor physicists. Note (§ 2.8.4) that, from an overall point of view, the firing of an externally accelerated proton with energy of the order of a GeV at a thick target made of heavy materials — tungsten, lead, bismuth, thorium, uranium, etc. — leads to a cascade of reactions followed by the emission of a considerable number of neutrons: about thirty per GeV in the proton[32].

The neutron source obtained in this way increases in intensity as the current of the proton accelerator increases, and can be of interest to reactor designers. In particular, C. Bowman introduced the idea of a *subcritical* (fission) reactor fed by this source, known as *ADS (accelerator-driven system),* in which the neutrons produced by spallation can provoke fission, leading to the emission of new neutrons that can in turn cause fission reactions according to a converging chain reaction (the system is designed to have a multiplication factor k less than one). (At equilibrium, we can use equation (18.1) from § 18.2.1:

$$\Phi \simeq \frac{\Phi_s}{1-k}, \qquad (18.3)$$

where Φ_s is the neutron flux without multiplication and Φ is the flux of all neutrons. This equation shows that neutrons from the source are *amplified* by a factor $1/(1-k)$ which increases as the system approaches criticality.)

The ADS concept was taken up again a few years ago by Carlo Rubbia and his team. Several designs have been studied. Below we present the overall configuration of the latest design (1995) in order to give an example of the hybrid–spallation concept.

Carlo Rubbia's energy amplifier

This project attempts to combine several key aspects that would not necessarily need to be implemented all at the same time. The main ones are as follows:

30 This blanket is obviously subcritical and the chain reaction taking place in it is limited.

31 During the Cold War this seemed an attractive concept, but it could equally well be used for the production of fuel for peaceful purposes.

32 For a uranium target; a little less for lead and tungsten.

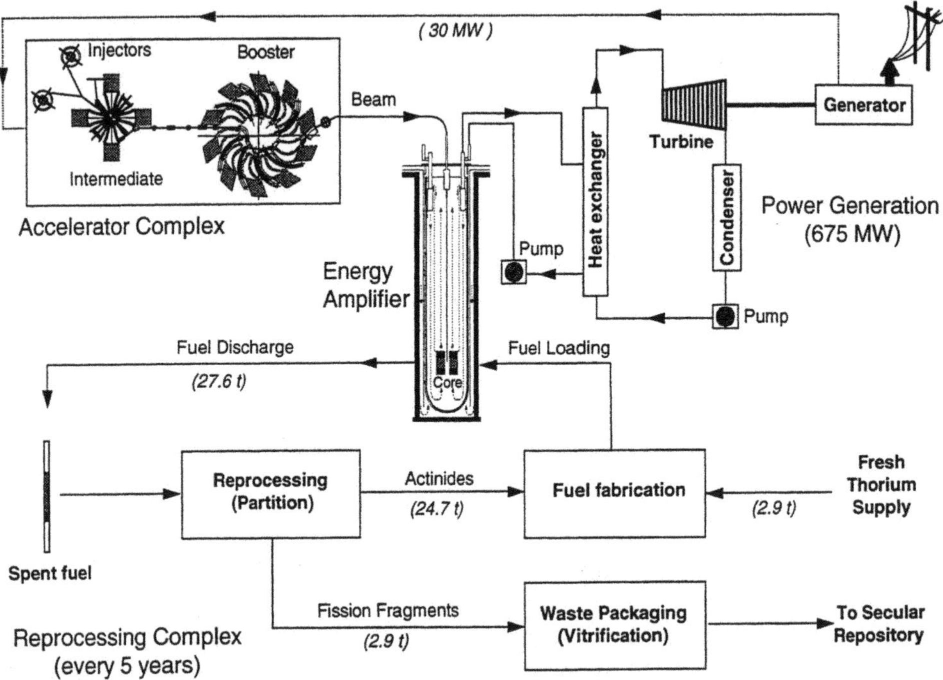

Figure 18.5. Flow diagram of the "Energy Amplifier" complex.

- Not only the reactor, but most of the *fuel cycle* — particularly the stages that involve a risk of theft of fissile materials[33] — is together on one site as shown in Figure 18.5, which is taken from publications by the authors[34];
- The reactor *core* is a "fast neutron"-type core and is *subcritical*. The essential factor in this choice is safety-related; this type of core does not need to be controlled using control rods, and the subcriticality eliminates the risk of a criticality accident because the chain reaction can be halted immediately by shutting off the proton beam. (Note, however, that the problem of residual power remains, as in critical reactors.)
- The shape of this core is annular, and the core is fed from inside by neutrons from the *source* placed on the axis of the system; (see Figure 18.6).
- This source is the result of *spallation reactions* induced on *lead* by protons brought from the accelerator *via* the central tube;
- The *fuel* is a mixture of a few percent *uranium 233* with *thorium 232*; Since the core is almost self-generating, a fairly long irradiation (of the order of 100 000 MWd/t) could be envisaged;

[33] In the thorium 232 - uranium 233 cycle, the uranium can be separated from the thorium by chemical means; because it contains the 233 isotope almost exclusively, it could be an excellent basic material for a weapon.

[34] Carlo Rubbia *et al.*, *Conceptual Design of a Fast Neutron Operated High Power Energy Amplifier*, CERN/AT/95-44 (ET).

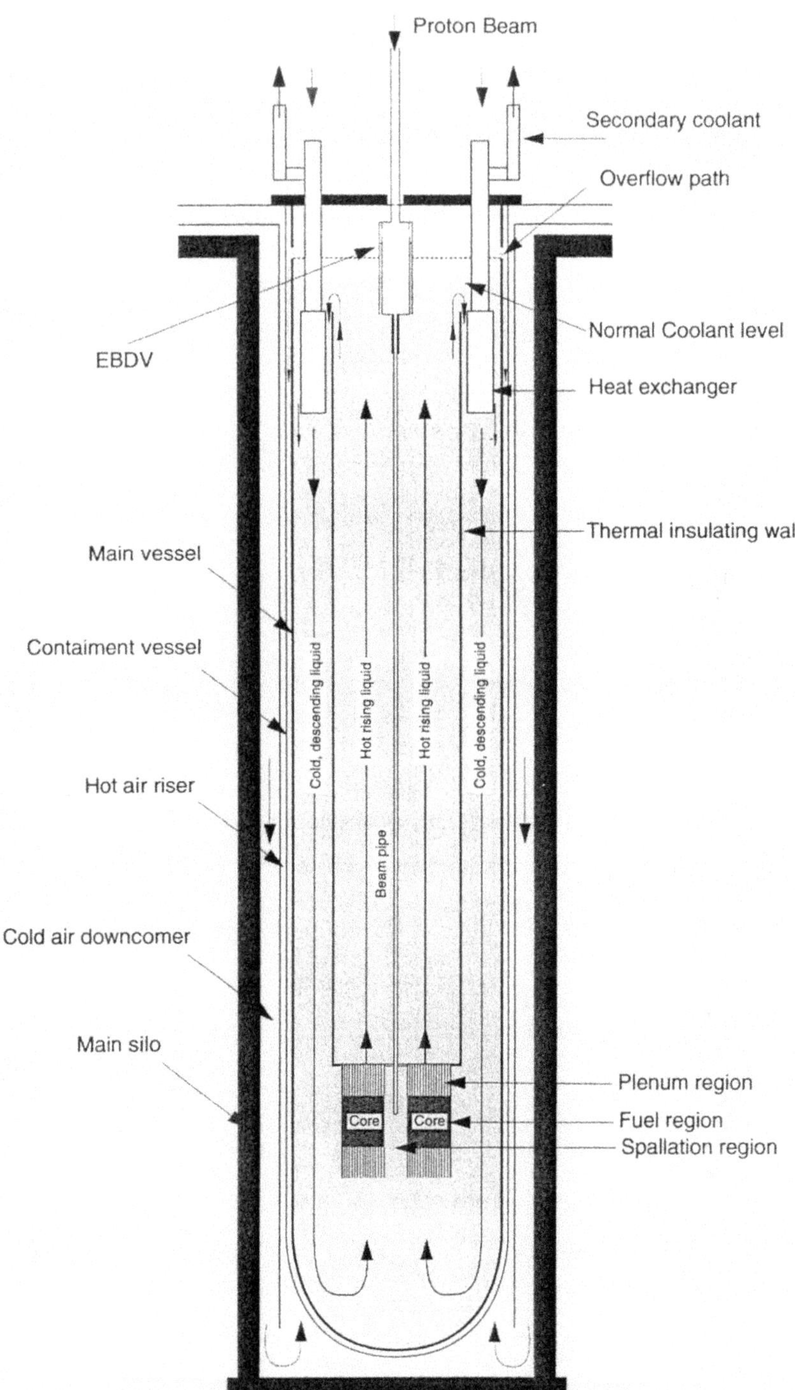

Figure 18.6. Main part of the "energy amplifier". (EBDV: emergency beam dump volume)

- The essential argument in favour of this cycle, however, is the lower production of minor actinides than in the uranium 238-plutonium 239 cycle, since the basic fuel material (thorium 232) is composed of nuclei with six fewer nucleons than in the usual case (uranium 238)[35];
- The irradiated fuel is *reprocessed on-site,* and the energetic material is recycled[36];
- The *lead* is not only a spallation target, but also a *coolant.* The very tall stack planned (over 30 metres) should allow cooling by *natural circulation*, which can be another safety argument;
- The proton *accelerator* could be a cyclotron (as in Figure 18.5) or a linear accelerator. The energies required for spallation (of the order of a GeV) are easy to obtain, but the currents that would be required (several tens of milliamperes) are not. In addition to the technological leap with respect to the current that technicians currently know how to produce, the supply of a beam that would be perfectly stable for long periods would present a tremendous technical challenge.

Developing formulae for the ADS

Let:

- n be the number of neutrons produced by spallation reactions per proton hitting the target;
- ω be the probability of a neutron placed in the system inducing fission there[37];
- ν be the average number of neutrons emitted by a fission event;
- $k = \omega\nu$ be the multiplication factor (§ 1.4).

For a proton injected into the system, the number of fissions induced by spallation neutrons and their descendants is:

$$f = n\omega + n\omega\nu\omega + n\omega\nu\omega\nu\omega + \cdots = \frac{n\omega}{1 - \nu\omega} = \frac{n}{\nu}\frac{k}{1 - k}. \qquad (18.4)$$

Let:

- E_p be the energy of a proton reaching the target;
- E_f be the energy produced by a fission;
- $\eta_a = E_p/E_c$ be the efficiency of the accelerator, i.e. the ratio of the energy imparted to the proton to the (electrical) energy consumed to accelerate it;

---

35 In another publication, the authors show that this reactor could start up with plutonium (on a thorium substrate) instead of uranium 233 and thus contribute to the incineration of the plutonium.

36 The principle of reprocessing of this fuel is known, but the technology would need to be developed. It is known that, even though this cycle produces fewer minor actinides, it involves other nuclides that lead to tricky radiation protection problems, in particular thallium 208 descending from uranium 232.

37 Strictly speaking, the probability for a neutron produced by spallation should be distinguished from the probability for a neutron produced by fission.

– η_e be the efficiency of the conversion of the heat produced in the core to electricity.

The thermal energy obtained per injected proton is the sum of the energy E_p of this proton (which will be dissipated in the target) and the energy produced by fissions:

$$E_{th} = E_p + fE_f,$$

and the electrical energy that could be obtained from this thermal energy will be:

$$E_e = \eta_e E_{th} = \eta_e(E_p + fE_f).$$

By comparing this to the electrical energy used by the accelerator, we see that the machine consumes an amount of electricity representing the following proportion:

$$c = \frac{E_c}{E_e} = \frac{E_p/\eta_a}{\eta_e(Ep + fE_f)} = \frac{1}{\eta_a\eta_e\left(\frac{n}{\nu}\frac{k}{1-k}\frac{E_f}{E_p} + 1\right)}, \tag{18.5}$$

of the electricity it produces. The inverse of this ratio is the *gain g* of the energy amplifier.

We have seen that, for a proton, we obtain n spallation neutrons, f fissions, and therefore νf fission neutrons. If we divide these numbers by f to normalise to *one fission event*: for this fission, we obtain:

$$\mu = \frac{n}{f} = \frac{1 - \nu\omega}{\omega} = \frac{\nu(1 - k)}{k}, \tag{18.6}$$

spallation neutrons and ν fission neutrons. It could be said that the presence of the external neutron source made necessary by the subcriticality *increases* by μ the average number ν of neutrons emitted by fission. Of these neutrons, one will induce the next fission; a certain number, which we shall call α, will be lost in the inevitable sterile captures in internal and external structures; the remainder can be considered to be *useful neutrons*, i.e. neutrons liable to be converted, for example, from fertile matter to fissile matter, or to incinerate radioactive waste. This remainder increases from $\nu - 1 - \alpha$ to $\nu - 1 - \alpha + \mu$ when the system goes from critical to subcritical. Independently of any safety-related arguments, this increase in number of useful neutrons is the main interest of *ADS*.

Here are some approximate values we can count on obtaining in this type of *ADS*:

– $n = 30$ for $E_p = 1$ GeV;

– $\nu = 2.5$; $E_f = 200$ MeV;

– $\eta_a = 1/2$; $\eta_e = 1/3$;

– $\alpha = 0.7$.

According to the level of subcriticality chosen, the values obtained for the main parameters are collected in Table 18.8; with regard to the number of useful neutrons available, the values of μ should be compared to $\nu - 1 - \alpha = 0.8$.

At present, this type of machine is of interest mainly for the purpose of incinerating nuclear waste — the subject of the next section — rather than for energy production. In this context, a fairly low multiplication factor will be sought.

Table 18.8. Variation of the main parameters characterising an ADS according to the level of subcriticality adopted (the line $k = 0.676$ corresponds to the energetic autarchy).

k	f	c	g	μ
1	∞	0	∞	0
0.95	228	0.129	7.77	0.132
0.9	108	0.265	3.77	0.278
0.8	48	0.566	1.77	0.625
0.7	28	0.909	1.1	1.071
0.676	25	1	1	1.2
0.6	18	1.304	0.767	1.667
0.5	12	1.765	0.566	2.5

18.3.6. The problem of waste management

Note that waste is sorted according to half-life, energy, and decay type, as well as the physical and chemical properties that determine how likely they are to migrate in the environment in the event of a spill and how they would be metabolised in the event of being inhaled or swallowed.

Waste with a low or medium level of activity and a half-life not exceeding 30 years[38] is placed in surface storage (in France, it used to be stored at La Hague, but is now stored at Soulaines-Dhuys); after three centuries (at least ten half-lives), during which time monitoring can realistically be ensured, the radioactivity will be negligible, and the site can be declared fit for other purposes.

Waste with a high level of activity and/or long half-life creates a different problem because it is a longer-term problem. Such waste is covered by French Law No. 91-1381 dated 30 December 1991 initiating a fifteen-year research programme before the subject comes up again for discussion in Parliament[39]. This programme explores three avenues: deep storage, long-term storage, and separation-transmutation. Only this last aspect is directly related to neutron physics.

This waste, essentially composed of fission products and minor actinides, is not currently separated and is stored in silos (in France, at Marcoule and La Hague). The third approach studied by this research programme is *separation* in view of special processing suited to each specific case (this theme with regard to chemists will not be dealt with here); for certain products, *transmutation* by neutron flux could be considered. The term *transmutation* harks back to the ancient dream of alchemists: to transform vulgar metals into noble metals. Nuclear physicist adopted this term to describe transformations of the atomic nucleus, particularly those inducing a change of element. In the context of radioactive waste management, only certain nuclear reactions are of interest, and here the term is used to denote any transformation, by one or more nuclear reactions, of a long half-life radionuclide to a stable nuclide or short half-life radionuclide[40] that decays to a stable nucleus. In this context, we also speak of the "incineration" of nuclear waste.

[38] Of the fission products obtained in significant quantities, the one with the longest half-life (30 years) is caesium 137, which is why this figure was chosen as a reference.

[39] In 2006, this matter was again discussed in Parliament, resulting in a law (dated 28 June 2006) extending the research programme.

[40] Short in the context of nuclear waste management, i.e. not exceeding 30 years.

Clearly, the easiest nuclear reactions to carry out on a large scale are those induced by neutrons. The main reactions of interest for transmutation are neutron capture, (n,2n) reactions, and fission. Fission in particular transforms an actinide, generally an alpha radioactive one with a long half-life, into beta radioactive fission products, almost all with short half-lives. For these reactions, we might consider using the available neutrons from a critical reactor or, even better, from a subcritical reactor. Whether critical or subcritical, the number of neutrons available is greater in a fast neutron reactor than a thermal neutron reactor. On the other hand, the cross-sections are smaller and the levels of incineration flux must therefore be higher.

Irradiation can be performed either outside the core, in the reflectors or blankets, or in the core. If it occurs in the core, the matter to be incinerated can either be distributed in the fuel at the rate of a few percent by mass ("homogeneous" recycling), or it can be placed in dedicated targets ("heterogeneous" recycling, which is obviously the only possible solution if irradiation occurs outside the core).

The most appropriate spectrum — fast or thermal — and the recycling mode must be chosen for each radionuclide to be incinerated. There does not seem to be any single mode that is preferable in general; for example, the incineration of neptunium 237 tends to be envisaged in homogeneous mode, but the incineration of americium could be performed in homogeneous or heterogeneous mode. In both cases, we can consider using FNRs or PWRs adapted for incineration.

In reactors primarily intended for energy production, it is important to take into account the effect of these products on the performance (cycle duration in particular) and the characteristics (reactivity coefficients in particular) of the reactors used. In dedicated incineration reactors, the transmuted quantities should be optimised, as energy production is only a secondary objective.

The anticipated incineration rates only become significant, i.e. at least 50%, after significant periods of irradiation, i.e. at least several years. This means that multirecycling (homogeneous case) or irradiation over several cycles (heterogeneous case) must generally be envisaged.

Clearly, a complete balance must be performed on a case-by-case basis and must take into account not only incineration, but also the production of new quantities, either *via* reactions related to plant operation, or from other products that we wish to incinerate.

This balance is often performed in terms of *radiotoxicity*: this is defined as the noxiousness of the product concerned, taking into account the characteristics of the radiation, and metabolisation in the case of ingestion by a person. This radiotoxicity is obviously a function of time because of transformations by radioactive decay. There are, therefore, many constraints on the incineration of nuclear waste.

Under these circumstances, which products are liable to be transmuted?

In a first category we can include the *energetic matter* that can be recovered by the reprocessing of irradiated fuels: uranium, thorium, and plutonium; not because they are waste but because they are governed by the same transmutation physics. If the irradiated fuel is not reprocessed, but stored as is, these materials are by far the main component of radiotoxicity. For example, for the irradiated fuel of PWRs, approximately 90% of the radiotoxicity comes from plutonium.

Table 18.9. Inventory of materials, in kg/year, at inlet and outlet of a 900 MWe pressurised water reactor.

Nuclide	Period (years)	Uranium fuel		MOX fuel	
		Inlet	Outlet	Inlet	Outlet
^{235}U	$7.08 \cdot 10^8$	751	221	11.1	5.8
^{236}U	$2.34 \cdot 10^7$		88		1.2
^{238}U	$4.47 \cdot 10^9$	20 734	20 204	4478	4261
^{238}Pu	88		3.3	11.3	12.4
^{239}Pu	24 100		123.1	209	105.2
^{240}Pu	6570		47.5	98.5	87.4
^{241}Pu	14		25.4	44.5	40.7
^{242}Pu	370 000		10.5	31.0	35.8
^{237}Np	2 140 000		8.8		0.8
^{241}Am	432		4.4		14.4
^{243}Am	7380		2.2		9.0
^{244}Cm	18		0.5		4.0
^{245}Cm	8500		0.1		0.5
^{93}Zr	1 500 000		15.5		3.0
^{99}Tc	210 000		17.7		4.8
^{107}Pd	6 500 000		4.4		3.2
^{126}Sn	100 000		0.4		0.2
^{129}I	15 700 000		3.9		1.3
^{135}Cs	2 000 000		7.7		4.8

Reducing the radiotoxicity in the short-term and long-term is therefore another argument in addition to the reprocessing and recycling of these materials. Apart from plutonium 241, a beta emitter with a fairly short half-life (14.4 years), and plutonium 238, an alpha emitter with a half-life of 87.7 years, these nuclides are long-half-life alpha emitters.

In a second category we can place all other actinides without any particular interest in energetic terms which are, for this reason and because the masses are lower, called *"minor actinides"* (MA for short). The main ones are neptunium 237, the americiums (241 and 243) and the curiums (244 and 245) (see evolution chain, Figure 12.2, § 12.1.1). Except for curium 244, these products are all alpha emitters with relatively long or very long half-lives.

Finally, in the third category, we can place the rare long-half-life fission products (LLFP or "long-lived fission products")[41], beta emitters like all radioactive fission products. The main ones, in view of the masses and half-lives, are zirconium 93, technetium 99, palladium 107, tin 126, iodine 129, and caesium 135.

[41] The other fission products can cause problems because of their high level of activity, but it is not of interest to consider incinerating them, because that would take longer than the spontaneous radioactive decay.

Table 18.9[42] gives a few values for a PWR-900 operating with an average load factor of 70%: the "Uranium fuel" column covers a complete core and fuel irradiation of 33 000 MWd/t; in the "MOX fuel" column, the values concern the part of the core loaded with MOX, i.e. 30%, and this fuel is irradiated at 43 500 MWd/t. In both cases, the masses of the fission products are calculated after three years of cooling.

The transmutation of long-lived fission products is possible only if there are sufficient neutron capture reactions, which there are with the nuclides mentioned above. With regard to actinides, in practice it is necessary to proceed until fission of the radionuclide to be incinerated or one of its descendants.

Note that, when discussing the question of waste transmutation, we must also consider any other isotopes of the element concerned in order to avoid creating new radionuclides that would cause problems. It does not seem feasible to perform isotope separation before irradiation.

To carry out a relevant assessment of the benefits of transmutation, two essential parameters can be introduced[43]:

- *Neutron consumption, D,* for an actinide until fission and for a fission product until a stable nucleus is obtained: this is the average number of neutrons that will be required to incinerate the nucleus concerned, calculated with all possible successive reactions and the branching ratios between them taken into account.

  This "consumption" can be negative (i.e. it is actually production), since the values per reaction are 1 for a capture, 0 for radioactive decay, −1 for an (n,2n) reaction, and $1 - \nu$ for a fission. The branching ratios and therefore the numerical values of average consumption depend on the irradiation conditions (spectrum and neutron flux level); Table 18.10 gives a few examples for actinides under typical irradiation conditions in a FNR and a PWR.

  Note that the "consumptions" are all negative for the FNR case. In this type of spectrum, all heavy nuclei are more or less fissile;

- The *transmutation half-life*, T_{transm}: this is the required irradiation time for half of the irradiated nuclei to have captured a neutron (assuming that one and only one capture is necessary to eliminate each nucleus from the nuclear waste concerned). This half-life obviously also depends on the irradiation conditions.

  Table 18.11 gives a few examples for LLFPs: note that these nuclides can be considered as "transmutable" in that the transmutation half-life is far shorter than the radioactive half-life[44].

In conclusion, note the following points:

- The cross-sections involved in these processes are often not well known: to carry out transmutation would require a demanding programme of nuclear measurements and qualification;

42 Source: Stéphanie Sala thesis, Université de Provence, 1995, cited by Massimo Salvatorès, *La transmutation, École nationale supérieure de chimie*, 2000.

43 For further details, refer to M. Salvatorès; the numerical values given in the following two tables (Tables 18.10 and 18.11) are taken from this document.

44 If not, then it is obviously simpler to allow the time for events to take their course. (For tin 126, the interest of a transmutation requiring several millennia is obviously debatable.)

Table 18.10. Neutron consumption per fission for the main actinides in FNR spectrum and PWR spectrum.

Nuclide	FNR conditions	PWR conditions
^{232}Th	−0.38	−0.24
^{238}U	−0.62	+0.07
^{238}Pu	−1.36	+0.17
^{239}Pu	−1.46	−0.67
^{240}Pu	−0.96	+0.44
^{241}Pu	−1.24	−0.56
^{242}Pu	−0.44	+1.76
^{237}Np	−0.59	+1.12
^{241}Am	−0.62	+1.12
^{243}Am	−0.60	+0.82
^{244}Cm	−1.39	−0.15
^{245}Cm	−2.51	−1.48

Table 18.11. Transmutation period examples (years).

Nuclide	FNR conditions	PWR conditions	Radioactive half-life
^{93}Zr	730	790	1 500 000
^{99}Tc	110	51	210 000
^{107}Pd	44	730	6 500 000
^{126}Sn	4400	4400	100 000
^{129}I	160	51	15 700 000
^{135}Cs	310	170	2 000 000

– the transmutation of waste will never be complete because the decay of the mass under irradiation is approximately exponential; it is hard to imagine reducing the masses, depending on the specific example, beyond about 90% to 99% of the initial mass. This means that residual quantities will always need to be stored. Incineration simplifies but does not eliminate the problem of permanent storage of products with a long half-life;

– A fast spectrum is often preferable to a thermal spectrum for incineration (smaller cross-sections but more available neutrons);

– The number of available neutrons can be increased in a system dedicated to incineration (particularly a hybrid spallation-fission system) compared to a system whose priority is to produce energy. The incineration of actinides, however, always ends in a fission, which means that energetic autarchy is possible even in a dedicated system.

18.3.7. Nuclear energy and sustainable development

Even though the management of nuclear waste gives rise to a few tricky problems, then, it does not lead to any dead ends. Note also that the problem is greatly simplified by the fact that the masses to be managed are small compared to the mass of waste generated by classic forms of energy.

Moreover, the energetic potential of nuclear material, if properly exploited, is considerable. Nuclear energy appears to have all the advantages that would allow sustainable development and preserve natural resources and the environment.

With the known resources of uranium and thorium and with the use of breeder reactors, and if energy requirements continue in the present trend, fission energy can meet our needs for millennia. Thus, even if these resources are poorly used with today's technologies, it is important for them to be preserved for the generations to come.

In the longer term, perhaps fusion technology will be mastered. In a first stage, to gain at least an order of magnitude compared to fission, the D + T reaction will be implemented; i.e. the use of deuterium (abundant) and lithium (more limited reserves). In the even longer term, we can count on the D + D and then the H + H reactions being mastered in turn. This will open up almost unlimited possibilities.

Exercises

The problems of reactor core design and management bring to a natural conclusion this course on neutron physics, since they are the ultimate objective of this work. Here is also a convenient point to recall the aspects of reactor core design and management that have been covered previously.

- Optimisation of moderation (exercise **9.4**).
- Composition of the fissile material (exercise **9.3**).
- Problem of using boron in solution (exercise **9.5**).
- Pile with a uniform power (exercise **6.8**).
- Pile with two multiplying zones (exercise **6.5**).
- Efficiency of an absorber (exercises **6.10** and **10.4**).
- Sensitivity of a core to power gradients (exercise **6.11**).
- Spatial instability (exercises **11.7** and **11.8**).

The exercises provided in this final chapter are either complementary to the preceding ones, or involve other aspects of the core design and management.

Exercise 18.1: the diameter of the fuel pellets

For any given type of reactor, the three main parameters to determine are the size of the fuel elements, the moderation ratio, and the composition of the fissile material. These last two choices are essentially linked to neutron physics criteria. However, the fuel element size is mainly controlled by thermal criteria. This point will next be examined in the following example for a pressurized water reactor.

a) Using the classical four-factor theory, and neglecting the possible effect on the fast fission factor, examine by which term(s) and in which direction(s) the infinite multiplication factor k_∞ may change when the fuel pellet diameter is changed, while keeping the moderation ratio and the fissile material composition unchanged.

b) For a reactor with a constant specific power by unit of core volume, and with a constant temperature T_s at the surface of the pellets, how does the temperature T_c at the centre of the pellets vary as a function of their diameter?

c) How does the manufacturing cost of the fuel vary for this reactor?

d) Consequently, in what way do the above factors affect the size of fuel pellets?

Exercise 18.2: effective leakage

For a bare, homogeneous core, the "leakage" is defined as the ratio between the number of neutrons escaping from the core (per unit of time) and the number of neutrons absorbed in the core (per unit of time). Naturally, the same definition can be used even for any heterogeneous core. However, to avoid ambiguity, the term "effective leakage" will be used instead. Two very simple examples follow.

a) *Axial case:* apply one-group diffusion theory to compare the leakage from a homogeneous critical core with the effective leakage from a radially infinite core (i.e. a slab of thickness H) of equal size comprising two zones differing only by $\nu\Sigma_f$ where

- in zone 1 for $-a < z < +a$, $k_\infty = 1$;
- in zone 2 for $-H/2 < z < -a$, and for $+a < z < +H/2$, $k_\infty > 1$.

Neglect the extrapolation distances.

b) *Radial case:* repeat part **(a)** comparing now a homogeneous critical core with an axially infinite core of equal size comprising two zones differing only by $\nu\Sigma_f$ where

- in zone 1 for $\rho < a$, $k_\infty = 1$;
- in zone 2 for $a < \rho < R$, $k_\infty > 1$.

Again, neglect the extrapolation distances.

Exercise 18.3: core partition and fissile content selection

Consider a 900 MWe-UO_2 PWR core. Calculations show that, when operated in three batches, the fuel yields about 33.5 GWd/t when an initial uranium-235 concentration by mass of 3.25% is used, or 44.2 GWd/t with an initial uranium-235 concentration by mass of 4.20%. Assuming

- that the average energy yield varies linearly with the initial uranium-235 concentration t by mass,
- and that this yield varies with the number of batches n according to the formula

$$E_n = \frac{2n}{n+1} E_1,$$

examine how the natural uranium consumption and the mass of fuel to reprocess per unit of energy produced vary with t and n. The concentration of uranium-235 by mass in natural uranium is 0.71%, and is 0.25% in depleted fuel. All the irradiated fuel is reprocessed.

Exercise 18.4: cycle extension

a) The end of a cycle is defined normally as the instant where the boron concentration reaches zero at the nominal power level. A cycle extension is obtained by decreasing the mean moderator temperature, and hence the power level. Explain how the decrease the moderator temperature permits this extension. Assume that the thermal power does not vary, and that the electric output is linked to the mean moderator temperature by the ideal Carnot efficiency. It can also be assumed that the reactivity varies linearly with the mean moderator temperature, and with the mean fuel consumption. *Data:* nominal temperature of the hot source = 600 K; temperature of the cold source = 400 K; reactivity parameters are -50 pcm K^{-1} (moderator), and -1000 pcm/month at the nominal thermal power (fuel consumption rate).

b) Under the present three-batch scheme, following a cycle extension, how do the lengths of the subsequent cycles vary on the basis that they are neither shortened nor extended.

Exercise 18.5: evolution of a burnable poison

a) A poison of infinite capture cross-section uniformly occupies a sphere of radius a. Neglecting neutron scattering in this sphere, examine how the concentration of the poison decays when it is irradiated by a neutron flux possessing spherical symmetry.

b) Use the methodolgy in exercise **6.10** to examine the variation of the efficiency of the poison with time to first order, when the sphere is located at the centre of a spherical reactor. Assume that the external medium does not vary, that the power of the reactor is constant, and that the zone burned in the poison sphere is replaced with the external medium.

Numerical data

- initial radius of the poison = 5 mm,
- atomic concentration of the absorber in the poison sphere 0.6×10^{27} m^{-3},
- neutron current per unit area entering the poisoned zone = 10^{17} m^{-2} s^{-1}.

How long does it take for the poison to disappear?

Exercise 18.6: uranium-plutonium interface

a) This exercise examines the variation of the flux near a planar interface between two infinite, homogeneous media. The first medium is a mixture of a matrix material and uranium-235 with a concentration N_1. The second medium is a mixture of the same matrix material and plutonium with a concentration N_2. The thermal flux is calculated by the diffusion equation assuming that the diffusion coefficient D is the same in both media, and that the source term Q — slowed-down, thermal neutrons — is equal and uniform in both media. Let Σ_c be the macroscopic capture cross-section of the matrix, and, following the usual notations, $\sigma_{a,i}$ and η_i are the characteristics of the fissile material, uranium-235 and plutonium. Find the flux distribution Φ and of the quantity $P = \eta\sigma_a N\Phi$ — assumed to be proportional to the power — in each part. In particular, give the asymptotic and interface values.

Material parameters

$$N_1\sigma_{a,1} = \frac{5}{3}\Sigma_c, \qquad N_2\sigma_{a,2} = \frac{13}{3}\Sigma_c,$$

$$\eta_1 = 2, \qquad \eta_2 = \frac{20}{13}.$$

b) Next, replace the constant concentrations of the fissile materials in the matrix with variable concentrations $N_1(x)$ and $N_2(x)$ that depend on the distance x to the interface. What form must these functions take in order to produce a critical situation with a uniform power? Assume that $\varepsilon p = 0.8$ in both regions.

Solutions

Exercise 18.1: the diameter of the fuel pellets

a) Three effects can be observed when the radius a of the pellets increases without changing the moderation ratio, owing to a homothetic transformation of the lattice's dimensions.

1. There is a decrease of the surface/volume ratio, and consequently of the equivalent dilution cross-section, leading to more selfshielding, a decrease of the effective integral, an increase of the factors p and k_∞.

2. There is a decrease of the Dancoff factor — the neutrons have a greater thickness of moderator to traverse in order to reach the neighbouring fuel pins without collision. This effect is in the opposite sense from the previous one, and partially compensates it.

3. There is a greater heterogeneity which increases the disavantage factor, and decreases f and k_∞.

The first effect is the most important one; therefore, k_∞ improves when the dimensions increase.

b) When the thermal conductivity k and the heat source are constant, the Fourier equation,

$$k\,\Delta T + Q = 0,$$

gives

$$T(\rho) = T(a) + \frac{Q(a^2 - \rho^2)}{4k}.$$

Therefore,

$$T_c = T_s + \frac{Qa^2}{4k},$$

where s and c mean surface and centre, respectively.

Within the present hypotheses, Q (proportional to the power per unit of volume of core) and T_s do not change. Hence, the temperature gradient from the surface to the pellet centre is proportional to the square of the pellet radius.

c) The unit cost of a pellet varies little with its size; therefore, the cost per unit of mass is inversely proportional to the square of the radius.

d) Neutron physics and the cost analysis both provide incentives to increase the size of the pellets. However, the temperature at the centre of the pellets must not exceed a limit value, i.e. the fusion temperature of the oxide minus a security margin. Thus, the size used corresponds to this limit.

Exercise 18.2: effective leakage

a) *Axial study:* consider the upper half-pile — the lower half-pile is symmetrical. The flux is

$$\Phi(z) = A \qquad (0 < z < a);$$

$$\Phi(z) = A \sin\left(\frac{H/2 - z}{H - 2a}\pi\right) \qquad \left(a < z < \frac{H}{2}\right).$$

The leakage rate is

$$-D\Phi'\left(\frac{H}{2}\right) = DA\frac{\pi}{H - 2a},$$

and the absorption rate is

$$\Sigma_a \int_0^{H/2} \Phi(z)\, dz = \Sigma_a A \left(a + \frac{H-2a}{\pi} \right).$$

Let $M^2 B_{\text{eff}}^2$ be the ratio between both these terms; then $M^2 = D/\Sigma_a$, and consequently

$$B_{\text{eff}}^2 = \frac{\pi^2}{(H-2a)(H-2a+\pi a)}.$$

This can be compared with the geometrical buckling $B_g^2 = \pi^2/H^2$. Using $a = \lambda H/2$, gives

$$\frac{B_{\text{eff}}^2}{B_g^2} = \frac{1}{(1-\lambda)(1-\lambda+\pi\lambda/2)}.$$

In particular, $B_{\text{eff}}^2 = B_g^2$ when λ goes to zero.

b) *Radial study:* the flux is

$$\Phi(\rho) = A \qquad (0 < \rho < a);$$

$$\Phi(\rho) = A\,\text{J}_0(\beta\rho) + B\,\text{Y}_0(\beta\rho) \qquad (a < \rho < R),$$

where the constants B and β are determined by the boundary conditions $\Phi'(a) = 0$ and $\Phi(R) = 0$. The parameter A is an arbitrary normalisation constant. Taking into account the identity

$$\text{J}_1(u)\,\text{Y}_0(u) - \text{J}_0(u)\,\text{Y}_1(u) = \frac{2}{\pi u},$$

this gives

$$B_{\text{eff}}^2 = \frac{\beta^2}{1 - (\beta a/2)\,\text{Y}_0(\beta R)/\text{Y}_1(\beta R)}.$$

When a goes to zero, β goes to j/R, and the denominator goes to one. This gives the usual formula for the geometrical radial buckling factor.

Exercise 18.3: core partition and fissile content selection

- The formula for the output thermal energy E (GWd/t), according to the uranium-235 content t (%), and the number of batches n is

$$E = \frac{4n}{3(n+1)}(11.263\,t - 3.105).$$

- The formula for the enrichment process is

$$\frac{M_1}{M_0} = \frac{t_0 - t_2}{t_1 - t_2},$$

where the label 0 denotes natural uranium, 1 is enriched uranium, and 2 is depleted uranium.

Therefore,

- the mass M_1 to load into the core, and to reprocess after irradiation, in t/GWd is

$$\frac{1}{M_1} = E = \frac{4n}{3(n+1)}\,(11.263\,t_1 - 3.105)\,;$$

- and the mass M_0 of natural uranium needed to produce this enriched fuel, in t/GWd is

$$\frac{1}{M_0} = \frac{M_1/M_0}{M_1} = \frac{4n}{3(n+1)}\,(11.263\,t_1 - 3.105)\,\frac{t_0 - t_2}{t_1 - t_2}.$$

When n is increased, there is a gain in both these masses; however, the reactor must be shut down more frequently, thereby reducing its availability.

When t_1 is increased, the mass to reprocess is smaller, and the consumption of natural uranium is increased slightly.

Below, are some numerical values of the masses M_0 and M_1, respectively, in t/GWd.

	$n = 3$	$n = 4$	$n = 5$
$t = 3.25$	0.19468	0.18251	0.17521
	0.02985	0.02799	0.02687
$t = 4.20$	0.19428	0.18213	0.17485
	0.02262	0.02121	0.02036

Exercise 18.4: cycle extension

a) Let T_n be the nominal temperature of the moderator, $T(t)$ its time dependence over the extension period, θ the temperature of the cold source and $\tau(t)$ the additional fuel consumption owing to the extension. Using the reactivity coefficients α and β — which are both negative — associated with the moderator temperature and to the mass fuel consumption, the conservation of the criticality during the extension gives the first equation:

$$\Delta\rho(t) = \alpha\,\left[\,T(t) - T_n\,\right] + \beta\,\tau(t) = 0.$$

Secondly, when it is assumed that the reactor output continues at the nominal thermal power, the electrical power varies as the Carnot thermodynamic efficiency, i.e.

$$r = 1 - \frac{\theta}{T}.$$

Therefore,

$$p(t) = \frac{P(t)}{P_n} = \frac{1 - \theta/T(t)}{1 - \theta/T_n}.$$

Thirdly, the variation in fuel consumption can be linked to the constant thermal power:

$$d\tau = \frac{P_{\text{th}}\,dt}{m}, \qquad \tau = \frac{P_{\text{th}}\,t}{m},$$

where m is the mass of the fuel in the core.

Then, the expression for the electrical power as a function of the fuel consumption can be deduced, or of the additional elapsed time. To find P as a function of t, replace τ with t, and express β per unit af time instead of unit of fuel used:

$$p(t) = \frac{P(t)}{P_n} = \frac{T_n}{T_n - \theta}\,\frac{T_n - \theta - \beta\tau/\alpha}{T_n - \beta\tau/\alpha}.$$

Numerical example

$$p(t) = 3\,\frac{10 - t}{30 - t},$$

where the time t is expressed in months. For example,

- $p = 0.931$ after one month,
- $p = 0.857$ after two months,
- $p = 0.778$ after three months.

Normally, the maximum extension period is six weeks, after which the electrical power output falls to about 90% of its nominal value.

b) Let L be the normal length of a cycle expressed, for instance, in GWd/t. Now assume that, at the start of the n-th cycle, the oldest batch of fuel has been irradiated for a period $2L + a_n$, and the next oldest for a period $L + b_n$. Using linear formulae, the duration $L + \varepsilon_n$ of the n-th cycle will be such that the average irradiation at the end of the n-th cycle is equal to its nominal value $2L$, when the reactivity is zero. Hence,

$$\frac{1}{3}\left[(2L + a_n + L + \varepsilon_n) + (L + b_n + L + \varepsilon_n) + (L + \varepsilon_n)\right] = 2L.$$

Therefore,

$$\varepsilon_n = -\frac{a_n + b_n}{3}.$$

From this the following recurrence formulae can be deduced:

$$a_{n+1} = b_n + \varepsilon_n = \frac{2b_n - a_n}{3}; \qquad b_{n+1} = \varepsilon_n = -\frac{a_n + b_n}{3}.$$

When it is the first cycle $n = 1$ being extended, and the additional fuel burn-up is τ, the initial values are $a_2 = \tau$ and $b_2 = \tau$. On subsequent cycles, the values of a_n and b_n are as shown in the following table, in units of τ.

n	*a*	*b*	ε
2	1	1	−2/3
3	+1/3	−2/3	+1/9
4	−5/9	+1/9	+4/27
5	+7/27	+4/27	−11/81
6	+1/81	−11/81	+10/243
7	−23/243	+10/243	+13/729
8	−43/729	+13/729	+30/2187

The total extra fuel burn-up for the extended cycle and all the following ones is $\tau + \frac{1}{2}\tau = \frac{3}{2}\tau$.

Exercise 18.5: evolution of a burnable poison

a) Any neutron entering through the surface of the poisoned region will be absorbed, and will destroy an atom near the surface of the poison. Let N be the concentration by volume of the poison atoms. Thus, per unit of surface,

$$| N\,dr | = | J_-\,dt |; \qquad r = r_0 - \frac{J_- t}{N}.$$

The neutron irradiation progressively ablates the surface of the absorbing material. This is called "onion peeling".

b) Using the formula obtained in **6.10.d** with $\gamma = 1$, it can be seen that the negative reactivity of the poison is

$$-\rho = \frac{2\chi(k_\infty - 1)}{k_\infty} \frac{(r_0 - J_- t/N)^2}{2D + r_0 - J_- t/N}.$$

It decreases continuously from its initial value until the complete consumption of the poison ($r = 0$). This occurs after being irradiated over a period $T = r_0 N/J_-$ in duration.

Numerical example: $T = 3.10^7$ s, i.e. about one cycle.

Exercise 18.6: uranium-plutonium interface

a) The equations to solve are

$$D\,\Delta\Phi - (N_i\sigma_{a,i} + \Sigma_c)\,\Phi + Q = 0,$$

with $i = 1$ for $x < 0$, and $i = 2$ for $x \geq 0$. The solution is obtained by taking into account the continuity of the flux and of the current at the interface:

$$\Phi_i(x) = \frac{Q}{\Sigma_i}\left[1 + \frac{\kappa_i - \kappa_j}{\kappa_j}\exp(-\kappa_i\,|\,x\,|)\right],$$

where

$$\Sigma_i = N_i\sigma_{a,i} + \Sigma_c, \qquad \kappa_i^2 = \frac{\Sigma_i}{D},$$

and $j = 2$ when $i = 1$, and $j = 1$ when $i = 2$.

The rate of neutron production — which is approximately proportional to the power — is obtained from the flux using

$$P_i(x) = (\eta\sigma_a N)_i\,\Phi_i(x).$$

When these functions are normalised to their asymptotic values for $x = -\infty$ (uranium zone), the result is as shown in the following table.

x	$-\infty$	-0	$+0$	$+\infty$
Φ	1	$\sqrt{2}/2$	$\sqrt{2}/2$	1/2
P	1	$\sqrt{2}/2$	$\sqrt{2}$	1

Notice the discontinuity of the power distribution due to the discontinuity of the production cross-section (by a factor of two here). Consequently there is a power peak in the plutonium region close to the interface.

Notice also that when $\varepsilon p = 0.8$ the system is critical.

b) The concentrations now are such that the absorption rates of the fissile material, and consequently the power levels, are constant in each region. For the power, the same value is required. When $\sigma_a N\Phi = C^t$ is substituted into each equation, and the continuity of the flux and current taken into account, the fluxes in both regions are obtained. Then the concentrations can be deduced:

$$N_i(x) = \frac{\Sigma_c/\sigma_{a,i}}{p\eta_i - 1 + \frac{1}{2}\left(1 - \eta_i/\eta_j\right)\exp\left(-\alpha \mid x \mid\right)},$$

where $\alpha = \sqrt{\Sigma_c/D}$ is used.

Normalising these functions, as done previously, to their asymptotic values for $x = -\infty$ (uranium region), the result is as shown in the following table.

x	$-\infty$	-0	$+0$	$+\infty$
Φ	1	3/4	3/4	1/2
N	1	4/3	13/15	13/10

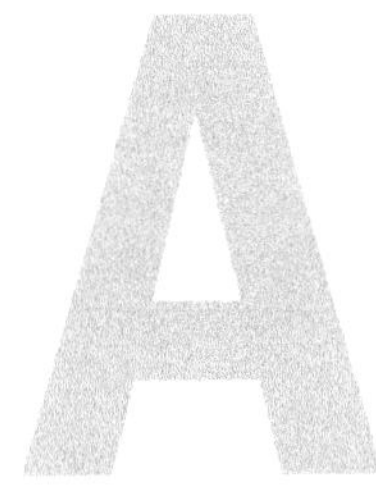

Annotated Bibliography

The following volumes have already been published by EDP Sciences in their "Génie Atomique" (Nuclear Engineering) series. The original (French) version of this book comes from this series:

- *Précis de neutronique,* Paul Reuss
- *Sciences des matériaux pour le nucléaire,* Clément Lemaignan
- *Le cycle du combustible nucléaire,* coordinated by Louis Patarin
- *L'économie de l'énergie nucléaire,* Evelyne Bertel and Gilbert Naudet
- *Exercices de neutronique,* Paul Reuss
- *La chaudière des réacteurs à eau sous pressions,* Pierre Coppolani, Nathalie Hassenboehler, Jacques Joseph, Jean-François Petetrot, Jean-Pierre Py, Jean-Sébastien Zampa.
- *Radioprotection et ingénierie nucléaire,* coordinated by Henri Métivier
- *L'épopée de l'énergie nucléaire,* Paul Reuss.

* * *

The following pages list bibliographic references to supplement the present book. These references are grouped according to theme.

Apart from some exceptions — notably with regard to certain publications by the author, which can be obtained on request from the INSTN[1] — this list is limited to easily obtainable documents[2]. In particular, many theses on the relevant topics are referenced. These are available from the universities concerned.

Contrary to common practice, we have given the full names of the authors wherever possible rather than just their initials and surname. We felt it was important in an educational document to acquaint the reader as much as possible with those who contributed to the developments being discussed.

[1] INSTN/UEIN, CEA/SACLAY, F - 91191 GIF-SUR-YVETTE CEDEX. Telephone: (33) 1 69 08 35 17.

[2] Note, however, that some of the works mentioned are out of print and can now only be found in libraries.

In general, we have used a chronological numbering system in each section or subsection. There are two reasons for this: firstly, to avoid favouring one of the authors if there are several, and secondly, to highlight the progression of developments. A brief presentation of the documents cited, *printed in italics and with square brackets,* directs the reader to sources of additional information.

The following letters are used to identify the different subject categories:

I - Introduction to nuclear energy and context.
G - General works on reactor physics.
N - Works by the present author on neutron physics.
Q - Nuclear physics, nuclear data and qualification.
C - Reactor kinetics.
R - Slowing down, thermalisation and resonant absorption of neutrons.
S - Neutron spectrum: multigroup theory.
T - Transport operator.
M - Monte Carlo method.
E - Equivalence, homogenisation and calculation of reflectors.
F - Neutron leakage.
P - Perturbation theory.
A - Mathematical methods and numerical analysis.
L - Software development.
V - Validation of software and calculation schemes.
D - Design and applied research.

A.1. Introduction to nuclear energy and context

A.1.1. General texts

[Two short books written by French specialists.]

[I-1] Colette LEWINER, *Les Centrales nucléaires,* "Que sais-je?" series, No. 1037, PUF, 1991.

[I-2] Rémy CARLE, *L'Électricité nucléaire,* "Que sais-je?" series, No. 2777, PUF, 1995.

[The first work cited below gives the American point of view, and the other two give the French point of view. The French books are intended for the general public and are particularly concerned with physical aspects and risk, safety, and environmental aspects respectively.]

[I-3] David BODANSKY, *Nuclear Energy: Principles, Practices, and Prospects,* American Institute of Physics, Woodbury, NY, USA, 1996.

[I-4] Paul REUSS, *L'Énergie nucléaire,* "Que sais-je?" Series, No. 317, PUF, 2006.

[I-5] Bernard WIESENFELD, *L'Atome écologique,* EDP - Sciences, 1998.

A.1.2. Historical aspects

[I-6] S. WEART, *La Grande Aventure des atomistes français*, Fayard, 1980.

[I-7] Emilio SEGRÉ, *Les Physiciens modernes et leurs découvertes*, Fayard, 1984.

[I-8] J.-C. DEBEIR, J.-P. DELÉAGE and D. HÉMERY, "Les servitudes de la puissance", *Une histoire de l'énergie*, Flammarion, 1986.

[I-9] Jacques LECLERCQ, *L'Ère nucléaire*, Chêne/Hachette, 1986.

[I-10] Bertrand GOLDSCHMIDT, *Pionniers de l'atome*, Stock, 1987.

[I-11] Georges LE GUELTE, *Histoire de la menace nucléaire*, Hachette, 1987.

[I-12] Pierre RADVANYI and Monique BORDRY, *Histoires d'atomes*, Belin, 1988.

[I-13] Paul REUSS, *L'épopée de l'énergie nucléaire, une histoire scientifique et industrielle,* EDP Sciences, "Génie atomique" series, 2007.

A.1.3. Risks, safety and accidents

[Two points of view on accidents.]

[I-14] Jean-Pierre PHARABOD and Jean-Paul SCHAPIRA, *Les Jeux de l'atome et du hasard : Les Grands Accidents nucléaires*, Calmann-Lévy, 1988.

[I-15] J.-L. NICOLET, Annick CARNINO and J.-C. WANNER, *Catastrophes? Non merci ! : La Prévention des risques technologiques et humains*, Masson, 1989.

[An eye-witness account.]

[I-16] G. MEDVEDEV, *La Vérité sur Tchernobyl*, Albin Michel, 1990.

[A simple book and a reference book about safety.]

[I-17] Daniel BLANC D., *La Sûreté de l'énergie électronucléaire*, "Que sais-je ?" series, No. 2032, PUF, 1991.

[I-18] Jacques LIBMANN, *Éléments de sûreté nucléaire*, EDP - Sciences, 1996.

A.1.4. Communication

[One of many books on the subject.]

[I-19] E. PARKER, *La Bombe à neurones : Désinformations en chaînes*, PUF, 1988.

A.1.5. Fuel cycle

[I-20] M. CUNEY, J. LEROY and M. PAGEL, *L'Uranium*, "Que sais-je?" series, No. 1070, PUF, 1992.

[I-21] Jean TEILLAC, *Les Déchets nucléaires*, "Que sais-je?" series, No. 2385, PUF, 1988.

[I-22] Armand FAUSSAT, *Les Déchets nucléaires : Les connaître, nous en protéger*, Stock, 1997.

[I-23] E. SURAUD, Ed., *Production d'énergie nucléaire et traitement des déchets*, EDP - Sciences, 2000.

A.1.6. Nuclear defence and risk of proliferation

[I-24] Henri PAC, *Le Droit de la défense nucléaire*, "Que sais-je?" series, No. 2472, PUF, 1989.

[I-25] François GÉRÉ, *La Prolifération nucléaire*, "Que sais-je?" series, No. 2978, PUF, 1995.

[I-26] Alain DURET, *La Nouvelle menace nucléaire*, Le Monde-Éditions, 1996.

A.1.7. Nuclear fusion

[I-27] J. ADAM, *La Fusion nucléaire, une source d'énergie pour l'avenir?*, "Pour la Science" series, Belin, 1993.

[I-28] Paul-Henri REBUT, *L'Énergie des étoiles : La Fusion nucléaire contrôlée*, Odile Jacob, 1999.

[I-29] Joseph WEISSE, *La Fusion nucléaire*, "Que sais-je?" series, No. 3659, PUF, 2003.

A.1.8. The Oklo phenomenon

[I-30] Roger NAUDET, *Oklo : des réacteurs nucléaires fossiles : Étude physique*, série Synthèses, "CEA" series, Eyrolles, 1991.

A.1.9. The world needs of energy and the possible contribution of the nuclear energy

[I-31] Pierre BACHER, *L'énergie en 21 questions*, Odile Jacob, 2007.

A.2. General works on reactor physics

[G-1] Alvin M. WEINBERG and Eugene P.WIGNER, *The Physical Theory of Neutron Chain Reactors*, University of Chicago Press, 1958.

[G-2] Samuel GLASSTONE and Milton C. EDLUND, *The Elements of Nuclear Reactor Theory*, D. Van Nostrand Company, New York, 1960.

[G-3] A.-F. HENRY, *Nuclear Reactor Analysis*, MIT Press, Cambridge, 1975.

[G-4] James J. DUDERSTADT and Louis J. HAMILTON, *Nuclear Reactor Analysis*, John Wiley, New York, 1976.

[G-5] Jean BUSSAC and Paul REUSS, *Traité de Neutronique* (see reference **[N-1]** in the next section).

[G-6] Robert BARJON, *Physique des réacteurs nucléaires*, Institut des Sciences Nucléaires, Grenoble, 1992.

[G-7] Samuel GLASSTONE and A. SESONSKE, *Nuclear Reactor Engineering*, 2 vol., Chapman & Hall, New York, 1994.

[G-8] Several authors, *Génie nucléaire*, 2 vol., Techniques de l'Ingénieur, 1996.

[G-9] Jacques LIGOU, *Introduction au génie nucléaire,* Presses polytechniques et universitaires romandes, Lausanne, 1997.

[G-10] Weston M. STACEY, *Nuclear Reactor Physics*, Wiley-VCH, 2007.

A.3. Works by the present author on neutron physics

[The first reference book in French; a bit old now.]

[N-1] Jean BUSSAC and Paul REUSS, *Traité de neutronique*, Hermann, 1985.

[A document summarising the basics; intended for students.]

[N-2] Paul REUSS, *Éléments de neutronique*, INSTN, 1986.

[An introduction to neutron physics.]

[N-3] Paul REUSS, "Au cœur des réacteurs : la neutronique"; "Réacteurs à eau sous pression et combustibles au plutonium", *Clefs CEA,* No. 11 (1988) and No. 20 (1991).

[Copy of figures and handbook.]

[N-4] Paul REUSS, "Neutronique des réacteurs à eau sous pression", documents accompanying the INSTN session on neutron physics in PWRs, updated regularly.

[A collection of problems with solutions.]

[N-5] Paul REUSS, *Clefs pour la neutronique des réacteurs à eau sous pression*, INSTN, 1990.

[Lecture notes for the training sessions given for EdF, followed by examples from other contexts.]

[N-6] Paul REUSS, *Neutronique: lectures*, INSTN, 1994.

[N-7] Paul REUSS, *Théorie du transport des neutrons*, INSTN, 1998.

[N-8] Paul REUSS, *L'Absorption neutronique*, INSTN, 1999.

[Introduction intended for the layman.]

[N-9] Paul REUSS, *La Neutronique*, "Que sais-je?" series, No. 3307, PUF, 1998.

[French edition of the present book.]

[N-10] Paul REUSS, *Précis de neutronique*, EDP Sciences, "Génie atomique" series, 2003.

[N-11] Paul REUSS, *Exercices de neutronique*, EDP Sciences, "Génie atomique" series, 2004.

A.4. Nuclear physics, nuclear data and qualification

A.4.1. General information on nuclear physics

[Q-1] W.-E. MEYERHOF, *Éléments de physique nucléaire*, Dunod, 1970.

[Q-2] Daniel BLANC, *Physique nucléaire*, Masson, 1973.

[Q-3] Luc VALENTIN, *Physique subatomique : noyaux et particules*, Hermann, 1982.

A.4.2. Radioactivity and radiation protection

[Q-4] Daniel BLANC, *La Physique nucléaire*, "Que sais-je?" Series, No. 2139, PUF, 1984.

[Q-5] Alain BAUR, *Protection contre les rayonnements : aspects physiques et méthodes de calcul*, Commissariat à l'Energie Atomique, 1985.

[Q-6] M. TUBIANA and M. BERTIN, *Radiobiologie-Radioprotection*, "Que sais-je?" series, No. 1070, PUF, 1992.

[Q-7] Colette CHASSARD-BOUCHAUD, *Environnement et radioactivité*, "Que sais-je?" series, No. 2797, PUF, 1993.

[Q-8] Pierre RADVANYI, *Les Rayonnements nucléaires*, "Que sais-je ?" series, No. 844, PUF, 1995.

[Q-9] Maurice TUBIANA and Robert DAUTRAY, *La Radioactivité et ses applications*, "Que sais-je?" series, No. 33, PUF, 1996.

[Q-10] Jacques FOOS, *Manuel de radioactivité à l'usage des utilisateurs*, three volumes, Formascience, 1993, 1994 and 1995.

[Q-11] Yves CHELET, *La radioactivité, manuel d'initiation*, NucléoN/EDP Sciences, 2006.

A.4.3. Fission

[The fundamental reactor phenomenon presented by top French specialists.]

[Q-12] André MICHAUDON, "La Fission nucléaire", *la Recherche*, No. 136, p. 990, September 1982.

[Q-13] Jean-François BERGER, Jacques DECHARGÉ, Michel GIROD, Gérard SIMON and Jean TROCHON, "Les secrets du mécanisme de la fission nucléaire", *Clefs CEA*, No. 17, 1990.

A.4.4. Nuclear Physics for Use in Neutron Physics

[The basics in less than 100 pages.]

[Q-14] Paul REUSS, *Éléments de physique nucléaire à l'usage du neutronicien*, INSTN, 1981 and 1987.

[This book was the product of a graduate course in Nuclear Reactor Physics.]

[Q-15] Henry TELLIER, *Réactions nucléaires induites par les neutrons*, INSTN, 1989.

A.4.5. Measurement of nuclear data

[Three recent examples of work in this field.]

[Q-16] Caroline BRIENNE-RAEPSAET, *Nouvelle détermination expérimentale des paramètres de résonances neutroniques du technétium 99*, thesis, Aix-Marseille I, 29 October 1998.

[Q-17] Jean GALY, *Mesure de la distribution en masse et en charge des produits de la fission rapide de l'uranium 233*, thesis, Aix-Marseille I, 21 September 1999.

[Q-18] Vincent GRESSIER, *Nouvelle détermination expérimentale des paramètres de résonances neutroniques du neptunium 237 en dessous de 500 eV*, thesis, Orsay, 13 October 1999.

A.4.6. Compilation and processing of nuclear data

[Q-19] S. F. MUGHABGHAB, *Neutron Cross Sections: Neutron Resonance Parameter and Thermal Cross Sections*, Academic Press, 1980.

[Q-20] R. E. MacFARLANE and D. W. MUIR, *The NJOY Nuclear Data Processing System*, LA-12740-M, October 1994.

[Q-21] JEF-PC, *A Personal Computer Program for Displaying Nuclear Data from the Joint Evaluated File Library*, OCDE/AEN, 1997.

A.4.7. Integral measurements and their use for the qualification of nuclear data

[Analysis of the main experimental methods and recent improvements to them.]

[Q-22] Jean-Pascal HUDELOT, *Développement, amélioration et calibration des mesures de taux de réaction neutroniques : élaboration d'une base de techniques expérimentales*, thesis, Grenoble I, 19 June 1998.

[Examples of integral experiments and their interpretation; note that one series of experiments — Muse III in this case — can give rise to several theses.]

[Q-23] Véronique ZAMMIT-AVERLANT, *Validation intégrale des estimations du paramètre bêta effectif pour les réacteurs MOX et incinérateurs*, thesis, Aix-Marseille I, 19 November 1998.

[Q-24] Cécile-Aline BOMPAS, *Contribution à la validation expérimentale du couplage entre un accélérateur et un massif sous-critique : expériences Muse III et Muse IV*, thesis, Grenoble I, 1 December 2000.

[Q-25] Gerardo ALIBERTI, *Caractérisation neutronique des systèmes hybrides en régimes stationnaire et transitoire*, thesis, Strasbourg, 5 October 2001.

[Some new developments in the field of experimental neutronics.]

[Q-26] Grégory PERRET, *Amélioration et développement des méthodes de détermination de la réactivité ; maîtrise des incertitudes associées*, thesis, Grenoble I, 15 October 2003.

[Q-27] Pierre LECONTE, *Développement et optimisation de techniques de mesure par spectrométrie gamma ; maîtrise et réduction des incertitudes associées*, thesis, Clermont-Ferrand II, 25 October 2006.

[Q-28] Benoit GESLOT, *Contribution au développement d'un système de mesure multi-mode pour des mesures neutroniques dynamiques et traitement des incertitudes associées*, thesis, Strasbourg I, 14 November 2006.

A.4.8. General approach to the qualification of nuclear data

[Formalised presentation of integral experiments to complement existing knowledge of nuclear data. This method was later extended to take covariances into account. Both of the following references are recent examples of this type of qualification approach. The associated bibliography also contains a broad spectrum of similar work carried out elsewhere.]

[Q-29] Paul REUSS, *La Méthode de recherche de tendances*, note CEA-N-2222, 1981.

[Q-30] Patrick BLAISE, *Mise au point d'une méthode d'ajustement des paramètres de résonance sur des expériences intégrales*, thesis, Aix-Marseille I, 13 February 1997.

[Q-31] Jean-Marc PALAU, *Corrélations entre données nucléaires et expériences intégrales à plaques : le cas du hafnium*, thesis, Aix-Marseille I, 22 October 1997.

[Example of sensitivity calculations: evaluating the effect of uncertainties on nuclear data and of technological uncertainties.]

[Q-32] David BERNARD, *Détermination des incertitudes liées aux grandeurs neutroniques d'intérêt des réacteurs à eau pressurisée à plaques combustibles et application aux études de conformité*, thesis, Clermont-Ferrand II, 18 December 2001.

A.4.9. Database of integral experiments

[Experiments concerning reactors.]

[Q-33] Patrick BLAISE, *Conception et réalisation de la base de données expérimentales*, technical note CEA/DEN/SPEx/LPE/01-042,2001.

[Experiments related to criticality.]

[Q-34] *ICSBEP: International Criticality Safety Benchmark Evaluation Project*, OCDE/-AEN/NSC/DOC *International Handbook Evaluated Criticality Benchmark Experiments*, 1995.

[How can these data be used relevantly?]

[Q-35] Emmanuel GAGNIER, "Contribution à la qualification du schéma de calcul de criticité CRISTAL", *Élaboration d'un système de caractérisation des configurations neutroniques*, second part, thesis, Aix-Marseille I, 24 June 1999.

A.5. Reactor kinetics

[A book that has remained a classic.]

[C-1] G. Robert KEEPIN, *Physics of Nuclear Kinetics*, Addison-Wesley Publishing Company, 1965.

[More than just an introduction! And it has CANDU examples.]

[C-2] Daniel ROZON, *Introduction à la cinétique des réacteurs nucléaires*, Éditions de l'École Polytechnique de Montréal, 1992.

[The basics. This book was the product of a graduate course in Nuclear Reactor Physics.]

[C-3] Henry TELLIER, *Cinétique des réacteurs nucléaires*, INSTN, 1994.

[A brief study showing the stochastic aspect of startup.]

[C-4] Paul REUSS, "Démarrage d'une chaîne de fissions", *Compléments de neutronique*, INSTN, booklet No. 13, 1992.

A.6. Slowing down, thermalisation and resonant absorption of neutrons

A.6.1. General information on resonant absorption

[Still a classic.]

[R-1] Lawrence DRESNER, *Resonance Absorption in Nuclear Reactors*, Pergamon Press, 1960.

[Summaries for teaching purposes.]

[R-2] Paul REUSS, *Théorie de l'absorption résonnante des neutrons*, note CEA-N-2679, 1991.

[R-3] Paul REUSS, *Historical Perspective on the Development of Methods for Reactor Analysis: from Elementary Theory to State-of-the-Art Methods Applied to Complex Reactor Lattices*, Frédéric Joliot/Otto Hahn Summer School, Cadarache, August 21-30 2000.

[R-4] Mireille COSTE et Paul REUSS, *Development of Computational Models Used in France for Neutron Resonance Absorption in Light Water Reactors*, Progress in Nuclear Energy, Vol. 1, No. 3, pp. 237-282, 2003.

A.6.2. The method of effective reaction rates

[The basic documents.]

[R-5] Françoise JEANPIERRE, *Méthode de calcul de sections effectives de corps résonnants lourds en géométrie hétérogène ; application à l'uranium 238*, thesis, Orsay, 15 September 1969.

[R-6] Françoise JEANPIERRE and Michel LIVOLANT, *Autoprotection des résonances dans les réacteurs nucléaires ; application aux isotopes lourds*, report CEA-R-4533, 1974.

[A few documents about developments.]

[R-7] Paul REUSS, "A Generalization of the Livolant-Jeanpierre Theory for Resonance Absorption Calculation", *Nuclear Science and Engineering*, No. 92, pp. 261–266, 1986.

[R-8] Mireille COSTE, *Absorption résonnante des noyaux lourds dans les réseaux hétérogènes : I - Formalisme du module d'autoprotection d'APOLLO-2*, note CEA-N-2746, 1994.

[R-9] Sophie PERRUCHOT-TRIBOULET, *Validation et extensions du module d'autoprotection du code de transport neutronique multigroupe APOLLO-2*, thesis, Aix-Marseille I, 7 October 1996.

[R-10] Mireille COSTE-DESCLAUX, *Modélisation du phénomène d'autoprotection dans le code de transport APOLLO2*, thesis, Conservatoire National des Arts et Métiers, 7 March 2006, Report CEA-R-6114, 2006.

[R-11] Noureddine HFAIEDH, *Nouvelle méthodologie de calcul de l'absorption résonnante*, thesis, Strasbourg I, 21 September 2006.

A.6.3. The method of probability tables

[General principles.]

[R-12] Pierre RIBON and Jean-Marie MAILLARD, *Les Tables de probabilité : application au traitement des sections efficaces pour la neutronique*, note CEA-N-2485, 1986.

[Extension to non-statistical cases.]

[R-13] Oum Keltoum BOUHELAL, *Prise en compte du ralentissement pour le calcul des sections efficaces : formalisme des tables de probabilité*, thesis, Rabat, Maroc, 2 November 1990.

[Recent developments.]

[R-14] Alain HÉBERT, "Advances in the Development of a Subgroup Method for the Self-Shielding of Resonant Isotopes in Arbitrary Geometries", *Nuclear Science and Engineering*, No. 126, pp. 245–263, 1997.

A.6.4. Doppler effect

[R-15] Mohamed OUISLOUMEN, "Noyau de l'opérateur de ralentissement tenant compte de l'agitation thermique ; effet sur les transferts en énergie", *Contribution aux développements du code de transport des neutrons APOLLO-2*, first part, thesis, Orsay, 22 March 1989.

[R-16] Claude MOUNIER, *Contribution à l'étude du coefficient de température des réacteurs à eau ordinaire*, thesis, Orsay, 13 December 1993.

[R-17] Dimitri NABEREJNEV, *Étude de l'influence des liaisons chimiques sur l'absorption et la diffusion des neutrons aux énergies de résonances,* thesis, Aix-Marseille I, 30 November 1998.

A.6.5. Validation and qualification

[Often-used measurements.]

[R-18] Eric HELLSTRAND, "Measurements of the Effective Resonance Integral in Uranium Metal and Oxide in Different Geometries", J. Appl. Phys. 28, No. 12, 1493, 1957.

[R-19] E. HELLSTRAND, P. BLOMBERG and S. HÖRNER, "The Temperature Coefficient of the Resonance Integral for Uranium Metal and Oxide", *Nuclear Science and Engineering,* No. 8, p. 497, 1960.

[Need for a multilevel formalism for uranium 238.]

[R-20] Henry TELLIER, Marc GRANDOTTO-BIETTOLI and Jacqueline VANUXEEM, *Une étude du désaccord entre les valeurs calculée et mesurée de l'intégrale effective de capture de l'uranium 238,* note CEA-N-2078, 1979.

[Qualification of the Livolant-Jeanpierre theory.]

[R-21] Henry TELLIER, Jean GONNORD, Catherine VAN DER GUCHT and Jacqueline VANUXEEM, *Dépendance spatiale et énergétique de la capture résonnante de l'uranium 238 dans un réseau hétérogène,* note CEA-N-2398, 1984.

[R-22] Henry TELLIER, Mireille COSTE, Caroline RAEPSAET and Catherine VAN DER GUCHT, *Absorption résonnante des noyaux lourds dans les réseaux hétérogènes : II - Qualification physique,* note CEA-N-2701, 1992.

[R-23] Henry TELLIER, Mireille COSTE, Caroline RAEPSAET and Catherine VAN DER GUCHT, "Heavy Nucleus Resonant Absorption Calculation Benchmarks", *Nuclear Science and Engineering,* No. 113, pp. 23–30, 1993.

[Validation by fine multigroup calculations. See also references **S-7** *and* **8***.]*

[R-24] Olivier BOULAND, *Amélioration du calcul de l'autoprotection des résonances résolues par un traitement quasi-exact du ralentissement des neutrons,* thesis, Orsay, 22 February 1994.

A.6.6. Thermalisation of neutrons

[R-25] C. H. WESTCOTT *et al.*, "Effective Cross-Sections and Cadmium Ratios for the Neutron Spectra of Thermal Reactors", Proc. Geneva Conference, 16, 202, p. 70, 1958.

[R-26] Michel CADILHAC, *Méthodes théoriques pour l'étude de la thermalisation des neutrons dans les milieux absorbants infinis et homogènes,* Doctoral thesis, Orsay, 15 November 1963; report CEA-R-2368, 1964.

[R-27] K. H. BECKURTS and K. WIRTZ, *Neutron Physics,* Springer-Verlag, 1964.

[R-28] M. M. R. WILLIAMS, *The Slowing Down and Thermalization of Neutrons,* North Holland Publishing Co, 1966.

[R-29] I. I. GUREWICH and L. V. TARASOV, *Low Energy Neutron Physics*, North Holland Publishing CO, 1968.

[R-30] P. A. EGELSTAFF and M. J. POOLE, *Experimental Neutron Thermalisation*, Pergamon Press, 1969.

[See also references **R-15** *and* **16**.*]*

A.7. Processing the neutron spectrum – multigroup theory

[The multigroup formalism is presented in all neutron physics books. For example, the following document sets out examples.]

[S-1] Xavier WARIN, "Méthodes déterministes de résolution de l'équation intégrodifférentielle du transport neutronique", EdF/DER, Service Informatique et Mathématiques Appliquées, HI-72/93/066, 1993.

[Below are the references for a few theses devoted either to a discussion of the optimisation of a multigroup breakdown or the qualification of a given breakdown.]

[S-2] Mohamed OUISLOUMEN, "Critères de choix du maillage énergétique multigroupe", *Contribution aux développements du code de transport des neutrons APOLLO-2*, second part, thesis, Orsay, 22 March 1989. *[See* **R-15** *for the first part.]*

[S-3] Gilles MATHONNIERE, *Bibliothèque neutronique à nombre de groupes restreint pour le calcul des réacteurs à eau*, thesis, Orsay, 27 October 1980.

[S-4] Jean-Yves DORIATH, *Méthodes numériques adaptatives pour les problèmes de transport dans les réacteurs nucléaires de sûreté par l'utilisation de signatures et de procédés de perturbation*, thesis, Aix-Marseille I, 6 May 1983.

[S-5] Philippe FOUGERAS, *Qualification des schémas de calcul pour le recyclage du plutonium dans les réacteurs à eau sous pression : expérience ÉPICURE*, thesis, Orsay, 10 November 1992.

[S-6] Bénédicte ROQUE, *Développement et qualification d'un formulaire de calcul de criticité*, thesis, Orsay, 10 November 1994.

[S-7] Stéphane MENGELLE, *APOLLO-2 : calculs de référence utilisant un maillage énergétique fin*, thesis, Orsay, 16 May 1995.

[S-8] Alain AGGERY, *Calculs de référence avec un maillage multigroupe fin sur les assemblages critiques par APOLLO-2*, thesis, Aix-Marseille I, 25 May 1999.

A.8. Transport operator

A.8.1. General publications

[The first classic book...still a classic today. Unfortunately, Volume II was never published.]

[T-1] K. M. CASE, F. de HOFFMANN and G. PLACZEK, *Introduction to the Theory of Neutron Diffusion*, Los Alamos Scientific Laboratory, 1953.

[The following titles are the other classics on the analytical approach to transport theory. See also the titles mentioned in the section on general books.]

[T-2] B. DAVISON, *Neutron Transport Theory*, Oxford University Press, London, 1957.

[T-3] G. M. WING, *An Introduction to Transport Theory*, Wiley, New York, 1962.

[T-4] Richard K. OSBORN, Sidney YIP, *The Foundations of Transport Theory*, Gordon & Breach, New York, 1966.

[T-5] K. M. CASE and P. F. ZWEIFEL, *Linear Transport Theory*, Addison-Wesley, Reading, Mass., 1967.

[The following title is more focused on numerical methods.]

[T-6] H. GRENNSPAN, C. N. KELBER and D. OKRENT, *Computing Methods of Reactor Physics*, Gordon & Breach, New York, 1968.

[The following two books are the most recent titles on transport theory and solution methods.]

[T-7] J. J. DUDERSTADT and W. R. MARTIN, *Transport Theory*, Wiley, New York, 1979.

[T-8] E. E. LEWIS, and W. F. MILLER Jr., *Computational Methods of Neutron Transport*, Wiley, New York, 1984.

[Finally, here is a comprehensive summary document on all techniques for solving the transport equation. This article also includes a complete list of original references.]

[T-9] R. SANCHEZ and N. J. McCORMICK, "A Review of Neutron Transport Approximations", *Nuclear Science and Engineering*, No. 80, pp. 481–535, 1982.

A.8.2. A few references about the French approach

[The following document has a historical interest; the basics of the diffusion approximation problem are explained in it.]

[T-10] M. BARBERGER, "Approximations de la diffusion et du transport en théorie multigroupe", *Bulletin du Centre de Recherches et d'Essais de Chatou*, No. 13, 1965.

[For the French approach to transport theory, refer to the lectures of P. Benoist on transport theory, integral in particular, and the documentation for the APOLLO code presented in Part **L**.*]*

[T-11] Pierre BENOIST, *Cours de troisième cycle, 2e année, option physique des réacteurs : Théorie du transport*, CEA-R-4778, 1976.

[T-12] Pierre BENOIST, *Lectures on Neutron Transport Theory*, CEA-N-2472, 1986.

[Here are references for a few theses (or associated publications) that expand on certain aspects. This list is not exhaustive.]

[T-13] Alain KAVENOKY, *La méthode C_N de résolution de l'équation du transport*, Doctorat ès Sciences thesis, Orsay, 16 November 1973; CEA-N-1710, 1974.

[T-14] Richard SANCHEZ, *Schémas approchés de résolution de l'équation intégrale du transport à deux dimensions*, Doctorat ès Sciences thesis, Orsay, 5 April 1979; CEA-N-2166, 1980.

[T-15] Richard SANCHEZ, "Approximate Solutions of the Two-Dimensional Transport Equation by Collision Probability Method", *Nuclear Science and Engineering*, No. 64, p. 384, 1977.

[T-16] Paul Marie GRANDJEAN, *Méthodes d'approximation de l'équation du transport : généralisation de la méthode C_N et comparaison avec la méthode F_N*, Doctorat ès Sciences thesis, Orsay, 22 February 1983; CEA-N-2388, 1984.

[T-17] Maurice MORDANT, *Résolution de l'équation du transport en géométrie x-y ou r-z à deux dimensions par des méthodes d'éléments finis sur l'espace des phases*, doctoral thesis, Orsay, 18 June 1985.

[T-18] Naïma LYOUSSI-CHARRAT, *Calcul de transport neutronique dans le code APOLLO-2 par une méthode des probabilités de collision dans une géométrie cartésienne générale*, thesis, Clermont-Ferrand II, 22 March 1994.

A.9. Monte Carlo method

A.9.1. General works

[General introductions to the Monte Carlo method.]

[M-1] N. P. BUSLENKO *et al.*, *The Monte Carlo Method: the Method of Statistical Trials*, Fizmatgiz, Moscow, 1962; English edition (translated by G. J. TEE): Pergamon Press, 1966.

[M-2] J. M. HAMMERSLEY and D. C. HANDSCOMB, *Les méthodes de Monte-Carlo*, French translation by F. ROSTAND, Dunod, 1967.

[Application to neutrons presented by two internationally renowned specialists.]

[M-3] J. SPANIER and E. M. GELBARD, *Monte Carlo Principles and Neutron Transport Problems*, Addison Wesley, 1969.

[A slightly old summary.]

[M-4] J. H. HALTON, "A Retrospective and Prospective Survey of the Monte Carlo Method", *SIAR Review*, vol. 12, No. 1, Jan. 1970.

[A general, more recent presentation by an Israeli author.]

[M-5] Reuven Y. RUBINSTEIN, *Simulation and the Monte Carlo Method*, John Wiley & Sons, 1981.

[The first reference to suggest to anyone looking for a complete and instructional presentation, with application to the transport of neutral particles. The first 150 pages give a solid introduction, and the rest of the book is mainly intended for specialists. The authors are research scientists at the Budapest Institute of Physics.]

[M-6] Ivan LUX and Laszlo KOBLINGER, *Monte Carlo Particle Transport Methods: Neutron and Photon Calculations*, CRC Press, Boston, 1991.

A.9.2. Examples of Monte Carlo codes

[The main French Monte Carlo code (but we could mention others, such as MORET used for criticality studies) is TRIPOLI, suitable for handling neutrons, gamma photons and their coupling. The first reference is complete but old, and the second presents the latest developments.]

[M-7] Jean-Claude NIMAL *et al.*, *Programme de Monte-Carlo polycinétique à trois dimensions TRIPOLI - 01*, notes CEA-N-1919 (seven volumes), 1976.

[M-8] Jean-Pierre BOTH, Hélène DERRIENNIC, Benjamin MORILLON and Jean-Claude NIMAL, *A Survey of TRIPOLI-4*, 8th International Conference on Radiation Shielding, Arlington, Texas, April 24-28 1994.

[TRIPOLI code has many "competitors" worldwide; MCNP is undoubtedly the most widely used one.]

[M-9] J. F. BRIESMEISTER, *MCNP: A General Monte Carlo Code for Neutron and Photon Transport*, LA-7396-M, 1986.

A.9.3. A few French developments

[Application of the Monte Carlo method to criticality studies.]

[M-10] Ali NOURI, *Contribution à la qualification du code Monte-Carlo TRIPOLI sur des expériences critiques et à l'étude de l'interaction neutronique entre unités fissiles*, thesis, Orsay, 21 January 1994.

[General introduction and analysis of biasing techniques.]

[M-11] Benjamin MORILLON, *Méthode de Monte-Carlo non analogue; application à la simulation des neutrons*, thesis, Orsay, 20 October 1995.

[Study of anisotropies and other analyses of biasing and perturbation methods.]

[M-12] Angélique LE COCQ, *Contributions au développement des méthodes de Monte-Carlo pour les études de criticité : anisotropie de la diffusion ; méthodes de perturbation*, thesis, Orsay, 23 November 1998.

[M-13] Li MAO, *Contribution à la résolution de l'équation de Boltzmann en multigroupe par les méthodes déterministes et Monte-Carlo*, thesis, Aix-Marseille I, 1998.

[M-14] François-Xavier GIFFARD, *Développements utilisant des méthodes stochastiques et déterministes pour l'analyse de systèmes nucléaires complexes*, thesis, Évry, 19 May 2000.

A.10. Equivalence, homogenisation and calculation of reflectors

A.10.1. Homogenisation

[The so-called SELENGUT approach.]

[E-1] D. S. SELENGUT, "Diffusion Coefficients for Heterogeneous Systems", *Trans. Am. Nucl. Soc.*, No. 3, p. 398, 1960.

[A few early French works.]

[E-2] Jacques MONDOT, *Mise en oeuvre de méthodes de calcul des poisons consommables dans les réacteurs à eau naturelle*, thesis, Orsay, 18 February 1973.

[E-3] Michel SOLDEVILA, *Contribution à l'étude du problème de l'équivalence transport-diffusion*, thesis, Orsay, 23 June 1978.

[The "ÉQUIPAGE" equivalence.]

[E-4] Jacques MONDOT, *Détermination des constantes équivalentes pour les calculs de diffusion aux différences finies*, AIEA, Comité technique en physique des réacteurs, Würenlingen, Switzerland, 1978 ; Proc. Specialists Mtg on Homogenization Methods in Reactor Physics, AIEA, Lugano, 1978.

[The poison problem.]

[E-5] Patrick CHAUCHEPRAT, *Qualification du calcul des poisons consommables au gadolinium dans les réacteurs à eau*, doctoral thesis, Orsay, 21 April 1988.

[The "EdF" equivalence.]

[E-6] Claude GARZENNE, *Équivalence transport-diffusion : présentation des méthodes utilisées au CEA et à EdF*, report EDF/DER/RNE/ PhR, HT-12/92 018 B, 1992.

[The "SPH" equivalence...]

[E-7] Alain KAVENOKY, *The SPH Homogenization Method*, AIEA, Comité technique en physique des réacteurs, Würenlingen, Switzerland, 1978; Proc. Specialists Mtg on Homogenization Methods in Reactor Physics, AIEA, Lugano, 1978.

[E-8] Alain HÉBERT, *Développement de la méthode SPH : homogénéisation de cellules dans un réseau non uniforme et calcul des paramètres de réflecteur*, thesis, Orsay, 18 December 1980; note CEA-N-2209, 1981.

[E-9] Alain HÉBERT and Pierre BENOIST, "A Consistent Technique for the Global Homogenization of a PWR Assembly", *Nuclear Science and Engineering*, No. 109, p. 360, 1991.

[E-10] Alain HÉBERT, "A Consistent Technique for the Pin-by-Pin Homogenization of a PWR Assembly", *Nuclear Science and Engineering*, No. 113, p. 327, 1993.

[... and its qualification.]

[E-11] Philippe FOUGERAS, *Qualification des schémas de calcul pour le recyclage du plutonium dans les réacteurs à eau sous pression : expérience ÉPICURE*, thesis, Orsay, 10 November 1992.

[E-12] Rachid SEKKOURI, *Analyse des techniques d'homogénéisation et des schémas de calcul pour les réacteurs à eau*, thesis Orsay, 7 April 1994.

[E-13] Pavel KLENOV, *Validation expérimentale des schémas de calcul relatifs aux absorbants et poisons consommables dans les REP*, thesis, Aix-Marseille I, 24 October 1995.

[E-14] Philippe MAGAT, *Analyse des techniques d'homogénéisation spatiale et énergétique dans la résolution de l'équation du transport des neutrons dans les réacteurs nucléaires*, thesis, Aix-Marseille I, 30 April 1997.

A.10.2. Reflector calculation

[The so-called "EdF" method.]

[E-15] Jean-Claude LEFEBVRE and Ph. LEBIGOT, *Tranches PWR- Études de cœur - Nouveau mode de calcul des réflecteurs*, report EdF/ DE-SEPTEN, E-SE-TB-78-02, 1978.

[Other methods, such as "Reuss-Nisan" and the "Beta" method of Jacques Mondot, were developed at the CEA. Presentations and extensions of these, accompanied by tests, can be found in the following publications.]

[E-16] Blaise MAIDOU, *Étude de l'interface cœur-réflecteur : application au calcul du réflecteur lourd*, thesis, Orsay, 24 June 1993.

[E-17] Edwige RICHEBOIS, *Calculs de cœurs REP en transport 3D*, thesis, Aix-Marseille I, 27 May 1999.

A.11. Neutron leakage

[The following document (an internal report) is of historical interest; it introduces the concept of the fundamental mode in a clear way.]

[F-1] Roger NAUDET, "Définition rigoureuse du bilan neutronique en milieu homogène", *Cours de troisième cycle de Physique des Réacteurs*, report CEA/SPM 966, 1967; "Définition du bilan et du coefficient de fuite dans un réseau", report CEA/SPM 1044, 1967.

[A work introducing the problem of how to define the diffusion coefficient.]

[F-2] Valentine C. DENIZ, *A New Consistent Definition of the Homogenized Diffusion Coefficient in a Lattice*, Proc. Specialists Mtg on Homogenization Methods in Reactor Physics, AIEA, Lugano, 1978.

[The concept of a diffusion coefficient on a lattice 1 - earlier works,]

[F-3] D.-J. BEHRENS, "The Effects of Holes in a Reacting Material on the Passage of Neutrons", *Proc. of Phys. Soc.*, 62, 10, 358 A, 1949.

[2 - The thesis by P. Benoist, basis of diffusion coefficient theory.]

[F-4] Pierre BENOIST, *Théorie du coefficient de diffusion des neutrons dans un réseau comportant des cavités*, doctoral thesis, Paris, 21 January 1964; report CEA-R-2278, 1964.

[A few milestones among the works that followed.]

[F-5] Michel LAM-HIME, *Homogénéisation : résolution de l'équation de transport en mode fondamental ; définition et calcul des coefficients de diffusion de cellules unidimensionnelles,* thesis, Orsay, 1980; note CEA-N-2223, 1981.

[F-6] Pierre BENOIST, *Homogenization Theory in Reactor Lattice,* note CEA-N-2471, 1986.

[F-7] Ivan PETROVIC, *Amélioration du modèle de fuites de neutrons dans le schéma de calcul des conditions critiques et des paramètres homogénéisés d'un réacteur nucléaire,* thesis, Orsay, 1993.

[F-8] Pierre BENOIST, Jacques MONDOT and Ivan PETROVIC, "Calculational and Experimental Investigations of Void Effect - A Simple Theoretical Model for Space-Dependant Leakage Treatment of Heterogeneous Assemblies", *Nuclear Science and Engineering,* No. 118, p. 197, 1994.

[F-9] Ivan PETROVIC, Pierre BENOIST and Guy MARLEAU, "A Quasi-Isotropic Reflecting Boundary Condition for the TIBERE Heterogeneous Leakage Model", *Nuclear Science and Engineering,* No. 122, p. 151, 1996.

[F-10] Xavier WOHLEBER, *Modélisation des fuites hétérogènes de neutrons dans un réacteur nucléaire,* thesis, Aix-Marseille I, 17 November 1997.

A.12. Perturbation calculations

[General works on the subject.]

[P-1] J. LEWINS, *Importance, the Adjoint Function,* Pergamon Press, 1965.

[P-2] M.-L. WILLIAMS, *Perturbation Theory for Nuclear Reactor Analysis,* vol. 3, CRC Handbook of Nuclear Reactor Calculations, Rogen, 1986.

[Summary document, many references.]

[P-3] Massimo SALVATORES, "La théorie des perturbations et les analyses de sensibilité", INSTN, *Compléments de neutronique,* vol. 8, 1987 ; EDF, *Bulletin de la Division des Études et Recherches* - Series A, No. 1, 1988.

[The ambiguity of the concept of adjoint flux.]

[P-4] Paul REUSS, "Définition et calcul du flux adjoint", INSTN, *Compléments de neutronique,* vol. 12, 1991.

A.13. Mathematical methods and numerical analysis

[General information on mathematical methods.]

[A-1] André ANGOT, *Compléments de mathématiques,* Éditions de la revue d'optique, 1961.

[General information on numerical methods; many references.]

[A-2] Franck JEDRZEJEWSKI, *Introduction aux méthodes numériques*, Springer, 2001.

[Various references, in English, on the main aspects of use in neutron physics, selected by E. E. Lewis and W. F. Miller Jr. (cf. **T-8**).]

[A-3] R. COURANT and D. HILBERT, *Methods of Mathematical Physics*, vol. 1, Interscience, New York, 1953.

[A-4] D. M. MORSE and H. FESCHBACH, *Methods of Theoretical Physics*, parts 1 and 2, McGraw-Hill, New York, 1953.

[A-5] M. J. LIGHTHILL, *Introduction to Fourier Analysis and Generalized Functions*, Cambridge University Press, London, 1958.

[A-6] R.-S. VARGA, *Matrix Iterative Analysis*, Prentice-Hall, Englewood Cliffs, N. J., USA, 1962.

[A-7] M. ABRAMOWITZ and I. STEGUN, *Handbook of Mathematical Functions*, Dover, New York, 1965.

[A-8] E. L. WACHSPRESS, *Iterative Solution of Elliptic Systems and Applications to the Neutron Diffusion Equation of Reactor Physics*, Prentice-Hall, Englewood Cliffs, N.J., USA, 1966.

[A-9] O. C. ZIENKIEWICZ, *The Finite Element Method in Engineering Science*, 2nd ed., McGraw-Hill, New York, 1971.

[A-10] W. C. BICKLEY and J. NAYLOR, "A Short Table of the Functions *Kin*(*x*) from $n = 1$ to 16", *Phil. Mag.*, No. 20, p. 343, 1935.

[The following book is the latest work published in French on mathematical methods in neutron physics. It essentially concerns the diffusion equation. It contains other references.]

[A-11] Jacques PLANCHARD, *Méthodes mathématiques en neutronique*, coll. "Direction des Études et Recherches d'Électricité de France", Eyrolles, No. 90, 1995.

[Two publications from Électricité de France on methods of solving the transport equation the first was also cited in **S-1**.*]*

[A-12] Xavier WARIN, "Méthodes déterministes de résolution de l'équation intégrale du transport neutronique", EdF/DER, Service Informatique et Mathématiques Appliquées, HI-72/93/081, 1993.

[A-13] Xavier WARIN, "Étude de quelques méthodes **SN** nodales en transport neutronique", EdF/DER, Service Informatique et Mathématiques Appliquées, HI-72/94/071/0, 1994, and HI-72/95/017/0, 1996.

A.14. Software development

[Every laboratory involved in nuclear energy has developed software, which means that an exhaustive bibliography would be extremely long. We have therefore decided to limit our list to French works on the subject.]

A.14.1. Calculation of first collision probabilities

[See also references **T-14,15** *and* **18** *above.]*

[L-1] Alain KAVENOKY, *Calcul et utilisation des probabilités de première collision pour les milieux hétérogènes à une dimension*, note CEA-N-1077, 1969.

A.14.2. The APOLLO 1 and APOLLO 2 codes

[L-2] Alain HOFFMANN, Françoise JEANPIERRE, Alain KAVENOKY, Michel LIVOLANT and Henri LORAIN, *APOLLO : code multigroupe de résolution de l'équation du transport pour les neutrons thermiques et rapides*, note CEA-N-1610, 1972.

[L-3] Mireille COSTE, Alain HÉBERT, Richard SANCHEZ, Zarko STANKOVSKI and Igor ZMIJAREVIC, *APOLLO-2 : notice théorique de la version 2.5*, report CEA/DMT/SERMA/LENR/RT/99-2719/A, 1999.

A.14.3. Simplified transport core calculations

[The method of simplified spherical harmonics.]

[L-4] C. G. POMRANING, "Asymptotic and Variational Derivations of the Simplified P_N Equations", *Ann. Nucl. Energy*, Vol. 20, No. 9, pp. 623–637, 1993.

[A few references concerning developments in ERANOS.]

[L-5] C.-B. CARRICO, E.-E. LEWIS and G. PALMIOTTI, "Three Dimensional Variational Nodal Transport Methods for Cartesian, Triangular, and Hexagonal Criticality Calculations", *Nuclear Science and Engineering*, No. 111, p. 168, 1992.

[L-6] G. PALMIOTTI, C. B. CARRICO and E. E. LEWIS, "Variational Nodal Transport Methods with Anisotropic Scattering", *Nuclear Science and Engineering*, No. 115, p. 233, 1993.

[L-7] E. E. LEWIS, C. B. CARRICO and G. PALMIOTTI, "Variational Nodal Formulation of the Spherical Harmonics Equations", *Nuclear Science and Engineering*, No. 122, p. 194, 1996.

[L-8] E. E. LEWIS and G. PALMIOTTI, "Simplified Spherical Harmonics in the Variational Nodal Method", *Nuclear Science and Engineering*, No. 126, p. 48, 1997.

[L-9] J. M. RUGGIERI, F. MALVAGI and R. BOYER, *TGV: a Coarse Mesh 3-Dimensional Diffusion-Transport Module for the CCRR/ERANOS Code System*, technical note CEA/SPRC/LEPh/93-209, 1993.

[L-10] Alexandre MAKARENKO, *Parallélisation de la méthode nodale variationnelle pour l'équation du transport neutronique*, thesis, Aix-Marseille I, October 1997.

[A few references concerning developments in CRONOS.]

[L-11] B. AKHERRAZ, *Résolution par la méthode des éléments finis de l'équation de transport en flux pair avec choc anisotrope*, thesis, Paris VI, December 1994.

[L-12] B. AKHERRAZ, C. FEDON-MAGNAUD, J.-J LAUTARD and R. SANCHEZ, "Anisotropic Scattering Treatment for the Neutron Transport Equation with Primal Finite Elements", *Nuclear Science and Engineering*, No. 120, pp. 187–198, 1995.

[L-13] Christine FEDON-MAGNAUD, *Résolution de l'équation de transport dans le code CRONOS*, note CEA-N-2751, 1994.

[L-14] Jean-Jacques LAUTARD, *La Méthode nodale de CRONOS : MINOS, approximation par des éléments mixtes duaux*, note CEA-N-2763, 1994.

A.14.4. Deterministic calculations for radiation protection

[It is important to note that the problem of solving the transport equation for neutrons, and particularly gamma photons, also arises in astrophysics and in calculations for radiation protection (the latter often using Monte Carlo calculations). The first reference cited below reviews the problems arising in astrophysics, and the second presents the problem of protection. The others are a few examples of recent developments concerning specific methods for deterministic protection calculations.]

[L-15] Subrahmanyan CHANDRASEKHAR, *Radiative Transfer*, Dover, New York, 1960.

[L-16] H. GOLDSTEIN, *Fundamental Aspects of Reactor Shielding*, Addison-Wesley, Reading Mass., USA, 1959.

[L-17] Ali ASSAD, *Amélioration de la modélisation du transport des gamma suivant la méthode de l'atténuation en ligne droite*, thesis, Orsay, 20 October 1995.

[L-18] Emmanuel LEFEVRE, *Mise au point et validation d'un nouveau formulaire adapté au calcul des protections neutroniques des réacteurs à neutrons rapides*, thesis, Aix-Marseille I, 19 June 1996.

[L-19] Olivier LITAIZE, *Contribution à la détermination de l'albédo doublement différentiel en angle et en énergie des neutrons ; application à la propagation dans les milieux lacunaires*, thesis, Strasbourg, 12 February 1999.

A.15. Validation of software and calculation schemes

[Software validation and the development of calculation schemes are closely related to the development of calculation methods and codes; for this reason, they are quite well illustrated in the above references (see, for example, **E-14** *and* **E-17** *for PWRs). To a large extent they also depend on the problems to be examined, and are therefore related to the applied studies covered in the next section. Here we give only a few references as examples (experimental reactors, PWRs and BWRs).]*

[V-1] Stéphane RAUCK, *Modélisation des phénomènes physiques dans les réacteurs de recherche à l'aide de développements réalisés dans les méthodes de calcul de transport et qualification*, thesis, 16 September 1999.

[V-2] Patrick BLANC-TRANCHANT, *Élaboration et qualification de schémas de calcul de référence pour les absorbants dans les réacteurs à eau pressurisée*, thesis, Aix-Marseille I, 8 November 1999.

[V-3] Frédéric BOUVERET, *Modélisation des phénomènes physiques spécifiques aux réacteurs à eau bouillante, notamment le couplage neutronique-thermohydraulique*, thesis, Grenoble, 5 December 2000.

A.16. Design and applied research

[System design and, more generally, applied research are the culmination of all developments in the fields of physics, mathematics, and code-writing. There is a wide variety of such activities, and it would not be possible to compile an exhaustive list of them. We therefore limit our list to a few examples by way of illustration.]

[Specific research is required for each reactor type or sub-type. The following small volume presents fast neutron reactors, and was written by the top specialist at the CEA in an era when these reactors were still an active subject of discussion.]

[D-1] Georges VENDRYES, *Les Surgénérateurs*, "Que sais-je ?" series, No. 2362, PUF 1987.

[Plutonium recycling, which has come about more in water reactors than fast neutron reactors, has given rise to many physical and engineering studies; here is just one example.]

[D-2] Paul REUSS, *Étude physique du recyclage du plutonium dans les réacteurs à eau*, doctoral thesis, Orsay, 21 May 1979.

[Here are a few other examples concerning recent work on standard reactors.]

[D-3] Dietrich KNOCHE, *Analyse des capacités des réacteurs à eau bouillante à recycler du combustible 100% MOX*, thesis, Grenoble, 23 June 1999.

[D-4] Stefano BALDI, *Détermination de schémas d'empoisonnement pour le contrôle de la réactivité de combustibles innovants. Application aux CERMET au plutonium*, thesis, Évry, 31 March 2000.

[D-5] Frédéric DAMIAN, *Analyse des capacités des réacteurs à haute température sous l'aspect de l'utilisation des matières fissiles*, thesis, Grenoble, 2 February 2001.

[Waste management has been the subject of much research since the law dated 1991. Most of these studies refer to the following thesis, where flux estimates for the materials concerned are given.]

[D-6] Stéphanie SALA, *Réduction de la radiotoxicité des déchets nucléaires à vie longue : études théoriques et stratégiques de la transmutation des actinides mineurs et des produits de fission dans les réacteurs électronucléaires*, thesis, Aix-Marseille I, 26 June 1995.

[Examples of transmutation research.]

[D-7] Myriam VALADE, *Étude de l'incinération des transuraniens en réacteur à sel fondu*, thesis, Strasbourg, 27 October 2000.

[D-8] David LECARPENTIER, *Contribution aux travaux sur la transmutation des déchets nucléaires. Voie des réacteurs à sel fondu : le concept AMSTER, aspects physiques et sûreté*, thesis, CNAM, 29 June 2001.

[After the publications of C. Bowman and C. Rubbia (first two references below), a certain number of studies on hybrid systems have been initiated. Here are a few examples in France.]

[D-9] Charles BOWMAN *et al.*, "Nuclear Energy Generation and Waste Transmutation Using Accelerator Driven Intense Thermal Neutron Source", LA-UR-91-2601, 1991 *or Nuclear Instruments and Methods in Physics Research*, A320, 336-367, 1992.

[D-10] Carlo RUBBIA *et al.*, *Conceptual Design of a Fast Neutron Operated High Power Energy Amplifier*, CERN/AT/95-44 (ET), 1995.

[D-11] Véronique BERTHOU, *Contribution à une proposition d'un développement à long terme de l'énergie nucléaire : le concept TASSE* (thorium based accelerator driven system with simplified fuel cycle for long term energy production), thesis, Évry, 30 October 2000.

[D-12] Denis KERDRAON, *Optimisation par simulation du couplage entre un réacteur sous-critique et sa source de spallation. Application à un démonstrateur*, thesis, Grenoble, 26 October 2001.

B Physical tables and constants

This appendix includes a periodic table of the elements, the main physical constants used in neutron physics, and some generally useful nuclear data: resonance integrals, cross-sections at 2200 m/s, average number of neutrons emitted per fission. Obviously, the nuclear data used in codes are too numerous to list. If necessary, these can be found in some of the existing databases (see § 2.12 and reference **Q-21**).

B.1. Table of the elements

See Figure B.1. The names of elements 101 to 109 were officially adopted by the *International Union of Pure and Applied Chemistry* (IUPAC) at its general assembly of August 1997 in Geneva.

B.2. Physical constants

Source: Handbook of Chemistry and Physics, 81st Ed., CRC Press, 2000.

B.2.1. Units used in nuclear physics

- Unit of surface area *(barn)*: 1 b = 10^{-28} m^2.
- Unit of atomic mass: 1 u = 1.660539 × 10^{-27} kg (Note: ${}^{12}_{6}\mathbf{C}$ = 12).
- Unit of energy *(electron volt)*: 1 eV = 1.6021765 × 10^{-19} J.
- Speed of light: c = 299 792 458 m/s.
- Mass-energy equivalence: 1 u = 931.4940 MeV.

B.2.2. Main physical constants used in neutron physics

- Mass of an electron: 0.0005485799 u = 0.511 MeV.
- Mass of a proton: 1.0072765 u = 938.282 MeV.

Legend: 92 **U** Uranium — Atomic number / Symbol / Name

1 **H** Hydrogen																	2 **He** Helium
3 **Li** Lithium	4 **Be** Beryllium											5 **B** Boron	6 **C** Carbon	7 **N** Nitrogen	8 **O** Oxygen	9 **F** Fluorine	10 **Ne** Neon
11 **Na** Sodium	12 **Mg** Magnesium											13 **Al** Aluminium	14 **Si** Silicon	15 **P** Phosphorus	16 **S** Sulphur	17 **Cl** Chlorine	18 **A** Argon
19 **K** Potassium	20 **Ca** Calcium	21 **Sc** Scandium	22 **Ti** Titanium	23 **V** Vanadium	24 **Cr** Chromium	25 **Mn** Manganese	26 **Fe** Iron	27 **Co** Cobalt	28 **Ni** Nickel	29 **Cu** Copper	30 **Zn** Zinc	31 **Ga** Gallium	32 **Ge** Germanium	33 **As** Arsenic	34 **Se** Selenium	35 **Br** Bromine	36 **Kr** Krypton
37 **Rb** Rubidium	38 **Sr** Strontium	39 **Y** Yttrium	40 **Zr** Zirconium	41 **Nb** Niobium	42 **Mo** Molybdenum	43 *Tc* Technetium	44 **Ru** Ruthenium	45 **Rh** Rhodium	46 **Pd** Palladium	47 **Ag** Silver	48 **Cd** Cadmium	49 **In** Indium	50 **Sn** Tin	51 **Sb** Antimony	52 **Te** Tellurium	53 **I** Iodine	54 **Xe** Xenon
55 **Cs** Caesium	56 **Ba** Barium	57 to 71 Lanthanides	72 **Hf** Hafnium	73 **Ta** Tantalum	74 **W** Tungsten	75 **Re** Rhenium	76 **Os** Osmium	77 **Ir** Iridium	78 **Pt** Platinum	78 **Au** Gold	80 **Hg** Mercury	81 **Tl** Thallium	82 **Pb** Lead	83 **Bi** Bismuth	84 **Po** Polonium	85 **At** Astatine	86 **Rn** Radon
87 **Fr** Francium	88 **Ra** Radium	89 à 103 Actinides	104 *Rf* Rutherfordium	105 *Db* Dubnium	106 *Sg* Seaborgium	107 *Bh* Bohrium	108 *Hs* Hassium	109 *Mt* Meitnerium	110 ? ?	111 ? ?	112 ? ?	113 ? ?	114 ? ?	115 ? ?	116 ? ?	117 ? ?	118 ? ?

Lanthanide series (rares earths)	57 **La** Lanthanum	58 **Ce** Cerium	59 **Pr** Prasoedymium	60 **Nd** Neodymium	61 *Pm* Promethium	62 **Sm** Samarium	63 **Eu** Europium	64 **Gd** Gadolinium	65 **Tb** Terbium	66 **Dy** Dysprosium	67 **Ho** Holmium	68 **Er** Erbium	69 **Tm** Thulium	70 **Yb** Ytterbium	71 **Lu** Lutetium
Actinide series	89 **Ac** Actinium	90 **Th** Thorium	91 **Pa** Protactinium	92 **U** Uranium	93 *Np* Neptunium	94 *Pu* Plutonium	95 *Am* Americium	96 *Cm* Curium	97 *Bk* Berkelium	98 *Cf* Californium	99 *Es* Einsteinium	100 *Fm* Fermium	101 *Mv* Mendelevium	102 *No* Nobelium	103 *Lw* Lawrencium

Figure B.1. Mendeleev table (classification of the elements). In *italics* : artificial elements.

- Mass of a neutron: 1.0086649 u = 939.565 MeV.
- Unit electrical charge: $e = 1.6021765 \times 10^{-19}$ C.
- Planck's constant: $h = 6.626068 \times 10^{-34}$ J.s.
- Avogadro's number: $N = 6.022142 \times 10^{23}$ mol-1.
- Boltzmann constant: $k = 1.380650 \times 10^{-23}$ J/K.
- Absolute zero: −273.15 °C.

B.3. Selection of nuclear data

Source: Joint Evaluated File: JEF - 2.2 and *Handbook of Chemistry and Physics.*

B.3.1. Data concerning the Elements

Table B.1 gives the following information for the *natural elements* with their average isotopic composition:

- Atomic number Z,
- Atomic symbol (or, for some of the more important materials in reactor physics, the chemical formula),
- Mass in units of atomic mass,
- Density ρ in kg/m^3 (under normal conditions, for liquid or solid materials),
- Number of atoms (or molecules) in 10^{28} per m^3,
- Microscopic absorption cross-section at 2200 m/s σ_a in barns,
- Absorption resonance integral I_a in barns,
- Microscopic scattering cross-section section at 2200 m/s σ_s in barns,
- Macroscopic absorption cross-section at 2200 m/s Σ_a in m^{-1},
- Macroscopic scattering cross-section at 2200 m/s Σ_s in m^{-1},
- Total macroscopic cross-section at 2200 m/s Σ_t in m^{-1}.

The density, concentration and macroscopic cross-sections are given only for solid or liquid materials under normal conditions.

Table B.1. Cross-sections of the natural elements.

Element	Mass (u)	ρ (kg/m^3)	N ($10^{28}/m^3$)	σ_a (barns)	I_a (barns)	σ_s (barns)	Σ_a (m^{-1})	Σ_s (m^{-1})	Σ_t (m^{-1})
(1) **H**	1.00794	-	-	0.322	0.149	30.3	-	-	-
H_2O	18.0153	998	3.34	0.644	0.299	107	2.15	357	359
D_2O	20.0276	1105	3.32	0.00133	$\simeq 0$	10.54	0.0044	35.0	35.0
(2) **He**	4.002602	-	-	0.0073	0.0033	0.856	-	-	-
(3) **Li**	6.941	534	4.63	70.57	31.7	1.10	327.0	5.1	332.1
(4) **Be**	9.012182	1850	12.36	0.0076	0.0048	6.34	0.094	78.4	78.5
BeO	25.0116	3010	7.25	0.00779	0.0049	10.10	0.056	73.2	73.3
(5) **B**	10.811	2340	13.03	764.9	342.7	4.51	9971	58.7	10030
(6) **C**	12.0107	1600	8.02	0.00337	0.00174	4.94	0.0270	39.7	39.7
(7) **N**	14.00674	-	-	1.959	0.848	10.29	-	-	-
(8) **O**	15.9994	-	-	0.000191	0.00009	3.76	-	-	-
(9) **F**	18.998403	-	-	0.0095	0.0214	4.11	-	-	-
(10) **Ne**	20.1797	-	-	0.039	0.018	2.42	-	-	-
(11) **Na**	22.989770	940	2.54	0.532	0.312	3.09	1.35	7.9	9.3
(12) **Mg**	24.3050	1740	4.31	0.063	0.034	3.41	0.27	14.7	15.0
(13) **Al**	26.981538	2700	6.03	0.213	0.133	1.45	1.28	8.7	10.0
(14) **Si**	28.0855	2329	4.99	0.172	0.085	2.10	0.86	10.5	11.4
(15) **P**	30.973761	1823	3.54	0.166	0.081	3.19	0.59	11.3	11.9
(16) **S**	32.066	2070	3.89	0.514	0.243	1.04	2.00	4.0	6.0
(17) **Cl**	35.4527	-	-	3.35	13.7	15.8	-	-	-
(18) **A**	39.948	-	-	0.675	-	0.65	-	-	-
(19) **K**	39.0983	890	1.37	2.10	1.21	2.23	2.88	3.1	6.0
(20) **Ca**	40.078	1540	2.31	0.430	0.223	3.03	1.00	7.0	8.0
(21) **Sc**	44.9559	2990	4.01	27.2	12	22.4	108.9	89.7	198.6
(22) **Ti**	47.867	4506	5.67	6.11	2.45	4.45	34.7	25.2	59.9
(23) **V**	50.9415	6000	7.09	5.06	2.54	5.05	35.9	35.8	71.7
(24) **Cr**	51.9961	7150	8.28	3.07	1.55	3.42	25.5	28.3	53.8
(25) **Mn**	54.938049	7300	8.00	13.31	15.33	1.77	106.5	14.1	120.6
(26) **Fe**	55.845	7870	8.49	2.59	1.36	11.35	22.0	96.3	118.3
(27) **Co**	58.933200	8860	9.05	37.24	75.52	6.01	337.2	54.4	391.6
(28) **Ni**	58.6934	8900	9.13	4.44	2.12	17.75	40.5	162.1	202.6
(29) **Cu**	63.546	8960	8.49	3.79	4.51	8.68	32.2	73.7	105.9
(30) **Zn**	65.39	7140	6.58	1.11	2.00	4.08	7.3	26.8	34.1
(31) **Ga**	69.723	5910	5.10	2.98	7.63	6.55	15.2	33.4	48.6
(32) **Ge**	72.61	5320	4.41	2.3	6.0	8.37	10.1	36.9	47.0
(33) **As**	74.92160	5750	4.62	4.30	60.2	1.89	19.9	8.7	28.6
(34) **Se**	78.96	4810	3.67	12.03	13.74	1.76	44.1	6.4	50.5
(35) **Br**	79.904	3103	2.34	7.00	96.1	3.29	16.4	7.7	24.1

Element	Mass (u)	ρ (kg/m³)	N (10^{28}/m³)	σ_a (barns)	I_a (barns)	σ_s (barns)	Σ_a (m^{-1})	Σ_s (m^{-1})	Σ_t (m^{-1})
(36) **Kr**	83.80	-	-	2.77	47.0	7.46	-	-	-
(37) **Rb**	85.4678	1530	1.08	0.377	5.20	5.23	0.41	5.6	6.0
(38) **Sr**	87.62	2640	1.81	1.41	8.87	3.47	2.6	6.3	8.9
(39) **Y**	88.90585	4470	3.03	1.29	0.92	7.73	3.9	23.4	27.3
(40) **Zr**	91.224	6520	4.30	0.184	0.932	6.49	0.79	27.9	28.7
(41) **Nb**	92.90638	8570	5.56	1.147	9.56	6.07	6.3	33.7	40.1
(42) **Mo**	95.94	10200	6.40	2.57	24.5	5.56	16.5	34.6	52.1
(44) **Ru**	101.07	12100	7.21	2.84	44.4	5.22	20.5	37.6	58.1
(45) **Rh**	102.90550	12400	7.26	146.4	1035	3.43	1062	24.9	1087
(46) **Pd**	106.42	12000	6.79	7.01	70.5	4.48	47.6	30.4	78.0
(47) **Ag**	107.8682	10500	5.86	63.2	766	5.04	371	29.5	400
(48) **Cd**	112.411	8690	4.66	2538	68.1	7.63	11 810	35.5	11 850
(49) **In**	114.818	7310	3.83	194	3097	2.61	744	10.0	754
(50) **Sn**	118.710	5769	2.93	0.681	6.24	4.27	2.0	12.5	14.5
(51) **Sb**	121.760	6680	3.30	5.24	173	3.79	17.3	12.5	29.8
(52) **Te**	127.60	6240	2.94	4.60	57.0	4.14	13.5	12.2	25.7
(53) **I**	126.90447	4930	2.34	6.20	144	3.57	14.5	8.3	22.8
(54) **Xe**	131.29	-	-	2.35	260	8.44	-	-	-
(55) **Cs**	132.90545	1930	0.875	29.1	439	3.89	25.5	3.4	28.9
(56) **Ba**	137.327	3620	1.58	1.29	8.04	6.29	2.1	10.0	12.1
(57) **La**	138.9055	6150	2.67	8.93	11.93	10.54	23.8	28.1	51.9
(58) **Ce**	140.116	6770	2.91	0.609	0.411	2.90	1.8	8.4	10.2
(59) **Pr**	140.90765	6770	2.89	11.5	18.0	2.60	33.3	7.5	40.8
(60) **Nd**	144.24	7010	2.93	49.2	40.3	14.80	144.0	43.3	187.3
(62) **Sm**	150.36	7520	3.01	5658	1432	35.8	17 040	107.9	17 150
(63) **Eu**	151.964	5240	2.08	4539	2362	5.16	9430	10.7	9440
(64) **Gd**	157.25	7900	3.03	48637	392	287	147 100	868	148 000
(65) **Tb**	158.92534	8230	3.12	23.2	415	6.57	72.4	20.5	92.9
(66) **Dy**	162.50	8550	3.17	907	1438	115	2870	365	3240
(67) **Ho**	164.93032	8800	3.21	66.5	763	3.29	214	10.6	224
(68) **Er**	167.26	9070	3.27	166	742	1.47	541	4.8	546
(69) **Tm**	168.93421	9320	3.32	105	1720	6.3	349	20.9	370
(70) **Yb**	173.04	6900	2.40	35.5	154	23.4	85.2	56.2	141.4
(71) **Lu**	174.967	9840	3.39	76.0	655	5.22	258	17.7	275
(72) **Hf**	178.49	13300	4.49	102.8	1990	8.58	461	38.5	500
(73) **Ta**	180.9479	16400	5.46	21.15	740	6.17	115.4	33.7	149.1
(74) **W**	183.84	19300	6.32	18.09	362	4.99	114.4	31.6	146.0
(75) **Re**	186.207	20800	6.73	89.6	833	6.14	603	41.3	644

Element	Mass (u)	ρ (kg/m³)	N (10^{28}/m³)	σ_a (barns)	I_a (barns)	σ_s (barns)	Σ_a (m^{-1})	Σ_s (m^{-1})	Σ_t (m^{-1})
(76) **Os**	190.23	22590	7.15	16	180	15	114	107	222
(77) **Ir**	192.217	22500	7.05	425	2150	14.2	3000	100	3100
(78) **Pt**	195.078	21500	6.64	10.3	140	12.4	68	82	151
(79) **Au**	196.96655	19300	5.90	98.76	1563	6.86	583.0	40.5	623.3
(80) **Hg**	200.59	13530	4.06	372	73	26.5	1510	108	1620
(81) **Tl**	204.3833	11800	3.48	3.43	12.9	10.01	11.9	34.8	46.7
(82) **Pb**	207.2	11300	3.28	0.178	0.114	11.22	0.6	36.9	37.5
(83) **Bi**	208.98038	9790	2.82	0.0372	0.239	9.33	0.1	26.3	26.4
(90) **Th**	232.038	11700	3.04	7.40	85.5	11.84	22.5	36.0	58.5
(92) **U**	238.0289	19100	4.83	7.61	281	9.42	36.8	45.5	82.3
UO_2	270.0277	10970	2.45	7.61	281	16.94	18.6	41.4	60.1

B.3.2. Data concerning nuclides

Table B.2 gives the following information for some of the most important nuclides:

- Symbol,
- Mass in units of atomic mass,
- Microscopic absorption cross-section at 2200 m/s σ_a in barns,
- Absorption resonance integral I_a in barns,
- Microscopic fission cross-section at 2200 m/s σ_f in barns,
- Fission resonance integral I_f in barns,
- Microscopic scattering cross-section at 2200 m/s σ_s in barns,
- Number ν of neutrons emitted by a thermal neutron-induced fission,
- Number η of neutrons emitted for one thermal neutron absorption.

Table B.2. Nuclear data for a few nuclides.

Nuclide	Mass (u)	σ_a (barns)	I_a (barns)	σ_f (barns)	I_f (barns)	Σ_s (barns)	ν	η
^{1}**H**	1.007 825 032	0.332	0.149	0	0	30.33	-	-
^{2}**H**	2.014 101 778	0.00051	0.00026	0	0	4.25	-	-
^{10}**B**	10.0129370	3844	1722	0	0	2.25	-	-
^{11}**B**	11.0093055	0.0055	0.0028	0	0	5.07	-	-
^{232}**Th**	232.038050	7.40	85.5	0	0	11.83	-	-
^{231}**Pa**	231.035879	227	607	0.01	5.1	8.46	2.296	0.0001
^{233}**Pa**	233.04	41.5	856	0	0	8.43	-	-

Nuclide	Mass (u)	σ_a (barns)	I_a (barns)	σ_f (barns)	I_f (barns)	Σ_s (barns)	ν	η
^{232}U	232.03713	149.7	746	77.1	429	7.54	3.131	1.613
^{233}U	233.039627	571.1	897	525.2	762	14.42	2.498	2.297
^{234}U	234.040946	103.5	667	0.46	6.5	12.33	2.352	0.011
^{235}U	235.043923	681.5	411	582.6	279	15.09	2.439	2.085
^{236}U	236.045561	5.21	355	0.047	7.8	8.09	2.317	0.021
^{238}U	238.050783	2.719	280	0.000012	2.0	9.38	1.644	0.00001
^{237}Np	237.048167	181.0	666	0.018	6.4	14.77	-	-
^{238}Pu	238.049553	563.4	176	17.3	33	19.90	2.889	0.089
^{239}Pu	239.052157	1017.7	483	747.3	298	8.82	2.874	2.110
^{240}Pu	240.053808	288.8	8459	0.068	10.1	1.51	2.783	0.0007
^{241}Pu	241.05	1375.3	750	1012.3	580	11.19	2.939	2.163
^{242}Pu	242.058737	18.5	1122	0.014	6.3	8.32	2.808	0.002
^{244}Pu	244.064198	1.83	106	0	5	9.65	-	-
^{241}Am	241.056822	618	1463	3.2	16	12.09	3.337	0.017
^{242m}Am	242.06	8694	1904	6885	1642	13.75	-	-
^{243}Am	243.061374	76	1821	0.050	6.5	7.13	3.061	0.002
^{242}Cm	242.058831	21.5	128	5.0	12	10.11	3.161	0.734
^{243}Cm	243.061362	545	2070	432	1786	6.80	3.393	2.688
^{244}Cm	244.062747	15.5	653	1.0	20	7.59	3.238	0.217
^{245}Cm	245.065484	2480	884	2131	760	10.45	-	-

The dashes indicate quantities that are not applicable or where the values are unknown. Masses not given by the Handbook of Chemistry and Physics *were taken from* Nuclides and Isotopes *Fifteenth Edition (1996), Lockheed Martin, GE Nuclear Energy.*

B.3.3. Energy released by fission

Table B.3 gives the values for energy released by neutron-induced fission, in MeV, for the main heavy nuclei (source: **CEA/ SERMA**).

Table B.3. Average energy released per induced fission for a few heavy nuclei (without the energy of antineutrinos).

Fissioned nucleus	Energy (MeV)
Thorium 232	187.0
Uranium 233	190.7
Uranium 235	193.7
Uranium 238	197.0
Plutonium 239	202.0
Plutonium 241	204.4
Americium 242	207.0

Note that these values:

- Exclude the energy taken away by antineutrinos (a little less than 10 MeV) because it will not be recovered,
- Exclude the energy deposited by gamma photons emitted during (fissionless) neutron capture reactions: users of APOLLO code must *add 8 MeV to the values given in this table* to calculate the burnup.

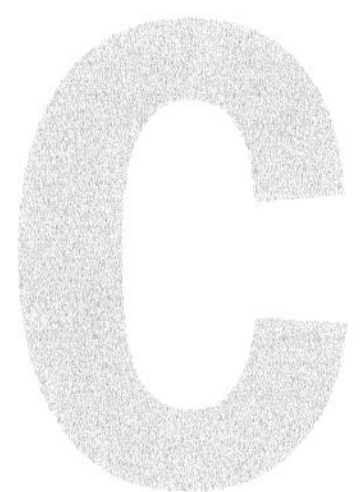

Mathematical supplement

This appendix contains the main mathematical elements used in neutron physics. Most of these elements are used in many other branches of physics, and can therefore be found scattered throughout many other books; some, such as the functions expressing the kernel of the Peierls operator, are probably less well known:

- The *Dirac distribution*, useful in representing the elementary sources of neutrons.
- The general properties of *linear operators*, which apply to the Boltzmann equation in particular.
- The *Fourier transform*, which facilitates the study of these operators when they are translation-invariant, which is the case for the Boltzmann equation in an infinite homogeneous medium, i.e. in fundamental mode.
- *Spherical harmonics*, eigenfunctions of the rotation-invariant operators, in particular the neutron scattering operator.
- The *eigenfunctions of the Laplace operator*, useful for studying flux in a bare homogeneous pile and, more generally, in fundamental mode.
- *Bessel functions*, used to express the solutions of the diffusion equation in cylindrical geometries; those most often used in reactors.
- Expressions for the *streaming operator* (involved in the differential expression for the transport operator) in the main coordinate systems.
- Expressions for the *Peierls operator* (integral expression for the tranport operator with the assumption of isotropic emission) for the main coordinate systems.
- The *integral exponential functions* involved in the expression for this operator in **x** planar geometry.
- The *Bickley-Naylor functions* involved in the expression for this operator in **x-y** geometry.
- The *quadrature formulae*, notably the *Gauss formulae*, very often chosen for the numerical calculation of integrals, for example in the $\mathbf{S_N}$ method or the method of probability tables.

C.1. Dirac distribution

The term *distribution* is a reminder that the Dirac "function" is not a "true" function, but a density function (§ 3.2.2).

C.1.1. Intuitive definition

The Dirac distribution $\delta(x)$ can be imagined as the limit of the distribution of a unit quantity placed on an axis x when it is completely concentrated at the origin. For example, the limit of a normalised Gaussian distribution:

$$\frac{1}{\sigma\sqrt{2\pi}} \exp\left(-\frac{x^2}{2\sigma^2}\right),$$

when σ tends to zero. This intuitive definition can be extended to 2- or 3-dimensional spaces.

C.1.2. Mathematical definitions

The Dirac distribution $\delta(x)$ is mathematically defined as a *functional* associating the number $f(0)$, i.e. the value at the origin, with any function $f(x)$ that is continuous at the origin:

$$\int_D f(x)\,\delta(x)\,dx = f(0),$$

whatever the interval D, provided that it contains the origin (if not, the integral is zero).

An equivalent definition, as shown by an integration by parts, is the derivative of the unit step:

$$\delta(x) = \frac{d}{dx}\Upsilon(x).$$

The definition as a functional can be generalised to n-dimensional space:

$$\iint \ldots \int_D f(x_1, x_2, \ldots, x_n)\,\delta(x_1)\,\delta(x_2)\ldots\delta(x_n)\,dx_1\,dx_2\ldots dx_n = f(0, 0, \ldots, 0),$$

(if D contains the origin; otherwise the integral is zero). This can be written more compactly:

$$\int_D f(\vec{r})\,\delta(\vec{r})\,d^n r = f(\vec{0}).$$

C.1.3. Definitions as Laplace operators

The following three formulae, where we have set $\rho^2 = x^2 + y^2$ et $r^2 = x^2 + y^2 + z^2$, are useful in diffusion equation calculations:

$$\delta(x) = \frac{1}{2}\Delta|x|, \quad \delta(\vec{\rho}) = \frac{1}{2\pi}\Delta(\ln\rho), \quad \delta(\vec{r}) = -\frac{1}{4\pi}\Delta\frac{1}{r}.$$

C.1.4. Generalisations

- The Dirac distribution placed at any point $\vec{r}_0$ is defined via a change of origin as follows:
$$\int_D f(\vec{r})\,\delta(\vec{r}-\vec{r}_0)\,d^n r = f(\vec{r}_0),$$
(if f is continuous at $\vec{r}_0$ and if D contains the point $\vec{r}_0$).

- By a change of variable, we obtain:
$$\delta(\lambda x) = \frac{\delta(x)}{|\lambda|},$$
and, more generally:
$$\delta[g(x)] = \sum_i \frac{\delta(x-x_i)}{|g'(x_i)|},$$
where the x_i are the zeros of g; this distribution $\delta[g(x)]$ *does not exist* if the derivative of g is zero for one of the x_i.

- By integration by parts, the derivatives of the Dirac distribution can be defined as functionals:
$$\int_D f(x)\,\delta'(x)\,dx = -f'(0), \qquad \int_D f(x)\,\delta^{(m)}(x)\,dx = (-)^m\, f^{(m)}\,(0),$$
(if D contains the origin and if the derivatives of f are continuous at the origin).

C.2. Linear operators

C.2.1. Definition

An operator A is a mathematical entity associating a function g with a function f, written as: $g = Af$. These functions depend on a scalar or vectorial variable x belonging to a certain set (we shall assume that the functions f and g belong to the same set and that the arguments of f and g belong to the same set, but these assumptions are not necessarily required).

An operator is "linear" if it observes the linear combinations:
$$A(\alpha_1 f_1 + \alpha_2 f_2) = \alpha_1 A f_1 + \alpha_2 A f_2,$$
where α_1 and α_2 are numbers.

C.2.2. Any linear operator is integral

An integral operator is an operator of the following form:
$$g(x) = (Af)(x) = \int k(x,x')f(x')dx',$$

where $k(x, x')$ is a given function of the variables x and x' called the *kernel of the operator*; the integral applies to the domain where the argument of f is defined. It is obvious that an integral operator is linear.

It can be shown that, conversely, any linear operator can be put in integral form; the kernel is:

$$k(x, x_0) = A\delta(x - x_0),$$

where the Dirac distribution is considered as a function of x.

C.2.3. Adjoint operator

The *scalar product* of two functions u and v is defined by:

$$\langle u, v\rangle = \int u(x)\, v(x)\, dx,$$

where the integral applies to the domain in which x is defined. (Here we limit ourselves to functions with real values; otherwise, replace the function placed as the first factor of the scalar product by its complex conjugate.)

The adjoint operator A^+ of a linear operator A is defined by the equation:

$$\langle u, Av\rangle = \langle A^+u, v\rangle,$$

satisfied for any u and v. For example, the adjoint of the operator d/dx is $-d/dx$; the adjoint of the Laplace operator Δ is Δ (an operator that is identical to its adjoint is called "self-adjoint").

It is easy to show that, in integral form, an operator can be converted to its adjoint by *permutation of the arguments x and x' of the kernel.*

C.2.4. Eigen elements of an operator

A function f is an *eigenfunction* of an operator A if:

$$Af = \mu f,$$

where μ is the associated *eigenvalue*.

The spectrum of eigenvalues can be discrete, continuous or a mixture of the two.

If there are several linearly independent functions associated with one eigenvalue, the eigenvalue is said to be *"degenerate"*; the order of degeneracy is the maximum number of associated linearly independent eigenfunctions.

An operator and its adjoint have the same spectrum of eigenvalues.

Any eigenfunction $f_i^+$of A^+ associated with μ_i is orthogonal to any eigenfunction f_j of A associated with μ_j, i.e. their scalar product is zero, *if μ_i and μ_j are different*:

$$\langle f_i^+, f_j\rangle = 0 \quad \text{si} \quad \mu_i \neq \mu_j.$$

We assume the spectrum of eigenvalues to be discrete. The following properties appear:

- The eigenfunctions f_i^+ of A^+ and the eigenfunctions f_j of A associated with the *same* degenerate eigenvalue can be chosen by pairs to be orthogonal:

$$\langle f_i^+, f_j \rangle = 0;$$

- Because all eigenfunctions are defined only to within a factor, these functions can be normed so that:

$$\langle f_i^+, f_i \rangle = 1,$$

- Under these conditions, the set is *orthonormed*:

$$\langle f_i^+, f_j \rangle = \delta_{ij},$$

where δ_{ij} is the Kronecker delta. If the eigenfunctions are orthonormed in this way and if we assumed that they form a complete basis, we can expand any function $\varphi(x)$:

$$\varphi(x) = \sum_n \varphi_n f_n(x)$$

with:

$$\varphi_n = \langle f_n^+, \varphi \rangle.$$

There are the following two properties in particular:

- The *kernel of the operator* can be written as:

$$k(x, x') = \sum_n \mu_n f_n^+(x') f_n(x);$$

- The Dirac distribution can be expanded on the eigenfunctions (like any distribution):

$$\delta(x - x') = \sum_n f_n^+(x') f_n(x).$$

This relation is the *closure relation*.

Note that, if the operator is self-adjoint (e.g. the Laplace operator), all of these relations can be written without the + sign.

C.3. Fourier transform

C.3.1. Translation-invariant operators

In neutron physics, we are interested in the case of the homogeneous, infinite medium known as the "fundamental mode". This case is characterised by translation invariance; if, for example, the neutron sources are displaced by a certain distance, the resulting flux undergoes the same displacement. More generally, if a function f undergoes a displacement x_0 — i.e. its argument x is replaced by $(x - x_0)$ — the function g related to it by a linear operator A undergoes the same displacement — i.e. $g(x)$ is replaced by $g(x - x_0)$. It can be shown that a consequence of this translation invariance is that the kernel $k(x, x')$ of A depends only on the *difference* $x - x'$; and, conversely, an operator whose kernel is a function of $x - x'$ only is translation invariant.

The general form of a translation-invariant operator is therefore:

$$g(x) = \int k(x - x')f(x')dx',$$

where the integral applies to the entire space under consideration, whether one-dimensional or multidimensional. This type of integral is called a *convolution product.* Using the symbol $*$ to denote this product, a translation-invariant operator can be expressed by the following formula:

$$g = k * f.$$

(This type of operator is also called a *convolution operator.*)

Notes:

1/ *The Dirac distribution* δ *is the unit of the convolution product.*

2/ *The convolution product is commutative.*

It is easy to show that the eigenfunctions of these operators are the exponential functions $e^{-a.x}$ where a is a constant (if x is an n-dimensional vector, the constant a must also be vectorial with dimension n and the point must be interpreted as a scalar product of the two vectors). If we limit the discussion to bounded eigenfunctions, we take a to be a pure imaginary number, i.e. $a = ib$ where b is real. The translation-invariant operators therefore *all have the same eigenfunctions,* let us say the functions $e^{-ib.x}$.

The only difference between them is in the associated eigenvalues. By replacing f by $e^{-ib.x}$ and g by λf, we can verify that the eigenvalue associated with $e^{-ib.x}$ is:

$$\lambda = \int k(x)\, e^{ib.x}\, dx.$$

This integral is the Fourier transform $\hat{k}(b)$ of the kernel $k(x)$.

Finally, we can see that the study of the fundamental mode is closely connected to the Fourier transform; that is why we shall review a few points on this subject.

C.3.2. Definitions of the Fourier transform of a function

The Fourier transform of a function $\varphi(x)$ exists if the function is square summable, which we assume here.

In the discussion that follows, x denotes a point in "physical" n-dimensional space, and b denotes a vector with n cartesian components in "dual" space. The integrals apply to all space considered as "physical" or "dual". dx and db are the respective volume elements.

There is a great deal of symmetry between these two spaces; the inversion formula, i.e. the formula giving the original function from its transform, is very similar to the formula defining the transform. (This inversion formula can be derived after the association between the constant function and the Dirac distribution has been established. The association is found by choosing a Gaussian distribution whose transform is also a Gaussian distribution and taking it to the limit.)

In the definitions there are several variants, which describe this symmetry to a greater or lesser extent.

- **Convention a**

  Direct formula:

$$\hat{\varphi}(b) = \int \varphi(x)\, e^{ib.x}\, dx.$$

  Inverse formula:

$$\varphi(x) = \frac{1}{(2\pi)^n} \int \hat{\varphi}(b) e^{-ib.x}\, db.$$

- **Convention b**

  Direct formula:

$$\hat{\varphi}(b) = \frac{1}{(2\pi)^{n/2}} \int \varphi(x)\, e^{ib.x}\, dx.$$

  Inverse formula:

$$\varphi(x) = \frac{1}{(2\pi)^{n/2}} \int \hat{\varphi}(b)\, e^{-ib.x}\, db.$$

- **Convention c**

  Direct formula:

$$\hat{\varphi}(b) = \int \varphi(x) e^{2\pi ib.x}\, dx.$$

  Inverse formula:

$$\varphi(x) = \int \hat{\varphi}(b)\, e^{-2\pi ib.x}\, db.$$

C.3.3. Fourier transform of a convolution product

A convolution product is transformed into a simple product: this is undoubtedly the property of the Fourier transform that has the most practical interest.

In particular, it provides a very simple way of finding the previous results on the eigen elements of translation-invariant operators.

C.3.4. One-dimensional examples (convention a)

A Gaussian is associated with a Gaussian. If the first is "wide", the second is "narrow", and vice-versa

$$\exp(-\alpha x^2) \Longrightarrow \sqrt{\pi/\alpha}\exp\left(-\frac{b^2}{4\alpha}\right).$$

If the parameter α tends to zero, we see that:

$$1 \Rightarrow 2\pi\delta(b).$$

Conversely, by definition of the Dirac distribution, we see that:

$$\delta(x) \Longrightarrow 1.$$

By differentiating the inversion formula with respect to x and then applying a recurrence method, we establish that the Fourier transforms of a function and its successive derivatives are connected by:

$$\varphi(x) \Longrightarrow \hat{\varphi}(b),$$
$$\frac{d\varphi(x)}{dx} \Longrightarrow -ib\hat{\varphi}(b),$$
$$\frac{d^2\varphi(x)}{dx^2} \Longrightarrow -b^2\hat{\varphi}(b),$$
$$\frac{dx^3\varphi(x)}{dx^3} \Longrightarrow +ib^3\hat{\varphi}(b),$$
$$\frac{d^4\varphi(x)}{dx^4} \Longrightarrow +b^4\hat{\varphi}(b),$$

and so forth.

From this, we conclude that the derivative of a convolution product $h = f * g$ is given by one or other of the expressions (not their sum):

$$\frac{dh(x)}{dx} = \frac{df(x)}{dx} * g(x) = f(x) * \frac{dg(x)}{dx}.$$

C.3.5. Fourier transforms in two- or three-dimensional space

The formulae for the derivatives can be generalised to multidimensional cases (we use the notation $\vec{r}$ instead of x to emphasise this, and remain with convention **a**):

$$\varphi(\vec{r}) \Longrightarrow \hat{\varphi}(\vec{b}),$$
$$\overrightarrow{\text{grad}}\,\varphi(\vec{r}) \Longrightarrow -i\vec{b}\hat{\varphi}(\vec{b}),$$
$$\Delta\,\varphi(\vec{r}) \Longrightarrow -b^2\hat{\varphi}(\vec{b}).$$

The Fourier transforms of functions factorised along the cartesian directions are themselves factorised along the cartesian directions, and each factor is given by the *one*-dimensional transformation formula.

A similar comment can be made for $[(x, y), z]$ or analogous factorised functions.

C.3.6. Fourier transforms of symmetric functions (convention a)

The transformation formulae are simpler if we consider functions that are symmetric about the origin, i.e. even (one dimension), with a rotational symmetric (two dimensions) or a spherical symmetry (three dimensions); note also that the transforms have the same symmetry. With convention **a**, and after integration on the angles in the two- and three-dimensional cases, the formulae are written as:

– *One dimension:*

$$\hat{\varphi}(b) = \int_0^{\infty} \varphi(x)\cos(bx)2dx.$$

– *Two dimensions:*

$$\hat{\varphi} = \int_0^{\infty} \varphi(\rho) J_0(b\rho)2\pi\rho\, d\rho.$$

– *Three dimensions:*

$$\hat{\varphi}(b) = \int_0^{\infty} \varphi(r)\frac{\sin(br)}{br}4\pi r^2\, dr.$$

The inverse formulae are analogous; the names of the variables are simply interchanged and the factors $1/(2\pi)^n$ are added in front of the integrals.

By expanding these Fourier transforms in powers of b, we see that the expansion coefficients are the *successive even moments* of the space variable, to within a numerical factor:

– *One dimension:*

$$\hat{\varphi}(b) = \left[1 - \frac{b^2}{2}\langle x^2\rangle + \frac{b^4}{24}\langle x^4\rangle + \ldots\right]\int_0^{\infty} \varphi(x)2\, dx.$$

– *Two dimensions:*

$$\hat{\varphi}(b) = \left[1 - \frac{b^2}{4}\langle x^2\rangle + \frac{b^4}{64}\langle x^4\rangle + \ldots\right]\int_0^{\infty} \varphi(\rho)2\pi\rho\, d\rho.$$

– *Three dimensions:*

$$\hat{\varphi}(b) = \left[1 - \frac{b^2}{6}\langle x^2\rangle + \frac{b^4}{120}\langle x^4\rangle + \ldots\right]\int_0^{\infty} \varphi(r)4\pi r^2\, dr.$$

C.3.7. Poisson summation formula

This formula expresses the fact that the two series obtained by summing the values of a function with integer abscissas and its Fourier transform (note: with convention **c**), also with integer abscissas, have the same sum:

$$\sigma = \sum_{m=-\infty}^{+\infty} \varphi(m) = \sum_{m=-\infty}^{+\infty} \hat{\varphi}(m).$$

This theorem transforms a slowly converging series into a rapidly converging series and vice-versa. This can be shown by introducing the following function:

$$e(x) = \sum_{m=-\infty}^{+\infty} \delta(x - m),$$

which is equal to its Fourier transform, as we can see by comparing the Fourier transform to the Fourier series representation.

The Poisson summation formula can be applied more generally to double and triple series, etc.

C.3.8. Eigenvalues of translation-invariant operators

We have seen that the eigenvalue of a translation-invariant operator associated with the eigenfunction $e^{-ib.x}$ is the Fourier transform $\hat{k}$ of the kernel k taken for the argument b.

In general, however, it turns out to be awkward to use this approach to the calculation, i.e. writing k and then calculating $\hat{k}$; it is preferable to apply the operator directly to the function $e^{-ib.x}$ and observe the result. For example for the Laplace operator, simply calculating the derivatives shows that:

$$\Delta e^{-i\vec{b}\cdot\vec{r}} = -b^2 \, e^{-i\vec{b}\cdot\vec{r}},$$

and therefore:

$$\lambda = \hat{k}(b) = -b^2.$$

(The kernel of this operator is the Laplace operator of the Dirac distribution and is quite difficult to imagine.)

C.3.9. Linear operators on an infinite, regular lattice

a) Translation invariance on a lattice

An infinite, regular lattice has a *periodicity*, which can be one-, two-, or three-dimensional. In this case, we limit the discussion to a two-dimensional lattice, which is the type most often created (approximately) in reactors. Seen in cross-section, the lattice is a regular, infinite tiling of the plan by identical mesh elements. The one-dimensional case (approximately created in plate reactors) and the three-dimensional case are very similar.

Like the infinite homogeneous medium, an infinite, regular lattice is characterised by the property of translation invariance, but *only* for translations of a *whole number of mesh elements*, i.e. translations ending in a point similar to the starting point, but located in a different mesh element. In a two-dimensional lattice, such translations are defined by the set of vectors of the following form:

$$\vec{t} = m\vec{p} + n\vec{q},$$

where $\vec{p}$ and $\vec{q}$ are the two vectors characterising unit displacements of a mesh element to a neighbouring mesh element, and m and n are integers that can be positive, negative or zero.

b) Translation-invariant operators on a lattice

A linear operator A on an infinite, regular lattice will be characterised by the invariance for these translations (for example, the displacement of neutron sources on a lattice: it leads to the same displacement of the flux). If the operator is written in integral form, its kernel k will be characterised by:

$$k(\vec{r} + m\vec{p} + n\vec{q}; \vec{r'} + m\vec{p} + n\vec{q}) = k(\vec{r}; \vec{r'}).$$

c) Eigen elements of translation-invariant operators on a lattice

By substitution, we verify that the eigenfunctions are *factorised* functions:

$$F(\vec{r}) = f(\vec{r}).e^{-i\vec{b}\cdot\vec{r}}.$$

The second factor (which can also be written as $e^{-\vec{a}\cdot\vec{r}}$ when the integrals converge) is the same as in the homogeneous case.

The first factor is a *periodic function* (with the periodicity of the lattice) which appears as an eigenfunction:

$$\lambda f(\vec{r}) = \int_{\text{One mesh}} \hat{k}(\vec{r}; \vec{r'}; \vec{b})\, f(\vec{r'})\, d^2 r',$$

of the operator whose kernel is defined by:

$$\hat{k}(\vec{r}; \vec{r'}; \vec{b}) = \sum_{m,n} k(\vec{r}; \vec{r'} + m\vec{p} + n\vec{q}) \exp\left[i\vec{b}\cdot(\vec{r} - \vec{r'} - m\vec{p} - n\vec{q})\right],$$

with the appropriate eigenvalue λ (depending on the choice of $\vec{b}$).

(Note that we can return to the case of the infinite, homogeneous medium by reducing the mesh to one point.)

As before, it is usually simpler to substitute this form for F in the equations as they appear and to examine the value of λ that satisfies them.

Note that the "macroscopic" exponential $e^{-i\vec{b}\cdot\vec{r}}$ and the periodic function f are both functions with complex values. The 'fine periodic structure' therefore has a symmetric component and an antisymmetric component.

C.4. Spherical harmonics

C.4.1. Rotation invariance

The trigonometric functions sine and cosine are used to represent the periodic functions; this is what happens in a Fourier expansion. If, where necessary, the period is adjusted to 2π by a change of variable, a periodic function can be considered as a function of the *direction in a plane* identified by a *longitude* φ. The functions $e^{in\varphi}$ (where n is an integer), whose real and imaginary parts are $\cos n\varphi$ and $\sin n\varphi$, are the *eigenfunctions of the rotation-invariant linear operators,* i.e. the operators (associating a function g with a function f) expressed, when written in integral form, by:

$$g(\varphi) = \int_{(2\pi)} k(\varphi - \varphi')\, f(\varphi')\, d\varphi',$$

with a kernel k that depends only on the difference $\varphi - \varphi'$, and is thus rotation invariant. This can be verified by taking $f(\varphi) = e^{in\varphi}$ under the integral; calculations show that the associated eigenvalue is given by the following formula:

$$\lambda_n = \int_{(2\pi)} k(\psi) e^{-in\psi}\, d\psi.$$

Rotation-invariant operators in a plane therefore have all the same eigenfunctions and differ only in their eigenvalues. A Fourier expansion is an expansion on these eigenfunctions. The choice of eigenfunctions to be used in the expansion can simplify all problems with this type of rotation invariance.

In three-dimensional space, spherical harmonics play the same role as the functions $\cos n\varphi$ and $\sin n\varphi$ in the plane. These functions are useful in neutron physics because the *cross-sections are rotation invariant* (in practice, materials are isotropic on the scale of the mean free path of neutrons). In particular, they can simplify the scattering operator.

Two angles are required to identify a direction in space (Figure 3.1): we often use the latitude (measured from the equator) or the colatitude θ (measured from the North Pole) and the longitude φ (defined in the same way as in the planar problem because it identifies the direction of the *projection* onto the equatorial plane of the direction in space).

Note that the three cartesian coordinates of a unit vector $\vec{\Omega}$ are (taking the longitude from the $\vec{x}$ direction):

$$\Omega_x = \sin\theta \cos\varphi,$$
$$\Omega_y = \sin\theta \sin\varphi,$$
$$\Omega_z = \cos\theta,$$

and that the solid angle element is expressed by:

$$d^2\Omega = \sin\theta\, d\theta\, d\varphi.$$

A rotation-invariant operator has a kernel k that depends only on the angle between $\vec{\Omega}$ and $\vec{\Omega'}$, the only parameter that is invariant for any rotation. It is often more convenient to use the cosine of this angle:

$$\nu = \vec{\Omega} \cdot \vec{\Omega'}.$$

Spherical harmonics are the eigenfunctions of rotation-invariant linear operators. They are usually introduced and studied using a *particular* (rotation invariant) operator, the angular part E of the Laplace operator Δ. This appears when Δ is written in spherical coordinates:

$$\Delta\cdot = \frac{\partial^2\cdot}{\partial r^2} + \frac{2}{r}\frac{\partial\cdot}{\partial r} + \frac{1}{r^2}E\cdot,$$

with:

$$E\cdot = \frac{1}{\sin\theta}\frac{\partial}{\partial\theta}\left(\sin\theta\frac{\partial\cdot}{\partial\theta}\right) + \frac{1}{\sin^2\theta}\frac{\partial^2}{\partial\varphi^2}.$$

Eigenfunctions that are independent of φ can be found. They are useful in problems that are invariant with respect to φ i.e. rotation invariant about the $\vec{z}$ axis: these are *Legendre* polynomials. *Spherical harmonics* are the general eigenfunctions that are dependent on both θ and φ.

The equation defining spherical harmonics is written by setting $\mu = \cos\theta$ and by using $-\lambda$ to denote the eigenvalue:

$$(1-\mu^2)\frac{\partial^2 Y}{\partial\mu^2} - 2\mu\frac{\partial Y}{\partial\mu} + \frac{1}{1-\mu^2}\frac{\partial^2 Y}{\partial\varphi^2} + \lambda Y = 0.$$

The equation defining functions that are independent of φ is reduced to:

$$(1-\mu^2)\frac{d^2 Y}{d\mu^2} - 2\mu\frac{dY}{d\mu} + \lambda Y = 0.$$

C.4.2. Legendre polynomials

A solution in the form of an expansion in powers of μ can be sought. In this way, we find a series with radius of convergence 1, which means that it diverges for at least one of the values $\mu = +1$ or $\mu = -1$, and therefore on at least one of the poles of the sphere, *unless* the series contains only a finite number of terms, i.e. is reduced to a polynomial. This is the case if and only if the eigenvalue has the following form (except for its sign):

$$\lambda = n(n+1).$$

These polynomial solutions, suitably normalised, are the Legendre polynomials, written as $P_n(\mu)$. The main properties are as follows:

- $P_n(\mu)$ is an n-degree polynomial;
- with parity $(-)^n$;
- is normalised by $P_n(1) = 1$;
- these polynomials are orthogonal in pairs but not normed:

$$\int_{-1}^{+1} P_k(\mu)P_n(\mu) = \delta_{kn}\frac{2}{2n+1};$$

- the Rodrigues formula:

$$P_n(\mu) = \frac{1}{2^n\, n!}\left(\frac{d}{d\mu}\right)^{(n)}(\mu^2-1)^n,$$

  could also be a definition of these polynomials;
- There is a recurrence relation between three successive polynomials:

$$(n+1)P_{n+1}(\mu) - (2n+1)\mu P_n(\mu) + nP_{n-1}(\mu) = 0,$$

which makes it possible to calculate them from $P_0(\mu) = 1$ and $P_1(\mu) = \mu$. The first thirteen are as follows:

$$P_0(\mu) = 1,$$
$$P_1(\mu) = \mu,$$
$$P_2(\mu) = (3\mu^2 - 1)/2,$$
$$P_3(\mu) = (5\mu^3 - 3\mu)/2,$$
$$P_4(\mu) = (35\mu^4 - 30\mu^2 + 3)/8,$$
$$P_5(\mu) = (63\mu^5 - 70\mu^3 + 15\mu)/8,$$
$$P_6(\mu) = (231\mu^6 - 315\mu^4 + 105\mu^2 - 5)/16,$$
$$P_7(\mu) = (429\mu^7 - 693\mu^5 + 315\mu^3 - 35\mu)/16,$$
$$P_8(\mu) = (6435\mu^8 - 12012\mu^6 + 6930\mu^4 - 1260\mu^2 + 35)/128,$$
$$P_9(\mu) = (12155\mu^9 - 25740\mu^7 + 18018\mu^5 - 4620\mu^3 + 315\mu)/128,$$
$$P_{10}(\mu) = (46189\mu^{10} - 109395\mu^8 + 90090\mu^6 - 30030\mu^4 + 3465\mu^2 - 63)/256,$$
$$P_{11}(\mu) = (88179\mu^{11} - 230945\mu^9 + 218790\mu^7 - 90090\mu^5 + 15015\mu^3 - 693\mu)/256,$$
$$P_{12}(\mu) = (676039\mu^{12} - 1939938\mu^{10} + 2078505\mu^8 - 1021020\mu^6 + 225225\mu^4 - 18018\mu^2 + 231)/1024;$$

(In a computer calculation, it is better to use this recurrence relation than these formulae.)

- There is a generating function:

$$\frac{1}{\sqrt{1 - 2z\mu + z^2}} = \sum_{n=0}^{\infty} P_n(\mu) z^n;$$

(Note that the first term is the inverse of the distance between two points located at distances 1 and z from the origin in the directions forming an angle $\theta = \text{Arccos}\,\mu$.)

- Characteristics of the zeros of Legendre polynomials:
  - the polynomial P_n has n distinct, real zeros:
  - they are between −1 and +1,
  - the zeros of P_n are interspersed with those of P_{n+1};
- the essential property of these polynomials is that they constitute a complete basis, i.e. they enable the functions of μ to be expanded:

$$f(\mu) = \sum_{n=0}^{\infty} f_n P_n(\mu), \qquad f_n = \frac{2n+1}{2} \int_{-1}^{+1} f(\mu) P_n(\mu)\, d\mu.$$

C.4.3. Spherical harmonics

Spherical harmonics are constructed in a similar way. In addition to the constraint of convergence of the series for all values of μ, there is the constraint of φ-periodicity. This periodicity is obtained by seeking factorised solutions, where one of the factors is $\exp(im\varphi)$ and the other is a function of θ; to within a factor, this is one of the *Legendre functions associated with the polynomials* having the following main properties:

- Definition:

$$P_n^m(\mu) = (1-\mu^2)^{m/2}\left(\frac{d}{d\mu}\right)^{(m)} P_n(\mu), \quad (0 \leqslant m \leqslant n);$$

- Orthogonality and norm:

$$\int_{-1}^{+1} P_k^m(\mu)P_n^m(\mu)d\mu = \delta_{kn}\frac{2}{2n+1}\frac{(n+m)!}{(n-m)!};$$

- Recurrence relations:

$$\begin{aligned}
(2n+1)\mu P_n^m(\mu) &= (n+m)P_{n-1}^m(\mu) + (n-m+1)P_{n+1}^m(\mu),\\
(2n+1)\sqrt{1-\mu^2}P_n^m(\mu) &= P_{n+1}^{m+1}(\mu) - P_{n-1}^{m+1}(\mu),\\
(2n+1)\sqrt{1-\mu^2}P_n^m(\mu) &= (n+m-1)(n+m)P_{n-1}^{m-1}(\mu)\\
&\quad - (n-m+1)(n-m+2)P_{n+1}^{m-1}(\mu).
\end{aligned}$$

Spherical harmonics are obtained by renorming to one the products of $\exp(im\varphi)$ by the associated functions $P_n^m(\mu)$. Here are the main formulae concerning the usual normalising conventions and the most important properties:

$$\begin{aligned}
Y_n^m(\vec{\Omega}) &= (-)^m\sqrt{\frac{2n+1}{4\pi}\frac{(n-m)!}{(n+m)!}}P_n^m(\mu)e^{im\varphi}, && (0 \leqslant m \leqslant n),\\
Y_n^m(\vec{\Omega}) &= (-)^m Y_n^{*-m}(\vec{\Omega}), && (m < 0),
\end{aligned}$$

where the asterisk denotes the complex conjugate function. (Other conventions exist; most notably to obtain real functions.)

- These functions are orthonormed: for two functions characterised by the same number m, but differing by the number n, the orthogonality is the result of that of the Legendre functions; for two functions that differ by the number m, the orthogonality is the result of that of the exponential factors; renormalisation is performed in view of the formulae giving the norm of the Legendre functions;
- The eigenvalues depend *only* on the number n: $\lambda_n = -n(n+1)$;
- This means that the concept of *eigen spaces* is more significant here than that of eigenfunctions. These are the functional spaces created by the $2n+1$ spherical harmonics of a given order n; i.e. *every* linear combination of *n-th* order spherical harmonics is an eigenfunction of the operator E with the eigenvalue $\lambda_n = -n(n+1)$;

- The *addition formula* can be considered as a special case of this type of combination:

$$\sum_{m=-n}^{m=+n} Y_n^m(\vec{\Omega})Y_n^{*m}(\vec{\Omega}') = \frac{2n+1}{4\pi}P_n(\vec{\Omega}\cdot\vec{\Omega}');$$

(Note: The asterisk denotes the complex conjugate function.)

- The recurrence formulae on the spherical harmonics are deduced from those on the Legendre functions; they make it possible to express the products of a spherical harmonic with each of the components of the vector $\vec{\Omega}$ as a function of the neighbouring spherical harmonics;

- The spherical harmonics constitue a *complete basis*: that is their essential property. In practice, this means that the "functions" of $\vec{\Omega}$ used in physics can be expanded in terms of spherical harmonics:

$$f(\vec{\Omega}) = \sum_{n=0}^{\infty}\sum_{m=-n}^{m=+n} f_n^m\, Y_n^m(\vec{\Omega}),$$

with:

$$f_n^m = \int_{(4\pi)} f(\vec{\Omega})Y_n^{*m}(\vec{\Omega})d^2\Omega;$$

- This applies in particular to the Dirac distribution:

$$\sum_{n=0}^{\infty}\sum_{m=-n}^{m=+n} Y_n^m(\vec{\Omega})Y_n^{*m}(\vec{\Omega}') = \delta(\vec{\Omega}-\vec{\Omega}').$$

The existence of this formula, known as the *closure relation*, is equivalent to the completeness of the basis.

C.4.4. Rotation-invariant operators

As has been said, the operator E, which was used to construct the spherical harmonics, is only a rotation invariant particular operator. The *properties common to all rotation invariant operators* are as follows:

- Definition: an operator A is rotation invariant if:

$$AR = RA,$$

whatever the rotation R (in three-dimensional space, a rotation is defined by *three* angles, for example, both angles [colatitude and longitude] that define the axis of rotation and the angle of rotation about this axis); Note that turning a function means taking the function at the point obtained by the inverse rotation: $Rf(\vec{\Omega}) = f(R^{-1}\vec{\Omega})$. (This is obviously true for any geometric transformation.)

- Assuming A to be linear, expressing it in integral form and writing $AR = RA$, we can see that the consequence of rotation invariance is that the kernel must be dependent on the scalar product $\vec{\Omega} \cdot \vec{\Omega}'$ only (and vice-versa):

$$(Af)(\vec{\Omega}) = \int_{(4\pi)} k(\vec{\Omega} \cdot \vec{\Omega}') f(\vec{\Omega}') d^2\Omega'.$$

  In other words, the kernel depends only on the parameter related to $\vec{\Omega}$ and $\vec{\Omega}'$ that is invariant by any rotation, i.e. the angle between these two vectors (or its cosine);

- By expanding the kernel k in Legendre polynomials and using the spherical harmonic addition formula, we note that spherical harmonics are eigenfunctions of A and that the eigenvalue is independent of the number m.

  Rotation invariant operators therefore all have the same eigen spaces; those of the operator E, i.e. the functional spaces generated by the spherical harmonics of a fixed order *n;*

- The *eigenvalue* for n-th order spherical harmonics is:

$$\lambda_n = 2\pi \int_{-1}^{+1} k(\mu)\, P_n(\mu)\, d\mu$$

C.5. Eigenfunctions of the Laplace operator

C.5.1. Definition

An eigenfunction f of the Laplace operator Δ is a function transformed into itself by the action of this operator, to within a factor (eigenvalue); we shall write this factor as $-\mu$. The equation giving the eigenfunctions and eigenvalues will therefore be written as:

$$\Delta f + \mu f = 0.$$

C.5.2. Laplace operator

- **Cartesian coordinates**: x, y and z,

$$\Delta \cdot = \frac{\partial^2 \cdot}{\partial x^2} + \frac{\partial^2 \cdot}{\partial y^2} + \frac{\partial^2 \cdot}{\partial z^2}.$$

- **Cylindrical coordinates**: ρ, φ and z,

$$\Delta \cdot = \frac{\partial^2 \cdot}{\partial \rho^2} + \frac{1}{\rho}\frac{\partial \cdot}{\partial \rho} + \frac{1}{\rho^2}\frac{\partial^2 \cdot}{\partial \varphi^2} + \frac{\partial^2 \cdot}{\partial z^2}.$$

- **Spherical coordinates**: r, θ and φ,

$$\Delta \cdot = \frac{\partial^2 \cdot}{\partial r^2} + \frac{2}{r}\frac{\partial \cdot}{\partial r} + \frac{1}{r^2} E \cdot,$$

  with:

$$E \cdot = \frac{1}{\sin\theta}\frac{\partial}{\partial \theta}\left(\sin\theta \frac{\partial \cdot}{\partial \theta}\right) + \frac{1}{\sin^2\theta}\frac{\partial^2 \cdot}{\partial \varphi^2}.$$

C.5.3. Factorised general solutions

- **Notation**

$$\mu = \alpha + \beta + \gamma,$$

$$a = \sqrt{|\alpha|}, \quad b = \sqrt{|\beta|}, \quad c = \sqrt{|\gamma|}.$$

- If α (or β, γ respectively) is *positive*:

$$T = \sin \text{ or } \cos, \quad A = \sin \text{ or } \cos, \quad B = J \text{ or } Y.$$

- If α (or β, γ respectively) is *negative*:

$$T = \sinh \text{ or } \cosh, \quad A = \sin \text{ or } \cos, \quad B = I \text{ or } K.$$

(*Note:* J, Y, I and K denote Bessel functions; see below.)

- **Cartesian coordinates**: x, y and z,

$$f = T(ax)\,T(by)\,T(cz).$$

- **Cylindrical coordinates**: ρ, φ and z (n is an integer),

$$f = B_n(a\rho)A(n\varphi)T(cz).$$

- **Spherical coordinates**: r, θ and φ (n and m are integers),

$$f = \frac{B_{n+1/2}(ar)}{\sqrt{r}} Y_n^m(\theta, \varphi).$$

If n is zero (spherical symmetry problem), this can be written more simply:

$$f = \frac{T(ar)}{r}.$$

C.5.4. Eigenfunctions of the Laplace operator vanishing at the surface of a domain

In the formulae below, N denotes the norm of f (square root of the integral of f^2 over the domain considered) and $j_{\nu m}$ is the mth zero of J_ν.

a) Geometries depending on only one space variable

(see Figure C.1.)

- Infinite plate between abscissas 0 and a:

$$f = \sin\left(n\pi\frac{x}{a}\right), \qquad \mu = \frac{n^2\pi^2}{a^2}, \qquad N = \sqrt{a/2}.$$

– Infinite cylinder of radius R with rotational symmetry:

$$f = J_0\left(j_{0n}\frac{\rho}{R}\right), \quad \mu = \frac{j_{0n}^2}{R^2}, \quad N = \sqrt{\pi R^2} J_1(j_{0n}).$$

– Sphere of radius R with spherical symmetry:

$$f = \frac{\sin\left(n\pi\frac{r}{R}\right)}{r}, \qquad \mu = \frac{n^2\pi^2}{R^2}, \qquad N = \sqrt{2\pi R}.$$

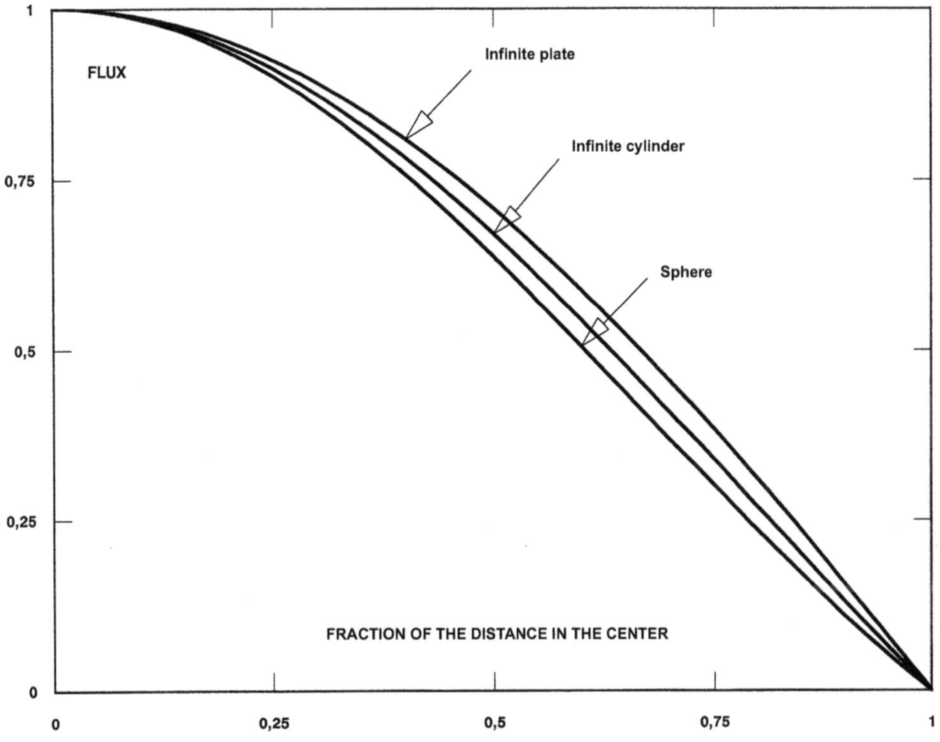

Figure C.1. Fundamental eigenfunction of the Laplace operator in the three geometries described by only one space variable (flux in the corresponding bare piles).

b) Geometry depending on only two space variables

– Infinite cylinder of radius R

$$f = J_m\left(j_{mn}\frac{\rho}{R}\right) e^{im\varphi}, \quad \mu = \frac{j_{mn}^2}{R^2}, \quad N = \sqrt{\pi R^2} J'_m(j_{mn}).$$

c) Geometries depending on the three space variables

- Rectangular parallelepiped with edges a, b and c:

$$f = \sin\left(\ell\pi\frac{x}{a}\right)\sin\left(m\pi\frac{y}{b}\right)\sin\left(n\pi\frac{z}{c}\right),$$

$$\mu = \frac{\ell^2\pi^2}{a^2} + \frac{m^2\pi^2}{b^2} + \frac{n^2\pi^2}{c^2}, \qquad N = \sqrt{abc/8}.$$

- Cylinder of radius R and height H

$$f = J_m\left(j_{mn}\frac{\rho}{R}\right)e^{im\varphi}\sin\left(\ell\pi\frac{z}{H}\right),$$

$$\mu = \frac{j_{mn}^2}{R^2} + \frac{\ell^2\pi^2}{H^2}, \qquad H = \sqrt{\pi R^2 H/2}J'_m(j_{mn}).$$

- Sphere of radius R (for n zero, we find the above functions to within a factor):

$$f = \frac{J_{n+1/2}\left(j_{n+1/2,\ell}\frac{r}{R}\right)}{\sqrt{r}}Y_n^m(\theta, \varphi),$$

$$\mu = \frac{j_{n+1/2,\ell}^2}{R^2}, \qquad N = \frac{R}{\sqrt{2}}J'_{n+1/2}(j_{n+1/2,\ell}).$$

C.6. Bessel functions

C.6.1. Bessel equations

Bessel functions are the solutions of the following second-order differential equations where ν is a real parameter; there are functions of the first kind which are regular at the origin, and functions of the second kind which are singular at the origin.

- ν-th order Bessel equation

$$\frac{d^2y}{dx^2} + \frac{1}{x}\frac{dy}{dx} + \left(1 - \frac{\nu^2}{x^2}\right)y = 0.$$

General solution: linear combination of $J_\nu(x)$, regular at the origin, and $Y_\nu(x)$, singular at the origin, sometimes written $N_\nu(x)$:

$$Y_\nu(x) \equiv N_\nu(x) = \frac{\cos\pi\nu J_\nu(x) - J_{-\nu}(x)}{\sin\pi\nu}.$$

- ν-th order modified Bessel equation ("-" sign in front of the y term).

$$\frac{d^2y}{dx^2} + \frac{1}{x}\frac{dy}{dx} - \left(1 + \frac{\nu^2}{x^2}\right)y = 0.$$

General solution: linear combination of $I_\nu(x) = i^{-\nu}J_\nu(x)$, regular at the origin, and $K_\nu(x)$, singular at the origin:

$$K_\nu(x) = \frac{\pi}{2}\frac{I_{-\nu}(x) - I_\nu(x)}{\sin\pi\nu}.$$

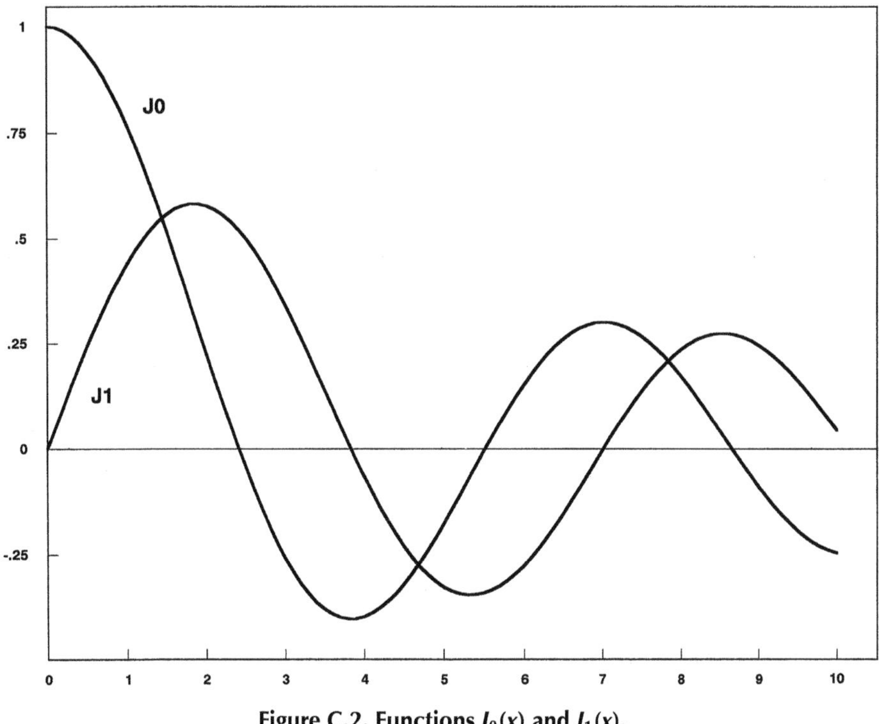

Figure C.2. Functions $J_0(x)$ and $J_1(x)$.

Figures C.2 to C.5 give the curves representing 0th and 1st order functions, which are the most useful types in neutron physics.

C.6.2. Expansions at the origin

$$J_n(x) = \left(\frac{x}{2}\right)^n \sum_{r=0}^{\infty} \frac{(-)^r}{r!(r+n)!} \left(\frac{x}{2}\right)^{2r},$$

$$Y_n(x) = \frac{2}{\pi}\left(\gamma + \ln\frac{x}{2}\right) J_n(x) - \frac{1}{\pi}\sum_{r=0}^{n-1} \frac{(n-r-1)!}{r!}\left(\frac{x}{2}\right)^{2r-n}$$

$$-\frac{1}{\pi}\sum_{r=0}^{\infty}(-)^r\left[1+\frac{1}{2}+\cdots+\frac{1}{r}+1+\frac{1}{2}+\cdots+\frac{1}{r+n}\right]\frac{(x/2)^{2r+n}}{r!(r+n)!},$$

$$I_n(x) = \left(\frac{x}{2}\right)^n \sum_{r=0}^{\infty} \frac{1}{r!(r+n)!} \left(\frac{x}{2}\right)^{2r},$$

$$K_n(x) = (-)^{n+1}\left(\gamma + \ln\frac{x}{2}\right) I_n(x) + \frac{1}{2}\sum_{r=0}^{n-1}(-)^r \frac{(n-r-1)!}{r!}\left(\frac{x}{2}\right)^{2r-n}$$

$$(-)^n\frac{1}{2}\sum_{r=0}^{\infty}(-)^r\left[1+\frac{1}{2}+\cdots+\frac{1}{r}+1+\frac{1}{2}+\cdots+\frac{1}{r+n}\right]\frac{(x/2)^{2r+n}}{r!(r+n)!}.$$

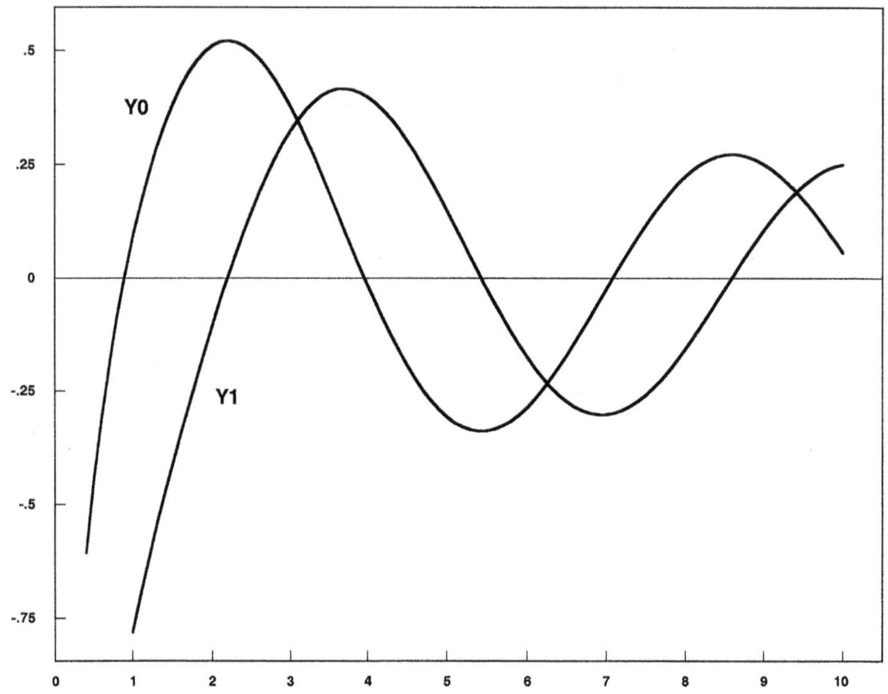

.5
.25
0
-.25
-.5
-.75
Y0
Y1
0
1
2
3
4
5
6
7
8
9
10

Figure C.3. Functions $Y_0(x)$ and $Y_1(x)$.

5
4.5
4
3.5
3
2.5
2
1.5
1
.5
0
I0
I1
0
.25
.5
.75
1
1.25
1.5
1.75
2
2.25
2.5
2.75
3

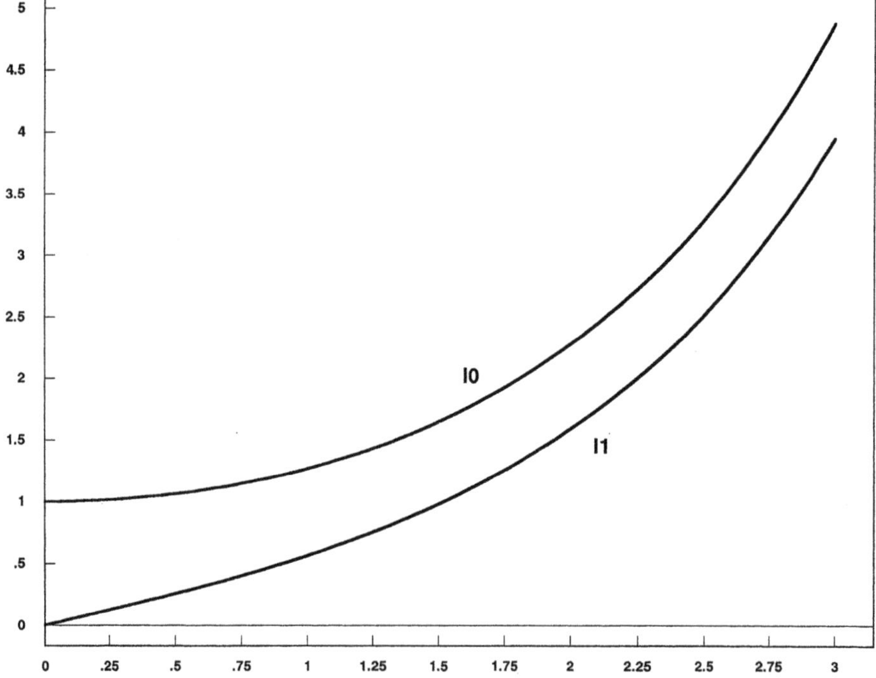

Figure C.4. Functions $I_0(x)$ and $I_1(x)$.

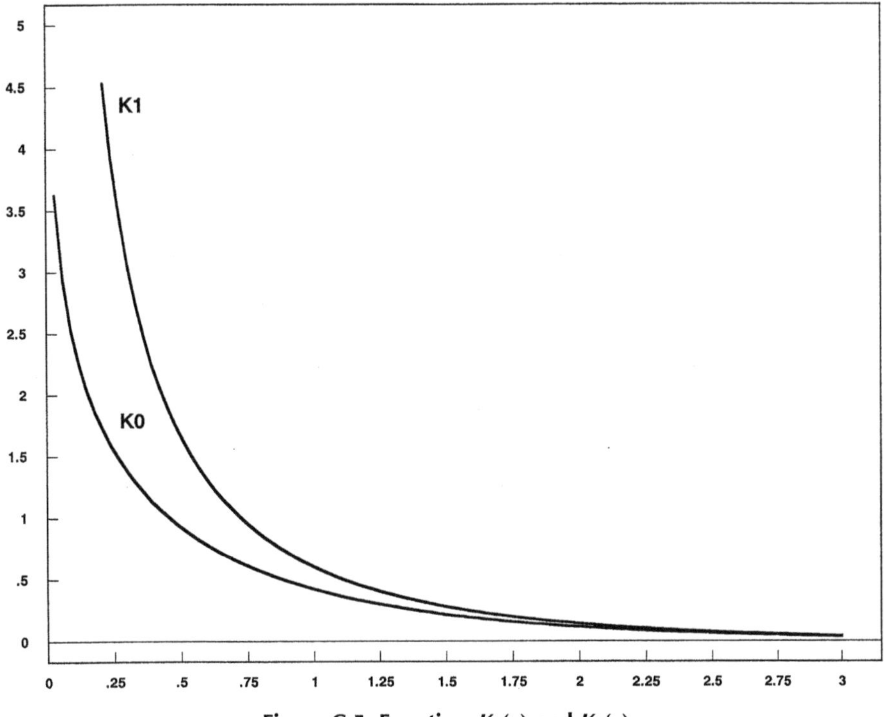

Figure C.5. Functions $K_0(x)$ and $K_1(x)$.

Notes:

- $\gamma = 0.577215665\ldots$ is Euler's constant.
- If the argument m of the factorials is not an integer, replace $m!$ with $\Gamma(m+1)$.
- If r is zero, replace the term in square brackets with $[1 + 1/2 + \cdots + 1/n]$.

Special cases:

$$J_0(x) = 1 - \frac{x^2}{4} + \frac{x^4}{64} - \frac{x^6}{2304} + \cdots,$$

$$J_1(x) = \frac{x}{2} - \frac{x^3}{16} + \frac{x^5}{384} - \frac{x^7}{18432} + \cdots,$$

$$Y_0(x) = \frac{2}{\pi}\left[\left(\gamma + \ln\frac{x}{2}\right)J_0(x) + \frac{x^2}{4} - \frac{3x^4}{128} + \frac{11x^6}{13824} + \cdots\right],$$

$$Y_1(x) = \frac{2}{\pi}\left[\left(\gamma + \ln\frac{x}{2}\right)J_1(x) - \frac{1}{x} - \frac{x}{4} + \frac{5x^3}{64} - \frac{5x^5}{1152} + \frac{47x^7}{442368} - \cdots\right],$$

$$I_0(x) = 1 + \frac{x^2}{4} + \frac{x^4}{64} + \frac{x^6}{2304} + \cdots,$$

$$I_1(x) = \frac{x}{2} + \frac{x^3}{16} + \frac{x^5}{384} + \frac{x^7}{18432} + \cdots,$$
$$K_0(x) = -\left(\gamma + \ln\frac{x}{2}\right) I_0(x) + \frac{x^2}{4} + \frac{3x^4}{128} + \frac{11x^6}{13824} + \cdots,$$
$$K_1(x) = \left(\gamma + \ln\frac{x}{2}\right) I_1(x) + \frac{1}{x} - \frac{x}{4} - \frac{5x^3}{64} - \frac{5x^5}{1152} - \frac{47x^7}{442368} - \cdots.$$

C.6.3. Asymptotic expansions

$$J_\nu(x) = \left[P_\nu(x)\cos\varphi - Q_\nu(x)\sin\varphi\right]\sqrt{\frac{2}{\pi x}},$$
$$Y_\nu(x) = \left[P_\nu(x)\sin\varphi + Q_\nu(x)\cos\varphi\right]\sqrt{\frac{2}{\pi x}},$$

with:

$$\varphi = x - \pi\frac{\nu + 1/2}{2},$$
$$P_\nu(x) = 1 - \frac{(4\nu^2 - 1^2)(4\nu^2 - 3^2)}{2!(8x)^2} + \frac{(4\nu^2 - 1^2)(4\nu^2 - 3^2)(4\nu^2 - 5^2)(4\nu^2 - 7^2)}{4!(8x)^4} - \cdots,$$
$$Q_\nu(x) = \frac{4\nu^2 - 1^2}{1!8x} - \frac{(4\nu^2 - 1^2)(4\nu^2 - 3^2)(4\nu^2 - 5^2)}{3!(8x)^3} + \cdots,$$

and:

$$I_\nu(x) = \frac{e^x}{\sqrt{2\pi x}}\left[1 - \frac{4\nu^2 - 1^2}{1!8x} + \frac{(4\nu^2 - 1^2)(4\nu^2 - 3^2)}{2!(8x)^2} - \cdots\right],$$
$$K_\nu(x) = \frac{\sqrt{\pi e^{-x}}}{\sqrt{2x}}\left[1 + \frac{4\nu^2 - 1^2}{1!8x} + \frac{(4\nu^2 - 1^2)(4\nu^2 - 3^2)}{2!(8x)^2} - \cdots\right].$$

C.6.4. Recurrence relations

$$xJ'_\nu(x) = \nu J_\nu(x) - xJ_{\nu+1}(x),$$
$$xJ'_\nu(x) = -\nu J_\nu(x) + xJ_{\nu-1}(x),$$
$$2\nu J_\nu(x) = xJ_{\nu-1}(x) + xJ_{\nu+1}(x),$$
$$2J'_\nu(x) = J_{\nu-1}(x) - J_{\nu+1}(x),$$
$$4J''_\nu(x) = J_{\nu-2}(x) - 2J_\nu(x) + J_{\nu+2}(x).$$

$$xY'_\nu(x) = \nu Y_\nu(x) - xY_{\nu+1}(x),$$
$$xY'_\nu(x) = -\nu Y_\nu(x) + xY_{\nu-1}(x),$$
$$2\nu Y_\nu(x) = xY_{\nu-1}(x) + xY_{\nu+1}(x),$$
$$2Y'_\nu(x) = Y_{\nu-1}(x) - Y_{\nu+1}(x),$$
$$4Y''_\nu(x) = Y_{\nu-2}(x) - 2Y_\nu(x) + Y_{\nu+2}(x).$$

$$\begin{aligned}
xI'_\nu(x) &= \nu I_\nu(x) + xI_{\nu+1}(x),\\
xI'_\nu(x) &= -\nu I_\nu(x) + xI_{\nu-1}(x),\\
2\nu I_\nu(x) &= xI_{\nu-1}(x) - xI_{\nu+1}(x),\\
2I'_\nu(x) &= I_{\nu-1}(x) + I_{\nu+1}(x),\\
4I''_\nu(x) &= I_{\nu-2}(x) + 2I_\nu(x) + I_{\nu+2}(x).
\end{aligned}$$

$$\begin{aligned}
xK'_\nu(x) &= \nu K_\nu(x) - xK_{\nu+1}(x),\\
xK'_\nu(x) &= -\nu K_\nu(x) - xK_{\nu-1}(x),\\
2\nu K_\nu(x) &= -xK_{\nu-1}(x) + xK_{\nu+1}(x),\\
2K'_\nu(x) &= -K_{\nu-1}(x) - K_{\nu+1}(x),\\
4K''_\nu(x) &= K_{\nu-2}(x) + 2K_\nu(x) + K_{\nu+2}(x).
\end{aligned}$$

Particular cases:

$$\begin{aligned}
J'_0(x) &= -J_1(x),\\
Y'_0(x) &= -Y_1(x),\\
I'_0(x) &= I_1(x),\\
K'_0(x) &= -K_1(x).
\end{aligned}$$

C.6.5. Integrals

$$\begin{aligned}
&\int x^n J_{n-1}(x)dx = x^n J_n(x), && \int x^{-n} J_{n+1}(x)dx = -x^{-n} J_n(x),\\
&\int x^n Y_{n-1}(x)dx = x^n Y_n(x), && \int x^{-n} Y_{n+1}(x)dx = -x^{-n} Y_n(x),\\
&\int x^n I_{n-1}(x)dx = x^n I_n(x), && \int x^{-n} I_{n+1}(x)dx = x^{-n} I_n(x),\\
&\int x^n K_{n-1}(x)dx = -x^n K_n(x), && \int x^{-n} K_{n+1}(x)dx = -x^{-n} K_n(x).
\end{aligned}$$

Lommel Integrals

$$\begin{aligned}
\int J_\nu(kx)J_\nu(\ell x)x\,dx &= \frac{x}{k^2-\ell^2}[kJ_\nu(\ell x)J_{\nu+1}(kx) - \ell J_\nu(kx)J_{\nu+1}(\ell x)],\\
&= \frac{x}{k^2-\ell^2}[\ell J_{\nu-1}(\ell x)J_\nu(kx) - kJ_{\nu-1}(kx)J_\nu(\ell x)],\\
\int [J_\nu(kx)]^2 x\,dx &= \frac{x^2}{2}\left\{\left(1-\frac{\nu^2}{k^2x^2}\right)[J_\nu(kx)]^2 + [J'_\nu(kx)]^2\right\},
\end{aligned}$$

and analogous relationships with the other functions.

C.6.6. Wronskians

$$J_{\nu+1}(x)J_{-\nu}(x) + J_\nu(x)J_{-(\nu+1)}(x) = -\frac{2\sin\pi\nu}{\pi x},$$
$$J_{\nu+1}(x)Y_\nu(x) - J_\nu(x)Y_{\nu+1}(x) = \frac{2}{\pi x},$$
$$I_{\nu+1}(x)I_{-\nu}(x) - I_\nu(x)I_{-(\nu+1)}(x) = \frac{2\sin\pi\nu}{\pi x},$$
$$I_{\nu+1}(x)K_\nu(x) + I_\nu(x)K_{\nu+1}(x) = \frac{1}{x}.$$

C.6.7. J_n generating function

$$e^{ix\sin\varphi} = \sum_{n=-\infty}^{+\infty} J_n(x)e^{in\varphi}. \quad \text{Note: } J_{-n}(x) = (-)^n J_n(x).$$

C.6.8. Representations by a definite integral

$$J_n(x) = \frac{1}{2\pi i^n}\int_{(2\pi)} \exp(in\varphi + ix\cos\varphi)\,d\varphi = \frac{1}{\pi}\int_0^\pi \cos(n\varphi - x\sin\varphi)\,d\varphi,$$
$$I_n(x) = \frac{(-1)^n}{2\pi}\int_{(2\pi)} \exp(in\varphi + x\cos\varphi)\,d\varphi = \frac{(-1)^n}{\pi}\int_0^\pi \cos(n\varphi)\exp(-x\cos\varphi)\,d\varphi,$$
$$K_n(x) = \frac{1}{2}\left(\frac{x}{2}\right)^n \int_0^\infty \exp\left(-t - \frac{x^2}{4t}\right)\frac{dt}{t^{n+1}}.$$

C.6.9. Addition formulae

Let there be a triangle with sides ω, x and X (with $X \geqslant x$); we use φ to denote the angle opposite ω, and ψ to denote the angle opposite x. The formula for the triangle gives:

$$\omega = \sqrt{X^2 + x^2 - 2xX\cos\varphi},$$

and we have:

$$J_\nu(\omega)e^{i\nu\psi} = \sum_{m=-\infty}^{+\infty} J_{\nu+m}(X)J_m(x)e^{im\varphi},$$
$$Y_\nu(\omega)e^{i\nu\psi} = \sum_{m=-\infty}^{+\infty} Y_{\nu+m}(X)J_m(x)e^{im\varphi},$$
$$I_\nu(\omega)e^{i\nu\psi} = \sum_{m=-\infty}^{+\infty} (-)^m I_{\nu+m}(X)I_m(x)e^{im\varphi},$$
$$K_\nu(\omega)e^{i\nu\psi} = \sum_{m=-\infty}^{+\infty} K_{\nu+m}(X)I_m(x)e^{im\varphi}.$$

C.6.10. Complete basis of functions defined in the interval [0,1]

The series of functions:

$$f_i(x) = \sqrt{x} J_n(j_{n,i} x),$$

where $j_{n,i}$ are the successive zeros of J_n, constitutes a complete basis of functions defined on [0,1]. These functions are orthogonal by pairs:

$$\int_0^1 f_i(x) f_j(x)\, dx = \frac{1}{2} \left[J'_n(j_{n,i}) \right]^2 \delta_{ij}.$$

C.6.11. Fourier transform in a plane

The Fourier transform in a plane is defined by:

$$\hat{g}(\vec{b}) = \int_{(\infty)} g(\vec{\rho}) e^{i\vec{b}.\vec{\rho}}\, d^2\rho,$$

where $\vec{\rho}$ and $\vec{b}$ are vectors with two components. If we write them in polar coordinates as (ρ, φ) and (b, α), respectively and then expand them in a Fourier series:

$$g(\vec{\rho}) = \sum_{n=-\infty}^{+\infty} g_n(\rho) e^{in\varphi}, \quad \hat{g}(\vec{b}) = \sum_{n=-\infty}^{+\infty} \hat{g}_n(b) e^{in\alpha},$$

we obtain the following formulæ:

$$\hat{g}_n(b) = i^n \int_0^\infty g_n(\rho) J_n(b\rho) 2\pi\rho\, d\rho,$$

$$g_n(\rho) = \frac{1}{4\pi^2} \frac{1}{i^n} \int_0^\infty \hat{g}_n(b) J_n(b\rho) 2\pi b\, db.$$

C.6.12. Half-order functions

Half-order Bessel functions are expressed analytically. Those of order ±1/2 are as follows:

$$J_{1/2}(x) = \sqrt{2/(\pi x)} \sin x, \qquad J_{-1/2}(x) = \sqrt{2/(\pi x)} \cos x,$$

$$Y_{1/2}(x) = -\sqrt{2/(\pi x)} \cos x, \qquad Y_{-1/2}(x) = \sqrt{2/(\pi x)} \sin x,$$

$$I_{1/2}(x) = \sqrt{2/(\pi x)} \operatorname{sh} x, \qquad I_{-1/2}(x) = \sqrt{2/(\pi x)} \operatorname{ch} x,$$

$$K_{1/2}(x) = \sqrt{2/(\pi x)} e^{-x}, \qquad K_{-1/2}(x) = \sqrt{2/(\pi x)} e^{-x}.$$

The others can be deduced using the recurrence formulæ.

Table C.1.

i	$j_{0,i}$	$J_1(j_{0,i})$	$j_{1,i}$	$J_0(j_{1,i})$
1	2.40483	+0.51915	3.83171	−0.40276
2	5.52008	−0.34026	7.01559	+0.30012
3	8.65373	+0.27145	10.17347	−0.24970
4	11.79153	−0.23246	13.32369	+0.21836
5	14.93092	+0.20655	16.47063	−0.19647

C.6.13. A few numerical values

Table C.1 shows the values of the first zeros of the function J_0, the values of J_1 at these abscissas, the first zeros of J_1 and the values of J_0 at these abscissas.

C.7. Streaming operator

The streaming operator $\text{div}[\vec{\Omega}\Phi]$ or $\vec{\Omega}.\overrightarrow{\text{grad}}\,\Phi$ expresses the transport of particles in a vacuum.

If the vector $\vec{\Omega}$ is identified by its colatitude θ (angle with respect to the direction of the north pole) and its longitude φ (angle between the meridian passing through $\vec{\Omega}$ and a meridian of origin), its cartesian coordinates are:

$$\kappa = \sin\theta\cos\varphi, \quad \lambda = \sin\theta\sin\varphi, \quad \mu = \cos\theta.$$

C.7.1. Cartesian coordinates

In cartesian coordinates x, y, z, the north pole is generally placed in the direction of the $\vec{z}$ axis, and the meridian taken as the origin is the one containing the $\vec{x}$ axis. Under these conditions, the streaming operator is written as:

$$\text{div}[\vec{\Omega}\Phi] = \kappa\frac{\partial\Phi}{\partial x} + \lambda\frac{\partial\Phi}{\partial y} + \mu\frac{\partial\Phi}{\partial z}$$

C.7.2. Cylindrical coordinates

In cylindrical coordinates, the point in $\vec{r}$ space is identified by the distance ρ to the $\vec{z}$ axis, by the longitude α, angle between the projection $\vec{\rho}$ of $\vec{r}$ on the (x, y) plane and the $\vec{x}$ axis, and by the dimension z.

It is convenient to identify the direction $\vec{\Omega}$ of the neutron in a *local* coordinate system: colatitude θ measured from $\vec{z}$, but longitude φ measured from the meridian containing $\vec{\rho}$.

Under these conditions, the streaming operator is written as:

$$\text{div}[\vec{\Omega}\Phi] = \frac{\kappa}{\rho}\frac{\partial}{\partial\rho}(\rho\Phi) + \frac{\lambda}{\rho}\frac{\partial\Phi}{\partial\alpha} + \mu\frac{\partial\Phi}{\partial z} - \frac{1}{\rho}\frac{\partial}{\partial\varphi}(\lambda\Phi).$$

This formula, like its equivalent in spherical geometry, is obtained by noting that $\text{div}(\vec{\Omega}\Phi)$ is the derivative $d\Phi/ds$ along the path of the neutron.

The formula is established by expressing this total derivative as the sum of the products $(\partial\Phi/\partial\xi)(\partial\xi/\partial s)$ where the ξ are the variables of the streaming problem; the spatial coordinates, obviously, but also the variables defining $\vec{\Omega}$ if they change along the neutron path because of the local nature of the coordinate system in curvilinear coordinates.

If one or two of the coordinates expressing $\vec{r}$ are not involved in the problem, the formula is simplified accordingly, but both coordinates expressing $\vec{\Omega}$ remain necessary in any case. For example, for a geometry that is solely dependent on ρ, we obtain:

$$\mathrm{div}[\vec{\Omega}\Phi] = \frac{\kappa}{\rho}\frac{\partial}{\partial\rho}(\rho\Phi) - \frac{1}{\rho}\frac{\partial}{\partial\varphi}(\lambda\Phi),$$

which can also be written as:

$$\mathrm{div}[\vec{\Omega}\Phi] = \kappa\frac{\partial\Phi}{\partial\rho} - \frac{\lambda}{\rho}\frac{\partial\Phi}{\partial\varphi}.$$

C.7.3. Spherical coordinates

In spherical coordinates, the point in $\vec{r}$ space is identified by the distance r to the origin, the colatitude β (measured from $\vec{z}$) and the longitude α (meridian of origin on $\vec{x}$) of the vector $\vec{r}/r$.

Here again, it is convenient to identify the direction $\vec{\Omega}$ of the neutron in a *local* coordinate system: colatitude θ measured from the "local vertical" $\vec{r}/r$, longitude φ measured from the meridian containing $\vec{z}$.

Under these conditions, and replacing the variable θ by its cosine $\mu = \cos\theta$, the streaming operator is written as:

$$\mathrm{div}[\vec{\Omega}\Phi] = \frac{\mu}{r^2}\frac{\partial}{\partial r}(r^2\Phi) + \frac{\kappa}{r\sin\beta}\frac{\partial}{\partial\beta}(\sin\beta\Phi) + \frac{\lambda}{r\sin\beta}\frac{\partial\Phi}{\partial\alpha} + \frac{1}{r}\frac{\partial}{\partial\mu}[(1-\mu^2)\Phi]$$
$$- \frac{\mathrm{cotg}\,\beta}{r}\frac{\partial}{\partial\varphi}(\lambda\Phi).$$

If one or two of the coordinates expressing $\vec{r}$ are not involved in the problem, the formula is simplified accordingly; if the problem depends on r only, then only the θ coordinate is involved with respect to $\vec{\Omega}$ (θ: angle between the direction of the neutron and the vector joining the origin to the current point; $\mu = \cos\theta$):

$$\mathrm{div}[\vec{\Omega}\Phi] = \frac{\mu}{r^2}\frac{\partial}{\partial r}(r^2\Phi) + \frac{1}{r}\frac{\partial}{\partial\mu}[(1-\mu^2)\Phi],$$

which can also be written as:

$$\mathrm{div}[\vec{\Omega}\Phi] = \mu\frac{\partial\Phi}{\partial r} + \frac{(1-\mu^2)}{r}\frac{\partial\Phi}{\partial\mu}.$$

C.8. Peierls operator

The Peierls operator is the integral operator giving the flux Φ of particles from the emission density Q if this emission density is assumed to be *isotropic*.

C.8.1. *Three-Dimensional Geometry* (x, y, z)

The general expression for the Peierls operator is:

$$\Phi(\vec{r}) = \int_{(\infty)} dx'\, dy'\, dz' \frac{\exp(-\tau)}{4\pi R^2} Q(\vec{r'}),$$

with:

$$R^2 = \left|\vec{r} - \vec{r'}\right|^2 = (x - x')^2 + (y - y')^2 + (z - z')^2.$$

The optical path τ is the integral of the total cross-section along the straight line segment connecting the particle starting point $\vec{r'}$ to the point $\vec{r}$ where it is observed.

C.8.2. *Two-Dimensional Geometry* (x, y)

In a problem invariant under translation parallel to the $\vec{z}$ axis, *i.e.* involving only the variables x and y (or $\vec{\rho}$), it is wise to begin by integrating over z':

$$\Phi(\vec{\rho}) = \int_{(\infty)} dx'\, dy' Q(\vec{\rho'}) \int_{(\infty)} dz' \frac{\exp(-\tau)}{4\pi R^2}.$$

The integral over z' can be written more simply with the variable θ, the angle between the direction of the particle path $\vec{R} = \vec{r} - \vec{r'}$ and the $\vec{z}$ axis, noting that the true optical path τ is also written as $\tau_{xy}/\sin\theta$, where τ_{xy} is its projection on the (x, y) plane:

$$\Phi(\vec{\rho}) = \int_{(\infty)} dx'\, dy' \frac{Ki_1(\tau_{xy})}{2\pi R_{xy}} Q(\vec{\rho'}),$$

with:

$$R_{xy}^2 = \left|\vec{\rho} - \vec{\rho'}\right|^2 = (x - x')^2 + (y - y')^2, \quad R_{xy} = R \sin\theta = z' \mathrm{tg}\,\theta,$$

where the function Ki_1 is one of the Bickley-Naylor functions (see below).

Note (Figure 14.3) that the variables Φ, R, t and t' are used to calculate the first collision probabilities in problems with an $(x - y)$ geometry.

C.8.3. *One-dimensional geometry* (x)

In a problem that is translation invariant parallel to the $\vec{y}$ and $\vec{z}$ axes, *i.e.* involving the x variable only, it is likewise advantageous to start by integrating along y' and z':

$$\Phi(x) = \int_{(\infty)} dx' Q(x') \int_{(\infty)} dy'\, dz' \frac{\exp(-\tau)}{4\pi R^2}.$$

The integral over y' and z' can be written more simply with the variable θ, which is the angle between the direction of the particle path $\vec{R} = \vec{r} - \vec{r'}$ and the $\vec{x}$ axis, noting that the true optical path τ is also written $\tau_x/\cos\theta$, where τ_x is its projection on the $\vec{x}$ axis:

$$\Phi(x) = \int_{(\infty)} dx' \frac{1}{2} E_1(\tau_x) Q(x'),$$

where the function E_1 is one of the integral exponential functions (see below).

C.9. Integral exponential functions

C.9.1. Definitions

Integral exponential functions are defined by the following formulae, which are equivalent, with the variable changes $\cos\theta = \sin\zeta = \mu = 1/u$:

$$E_n(x) = \int_1^{\infty} \exp(-xu)\frac{du}{u^n} = \int_0^1 \exp\left(-\frac{x}{\mu}\right)\mu^{n-2}\,d\mu$$
$$= \int_0^{\pi/2} \exp\left(-\frac{x}{\cos\theta}\right)\cos^{n-2}\theta\sin\theta\,d\theta = \int_0^{\pi/2} \exp\left(-\frac{x}{\sin\zeta}\right)\sin^{n-2}\zeta\cos\zeta\,d\zeta,$$
$$E_0(x) = \frac{e^{-x}}{x}.$$

C.9.2. Recurrence relations

$$E_n(x) = \frac{e^{-x}}{(n-1)!}\left[\sum_{m=0}^{n-2}(n-m-2)!(-x)^m + (-x)^{n-1}E_1(x)\right],$$
$$E_n(x) = \int_x^{\infty} E_{n-1}(x')\,dx', \quad E_n'(x) = -E_{n-1}(x),$$
$$E_n(x) = \frac{1}{n-1}\left[e^{-x} - xE_{n-1}(x)\right], \quad (n>1).$$

C.9.3. Expansions at the origin

$$E_n(x) = (-)^n\frac{x^{n-1}}{(n-1)!}(\ln x - A_n + \gamma) + \sum_{m=0;m\neq n-1}^{\infty}\frac{(-x)^m}{m!(n-1-m)},$$

with:

$$A_0 = 0, \quad A_n = \sum_{m=1}^{n-1}\frac{1}{m}, \quad (n>1),$$

and $\gamma = 0.557215665\ldots$ (Euler's constant). We note that E_0 and E_1 are infinite at the origin, and that:

$$E_n(0) = \frac{1}{n-1}, \quad (n>1).$$

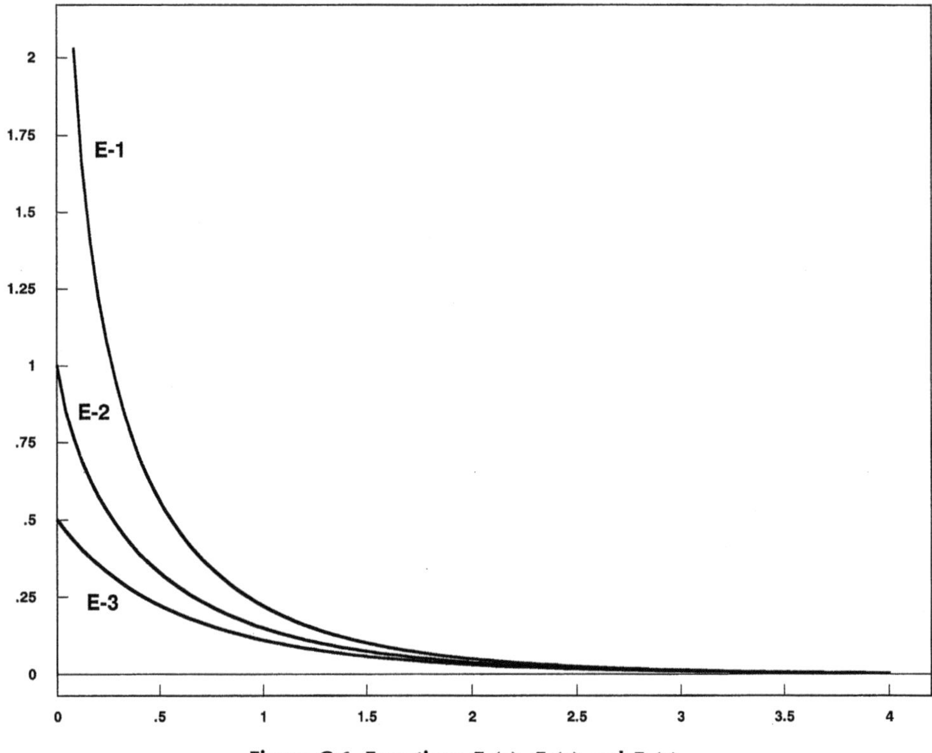

Figure C.6. Functions $E_1(x)$, $E_2(x)$ and $E_3(x)$.

C.9.4. Asymptotic expansions

$$E_n(x) = \frac{x^{-x}}{x}\left[1 - \frac{n}{x} + \frac{n(n+1)}{x^2} - \frac{n(n+1)(n+2)}{x^3} + \cdots\right],$$

or:

$$E_n(x) = \frac{x^{-x}}{x+n}\left[1 + \frac{n}{(x+n)^2} + \frac{n(n-2x)}{(x+n)^4} + \frac{n(6x^2 - 8nx + n^2)}{(x+n)^6} + \cdots\right].$$

C.9.5. Fourier cosine transforms

Let us define the Fourier transform by:

$$\hat{E}_n(b) = \int_0^\infty \cos(bx) E_n(x)\, dx.$$

The transform of E_0 does not exist; the transforms of the following functions are:

$$\hat{E}_1(b) = \frac{1}{b}\text{Arctan}\, b,$$

$$\hat{E}_2(b) = \frac{1}{2b^3}\ln(1+b^2),$$

$$\hat{E}_3(b) = \frac{1}{b^2} - \frac{1}{b^3}\text{Arctan}\, b.$$

C.10. Bickley-Naylor functions

Bickley-Naylor functions are defined by the following integrals:

$$Ki_n(x) = \int_0^{\pi/2} \exp\left(-\frac{x}{\sin\zeta}\right)\sin^{n-1}\zeta\, d\zeta.$$

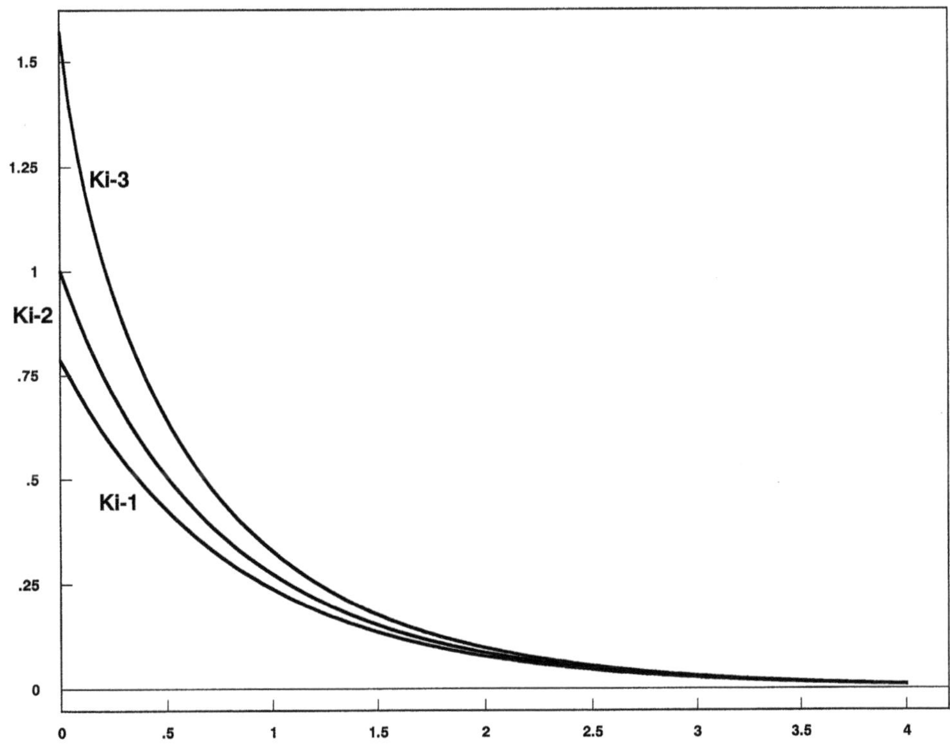

Figure C.7. Functions $Ki_1(x)$, $Ki_2(x)$ and $Ki_3(x)$.

Like the integral exponential functions, they are interconnected:

$$Ki_n(x) = \int_x^{\infty} Ki_{n-1}(x')\, dx', \quad Ki_n'(x) = -Ki_{n-1}(x),$$

$$(n-1)Ki_n(x) = (n-2)Ki_{n-2}(x) + x[Ki_{n-3}(x) - Ki_{n-1}(x)].$$

The values at the origin are given by the following formulae:

$$Ki_{2n}(0) = \frac{(n-1)!}{(2n-1)!} 2^{n-1},$$

$$Ki_{2n+1}(0) = \frac{(2n-1)!}{n!} 2^{-(n+1)} \pi.$$

and the asymptotic behaviour is described by:

$$Ki_n(x) \sim a_n \frac{e^{-x}}{\sqrt{x}},$$

where a_n is a constant.

C.11. Quadrature formulae

C.11.1. General information

A *numerical integration formula,* or *quadrature formula,* is used to evaluate an integral when it is not expressed analytically or by pretabulated functions or functions represented by approximate expressions.

There are many quadrature formulae, and all of them[1] take the following form:

$$\int_a^b f(x)dx \simeq \sum_{i=1}^{I} w_i f(x_i),$$

where x_i are the abscissas where the function f is known or can be calculated, and w_i are *"weights"*. The trapezoidal rule, for example, which involves calculating the integral by replacing the function f with a straight line segment in each interval $[x_i, x_{i+1}]$, belongs to this type.

In certain cases, the abscissas x_i are imposed or the choice is made *a priori*; then all that remains is to choose the weights w_i. We shall discuss this in the first part. For the sake of simplicity, the discussion will be limited to the case where the abscissas are arranged regularly in the integration interval.

If we also have the freedom to choose the abscissas, we can try to optimise not only the w_i, but also the x_i. We shall look at this case in the second part. (The procedure that involves setting the w_i and selecting the best possible x_i will only be mentioned briefly.)

To develop a quadrature formula, a "best choice" criterion must be specified. In practice, we shall try to obtain the highest possible order of precision, defining the order of precision as follows:

A quadrature formula is said to be precise to order k if it is exact for the mononomials $1, x, x^2, ...x^k$, and therefore for all the polynomials of order less than or equal to k.

We can also say that *a formula that is exact to order k commits an error of order* h^{k+1}, which means "approximately proportional to h^{k+1}". Therefore, if the integration step is reduced by a factor of λ, the error of the quadrature formula is reduced approximately by a factor of λ^{k+1}.

[1] For the sake of simplicity, here we limit the discussion to simple integrals and we will not discuss the Monte Carlo method. Multiple integrals can be calculated in a very similar way; in the Monte Carlo method, the abscissas are "drawn at random" and the weights are calculated according to the abscissa obtained.

C.11.2. Constant-step formulae

Let us assume that the interval $[a, b]$ is divided into N intervals:

$$h = \frac{b-a}{N}.$$

We take $x_0 = a$ and $x_N = b$. The quadrature formula that we now rewrite in the following form:

$$\int_a^b f(x)dx \simeq \sum_{i=0}^{N} w_i f(x_i),$$

will then be called an *"N-step formula"* (constant steps in this case).

For the various formulae that are commonly used, N is never very large, and there might be concern about the step h being too wide and preventing the required accuracy from being achieved. If so, we can cut up $[a, b]$ into $I = MN$ elementary intervals and use an N-step formula in each of the M macro-intervals of N steps. Note that this leads to *doubling* the coefficients w_{mN} associated with the limit abscissas of the macro-intervals, except w_0 and w_I.

The weight values for an integration interval of length N are given below. In practice, the length of the macro-interval needs to be normalised to this length N by a change of variable.

a) Elementary formulae

The simplest formula is the "trapezoidal rule", so-called because it amounts to replacing the curve to be integrated with a series of trapezia that preserve the values at the discretised abscissas.

The other "elementary formulae" indicated below are constructed using the trapezoidal rule: we write (if possible) the trapezoidal rule for the elementary step and for one or more multiples of this elementary step, and then we combine these formulae in order to make the h^2 terms disappear, or possibly the h^4 terms, etc.

• *Trapezoidal rule*: this corresponds to $N = 1$ and the *error is of the order* h^2. The coefficients are obviously:

Table C.2.

Abscissa number	Weight
0	1/2
1	1/2

• *Simpson's rule:* let us take $N = 2$, i.e. I even. We can then write the trapezoidal rule for the elementary step h and the step $2h$. By combining the two formulae with appropriate coefficients, we can make the remainder of order h^2 vanish and thus obtain a more accurate formula, which in this case has an *error of order* h^4. This is Simpson's rule. Its coefficients are given in Table C.3.

Table C.3.

Abscissa number	Weight
0	1/3
1	4/3
2	1/3

• *Modified Simpson's rule:* let us take $N = 4$, i.e. I a multiple of 4. We can then write the trapezoidal rule for the elementary steps h, $2h$ and $4h$. By combining the three formulae with suitable coefficients, we can make not only the h^2 terms vanish, but also the h^4 terms, thus obtaining a more precise formula which, in this case, has *error of order* h^6. The coefficients are as follows:

Table C.4.

Abscissa number	
0	14/45
1	64/45
2	24/65
3	64/45
4	14/45

• *Weddle's rule:* let us take $N = 6$, i.e. I a multiple of 6. We can now combine formulae for the elementary step h, for the step $2h$ and for the step $3h$ to make the h^2 and h^4 terms vanish in the expression for the remainder. We thus obtain Weddle's rule, with *error of order* h^6. The coefficients are remarkably simple (especially if 3/10 is added as a factor): see Table C.5.

• *Modified Weddle's rule:* still using $N = 6$, i.e. I a multiple of 6, we can combine not only formulae for the steps h, $2h$ and $3h$, but also the formula with step $6h$: we thus improve the formula, because the *error is then of order* h^8, but the coefficients are not as simple:

Table C.5.

Abscissa	Weight of number Weddle's rule:	
	standard	modified
0	3/10	41/140
1	15/10	216/140
2	3/10	27/140
3	18/10	272/140
4	3/10	27/140
5	15/10	216/140
6	3/10	41/140

• *Other elementary formulae*: this type of reasoning can obviously be pursued: here are the coefficients of three 12-step formulae, whose *errors* are of order h^8, h^{10} and h^{12} respectively (see Table C.6).

Table C.6.

Abscissa number	Weight of 12-interval formulae		
0	10/35	49/175	41833/150150
1	56/35	288/175	248832/150150
2	0	–27/175	–29160/150150
3	80/35	448/175	395264/150150
4	–4/35	–63/175	–63909/150150
5	56/35	288/175	248832/150150
6	24/35	134/175	118416/150150
7	56/35	288/175	248832/150150
8	–4/35	–63/175	–63909/150150
9	80/35	448/175	395264/150150
10	0	–27/175	–29160/150150
11	56/35	288/175	248832/150150
12	10/35	49/175	41833/150150

b) Newton-Cotes formulae

The Newton-Cotes formulae are obtained by systematically seeking the "best choice" of weights when N has been chosen. With N unknowns w_i to be determined, we can write N equations by writing out the precision for the successive mononomials up to x^{N-1}. We thus write out exact formulae to the order $N - 1$, with error of order h^{N+1} if N is even. We thus find, respectively, the trapezoidal rule, Simpson's rule, the modified Simpson's rule, and the modified Weddle's rule, for the values 1, 2, 4 and 6 of N. These formulae therefore turn out to be the best possible constant-step formulae for these values of N. For even values of N beyond 6, we can find formulae of order h^{N+2} higher than that of the elementary formulae. The odd values of N are less interesting because, for reasons of symmetry, the formulae with $2n$ steps and $2n + 1$ steps are of the same order.

C.11.3. Gauss formulae

As we can see from the tables (tables C.2 to C.6), the weights w_i are increasingly dispersed in orders of magnitude as N increases. There is one disadvantage to this *dispersion*: it *increases the sensitivity to numerical errors*. The advantage of moving by one order of precision thus ends up being counterbalanced by the increase in numerical errors, and so it is pointless to try to develop formulae with even larger values of N.

This observation led Chebyshev to construct the least sensitive formulae possible — i.e. *with weights w that are all equal* — and the most exact formulae possible, by adjusting the choice of abscissas x_i. This route, however, soon leads to a dead end, because beyond $N = 8$, the polynomial equation giving the values for x_i has complex roots.

In practice, the Chebyshev formulae are of little benefit here, because the Gauss formulae are almost as insensitive to numerical errors as the Chebyshev formulae, but have a higher order of precision.

For the Gauss formulae, *both the* abscissas x_i and the weights w_i are chosen to obtain the highest possible order of precision. Because we note that the boundaries are no longer a part of the abscissas defined in this way, the quadrature formula must be rewritten as:

$$\int_a^b f(x)dx \simeq \sum_{i=1}^{N} w_i f(x_i).$$

We are now dealing with an N-point formula rather than an N-interval formula.

As before, we could introduce several macro-intervals if necessary, and use an N-point formula in each one.

An N-point formula is thus characterised by $2N$ parameters. They can be determined by $2N$ equations by writing down the precision for the first $2N$ mononomials: we thus see that *the N-point formula can be exact to the order of* $2N-1$ or can be characterised by an *error of the order* h^{2N} if h is defined as the fraction $1/N$ of the integration interval.

Tables C.7 give the coefficients for the first twelve formulae (we give them to fifteen decimal places, because it is preferable to perform "double precision" calculations in order to take full advantage of the great precision of the Gauss formulae). For the sake of convenience, particularly in order to use the symmetry with respect to the centre of the interval, these coefficients x_i and w_i are given for the integration interval $[-1,+1]$. The necessary changes of integration variable need to be performed in order to produce this situation.

Note: The abscissas x_i for the N-point formula are the zeros of the Legendre polynomial P_N.

Table C.7. **Continued on next page.**

N = 2	Abscissa	Weight
	±0.577350269189627	1.000000000000000
N = 3	Abscissa	Weight
	±0.774596669241483	0.555555555555556
	0.000000000000000	0.888888888888889
N = 4	Abscissa	Weight
	±0.861136311594053	0.347854845137454
	±0.339981043584856	0.652145154862546
N = 5	Abscissa	Weight
	±0.906179845938664	0.236926885056189
	±0.538469310105683	0.478628670499367
	0.000000000000000	0.568888888888889
N = 6	Abscissa	Weight
	±0.932469514203152	0.171324492379171
	±0.661209386466264	0.360761573048138
	±0.238619186083197	0.467913934572691
N = 7	Abscissa	Weight
	±0.949107912342758	0.129484966168870
	±0.741531185599395	0.279705391489277
	±0.405845151377397	0.381830050505119
	0.000000000000000	0.417959183673470

Table C.7. Continued.

$N = 8$	Abscissa	Weight
	±0.960289856497536	0.101228536290376
	±0.796666477413627	0.222381034453374
	±0.525532409916329	0.313706645877888
	±0.183434642495650	0.362683783378362
$N = 9$	Abscissa	Weight
	±0.968160239507626	0.081274388361575
	±0.836031107326635	0.180648160694858
	±0.613371432700590	0.260610696402935
	±0.324253423403809	0.312347077040002
	0.000000000000000	0.330239355001261
$N = 10$	Abscissa	Weight
	±0.973906528517171	0.066671344308689
	±0.865063366688985	0.149451349150580
	±0.679409568299024	0.219086362515983
	±0.433395394129247	0.269266719309996
	±0.148874338981631	0.295524224714753
$N = 11$	Abscissa	Weight
	±0.978228658146058	0.055668567116172
	±0.887062599768095	0.125580369464908
	±0.730152005574049	0.186290210927730
	±0.519096129206812	0.233193764591996
	±0.269543155952345	0.262804544510240
	0.000000000000000	0.272925086777908
$N = 12$	Abscissa	Weight
	±0.981560634246714	0.047175336386513
	±0.904117256370491	0.106939325995295
	±0.769902674194306	0.160078328543379
	±0.587317954286618	0.203167426723046
	±0.367831498998180	0.233492536538367
	±0.125233408511469	0.249147045813398

D Handbook

Various conference participants have suggested that the documentation accompanying the "Neutron Physics of Pressurised Water Reactors" session should be supplemented by a ready reference guide. It is indeed useful to be able to look up a constant, an order of magnitude, or a basic formula at any time, and that is the purpose of this "Neutron Physics Handbook".

For ease of use, this type of document must be short, but if it is short, it cannot be exhaustive. I have therefore tried to find a compromise by selecting the information that I deem should be readily available.

I hope that the readers of Neutron Physics: A Guide *will also find this handbook useful.*

D.1. Units and constants

D.1.1. Units

• Unit of length: **fermi**	**F** or **fm**	1 **F** = 10^{-15} m
• Unit of surface area: **barn**	**b**	1 **b** = 10^{-28} m^2 = 10^{-24} cm^2
• Unit of atomic mass:	**u**	1 **u** = 1.660539 × 10^{-27} kg
(1/12 of the mass of the carbon 12 atom)		
• Unit of energy: **electron volt**	**eV**	1 **eV** = 1.6021765 × 10^{-19} J
Sub-multiple and multiples:		1 meV = 10^{-3} eV
		1 keV = 10^{3} eV
		1 MeV = 10^{6} eV
		1 GeV = 10^{9} eV
		1 TeV = 10^{12} eV
• Unit of activity: **becquerel**	**Bq**	1 **Bq** = 1 decay/s
(Old unit: curie	Ci	1 Ci = 3.7 · 10^{10} Bq)
• Unit of dose: **gray**	**Gy**	1 **Gy** = 1 J/kg
(Old unit: rad	rad	1 rad = 10^{-2} Gy)
• Unit of equivalent dose: **sievert**	**Sv**	1 **Sv** = 1 J/kg
(Old unit: rem	rem	1 rem = 10^{-2} Sv)

D.1.2. A few physical constants

• Speed of light:	**c**	**c** = 299 792 458 m/s
• Mass-energy equivalence: $E = mc^2$		1 kg = 8.98755×10^{16} J
		1 **u** = 931.4940 MeV
• Mass of an electron:	$\mathbf{m}_e$	$\mathbf{m}_e$ = 0.0005485799 u
		$\mathbf{m}_e = 9.10938 \times 10^{-31}$ kg
		$\mathbf{m}_e$ = 0.51100 MeV
• Mass of a proton:	$\mathbf{m}_p$	$\mathbf{m}_p$ = 1.0072765 u
		$\mathbf{m}_p = 1.67262 \times 10^{-27}$ kg
		$\mathbf{m}_p$ = 938.27 MeV
		$\mathbf{m}_p$ = 1836.15 $\mathbf{m}_e$
• Mass of a neutron:	$\mathbf{m}_n$	$\mathbf{m}_n$ = 1.0086649 u
		$\mathbf{m}_n = 1.67493 \times 10^{-27}$ kg
		$\mathbf{m}_n$ = 939.57 MeV
		$\mathbf{m}_n$ = 1838.68 $\mathbf{m}_e$
• Unit electric charge:	**e**	**e** = $1.6021765 \times 10^{-19}$ C
• Planck's constant:	**h**	**h** = 6.626068×10^{-34} J.s
• Boltzmann constant:	**k**	**k** = 1.380650×10^{-23} J/K
		k = 8.61734×10^{-5} eV/K
• Avogadro's number:	**N**	**N** = 6.022142×10^{23} mol^{-1}

D.2. Nuclear physics

D.2.1. Characteristics of a (non-relativistic) particle

• Kinetic energy:	$E = \frac{1}{2}mv^2$
• Momentum:	$\vec{p} = m\vec{v}$ (a vector quantity)
• Wavelength:	$\lambda = h/p$

D.2.2. Constitution of an atom

• Element: characterised by the number **Z** of protons and electrons

• Isotopes: varieties of an element that vary according to the number **N** of neutrons

• Notation for nuclides: ${}^{A}_{Z}\mathbf{X}$

	X: Chemical symbol of the element
	Z: Number of protons
	N: Number of neutrons
	A = **Z** + **N**: number of nucleons

D.2.3. Binding energy in nuclei

• Mass defect:	$\Delta \mathbf{m} = (\mathbf{Zm}_p + \mathbf{Nm}_n) - \mathbf{m}_X$
• Binding energy:	$\mathbf{W} = \Delta \mathbf{mc}^2$
• Binding energy per nucleon:	$\mathbf{W/A}$

D.2.4. Radioactivity

• α decay:	${}^A_Z\mathbf{X} \Longrightarrow {}^{A-4}_{Z-2}\mathbf{Y} + {}^4_2\mathbf{He}$
• β^- decay:	${}^A_Z\mathbf{X} \Longrightarrow {}^{A}_{Z+1}\mathbf{Y} + \mathbf{e}^- + \bar{\nu}$
• β^+ decay:	${}^A_Z\mathbf{X} \Longrightarrow {}^{A}_{Z-1}\mathbf{Y} + \mathbf{e}^+ + \nu$
• Electron capture:	${}^A_Z\mathbf{X} + \mathbf{e}^- \Longrightarrow {}^{A}_{Z-1}\mathbf{Y} + \nu$
• γ decay:	${}^A_Z\mathbf{X}^* \Longrightarrow {}^A_Z\mathbf{X} + \gamma$
• Radioactive decay constant λ:	$d\mathbf{N} = -\lambda \mathbf{N} dt$
• Law of radioactive decay:	$\mathbf{N}(t) = \mathbf{N}(0)\exp(-\lambda t)$
• Radioactive half-life:	$\mathbf{T} = \ln 2/\lambda = 0.693/\lambda$
• Activity:	$\mathbf{a} = \lambda \mathbf{N}$

D.2.5. Fission

	Uranium 235	**Plutonium 239**
• Energy obtained by fission:	202 MeV	210 MeV
[Overall, 3.1×10^{10} fissions give 1 joule.]		
• Number of neutrons emitted per fission:	2.439	2.874
• Proportion of delayed neutrons (pcm):	679	224
• Average energy of prompt neutrons:	approximately 2 MeV	
• Fission spectrum of prompt neutrons:	$\chi(E) = C^t\sqrt{E}\exp(-E/\theta)$	
	$\theta = (2/3)E_{average}$	
• Average energy of delayed neutrons:	Approximately 0.5 MeV	

D.2.6. Reactions involving neutrons

Excitation energy of compound nucleus	= Neutron binding energy + Kinetic energy supplied by the neutron
• Total microscopic cross-section: [σ is generally expressed in barns.]	$\sigma_t = \sigma_a + \sigma_s$ (absorption + scattering)
• Microscopic absorption cross-section:	$\sigma_a = \sigma_f + \sigma_c$ (fission + sterile capture)
• Macroscopic cross-sections: [Σ is often expressed in cm^{-1}; here in m^{-1}.]	$\Sigma = N_1\sigma_1 + N_2\sigma_2 + \cdots$
• Calculation of concentrations: [N is often expressed in cm^{-3}; here in m^{-3}.]	$N = \rho \mathbf{N}_{Avogadro}/\mathbf{A}$ ρ: Density **A**: Mass number

D.3. Neutron diffusion

D.3.1. Multiplication factor

• Average number of neutrons emitted per fission for one neutron emitted by fission:

$$\mathbf{k} = \omega\nu$$

ω: probability that the neutron will provoke a fission
ν: average number of neutrons emitted by this fission.

D.3.2. Neutron paths

• Mean free path:	$\lambda = 1/\Sigma$
	Σ: Total macroscopic cross-section
• Average number of paths:	$\langle n \rangle = \Sigma/\Sigma_a$
	Σ_a: Macroscopic absorption cross-section

D.3.3. Neutron population

• Density:	$\mathbf{n}$	Neutrons per unit volume
		[Usual unit: cm^{-3}; here m^{-3}.]
• Flux:	$\Phi = \mathbf{nv}$	Neutrons per unit of surface area and time
		[Usual unit: $cm^{-2}.s^{-1}$; here $m^{-2}.s^{-1}$.]
• Reaction rate:	$R_k = \Sigma_k\Phi$	Reactions per unit volume and time
		[Usual unit: $cm^{-3}.s^{-1}$; here $m^{-3}.s^{-1}$.]

D.3.4. Transport equation

(monokinetic neutrons, isotropic collision, steady state)

$$\Phi(\vec{r}) = \int_{\text{Reactor}} Q(\vec{r'}) \frac{e^{-\tau}}{4\pi R^2} d^3 r'$$

$$Q(\vec{r'}) = S(\vec{r'}) + \Sigma_s(\vec{r'})\Phi(\vec{r'})$$

$$S(\vec{r'}) = \nu\Sigma_f(\vec{r'})\Phi(\vec{r'})$$

R: distance from $\vec{r'}$ to $\vec{r}$
τ : integral of Σ along the segment joining $\vec{r'}$ to $\vec{r}$

D.3.5. Diffusion approximation

(monokinetic neutrons, steady state)

$$D\Delta\Phi - \Sigma_a\Phi + S = 0$$

$D = 1/(3\Sigma_{tr})$: diffusion coefficient
$\Sigma_{tr} = \Sigma - \bar{\mu}\Sigma_s$: transport cross-section
$\bar{\mu} = \langle\cos\psi\rangle \simeq 2/(3A)$: average cosine of the scattering angle (*refer to the section on* "Slowing down")

- *Currents:*

$$J_+ = \frac{\Phi}{4} - \frac{D}{2}\frac{\partial\Phi}{\partial N} \quad J_- = \frac{\Phi}{4} + \frac{D}{2}\frac{\partial\Phi}{\partial N} \quad J_{\text{net}} = J_+ - J_- = -D\frac{\partial\Phi}{\partial N}$$

Vectorially *(Fick's law)*: $\vec{J} = -D\,\overrightarrow{\text{grad}}\,\Phi$

- *Black body extrapolation distance*: $d = 0.7104/\Sigma_{tr}$

D.3.6. One-group neutron theory

- *Sources*: $S = \nu\Sigma_f\Phi = k_\infty\Sigma_a\Phi$
- *Critical condition of bare homogeneous pile*:

$$k_{\text{eff}} = \frac{k_\infty}{1 + M^2B^2} = 1$$

$1/(1 + M^2B^2)$: non-leakage probability
B^2: geometric buckling
$M^2 = D/\Sigma_a$: migration area ($M^2 = \frac{1}{6}\langle R^2\rangle$)

Sphere:	$B^2 = \dfrac{\pi^2}{R^2}$	$\Phi = C^t \dfrac{\sin\frac{\pi r}{R}}{r}$
Cylinder:	$B^2 = \dfrac{j^2}{R^2} + \dfrac{\pi^2}{H^2}$	$\Phi = C^t J_0\left(\dfrac{j\rho}{R}\right)\sin\dfrac{\pi z}{H}$ ($j = 2.40483$)
Parallelepiped:	$B^2 = \dfrac{\pi^2}{a^2} + \dfrac{\pi^2}{b^2} + \dfrac{\pi^2}{c^2}$	$\Phi = C^t \sin\dfrac{\pi x}{a}\sin\dfrac{\pi y}{b}\sin\dfrac{\pi z}{c}$

(Dimensions include the extrapolation distance.)

D.4. Neutron spectrum

D.4.1. Infinite multiplication factor of thermal neutron reactors

$$k_\infty = \varepsilon p f \eta$$

- ε Fast fission factor: gain on the production resulting from fast fissions (essentially from uranium 238)
- p Resonance escape probability: probability of escaping from capture (essentially by uranium 238) during slowing down
- f Thermal utilisation factor: probability of absorption in the fuel for a thermal neutron
- η Reproduction factor: number of neutrons produced per fission for a thermal neutron absorption in the fuel

D.4.2. Elastic slowing down

- *Post-collision energy:*

$$\frac{E_{\text{after}}}{E_{\text{before}}} = \frac{A^2 + 2A\cos\theta + 1}{(A+1)^2}$$

A: mass of target nucleus/mass of neutron
θ: deflection angle in the centre of mass system

- *Relationship between the angles*:

$$\cos\psi = \frac{A\cos\theta + 1}{\sqrt{A^2 + 2A\cos\theta + 1}}$$

ψ: deflection angle in the laboratory system (reactor)

- *Isotropic case in the centre of mass:*

$\Longrightarrow$ $\cos\theta$ and E_{after} are uniform random variables

$\Longrightarrow$ E after is between E_{before} and αE_{before} with $\alpha = \left[\dfrac{A-1}{A+1}\right]^2$

$\Longrightarrow$ Average cosine of scattering angle: $\bar{\mu} = \langle\cos\psi\rangle = 2/(3A)$

$\Longrightarrow$ Maximum lethargy gain: $\epsilon = \ln(1/\alpha)$

$\Longrightarrow$ Average lethargy gain: $\xi = 1 - \alpha\epsilon/(1-\alpha)$

$\Longrightarrow$ Average number of impacts to slow down a neutron: $\langle n\rangle = \ln(E_{\text{initial}}/E_{\text{final}})/\xi$

D.4.3. Maxwell spectrum

• *Energy:*

$n(E)dE = C^t\sqrt{E}\exp(-E/E_0)dE$	with: $E_0 = kT$
For 20 °C:	$T = 293.16$ K
	$E_0 = 0.0253$ eV
	Associated speed $\frac{1}{2}m_n v_0^2 = E_0 : v_0 = 2200$ m/s
Most probable energy:	$E_p = \frac{1}{2}E_0$
Average energy:	$\langle E\rangle = \frac{3}{2}E_0$

• *Speed:*

$n(v)dv = C^{te}v^2\exp(-v^2/v_0^2)dv$	
Most probable speed:	$v_p = v_0$
Average speed:	$\langle v\rangle = \frac{2}{\sqrt{\pi}}v_0 = 1.128v_0$

D.5. Reactor kinetics

D.5.1. Reactivity

Definition:	$\rho = \frac{k-1}{k}$	k: "effective" multiplication factor
Units:	$\Longrightarrow$	**p.c.m.** = (per hundred thousand), also known as "millinile"
	$\Longrightarrow$	**dollar** = proportion of delayed neutrons (refer to the section on "Fission")

D.5.2. Supercritical situation with prompt neutrons

(reactivity ρ greater than one dollar)

$$n(t) \simeq n(0)\exp[(k_{prompt} - 1)/\ell]$$

with:	$k_{prompt} = k(1-\beta)$	prompt neutron multiplication factor β: proportion of delayed neutrons ℓ: lifetime of prompt neutrons (approximately 2×10^{-5} s for PWRs)

D.5.3. Evolution in the neighbourhood of criticality

$$n(t) \simeq n(0)\exp[(k-1)/\ell_{\text{eff}}]$$

with:	$\ell_{\text{eff}} = \ell + \sum_{j=1}^{G} \beta_j\tau_j$	"effective" lifetime of neutrons
		G: number of groups of delayed neutrons
		β_j: proportion of (delayed) neutrons emitted in group G
		$\lambda_j = 1/\tau_j$: radioactive decay constant of the jth precursors
		τ_j: average lifetime of jth precursors
		ℓ_{eff} in the region of 1/13 s for uranium 235

D.6. Reactor operation

D.6.1. Temperature effects

(For pressurised water reactors)

- **Doppler Effect**

$\Longrightarrow$	Related to temperature of fuel
$\Longrightarrow$	Instantaneous effect
$\Longrightarrow$	Always negative effect

- **Spectrum effect**

$\Longrightarrow$	Related to moderator temperature
$\Longrightarrow$	Delayed effect
$\Longrightarrow$	Negative effect for uranium 235 and positive effect for plutonium 239

- **Water expansion effect**

$\Longrightarrow$	Related to moderator temperature
$\Longrightarrow$	Delayed effect
$\Longrightarrow$	Negative effect on resonance escape probability
$\Longrightarrow$	Positive effect on thermal utilisation factor, increasing with higher boron concentration
$\Longrightarrow$	Note: optimum moderation is equivalent to a zero overall expansion effect.

D.6.2. Poisoning by fission products

- *Fission fragments:* nuclides that can be obtained directly by fission
- *Fission products:* fission fragments + possible derived products
- There are approximately 1000 known fission products
- The sum of the fission yields γ_i is 200%
- *Indepedent* yield γ_i: relative to a given nuclide *i*; *cumulative* yield: sum of independent yields for all nuclides of a given mass **A**.
- The curve plotting cumulative yields by mass has a humped shape, indicating generally asymmetric fission.
- Almost all fission fragments are β^- radioactive.

• **General evolution equations** (certain terms may be zero or negligible)

$dN_i/dt =$	change in concentration of product *i*
$+\gamma_i \Sigma_f \Phi$	production by fission
$-\sigma_{a,i} N_i \Phi$	destruction by neutron irradiation
$-\lambda_i N_i$	destruction by radioactive decay
$+\sigma_{c,j} N_j \Phi$	production by capture by nuclei *j*
$+\lambda_k N_k$	production by decay of a product *k*

• **Neutron poisoning** (approximately the reactivity effect, except for the sign)

$$\pi = \frac{\Sigma_{a,F.P.}}{\Sigma_{a,comb}} \simeq -\frac{\Delta k}{k}$$

• **Approximate values for pressurised water reactors**

— **Xenon 135**	Equilibrium: 2800 pcm	Peak after shutdown: 2000 pcm
— **Samarium 149**	Equilibrium: 700 pcm	Excess after shutdown: 300 pcm
— **Total**	End of irradiation: 15 000 pcm	

D.6.3. Evolution of heavy nuclei

• **Evolution equations**: analogous to those of fission products (except for the fission production term); certain terms can be zero or negligible

$dN_i/dt =$	change in concentration of product *i*
$-\sigma_{a,i} N_i \Phi$	destruction by neutron irradiation
$-\lambda_i N_i$	destruction by radioactive decay
$+\sigma_{c,j} N_j \Phi$	production by capture by nuclei *j*
$+\lambda_k N_k$	production by decay of a product *k*

- **Conversion**

— *uranium 238 / plutonium 239 series*

$$^{238}_{92}\mathbf{U} + \mathbf{n} \Longrightarrow {}^{239}_{92}\mathbf{U} \underset{23\ \text{min}}{\overset{\beta^-}{\Longrightarrow}} {}^{239}_{93}\mathbf{Np} \underset{2.3\ \text{days}}{\overset{\beta^-}{\Longrightarrow}} {}^{239}_{94}\mathbf{Pu}$$

— *thorium 232 / uranium 233 series*

$$^{232}_{90}\mathbf{Th} + \mathbf{n} \Longrightarrow {}^{233}_{90}\mathbf{Th} \underset{23\ \text{min}}{\overset{\beta^-}{\Longrightarrow}} {}^{233}_{91}\mathbf{Pa} \underset{27\ \text{days}}{\overset{\beta^-}{\Longrightarrow}} {}^{233}_{92}\mathbf{U}$$

— *Conversion factor*

C = (Number of fissile nuclei produced)/(Number of fissile nuclei destroyed)

— *Regeneration gain*

G = (**Net** number of fissile nuclei produced)/(Number of fissions)

- **Evolution measurements**

— *Fluence*

= **integral of flux over time**
Usual unit: neutron per kilobarn (n/kb)
Approximate value for PWRs: 2

— *Burnup*

= **energy produced / initial mass (of heavy nuclei)**
Usual units: terajoule/kilogramme (TJ/kg)
Approximate value for PWRs: 4
and: megawatt-day/tonne (MWd/t)
Approximate value for PWRs: 40 000

— *Burnup rate (fission)*

= **number of fissioned nuclei / initial number of heavy nuclei**
Usual unit: percent
Approximate value for PWRs: 4

Index and glossary

Below is a list of the main topics mentioned or discussed in this book, unless the exercises.

The terms listed in the index often comprise several words and, in this case, there is an entry for each important word in the expression. For example, 'Buffon needle' is listed under both 'Buffon' and 'needle'.

When compiling the index, we tried to distinguish between a simple reference to a concept (the 'index' function) and a passage where the concept is defined or discussed (the 'glossary' function). For glossary-type entries, the paragraph number is shown in bold print. '.0' means the introduction of the referenced chapter.

B

C

D

N

T

U

V

W

X

Y

Z

www.ingramcontent.com/pod-product-compliance
Ingram Content Group UK Ltd.
Pitfield, Milton Keynes, MK11 3LW, UK
UKHW050920270726
13994UKWH00011B/2450